Organic Reactions

Organic Reactions

VOLUME 23

JOHN WILEY & SONS, INC.
NEW YORK · LONDON · SYDNEY · TORONTO

Published by John Wiley & Sons, Inc.

Library of Congress Catalogue Card Number: 42-20265

ISBN 0-471-19624-x

Printed in the United States of America.

10 9 8 7 6 5 4 3 2

PREFACE TO THE SERIES

In the course of nearly every program of research in organic chemistry the investigator finds it necessary to use several of the better-known synthetic reactions. To discover the optimum conditions for the application of even the most familiar one to a compound not previously subjected to the reaction often requires an extensive search of the literature; even then a series of experiments may be necessary. When the results of the investigation are published, the synthesis, which may have required months of work, is usually described without comment. The background of knowledge and experience gained in the literature search and experimentation is thus lost to those who subsequently have occasion to apply the general method. The student of preparative organic chemistry faces similar difficulties. The textbooks and laboratory manuals furnish numerous examples of the application of various syntheses, but only rarely do they convey an accurate conception of the scope and usefulness of the processes.

For many years American organic chemists have discussed these problems. The plan of compiling critical discussions of the more important reactions thus was evolved. The volumes of *Organic Reactions* are collections of chapters each devoted to a single reaction, or a definite phase of a reaction, of wide applicability. The authors have had experience with the processes surveyed. The subjects are presented from the preparative viewpoint, and particular attention is given to limitations, interfering influences, effects of structure, and the selection of experimental techniques. Each chapter includes several detailed procedures illustrating the significant modifications of the method. Most of these procedures have been found satisfactory by the author or one of the editors, but unlike those in *Organic Syntheses* they have not been subjected to careful testing in two or more laboratories.

Each chapter contains tables that include all the examples of the reaction under consideration that the author has been able to find. It is inevitable, however, that in the search of the literature some examples will be missed, especially when the reaction is used as one step in an extended synthesis. Nevertheless, the investigator will be able to use the tables and their accompanying bibliographies in place of most or all of the literature search so often required.

Because of the systematic arrangement of the material in the chapters and the entries in the tables, users of the books will be able to find information desired by reference to the table of contents of the appropriate chapter. In the interest of economy the entries in the indices have been kept to a minimum, and in particular, the compounds listed in the tables are not repeated in the indices.

The success of this publication, which will appear periodically, depends upon the cooperation of organic chemists and their willingness to devote time and effort to the preparation of the chapters. They have manifested their interest already by the almost unanimous acceptance of invitations to contribute to the work. The editors will welcome their continued interest and their suggestions for improvements in *Organic Reactions*.

Chemists who are considering the preparation of a manuscript for submission to Organic Reactions are urged to write either secretary before they begin work.

CONTENTS

CHAPTER 1

REDUCTION AND RELATED REACTIONS OF α,β-UNSATURATED CARBONYL COMPOUNDS WITH METALS IN LIQUID AMMONIA

Drury Caine

Georgia Institute of Technology
Atlanta, Georgia

CONTENTS

INTRODUCTION

This chapter is concerned with the reduction and related reactions of α,β-unsaturated ketones by alkali and alkaline earth metals in liquid ammonia.* α,β-Unsaturated acids, esters, and aldehydes also can be reduced by metals in liquid ammonia; reduction of these classes of compounds is included as well. Upon treatment with a metal in liquid ammonia (usually containing an ether co-solvent) a simple α,β-unsaturated ketone **1** is converted into the metal enolate **2** corresponding to the saturated ketone **3**. The enolate is usually stable in liquid ammonia, but can be converted into **3** by treatment with a relatively acidic proton donor (*e.g.*, ammonium chloride). If proton donors such as alcohols of comparable acidity to **3** are present during the reduction or added at the end of the reaction when excess metal is present, equilibrium between **2** and **3** will be established, and **3** will be reduced to the saturated alcohol **4**.

$$\underset{\mathbf{1}}{>C{=}C{-}C{=}O} \xrightarrow{M/NH_3} \underset{\mathbf{2}}{>C(H){-}C{=}C{-}O^-M^+}$$

$$\mathbf{2} \underset{}{\overset{HA}{\rightleftharpoons}} \underset{\mathbf{3}}{>C(H){-}C(H){-}C{=}O} \underset{HA}{\xrightarrow{M/NH_3}} \underset{\mathbf{4}}{>C(H){-}C(H){-}C(H){-}OH}$$

Research during the first half of this century revealed that a variety of organic compounds undergo reduction with metals in liquid ammonia,[1,2] but preparative metal-ammonia reductions of α,β-unsaturated carbonyl compounds were apparently not described until 1951. In that year Wilds, working on steroid syntheses, reported that the tricyclic ketone **6** could be prepared by treatment of the keto enone **5** with lithium-ammonia-ethanol,† followed by oxidation with chromium trioxide.[3] At the same time he found that under similar conditions cholest-4-en-3-one could be converted into cholestanone in good yield.[4] A few months later groups at Syntex[5] and at Merck[6] reported simultaneously the conversions of steroidal 8(9)-en-11-ones into 11-ones having the natural $8\beta,9\alpha$-B/C-backbone by lithium-ammonia reduction (11α-hydroxy-$8\beta,9\alpha$-steroids were produced when the

* Metal-ammonia reductions are often referred to as Birch reductions in honor of Arthur J. Birch, who, with his collaborators, has applied the reaction extensively to a variety of organic compounds.

† This reduction medium is often referred to as Wilds's reagent.

[1] A. J. Birch, *Quart. Rev.* (London), **4**, 69 (1950).

[2] G. W. Watt, *Chem. Rev.*, **46**, 317 (1950).

[3] A. L. Wilds, Abstr. ACS Meeting, New York, Sept. 1951, p. 20M.

[4] A. L. Wilds, personal communication.

[5] F. Sondheimer, R. Yashin, G. Rosenkranz, and C. Djerassi, *J. Amer. Chem. Soc.*, **74**, 2696 (1952).

[6] E. Schoenewaldt, L. Turnbull, E. M. Chamberlin, D. Reinhold, A. E. Erickson, W. V. Ruyle, J. M. Chemerada, and M. Tishler, *J. Amer. Chem. Soc.*, **74**, 2696 (1952).

lithium-ammonia-ethanol combination was employed). Shortly after, the first examples of metal-ammonia reductions of α,β-unsaturated acids were reported by other workers at Merck.[7,8]

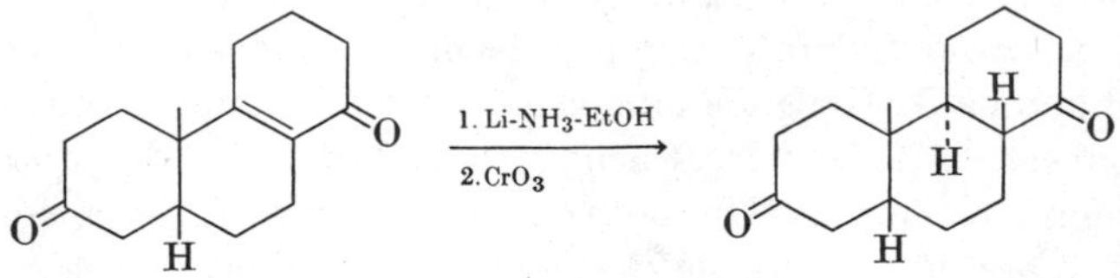

5 → (1. Li-NH_3-EtOH; 2. CrO_3) → 6

Metal-ammonia reduction of a large number of steroid and terpenoid enones in which the β-carbon atom was located at the fusion of two six-membered rings revealed that, in general, the reaction leads to the formation of the thermodynamically more stable isomer at that position.[9] However, using results of reductions in the octalone series, Stork and Darling pointed out that the more stable isomer is not always obtained, and that "the product will be the more stable of the two isomers (*cis* or *trans*) having the newly introduced β-hydrogen axial to the ketone ring."[10] This rule has been widely applied for correctly predicting the stereochemical outcome of a large number of metal-ammonia reductions, and in only a few cases in which exceptionally complex strain or other factors are involved does it appear to have been violated.

Generally, the conditions employed in the workup of metal-ammonia reduction reactions lead to products having the more stable configuration at the α-carbon atom, but products having the less stable configuration at this center have been obtained by kinetic protonation of enolate intermediates.[11, 12]

In addition to affording a remarkable degree of stereoselectivity, many metal-ammonia reductions of unsaturated carbonyl systems can be performed in the presence of a variety of functional and protective groupings, and the reaction is generally free of the rearrangements sometimes observed when other chemical reduction methods are employed.

These features have caused metal-ammonia reductions of unsaturated carbonyl compounds to be of immense value in synthetic organic chemistry. However, a new dimension was added to the reaction when it was

[7] G. E. Arth, G. I. Poos, R. M. Lukes, F. M. Robinson, W. F. Johns, M. Feurer, and L. H. Sarett, *J. Amer. Chem. Soc.*, **76**, 1715 (1954).

[8] L. H. Sarett, G. E. Arth, R. M. Lukes, R. E. Beyler, G. I. Poos, W. F. Johns, and J. M. Constantin, *J. Amer. Chem. Soc.*, **74**, 4974 (1952).

[9] D. H. R. Barton and C. H. Robinson, *J. Chem. Soc.*, **1954**, 3054.

[10] G. Stork and S. D. Darling, *J. Amer. Chem. Soc.*, **82**, 1512 (1960); **86**, 1761 (1964).

[11] A. J. Birch, H. Smith, and R. E. Thornton, *J. Chem. Soc.*, **1957**, 1339.

[12] H. E. Zimmerman, *J. Amer. Chem. Soc.*, **78**, 1168 (1956).

discovered that lithium enolates of unsymmetrical ketones generated in the reduction process undergo C-alkylation with alkyl halides and carbonation with carbon dioxide, either in liquid ammonia or after exchange of solvents.[13,14] These enolate trapping reactions thus allow regiospecific* introduction of groups at the α-carbon atoms of unsymmetrical ketones via the appropriate enone precursors. Reactions of reductively formed lithium enolates have been observed with other electrophilic reagents at carbon or at oxygen.

The unique features of metal-ammonia reductions and related reactions have led to their wide application in total synthesis and in transformations of steroids and terpenoids. More recently, interest has been aroused in the use of these reactions to achieve synthetic objectives in simple monocyclic and acylic systems. A number of reviews have appeared that include coverage of metal-ammonia reductions of α,β-unsaturated carbonyl compounds.[16–24]

MECHANISM

The Nature of Metal-Ammonia Solutions

With the exception of beryllium, all alkali and alkaline earth metals dissolve to some extent in liquid ammonia. Magnesium has very limited solubility, but solutions of it[25] as well as of beryllium[26] and certain other metals can be prepared by cathodic reduction of liquid ammonia solutions

* The term regiospecific is used as originally coined by A. Hassner.[15]

[13] G. Stork, P. Rosen, and N. L. Goldman, *J. Amer. Chem. Soc.*, **83**, 2965 (1961).

[14] G. Stork, P. Rosen, and N. L. Goldman, R. V. Coombs, and J. Tsuji, *J. Amer. Chem. Soc.*, **87**, 275 (1965).

[15] A. Hassner, *J. Org. Chem.*, **33**, 2684 (1968).

[16] A. J. Birch and H. Smith, *Quart. Rev.* (London), **12**, 17 (1958).

[17] A. J. Birch and G. Subba-Rao, *Advances in Organic Chemistry*, E. C. Taylor, Ed., Vol. 8, Wiley-Interscience Publishers, New York, 1972, p. 1.

[18] C. Djerassi, *Steroid Reactions*, Holden-Day, Inc., San Francisco, 1963, pp. 299–325.

[19] H. L. Drydon, Jr., Organic Reactions in Steroid Chemistry, J. Fried and J. A. Edwards, Eds., Vol. I, Van Nostrand Reinhold Co., New York, 1972, p. 1.

[20] H. O. House, *Modern Synthetic Reactions*, 2nd ed., Benjamin, Menlo Park, California, 1972, Chap. 3.

[21] F. Johnson, *Chem. Rev.*, **68**, 375 (1968).

[22] F. J. McQuillin, *Technique of Organic Chemistry*, A. Weissberger, Ed., Vol. XI, Part I, Interscience, New York, 1963, Chap. 9.

[23] H. Smith, *Organic Reaction in Liquid Ammonia, Chemistry in Non-Aqueous Ionizing Solvents*, Vol. I, Pt. 2, Wiley, New York, 1963.

[24] M. Smith, *Reduction*, R. L. Augustine, Ed., Marcel Dekker, New York, 1968, pp. 95–170.

[25] P. Angibeaund, H. Rivière, and B. Tchoubar, *Bull. Soc. Chim. Fr.*, **1968**, 2937.

[26] W. L. Jolly, *Progr. Inorg. Chem.*, **1**, 235 (1959).

of the corresponding metal salts. Dissolution of the alkaline earth metals and lithium is exothermic, whereas the remaining alkali metals have heats of solution near zero.[26] Metal-ammonia solutions are characteristically blue when dilute, but saturated solutions of the more soluble metals exhibit a bronze metallic luster.

The nature of metal-ammonia solutions has been the subject of a large number of physical chemical studies for over one hundred years. Although there is still not complete agreement concerning the precise nature of these solutions, most reviewers formulate different models according to the concentration.[26–31] Solutions which are 0.005 *M* or less are considered, as originally suggested,[32] to consist of essentially independent ammoniated metal cations and solvated electrons. The "free" electrons are thought to occupy cavities surrounded by ammonia molecules having their protons oriented toward the cavity. As the concentration is increased to the 0.005–1.0 *M* range, species of stoichiometry M, M^-, and M_2 appear to become important. The paramagnetic species M and the two diamagnetic ones, M^- and M_2, are thought to arise by simple electrostatic interactions of solvated electrons and cations. Indeed, on the basis of spectral studies it has been concluded that the characteristics of the electrons and cations are essentially the same in these ion pair species as they are in very dilute solutions.[33] More concentrated solutions (1.0 *M* or greater) are considered to be composed mainly of solvated cations held together by electrons, that is, the nature of these solutions is thought to be closely akin to that of the metallic state of the metal.[26]

Pure liquid ammonia has a very low tendency to react with dissolved metals.[34] Although the reaction of solvated electrons with ammonium ions is probably diffusion-controlled,[35] the low tendency of ammonia toward autoionization ($pK_a \simeq 34$) ensures that the ammonium ion concentration will be extremely low. Metals do react slowly with ammonia to form metal amides and hydrogen; this reaction is strongly catalyzed by transition metals such as iron, cobalt, and nickel and by ultraviolet light.* In

* The catalyzed reaction provides a convenient means of preparation of various metal amides which are used widely as basic catalysts in organic chemistry.[36]

[27] J. L. Dye, *Accounts Chem. Res.*, **1**, 306 (1968).

[28] W. L. Jolly, "Solvated Electron," *Advan. Chem. Ser.*, No. 50, American Chemical Society, Washington, D.C., 1965, pp. 27f.

[29] U. Schindewolf, *Angew. Chem., Int. Ed. Engl.*, **7**, 190 (1968).

[30] M. Szwarc, *Progr. Phys. Org. Chem.*, **6**, 323 (1968).

[31] M. C. R. Symons, *Quart. Rev.* (London), **13**, 99 (1959).

[32] C. A. Kraus, *J. Amer. Chem. Soc.*, **29**, 1557 (1907).

[33] M. Gold, W. L. Jolly, and K. S. Pitzer, *J. Amer. Chem. Soc.*, **84**, 2264 (1962).

[34] J. F. Dewald and G. LePoutre, *J. Amer. Chem. Soc.*, **76**, 3369 (1954).

[35] W. L. Jolly and L. Prizant, *Chem. Commun.*, **1968**, 1345.

[36] L. F. Fieser and M. Fieser, *Reagents for Organic Synthesis*, Wiley, New York, 1967.

general, metals of lower atomic weight yield more stable liquid ammonia solutions.

$$\mathrm{M + NH_3 \rightarrow MNH_2 + \tfrac{1}{2}H_2}$$

The general theory of chemical reduction provides the foundation for present-day consideration of mechanisms of chemical reduction of organic systems.[37,38] According to this idea, reversible addition of an electron to a vacant orbital of the substrate (S) can yield a radical anion. This species can then be protonated to give a radical which can either dimerize or accept another electron and a proton. Alternatively, stepwise or simultaneous reversible addition of two electrons to S can give a dianion that can accept two protons.

$$\mathrm{S} \underset{-e^-}{\overset{+e^-}{\rightleftharpoons}} \mathrm{S}^{\dot{-}} \xrightarrow{\mathrm{H^+}} \mathrm{HS\cdot} \xrightarrow{\mathrm{HS\cdot}} \mathrm{HS{-}SH}$$

$$\mathrm{S} \xrightarrow{-2e^-} \mathrm{S^{2-}} \qquad \mathrm{S}^{\dot{-}} \underset{+e^-}{\overset{-e^-}{\rightleftharpoons}} \mathrm{S^{2-}} \qquad \mathrm{HS\cdot} \xrightarrow{e^-} \mathrm{HS^-}$$

$$\mathrm{S^{2-}} \xrightarrow{\mathrm{H^+}} \mathrm{HS^-} \xrightarrow{\mathrm{H^+}} \mathrm{HSH}$$

The exact sequence and timing of these steps would be expected to be dependent upon factors such as the nature of the substrate, the homogeneity and reduction potential of the medium, and the presence and nature of proton donors in the medium.

Metal-ammonia solutions provide excellent media for homogeneous chemical reduction of a large number of organic substances. Many organic substances are soluble in liquid ammonia or can be brought into solution through the use of appropriate co-solvents, *e.g.*, diethyl ether, tetrahydrofuran, dioxane. The dipolar hydrogen-bonding character of liquid ammonia readily permits the formation of charged species such as radical anions and dianions as well as cations. Because ammonia provides a very low limit of proton acidity, many anionic species are stable in the medium. However, the more basic anionic species are protonated by ammonia. Moreover, the acidity of the medium can often be controlled by the addition of relatively acidic proton donors such as acidic hydrocarbons, alcohols, and water. These substances may accelerate reductions, alter their course, or prevent build-up of strongly basic metal amides. Reaction of such acids with the reducing system, according to the accompanying equations, is generally much slower than reduction because proton transfer to ammonia is strongly suppressed as the concentration of the base increases.[39, 40]

$$\mathrm{HA + NH_3 \rightleftharpoons A^- + NH_4^+}$$
$$\mathrm{NH_4^+} + e^- \rightarrow \mathrm{NH_3 + \tfrac{1}{2}H_2}$$

[37] R. Willstätter, F. Seitz, and E. Bumm, *Chem. Ber.*, **61**, 871 (1928).
[38] L. Michaeles and M. P. Schubert, *Chem. Rev.*, **22**, 437 (1938).
[39] R. R. Dewald and R. V. Tsina, *Chem. Commun.*, **1967**, 647.
[40] J. F. Eastham and D. R. Larkin, *J. Amer. Chem. Soc.*, **81**, 3652 (1959).

The Electron and Proton Transfer Steps

The metal-ammonia reduction of α,β-unsaturated carbonyl compounds involves a series of electron and proton transfers that follow the general pattern of chemical reduction outlined above. Although the details may vary considerably with the specific system and reduction conditions, sufficient information, particularly from studies of the stereochemistry of the proton transfer steps and from the reactions of various anionic intermediates, has been accumulated to develop a reasonably clear picture of the mechanism. In addition, studies on the reduction of α,β-unsaturated carbonyl compounds with metals (both in suspension and in solution) in other media, as well as by electrochemical methods, have shed light on the details of the pathways involved.

Unless proton donors of acidity greater than ammonia are present, alkyl-substituted α,β-unsaturated ketones **1** (p. 3) are reduced to metal enolates **2** by solutions of alkali metals in liquid ammonia.[11–14,41] As discussed later, reductions in which calcium is employed may lead directly to the alkoxide of the corresponding saturated alcohol.[42] Similar results have been observed in reductions with magnesium in liquid ammonia and to a lesser degree with concentrated solutions of lithium in liquid ammonia and extended reaction times.[43]

In 1954 it was proposed that the reduction involves β protonation of a dianionic intermediate such as **7**, which was thought to arise by transfer

$$\mathbf{1} \xrightarrow{2e} \mathrm{>\bar{C}-C=C-O^-}\ (\mathbf{7}) \qquad \mathrm{>\dot{C}-C=C-O^-}\ (\mathbf{8})$$

of two electrons from the solution to the conjugated system.[9] It was recognized that the formulation **7** did not exclude the coordination of metal cations with negative centers to form ion pairs or, perhaps, covalently bonded species that could be in equilibrium with the free dianion.[9] No attempt was made to specify whether the addition of two electrons occurred simultaneously or stepwise, *i.e.*, via the initial formation of a radical anion **8** which then added a second electron. Later, it was suggested that β protonation might take place at either the radical anion or the dianion stage; in the former case an enolate radical could be formed that would add a second electron to give the enolate.[10]

Recently, the polarographic reduction potentials for several alkyl- and

[41] H. E. Zimmerman, *Molecular Rearrangements*, Pt. I, P. de Mayo, Ed., Interscience Publishers, New York, 1963, pp. 345–406.

[42] P. Angibeaund and H. Rivière, *C.R. Acad. Sci., Ser. C*, **263**, 1076 (1966).

[43] A. Spassky-Pasteur, *Bull. Soc. Chim. Fr.*, **1969**, 2900.

aryl-substituted enones have been determined in dimethylformamide.[44] Aryl-substituted enones generally exhibit two reduction half-waves corresponding to the formation of radical anion **8** and dianion **7**, respectively, the second occurring at a potential 0.5–1.0 volt more negative than the first. Only one reduction wave could be observed for alkyl-substituted enones, and it was estimated that for these systems the second reduction wave corresponding to the formation of dianion **7** from radical anion **8** would require a potential more negative than −3.0 volts versus a saturated calomel electrode (sce). A potential of −3.0 volts is considerably more negative than the reduction potentials which have been measured for solutions of metals in liquid ammonia (vs sce).[45] These results make it appear highly unlikely that in metal-ammonia reductions the dianion **7** is formed by simultaneous addition of two electrons from the solution to the enone **1**. Additionally, the data suggest that *free dianionic* species are probably not involved in metal-ammonia reductions of aliphatic enones.[44]

The polarographic results indicate that the initial step in the reduction is a transfer of one electron from the solution to an antibonding π orbital of the conjugated system to produce a radical anion **8**.* This species may combine with a metal cation to produce an ion pair species **9**, which may

$$2\ {>}\dot{C}{-}\overset{|}{C}{=}\overset{|}{C}{-}O^- + 2\,M^+ \rightleftharpoons 2\ {>}\dot{C}{-}\overset{|}{C}{=}\overset{|}{C}{-}O^-M^+$$

8 **9**

$$\Updownarrow$$

$${>}\dot{C}{-}\overset{|}{C}{=}\overset{|}{C}{-}O^- \cdots M^+ \cdots {}^-O{-}\overset{|}{C}{=}\overset{|}{C}{-}\dot{C}{<} \quad (\text{bridged by two } M^+)$$

10

* The structure **8** appears to be a reasonable representation of the radical anion species. Electron spin resonance studies have been made on radical anions electrochemically generated in DMF solution from 2,2,6,6-tetramethyl-4-hepten-3-one[44] and several monocyclic ketones,[46] as well as by mixing dilute solutions of monocyclic enones in liquid ammonia with dilute solutions of sodium in liquid ammonia using a flow system.[47] The results showed that about 50 percent of the unpaired electron density is located on the β-carbon atom, very little at the α-carbon atom, and the remainder at the carbonyl group.

[44] K. W. Bowers, R. W. Giese, J. Grimshaw, H. O. House, N. H. Kolodny, K. Kronberger, and D. K. Roe, *J. Amer. Chem. Soc.*, **92,** 2783 (1970).

[45] H. Strehlow, *The Chemistry of Non-Aqueous Solvents*, Vol. 1. J. J. Lagowskii, Ed., Vol. 1, Academic Press, New York, 1966, pp. 129–172.

[46] G. A. Russell and G. R. Stevenson, *J. Amer. Chem. Soc.*, **93,** 2432 (1971).

[47] I. H. Elson, T. J. Kemp, and T. J. Stone, *J. Amer. Chem. Soc.*, **93,** 7091 (1971); I. H. Elson, T. J. Kemp, D. Greatorex, and H. D. B. Jenkins, *J. Chem. Soc. Faraday II*, **69,** 665, 1402 (1973).

be in equilibrium with the ion pair dimer **10**.[30, 48, 49] The exact nature of the association (solvent-separated ion pair, tight ion pair, or covalent bond) between the metal cation and the oxygen atom, and the importance of the various species in the equilibrium described above would expectedly depend upon the structure of the carbonyl system, the metal, the polarity of the medium as influenced by the presence of co-solvents, the concentration, and the temperature.

On the basis of present evidence, it appears that there are at least two possible, perhaps competing, pathways by which an initially formed radical anion may be converted into a metal enolate **2** (p. 3). These are shown in Eqs. 1 and 2, the radical anion being represented by the ion pair species **9**. Another pathway that has been fairly widely considered, but seems less likely than the other two, is shown in Eq. 3.

$$\underset{\mathbf{9}}{\underset{\beta\quad\alpha}{\dot{C}-C=C-O^-M^+}} \xrightarrow{e,M^+} \underset{\mathbf{11}}{\overset{M^+}{\bar{C}}-C=C-O^-M^+} \xrightarrow{H^+} \underset{\mathbf{2}}{\overset{H}{C}-C=C-O^-M^+} \qquad \text{(Eq. 1)}$$

$$\underset{\mathbf{9}}{\dot{C}-C=C-O^-M^+} \xrightarrow{H^+} \underset{\mathbf{12}}{\dot{C}-C=C-OH} \xrightarrow{e,M^+} \underset{\mathbf{13}}{\overset{M^+}{\bar{C}}-C=C-OH}$$

$$\mathbf{13} \xrightarrow{H^+} \underset{\mathbf{14}}{\overset{H}{C}-C=C-OH} \xrightarrow{B^-,M^+} \underset{\mathbf{2}}{\overset{H}{C}-C=C-O^-M^+} \qquad \text{(Eq. 2)}$$

$$\underset{\mathbf{9}}{\dot{C}-C=C-O^-M^+} \xrightarrow{H^+} \underset{\mathbf{15}}{\overset{H}{C}-C=C-\dot{O} + M^+} \xrightarrow{e} \underset{\mathbf{2}}{\overset{H}{C}-C=C-O^-M^+} \qquad \text{(Eq. 3)}$$

The route shown in Eq. 1 is similar to the original proposal (p. 8)[9] if one considers that the dianionic species arises by stepwise addition of electrons to the enone **1** and that the dianionic species undergoing β protonation, *i.e.*, **11**, is strongly coordinated with a metal cation at oxygen and, perhaps, associated in some way with a second metal cation at the β position. Additional work has provided evidence that stable dianionic species can be produced by treatment of aryl-substituted enones with metals in liquid ammonia.[50] Treatment of benzalacetophenone (**16**) with potassium in

[48] N. Hirota, in *Radical Ions*, E. T. Kaiser and L. Kevan, Eds., Interscience Publishers, New York, 1968, p. 57.

[49] M. Szwarc, *Accounts Chem. Res.*, **2**, 87 (1969).

[50] P. J. Hamrick and C. R. Hauser, *J. Amer. Chem. Soc.*, **81**, 493 (1959).

liquid ammonia followed by addition of 1 equivalent of benzyl chloride and then excess ammonium chloride gave the β-benzylated ketone **17** in 73% yield. On the other hand, addition of 2 equivalents of benzyl chloride to the reduction mixture gave the α,β-dibenzylated ketone **18** in 76% yield. The high yield of each of the products and the apparent absence of coupling

$$\underset{\mathbf{16}}{(H)(C_6H_5)C{=}C(H)C({=}O){-}C_6H_5} \xrightarrow[NH_3,Et_2O]{2\,K} \underset{\mathbf{19}}{\overset{K^+}{C_6H_5\bar{C}H}{-}CH{=}\overset{\bar{O}K^+}{C}{-}C_6H_5}$$

$$\mathbf{19} \xrightarrow[2.\ NH_4Cl]{1.\ 1\ eq.\ C_6H_5CH_2Cl} \underset{\mathbf{17}}{C_6H_5CH(CH_2C_6H_5)CH_2COC_6H_5}$$

$$\mathbf{19} \xrightarrow{2\ eq.\ C_6H_5CH_2Cl} \underset{\mathbf{18}}{C_6H_5CH(CH_2C_6H_5)CH(CH_2C_6H_5)COC_6H_5}$$

products derived from reaction of benzyl chloride with potassium in liquid ammonia[51] suggest that the dipotassium dianion **19** is formed in high concentration in the reaction mixture and that it undergoes mono-benzylation at the more nucleophilic β position to give the potassium enolate of **17**. The latter can either react with a proton donor to give the β-benzylated ketone **17** or with additional benzyl chloride to give the α,β-dibenzylated ketone **18**.

Recently, the preceding observations have been confirmed by studies involving the β-alkylation of benzalacetophenone (**16**) and some of its derivatives with a variety of alkylating agents.[52,53] The interesting observation was made that, unlike the dipotassium dianion, the related dilithium dianion is apparently rather rapidly protonated at the β position in liquid ammonia. Only α-alkylation products such as **20** were formed when benzalacetophenone and its derivatives were treated with 2 equivalents of lithium in liquid ammonia followed by addition of an alkylating agent.

$$\underset{\mathbf{16}}{C_6H_5CH{=}CHCOC_6H_5} \xrightarrow[2.\ RX^-]{1.\ 2\ Li/NH_3\cdot Et_2O} \underset{\mathbf{20}}{C_6H_5CH_2CHRCOC_6H_5}$$

The behavior of the dimetal dianions derived from benzalacetophenone appears to parallel that observed for the corresponding species derived

[51] C. B. Wooster and N. W. Mitchell, *J. Amer. Chem. Soc.*, **52**, 688 (1930).
[52] J. A. Gautier, M. Miocque, and J. P. Duclos, *Bull. Soc. Chim. Fr.*, **1969**, 4348.
[53] J. A. Gautier, M. Miocque, and J. P. Duclos, *Bull. Soc. Chim. Fr.*, **1969**, 4356.

$$(C_6H_5)_2C{=}O \xrightarrow[NH_3]{2\,M} M^{+}\,{}^{-}C(C_6H_5)_2{-}O^{-}M^{+} \xrightarrow{RX} R{-}C(C_6H_5)_2{-}O^{-}M^{+}$$

21 **22**

$$\mathbf{22} \xrightarrow{NH_3} H{-}C(C_6H_5)_2{-}O^{-}M^{+}$$

from benzophenone (**21**). It has been reported that the dipotassium dianion **22** (M = K) is rapidly formed and undergoes C-alkylation in liquid ammonia,[50, 54] but it has also been observed that, when lithium is substituted for potassium, C-alkylation products are not formed on addition of alkylating agents.[54] This suggests that dianion **22** (M = Li), like the related species derived from benzalacetophenone, undergoes protonation by ammonia.

While it seems clear that the formation of dianionic species such as **11** (p. 10) is possible for α,β-unsaturated systems having substituents capable of delocalizing negative charge, *e.g.*, benzalacetophenone, the question whether or not the reducing power of metal-ammonia solutions is sufficiently high to produce similar intermediates in reductions of simple alkyl-substituted enones has not been answered. Although the formation of free dianions is likely to be out of the range of the reduction potential of metal-ammonia solutions, it is possible that the formation of tight ion pairs (or perhaps covalent bonds) between the metal cation and oxygen in a species such as **9** could cause sufficient neutralization of negative charge to allow the addition of a second electron. Indeed, there is polarographic evidence to indicate that in reduction of aromatic ketones in aprotic media the second electron transfer step occurs at a less negative potential in the presence of metal cations.[55] The formation of tightly associated ion pairs should be more favorable when small cations such as lithium are employed[44, 56] and when the polarity of the medium is relatively low. The disproportionation of radical anions also provides a possible pathway for dianion formation.[30]

The second step in the reduction mechanism shown in Eq. 1 involves protonation at the β position of the dianionic intermediate **11**. There is excellent evidence that in reductions of alkyl substituted α,β-unsaturated

[54] W. S. Murphy and D. J. Buckley, *Tetrahedron Lett.*, **1969**, 2975.
[55] C. L. Perrin, *Progr. Phys. Org. Chem.*, **3**, 165 (1965).
[56] B. R. Eggins, *Chem. Commun.*, **1969**, 1267.

ketones the hydrogen introduced at the β position is derived from a proton and not a hydrogen atom donor and, in the absence of more acidic species, ammonia can serve as the proton donor. This is also true for reductions of aryl-substituted enones when lithium is employed as the metal.[52,53]. For example, the lithium-ammonia reduction of 2,2,6,6-tetramethyl-*trans*-4-hepten-3-one (**23**) and isophorone (**24**) was carried out in the presence of the deuterium atom donor 2-deuterio-2-propanol, but it was found that the reduction products contained no deuterium at the β position.[44,57] It had been shown earlier that, when 1(9)-octalin-2-one (**25**) was reduced

t-C_4H_9(H)C=C(H)COC_4H_9-t (**23**) $\xrightarrow[\text{2. } H_2O, CrO_3]{\text{1. } Li/NH_3, (CH_3)_2CDOH, -78°}$ t-$C_4H_9CH_2CH_2COC_4H_9$-t (57%)

24 $\xrightarrow[\text{2. } H_2O, CrO_3]{\text{1. } Li/NH_3, (CH_3)_2CDOH, -78°}$ (84%)

with lithium in ammonia, the ammonia was replaced by dry benzene, and deuterium oxide was added, only α-deuterated *trans*-2-decalone was isolated.[14]

25 $\xrightarrow{\text{1. } Li/NH_3;\ \text{2. } -NH_3, +C_6H_6;\ \text{3. } D_2O}$ (H, H, D)

Additional results favoring protonation at the β position by ammonia (or an amine) are provided by work that showed the reduction of several α,β-unsaturated ketones with lithium in dideuterated propylamine led to the introduction of deuterium at the β position.[58] Other evidence[59,60] involving the introduction of deuterium at the β position in reductions of unsaturated ketones with metals in trideuterated ammonia also bears upon this point, but is somewhat less rigorous since in at least one reaction the trideuterated ammonia probably contained deuterium oxide.[60]

[57] H. O. House, R. W. Giese, K. Kronberger, J. P. Kaplan, and J. P. Simeone, *J. Amer. Chem. Soc.*, **92,** 2800 (1970).

[58] M. Fetizon and J. Gore, *Tetrahedron Lett.*, **1966,** 471.

[59] J. Karliner, H. Budzikiewicz, and C. Djerassi, *J. Amer. Chem. Soc.*, **87,** 580 (1965).

[60] D. H. Williams, J. M. Wilson, H. Budzikiewicz, and C. Djerassi, *J. Amer. Chem. Soc.*, **85,** 2091 (1963).

The observation that ammonia, a relatively weakly acidic substance, can serve as the proton donor in metal-ammonia reductions of α,β-unsaturated systems suggests that the β position becomes highly basic during the reduction. Indeed, the available evidence concerning the stereochemistry of metal-ammonia reductions of fused-ring α,β-unsaturated ketones, which is discussed in detail later, suggests the involvement of tetrahedral β-carbanionic intermediates[20,57,61,62] and that these species undergo rapid protonation with retention of configuration.[20,57,63,64] Several groups have pointed out that a dianionic species such as **11** (p. 10), derived from an alkyl-substituted enone, should have a highly basic β-carbon atom.[9,10,14,41]

It has been suggested that in a species such as dianion **11** there should be a relatively small amount of overlap between the electron pair at the β position and the adjacent enolate system which already bears a negative charge.[*,41] To the extent that this is true the basicity of this carbanionic center should approach that of an alkyl carbanion, which should remove a proton rapidly from ammonia.[65] It also seems likely that the acidity of ammonia might be significantly increased by coordination with metal cations, particularly, Li^+, Mg^{2+}, and Ca^{2+}.[42,47] In fact, it has been argued that calcium-ammonia and magnesium-ammonia solutions are sufficiently acidic to bring about protonation of the corresponding metal enolates.[42,43] The very low solubility of lithium and calcium amides may also be, in part, responsible for the proton-donating ability of ammonia solutions of these metals as compared with solutions of metals such as potassium and sodium.[66]

On the basis of thorough studies on reduction of aliphatic unsaturated ketones, both electrochemically and with alkali metals under various conditions, a general mechanism for these reactions has been proposed, which is considered to be applicable to metal-ammonia reduction.[44,57] With the appropriate modification to allow for the formation of the enolate **3**, the proposed pathway is outlined in Eq. 2 (p. 10). In this case the initially formed radical anion **9**, rather than adding a second electron to form the dianion **11**, undergoes rate-limiting protonation to give

* This point was reinforced by molecular orbital calculations which revealed that a relatively small loss in π delocalization energy resulted from the localization of the unshared pair at the terminal position in a dianionic species such as **11**.[41]

[61] D. J. Cram, *Fundamentals of Carbanion Chemistry*, Academic Press, New York, 1965, pp. 47–84, 175–210.

[62] P. E. Verkade, K. S. DeVries, and B. M. Wepster, *Rec. Trav. Chim. Pays-Bas*, **84**, 1295 (1965).

[63] S. D. Darling, O. N. Devgan, and R. E. Cosgrove, *J. Amer. Chem. Soc.*, **92**, 696 (1970).

[64] W. H. Glaze, C. M. Selman, A. L. Ball, and L. E. Bray, *J. Org. Chem.*, **34**, 641 (1969).

[65] Ref. 61, Cram, p. 14.

[66] R. G. Harvey, *J. Org. Chem.*, **36**, 3306 (1971).

the hydroxyallyl radical **12**. This step is followed by rapid addition of a second electron to give a hydroxyallyl anion **13** which undergoes protonation (by ammonia or other proton donors) to give enol **14**. Transfer of a proton from this enol to a base before ketonization would lead to the specific enolate **2**.

The suggested rate-limiting protonation of radical anion **9** is based upon several findings. First, stable radical anions could be generated electrochemically from aliphatic enones in aprotic solvents such as dimethylformamide only if no hydrogen atoms were present at the α' and γ positions; this observation has been confirmed in work involving alicyclic systems.[46] Second, the relatively stable radical anion **26** (derived from the enone **23**) was found to be rapidly converted into the racemic dihydro dimer **27** in the presence of proton donors such as alcohols or in the presence of lithium cations. Neutralization of the negative charge on the oxygen of radical anion **26** either by protonation or tight ion pair forma-

t-C_4H_9, H, $\dot{C}$—C, H, C—C_4H_9-t, ^-O

26

t-C_4H_9, $CH_2COC_4H_9$-t, H, H, $CH_2COC_4H_9$-t, t-C_4H_9

27

tion with a lithium cation apparently allows rapid dimerization; in the latter case the process probably takes place via formation of an ion pair dimer analogous to **10** (p. 9).[44,48] Third, in reduction with metals in which proton donors were excluded, *i.e.*, conditions favoring long-lived radical anions, β,β dimerization of ion pairs was favored over reduction. However, under conditions which favor formation of a species such as the hydroxyallyl radical **12**, simple reduction and β,β dimerization were competitive processes. Also, in support of the mechanism outlined in Eq. 2, it has been noted that metal-ammonia reductions of unsaturated ketones conducted in the absence of at least 1 equivalent of a proton donor frequently led to the recovery of significant amounts of the starting material and to the formation of by-products of high molecular weight.[44,47] Processes that could account for these results are transfer of an acidic proton (α'- or γ-) from the enone to the radical anion **9** to produce the hydroxyallyl radical **12** and enolate **28** or **29**, or formation of aldol or Michael products from the reaction of these species and the starting material, or possible dimerization reactions of **9**. However, it has been widely observed that in many reactions high yields of simple reduction products are obtained in metal-ammonia reductions under conditions in which a proton donor is not added until the enolate stage is reached; and,

$$9 + \text{H}-\underset{\gamma}{\text{C}}-\text{C}=\text{C}-\text{C}(=\text{O})-\underset{\alpha'}{\text{C}}-\text{H} \rightarrow 12 + \text{H}-\text{C}-\text{C}=\text{C}-\text{C}(\text{O}^-)=\text{C}$$

28

or

$$\text{C}=\text{C}-\text{C}=\text{C}(\text{O}^-)-\text{C}-\text{H}$$

29

in the reduction-enolate alkylation sequence to be discussed later, products of alkylation of conjugate enolates of the starting material have been observed rarely.[67] In explanation it has been suggested that ammonia itself supplies the proton required for the conversion of the radical anion **9** into the hydroxyallyl radical **12**.[19] A high kinetic basicity of oxygen in the radical anion together with a high kinetic acidity of ammonia would favor the proton transfer.

At present there appears to be no firm evidence to allow one to rule in favor of the exclusive involvement of the mechanisms represented in either Eq. 1 or Eq. 2 in metal-ammonia reductions, and it appears quite likely that in most cases the two pathways are competitive. Formation of hydroxyallyl radicals such as **12** from radical anions such as **9** would be expected to be favored when a stronger proton donor than ammonia is present in the medium in significant concentration. Conditions under which proton donors other than ammonia are rigorously excluded and which favor tight ion pair formation (*i.e.*, relatively nonpolar co-solvents and the presence of small cations such as lithium) should favor the formation of dianionic species such as **11**.

A third mechanism which has been considered as a possibility for conversion of radical anions into metal enolates is outlined in Eq. 3 (p. 10).[10] This involves protonation of the radical anion **9** at the β position to produce the enolate radical **15** which adds an electron to give the enolate **3**. This pathway appears rather unlikely. Although no data are available on the basicity of species such as **9**, it seems unlikely that such intermediates would be sufficiently basic to abstract a proton from ammonia. The basicity of such radical anions would be expected to be roughly comparable to that of metal enolates like **3**, which are stable in liquid ammonia, at least when alkali metals are involved. Also, existing data[44,57] suggest that in the presence of relatively acidic proton donors radical anions, such as **9**

[67] M. J. Weiss, R. E. Schaub, G. R. Allen, Jr., J. F. Poletta, C. Pidacks, R. B. Conrow and C. J. Coscia, *Tetrahedron*, **20**, 367 (1964); *Chem. Ind.* (London), **1963**, 118.

undergo protonation preferentially on oxygen—a result that is expected in view of the predicted high electron density on this atom and the greater exothermal character of hydrogen-oxygen as compared with hydrogen-carbon bond formation.[19]

As noted in the introduction, α,β-unsaturated esters and acids are reduced to the corresponding α,β-dihydro derivatives by metals in liquid ammonia. The literature contains essentially no information pertaining specifically to the mechanistic details of reduction of these systems. It seems likely, however, that after formation of an initial radical anion, pathways analogous to those described in Eqs. 1 and 2 are involved. With α,β-unsaturated acids, carboxylate anion formation may well precede the initial electron transfer step.

α,β-Unsaturated aldehydes also presumably undergo metal-ammonia reduction by a mechanism similar to that of unsaturated ketones. The use of metal-ammonia solutions for the reduction of α,β-unsaturated aldehydes has been quite limited. Indeed, it appears that only a few examples of metal-ammonia reductions of these compounds have been reported in the literature. Aldehydes show a strong tendency to undergo imine formation in liquid ammonia.[23, 24] This feature probably had discouraged the use of metal-ammonia solutions for their reduction, but it is not clear that careful attempts have been made to find optimum conditions for converting α,β-unsaturated aldehydes into the corresponding dihydro derivatives by this method.

Intramolecular Processes Involving Reactive Intermediates

Metal-ammonia reductions of unsaturated ketones involve intermediates having carbanionic character at the β position. Therefore, as might be expected, intramolecular displacements, additions, and eliminations occur during the reduction of appropriate enones. For example, it was observed that the unsaturated keto tosylate **30**, when treated with lithium in liquid ammonia-tetrahydrofuran followed by addition of ammonium chloride, gave the tricyclic ketone **31**.[68] The conversion of the keto enone **32** into the cyclopropanol **33**[69] and of the diacetoxy enone **34** into the unconjugated enone **35**[70] are examples of reactions involving intramolecular addition and elimination, respectively. (Equations on p. 18).

Note that in the reduction of the enone **30**, displacement of the leaving group might take place at the radical anion or the dianion stage of the reaction.[68] The β position of a radical anion such as **9** (p. 9), though possibly

[68] G. Stork and J. Tsuji, *J. Amer. Chem. Soc.*, **83**, 2783 (1961).

[69] P. S. Venkataramani, J. E. Karoglan, and W. Reusch, *J. Amer. Chem. Soc.*, **93**, 269 (1971).

[70] T. A. Spencer, K. K. Schmiegel, and W. W. Schmiegel, *J. Org. Chem.*, **30**, 1626 (1965).

CH_2OTs

1. NH_3/Li, THF
2. NH_4Cl

30 → 31

1. Li/NH_3, ether
2. NH_4Cl

32 → 33

OAc OAc

1. Li/NH_3, THF
2. NH_4Cl

34 → 35

incapable of being protonated, is perhaps sufficiently nucleophilic to take part in intramolecular displacement as well as addition and elimination reactions in suitably arranged systems. However, as yet there seems to be no strong evidence to allow one to decide which of three possible reduction intermediates, the anion radical **9**, the dianion **11**, or the hydroxyallyl anion **13**, might be involved in these reactions. Indeed, the exact species may differ depending upon the structure of the system and the reaction conditions.

Dimerization Processes

Because of the intervention of radical anions such as **9** and/or hydroxyallyl radicals such as **12** in metal-ammonia reductions of conjugated carbonyl systems (see p. 10), dimerization processes involving these species may compete with simple reduction. The radical anion **9** may be used to illustrate the three types of dimerization products that may be produced. 1,6-Diketones such as **36** may be formed from coupling at the β position of two radical anions; unsaturated pinacols (**37**) may be produced if coupling occurs at the carbonyl carbon atoms; and unsaturated γ-hydroxy ketones such as **38** may arise if the β-carbon atom of one radical anion and the carbonyl carbon atom of another undergo coupling. Similar processes involving the hydroxyallyl radical **12** could lead directly to **36**–**38**. Also, pathways involving 1,4-addition of radical or charged species such as **9**,

9

36

37

38

11, **12**, or **13** to the starting unsaturated system followed by the appropriate electron and/or proton transfer steps would lead to the diketone **36**.

In general the metal-ammonia reduction of conjugated carbonyl systems has been applied when synthesis of the simple reduction product was desired, and relatively little attention has been devoted to determining the structures of the by-products of high molecular weight that are sometimes formed in these reactions. Although dimerization products of all three types have been observed in reductions with metals in other media,[71] only reports involving the isolation of 1,6-diketones and unsaturated pinacols in metal-ammonia reductions have appeared. For example, a small amount of the β,β-dimer **39** was isolated from the lithium ammonia reduction of isophorone,[72] and unsaturated pinacols have been obtained in

39

[71] J. Wiemann, S. Risse, and P.-F. Casals, *Bull. Soc. Chim. Fr.*, **1966**, 381.

[72] H. A. Smith, B. J. L. Huff, W. J. Powers, and D. Caine, *J. Org. Chem.*, **32**, 2851 (1967).

lithium-ammonia reductions of cholest-4-en-3-one[4, 73, 74] and the keto enone **5** (p. 4).[4, 73] In the last two reactions, dimerization was a serious problem when proton donors were absent.[4, 73]

Dimerization products of the types **36–38** are generally the major ones obtained in electrochemical reductions[44, 55, 75, 76] and in reductions at metal surfaces[44, 71] for which species such as the radical anion **9** or the hydroxyallyl radical **12** must diffuse to a surface for further electron transfer to take place. However, in metal-ammonia solutions simple reduction is generally favored over dimerization. These solutions provide high concentrations of electrons; therefore, collisions between species such as the radical anion **9** and an electron are favored. If the addition of a second electron to the radical anion **9** is slow, addition of proton donors will convert this species into the hydroxyallyl radical **12**, which should react rapidly with dissolved electrons. Also, the cation- and anion-solvating power of ammonia tends to prevent the formation of ion pair dimers such as **10** (p. 9) which are likely to be involved in dimerization processes.[44] For this reason, dimerization reactions usually become more important when high concentrations of nonpolar co-solvents are employed in metal-ammonia reductions.[77]

Stereochemistry of Metal-Ammonia Reductions

One of the most intriguing aspects of the mechanism of metal-ammonia reductions of α,β-unsaturated carbonyl systems is the stereochemical outcome of the β-protonation step. In a review of the results of reduction of a number of unsaturated steroid and terpene ketones, it was observed that, when the β position of the conjugated system was located at a junction of six-membered rings, the more thermodynamically stable reduction product was usually formed.[9] The reduction has been viewed as proceeding via a free dianionic intermediate, *i.e.*, **7** (p. 8). Although such a species should have only transient existence, it was considered that the β carbanion could achieve a definite, though easily invertible, tetrahedral configuration in which the lone pair occupied an orbital approximately the size of a C—H bond. It was considered that the more stable product would be formed as a result of protonation of the carbanion in its more stable configuration.

Later, using examples of reduction in the octalone series, it was demonstrated that the more stable of the two possible reduction products is not

[73] J. S. Jellinek, Ph.D. Dissertation, University of Wisconsin, 1955.

[74] T. A. Spencer, unpublished work; see ref. 20, House, footnote 72c.

[75] M. M. Baizer and J. P. Petrosrich, *Advan. Phy. Org. Chem.*, **7,** 189 (1970).

[76] D. Miller, L. Mandell, and R. A. Day, Jr., *J. Org. Chem.*, **36,** 1683 (1971).

[77] J. Fried and N. A. Abraham, *Tetrahedron Lett.*, **1964,** 1879.

always obtained.[10] For example, reduction of the octalone **40** with lithium-ammonia-ethanol followed by oxidation with chromic acid afforded the *trans*-2-decalone **41**; by contrast, it was pointed out that the *cis*-2-decalone related to **41** should be about 2 kcal/mol more stable than the *trans* isomer.

OCH_3 H H 1 Li/NH_3,Et_2O-EtOH 2. CrO_3 OCH_3 H H O O H

40 **41**

This result and related observations led to the proposal that, in reductions of octalones of the type **42**, only two (**43** and **44**) of three possible protonation transition states involving a half-chair conformation of ring A would be stereoelectronically allowed; in these two conformations the orbital of the developing C—H bond overlaps with the remainder of the π system of the enolate.[10] The alternative *cis* conformer **45** would not be allowed because it does not fulfill the overlap requirement. Thus, when the substituents R_2 and/or R_3 are larger than hydrogen, the *trans* transition state **43** would be more stable than the *cis* **44**, and the trans-2-decalone reduction product would be obtained in spite of the fact that the *cis* isomer having a conformation related to **45** should be more stable. (Formula on p. 22).

It was considered that β protonation might take place at the radical anion or dianion stage of the reduction.[10] However, in light of the foregoing discussion it seems likely that protonation occurs at either the dianion or the hydroxyallyl anion (*cf.* **13**, p. 10) stage. For simplicity of illustration in the subsequent discussion the species undergoing β protonation is represented as the dianion (*cf.* **11**).

On the basis of the preceding considerations, it was pointed out that, in order to predict the stereochemistry of a metal-ammonia reduction in the octalone series, one should consider the relative energies of the two isomers of the reduction product having a new C—H bond axial to the ketone-containing ring. The more stable of these, whether of the *cis* or *trans* configuration, would be expected to be the major product of the reduction.[10] Although this rule of "axial protonation" has been found to be widely applicable to metal-ammonia reductions of octalones, steroids, and other fused-ring systems, accurate predictions of the stereochemical outcome of reductions in complex systems are difficult for two reasons. First, the stereoselectivity of the reaction is often much greater than would be

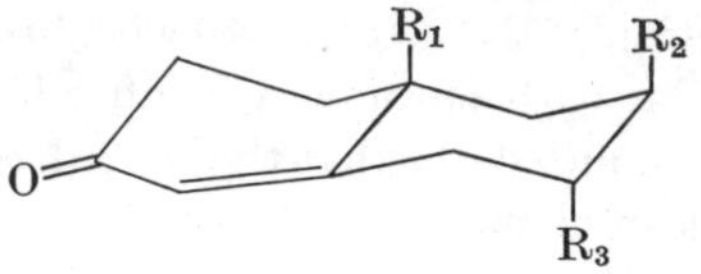

42

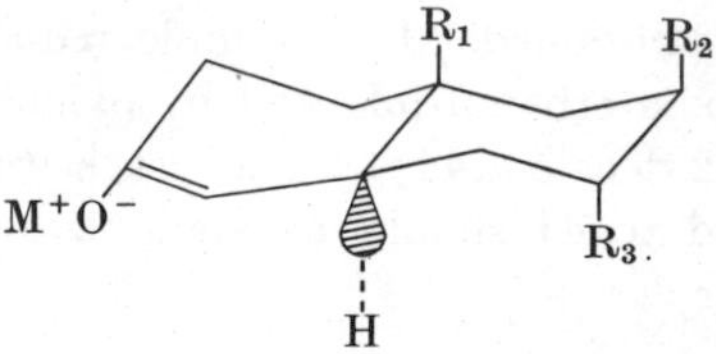

43

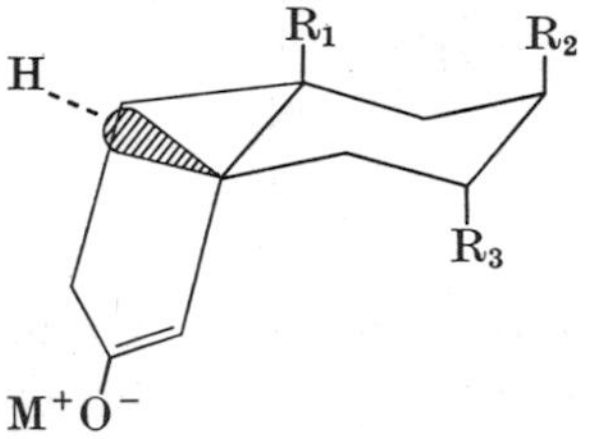

44

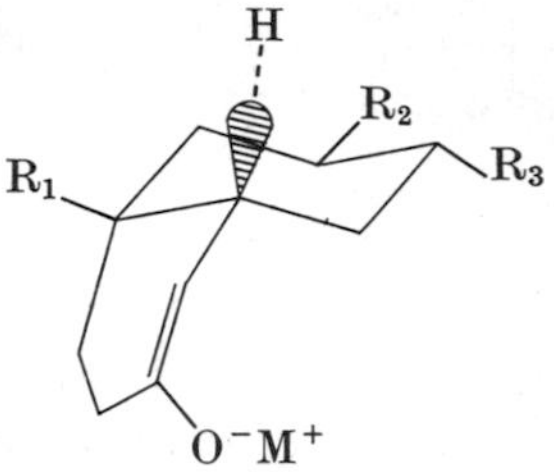

45

predicted from simple analysis of nonbonded interactions in the two stereoelectronically allowed reduction products; and, second, in systems in which a significant amount of strain must be introduced in order for protonation to occur axially, transition states resembling **45** in which the new C—H bond forms quasi-equatorially to the enolate ring may become important.

In connection with the first point, it may be noted that reductions of many 1(9)-octalin-2-ones yield *trans* products with a high degree of stereoselectivity.[78] For example, it was pointed out that 1(9)-octalin-2-one

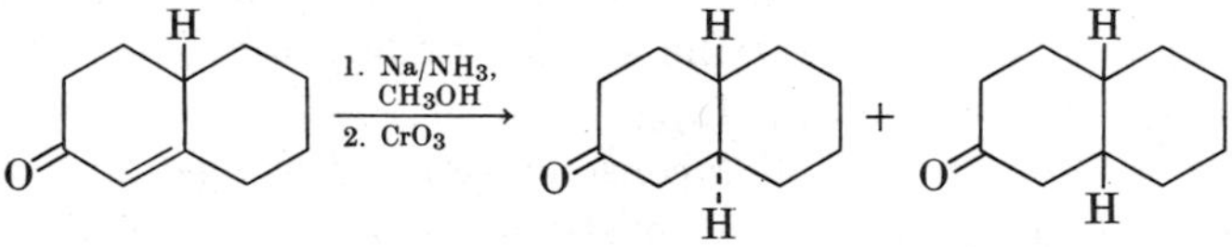

1. Na/NH_3, CH_3OH 2. CrO_3

25 → **46** (99%) + **47** (1%)

(**25**) yielded a 99/1 mixture of the *trans*- and *cis*-decalones **46** and **47** on reduction with sodium in liquid ammonia, whereas analysis of nonbonded interactions in the corresponding 1(2)-enolates **48** and **49** indicated that the

[78] M. J. T. Robinson, *Tetrahedron*, **21**, 2475 (1965).

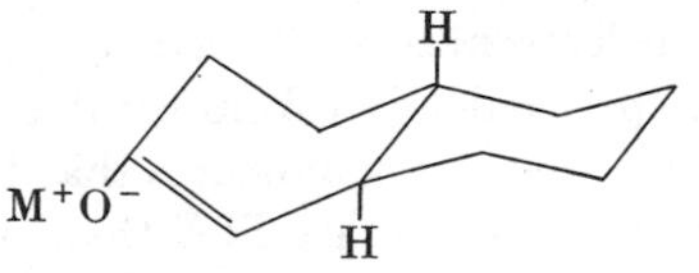

48 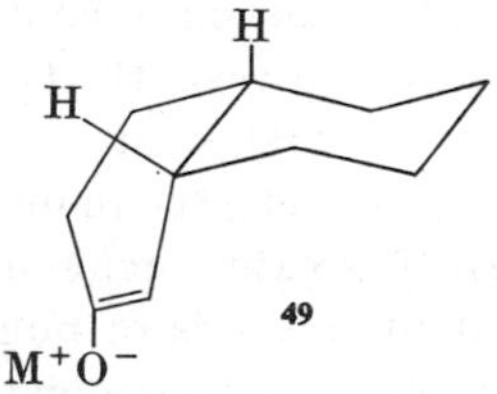 49

former should be favored only by about 1.0 kcal/mol, which would correspond to an approximately 80/20 *trans*/*cis* ratio.

Small variations in the *trans*/*cis* product ratio were observed when methyl substituents were present at the angular position and at various positions on ring B of the parent octalone, and it was suggested that the stereoselectivity of the reduction might be explained by assuming that the β-carbon atom is trigonal in the transition state for protonation.[78] Substituents were considered to influence the reduction stereochemistry by causing small changes in the position of the equilibrium involving the two half-chair conformations, **50A** (* = —) and **50B** (* = —), of ring A of the species undergoing protonation. However, it does not appear that a protonation transition state involving a trigonal β-carbon atom can adequately account for the high preference for formation of *trans* products in reductions of enones such as **25**. If a trigonal β carbon were involved, the stereochemistry of the reduction would be expected to be controlled largely by steric hindrance associated with the approach of the proton donor. Yet examination of models of conformations **50A** and **50B** reveals that for both species approach of the proton donor from the topside leading to a *cis* product should be at least as favorable as bottomside approach leading to a *trans* product. (Except in extreme cases[79] the stereochemistry of metal ammonia reductions does not appear to be significantly influenced by steric hindrance associated with the approach of the proton donor.[23])

The evidence concerning the stereoselectivity of metal-ammonia reductions is consistent with the view that tetrahedral β-carbanionic intermediates are involved[20, 57, 61, 62] and that protonation of the β carbanion

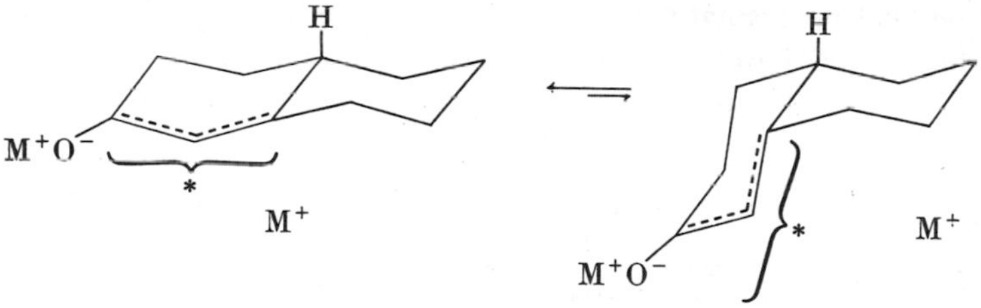

50A 50B

[79] R. E. Ireland and U. Hengartner, *J. Amer. Chem. Soc.*, **94**, 3653 (1972).

occurs with retention of configuration.[20, 57, 63, 64] However, if equilibrium is established between the two stereoelectronically allowed configurations (*cf.* **51A** and **51B**) of the dianionic species derived from octalone **25** and related compounds, the reduction stereochemistry indicates that there is a significantly greater preference for the *trans* species **51A** than would be predicted by analysis of nonbonded interactions. It is conceivable that charge repulsions and solvation factors could influence the position of equilibrium between dianions **51A** and **51B** in favor of the *trans* species **51A**.[80] However, it seems more likely that there is a kinetic preference for formation of dianion **51A** from the radical anion intermediate, and that this species undergoes protonation more rapidly than equilibration.

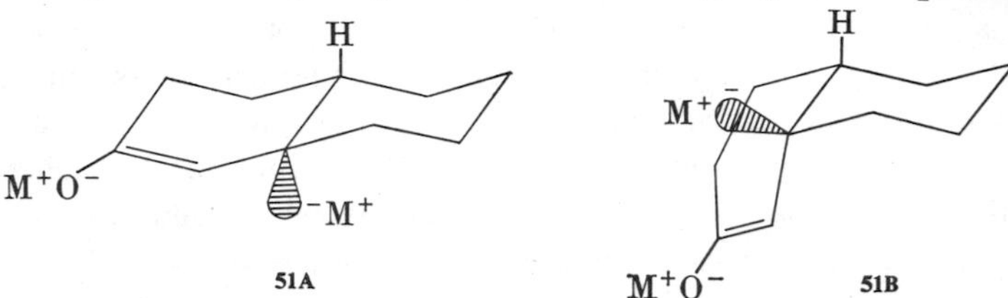

The suggestion that β protonation of species such as **51A** or **51B** is more rapid than equilibration in metal-ammonia reductions is supported in two ways. First, the reduction stereochemistry apparently is not affected by the presence or concentration of added proton donors.[19, 78] To the extent that the β carbanion has highly basic character resembling an alkyl carbanion,[9, 10, 14, 41] it would be expected to undergo protonation by relatively acidic proton donors such as alcohols at a diffusion-controlled rate.[61] Since the reduction stereochemistry seems to be unchanged when ammonia is the only proton donor, β protonation by ammonia or a metal cation-ammonia complex also may be quite rapid. On the other hand, the magnitudes of carbanion inversion energy barriers that have been calculated[81–83] and experimentally determined[84] indicate that inversion may be slower than a diffusion-controlled process. (The formation of cholestane from lithium-ammonia reduction of both 5α-chlorocholestane and 5β,6α-dibromocholestane often has been cited[9, 23, 41] as evidence that carbanionic intermediates undergo equilibration more rapidly than protonation in metal-ammonia reductions. However, recent evidence[85–87] on

[80] S. Y. Wong, Ph.D. Dissertation, University of Southern California, 1968; *Diss. Abstr. B*, **29**(10), 367 (1969).

[81] M. J. S. Dewar and M. Shanshal, *J. Amer. Chem. Soc.*, **91**, 3654 (1969).

[82] G. W. Koeppl, D. S. Sagatys, G. S. Krishnamurthy, and S. J. Miller, *J. Amer. Chem. Soc.*, **89**, 3396 (1967).

[83] M. Shanshal, *Z. Naturforsch. B*, **25**, 1065 (1970).

[84] M. Witanowski and J. D. Roberts, *J. Amer. Chem. Soc.*, **88**, 737 (1966).

[85] J. Jacobus and J. F. Eastham, *Chem. Commun.*, **1969**, 138.

[86] J. Jacobus and D. Pensak, *Chem. Commun.*, **1969**, 400.

[87] H. M. Walborsky, F. P. Johnson, and J. B. Pierce, *J. Amer. Chem. Soc.*, **90**, 5222 (1968).

the mechanism of metal-ammonia reductions of alkyl halides suggests that in the reduction of both of the above-mentioned halides a common radical intermediate may be formed which may adopt a configuration similar to that of cholestane and then undergo rapid addition of an electron and a proton with retention of configuration.)

Secondly, the octalone **25** has been reduced under conditions considered to favor establishment of equilibrium between dianions **51A** and **51B**.[63] By treating the enones with solutions of metals in various solvents containing trimesitylboron and adding proton donors after an extended time, mixtures of **46** and **47** containing significantly more of the *cis* isomer than is normally produced in metal-ammonia reductions were obtained. Although carbanion-metal association and solvation factors involved in these reactions may be quite different from those in metal-ammonia reductions, the *cis*/*trans* ratios agreed reasonably well with the calculated value[78] and an experimentally determined value[88] for the thermodynamic *cis*/*trans* ratio for the 1(9)-octalin system.

The stereoselective formation of *trans*-decalones in many octalone reductions is possibly explained by assuming that most radical anion intermediates strongly prefer conformation **50A** (* = ·). On addition of the second electron, torsional strain energy involving movement of atoms and energy changes associated with reorganization of the solvent shell would be minimized if the carbanion developed in the *trans* configuration **51A**.[89] Very rapid protonation of **51A** would lead to the observed *trans*-fused products.

Recent studies[90, 91] on reductions of octalones having 4α-methyl,[90, 91] ethyl,[91] and isopropyl[91] substituents are of interest. Here, because the two stereoelectronically allowed configurations of the dianion intermediates analogous to **51A** and **51B** appear quite close in energy, significant amounts of the *cis*-decalone reduction products would be expected. While this was found to be true for the methyl and ethyl compounds there was still a significant preference for the *trans* products.[90,91] However, reduction of 4α-isopropyl-1(9)-octalin-2-one yielded a 99/1 mixture of the corresponding *cis*- and *trans*-fused decalones.[91] It seems likely that in the latter case a conformation analogous to **50B** (* = ·) is strongly preferred by the radical anion intermediate. If this were so, then torsional strain and solvation energy changes would be minimized, assuming a *cis* dianion analogous

[88] S. K. Malhotra, D. F. Moakley, and F. Johnson, *J. Amer. Chem. Soc.*, **89**, 2794 (1967).

[89] R. E. Cosgrove, II, Ph.D. Dissertation, University of Southern California, 1968; *Diss. Abstr. B*, **29**(10), 3671 (1969).

[90] R. M. Coates and J. E. Shaw, *Chem. Commun.*, **1968**, 47; *J. Amer. Chem. Soc.*, **92**, 5657 (1970).

[91] E. Piers, W. M. Phillips-Johnson, and C. Berger, *Tetrahedron Lett.*, **1972**, 2915.

to **51B** developed initially on addition of a second electron. The finding that 10β-methyl-4α-isopropyl-1(9)-octalin-2-one yielded mainly the corresponding *trans*-decalone[91] is consistent with this idea. Because of the *gauche* methylisopropyl interaction in the radical anion analogous to **50B** (* = ·), it would be expected to be destabilized relative to the species in which the alkyl groups are *anti*.

The rule of "axial protonation" appears to be violated in reductions of 1(9)-octalin-2-ones having exceptionally bulky 6β and 7α substituents. For example, enone **52**, having a 6β-*t*-butyl group, yields almost exclusively the *cis*-decalone on reduction with sodium in liquid ammonia.[92] (In contrast, the enone related to **52** having a 6α-*t*-butyl group gives mainly the corresponding *trans*-decalone as expected.[92]) In the radical anion derived from **52**, ring B probably adopts a boat conformation in order to relieve the strong 1,3-diaxial methyl-*t*-butyl interaction that would be present in the chair form. On addition of the second electron, formation of either of the two stereoelectronically allowed dianion intermediates, *i.e.*, **53A** or **53B**, requires that ring B remain in a boat conformation. If these two configurations of the dianion were the only ones considered, the *trans* reduction product would be expected since **53A** should be much more stable than **53B**. However, if the β carbanion develops in conformation **53C**, in which it is quasi-equatorial with respect to ring A, both rings A and B can take up chair conformations with a minimum amount of motion. Protonation of the β carbanion in conformation **53C** would lead to the observed *cis* product. The formation of *cis*-fused products on

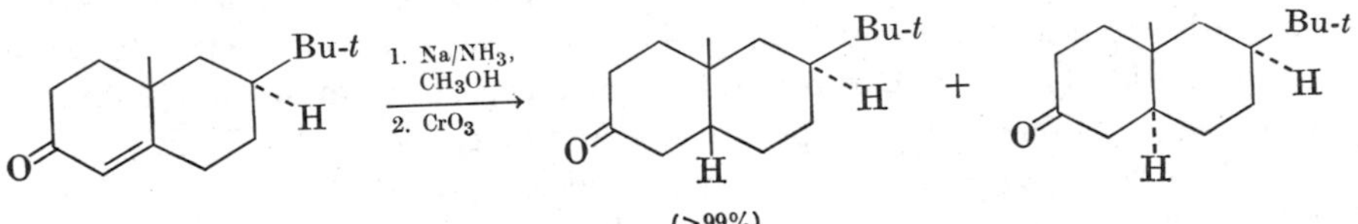

Bu-*t* H O **52** — 1. Na/NH_3, CH_3OH 2. CrO_3 → Bu-*t* H O H (>99%) + Bu-*t* H O H

NaO$^-$ (Na$^+$) Bu-*t* $^-$ Na$^+$ **53A**

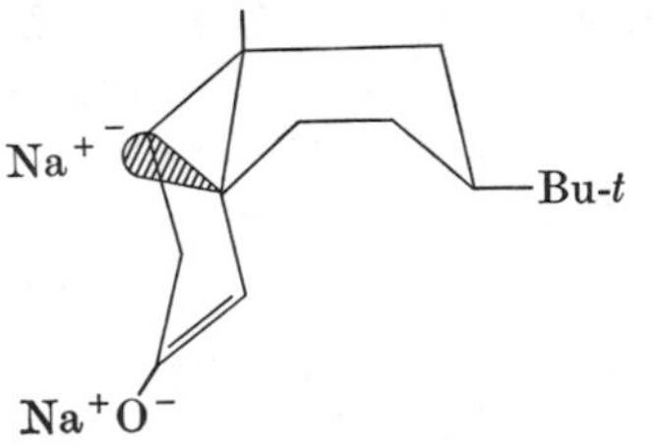

Na$^+$ $^-$ Bu-*t* Na$^+$O$^-$ **53B**

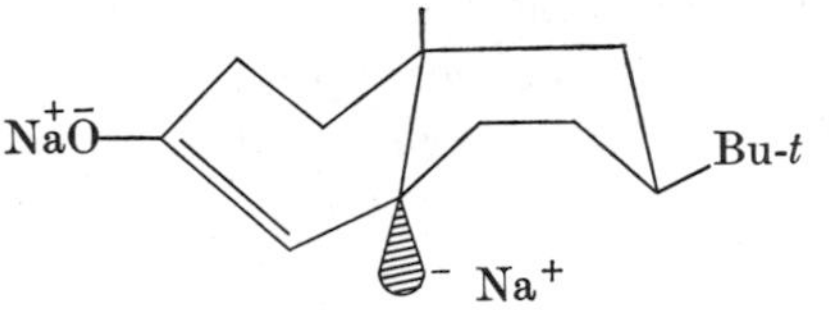

[92] M. J. T. Robinson, personal communication.

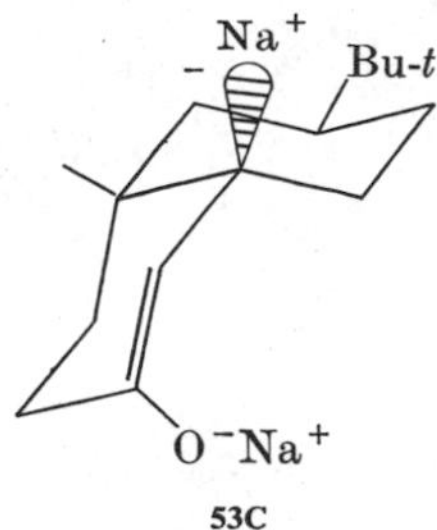

53C

lithium-ammonia reduction of tricyclic enones related to **52** but having a 6β, 7β-dimethyl-substituted carbon bridge can be explained similarly.[92a]

1(9)-Octalin-2-ones like **54a**, **54b**,[93] and **54c**[94, 95] (p. 27) give mainly *trans*-decalones on metal-ammonia reduction. However, reductions of the enones **54d** and **54e** have been found to give mixtures of the corresponding *cis* and *trans* isomers in 2/3 and 5/1 ratios, respectively.[96] The latter systems seem to be examples of cases in which protonation of the more stable stereoelectronically allowed dianion intermediate **55A** and the sterically favored one **(55B)** are competitive. Again, the bulky 7α-$C(CH_3)_2OH$ and $-C(CH_3)_2OAc$ groups probably cause ring B of the radical anions derived from **54d** and **54e** to prefer a boat conformation. Interestingly, the orbital overlap requirement seems to be sufficiently important to control the reduction stereochemistry when 7α-isopropenyl or isopropyl groups are present (cf. **54a–c**). The product ratios obtained in reductions of octalones such as **40** (p. 21), **52**, and **54** and rough estimates of energy differences between the more stable stereoelectronically allowed and non-allowed dianion intermediates derived from these enones indicate that the stereoelectronic factor may provide a stabilization energy of 3–5 kcal/mol.

The conversion of octalone **56** to the *trans*-decalone **57** provides another interesting example of a reaction from which the product having the less stable configuration at the β position (C–10) is obtained.[97] If the allowed configurations **58A** and **58B** of the dianion intermediate are considered, the latter appears to be more stable since $A^{1,3}$-strain involving the equatorial phenyl group and the $-O^-M^+$ ion pair (or —OH group) would be severe in the former.

[92a] F. Fringuelli, A. Taticchi, F. Fernandez, D. N. Kirk, and M. Scopes, *J. Chem. Soc., Perkin I*, **1974**, 1103.

[93] R. Howe and F. J. McQuillin, *J. Chem. Soc.*, **1956**, 2670.

[94] C. Djerassi, J. Burakevich, J. W. Chamberlin, D. Elad, T. Toda, and G. Stork, *J. Amer. Chem. Soc.*, **86**, 465 (1964).

[95] B. J. L. Huff, Ph.D. Dissertation, Georgia Institute of Technology, 1969; *Diss. Abstr. B*, **29** (12), 4589 (1969).

[96] P. Deslongchamps, personal communication.

[97] H. O. House and H. W. Thompson, *J. Org. Chem.* **28**, 360 (1963).

54

a, R_1 = isopropenyl; $R_2 = CH_3$
b, R_1 = isopropyl; $R_2 = CH_3$
c, R_1 = isopropyl; R_2 = H
d, $R = C(CH_3)_2OH$; $R_2 = CH_3$
e, $R = C(CH_3)_2OAc$; $R_2 = CH_3$

55A **55B**

1. Li/NH_3
2. NH_4Cl

56 **57**

58A **58B**

Lithium-ammonia reduction of (±)-5-*epi*-4-demethylaristolone **(59)** gives the *cis*-fused product **60**.[98] The formation of the stereoelectronically allowed *trans* dianion intermediate would introduce a severe nonbonded interaction between the angular methyl group and the β-methyl group of the cyclopropane ring; this interaction would not be present in the allowed *cis* dianion intermediate.

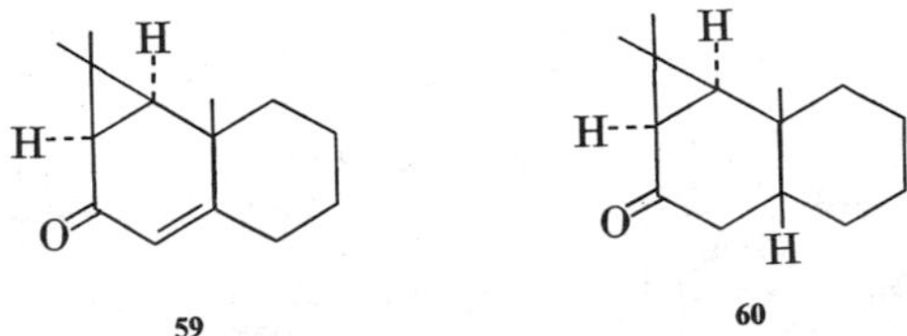

59 **60**

[98] E. Piers, W. deWaal, and R. W. Britton, *Chem. Commun.*, **1968,** 188; *Can. J. Chem.*, **47,** 4299 (1969).

Like 1(9)-octalin-2-ones, steroidal 4-en-3-ones are usually reduced to 5α products by metals in ammonia.[9, 18, 19, 23] However, in the reductions of 2α-cyano-2β-methyl-4-cholesten-3-one (**61**)[99] and the 1,2β-methylene steroid **62**,[100] the respective 5β products predominate. In each case it appears that the allowed β configuration of the C–5 carbanion which leads to the *cis* product is more stable than the alternative α configuration which would give the *trans* product.

C_8H_{17} NC O H H H

61

OAc H H O H H H

62

Recent evidence indicates that in 10α steroids the *cis*-A/B ring fusion is more stable than the *trans*.[101] Thus the reduction of the 10αH-4-en-3-one **63** to a 3/1 mixture of the corresponding A/B *trans* and *cis* products provides an example in the steroid series in which the stereoelectronic factor favors formation of the thermodynamically less stable product.[102–104] An example in which steric factors apparently override the overlap requirement is found in the reduction of the lumisterol derivative **64** which leads to the thermodynamically more stable *cis* dihydro ketone **65**.[105] The factors governing the stereochemistry of this reduction have been discussed in detail elsewhere.[10, 106, 107] In reductions of steroidal 8-en-11-ones and 8-en-7-ones, product stability and stereoelectronic factors favor the formation of "normal" *trans-anti-trans*-fused products.[5,6,18] However, it has been reported that the related resin acid enones **66** and **67** yield mainly the corresponding *trans-anti-cis* and *trans-syn-cis* products on lithium-ammonia reduction.[108] (Formulas **63–67** are on p. 30).

[99] P. Beak and T. L. Chaffin, *J. Org. Chem.*, **35**, 2275 (1970).

[100] R. Wiechert, O. Engelfried, U. Kerb, H. Laurent, H. Müller, and G. Schulz, *Chem. Ber.*, **99**, 1118 (1966).

[101] B. A. Shoulders, W. W. Kwie, W. Klyne, and P. D. Gardner, *Tetrahedron*, **21**, 2973 (1965).

[102] M. Debono, E. Farkas, R. M. Molloy, and J. M. Owen, *J. Org. Chem.*, **34**, 1447 (1969).

[103] E. Farkas, J. M. Owen, M. Debono, R. M. Molloy, and M. M. Marsh, *Tetrahedron Lett.*, **1966**, 1023.

[104] D. N. Kirk and M. P. Hartshorn, *Steroid Reaction Mechanisms*, Elsevier, New York, 1968, p. 197.

[105] P. A. Mayor and G. D. Meakins, *J. Chem. Soc.*, **1960**, 2800.

[106] D. H. R. Barton and G. A. Morrison, *Fortschr. Chem. Org. Naturstoffe*, **19**, 223–225, (1961).

[107] E. L. Eliel, N. L. Allinger, S. J. Angyal, and G. A. Morrison, *Conformational Analysis*, Interscience, New York, 1965, p. 319.

[108] W. Herz and J. J. Schmid, *J. Org. Chem.*, **34**, 3473 (1969).

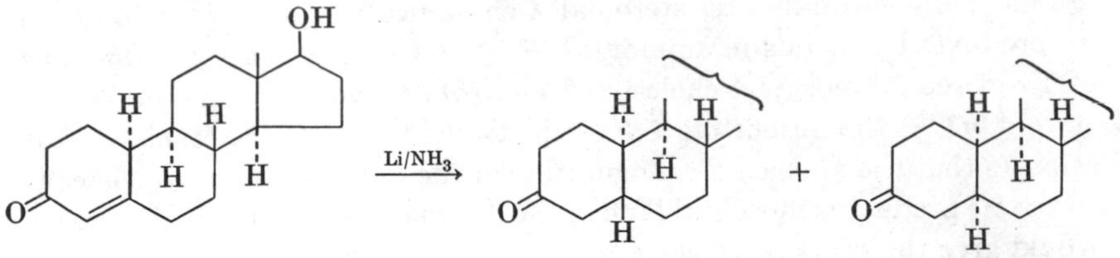

63 (60%) (20%)

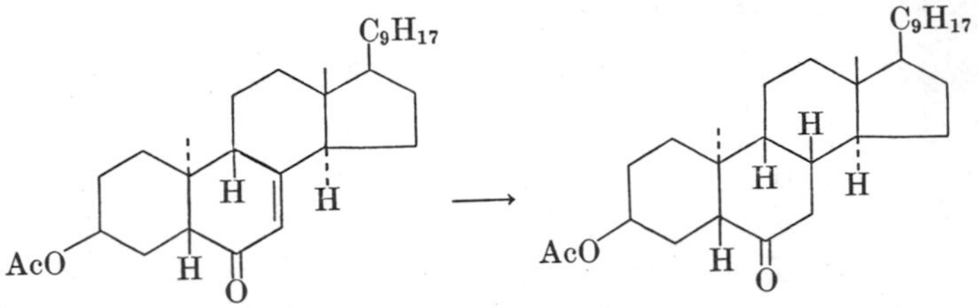

64 65

The course of reduction of enone **66** was explained by the observation that of the two configurations of the dianionic intermediate having the isopropyl group equatorial to ring C, the 8α configuration of the carbanion is more favorable stereoelectronically than the 8β configuration and thus the new C—H bond should form quasi-axially with respect to ring C.

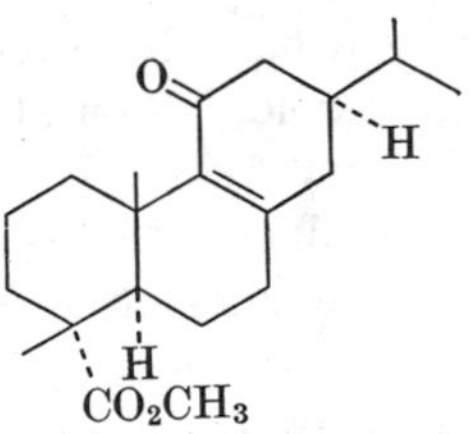

66 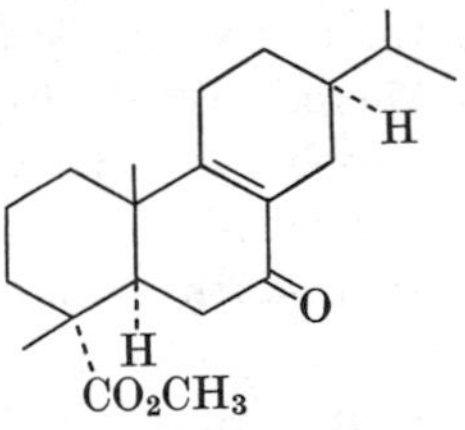 67

However, the formation of the *trans-syn-cis* product from enone **67** appears to violate the Stork-Darling rule. Here the reduction seems to proceed by way of the 9β carbanion having the lone pair quasi-equatorial to ring B, rather than via the stereoelectronically allowed 9α configuration, which is more sterically strained.

Metal-ammonia reductions of a series of 3,4-disubstituted cyclohex-2-enones led to some interesting stereochemical observations.[109] Reductions of the 3,4-dimethyl **(68a)** and 3-methyl-4-ethyl compounds **(68b)** gave mixtures composed of 84% of the *trans-* and 16% of the *cis*-3,4-dialkycyclohexanones. The diethyl enone **68c** afforded a mixture of 44% of the *cis*

[109] S. K. Malhotra, D. F. Moakley, and F. Johnson, *Tetrahedron Lett.*, **1967**, 1089.

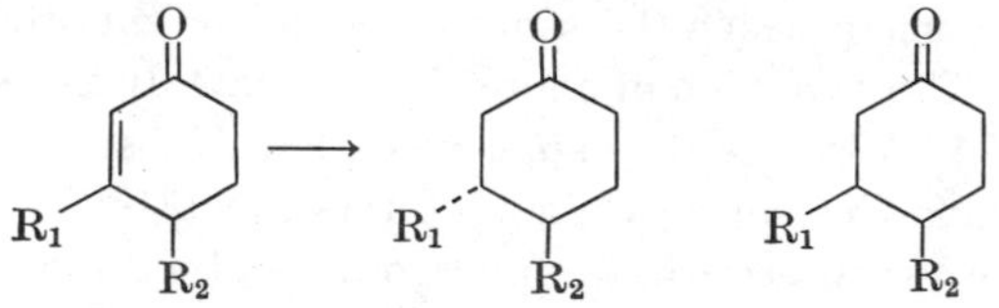

68

a, $R_1 = R_2 = CH_3$
b, $R_1 = CH_3$; $R_2 = C_2H_5$
c, $R_1 = R_2 = C_2H_5$
d, $R_1 = C_6H_5$; $R_2 = CH_3$
e, $R_1 = R_2 = C_6H_5$

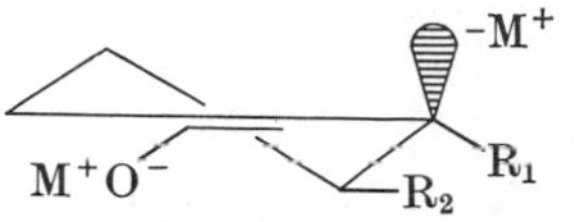

69A

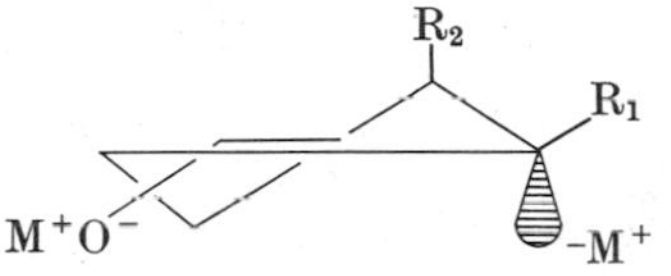

69B 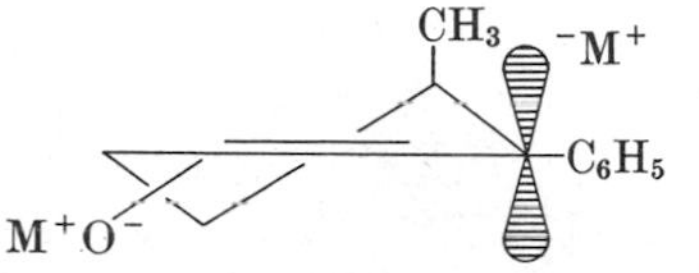 **70**

and 56% of the *trans* isomers. The two enones having phenyl groups at C-3, *i.e.*, **68d** and **68e**, gave reduction mixtures containing 94% and 98% of the corresponding *cis* products.

The results with the first two enones were explained by assuming that the β carbanion (*C*-3) achieves a tetrahedral geometry before protonation, and that the stereoelectronically allowed configuration **69A** is favored over the alternative one, **69B**, because nonbonded interactions involving R_1 and R_2 are minimized. It was pointed out that the formation of nearly equal amounts of *cis*- and *trans*-3,4-diethylcyclohexanones from **68c** was difficult to rationalize unless exceptionally large interactions between the two ethyl groups occurred in configuration **69A**.[21, 109] The results for the enones **68d** and **68e** have been explained by considering that delocalization of negative charge by the phenyl group causes the dianion intermediate to have a planar geometry. If so, a species derived from, for example, **68d** would be expected to prefer a conformation such as **70** in which $A^{1,2}$-strain is minimized. If the transition state for protonation also resembles this conformation, and/or, if because of the relatively low basicity of the β carbanion its protonation is subject to steric hindrance, attack on **70** from the α side leading to the *cis* product would be expected.[10, 21, 109]

Reductions of the 3-alkyl-substituted enones **68a** and **68b**,[109] the enone **71**,[57] and 3,5-dimethylcyclohex-2-enone (**72**)[57, 92, 110] provide examples of

[110] M. Anteunis, personal communication.

monocyclic systems in which the more stable reduction product is formed predominately. The reduction of enone **73** gives mainly the apparently less stable ketone **74**.[111] It has been suggested that $A^{1,2}$-strain involving the O^-M^+ ion pair and the isopropyl group forces the latter into a quasi-axial conformation in the intermediate undergoing protonation.[21]

t-Bu– =CHCOCH$_3$ $\xrightarrow[\text{THF, }t\text{-BuOH}]{\text{Li or Na/NH}_3}$ *t*-Bu– (H, CH$_2$COCH$_3$) (73–84%) + *t*-Bu– (CH$_2$COCH$_3$, H) (10–15%)

71

$\xrightarrow[\text{2. CrO}_3]{\text{1. M/NH}_3\text{, ROH}}$ +

72 (M = Li, Na, K) (91–92%) (8–9%)

$\xrightarrow{\text{Li/NH}_3}$ +

73 **74** (75%) (25%)

For the reasons discussed earlier, the cyclopentenone derivative **75**[10] and the cyclohexenone derivative **76**,[112] having a group capable of delocalizing negative charge at the β position, yield the less stable dihydro products on metal-ammonia reduction.

Metal-ammonia reductions of tricyclic ketones such as **77** (n = 1 and 2) in which the B ring is constrained in a boat conformation lead exclusively to the thermodynamically more stable *cis*-fused products.[80, 113] The tricyclic ketone **78** also gives largely the A/B *cis* product on metal-ammonia reduction.[114]

[111] C. Djerassi, J. Osiecki, and E. J. Eisenbraun, *J. Amer. Chem. Soc.*, **83**, 4433 (1961).

[112] F. E. Ziegler and M. E. Condon, *J. Org. Chem.*, **36**, 3707 (1971); *Tetrahedron Lett.*, **1969**, 2315.

[113] S. D. Darling and S. Y. Wong, Abstr. 155th National ACS Meeting, San Francisco, April 1968, p. 92.

[114] J. A. Marshall and S. F. Brady, *J. Org. Chem.*, **35**, 4068 (1970).

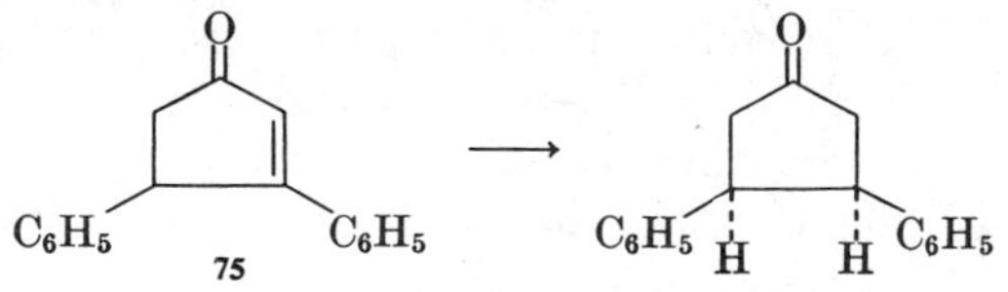

C_6H_5 C_6H_5 **75** → C_6H_5 H H C_6H_5

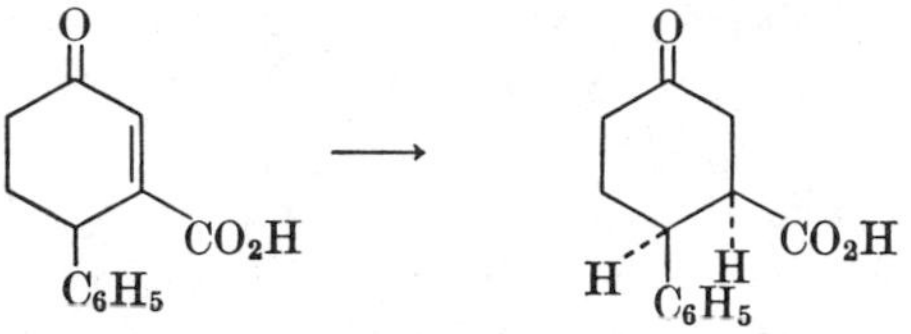

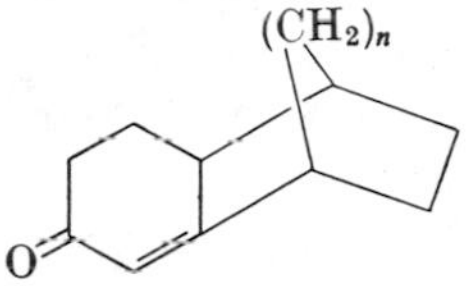

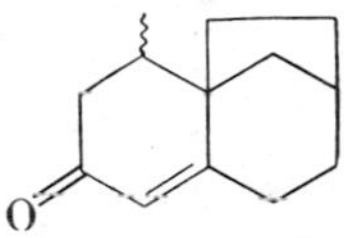

CO_2H C_6H_5 **76** → H H CO_2H C_6H_5

$(CH_2)_n$ O **77**

O **78**

The stereochemical course of metal-ammonia reductions of bi- and polycyclic compounds in which one or both of the rings involving the α,β-unsaturated system are not six-membered is in general determined by the thermodynamic stability of the reduction products. 5/5-Fused enones are reduced exclusively to *cis*-fused products.[115–118] For example, as a key step in the synthesis of cedrol, lithium-ammonia reduction of the diester enone **79** was employed to obtain the more stable dihydro product **80**.[118]

Several groups[14, 119–121] have reported that lithium-ammonia reduction of the hydrindenone **81a** forms a mixture of the *trans*- and *cis*-fused products in about an 85/15 ratio. By contrast similar reduction of the hydrindenone **82a** gave 99% of the *trans* isomer.[119] On the other hand, hydrindenones **81b** and **82b**, having angular methyl groups, have been found to yield

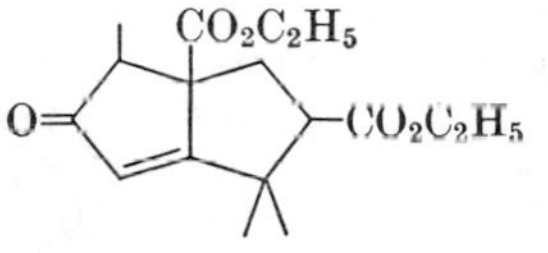

$CO_2C_2H_5$ O $CO_2C_2H_5$ **79**

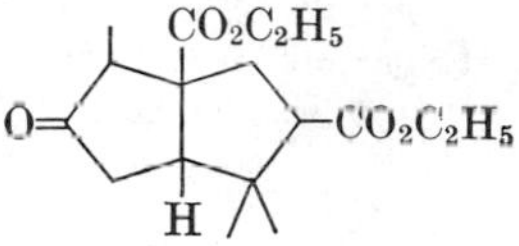

$CO_2C_2H_5$ O $CO_2C_2H_5$ H **80**

[115] W. G. Dauben, D. J. Ellis, and W. H. Templeton, *J. Org. Chem.*, **34,** 2297 (1969).
[116] P. T. Lansbury, N. Y. Wang, and J. E. Rhodes, *Tetrahedron Lett.*, **1971,** 1829.
[117] P. T. Lansbury and N. Nazarenko, *Tetrahedron Lett.*, **1971,** 1833.
[118] G. Stork and F. H. Clarke, *J. Amer. Chem. Soc.*, **77,** 1072 (1955); **83,** 3114 (1961).
[119] W. G. Dauben, personal communication.
[120] F. Granger, J. P. Chapat, and J. Crassous, *C.R. Acad. Sci., Ser. C*, **265,** 529 (1967).
[121] R. Granger, J. P. Chapat, J. Crassous, and F. Simon, *Bull. Chim. Soc. Fr.*, **1968,** 4265.

largely the *cis*-fused reduction products; the *cis/trans* ratios were 9/1 and 7/3, respectively.[119,122] *trans*-Fused hydrindanones related to **81a** and **82a** are apparently more stable than the *cis* isomers,[123] but heat of combustion data indicate that the *cis*-fused dihydro ketone derived from enone **82b** is thermodynamically more stable than the *trans* isomer.[124] It seems likely that this is also true for the corresponding dihydro derivative of **81b**. Energy calculations that have been recently reported provide support for these points.[125]

Derivatives of **81a** having a six-membered ring *trans*-fused to the B ring yield mainly *cis*-fused products on lithium-ammonia reduction.[126, 127] The stereochemistry of 4-substituents has been shown to have a significant effect upon the stereochemical course of metal-ammonia reduction of derivatives of **82b**.[128, 129] A-Norsteroids related to **82b** give mixtures of *cis*- and *trans*-fused products on reduction under similar conditions.[130, 131] Interestingly, the nature of the substituent at C–17 has an influence upon the stereochemistry of the reduction. For example, the ratio of the A/B-*cis* to the A/B-*trans* product was about 4/1 when A-norcholestenone was reduced with lithium in liquid ammonia,[130] but under the same conditions A-nortestosterone gave nearly equal amounts of the two products.[131]

Lithium-ammonia reductions of 8(14)-en-15-ones afford dihydro products having the natural $8\beta,14\alpha$ configurations.[132–134] However, it has been recently found that similar reduction of an 8(14)-en-7-one yields exclusively a product having an unnatural $8\beta,14\beta$ configuration.[134] This result again emphasizes the point that in complex systems metal-ammonia reductions often do not yield the more thermodynamically stable product. In two reductions of 14-en-16-ones, mixtures of products containing predominantly the *cis*-C/D isomers were obtained.[119] Tricyclic enones

[122] R. Fraisse-Jullien, C. Frejaville, V. Toure, and M. Derieux, *Bull. Soc. Chim. Fr.*, **1966,** 3725.

[123] W. G. Dauben and K. S. Pitzer, in *Steric Effects in Organic Chemistry*, Wiley, New York, 1956, p. 37.

[124] P. Sellers, *Acta Chem. Scand.*, **24,** 2453 (1970).

[125] N. L. Allinger and M. T. Tribble, *Tetrahedron*, **28,** 1191 (1972).

[126] M. J. Green, N. A. Abraham, E. B. Fleischer, J. Case, and J. Fried, *Chem. Commun.*, **1970,** 234.

[127] J. P. Kutney, J. Cable, W. A. F. Gladstone, H. W. Hanssen, E. J. Torupka, and W. D. C. Warnock, *J. Amer. Chem. Soc.*, **90,** 5332 (1968).

[128] F. Giarrusso and R. E. Ireland, *J. Org. Chem.*, **33,** 3560 (1968).

[129] R. E. Ireland, P. S. Grand, R. E. Dickerson, J. Bordner, and R. R. Rydjeski, *J. Org. Chem.*, **35,** 570 (1970).

[130] W. G. Dauben, G. A. Boswell, and W. H. Templeton, *J. Amer. Chem. Soc.*, **83,** 5006 (1961).

[131] F. L. Weisenborn and H. E. Applegate, *J. Amer. Chem. Soc.*, **81,** 1960 (1959).

[132] C. S. Barnes, D. H. R. Barton, and G. F. Laws, *Chem. Ind.* (London), **1953,** 616.

[133] D. H. R. Barton and G. F. Laws, *J. Chem. Soc.*, **1954,** 52.

[134] I. Midgley and C. Djerassi, *J. Chem. Soc., Perkin Trans.*, I, 155 (1973).

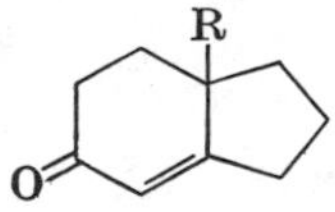

81 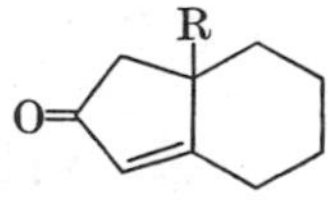 82

a, R = H
b, R = CH_3

related to **81**[135] and **82**[136] yield *cis*-fused products on lithium-ammonia reduction.

Two 5/7-fused enones reduced with lithium in liquid ammonia formed the *trans*-dihydro products with a high degree of stereoselectivity.[137] The hydroxy enone **83** has been converted into the *trans*-fused dihydro derivative **84** by treatment with lithium-ammonia-methanol followed by oxidation; **84** was further converted into 5-*epi*-α-bulnesene.[137]

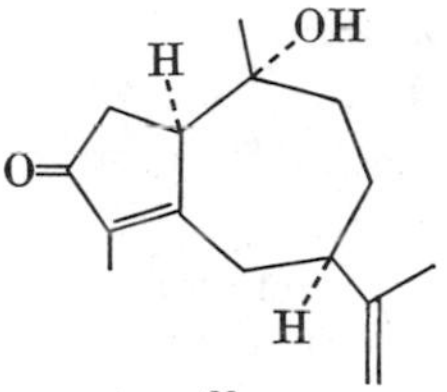

83 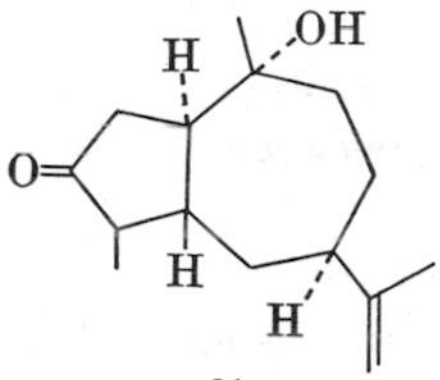 84

The possible influence of the nature of the metal on the stereochemistry of metal-ammonia reductions of α,β-unsaturated carbonyl compounds is an aspect of the reduction mechanism which has received little attention in the literature. The carbanionic species involved in these reductions would be expected to have the negative charge at the β position neutralized to some degree by association with the metal cation before and during the protonation step. The nature and extent of this association should depend upon the size of the cation and its degree of solvation in the reduction medium, and may range from essentially free ions, over a gamut of ion pair species, to essentially covalent bonds.[9, 106] Steric and solvation factors associated with the carbanion-metal interaction could favor the development of the carbanion in a particular geometry, or influence the populations of various configurations of the carbanion if equilibration is more rapid than protonation. Also, transition states of type **85** in which a proton is transferred to the carbanion from a solvent molecule already complexed with a metal cation should be favorable for the protonation step.[20, 57, 64, 68, 138] The nature and size of the metal and its accompanying

[135] D. Becker and H. J. E. Lowenthal, *J. Chem. Soc.*, **1965**, 1338.
[136] A. J. Birch and H. Smith, *J. Chem. Soc.*, **1956**, 4909.
[137] E. Piers and K. F. Cheng, *Can. J. Chem.*, **48**, 2234 (1970).
[138] D. N. Kirk and A. Mudd, *J. Chem. Soc., C*, **1969**, 968.

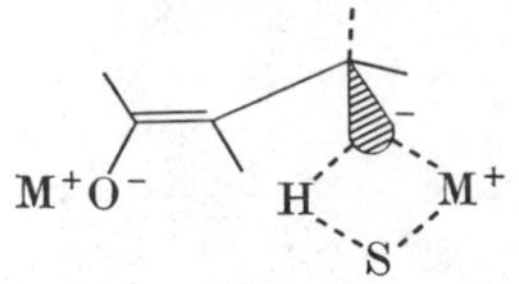

85

solvent shell could have an influence on relative energies of possible protonation transition states. These features could also influence the reduction stereochemistry if such transition states are in equilibrium.

Lithium has been used in carrying out the majority of metal-ammonia reductions of unsaturated carbonyl compounds. Sodium also has been employed but to a much more limited extent; the use of other metals is rare. In most reactions in which lithium or sodium is employed as the metal there does not seem to be a significant steric factor associated with formation of the new β-C—H bond. The literature contains a number of examples of reductions in which products having a severe 1,3-diaxial interaction between the new C—H bond and a bulky substituent are formed.[10, 93–95, 97, 139, 140]

Only a few studies have been made in which a careful examination of the influence of the nature of the metal on the reduction stereochemistry was undertaken. In a study in which the enone **72** was reduced with lithium, sodium, or potassium in liquid ammonia containing various alcohols, essentially identical percentages of *trans*- and *cis*-3,5-dimethylcyclohexanone were produced.[92] Some rather significant variations in the stereochemistry of the products of reduction of the 6/7-fused enone **86** were obtained upon changing the reducing metal.[121] It was considered that the

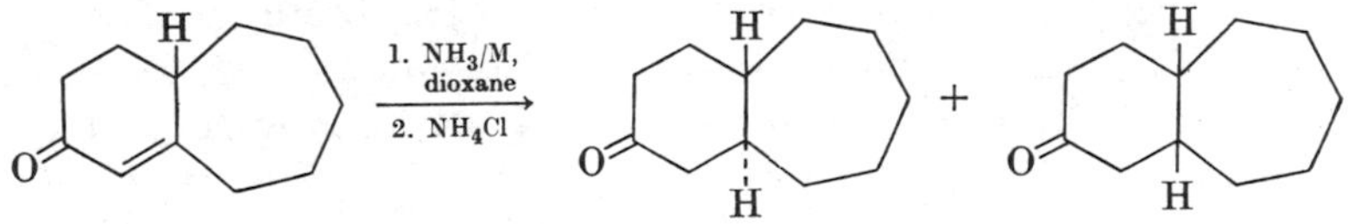

1. NH_3/M, dioxane
2. NH_4Cl

86

M	Product Ratio	
	trans	*cis*
Li	76	24
Na	87	13
Ca	56	44
Ba	44	56

stereoelectronically allowed transition states **87A** and **87B** were in equilibrium and that the increased preference for the *cis* product in going from sodium to calcium to barium represented an increased preference for the transition state **87B** in which nonbonded interactions involving the

[139] T. G. Halsall, D. W. Theobald, and K. B. Walshaw, *J. Chem. Soc.*, **1964**, 1029.
[140] F. J. McQuillin, *J. Chem. Soc.*, **1955**, 528.

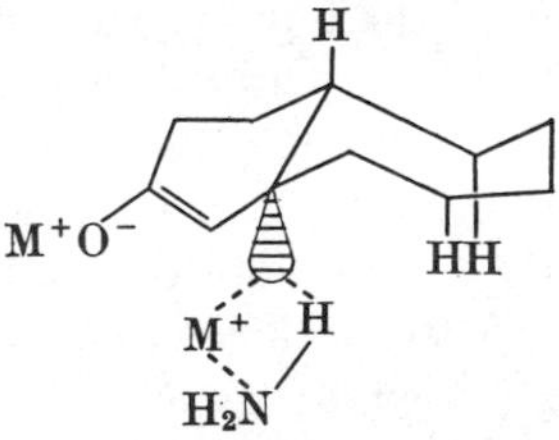

M⁺O⁻ H HH M⁺ H H₂N

87A

H₂N H M H M⁺O⁻ H H H

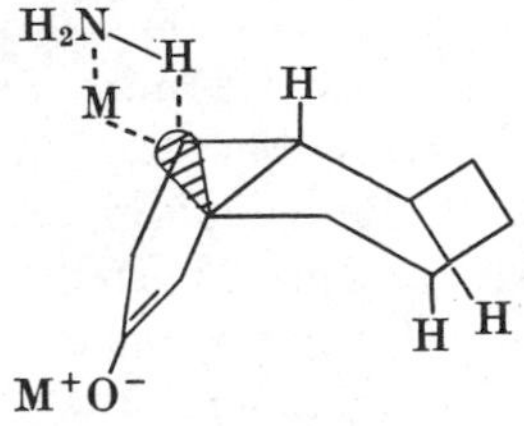

87B

indicated hydrogen atoms and the β carbanion are minimized. Steric effects associated with the kinetic development of the carbanion in a configuration similar to **87A** may also explain the results. The slight preference for the *cis* product with lithium as compared with sodium was attributed to more extensive solvation of the latter cation.

The most dramatic effect on the stereochemical course of a metal-ammonia reduction attributable only to a difference in the metal is found in the reduction of the α,β-unsaturated acid **88**.[7, 8] Reduction with

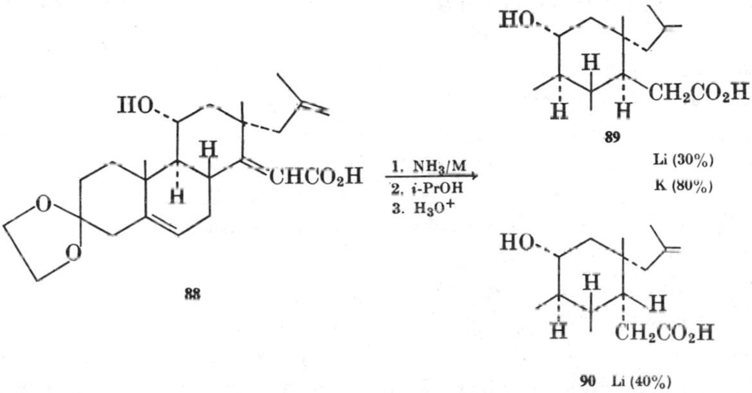

HO. H H O O =CHCO₂H

88

1. NH_3/M
2. *i*-PrOH
3. H_3O^+

HO. H H H CH₂CO₂H

89 Li (30%) K (80%)

HO. H H H CH₂CO₂H

90 Li (40%)

potassium in liquid ammonia followed by addition of 2-propanol gave isomer **89** having the acetic acid side chain β in 80% yield. None of the α epimer was found. Under the same conditions but using lithium in place of potassium a 30% yield of **89** and a 40% yield of **90** was obtained. Interpretation of these results is difficult. However, it should be pointed out that the metals were added to a suspension of the acid in liquid ammonia. It seems possible that the metals may have undergone dissolution at different rates so that reduction at a metal surface might have been more significant in one case than the other.

In some instances hydroxyl groups in proximity to the β position have been found to influence the stereochemistry of metal-ammonia reductions of α,β-unsaturated carbonyl compounds. However, present evidence allows only speculation concerning the role of these groups. It has been reported that a 2/1 mixture of the A/B *cis*- and *trans*-fused dihydro products was obtained on lithium-ammonia reduction of 19-hydroxytestosterone.[141] (Protection of the hydroxyl group as the tetrahydropyranyl derivative followed by reduction gave only the *trans*-fused product, as expected.[141]) It was suggested that the *cis* product arose as a result of internal protonation of the kinetically formed 5β-carbanionic intermediate.[141] In contrast, the related 10-hydroxymethyl-1(9)-octalin-2-one, has been reported to give predominantly the *trans* product on lithium-ammonia reduction.[89] As noted earlier, reduction of the octalone **54d** (p. 28) gives a 2/3 mixture of *cis* and *trans* reduction products.[96] However if the tertiary hydroxy group of **54d** was converted into the lithium alkoxide before reduction of the enone, the *cis/trans* ratio decreased to about 1/9.[96] These results suggest the possibility that the stabilization of the 9α-carbanion which leads to the *trans* reduction product may be enhanced by its coordination with a metal cation which is already bonded to the alkoxide ion of the side chain. If the assumption is made that the alkoxide ion formation took place prior to reduction, a similar factor may provide an explanation for the results obtained in the reduction of 19-hydroxytestosterone.

Although other explanations have been suggested,[142] cation bridging between the 11β-alkoxide ion and the carbanion center at position 14 also may account for the exclusive formation of the product **92**, having the acetic acid side chain α, on the reduction of the α,β-unsaturated acid **91** with lithium, sodium, or potassium in liquid ammonia.[7] A boat conformation of ring C would be required for such bridging, but the strong 1,3-diaxial interaction which would exist between the 11-alkoxide-metal ion pair and the 13-methyl group if the C ring were in a chair conformation might cause the energies of the two forms to be relatively close.

Lithium-ammonia reductions of 2,3-dialkyl-4-hydroxy-2-cyclopentenones have recently been shown to yield products having the 3-alkyl

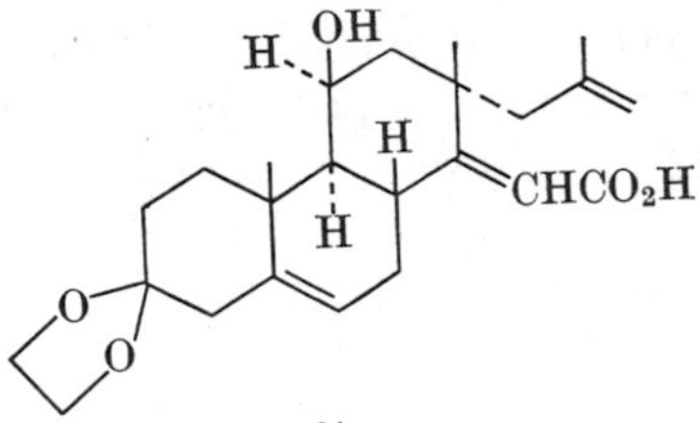

H. OH H H $CHCO_2H$

91

H. OH H H H CH_2CO_2H

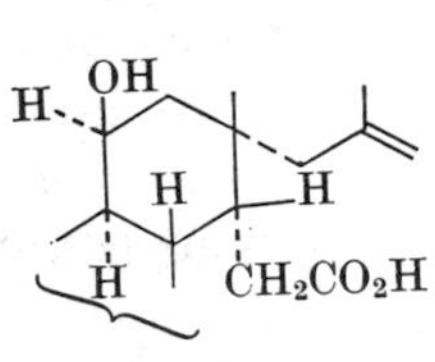

92

[141] L. H. Knox, E. Blossy, H. Carpio, L. Cervantes, P. Crabbé, E. Velarde, and J. A Edwards, *J. Org. Chem.*, **30**, 2198 (1965).

[142] G. W. Kenner, *Ann. Rep.*, **51**, 177 (1954).

group and the 4-hydroxy group *trans* with a high degree of stereoselectivity.[142a,142b] The occurrence of β protonation *cis* to the oxygen function suggests that cation bridging may be an important factor in these cases as well. Clearly, further examination of the influence of neighboring hydroxyl groups on the stereochemistry of enone reductions is needed.

Kinetic and Thermodynamic Control in the Protonation of Metal Enolate Intermediates

The metal enolates formed on metal-ammonia reduction of α,β-unsaturated carbonyl systems are converted into the saturated carbonyl compounds by addition of proton donors more acidic than ammonia. This step is subject to the usual factors controlling the stereochemistry of ketonization, and when an asymmetric center is generated at the α position the more or the less stable diastereoisomer may be produced initially. Usually, the product having the more stable configuration at the α position is isolated in metal-ammonia reductions. The reason is that, under the usual conditions employed in the quenching of enolates, *i.e.*, addition of the proton source to liquid ammonia containing the enolate in solution or suspension, and in the workup of reaction mixtures, the initial product is exposed to an acidic or basic medium for sufficient time to cause partial or complete equilibration. When inverse quenching conditions are employed or when the kinetic ketonization product is not very labile, the isolation of the product having the less stable configuration at the α position has been possible. For example, it has been shown that addition of a solution of the enolate **94** (produced by reduction of 2-methyl-3-phenylindanone **93**) to a saturated aqueous ammonium chloride solution gave, depending upon the exact conditions employed, mixtures containing up to 80% of the *cis* product **95**.[18] Normal decomposition by addition of solid ammonium chloride to the ammonia solution produced a mixture containing approximately 70% of the *trans* isomer **96**. This composition is close to the thermodynamic product ratio, which is 4/1 in favor of the *trans* product.[12] (Formulas **93–101** are on p. 40.)

The formation of the *cis*-phenanthrene derivative **98** from enone **97**[11] and the *trans*-decalone **57** from enone **56** (p. 28)[97] provide other examples of the isolation of less stable diastereoisomers in metal-ammonia reductions.

The reduction of 9-octalin-1-one **(99)** is of interest because it gave only *trans*-1-decalone **(100)** when reduced under the conditions employed for the preparation of the *cis* product **98** from **97**;[11, 16] with inverse workup only 42% of the *cis* product **101** could be obtained.[41]

[142a] P. DeClercq, D. Van Haver, D. Tavernier, and M. Vandewalle, *Tetrahedron*, **30**, 55 (1974).

[142b] F. Van Hulle, V. Sipido, and M. Vandewalle, *Tetrahedron Lett.*, **1973**, 2213.

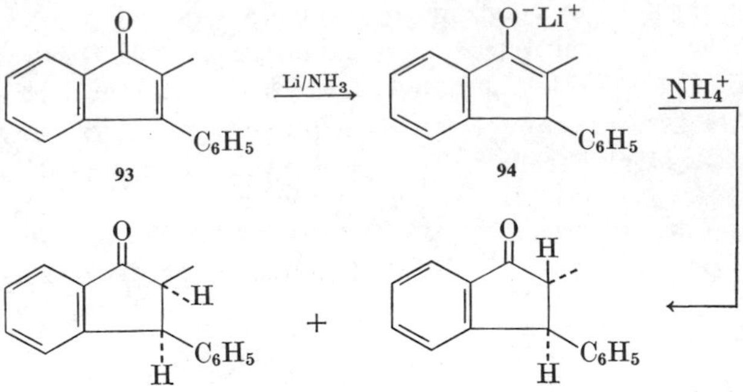

The ketonization of the enolate **103** derived from 1-benzoyl-2-phenylcyclohexanone **(102)** has been widely studied.[21, 25, 41, 143] When solutions of this enolate were prepared by reduction of the enone **102** with solutions of lithium, sodium, potassium, magnesium, or cesium in liquid ammonia and quenched under conditions favoring kinetic protonation, the ratio of the *cis* reduction product **104** to the *trans* product **105** was approximately 4/1, and was essentially independent of the nature of the metal.[25] The proportion of the more stable *trans* product was found to increase when **103** was quenched in increasingly basic media.[143]

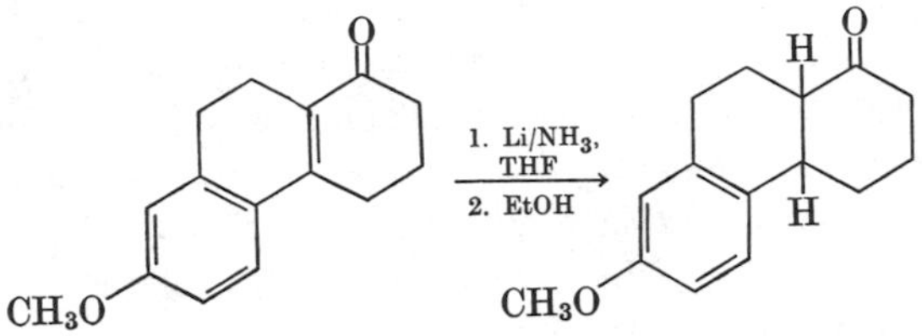

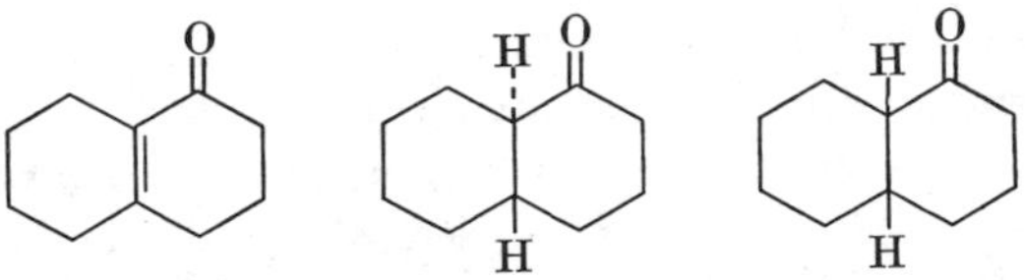

Detailed discussions of the factors controlling the stereochemistry of protonation of enolates (and enols) have been presented elsewhere.[21, 41] In summary it appears that for stereoelectronic reasons protonation occurs perpendicular to the plane of the enolate anion (or enol); that C-protonation of enolates as well as ketonization of enols is generally involved;[144] and that, when protonation is carried out with relatively

[143] P. Angibeaund, H. Rivière, and B. Tchoubar, *C.R. Acad. Sci., Ser. C*, **263**, 160 (1966).

[144] H. O. House, B. A. Tefertiller, and H. D. Olmstead, *J. Org. Chem.*, **33**, 935 (1968).

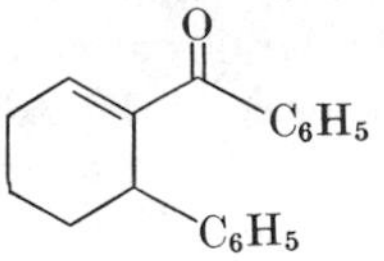

102

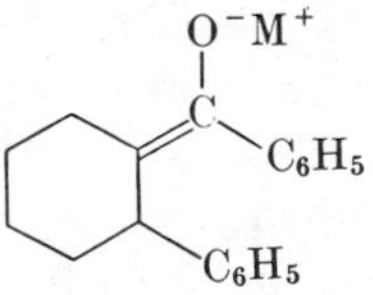

103

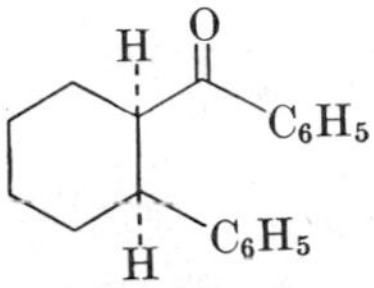

104

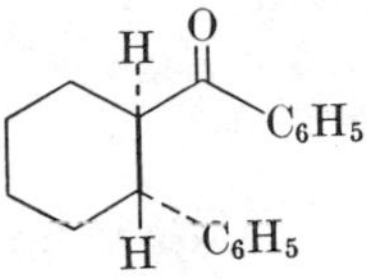

105

strong acids, the transition state closely resembles the enolate or enol, so that steric factors within the reactant exert a controlling influence upon the direction of proton addition. However, in accordance with the Hammond postulate,[145] it is expected that as the acidity of the medium employed for the protonation is decreased the transition state for protonation should have more product character; then product stability should become increasingly important.[21]

The Generation and Trapping of Specific Lithium Enolate Intermediates

In addition to the ketonization reactions described above, the observation of other reactions characteristic of metal enolates provides confirmation that these species are the initial products of metal-ammonia reduction of α,β unsaturated ketones. In 1961 the important observation was made that products derived from C-alkylation of the lithium enolate generated in the reduction were obtained if an alkylating agent rather than a proton donor was added to the reaction mixture at the end of the reduction.[13] For example, it was found that the 1-enolate prepared from lithium ammonia reduction of 1(9)-octalin-2-one (**25**) could be

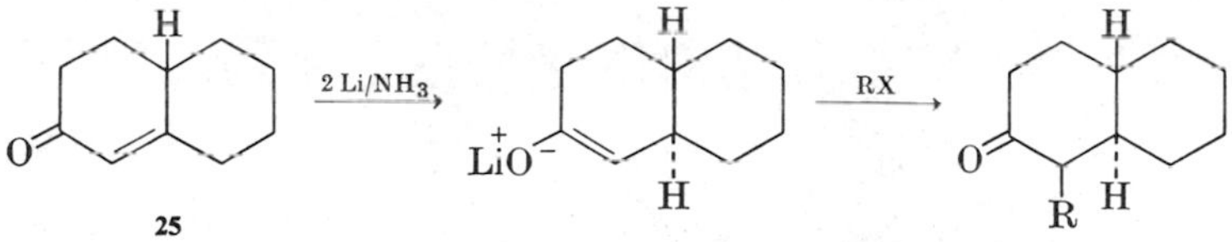

$\xrightarrow{2\ Li/NH_3}$ $\xrightarrow{RX}$

25

alkylated with alkyl halides in liquid ammonia to produce 1-alkyl-*trans*-2-decalones. Under the usual base-catalyzed conditions, alkylation of *trans*-2-decalones gives mainly 3-alkyl-substituted products via the 2-enolate. The integrity of the 1-enolate was not maintained if sodium or potassium

[145] G. S. Hammond, *J. Amer. Chem. Soc.*, **77**, 334 (1955).

was employed as the reducing metal or if the lithium enolate was transferred to dimethyl sulfoxide before alkylation; instead 3-alkyl-*trans*-2-decalones were the major products. The success of the reduction-alkylation method depends upon the now well-established fact that alkylation of lithium enolates (as opposed to other alkali metal enolates) of unsymmetrical ketones with relatively reactive alkylating agents occurs faster in a variety of solvents than does equilibration among the structurally isomeric enolates via proton transfer with the products of such alkylations.[13, 14, 146]

The Mechanism of Formation of Saturated Alcohols in Metal-Ammonia Reductions of α,β-Unsaturated Carbonyl Compounds

As discussed earlier, it often has been found useful to employ relatively acidic proton donors to increase yields in metal-ammonia reductions of α,β-unsaturated carbonyl compounds.[57, 72, 78, 109, 147] If an excess of a proton donor is present during the reduction or is added at the end of the reaction when the metal is still present in excess, partial or complete reduction of α,β-unsaturated ketones to saturated secondary alcohols results. While the initial reduction product, the metal enolate **2** (p. 3), resists reduction, the presence of a proton donor of comparable acidity to the parent ketone **3** allows equilibrium to be established between the enolate and the ketone, and the latter is readily reduced. Reductions of α-cyperone **(106)** illustrate how the course of the reaction can be influenced to produce the saturated ketone **107** or the saturated alcohol **108** by the choice of the proton donor employed at the end of the reaction.[148] In both products the methyl group has the more stable configuration. This is

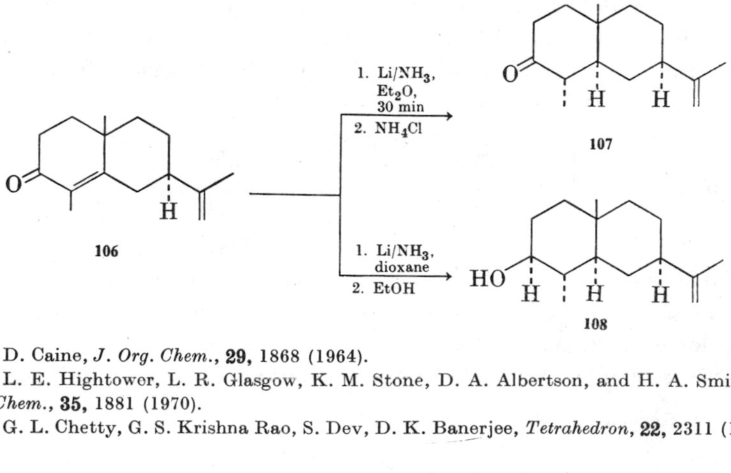

106 → (1. Li/NH_3, Et_2O, 30 min; 2. NH_4Cl) → 107

106 → (1. Li/NH_3, dioxane; 2. EtOH) → 108

[146] D. Caine, *J. Org. Chem.*, **29**, 1868 (1964).

[147] L. E. Hightower, L. R. Glasgow, K. M. Stone, D. A. Albertson, and H. A. Smith, *J. Org. Chem.*, **35**, 1881 (1970).

[148] G. L. Chetty, G. S. Krishna Rao, S. Dev, D. K. Banerjee, *Tetrahedron*, **22**, 2311 (1966).

generally the case for reductions leading to saturated alcohols when the enone has an α substituent.

The mechanism and stereochemistry of the reduction of ketones to secondary alcohols by dissolved metals has been widely investigated.[9, 16, 149–154] Although space does not permit a detailed treatment of this subject, some of the salient features of the ketone reduction process are discussed briefly. Transfer of an electron from the solution to the antibonding π orbital of the carbonyl group can give rise to a radical anion (or ketyl) **109** having the greater unpaired electron density on carbon.[155, 156] Many workers have considered that a radical anion such as **109** may add a

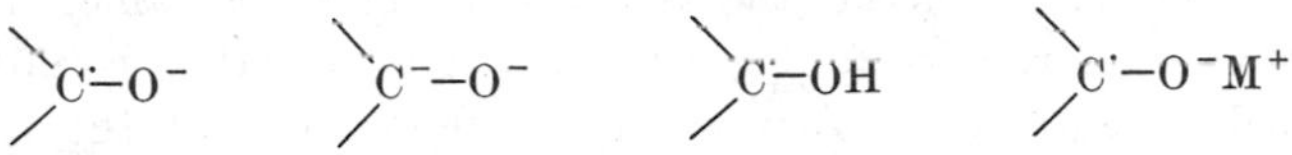

$$\gt\dot{C}{-}O^- \qquad \gt C^-{-}O^- \qquad \gt\dot{C}{-}OH \qquad \gt\dot{C}{-}O^-M^+$$

109 **110** **111** **112**

second electron to give a dianion **110**, which may be protonated on carbon and on oxygen by proton donors of the proper acidity to give the alcohol.[9, 22, 23, 154] Dianion formation can undoubtedly take place with diaryl ketones[50, 54] and, perhaps, with monoaryl ketones, but it is unlikely that the reduction potential of metal-ammonia solutions is sufficiently negative to form free dianions from aliphatic ketones.[20] However, species such as **111** and **112**, which may be formed by protonation of **109** or by tight ion pair formation with a metal cation, should be capable of adding a second electron. The formation of the hydroxy radical intermediates would, of course, be favorable when relatively acidic proton donors are present in the medium. When proton donors are not present, the formation of both the ion pair **112** and ion pair dimers derived from it would be expected, especially when small cations such as lithium and relatively nonpolar cosolvents are employed.[48] The latter conditions also appear to favor the formation of pinacols in ketone reductions.[77]

In general, metal-ammonia reductions of aliphatic ketones lead to the formation of the more stable alcohol. This is clearly illustrated in alicyclic

[149] A. Coulombeau, *Bull. Soc. Chim. Fr.*, **1970**, 4407.

[150] A. Coulombeau and A. Rassat, *Bull. Soc. Chim. Fr.*, **1970**, 4399, 4044; *Chem. Commun.*, **1968**, 1587.

[151] J. W. Huffman and J. T. Charles, *J. Amer. Chem. Soc.*, **90**, 6486 (1968).

[152] J. W. Huffman, D. M. Alabran, T. W. Bethea, and A. C. Ruggles, *J. Org. Chem.*, **29**, 2963 (1964).

[153] W. S. Murphy and D. F. Sullivan, *Tetrahedron Lett.*, **1971**, 3707; *J. Chem. Soc., Perkin Trans.*, **I**, 999 (1972).

[154] O. H. Wheeler, in *The Chemistry of the Carbonyl Group*, S. Patai, Ed., Interscience, New York, 1966, Chap. 11.

[155] B. Mile, *Angew. Chem., Int. Ed. Engl.*, **7**, 507 (1968).

[156] M. Steinberger and G. K. Frenkel, *J. Chem. Phys.*, **40**, 723 (1964).

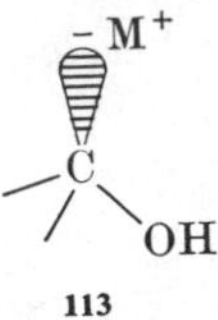

113

systems by reduction of cyclohexanones,[151] decalones,[148, 157, 158] and steroidal ketones.[18] Addition of a second electron to the radical **111** (or the related species **112**) should yield a carbanion intermediate which would be expected to achieve a tetrahedral geometry (*cf.* **113**). However, as in the case of β-carbanionic species involved in metal-ammonia reductions of enones, it is not clear whether the kinetically formed carbanion undergoes protonation with retention of configuration or whether equilibrium is established between the two possible configurations of the carbanion before protonation. Either view seems to provide an adequate explanation for the formation of equatorial alcohols in reductions of cyclohexanone derivatives. However, kinetic formation of a particular carbanion, whose geometry is controlled by torsional strain factors and by steric factors associated with the approach of a metal cation to the carbanionic center, seems to provide the more reasonable explanation for the stereochemistry of those reductions in hindered systems which often give the less stable alcohol.[149, 150, 152] The stereochemistry of reductions involving addition of an electron to an anion pair species such as **112** may also be influenced by the close proximity of functional groups which are capable of coordinating with the metal cation.[159]

SCOPE AND LIMITATIONS

Reduction and Reduction-Alkylation of α,β-Unsaturated Ketones

α,β-Unsaturated ketones of structural types ranging from simple acyclic compounds to complex alicyclic ones such as steroids, terpenoids, and alkaloids have been reduced to saturated ketones, usually in good yield, by solutions of metals in liquid ammonia. The reduction may be applied to compounds with any degree of substitution on the double bond; and alkyl and aryl groups, ring residues, and certain functional groups may be present at the α and/or β position.

Ethers such as diethyl ether, tetrahydrofuran, or dioxane are employed as co-solvents in most metal-ammonia reductions. Alkali metals, principally lithium and sodium, are generally the reducing metals. Although

[157] S. W. Pelletier, R. L. Chappell, and S. Prabhakar, *J. Amer. Chem. Soc.*, **90,** 2889 (1968).

[158] R. B. Turner, R. B. Miller, and J. J. Lin, *J. Amer. Chem. Soc.*, **90,** 6124 (1968).

[159] D. M. S. Wheeler, M. M. Wheeler, and M. Fetizon, *Tetrahedron,* **23,** 3909 (1967).

only 2 equivalents of these metals are required for the conversion of an enone to a saturated ketone, many workers have found it experimentally convenient to employ the metal in excess. For the reasons discussed in connection with the mechanistic aspects of the reaction, proton donors are often employed to reduce various competing side reactions such as dimerization. The presence of proton donors in the medium may lead to the conversion of the α,β-unsaturated ketone to a saturated alcohol. Of course, at least 4 equivalents of metal must be present for this type of reduction to take place. It is also dependent upon the nature of the proton donor and proton donor/enone ratio.

Alcohols such as methyl and ethyl alcohol lead to complete formation of saturated alcohols when present in excess during the reduction, and mixtures of ketone and alcohol are generally formed when one equivalent of these proton donors is employed.[147] These alcohols have acidity comparable to that of saturated ketones, and when they are present equilibrium can be established between the initially formed metal enolate and the saturated ketone. The latter is then reduced to the saturated alcohol (or the derived alkoxide). Such reduction generally does not occur to a very significant extent when 1 equivalent of *t*-butyl alcohol[72] or some less acidic proton donor, *e.g.*, triphenylcarbinol,[147] is employed. With excess *t*-butyl alcohol the results are less clear cut. For example, the formation of only the saturated ketone has been reported when reduction of an enone derived from podocarpic acid was carried out in the presence of a large excess of *t*-butyl alcohol.[160] In other reductions, product mixtures containing hydroxylic materials were obtained when a small excess of *t*-butyl alcohol was employed.[57] It appears that, under certain conditions, equilibrium between the metal enolate and the ketone is not established in the presence of *t*-butyl alcohol. The acidity of the ketone involved as well as the solubility of the metal enolate in the reaction medium may be of importance in determining whether alcohols are formed.

Even though the reduction conditions may lead to the formation of the metal enolate in high yield, further reduction may occur during the quenching step of the reaction. Alcohols such as methanol and ethanol convert metal enolates to saturated ketones much faster than they react with metals in ammonia,[40, 161] and quenching of reduction mixtures with these alcohols usually leads partially or completely to alcohols rather than saturated ketones. Rapid addition of an excess of solid ammonium chloride is the procedure most commonly employed in the quenching step if ketonic products are to be obtained. The reaction of ammonium ions with solvated electrons[35] apparently destroys the reducing system before reduction of

[160] R. H. Bible, Jr., and R. R. Burtner, *J. Org. Chem.*, **26**, 1174 (1961).

[161] H. O. House, *Rec. Chem. Progr.*, **28**, 98 (1967).

the ketone can take place. Alcohol formation can be avoided by adding any one of a number of reagents such as sodium benzoate,[162] ferric nitrate,[163, 164] sodium nitrite,[165] bromobenzene,[166] sodium bromate,[167] 1,2-dibromoethane,[19] and acetone[9] to destroy excess metal before the reaction mixture is neutralized.

If the reaction conditions employed lead to over-reduction and if the saturated ketone is the desired product, then the crude mixture may be treated with aqueous chromic acid in acetone (Jones' Reagent)[168] or other oxidizing agents before purification.

Assuming proton donors are not used in enone reductions, significant amounts of starting materials are sometimes recovered along with the α,β-dihydro derivative.[44, 57, 72, 78, 147] This is particularly troublesome in reduction of β-unsubstituted cyclohexenones,[72] acyclic enones,[147] and pregn-16-en-20-one derivatives.[19, 67, 169] The recovery of the starting material is likely to be the result of its being deactivated in some way toward reduction during the reaction, but being regenerated during the quenching step. Conjugate enolate formation by reaction of the enone with its derived radical anion or with lithium amide (the latter formed as the reduction proceeds) or the formation of 1,2 and/or 1,4 adducts of the enone with lithium amide, may explain the recovery of starting material.[72] Recent results indicate that 1,4 addition of lithium amide is an important side reaction in reduction of pregna-16-en-20-one derivatives.[19] However, evidence to allow the determination of the relative importance of various possible deactivation processes has not been obtained. The addition of 1 equivalent of *t*-butanol along with the enone to the metal solution usually provides an effective means of eliminating the recovery of unreacted starting material without causing a significant amount of saturated alcohol to be formed.[19, 72, 147]

The enone reduction-enolate alkylation sequence has been widely employed for the regiospecific alkylation of unsymmetrical ketones of a variety of structural types.[13,14,67,72,149,169–172] The procedure usually involves (1) the generation of a specific lithium enolate of an unsymmetrical ketone by reduction of the corresponding α,β-unsaturated

[162] A. P. Krapcho and A. A. Bothner-By, *J. Amer. Chem. Soc.*, **81**, 3658 (1959).
[163] D. C. Burke, J. H. Turnbull, and W. Wilson, *J. Chem. Soc.*, **1953**, 3237.
[164] I. N. Nazarov and I. A. Gurvich, *J. Gen. Chem. USSR*, **25**, 921 (1955).
[165] A. J. Birch, E. Pride, and H. Smith, *J. Chem. Soc.*, **1958**, 4688.
[166] G. Büchi, S. J. Gould, and F. Näf, *J. Amer. Chem. Soc.*, **93**, 2492 (1971).
[167] M. E. Kuehne, *J. Amer. Chem. Soc.*, **83**, 1492 (1961).
[168] K. Bowden, I. M. Heilbron, E. R. H. Jones, and B. C. L. Weedon, *J. Chem. Soc.*, **1946**, 39
[169] R. Deghenghi, C. Revesz, and R. Gaudry, *J. Med. Chem.*, **6**, 301 (1963).
[170] R. M. Coates and R. L. Sowerby, *J. Amer. Chem. Soc.*, **93**, 1027 (1971).
[171] R. Deghenghi and R. Gaudry, *Tetrahedron Lett.*, **1962**, 489.
[172] R. E. Schaub and M. J. Weiss, *Chem. Ind.* (London), **1961**, 2003.

$$\gt C{=}C{-}\overset{\overset{O}{\|}}{C}{-}\overset{|}{\underset{|}{C}}{-} \xrightarrow[\text{1 eq. ROH}]{\text{2 Li/NH}_3} H{-}\overset{|}{\underset{|}{C}}{-}\overset{|}{C}{=}\overset{\overset{Li^+}{O^-}}{C}{-}\overset{|}{\underset{|}{C}}{-} \xrightarrow{\text{RX}} H{-}\overset{|}{\underset{|}{C}}{-}\overset{|}{\underset{R}{C}}{-}\overset{\overset{O}{\|}}{C}{-}\overset{|}{\underset{|}{C}}{-} \quad \text{(Eq. 4)}$$

ketone with 2 equivalents of lithium in liquid ammonia containing no proton donor or 1 equivalent of a proton donor, and (2) reaction of this enolate with excess alkylating agent either in liquid ammonia or some other solvent system (Eq. 4). Co-solvents such as diethyl ether and tetrahydrofuran have often been employed.

High yields are generally obtained when methyl iodide,[14, 72, 170] other primary alkyl iodides,[14, 170] and such activated halides as allyl[14, 170] and benzyl[14] and halomethyl isoxazole derivatives[173] are employed as alkylating agents. Poorer yields and loss of selectivity are observed when alkylations are carried out with secondary halides such as isopropyl iodide.[170] Variable amounts of simple reduction products often are recovered from the reduction-alkylation sequence. These products are likely to arise from reaction of the metal enolate with an ammonium salt derived from reaction of the alkylating agent with ammonia.[72, 147] In the reduction-methylation of 1,10-dimethyl-1(9)-octalin-2-one to produce 1,1,10-trimethyl-*trans*-2-decalone it was found that the yield of the simple reduction product could be significantly reduced if the ammonia was removed and replaced by tetrahydrofuran before the addition of methyl iodide.[14] The lithium enolate intermediate is more soluble in tetrahydrofuran than in ammonia and apparently is more rapidly alkylated in the former solvent.[14] α-Nitro and α-cyano ketones have been prepared by reaction of reductively formed lithium enolates with amyl nitrate[174] and cyanogen chloride,[175] respectively.

Polyalkylation, as a result of base-catalyzed (either by lithium amide or conjugate bases of more acidic proton donors such as *t*-butyl alcohol) conversion of monoalkylated materials into their enolate anions and subsequent alkylation, is sometimes a serious side reaction in the reduction-alkylation sequence,[72, 147, 170] particularly when reactive ketones and alkylating agents are involved. Polyalkylation can generally be minimized through use of water as a proton donor[72, 170, 176] and brief reaction times for the alkylating step.[170, 176] The relatively weak base, lithium hydroxide, which is formed when water is employed as a proton donor, is apparently not effective in promoting enolization of alkylation products, and the use

[173] J. Pugach, Ph.D. Dissertation, Columbia University, 1964; *Diss. Abstr. B*, **25** (10), 5567 (1965).

[174] R. E. Schaub, W. Fulmor, and M. J. Weiss, *Tetrahedron*, **20,** 373 (1964).

[175] M. E. Kuehne and J. A. Nelson, *J. Org. Chem.*, **35,** 161 (1970).

[176] D. Caine, T. I. Chao, and H. A. Smith, accepted for publication in *Org. Syn.*

of water does not seem to cause a significant amount of over-reduction.

Elegant use of the reduction-alkylation sequence has been made in the total synthesis of natural products. The conversion of the tricyclic enone **114** into the reduction-methylation product **115** and the tetracyclic enone **116** into the reduction-allylation product **117** were important steps in the syntheses of *dl*-progesterone[177] and lupeol,[178] respectively.

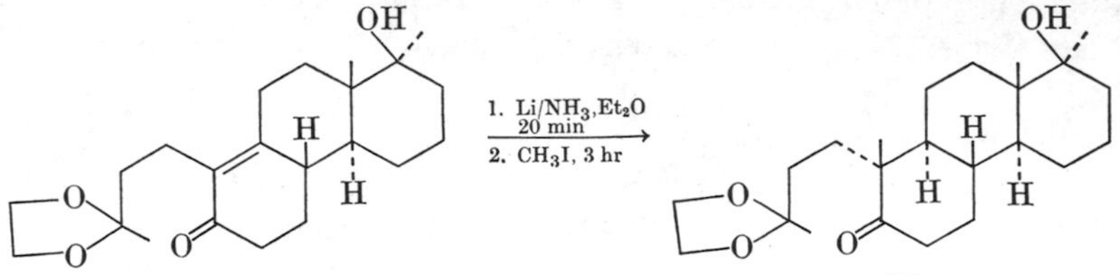

1. $Li/NH_3, Et_2O$, 20 min
2. CH_3I, 3 hr

114 → **115**

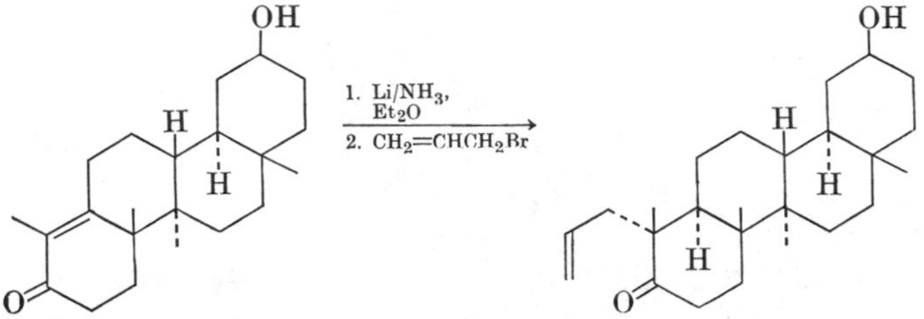

1. Li/NH_3, Et_2O
2. $CH_2{=}CHCH_2Br$

116 → **117**

The stereochemistry of methylation of reductively formed lithium 1- and 2-enolates in the *trans*-2-decalone series[179–181] and of the similarly prepared lithium 3-enolate from 4-methylcholest-4-en-3-one[182] has been established.

The enone reduction-enolate alkylation sequence originally demonstrated the unique behavior of lithium enolates (as opposed to other alkali metal enolates) toward alkylating agents.[13, 14] This led to the development of a number of new methods for generating such species.[67, 146, 183–185] In addition, new procedures for the base-promoted generation of kinetic and thermodynamic mixtures of lithium enolates of unsymmetrical ketones and conditions for alkylating them have been reported.[161, 183, 186]

177 G. Stork and J. E. McMurry, *J. Amer. Chem. Soc.*, **89,** 5464 (1967).

178 G. Stork, S. Uyeo, T. Wakamatsu, P. Grieco, and J. Labovitz, *J. Amer. Chem. Soc.*, **93** 4945 (1971).

179 R. S. Matthews, S. S. Girgenti, and E. A. Folkers, *Chem. Commun.*, **1970,** 708.

180 R. S. Matthews, P. K. Hyer, and E. A. Folkers, *Chem. Commun.*, **1970,** 38.

181 P. Lansbury and G. E. DuBois, *Tetrahedron Lett.*, **1972,** 3305.

182 G. P. Moss and S. A. Nicholaidis, *J. Chem. Soc., D*, **1969,** 1077.

183 H. O. House, L. J. Czuba, M. Gall, and H. D. Olmstead, *J. Org. Chem.*, **34,** 2324 (1969).

184 H. O. House and B. M. Trost, *J. Org. Chem.*, **30,** 2502 (1965).

185 G. Stork and P. F. Hudrlik, *J. Amer. Chem. Soc.*, **90,** 4464 (1968).

186 B. J. L. Huff, F. N. Tuller, and D. Caine, *J. Org. Chem.*, **34,** 3070 (1969).

If the ammonia is removed and replaced by anhydrous ether, reductively formed lithium enolates can be converted to β-keto esters in fair yield by carbonation, acidification, and treatment with diazomethane.[14, 187] The conversion of the octalone derivative **118** to the carbomethoxy decalone **119** illustrates this sequence.[187]

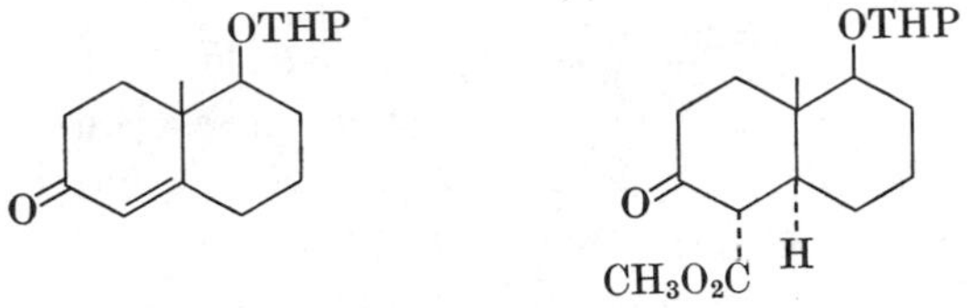

OTHP OTHP O O CH_3O_2C H

118 THP = tetrahydropyranyl group **119**

Specific enol acetates, enol benzoates, and enol carbonates have been prepared in good yield from reductively formed lithium enolates by replacing the ammonia with an ethereal solvent and adding acetic anhydride[95, 188] (or acetyl chloride[95]), benzoyl chloride,[95] or methyl chloroformate.[189] There has been described a particularly interesting route to specific olefins from reductively formed enolate anions.[190] For example, lithium-ammonia reduction of the octalone derivative **120**, phosphorylation of the resulting enolate with diethyl phosphorochloridate, and reduction of the enol phosphate product with lithium in ethylamine gave the 1-octalin **121** in 50% overall yield.

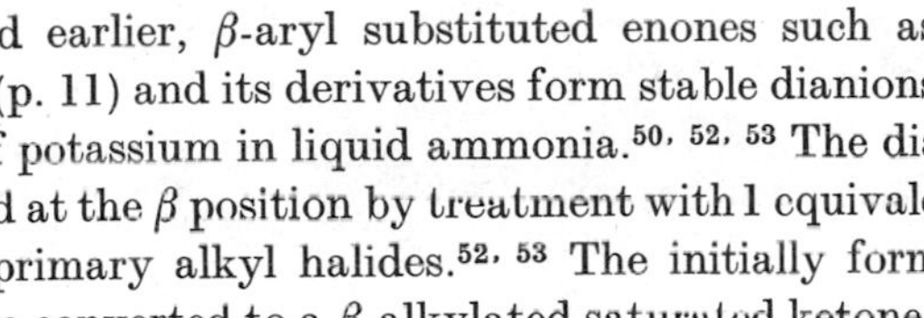

1. $Li/NH_3,Et_2O$
2. $(EtO)_2POCl,Et_2O$
3. $Li/EtNH_2$

120 **121**

As discussed earlier, β-aryl substituted enones such as benzalacetophenone **(16)** (p. 11) and its derivatives form stable dianions on treatment with 2 g-at. of potassium in liquid ammonia.[50, 52, 53] The dianions may be monoalkylated at the β position by treatment with 1 equivalent of benzyl,[50] allyl,[52, 53] or primary alkyl halides.[52, 53] The initially formed potassium enolate may be converted to a β-alkylated saturated ketone by addition of a proton donor, or into an α,β-dialkylated ketone by addition of a second

187 T. A. Spencer, T. D. Weaver, R. M. Villarica, R. J. Friary, J. Posler, and M. A. Schwartz, *J. Org. Chem.*, **33**, 712 (1968).

188 G. Stork, M. Nussim, and B. August, *Tetrahedron Suppl.*, **8**, 105 (1966).

189 T. A. Spencer, R. J. Friary, W. W. Schmiegel, J. F. Simeone, and D. S. Watt, *J. Org. Chem.*, **33**, 719 (1968).

190 R. E. Ireland and G. Pfister, *Tetrahedron Lett.*, **1969**, 2145.

equivalent of alkylating agent. The dianion **19** from benzalacetophenone (p. 11) reacts with benzophenone to form a γ-hydroxy ketone.[50]

Intramolecular Substitution, Addition, and Elimination Reactions Accompanying Metal-Ammonia Reductions of α,β-Unsaturated Carbonyl Compounds

Many α,β-unsaturated carbonyl compounds have structural features which allow intramolecular substitution, addition, and elimination reactions involving carbanionic intermediates to occur during metal-ammonia reduction. Since the original report that the unsaturated keto tosylate **30** (p. 18) and its 1-methyl derivative gave tricyclic ketones (*cf.* **31**) having a newly formed cyclopropane ring on lithium-ammonia reduction,[68] other examples have appeared of cyclopropane ring formation during reductions of 19-tosyloxy- and 19-halo-3-en-4-ones.[141,191,192] However, despite efforts to the contrary,[14] it has not been found possible to utilize this type of intramolecular substitution reaction for the formation of larger rings.

The isolation of cyclopropanol derivatives from lithium-ammonia reduction of keto enones having a suitably disposed carbonyl group (*cf.* **32**) has been reported recently.[69, 193] Earlier workers did not detect cyclopropanols on reduction of **32** under similar conditions.[164, 194, 195] The preparation of the bis-2,4-dinitrophenylhydrazone derivative of the reduction product, *trans*-10-methyl-2,5-decalindione, has been reported.[194] The cyclopropanol derivative **33** may have been converted into the decalindione on treatment with the 2,4-dinitrophenylhydrazine reagent.

Reductive intramolecular additions to carbonyl groups of esters also have been observed. For example, the enone ester **122** gave the perhydroindanedione **123** on reduction with lithium or potassium in liquid ammonia,[196] while the steroidal acetoxy dienone **124** yielded the stable hemiketal **125** on treatment with lithium in liquid ammonia.[197] Another interesting example of this type of reduction is found in the lithium-ammonia conversion of the bicyclic unsaturated triester **126** into the tricyclic keto diester **127**.[198]

[191] L. H. Knox, E. Velarde, and A. D. Cross, *J. Amer. Chem. Soc.*, **87**, 3727 (1965); **85**, 2533 (1963).

[192] S. Rakhit and M. Gut, *J. Amer. Chem. Soc.*, **86**, 1432 (1964).

[193] P. S. Venkataramani and W. Reusch, *Tetrahedron Lett.*, **1968**, 5283.

[194] C. B. C. Boyce and J. S. Whitehurst, *J. Chem. Soc.*, **1960**, 2680.

[195] V. Prelog and D. Zach, *Helv. Chim. Acta* **42**, 1862 (1959).

[196] R. G. Carlson and R. G. Blecke, *Chem. Commun.*, **1969**, 93.

[197] M. Tanabe, J. W. Chamberlin, and P. Y. Nishiura, *Tetrahedron Lett.*, **1961**, 601.

[198] B. M. Trost, Abstr. of Papers, Joint Conference CIC-ACS, Toronto, Canada, May 24–29, 1970, Organic Section, Paper No. 42.

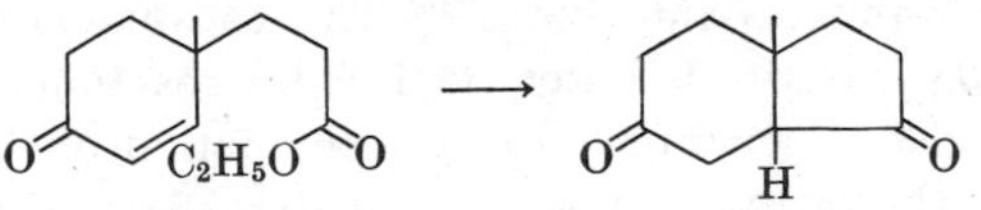

122 → 123

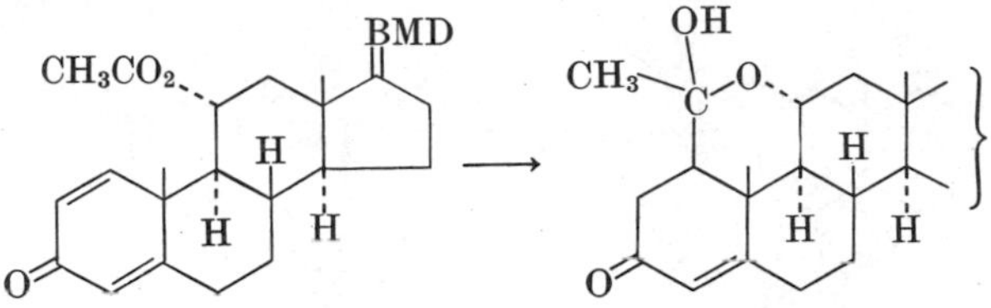

124 → 125

BMD = Bismethylenedioxy protected side chain

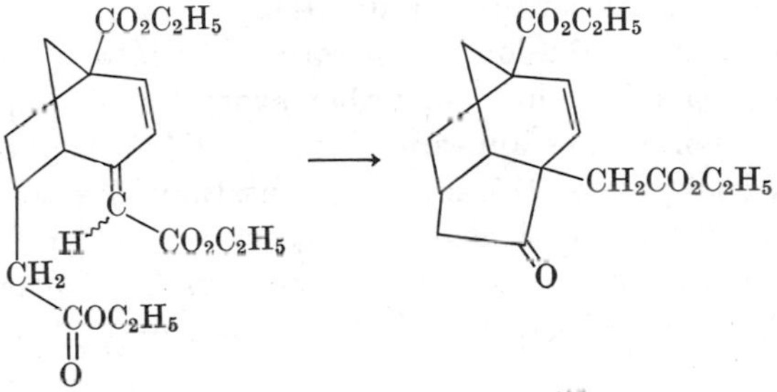

126 → 127

α,β-Unsaturated ketones having leaving groups at the γ position normally undergo reductive elimination to give initially metal dienolates on reaction with metals in ammonia (Eq. 5). Quenching these enolates with ammonium chloride allows the isolation of the β,γ-unsaturated ketone, but isomerization of the double bond into conjugation may be brought about by bases. Reductions have been reported that involve elimination of hydroxide ions,[199, 200] alkoxide ions,[201] and acetate ions,[70] as well as fission

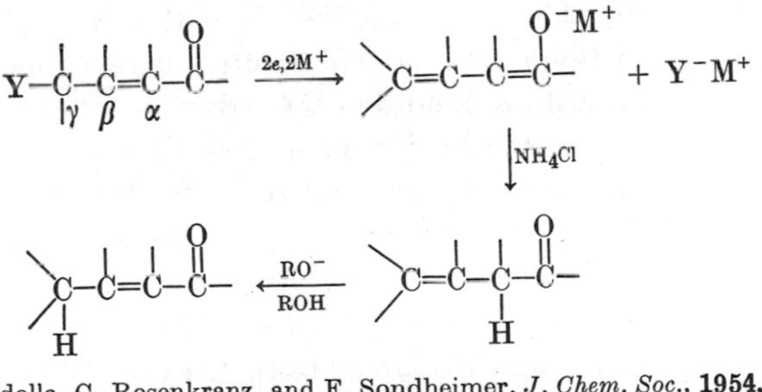

$$\text{Y—}\overset{|}{\underset{|\gamma}{\text{C}}}\text{—}\overset{|}{\underset{\beta}{\text{C}}}\text{=}\overset{|}{\underset{\alpha}{\text{C}}}\text{—}\overset{\text{O}}{\overset{\|}{\text{C}}}\text{—} \xrightarrow{2e,\,2M^+} \text{>C=}\overset{|}{\text{C}}\text{—}\overset{|}{\text{C}}\text{=}\overset{\text{O}^-\text{M}^+}{\overset{|}{\text{C}}}\text{—} + \text{Y}^-\text{M}^+$$

$$\downarrow \text{NH}_4\text{Cl}$$

$$\text{>}\underset{\text{H}}{\underset{|}{\text{C}}}\text{—}\overset{|}{\text{C}}\text{=}\overset{|}{\text{C}}\text{—}\overset{\text{O}}{\overset{\|}{\text{C}}}\text{—} \xleftarrow[\text{ROH}]{\text{RO}^-} \text{>C=}\overset{|}{\text{C}}\text{—}\underset{\text{H}}{\underset{|}{\overset{|}{\text{C}}}}\text{—}\overset{\text{O}}{\overset{\|}{\text{C}}}\text{—} \qquad \text{(Eq. 5)}$$

[199] C. Amendolla, G. Rosenkranz, and F. Sondheimer, *J. Chem. Soc.*, **1954**, 1226.

[200] T. Anthonsen, P. H. McCabe, R. McGrindle, and R. D. H. Murray, *Tetrahedron*, **25**, 2233 (1969).

[201] T. Masamune, A. Murai, K. Orito, H. Ono, S. Numata, and H. Suginome, *Tetrahedron*, **25**, 4853 (1969).

of lactone[202–204] and epoxide rings.[205] The conversion of solidagenone **(128)** to the β,γ-unsaturated ketone **129** by reaction with lithium in ammonia provides an example for the elimination of hydroxide ion.[200] The reaction of the 7α-hydroxy-8(14)-cholesten-15-one **(130)** with lithium

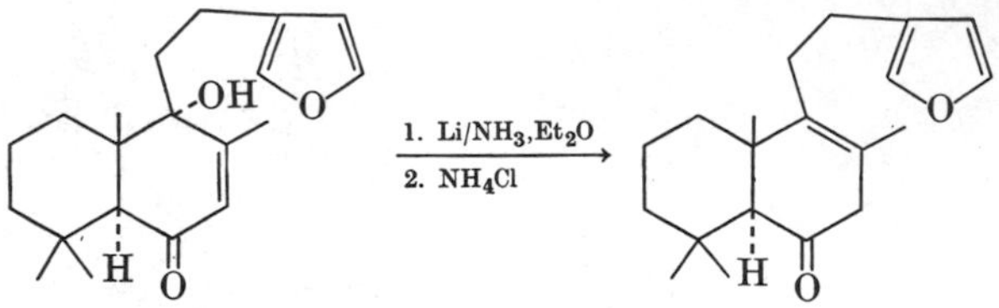

1. $Li/NH_3, Et_2O$
2. NH_4Cl

128 **129**

in liquid ammonia at −80° has been reported to lead to reduction of the 8,14 double bond rather than elimination of hydroxide ion.[133] In this case it appears that, at the low temperature of the reaction, elimination is slower than simple reduction. Other examples of simple reductions of γ-hydroxy enones have appeared recently.[142a,142b] The conversion of the diacetoxy enone **34** (p. 18) to the β,γ-unsaturated ketone **35** presumably involves reductive elimination of acetate ion from C-10 followed by elimination of acetate ion from C-5 of the resulting dienolate and reduction of the 1(9),5(10)-conjugated dienone product.[70]

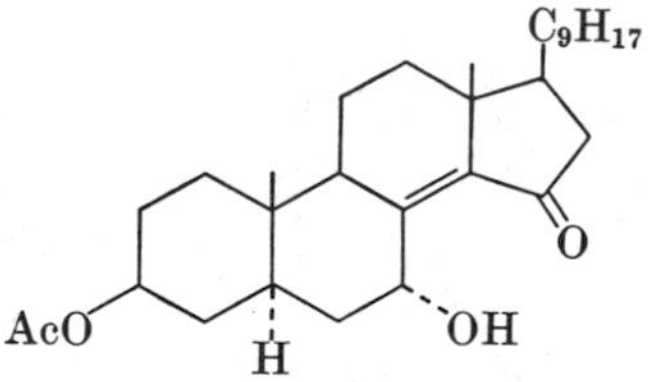

C_9H_{17} AcO OH H O

130

The reduction of santonin **(131)** with excess lithium in liquid ammonia yields the unsaturated keto acid **132**.[202–204] In this case, fission of the lactone ring and reduction of the less substituted double bond takes place despite the absence of proton donors in the reduction. Protonation of the initially formed trienolate **133** by ammonia may allow reformation of the cross-conjugated system which then undergoes further reduction of the less substituted double bond. Alternatively, reduction of **133** itself by excess metal may be involved.

[202] H. Bruderer, D. Arigoni, and O. Jeger, *Helv. Chim. Acta*, **39**, 858 (1956).
[203] R. Howe, F. J. McQuillin, and R. W. Temple, *J. Chem. Soc.*, **1959**, 363.
[204] K. S. Kulkarni and A. S. Rao, *Tetrahedron*, **21**, 1167 (1965).
[205] K. Irmscher, W. Beerstecher, H. Metz, R. Watzel, and K.-H. Bork, *Chem. Ber.*, **97**, 3363 (1964).

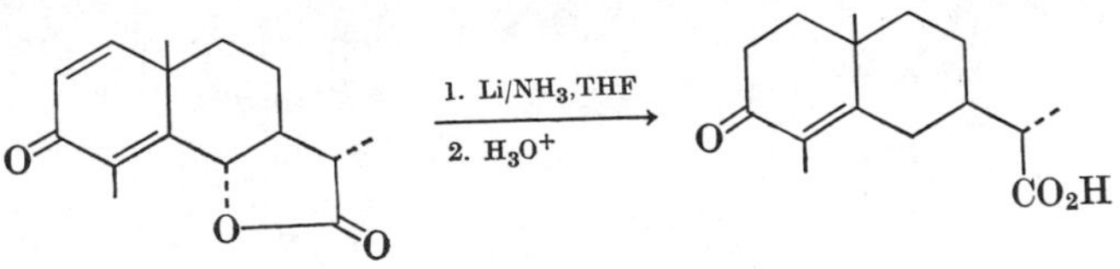

1. Li/NH_3,THF
2. H_3O^+

CO_2H

131 132

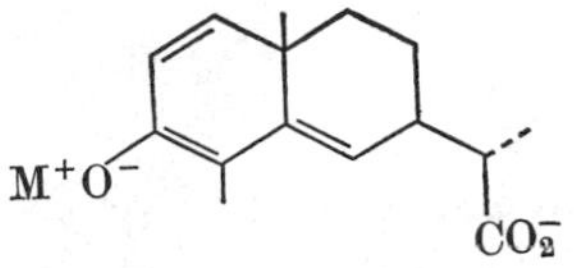

M^+O^-

CO_2^-

133

Cross-conjugated dienones having anionic leaving groups at the γ position undergo aromatization on treatment with metals in liquid ammonia.[197, 206–209] The reduction of the prednisone derivative **134** to the *seco*-steroid **135** and aromatizations of dienone rings in alkaloids of the proapophine group,[209] satularidine,[206] O-methylandrocymbine,[208] and erysodienone[207] provide examples of these reactions.

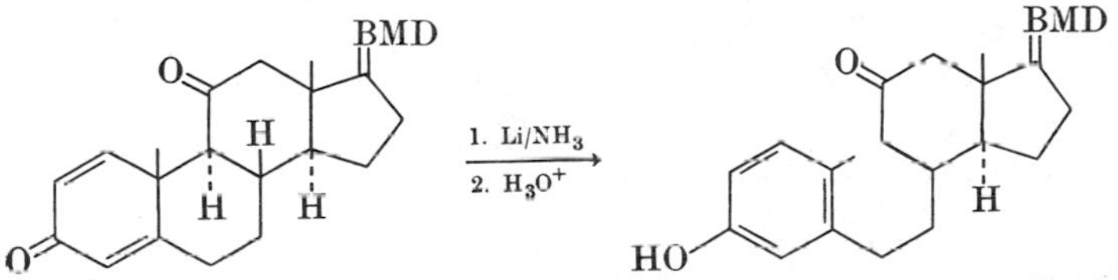

BMD BMD

1. Li/NH_3
2. H_3O^+

HO

BMD = bismethylenedioxy protected side chain

134 135

Reductions of α,β-Unsaturated Carbonyl Compounds Having Leaving Groups at the β Position

On metal-ammonia reduction α,β-unsaturated carbonyl compounds having leaving groups at the β position form metal enolates which may undergo elimination to give new α,β-unsaturated carbonyl compounds that are susceptible to further reduction (Eq. 6).[43, 170, 210–215] In a study of

206 D. H. R. Barton, D. S. Bhakuni, R. James, and G. W. Kirby, *J. Chem. Soc., C*, **1967**, 128.

207 D. H. R. Barton, R. B. Boar, and D. A. Widdowson, *J. Chem. Soc., C*, **1970**, 1208.

208 A. R. Battersby, R. B. Berbert, L. Pijewska, and F. Santavy, *Chem. Commun.*, **1965**, 228.

209 M. P. Cava, K. NoMura, R. H. Schlessinger, K. T. Buck, B. Douglas, R. F. Raffaud, and J. A. Woisbach, *Chem. Ind.* (London), **1964**, 282.

210 R. M. Coates and J. E. Shaw, *Tetrahedron Lett.*, **1968**, 5405; *J. Org. Chem.*, **35**, 2597 (1970).

211 R. M. Coates and J. E. Shaw, *J. Org. Chem.*, **35**, 2601 (1970).

212 R. E. Ireland and J. A. Marshall, *J. Org. Chem.*, **27**, 1615 (1962).

213 M. Vandewalle and F. Compernolle, *Bull. Soc., Chim. Belges*, **75**, 349 (1966).

214 M. Vandowalle and F. Compernolle, *Bull. Soc. Chim. Belges*, **76**, 43 (1967).

215 D. S. Watt, J. M. McKenna, and T. A. Sponcer, *J. Org. Chem.*, **32**, 2674 (1967).

$$\begin{array}{l} (Y)C{=}C{-}C({=}O){-} \xrightarrow[H^+]{2e,\,M^+} H{-}C(Y){-}C{=}C(O^-M^+){-} \xrightarrow{H^+} H{-}C(Y){-}C(H){-}C({=}O){-} \\ \qquad\qquad\qquad\qquad \downarrow -Y^- \\ \qquad\qquad\qquad (H)C{=}C{-}C({=}O){-} \xrightarrow[2H^+]{2e} H{-}C(H){-}C(H){-}C({=}O){-} \end{array} \qquad \text{(Eq. 6)}$$

reductions of 3-ethoxycyclohex-2-enone (**136a**) with lithium in liquid ammonia, it was found that the extent of double reduction (reduction, elimination and subsequent reduction) compared with that of simple reduction was strongly dependent upon the temperature and the reaction time before quenching with a proton donor.[215] By the use of low temperature (−78°) and short reaction times the double reduction process could be completely prevented; however, if the reaction was conducted at −33° and the mixture was stirred for 1 hour or more prior to quenching, the products of double reduction were the only ones observed. The formation of saturated alcohols was observed when ethyl alcohol rather than ammonium chloride was employed as the quenching agent. In a related study involving reduction of 3-isobutoxycyclohex-2-enone (**136b**), the elimination process was significantly retarded if magnesium rather than lithium was employed as the reducing metal.[43] The greater covalent character of the oxygen-metal bond of the initial enolate intermediate when magnesium is involved adequately accounts for this observation.

O
OR
136

a, R = C_2H_5
b, R = $(CH_3)_2CHCH_2$

β-Alkoxy α,β-unsaturated esters[210, 211] and acids[216] have been found to undergo double reduction. This procedure has been used in an elegant way to prepare the ester **138**, a key intermediate in the total synthesis of eremophilane sesquiterpenes, from the alkoxy ester **137**.[210] It should be noted that kinetic protonation of the ester enolate intermediate leads to the formation of the thermodynamically less stable product having the carbethoxy group in the axial orientation.[210]

n-Butylthiomethylene derivatives of ketones undergo double reduction with lithium and liquid ammonia to form methyl-substituted lithium enolates that can be alkylated with methyl or other alkyl halides.[170]

[216] J. E. Shaw and K. K. Knutson, *J. Org. Chem.*, **36**, 1151 (1971).

$$\mathbf{137} \xrightarrow[\text{2. } NH_4Cl, -70°]{\text{1. } Li/NH_3, Et_2O, -33°, 12 \text{ min}} \mathbf{138}$$

As illustrated in Eq. 7, this scheme provides a convenient method of selective geminal alkylation of methylene ketones. This sequence, like the sodium-ethanol-ammonia reduction of the *n*-butylthiomethylene

$$>CH-C(=O)-C(=CHSBu\text{-}n)- \xrightarrow[Et_2O, \text{ 2eq. } H_2O]{\text{6eq. } Li/NH_3} >CH-C(O^-Li^+)=C(CH_3)-$$

$$\xrightarrow{RX} >CH-C(=O)-C(R)(CH_3)- \qquad \text{(Eq. 7)}$$

derivative of 9-methyl-6-octalin-1-one to 2,9-dimethyl-6-octalin-1-ol reported earlier,[212] presumably proceeds via enolate elimination of the *n*-butylthio anion.[170,215] However, simple alkyl vinyl sulfides are known to undergo cleavage with sodium in liquid ammonia.[217]

Metal-Ammonia Reductions of Linear- and Cross-Conjugated Dienones

Both linear- and cross-conjugated dienones are reduced by solutions of metals in liquid ammonia. Heteroannular systems such as steroidal 4,6-dien-3-ones (**139**) and related compounds are reduced initially to 3,5-dienolates (**140**).[172, 218–225] As expected, δ-protonation occurs axially with respect to ring B.[218, 221] Addition of ammonium chloride to the enolate **140** leads to the formation of the 5-en-3-one **141**; [172, 222] addition of proton donors such as ethanol or water leads to the 4-en-3-one **142**.[219, 220] When the latter proton donors are employed, the initially formed β,γ-unsaturated ketone undergoes isomerization to the more stable α,β-unsaturated

[217] L. Brandsma, *Rec. Trav. Chim. Pays-Bas*, **89**, 593 (1970).

[218] J. A. Campbell and J. C. Babcock, *J. Amer. Chem. Soc.*, **81**, 4069 (1959).

[219] A. F. Daglish, J. Green, and V. D. Poole, *J. Chem. Soc.*, **1954**, 2627.

[220] F. Johnson, G. T. Newbold, and F. S. Spring, *J. Chem. Soc.*, **1954**, 1302.

[221] J. A. Marshall and H. Roebke, *J. Org. Chem.*, **33**, 840 (1968).

[222] M. Nussim, Y. Mazur, and F. Sondheimer, *J. Org. Chem.*, **29**, 1120 (1964).

[223] H. Van Kamp, P. Westerhof, and H. Niewind, *Rec. Trav. Chim. Pays-Bas*, **83**, 509 (1964).

[224] E. Wenkert, A. Afonso, J. B. Bredenberg, C. Kaneko, and A. Tahara, *J. Amer. Chem. Soc.*, **86**, 2038 (1964).

[225] P. Westerhof and E. H. Reerink, *Rec. Trav. Chim. Pays-Bas*, **79**, 771 (1960).

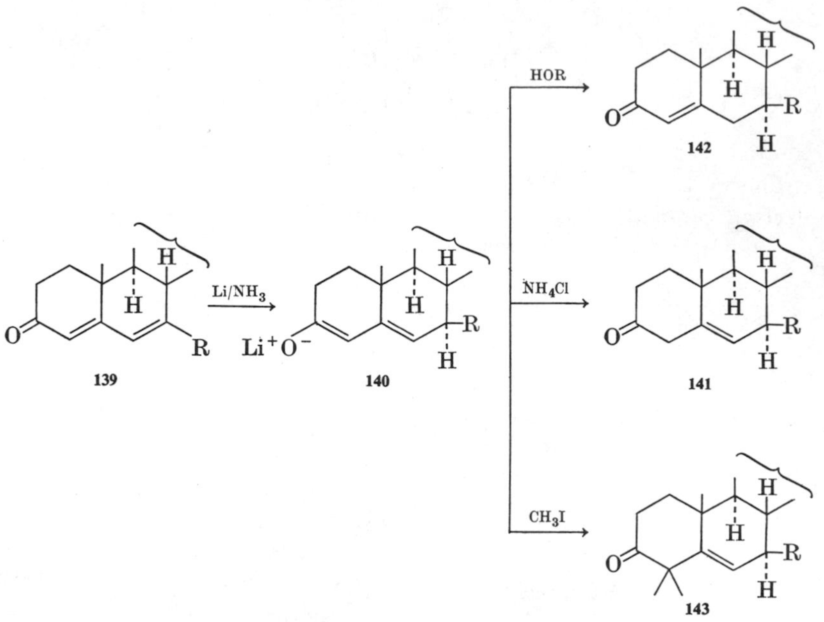

HOR

142

Li/NH$_3$

NH_4Cl

139

140

141

CH_3I

143

system. Treatment of the conjugate enolate **140** obtained from lithium-ammonia reduction of 6-dehydrotestosterone with excess methyl iodide gives the 4,4-dimethyl-5-en-3-one **143** in low yield.[172] Recently, improved yields of *gem*-dimethylated products derived from bicyclic and steroidal linearly conjugated dienones have been obtained by this method.[225a] A detailed explanation has been provided for the formation of *gem*-dimethyl products in the alkylation of enolates such as **140**.[226] In addition to compounds of the type **139** metal-ammonia reductions of a number of other alicyclic heteroannular dienones have been reported.[221,227–229] Linearly conjugated dienones may be completely reduced to saturated alcohols using excess lithium in liquid ammonia.[229]

Homoannular dienones also are reduced with metals in liquid ammonia.[230–232] Interestingly, treatment of the steroidal 14,16-dien-2-one **144** with lithium in liquid ammonia[230] or lithium in liquid ammonia containing

[225a] K. P. Dastur, *Tetrahedron Lett.*, **1973,** 4333.
[226] H. J. Ringold and S. K. Malhotra, *J. Amer. Chem. Soc.*, **84,** 3402 (1962).
[227] K. K. Pivnitsky, N. N. Gaidamovich, and I. V. Torgov, *Tetrahedron*, **22,** 2837 (1966).
[228] G. Stork, *Pure Appl. Chem.*, **1969,** 383.
[229] A. Zurcher, H. Heusser, O. Jeger, and P. Geistlich, *Helv. Chim. Acta*, **37,** 1562 (1954).
[230] G. Bach, J. Capitaine, and Ch. R. Engel, *Can. J. Chem.*, **46,** 733 (1968).
[231] K. Grimm, P. S. Venkataramani, and W. Reusch, *J. Amer. Chem. Soc.*, **93,** 270 (1971).
[232] H. Heusser, M. Roth, O. Rohr, and R. Anliker, *Helv. Chim. Acta*, **38,** 1178 (1955).

n-propanol[232] leads mainly to reduction of the 16,17 double bond; the 14-en-20-one **145**[230] and the related 20-hydroxy derivative **146** are obtained.[232]

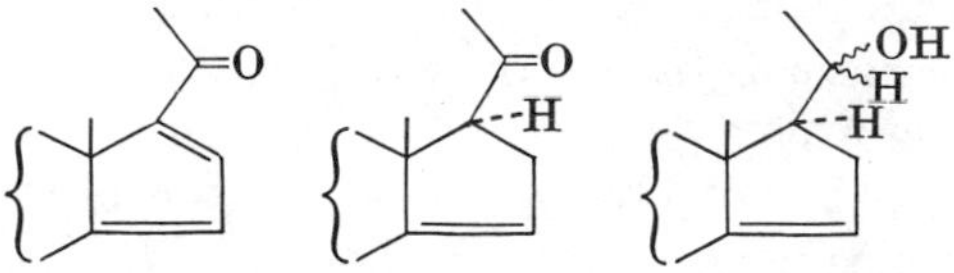

144 **145** **146**

The less substituted double bond of cross-conjugated dienones such as the steroidal 1,4-dien-3-one **147**[19, 172, 233, 234] or santonin **(131)** and related compounds[202–204, 235] is selectively reduced by lithium in liquid ammonia. This result is expected in light of the mechanism of the reduction of enones discussed earlier (p. 8). Dehydrofukinone **(148)** undergoes selective reduction of the exocyclic double bond on treatment with lithium in liquid ammonia.[236]

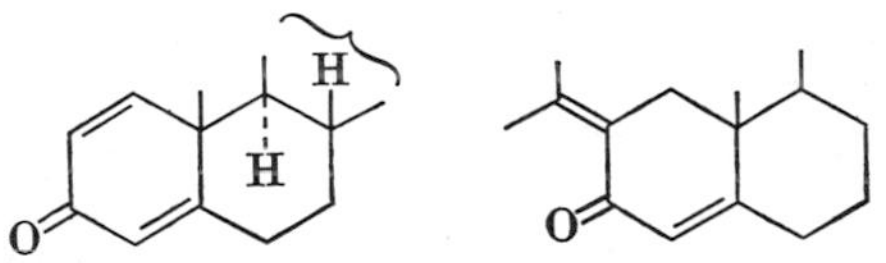

147 **148**

Selective Reduction of α,β-Unsaturated Carbonyl Compounds Containing Other Reducible Groups

Although a host of groups are reduced by metal ammonia solutions,[16, 18, 20–22, 24] it is often possible to reduce double bonds of α,β-unsaturated carbonyl systems without affecting other reducible groups. Selective reductions of conjugated enones in the presence of an internal isolated double bond generally present no difficulty.[237] Such double bonds, unless they have very low-lying antibonding orbitals[238] or special structural features which stabilize radical anion intermediates,[239] are normally stable to metal-ammonia solutions. However, nonconjugated terminal olefins may be reduced on extended exposure to metal-ammonia solutions containing alcohols as proton donors.[139] Although carbon-carbon double

[233] W. F. Johns, *J. Org. Chem.*, **36,** 711 (1971).

[234] E. Shapiro, T. Legatt, L. Weber, M. Steinberg, and E. P. Oliveto, *Chem. Ind.* (London), **1962,** 300.

[235] W. Cocker, B. Donnelly, H. Gobinsingh, T. B. H. McMurry, and N. A. Nisbet, *J. Chem. Soc.*, **1963,** 1262.

[236] K. Naya, I. Takagi, Y. Kawaguchi, and Y. Asada, *Tetrahedron,* **24,** 5871 (1968).

[237] C. Djerassi, D. Marshall, and T. Nakano, *J. Amer. Chem. Soc.*, **80,** 4853 (1958).

[238] B. R. Ortiz de Montellano, B. A. Loving, T. C. Shields, and P. D. Gardner, *J. Amer. Chem. Soc.*, **89,** 3365 (1967).

[239] D. J. Marshall and R. Deghenghi, *Can. J. Chem.* **47,** 3127 (1969).

bonds that are conjugated with other multiple bonds or aromatic rings are readily reduced by metal-ammonia solutions, selective reductions of enones containing such groups have been reported. Examples are the reduction-ethylation of 3-methoxy-6-methylpregna-3,5,16-trien-20-one **(149)** and the corresponding 3-acetoxy compound.[67] Thus conjugated ketones may be protected from reduction by conversion into the corresponding conjugated enol ethers or enol acetates.[16, 240] Also, certain styrenoid compounds have been found not to undergo reduction unless proton donors are present.[241] The selective reduction of the 4,5 double bond of the tetracyclic enone **150** has been effected by treating it with a limited amount of lithium in liquid ammonia at low temperature and excluding proton donors.[242] A small amount of a product resulting from

1. 2 eq. Li/NH_3, THF
2. EtI
3. H_3O^+

149 (17%)

1. Li/NH_3, dioxane-Et_2O, −70°
2. NH_4Cl

150 (53%)

cleavage of the aromatic ether was also isolated. Reductions of **150**[242] and the related compound having a methoxy group[243] at C-18 with a limited amount of lithium in ammonia-ethanol gave the corresponding saturated ketones, as well as the derived alcohols, having both the 4,5 and 11,12 double bonds reduced.

Mono- and polycyclic aromatic compounds undergo reduction with metals in liquid ammonia (Birch reduction)[1, 16, 20, 23, 24, 244, 245] but these

[240] J. S. Mills, H. J. Ringold, and C. Djerassi, *J. Amer. Chem. Soc.*, **80,** 6118 (1958).

[241] J. E. Cole, Jr., W. S. Johnson, P. A. Robins, and J. Walker, *J. Chem. Soc.*, **1962,** 244.

[242] W. Nagata, T. Terasawa, S. Hirai, and K. Takeda, *Tetrahedron Lett.*, No. 17, 27 (1960); *Chem. Pharm. Bull.* (Tokyo), **9,** 769 (1961).

[243] W. S. Johnson, E. R. Rogier, J. Szmuszkovicz, H. I. Hadler, J. Ackerman, B. K. Bhattacharyya, B. M. Bloom, L. Stalmann, R. A. Clement, B. Bannister, and H. Wynberg, *J. Amer. Chem. Soc.*, **78,** 6289 (1956).

[244] Ref. 18, Djerassi, Chap. 6.

reactions are usually slow unless proton donors are added. Thus it is possible to reduce α,β-unsaturated ketones selectively in the presence of aromatic rings,[242, 243, 246] even relatively easily reducible rings such as those containing electron-withdrawing carboxylate groups.[247] The reduction of the enone **151**[166] illustrates a selective reduction in the presence of a reducible indole ring.[248]

1. Li/NH_3 THF, 2 min
2. C_6H_5Br
3. H_2O

151 (47%) (25%)

The selective lithium-ammonia reduction of a 1(9)-octalin-2-one derivative containing a reducible unconjugated triple bond has been reported.[249] Ethynyl carbinols are reduced to allyl alcohols and eventually to olefins with metal-ammonia solutions containing proton donors.[250] However, by excluding proton donors, selective reduction of conjugated enones in the presence of ethynyl carbinol groups has been carried out.[164,251–253] Presumably the enone system undergoes reduction first and the metal amide formed as the reduction proceeds converts the ethynyl carbinol group into an alkoxide which is resistant to further reduction. Relatively low solubility of the enolate-alkoxide double salts, particularly those derived from steroidal enones,[19] may also be an important factor accounting for the slow reduction of the ethynyl group. Selective reductions of conjugated enones containing allylic alcohol groupings, which may undergo fission with metals in ammonia,[16, 23, 24] have also been reported.[164, 251, 253]

Carbon-halogen bonds of alkyl and vinyl halides are readily cleaved by metals in ammonia.[20, 23, 24] However, the 1,2 double bond of the steroid

[245] R. G. Harvey, *Synthesis*, No. 4, 161 (1970).

[246] W. F. Johns, *J. Org. Chem.*, **28**, 1856 (1963).

[247] W. S. Johnson, J. M. Cox, D. W. Graham, and H. W. Whitlock, Jr., *J. Amer.Chem.Soc.*, **89**, 4524 (1967).

[248] M. V. R. Koteswara Rao, G. S. Krishna Rao, and S. Dev, *Tetrahedron*, **22**, 1977 (1966).

[249] P. Lansbury and G. T. DuBois, *J. Chem. Soc., D*, **1971**, 1107.

[250] F. B. Colton, L. N. Nysted, B. Riegel, and A. L. Raymond, *J. Amer. Chem., Soc.* **79**, 1123 (1957).

[251] A. Bowers, H. J. Ringold, and E. Denot, *J. Amer. Chem. Soc.*, **80**, 6115 (1958).

[252] I. A. Gurvich, V. F. Kucherov, and T. V. Ilyakhina, *J. Gen. Chem. USSR*, **31**, 738 (1961).

[253] P. S. Venkataramani, J. P. John, V. T. Ramakrishnan, and S. Swaminathan, *Tetrahedron*, **22**, 2021 (1966).

derivative **152** was reduced in good yield without removal of fluorine by using lithium in liquid-ammonia and employing a short reaction time.[234] The resistance of alkyl fluorides to reduction by metals in liquid ammonia at $-33°$ has been noted.[85] The conjugated double bond of the bicyclic ketone **153** was reduced without cleavage of the vinyl halide through inverse addition.[254] Selective reductions of this type have potentially wide application in connection with ring annelation sequences. Under normal conditions both the enone and the vinyl halide groups are reduced.[254, 255]

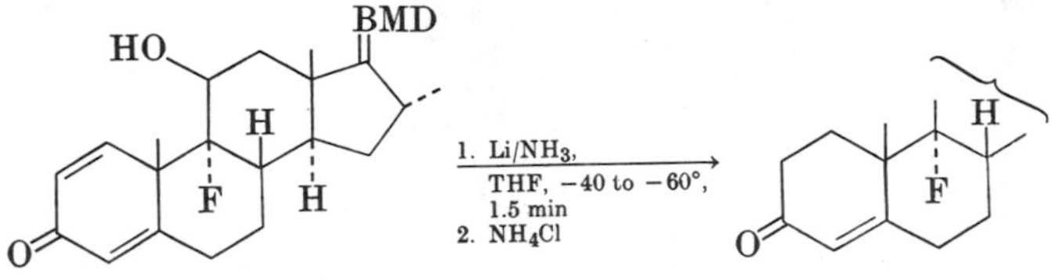

HO BMD H F H 1. Li/NH_3, THF, -40 to $-60°$, 1.5 min 2. NH_4Cl H F

152

BMD = bismethylenedioxy group protected side chain

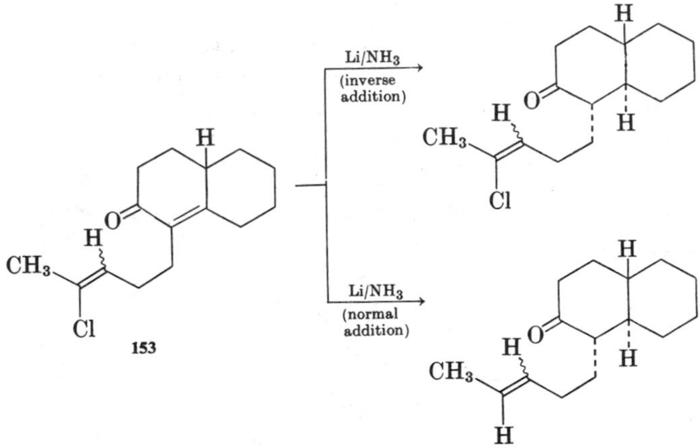

Li/NH_3 (inverse addition) H CH_3 H Cl H

153 H CH_3 H Cl

Li/NH_3 (normal addition) H CH_3 H H H

Although carbon-sulfur bonds of thioethers, thioketals, and hemithioketal are readily cleaved by metals in ammonia,[20, 23, 24] the double bond of steroidal 4-en-3-ones, such as **154**, containing a thioalkyl ether grouping at C-7 may be selectively reduced with lithium in liquid ammonia without carbon-sulfur bond cleavage.[256, 257] Carbon-sulfur bond cleavage did occur

[254] P. T. Lansbury, P. C. Briggs, T. R. Demmin, and G. E. DuBois, *J. Amer. Chem. Soc.*, **93**, 1311 (1971).

[255] D. H. R. Barton and D. Kumari, *Ann. Chem.*, **737**, 108 (1970).

[256] H. Kaneko, K. Nakamura, Y. Yamoto, M. Kurokawa, *Chem. Pharm. Bull.* (Tokyo), **17**, 11 (1969).

[257] R. E. Schaub and M. J. Weiss, *J. Org. Chem.*, **26**, 3915 (1961).

in those compounds in which the R groups of **154** were allyl, aryl, or benzyl.[256]

The reduction of the conjugated double bond of 2α-cyano-2β-methyl-4-cholesten-3-one (**61**, p. 29) provides an example of a selective enone

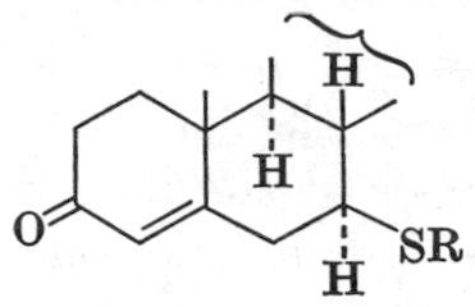

154

reduction in the presence of a reducible tertiary nitrile group.[99] Tertiary nitriles may be reduced to hydrocarbons by metal-ammonia solutions.[258] Because of salt formation, carboxylic acids are not reduced by metals in liquid ammonia; but acid derivatives such as esters and amides can be reduced in such media. Esters can be converted to saturated alcohols by acyl-oxygen fission or, in hindered systems, carboxylic acids may be formed by alkyl-oxygen fission.[259] Amides can be converted to aldehydes or alcohols depending upon the acid strength of the proton donor employed.[16, 23] However, metal-ammonia reductions of enones are more rapid than those of either amides or esters. Thus the double bond of the enone diester **79** (p. 33) is selectively reduced using a limited amount of lithium.[118] Also the keto amide **156** may be obtained by lithium-ammonia reduction of the enone **155** if a short reaction time is employed.[260] On extended exposure to lithium in liquid ammonia the amide group of **155** is

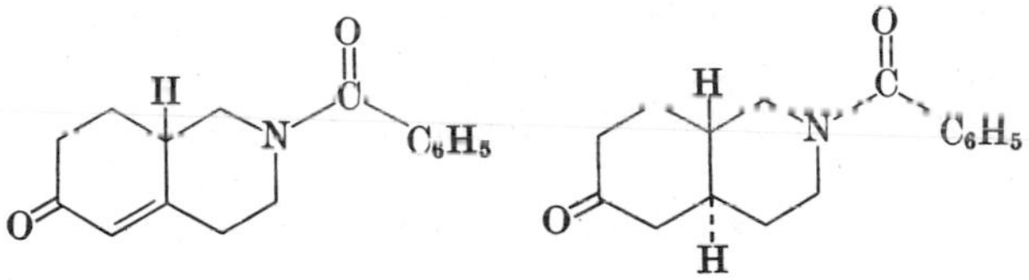

155 **156**

also reduced.[261] Other examples of selective reductions of α,β-unsaturated ketones in the presence of amide groups are found in the lithium ammonia reductions of the enone **151**[166] and of dihydroisojervine triacetate.[262]

[258] P. G. Arapakos, *J. Amer. Chem. Soc.*, **89**, 6794 (1967).

[259] E. Wenkert and B. G. Jackson, *J. Amer. Chem. Soc.*, **80**, 217 (1958).

[260] S. Dubé and P. Deslongchamps, *Tetrahedron Lett.*, **1970**, 101.

[261] R. L. Augustine, *Catalytic Hydrogenation*, Marcel Dekker, New York, 1965, p. 44–60, 60–66.

[262] W. G. Dauben, W. W. Epstein, M. Tanabe, and B. Weinstein, *J. Org. Chem.*, **28**, 293 (1963)

Since reduction potentials of α,β-unsaturated ketones are less negative than those of saturated ketones,[20] selective reductions of the former groups in the presence of the latter ones are probably possible. However, rather than attempting selective reductions, most workers have found it more convenient to protect saturated ketone groups as the ketals or enol ethers (which are not reducible unless the alkoxyl groups are allylic or benzylic),[16] or simply to reduce the conjugated and unconjugated ketone groups to saturated alcohols and to reoxidize both groups with Jones's Reagent.[168] Leaving groups such as halo, amino, acyloxyl, methoxyl, and hydroxyl that are α to carbonyl groups are generally eliminated by solutions of metals in liquid ammonia.[18, 263, 264] For example, the α-acetoxy dienone **157** gave the ketone **158** on reduction with calcium-ammonia-methanol followed by oxidation.[265] In this case the proton donor allows protonation of the initially formed enolate, derived either from reductive removal of the acetoxyl group or reduction of the carbon-carbon double bond of the conjugated enone, so that the second reduction step can take place.

1. Ca/NH_3, THF
2. MeOH
3. CrO_3

157 **158**

The conversion of $1\alpha,2\alpha$-epoxy-4,6-dien-3-ones into the corresponding $1\alpha,3\beta$-dihydroxy-5-ene sterols, intermediates in the synthesis of vitamin D_3 derivatives, with lithium in liquid ammonia and ammonium chloride provide interesting examples of reductions involving cleavage of α',β'-epoxides.[265a,265b]

α'-Hydroxyl groups of α,β-unsaturated ketones may be removed or retained depending upon the reduction conditions employed. For example, treatment of 3β-acetoxy-5β-hydroxylumista-7,22-diene-6-one with excess lithium in liquid ammonia for a 30-minute period led to reduction of the 7,8 bond and removal of the 5β-hydroxyl group,[105] while only the 7,8 bond was reduced when 3β-acetoxy-5α-hydroxyergosta-7,22-diene-6-one

[263] J. H. Chapman, J. Elks, G. H. Phillips, and L. J. Wyman, *J. Chem. Soc.*, **1956**, 4344.

[264] E. S. Rothman and M. E. Wall, *J. Amer. Chem. Soc.*, **79**, 3228 (1957).

[265] R. F. Church, R. E. Ireland, and D. R. Shridhar, *J. Org. Chem.*, **27**, 707 (1962).

[265a] D. H. R. Barton, R. H. Hesse, M. M. Pechet, and E. Rizzardo, *J. Amer. Chem. Soc.*, **95**, 2748 (1973).

[265b] D. H. R. Barton, R. H. Hesse, M. M. Pechet, and E. Rizzardo, *Chem. Commun.*, **1974**, 203.

was treated with lithium in ammonia for a short time.[9] These results suggest that in the first reaction the enolate derived from reduction of the double bond underwent protonation to give an α-hydroxy ketone which was further reduced to the saturated ketone.[266] Although selective reduction of α,β-unsaturated ketones having other leaving groups at the α' position may be possible, such reactions do not appear to have been reported.

Conjugated cyclopropyl ketones are readily reduced by solutions of metal in liquid ammonia.[20, 267] In the absence of proton donors, ketones conjugated with both a double bond and a cyclopropane ring are easily reduced to saturated cyclopropyl ketones.[98, 100, 122, 268–272] Reductions of ($\pm$)-4-demethylaristolone,[98] (+)-5-*epi*-4-demethylaristolone (**59**, p. 28),[98] and ($\pm$)-aristolone[270] to the corresponding dihydro products provide examples of these reactions.

Generally the presence of unprotected hydroxyl groups causes no complications during metal-ammonia reduction of enones. These groups may serve as internal proton donors and, for the reasons discussed earlier, lead to improved yields of reduction products. However, hydroxyl groups in close proximity to the β position of an enone may influence the reduction stereochemistry.[141] If protection of hydroxyl groups during enone reductions is desirable, the tetrahydropyranyl protecting group should be used.[273] This group is stable to metal-ammonia reduction, whereas other protecting groups such as acetate and benzoate are cleaved by metal-ammonia solutions.[18] The bismethylenedioxy group which is often used for protection of the C-17 side chain of cortisone and its derivatives is also stable to metal-ammonia reduction.[18]

Comparison with Other Methods of Reduction

Before the introduction of metal-ammonia solutions for the reduction of α,β-unsaturated carbonyl compounds,[3, 6, 8, 234, 274] sodium, sodium amalgam, or zinc in protic media were most commonly employed for this purpose. Some early examples of their use for the reduction of α,β-unsaturated ketones are the conversion of carvone to dihydrocarvone with zinc

[266] G. Just and K. St. C. Richardson, *Can. J. Chem.*, **42**, 456 (1964).

[267] S. W. Staley, *Selec. Org. Transform.*, **2**, 97 (1972).

[268] D. E. Evans, G. S. Lewis, P. J. Palmer, D. J. Weyell, *J. Chem. Soc., C*, **1968**, 1197.

[269] R. Fraisse-Jullien and C. Frejaville, *Bull. Soc. Chim. Fr.*, **1968**, 4449.

[270] E. Piers, R. W. Britton, and W. deWaal, *Can. J. Chem.*, **47**, 4307 (1969).

[271] M. J. Thompson, C. F. Cohen, and S. M. Lancaster, *Steroids*, **5**, 745 (1965).

[272] R. Wiechert and E. Kaspar, *Chem. Ber.*, **93**, 1710 (1960).

[273] R. L. Clarke and C. M. Martini, *J. Amer. Chem. Soc.*, **81**, 5716 (1959).

[274] F. Sondheimer, O. Mancera, G. Rosenkranz, and C. Djerassi, *J. Amer. Chem. Soc.*, **75**, 1282 (1953).

in acid or alkaline medium,[275] and the conversion of carvone to dihydrocarveol and of cholest-4-en-3-one to cholestanol with sodium in alcohol.[276, 277] Reductions using these earlier methods may be complicated by a variety of side reactions—such as over-reduction, dimerization, skeletal rearrangements, and acid- and base-catalyzed isomerization of the starting material—which can be avoided entirely or significantly minimized by the use of metal-ammonia reagents.

Reduction of α,β-unsaturated ketones with sodium in alcohols leads to reduction of both the carbon-carbon double bond and the carbonyl group, and, as illustrated by the conversion of the tricyclic enone **159** to the saturated alcohol **160** with sodium and *n*-pentyl alcohol, the reaction usually gives rise to the more thermodynamically stable product.[278, 279] As in metal-ammonia reductions, the conversion of the dienone **161** to the unsaturated alcohol **162** with sodium in ethyl alcohol shows that isolated double bonds are not reduced by metal-alcohol combinations.[280]

Na, n-$C_5H_{11}OH$, reflux

159 → **160**

Na, EtOH

161 → **162**

Clearly the use of sodium-alcohol combinations for enone reductions does not offer the flexibility afforded by metal-ammonia combinations, because with the latter reagents the reduction may be halted at the metal enolate or the saturated ketone stage. Also, the former reagents may cause base-catalyzed isomerizations at labile asymmetric centers of the starting material before reduction. Because reductions with metal in ammonia are very rapid and because strongly basic species are generated only as the reduction proceeds, such isomerizations normally may be avoided by using metal-ammonia reagents. Indeed, in some of the earliest examples of

[275] O. Wallach, *Ann.* **279,** 377 (1894).

[276] O. Wallach, *Ann.*, **275,** 111 (1893).

[277] O. Diels and E. Abderhalden, *Chem. Ber.*, **39,** 884 (1906).

[278] W. S. Johnson, *Chem. Ind.* (London), **1956,** 167.

[279] A. R. Pinder and R. Robinson, *J. Chem. Soc.*, **1955,** 3341.

[280] L. H. Zalkow, F. X. Markley, and C. Djerassi, *J. Amer. Chem. Soc.*, **82,** 6354 (1960).

the application of lithium-ammonia solutions for the reduction of α,β-unsaturated ketones, esters of 22a,5α-spirosten-3β-ol-8-en-11-one were converted into 22a,5α-spirostan-3β-ol-11-one in good yield;[5, 6, 274] but reductions of the same enones with metal-alcohol combinations gave intractable material, probably because partial isomerization at C-14 of the starting material occurred before reduction of the double bond.[274]

Base-catalyzed aldol condensations of starting materials and intermediate reduction products are also important side reactions in metal-alcohol reductions. Recently, it has been shown that certain α,β-unsaturated aldehydes[281] and ketones[282] that are especially prone to undergo base-catalyzed condensations give good yields of tetrahydro products when reduced with sodium in aqueous ammonia rather than sodium in alcohols. Sodium amalgam in basic media has been used to convert certain complex α,β-unsaturated ketones to saturated ketones, *e.g.*, the conversion of metathabainone to its dihydro derivative.[283] However, the utility of sodium amalgam does not extend to simple reduction of less complex enones since treatment of 4,4-dimethylcyclohex-2-enone with this reagent in a basic medium gives dimerization products exclusively.[71] Treatment of cholest-4-en-3-one with sodium amalgam in propanol-acetic acid has been used to prepare cholestene pinacol.[284]

Various metals in alcohols[7, 8] as well as sodium amalgam in a basic medium[285] are useful reagents for the reduction of α,β-unsaturated acids. For example, mixtures of the hydroxy acids **89** and **90** (p. 37) were obtained when the unsaturated acid **88** and the related 11-keto compound were treated with various alkali metals in ethanol or *n*-butanol.[7] Similar results were obtained using metal-ammonia combinations.[7]

163 —Zn/HOAc→ **164** —Zn/HOAc→ **165**

Reductions of conjugated carbonyl compounds have been effected with zinc in acidic media. β-Cyperone (**163**) is reduced to the saturated ketone **165** by way of the intermediate α,β-unsaturated ketone **164**.[93] However, treatment of the simpler monocyclic enone 4,4-dimethylcyclohex-2-enone with this reagent causes extensive dimerization, and only a small

[281] H. Kayahara, H. Ueda, K. Takeo, and C. Tatsumi, *Agr. Biol. Chem.* (Tokyo), **33,** 86 (1969).

[282] H. Ueda and S. Shimizu, *Agr. Biol. Chem.* (Tokyo), **23,** 524 (1959).

[283] M. Gates and D. A. Klein, *J. Med. Chem.*, **10,** 380 (1967).

[284] E. M. Squire, *J. Amer. Chem. Soc.*, **73,** 2586 (1951).

[285] W. E. Bachmann, W. Cole, and A. L. Wilds, *J. Amer. Chem. Soc.*, **61,** 974 (1939); **62,** 824 (1940).

amount of the simple reduction product is obtained.[71] On extended exposure to excess zinc dust in glacial acetic acid at room temperature, cholest-4-en-3-one gives 5α-cholest-3-ene in about 40% yield; mixtures of 5α- and 5β-3-enes were formed when other 4-en-3-ones were treated similarly.[286] In contrast, cholest-4-en-3-one under Clemmensen conditions, *i.e.*, reflux in toluene with amalgamated zinc and 7 *M* hydrochloric acid, gave 5β-cholest-3-ene in 48% yield.[287] As yet no adequate explanation for these different stereochemical results has appeared.

Carbon skeleton rearrangements have also been observed when α,β-unsaturated ketones are reduced under Clemmensen conditions.[287, 288] 2-Methylcyclohex-2-enone is converted to a mixture of 2-ethylcyclopentanone and 2-methylcyclohexanone, presumably through a cyclopropanol intermediate, by zinc amalgam in hydrochloric acid.[287, 288]

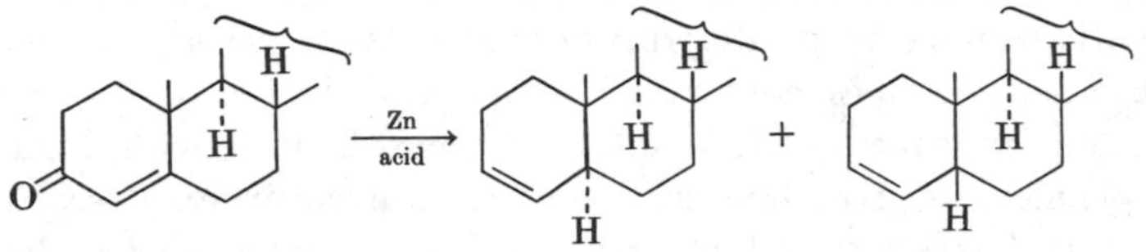

H H O —Zn/acid→ H H H + H H H

Because alkene formation and carbon skeleton rearrangements are not observed in metal-ammonia reductions, and because dimerization reactions are usually much less important in metal-ammonia than in zinc-acid reductions, the advantages of the former reagents for simple reduction of conjugated compounds are clear. Zinc in aqueous base has found utility for simple reduction of α,β-unsaturated ketones,[289] but for the reasons discussed in connection with sodium-alcohol reductions the basic conditions may cause complications.

Although solutions of lithium and, less extensively, sodium in primary amines of low molecular weight are powerful reducing agents which have been used quite widely for the reduction of organic compounds,[24] these reagents have not been applied extensively to reductions of α,β-unsaturated carbonyl systems. However, a very important use of these solutions has been reported whereby β-deuterio cyclohexanones and steroidal ketones could be obtained by treating the corresponding enones with solutions of lithium in dideuterated *n*-propylamine.[58] The yields were comparable to those reported in lithium-ammonia reductions of the corresponding compounds, and this method of β-deuteration offers distinct advantages over the use of lithium in trideuterated ammonia.[60]

[286] J. McKenna, J. K. Norymberski, and R. D. Stubbs, *J. Chem. Soc.*, **1959**, 2502.

[287] B. R. Davis and P. D. Woodgate, *J. Chem. Soc., C*, **1966**, 2006.

[288] B. R. Davis and P. D. Woodgate, *J. Chem. Soc.*, **1965**, 5943.

[289] M. Yoshida, *Chem. Pharm. Bull.* (Tokyo), **3**, 215 (1955).

Lithium-amine solutions have such powerful reducing action that even isolated double bonds may be reduced by them.[290] The lack of selectivity and the additional expense for amines as compared with liquid ammonia indicates that the use of metals in ammonia for reductions of α,β-unsaturated carbonyl compounds is clearly preferred in most instances.

Recently it has been found that α,β-unsaturated carbonyl compounds can be reduced with solutions of metals in hexamethylphosphoramide (often containing tetrahydrofuran as a co-solvent)[*, 44, 57, 63, 291–293] and by ethereal solutions of the radical anion derived from metals and trimesitylboron.[63, 89] Yields are generally similar to those obtained in metal-ammonia reductions, and these new reducing media offer certain practical advantages over metal-ammonia solutions: reduction can be carried out over a wider temperature range, including room temperature, and most substrates have higher solubility in hexamethylphosphoramide than in liquid ammonia. In metal-hexamethylphosphoramide reductions supplemental proton donors bring about protonation at the β position of α,β-unsaturated ketones,[44] but in the absence of such proton donors the hydrogen introduced at the β position is derived either from hydrogen atom[291–293] or from proton abstraction from the solvent.[44] Of course, proton donors must be added when metals in ethereal solvents containing trimesitylboron are employed for reductions.[63,89]

Perhaps the most interesting feature associated with the use of these new reducing agents is that the reduction stereochemistry is often different from that obtained in metal-ammonia reductions and frequently is influenced significantly by changing the metal, the solvent polarity, or the temperature. In a particularly striking case, reduction of the enone **71** (p. 32) at low temperature with excess sodium in hexamethylphosphoramide-tetrahydrofuran gave up to 70% of the less stable reduction product having the acetone side chain *cis* to the *t*-butyl group; metal-ammonia reduction of this ketone yields predominantly the more stable *trans* product.[57] Compared with metal-ammonia reductions, only small changes in the *trans*/*cis* ratio of reduction products were observed when the octalone **25** (p. 13) and its 10-methyl derivative were reduced with sodium in hexamethylphosphoramide under a variety of conditions.[57] A much greater amount of *cis*-2-decalone (**47**, p. 22) was obtained when octalone **25** was reduced with lithium in hexamethylphosphoramide.[63]

* Addition of tetrahydrofuran greatly enhanced the stability of sodium-hexamethylphosphoramide solutions and allowed reductions to be run at $-78°$ without freezing.[44]

290 A. P. Krapcho and M. E. Nadel, *J. Amer. Chem. Soc.*, **86**, 1096 (1964).

291 P. Angibeaud, M. Larchevêque, H. Normant, and B. Tchoubar, *Bull. Soc. Chim. Fr.*, **1968**, 595.

292 M. Larchevêque, *Ann. Chim.* (Paris), [14] **5**, 129 (1970).

293 H. Normant, *Angew. Chem., Int. Ed. Engl.*, **6**, 1046 (1967).

When the trimesitylboron radical anion was used as the reducing agent, different ratios of the decalones **46** and **47** were obtained, depending upon the period of time before addition of proton donors, the metal cation present, and the solvent.[63] On the basis of these results it was suggested that since steric hindrance to electron transfer is minimized in the formation of the *cis* dianion (*cf.* **51B**, p. 24), then this species is formed kinetically, but the alternative *trans*-fused dianion **51A** is usually favored thermodynamically. The results also indicated that the nature of the cation and the solvent exerts a significant influence on the **51A**/**51B** ratios. It has been pointed out that significant amounts of *cis*-decalones are formed on reductions of octalones under conditions favoring the formation of tight ion pairs or covalent bonds (lithium cations, nonpolar solvents, low temperature) between the β-carbanion and the metal in a species such as **51A** or **51B** or the related hydroxy anion.[20] In such cases the species with the *cis* configuration (*cf.* **51B**) would be expected to be favored because the metal and its attendant solvent shell are in the equatorial position with respect to one of the rings.[20]

Reduction-alkylation reactions may be carried out using metal-hexamethylphosphoramide solutions for the generation of reactive metal enolates.[292] Reductive alkylations of both α,β-unsaturated ketones and carboxylic acids have been reported.[292] However, regiospecific reduction-alkylations, which can be carried out using lithium enolates generated in liquid ammonia, are probably not possible when metals in hexamethylphosphoramide are used because enolate equilibration via proton transfer reactions is extremely rapid in such dipolar aprotic solvents.[13]

Cu cathode, dil H_2SO_4, EtOH → (~90%)

$$C_6H_5CH{=}CHCO_2^-Na^+ \xrightarrow[\text{2. } H_2SO_4]{\text{1. Hg cathode, } H_2O,\ Na_2SO_4} C_6H_5CH_2CH_2CO_2H \quad (80\text{–}90\%)$$

As illustrated by the conversion of carvone to dihydrocarvone[294] and of sodium cinnamate to hydrocinnamic acid,[295] electrolytic reduction is sometimes useful for the conversion of α,β-unsaturated carbonyl compounds to the corresponding dihydro derivatives. Electrochemical reduction has

[294] H. D. Law, *J. Chem. Soc.*, **101**, 1549 (1912).

[295] A. W. Ingersoll, *Org. Syn., Coll. Vol.*, **1**, 311 (1944).

been used to a rather limited extent for the simple reduction of α,β-unsaturated carbonyl compounds because it generally leads to significant amounts of bimolecular reduction products of the type **36** and **37** (p. 19). The conversion of cholest-4-en-3-one to cholestene pinacol (the same product obtained on metal-ammonia[4, 73, 74] and sodium amalgam reductions[284]) by reduction in ethyl alcohol containing sodium acetate using a mercury cathode,[296] and the formation of the β,β-dimer **27** (p. 15) by controlled potential electrolysis of the enone **23** (p. 13) in dimethylformamide containing tetra-*n*-propylammonium perchlorate with a mercury cathode followed by addition of water provide examples of the two types of dimerization processes normally observed.[44] Much work in recent years has been devoted to studies of the mechanism of electrochemical dimerization reactions and to a search for conditions engendering optimum yields of these products.[20, 75, 76] The electrochemical reduction of α,β-unsaturated ketones in aprotic media is of special interest since relatively stable radical anions, which can be studied by electron spin resonance spectroscopy, may be produced.[44, 46]

The value of catalytic hydrogenation for reducing organic compounds, including α,β unsaturated carbonyl compounds, is well known.[22, 261, 297–299] Catalytic hydrogenations are relatively easy to carry out and, unless the carbon-carbon double bond is highly hindered, α,β-unsaturated ketones, acids, and esters generally may be reduced to the corresponding saturated compounds in good yield. Reductions of carbon-carbon double bonds of α,β-unsaturated carbonyl compounds, particularly α,β-unsaturated ketones, are generally carried out using palladium on activated charcoal or calcium or strontium carbonate as the catalyst. α,β-Unsaturated aldehydes also can be reduced to the corresponding saturated aldehydes using hydrogen and catalysts, but those systems having aromatic β substituents or tri- or tetra-substituted double bonds are prone to undergo reduction to unsaturated alcohols.[300]

The stereochemical course of catalytic hydrogenation of alicyclic conjugated ketones is determined mainly by the accessibility of the reaction center to the catalytic surface. Thus catalytic reduction often provides a method of obtaining saturated ketones having the opposite configuration to those formed in metal-ammonia reductions. However, the stereochemistry of catalytic reduction of conjugated enones is often difficult to predict, and it may be strongly influenced by the degree and nature of

[296] P. Bladon, J. W. Cornforth, and R. H. Jaeger, *J. Chem. Soc.*, **1958**, 863.

[297] Ref. 20, House, pp. 1–34.

[298] H. J. E. Loewenthal, *Tetrahedron*, **6**, 269 (1959).

[299] P. N. Rylander, *Catalytic Hydrogenation over Platinum Metals*, Academic Press, New York, 1967.

[300] Ref. 299, Rylander, Chap. 14.

substitution of the double bond, the nature of neighboring and remote substituents, and by the reduction conditions. 1(9)-Octalin-2-one (**25**, p. 13) and steroidal 4-en-3-ones usually give mixtures of *cis*- and *trans*-fused products on catalytic hydrogenation. However, by modification of reduction conditions the *cis* product may be produced with a high degree of stereoselectivity. For example, a mixture containing 93% *cis*-2-decalone (**47**, p. 22) was obtained when 1(9)-octalin-2-one (**25**) was reduced with palladium on carbon in aqueous ethanolic hydrochloric acid.[301–303] 10-Methyl-1(9)-octalin-2-one and other C-1 unsubstituted derivatives are reduced primarily to *cis*-fused decalones with hydrogen and catalysts.[194, 304] In steroidal 4-en-3-one reductions, the use of solvents of high dielectric constant[305] and alkaline media[298] generally favors the formation of products having *cis*-fused A and B rings. Furthermore the stereochemistry of reduction of these compounds is strongly dependent upon remote structural features.[305, 306]

While metal-ammonia reductions of hydrindenones such as **81a** and **82a** (p. 35) give largely *trans*-fused products,[14, 119–121] the corresponding *cis* isomers are obtained on catalytic reduction.[261] Unfortunately, both metal-ammonia[59, 119, 122, 307] and catalytic reduction[119, 122, 307] of methyl-substituted hydrindenones such as **81b** and **82b** give largely *cis*-fused products.

Derivatives of 10-methyl-1(9)-octalin-2-one and the hydrindenone **81b**, having bulky groups at the 1 position, such as **166**[308] and **167**,[309] are catalytically reduced to *trans*-fused bicyclic ketones.

Metal-ammonia solutions offer distinct advantages over catalytic hydrogenation for the reduction of highly hindered carbon-carbon double bonds in α,β-unsaturated ketones. For example, the double bonds of 7-en-6-ones[9, 229, 310] and 9-en-12-ones[311] of steroidal systems as well as 12-en-11-ones[9, 312] of pentacyclic triterpenes are readily reduced with lithium in liquid ammonia. Catalytic hydrogenation of these compounds

[301] R. L. Augustine, *J. Org. Chem.*, **23**, 1853 (1958).

[302] R. L. Augustine and A. D. Broom, *J. Org. Chem.*, **25**, 802 (1960).

[303] R. L. Augustine, D. C. Migliorini, R. E. Foscante, C. S. Sodano, and M. J. Sisbarro *J. Org. Chem.* **34**, 1075 (1969).

[304] F. Sondheimer and D. Rosenthal, *J. Amer. Chem. Soc.*, **80**, 3995 (1958).

[305] M. G. Combe, H. B. Henbest, and W. R. Jackson, *J. Chem. Soc., C*, **1967**, 2467.

[306] F. J. McQuillin, W. Ord, and P. L. Simpson, *J. Chem. Soc.*, **1963**, 5996.

[307] C. B. C. Boyce and J. S. Whitehurst, *J. Chem. Soc.*, **1960**, 4547.

[308] G. Stork, S. Danishefsky, and M. Ohashi, *J. Amer. Chem. Soc.*, **89**, 5459 (1967).

[309] G. H. Douglas, J. M. H. Graves, D. Hartley, G. A. Hughes, B. J. McLoughlin, J. Siddall, and H. Smith, *J. Chem. Soc.*, **1963**, 5072.

[310] J. Jizba, L. Dolejš, V. Herout, F. Sorm, H.-W. Fehlhaber, G. Snatzke, R. Tschesche, and G. Wulff, *Chem. Ber.*, **104**, 837 (1971).

[311] Y. Mazur, N. Danieli, and F. Sondheimer, *J. Amer. Chem. Soc.*, **82**, 5889 (1960).

[312] J. Karliner and C. Djerassi, *J. Org. Chem.*, **31**, 1945 (1966).

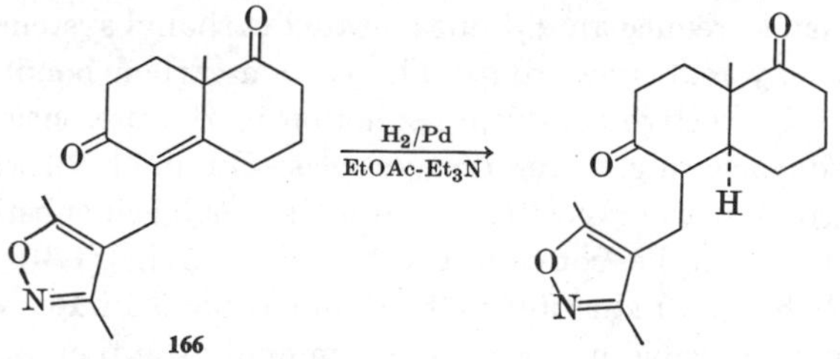

H_2/Pd, EtOAc-Et_3N

166

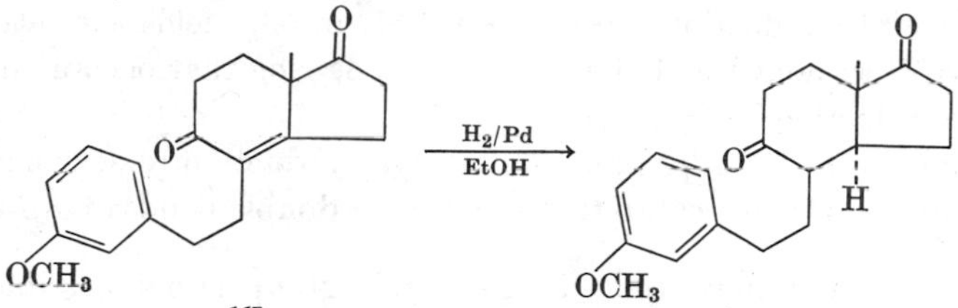

H_2/Pd, EtOH

167

leads either to the formation of unsaturated alcohols or to alkenes formed by hydrogenolysis.[9, 229, 298, 312] The double bonds of 8(9)-en-7-ones,[313] 8(9)-en-11-ones,[314, 315] and 14-en-16-one[316] can be reduced catalytically.

Most functional groups that are reduced by metal-ammonia solutions, as well as isolated double bonds, can be catalytically hydrogenated under the proper conditions.[22, 261, 299] However, catalytic reduction of conjugated enones frequently occurs with sufficient readiness, particularly in basic media, to permit selective reduction. For example, hydrogenation has

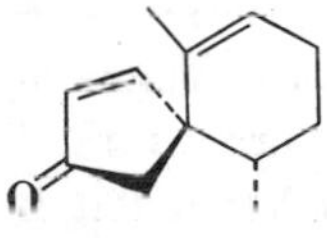

168 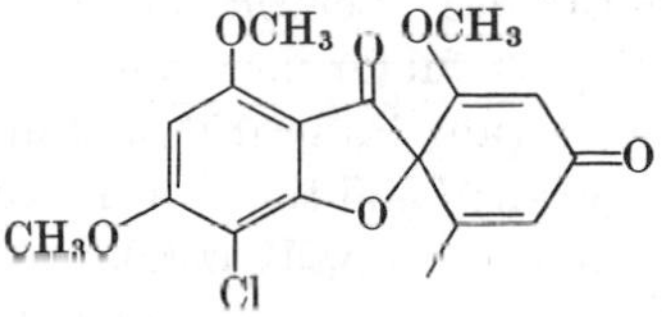 169 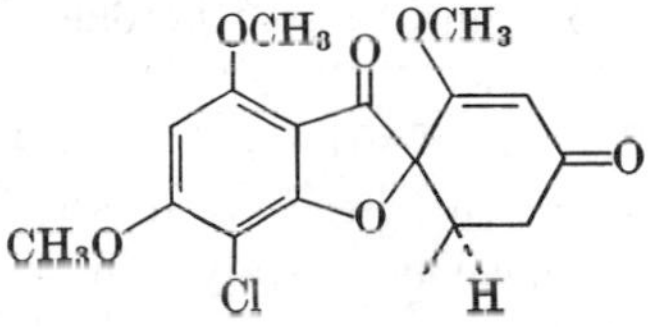 170

been reported for the conjugated double bond in α-cyperone (**106**, p. 42) and *epi*-α-cyperone (**54a**, p. 28),[317] the spiro dienone **168**,[318] and several steroidal enones containing isolated double bonds.[298] Selective catalytic reductions of conjugated enones containing saturated ketone and ester functions as well as aromatic groups have been carried out. Catalytic hydrogenation may offer advantages over metal-ammonia reduction when

[313] J. Elks, R. M. Evans, A. G. Long, and G. H. Thomas, *J. Chem. Soc.*, **1954,** 451.

[314] C. Djerassi, W. Frick, G. Rosenkranz, and F. Sondheimer, *J. Amer. Chem. Soc.*, **75,** 3496 (1953).

[315] C. Djerassi and G. H. Thomas, *J. Amer. Chem. Soc.*, **79,** 3835 (1957).

[316] P. Wieland, K. Heusler, H. Ueberwasser, and A. Wettstein, *Helv. Chim. Acta*, **41,** 74 (1958).

[317] R. Howe and F. J. McQuillin, *J. Chem. Soc.*, **1958,** 1194.

[318] J. A. Marshall and P. C. Johnson, *J. Org. Chem.*, **35,** 192 (1970).

it is necessary to reduce an α,β-unsaturated carbonyl system containing a relatively easily reducible group. The carbon-carbon bond of the enone **166** undergoes selective catalytic reduction in the presence of the easily reducible isoxazole ring,[308] and dehydrogriseofulvin **(169)**, which contains a leaving group at the γ position as well as a halogen substituent on the aromatic ring, can be converted into griseofulvin **(170)** by catalytic reduction.[319] Simple reduction of the double bond in **169** would not be expected to be possible using a metal-ammonia solution as the reducing agent. Catalytic reduction of santonin (**131**, p. 53) yields the tetrahydro derivative[320] without the fission of the lactone ring that occurs on metal-ammonia reduction.[202, 204]

Ethynyl carbinol groups undergo catalytic reduction to allylic alcohols rapidly; hence selective catalytic reduction of double bonds of α,β-unsaturated ketones containing the $-C{\equiv}C-\overset{|}{\underset{|}{C}}-OH$ group is not possible.[298, 321] However, as discussed on p. 59, metal-ammonia reductions of enones in the presence of ethynyl carbinol groups are possible.[164, 251–253]

There are various other less frequently used methods for reducing α,β-unsaturated carbonyl compounds. Some of them are: reduction with metal hydrides, which is useful for the synthesis of allylic alcohols but often causes the formation of saturated ketones and/or saturated alcohols;[322] the use of diimide, which under the proper conditions reduces carbon-carbon double bonds of α,β-unsaturated acids and esters but not of α,β-unsaturated ketones;[323] the use of chromium(II) salts in aqueous ammonia,[324] these being especially useful for the reduction of enediones;[325] the use of lithium biphenyl in tetrahydrofuran[326] and of suspensions of metals in ether[327] or tetrahydrofuran;[44] and the exciting new procedure of homogeneous catalytic hydrogenation, usually employing tris(triphenylphosphine)chlororhodium.[328,329] Synthetically useful reductions of α,β-unsaturated carbonyl compounds to the corresponding α,β-dihydro

[319] D. Taub, C. H. Kuo, H. L. Slates, and N. L. Wendler, *Tetrahedron*, **19**, 1 (1963).

[320] W. Cocker and T. B. H. McMurry, *J. Chem. Soc.*, **1956**, 4549.

[321] E. B. Hershberg, E. P. Oliveto, C. Gerold, and L. Johnson, *J. Amer. Chem. Soc.*, **73**, 5073 (1951).

[322] Ref. 20, House, pp. 89–96.

[323] Ref. 20, House, 252–253.

[324] K. D. Kopple, *J. Amer. Chem. Soc.*, **84**, 1586 (1962).

[325] J. R. Hanson and E. Premuzic, *J. Chem. Soc., C*, **1969**, 1201.

[326] P. Wieland and G. Anner, *Helv. Chim. Acta*, **51**, 1698 (1968).

[327] J. Wiemann and A. Jacquet, *C.R. Acad. Sci., Ser. C*, **263**, 546 (1966).

[328] M. Fieser and L. F. Fieser, *Reagents for Organic Synthesis*, Vol. 2, Wiley, New York, 1969, p. 448.

[329] J. E. Lyons, L. E. Rennick, and L. E. Burmeister, *Ind. Eng. Chem., Prod. Res. Develop.*, **9**, 2 (1970).

derivatives by "ate" complexes of copper(I) hydride have been reported.[329a,329b]

A unique method has appeared for the reduction of α,β-unsaturated ketones.[88] The procedure involves converting the enone to its Schiff base with benzylamine, treatment of this material with base to bring about isomerization to a vinyl imine of benzaldehyde, and acid-catalyzed hydrolysis to the saturated ketone and benzaldehyde. This procedure may be applied to determine thermodynamic stabilities of various olefins, *e.g.*, 1(9)-octalins.[88]

$$C_6H_5CH_2NH_2 + -\overset{\overset{O}{\|}}{C}-\overset{|}{C}=\overset{|}{C}- \longrightarrow C_6H_5CH_2N=\underset{|}{C}-\overset{|}{C}=\overset{|}{C}-$$

$$\xrightarrow{\text{Base}} C_6H_5CH=N\underset{|}{C}=\overset{|}{C}-\overset{|}{\underset{\underset{H}{|}}{C}}- \xrightarrow[\text{AcOH}]{\text{Dil.}} C_6H_5CHO + -\overset{\overset{O}{\|}}{C}-\overset{|}{\underset{\underset{H}{|}}{C}}-\overset{|}{\underset{\underset{H}{|}}{C}}-$$

EXPERIMENTAL CONDITIONS

Handling Liquid Ammonia

Anhydrous liquid ammonia may be obtained commercially in steel cylinders. The solvent may be taken from these cylinders in gaseous or liquid form. Reactions involving the use of liquid ammonia should be conducted in an efficient hood because the gas is toxic and has a pungent odor. Some commercial cylinders are provided with eductor tubes by means of which the liquid may be taken from the upright cylinder. Cylinders not so equipped may be secured in a suitable wooden or metal cradle, constructed so that the outlet valve is inclined below the body of the cylinder. In this position the pressure forces the liquid ammonia out when the valve is opened. The liquid or gaseous material may be transferred through rubber or Tygon tubing. The concentration of iron in undistilled commercial liquid ammonia may be as high as 1 part per million. Iron catalyzes the reaction of metals (particularly sodium and potassium) with ammonia and with proton donors such as alcohols.[330] Thus, when sodium or potassium is to be employed in metal-ammonia reductions, it is desirable that the ammonia be distilled. A rather widespread practice has been to introduce ammonia as a liquid or gas into an intermediate vessel, to dry it by adding

[329a] R. K. Boeckman, Jr. and R. Michalak, *J. Amer. Chem. Soc.*, **96,** 1623 (1974).

[329b] S. Masamune, G. S. Bates, and P. E. Georghiou *J. Amer. Chem. Soc.*, **96,** 3686 (1974).

[330] H. L. Dryden, Jr., G. M. Webber, R. P. Burtner, and J. A. Cella, *J. Org. Chem.*, **26,** 3237 (1961).

a small amount of sodium, and then to distill into the reaction flask. This procedure ensures that colloidal iron will be removed from the ammonia and that it is strictly anhydrous. However, in the majority of metal-ammonia reductions of α,β-unsaturated carbonyl compounds, lithium is employed as the reducing metal, and the concentration of iron in commercial ammonia is not sufficient to interfere with the reducing action of this metal.[330] The author has encountered no difficulty in carrying out lithium-ammonia reductions using ammonia which has been introduced as the liquid directly into the reaction flask. Ammonia which has not been predried as described above may be dried by the addition of small pieces of metal until the blue color persists. Even when proton donors such as water are employed, drying the ammonia and other solvents is desirable so that the metal-ammonia solution is not exposed to water during preparation and so that the amount of proton donor can be accurately controlled.

Liquid ammonia has a low boiling point (−33.4°) and a high heat of vaporization (5.58 kcal/mol); therefore an efficient condenser is required. A Dewar condenser containing a slurry of solid carbon dioxide (dry ice) and acetone, ethanol, or isopropanol may be used for condensing ammonia. Normally less frothing is observed with the latter liquid. Reaction mixtures may be protected from atmospheric moisture and carbon dioxide by using a soda lime guard tube. However, many workers, including the author, have found it convenient to conduct metal-ammonia reductions under a nitrogen or argon atmosphere. A static atmosphere device having the gas source connected at the top of the Dewar condenser is convenient for this purpose.[331] Using this unit, an inert atmosphere may be maintained in the system during the reaction and a slight positive inert gas flow may be maintained while the system is open for the addition of reagents.

Metals

The metals employed in metal-ammonia reduction should have low transition metal content, particularly of iron. There are numerous suppliers of metals of sufficient purity to be used directly. The metals should be stored under high-boiling hydrocarbon solvents, freshly cut under the solvent, and washed with a low-boiling unreactive solvent such as ether or pentane before use. Metals may be freed of oxide coatings and cut into small pieces with a stainless steel knife or scissors, but care must be taken to avoid contamination with iron during the cutting. For further details concerning the handling of various metals one should consult the

[331] W. S. Johnson and W. P. Schneider, *Org. Syn., Coll. Vol.*, **IV**, Wiley, 1953, p. 132.

appropriate sections of *Reagents for Organic Synthesis*.[36] Lithium is available in the form of ⅛-in. wire which is very convenient for handling.

Lithium is by far the most widely used metal in metal-ammonia reductions. It has a higher solubility (g-at./100 g NH_3) and reduction potential in liquid ammonia than sodium or potassium,[24,332] and solutions of lithium in liquid ammonia are stable in the presence of a fairly high concentration of iron.[330] Lithium must be employed as the reducing metal if regiospecific alkylation and other reactions of metal enolate intermediates are to be carried out.[14] Sodium and potassium have been employed, though much less frequently than lithium, for reduction of α,β-unsaturated carbonyl compounds. Sodium is the most inexpensive of the alkali metals and for simple reductions it is normally adequate. Potassium rather than lithium must be used if alkylation or other reactions are to be carried out at the β position of α,β-unsaturated systems having electron-withdrawing β substituents.[52,53]

Barium and calcium have been used in metal-ammonia reductions. With calcium, reduction of conjugated enones to saturated alcohols takes place even in the absence of proton donors.[42] Apparently calcium-ammonia solutions are sufficiently acidic to protonate calcium enolate intermediates.

Solutions of magnesium in liquid ammonia may be prepared by electrolysis of solutions of magnesium salts.[25, 43] However, studies involving the use of these solutions have been rare.

Presently there is insufficient evidence to allow firm predictions concerning the possible effect of a change of the metal on the stereochemical course of the reduction of a particular α,β-unsaturated carbonyl system. However, a sufficient number of examples of reductions in which the stereochemistry has been influenced by a change in the metal have been reported to warrant an investigation of this factor if a particular stereochemical result is desired.[7, 121]

No detailed studies concerning the influence of metal concentration on yields have been reported. Most workers prefer to use dilute solutions, usually in the range 0.1–0.5 g/100 ml. The solubilities (g/100 g NH_3) of some commonly used metals in liquid ammonia are as follows: Li, 10.9 g (−33.2°); Na, 24.6 g (−33°); K, 46.5 g (−33.2°); Ca, 33.6 g (−35°).[333]

Proton Donors and Co-Solvents

The possible role of proton donors in influencing yields of reduction products and the dependence of the extent of over-reduction on the nature and concentration of the proton donor have been discussed in detail

[332] A. L. Wilds and N. A. Nelson, *J. Amer. Chem. Soc.*, **75**, 5360 (1953).

[333] J. Jander, *Anorganische und Allgemeine Chemie in Flussigem Ammoniak*, Chemistry in Nonaqueous Ionizing Solvents, Vol. 1, Pt. 1, Wiley, New York, 1966, p. 242.

previously. Ethanol, methanol, and *t*-butanol as well as water have been most frequently used as proton donors, but other less acidic substances have been employed.[147] Substances more acidic than water are normally not used because they react rapidly with the dissolved metals to produce hydrogen.[24, 244]

Protic substances are occasionally used as both the proton donor and the co-solvent, but more frequently aprotic solvents such as diethyl ether, tetrahydrofuran, 1,2-dimethoxyethane, dioxane, or dioxane-ether mixtures are employed to increase the solubility of nonpolar substrates in the reduction medium. A few examples of use of nonethereal co-solvents have been reported. A 1:1:2 toluene-tetrahydrofuran-ammonia mixture is a good medium for metal-ammonia reductions.[19]

All co-solvents should be freed of peroxides and other impurities before use. Purification of the lower-boiling ethers such as diethyl ether, tetrahydrofuran, and 1,2-dimethoxyethane can be accomplished by refluxing them over lithium aluminum hydride for a short period and then distilling them from the metal hydride just prior to use.* Dioxane can be purified by reflux over freshly cut sodium and distillation prior to use.

Normally the volume of co-solvent should be kept below 20% of the total reaction volume. If the polarity of the reduction medium is sufficiently lowered by the addition of nonpolar co-solvents, dimerization may become an important side reaction.[77]

It appears that diethyl ether is an acceptable co-solvent for reductions involving substrates of low as well as high molecular weight. Tetrahydrofuran has been rather widely used when the enone reduction-enolate alkylation sequence was employed.[14]

Apparatus

For most reductions a three-necked round-bottomed flask fitted with a mechanical stirrer, an addition funnel with a pressure-equalizing side arm, and a Claisen adapter fitted with a Dewar condenser in the offset neck may be employed. A soda-lime guard tube or a gas source, if the reaction is to be carried out in an inert atmosphere, should be attached at the top of the Dewar condenser. Ammonia from a cylinder (or from an intermediate container if predrying and distillation of ammonia are carried out) may be introduced via an adapter fitted in the straight neck of the Claisen adapter. The Dewar condenser is about one-fourth filled with acetone, ethanol,

* *Caution!* Anhydrous reagent grade ether and reagent grade tetrahydrofuran and 1,2-dimethoxyethane may be treated directly with lithium aluminum hydride. Lower grade solvents should be predried before adding the metal hydride. Under no circumstances should distillations from lithium aluminum hydride be conducted to dryness as the metal hydride may undergo explosive decomposition at temperatures above 120°

or isopropanol. Ammonia is introduced for about 1 minute, the Dewar condenser is then carefully filled with powdered dry ice, and the ammonia flow is increased to the desired rate. Additional dry ice should be carefully added to the condenser from time to time. After the ammonia has been added, the inlet adapter should be replaced by a ground glass stopper. The metal also may be added through this neck of the Claisen adapter. Metal-ammonia reductions are normally conducted at the reflux temperature of liquid ammonia or at the temperature (approximately −78°) obtained by applying a dry ice-liquid coolant to the outside of the flask. A low-temperature thermometer with a proper adapter may be placed in the straight neck of the adapter if accurate measurement of the temperature is desired. Unless a vacuum-jacketed flask is used, the reaction flask will become coated with a layer of frost. This acts as a good insulator, and may be washed off with acetone or ethanol if one wishes to observe the reaction mixture.

Although mechanical stirrers are recommended, magnetic stirrers may be used. Glass stirrer blades or glass- or polyethylene-coated magnets are recommended. The use of Teflon stirrer blades or Teflon-coated magnets is less desirable because a slow reaction of the polymer occurs with metal-ammonia solutions.

An apparatus has been described that is useful for small-scale metal-ammonia reductions.[334] This equipment with minor modifications has been discussed in detail by other authors.[24,244]

Reaction Procedures

Generally, procedures similar to those employed in the Birch reduction of aromatic compounds[24, 244] have been used for the reduction of α,β-unsaturated carbonyl compounds. They are: (1) addition of the substrate, which is usually dissolved in an ethereal solvent (perhaps containing a proton donor), to a well-stirred solution of the metal in liquid ammonia; (2) addition of the metal, cut into small pieces, to a solution of the substrate in liquid ammonia (perhaps containing a co-solvent and/or a proton donor); and (3) addition of the substrate to an ammonia solution containing an excess of the metal, followed by a period of stirring and then addition of an excess of a proton donor.

The first procedure mentioned above is by far the most commonly employed. For relatively small-scale reductions, a solution containing *ca.* 5 g of conjugated carbonyl compound/100 ml of solvent is employed. This is added to a solution of at least 2 equivalents of the metal (because of salt formation at least 3 equivalents of metal must be used in reductions of α,β-unsaturated acids) in liquid ammonia. For simple reductions most

[334] A. Sandoval, *Chem. Ind.* (London), **1960,** 1082.

workers employ an excess of the metal, while slightly more than 2 equivalents are used if reduction-alkylation or related reactions are to be carried out. A sufficient volume of the metal-ammonia solution (usually containing 0.1–0.5 g of metal/100 ml of ammonia) should be used so that the final substrate/solvent concentration is *ca.* 1 g/100 ml. Large-scale reductions may be performed using more concentrated solutions, but it should be realized that dimerization and other side reactions are likely to become more important as the concentration is increased.

In general, reductions of α,β-unsaturated ketones, esters, and acids are quite rapid; therefore the substrate may be added quickly to the metal-ammonia solution and the reaction mixture quenched immediately after the addition. If the procedure of adding the substrate to the metal-ammonia solution is employed and if other reducible groups are present, it is particularly important to use short reaction times. Otherwise, metal enolate intermediates are usually stable in liquid ammonia unless relatively acidic proton donors are added. Thus, procedures involving the slow addition of the unsaturated compound followed by stirring periods of 0.5 to 1 hour are quite common.

Reductions are normally carried out at the reflux temperature of the liquid ammonia solution but, if selective reductions are being attempted or if the metal enolate intermediate may undergo elimination reactions, it is advisable to carry out the reduction at $-78°$. If an excess of proton donors such as methanol or ethanol is added along with the substrate to the metal-ammonia solution, over-reduction to a saturated alcohol will occur.

If over-reduction is to be avoided, as in reduction-alkylation, 1 equivalent of a proton donor such as *t*-butanol, other alcohols, or water may be added. This leads to the formation of the lithium enolate and the lithium alkoxide or hydroxide.

For simple reductions, reaction mixtures are normally quenched with ammonium chloride, which rapidly destroys the excess metal and converts the metal enolate to the saturated ketone. Alcohols or water may be employed to protonate enolate intermediates, but unless the excess metal is removed before the addition of these reagents over-reduction is likely. Sodium benzoate[162] and a number of other reagents, including ferric nitrate,[163, 164] sodium nitrite,[165] bromobenzene,[166] sodium bromate,[167] 1,2-dibromoethane,[19] and acetone,[9] may be used to destroy excess metal.

If alkylations or other reactions of lithium enolates are to be carried out, the reagent, perhaps with an additional quantity of co-solvent, can be added directly to the ammonia solution or the ammonia may be removed and replaced by other solvents.

After the quenching agent has been added, the ammonia is allowed to evaporate by removing the Dewar condenser and replacing it with a soda-lime guard tube. This process may be hastened by warming the flask with water or a heat gun and/or by passing a stream of inert gas through the solution. After removal of the ammonia, water and ether are added and the ether layer is separated and washed with dilute acid and water. After drying the solution, the solvent is evaporated and the product is purified and characterized by the usual methods.

Frequently the procedure of adding metal to the substrate in liquid ammonia is employed. In the majority of cases the order of addition seems to have little influence on yields, but the metal should be introduced last when reductions of α,β-unsaturated carbonyl compounds containing other reducible groups are being attempted. Using this method, the concentration of metal in solution may be kept low at all times. The technique of dropwise addition of a solution of the metal in liquid ammonia to an ammonia solution of the substrate affords greater control of the metal concentration than direct addition of the metal.[254]

The procedure of using excess metal and adding an alcohol at the end of the reduction, employed quite successfully in Birch reductions,[24, 244, 332] has often been utilized in reductions of α,β-unsaturated ketones and α,β-unsaturated esters. Under these conditions metal enolate intermediates are protonated and the resulting saturated ketones or esters are further reduced to saturated alcohols. This method is useful for the direct conversion of unsaturated carbonyl compounds to saturated alcohols, which may be oxidized to the corresponding saturated carbonyl compounds if desired.

Experimental Procedures

4,5α-Dihydro-17α-ethynyl-19-nortestosterone (Enone Reduction in Aprotic Co-solvent).[251] A solution of 17α-ethynyl-19-nortestosterone (15 g) in dioxane-ether (1:1, 250 ml) was added rapidly to a well-stirred solution of lithium (2.25 g) in liquid ammonia (1.5 l.). Ammonium chloride (30 g) then was added and the ammonia allowed to evaporate. Isolation with methylene chloride afforded a product that was dissolved in 500 ml of 1:1 benzene-hexane and chromatographed on 700 g of alumina. Elution with benzene (1.5 l.) gave 4,5α-dihydro-17α-ethynyl-19-nortestosterone (10.8 g, 72%), mp 195–215°, raised by crystallization from acetone-hexane to 222–223°; $[\alpha]_D + 6°$; λ_{max}^{KBr} 3330 (OH), 2750 (—C≡H) and 1700 cm^{-1} (C═O).

***trans*-7β-Isopropyl-10α-methyl-2-decalone (Enone Reduction in Aprotic Co-solvent Containing One Equivalent of Proton Donor).**[95] A three-necked flask fitted with a mechanical stirrer, a dropping funnel, and a Dewar condenser was flame-dried under positive nitrogen flow, and the system was maintained under nitrogen throughout the experiment.

Anhydrous liquid ammonia (1 l.) was introduced into the reaction flask by distillation from sodium, the stirrer was started, 0.462 g (0.066 g-at.) of freshly cut lithium wire was added, and stirring was continued for 15 minutes. To the lithium-ammonia solution was added dropwise a solution of 6.18 g (0.030 mol) of 7β-isopropyl-10α-methyl-1(9)-octalin-2-one and 2.22 g (0.030 mol) of *t*-butanol in 200 ml of anhydrous diethyl ether over 30 minutes. After 30 minutes 6.42 g (0.12 mol) of ammonium chloride was added, the Dewar condenser was removed, and the ammonia was allowed to evaporate. A mixture of 200 ml of ether and 200 ml of water was added, the ether layer was separated, and the aqueous layer was saturated with sodium chloride and extracted with ether. The combined ethereal solutions were washed with 50 ml of 5% hydrochloric acid and 50 ml of saturated aqueous sodium chloride, and dried over magnesium sulfate. Removal of the ether under reduced pressure and distillation of the residue gave 4.19 g (67%) of *trans*-7β-isopropyl-10α-methyl-2-decalone; bp 115–122° (0.2 mm); mp 101–103°; ir (CCl_4) 1720 cm^{-1}; nmr (CCl_4) 2.50–1.10 (broad adsorption, 15*H*), 1.05 (*s*, 3*H*, C-10α CH_3), and 0.89 ppm [(*d*, 6*H*, $J = 6$ Hz, $CH(CH_3)_2$]. The material showed only one component on several gas chromatography columns.

(+)-Dihydrocarvone (Enone Reduction in Alcohol Co-solvent Followed by Oxidation of Alcohol Product).[335] A solution of 100 g (0.67 mol) of (−)-carvone in 830 ml of anhydrous diethyl ether was added over 3 hours to a solution of 18.7 g of lithium in 3 l. of distilled ammonia, after which 0.2 l. of absolute ethanol was added over 3.5 hours. After the addition of *ca.* 170 g of ammonium chloride, the ammonia was allowed to evaporate. The yellow residue was isolated with ether, affording 96.4 g (95%) of a mixture of dihydrocarvone and dihydrocarvol. The crude mixture was dissolved in 700 ml of acetone (distilled from potassium permanganate), cooled to 0°, and treated over a 45-minute period with Jones's reagent[168] until a persistent red color developed (*ca.* 80 ml). 2-propanol was added to destroy the excess oxidizing agent, and the solution was neutralized with sodium bicarbonate. The mixture was filtered and the solid washed well with ether. Most of the acetone was removed from the filtrate at reduced pressure and the product was isolated with ether. Distillation gave 66.2 g (66%) of (+)-dihydrocarvone: bp 87–90° (5 mm), λ_{max}^{film} 5.84 (ketone CO) and 6.08 μ (C=C).

4-Ethylcholestan-3β-ol (Reduction of an Enone to an Alcohol).[336] 4-Ethylcholest-2-en-3-one (200 mg) in 15 ml of dioxane was added to a solution of 150 mg of lithium in 50 ml of liquid ammonia. After stirring

[335] J. A. Marshall, W. I. Fanta, and H. Roebke, *J. Org. Chem.*, **31**, 1016 (1966).
[336] C. Djerassi, M. Cais, and L. A. Mitscher, *J. Amer. Chem. Soc.*, **81**, 2386 (1959).

for 1 hour, methanol was added until all the lithium had reacted, and a new portion of lithium metal was added to maintain the blue color for 1 hour. Ammonium chloride was added, and the ammonia was allowed to evaporate. Extraction with ether and removal of the solvent yielded 190 mg of crystals, mp 135–143°; one recrystallization from pentane afforded 4α-ethylcholestan-3β-ol, mp 140–143°, undepressed on admixture with an authentic sample.

2-Methyl-3-phenylpropanoic Acid (Reduction of an α,β-Unsaturated Acid).[216] A solution of 25 mmol of 2-methyl-3-phenylpropenoic acid in 75 ml of anhydrous diethyl ether was rapidly added to a magnetically stirred solution of 695 mg (0.100 g-at.) of lithium in 175 ml of anhydrous ammonia under nitrogen. After 30 minutes at the liquid ammonia boiling point (−33°), the blue reaction mixture was carefully quenched by the slow addition of 20 g of ammonium chloride. Then 125 ml of ether was added, and the dry ice-2-propanol condenser was replaced by a sodium hydroxide drying tube. After evaporation of the liquid ammonia overnight, the reaction mixture was acidified with 6*N* hydrochloric acid, more water was added to bring the inorganic salts into solution, and the solution was extracted twice with ether.

The combined ether extracts (250 ml) were washed three times with 40-ml portions of saturated sodium chloride solution and then dried over sodium sulfate. Evaporation of the ether under reduced pressure gave crude 2-methyl-3-phenylpropanoic acid which, on recrystallization from ether-pentane, gave 3.90 g (95%) of product; mp 36–38°; ir ($CHCl_3$) 1705, 1200, and 720 cm^{-1}.

Ethyl 8β,8aβ-Dimethyl-1,2,3,4,5,6,7,8,8a-octahydro-2β-naphthoate (Enol Ether Hydrogenolysis).[210] A solution of 5.70 g of the crude methoxymethyl enol ether of ethyl 8β,8aβ-dimethyl-1-oxo-1,2,3,4,6,7,8,8a-octahydro-2-naphthoate in 105 ml of ether was added rapidly to a magnetically stirred, dark-blue solution of 750 mg (0.108 g-at.) of lithium in 330 ml of anhydrous ammonia under argon. Powdered dry ice was used to cool the reaction flask while the addition was made. After stirring for 12 minutes at the liquid ammonia boiling point (−33°), the reaction flask was again cooled with powdered dry ice for 10 minutes, and then 30 g of ammonium chloride was added essentially all at once to quench. The blue solution actually faded 2 minutes before quenching. After 250 ml of ether was added, the dry ice-2-propanol condenser was removed and replaced by a sodium hydroxide drying tube. The reaction mixture was allowed to stand at room temperature until the ammonia had evaporated. The mixture was then filtered, and the inorganic salts were crushed and washed three times with ether.

The ether solution was evaporated under reduced pressure to give 4.65 g of a slightly yellow oil. Glpc analysis (6 ft × $\frac{3}{8}$ in. column of 20% SE-30 on 60–80 mesh Chromosorb *W*, 165°, 200 ml/min helium flow) of this oil revealed that there was essentially only one volatile component. Chromatography of the crude product on 80 g of Woelm neutral alumina (Activity II) and elution with 0–10% ether in petroleum ether (bp 30–60°) gave 2.85 g of ethyl 8β,8aβ-dimethyl-1,2,3,4,6,7,8,8a-octahydro-2β-naphthoate as a colorless oil. The analytical sample obtained by preparative glpc on the column described above showed: n_D^{25} 1.4905; ir 1729 (C=O), 1460, 1375, 1200, 1176, 1100, 1063, and 1045 cm^{-1}; nmr τ 4.75 (*m*, 1*H*), 5.90 (quartet, 2*H*, $J = 7.0$ Hz), 8.74 (*t*, 3*H*, $J = 7.0$ Hz), 9.12 (*d*, 3*H*, $J = 6.0$ Hz), and 9.18 (*s*, 3*H*).

2-Allyl-3-methylcyclohexanone (Enone Reduction Followed by Alkylation).[176] A 2-l., three-necked flask was fitted with a mechanical stirrer, a pressure-equalizing dropping funnel, and a two-necked Claisen adapter holding a Dewar condenser (in the offset neck) and an adapter with stopcock for introduction of ammonia. The entire apparatus was flame-dried under a positive nitrogen flow which entered the top of the Dewar condenser, and was maintained under nitrogen throughout the experiment. Commercial anhydrous ammonia (1 l.) was introduced into the flask in liquid form. The condenser was filled with a dry ice-acetone slurry and the inlet adapter was removed and replaced by a glass stopper. The stirrer was started and the ammonia was dried by the addition of minimal pieces of lithium metal, *ca.* 5 mg, until the blue color persisted. Freshly cut 3-mm lithium wire, 2.77 g (0.40 g-at.), was introduced into the flask with stirring, and stirring was continued for 20 minutes to allow dissolution of the lithium. While stirring was continued, a solution prepared from 20.00 g (0.182 mol) of 3-methylcyclohex-2-enone, 3.27 g (0.182 mol) of water, and 400 ml of anhydrous diethyl ether was added dropwise over 60 minutes. Stirring was contined for 10 minutes and a solution of 65 g (0.54 mol) of allyl bromide in 150 ml of anhydrous ether was added from the dropping funnel in a stream over 60 seconds. The reaction mixture was stirred for 5 minutes and 30 g of solid ammonium chloride was added as rapidly as possible. The Dewar condenser was removed and the ammonia was allowed to evaporate. A mixture of 300 ml of ether and 300 ml of water was added and the reaction mixture was transferred to a separatory funnel, shaken, and the ether layer was separated. The water layer was saturated with sodium chloride and extracted with two 100-ml portions of ether. The combined ether extracts were washed with 100 ml of 5% aqueous hydrochloric acid, 100 ml of saturated aqueous sodium chloride, and dried over anhydrous magnesium sulfate. The drying agent was removed by filtration and the ether was removed using a rotary evaporator. The residual oil was distilled through a 37-cm column packed with 6-mm Raschig rings and equipped with a resistance heater for thermal balance. After a fore-run, bp 74–80° (12 mm), consisting principally of 3-methylcyclohexanone, 12.31–13.63 g (44.5–49.3%) of about a 20:1 mixture of *trans*- and *cis*-2-allyl-3-methylcyclohexanone, bp 99–102° (12 mm), n_D^{24} 1.4680–1.4683, was collected.

The distillate was analyzed by gas chromatography using a 3 mm × 2 m column containing 10% Carbowax 20 M on HMDS Chromosorb *W* (60/80 mesh). Using a column temperature of 104° and a carrier gas flow rate of 30 ml/minute, the retention times for the *trans* and *cis* isomers were 9.0 minutes and 11.2 minutes, respectively. The product contained greater than 98% of the mixture of 2-allyl-3-methylcyclohexanones. A small amount of the product was refluxed with methanolic sodium methoxide to convert it into the thermodynamic mixture of *trans* (~65%) and *cis* (35%) isomers. The same thermodynamic mixture of isomers was prepared independently on lithium-ammonia reduction[72] of α-allyl-3-methylcyclohex-2-enone[337] followed by equilibration with methanolic sodium methoxide.

***trans*-1-Butyl-2-decalone (Enone Reduction Followed by Alkylation).**[13] A 500-ml, three-necked flask fitted with a stirrer, Hirshberg dropping funnel, and dry ice condenser was heated for 15 minutes with a free flame. Dry nitrogen was swept through the system while heating and for several additional minutes. The system was then evacuated and filled with nitrogen. Ammonia (250 ml) was distilled into the reaction flask from an ammonia-sodium solution and 0.70 g (0.101 g-at.) of lithium was added. 1(9)-Octalin-2-one (5.05 g, 0.0337 mol) was then added and the reaction mixture was stirred for 1 hour. *n*-Butyl iodide (37 g, 0.2 mol) was then added dropwise and the medium soon turned white. After 30 minutes more, the dry ice condenser was replaced by a water condenser and the ammonia was allowed to evaporate overnight. The remaining salts were dissolved in 100 ml of water, and the mixture was made acid by the addition of 10% hydrochloric acid. The organic material was taken up in ether, the ether was washed, dried, and concentrated. The yield was 3 g (47%) of *trans*-1-butyl-2-decalone, bp 90–91° (0.05 mm). Gas chromatography indicated 99% purity. The 2,4-dinitrophenylhydrazone showed mp 125–126.5, which was undepressed when mixed with an authentic sample.

A higher-boiling fraction (0.8 g) of polyalkylated material was also obtained, bp 115–122° (0.05 mm).

[337] J.-M. Conia and F. Rouessac, *Bull. Soc. Chim. Fr.*, **1963,** 1925.

Tetrahydropyranyl Derivatives of *trans*-1-Carbomethoxy-5-hydroxy-10-methyl-2-decalone (Enone Reduction Followed by Carboxylation).[187] To a 2-l., three-necked flask equipped with a mechanical stirrer and a reflux condenser was added 800 ml of liquid ammonia. To this was added 2.8 g (0.4 g-at.) of lithium wire (47.5 cm) which had been cut into 2- to 3-in. lengths, quickly rinsed with hexane to remove mineral oil, and dried quickly with a towel. The blue mixture was stirred for 10 minutes and then a solution of 26.4 g (0.1 mol) of a mixture of tetrahydropyranyl derivatives of 5-hydroxy-10-methyl-1(9)-octalin-2-one in 300 ml of anhydrous diethyl ether was added rapidly (4–5 minutes) from a pressure-equalizing dropping funnel while very vigorous stirring was maintained. As soon as the addition was complete, a steam bath was applied to the flask and the ammonia was evaporated as quickly as possible through the condenser (15–20 minutes). When the coating of ice around the flask melted, 500 ml of anhydrous diethyl ether was added and a Drierite (anhydrous calcium sulfate) drying tube was attached to the condenser. The mixture was refluxed for 15 minutes to drive off any residual ammonia and was then cooled to dry ice-acetone temperature.

During this cooling period a piece of dry ice was chipped on all sides to about 200 g, and then pulverized inside a cloth bag inside a large, dry plastic bag. This fine powder was then added to the cold reaction mixture through a powder funnel which was also encased in a larger plastic bag. Care was taken to exclude moisture. The mixture was removed from the dry ice-acetone bath and stirred for 30 minutes, and then for 30 minutes more in a room-temperature water bath. Then the mixture was recooled in a dry ice-acetone bath and 500 g of powdered dry ice was added, followed by 500 ml of cold distilled water. The contents of the flask were transferred to a separatory funnel, and the reaction flask was rinsed with cold water that was added to the funnel. The ether layer was separated and set aside; subsequent evaporation of this layer yielded 7.5 g of oil.

The aqueous layer was mixed with 500 ml of cold ether, cooled with stirring in a dry ice-acetone bath, and carefully acidified with a mixture of 50 ml of concentrated hydrochloric acid and 50 g of ice. The aqueous layer turned cloudy and then clear as the freed acid dissolved in the ether layer. The layers were separated, and the water layer was extracted with two 250-ml portions of cold ether. The combined, very cold, ether layers were washed with two 250-ml portions of very cold brine, and then were filtered into an ethereal solution of diazomethane. After 30 minutes just enough acetic acid was added to dispel *partially* the yellow diazomethane color, and the solvent was evaporated at aspirator pressure. Upon trituration of the residue with 100 ml of ether, followed by evaporation, there was obtained 21.8 g (68%) of a mixture of tetrahydropyranyl derivatives of *trans*-1-carbomethoxy-5-hydroxy-10-methyl-2-decalone as a creamy solid, mp 102–122°. This product was of sufficiently good quality for use without further purification.

TABULAR SURVEY

For reductions, α,β-unsaturated ketones are grouped in Tables I to VI according to whether the functional group is contained in an acyclic chain, a monocyclic system, a bicyclic fused-ring system, etc. Within each table the compounds are arranged in order of increasing complexity of molecular formula using the *Chemical Abstracts* convention. A similar system is followed for Tables VIII to XII covering reduction-alkylation and related reactions of α,β-unsaturated ketones. Table VII contains examples of reductions of α,β-unsaturated acids, esters, and aldehydes, and the arrangement within this table is the same as that described above.

In column 2 of the tables the notation M/NH_3 indicates that the substrate was added to the metal-ammonia solution, while NH_3/M indicates that the metal was added last. Unless otherwise noted, it should be assumed that the metal, the proton donor, and the quenching agent (column 3) were employed in excess. The reaction time has been indicated where reported and is the total of the time of addition of the substrate and the reaction period before addition of other reagents or quenching. The reaction temperature is indicated when it differed from the reflux temperature of liquid ammonia. Column 3 shows the quenching agent and other reactants required to convert the reduction product into the material actually isolated. The structures (or partial structures) of products are shown in column 4, yields being given in parentheses. In some cases approximate yields have been calculated by the author from published data. Numbers not in parentheses indicate product ratios when percentage yields were not reported.

The literature has been reviewed through 1973, although references from readily available journals in 1974 are also included.

A list of abbreviations employed in the tables follows.

Ac	acetyl	Ms	methanesulfonyl
Bu	butyl	Me	methyl
BMD	bismethylenedioxy	Pr	propyl
Bz	benzoyl	Py	pyridine
DME	1,2-dimethoxyethane	PTSA	*p*-toluenesulfonic acid
DMSO	dimethyl sulfoxide	THF	tetrahydrofuran
DMF	dimethylformamide	THP	tetrahydropyranyl
Et	ethyl	Ts	*p*-toluenesulfonyl
Hex	hexyl	Trityl	triphenylmethyl
LAH	lithium aluminum hydride		

TABLE I. Reduction of Acyclic α,β-Unsaturated Ketones

	Reactant	Reduction Conditions	Quenching Agent	Products (% Yield)	Refs.
C_6	$(CH_3)_2C{=}CHCOCH_3$	2 eq. Li/NH_3, Et_2O, 30 min	NH_4Cl	$(CH_3)_2CHCH_2COCH_3$ + $(CH_3)_2CHCH_2CHOHCH_3$ + starting material 42 38 20	147
		2 eq. Li/NH_3, Et_2O, 1 eq. t-BuOH, 30 min	"	" (37) + " (3)	
		2 eq. Li/NH_3, Et_2O, 1 eq. $(C_6H_5)_3COH$, 30 min	"	" (49) + " (2)	
		2 eq. Li/NH_3, Et_2O, 1 eq. MeOH, 30 min	"	" 78 + " 22	
		2 eq. Li/NH_3, Et_2O, 1 eq. pyrrole, 30 min	"	" 49 + " 51	
		2 eq. Li/NH_3, Et_2O, 1 eq. $(C_6H_5)_2NH$, 30 min	"	" 72 + " 28	
C_{10}	$C_6H_5CH{=}CHCOCH_3$ (*trans*)	Li/NH_3, Et_2O, 1 eq. t-BuOH, 30 min	NH_4Cl	$C_6H_5CH_2CH_2COCH_3$ (66)	147

	Reactant	Conditions	Quench	Product(s) and Yield(s) (%)	Refs.
		Li/NH_3, Et_2O, 1 eq. MeOH 30 min	"	" 87 + $C_6H_5CH_2CH_2CHOHCH_3$ 13	
		Li/NH_3, Et_2O, 1 eq. $(C_6H_5)_3COH$ 30 min	"	" (54)	
C_{11}	$(H)(C_6H_5)C{=}C(COCH_3)(CH_3)$	Li/NH_3, Et_2O, 1 eq. *t*-BuOH, 30 min	NH_4Cl	$C_6H_5CH_2CH(CH_3)COCH_3$ (59) + $C_6H_5CH_2CH(CH_3)CHOHCH_3$ (2)	
	t-$C_4H_9CH{=}CHCOC_4H_9$-*t* (*trans*)	3 eq. Li/NH_3, THF, 3 eq. *t*-BuOH, 2 hr	H_2O, H_2CrO_4	*t*-$C_4H_9CH_2CH_2COC_4H_9$-*t* (31) + *t*-$C_4H_9CHCH_2COC_4H_9$-*t* | *t*-$C_4H_9CHCH_2COC_4H_9$-*t* (—)	44
		Li/NH_3, 3 eq. $(CH_3)_2CDOH$, −78°, 10 min	"	" (57) + starting material (41)	
C_{15}	$C_6H_5CH{=}CHCOC_6H_5$ (*trans*)	3 eq. K/NH_3, Et_2O, 20 min	NH_4Cl	$C_6H_5CH_2CH_2COC_6H_5$ (—)	53
		2 eq. K/NH_3, Et_2O, 5 min	"	" (98)	50

Note: References 338–543 are on pp. 253–258.

TABLE II. Reductions of Monocyclic α,β-Unsaturated Ketones

Reactant	Reduction Conditions	Quenching Agent	Product(s) (% Yield)	Refs.
C_7 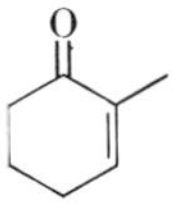	2.2 eq. Li/NH_3, Et_2O, 1 eq. *t*-BuOH, 30 min	NH_4Cl	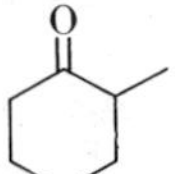 (50)	72
	2.2 eq. Li/NH_3, Et_2O, 30 min	,,	,, (47) + starting material (15)	
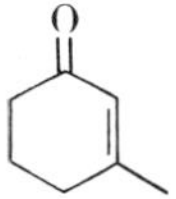	2.2 eq. Li/NH_3, Et_2O, 1 eq. *t*-BuOH, 30 min	NH_4Cl	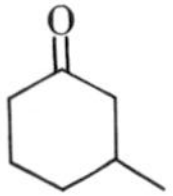 (62)	72
	2.2 eq. Li/NH_3, EtO_2, 30 min	,,	,, (66)	
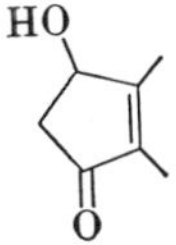 HO	NH_3/Li, EtOH, THF 45 min	NH_4Cl, H_3O^+, Ac_2O	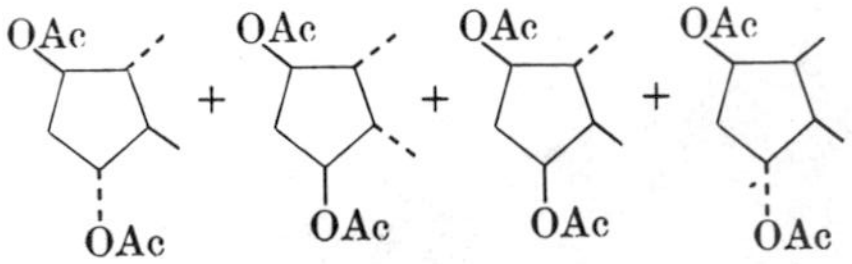 OAc OAc 60 + OAc OAc 35 + OAc OAc 4.5 + OAc OAc 0.5	142a
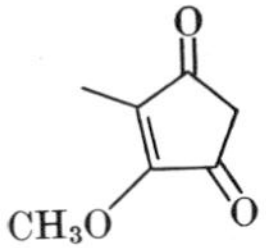 CH_3O	4 eq. K/NH_3, Et_2O	NH_4Cl	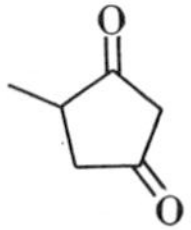 (41)	213
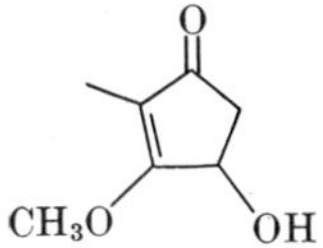 CH_3O OH	4 eq. K/NH_3, Et_2O	NH_4Cl	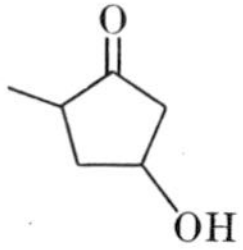 OH (14) 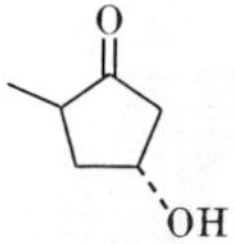 + OH (26) 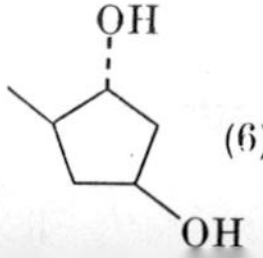 OH OH (6) 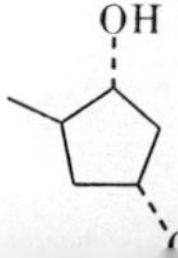 + OH OH (7)	214

	Conditions		Product(s) and Yield(s) (%)	Refs.
C_8 	2.2 eq. Li/NH_3, Et_2O, 1 eq. t-BuOH, 30 min	NH_4Cl	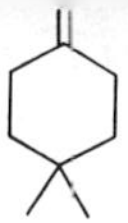 (62)	147
	2.2 eq. Li/NH_3, Et_2O, 30 min	,,	,, 60 + starting material 40	
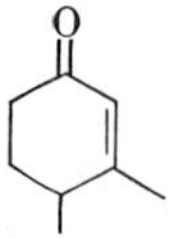	Na/NH_3, dioxane	—	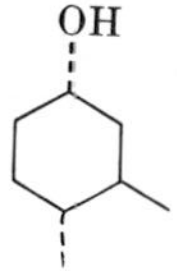 OH (25)[a] + OH (—) 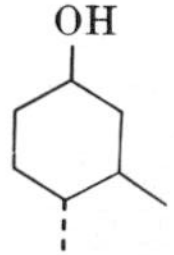	338
	Excess Li/NH_3, EtOH	CrO_3	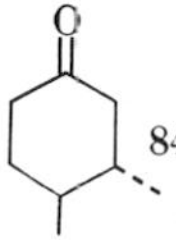 84 + 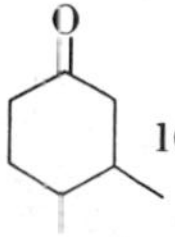 16	109
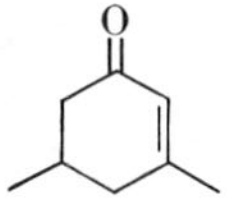	Li/NH_3, THF, t-BuOH	H_2O, H_2CrO_4	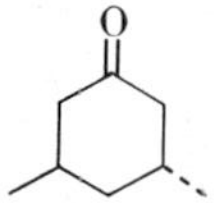 I (10) 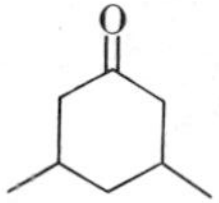 + II (70)	57
	Li/NH_3, THF	,,	I (4) + II (46) + starting material (3)	
	NH_3/Li,[b] t-BuOH, THF	,,	I (4) + II (61)	
	2.4 eq. Li/NH_3, THF	—	I (8) + II (75) + starting material (13)	110
	Li/NH_3, MeOH	CrO_3	I ~8 + II ~92	92

Note: References 338–543 are on pp. 253–258.

[a] The yield is based upon amount of saturated ketone obtained upon oxidation.
[b] A solution of lithium in liquid ammonia was added to the solution of the enone.

TABLE II. REDUCTIONS OF MONOCYCLIC α,β-UNSATURATED KETONES (*Continued*)

Reactant	Reduction Conditions	Quenching Agent	Product(s) (% Yield)	Refs.
C_8 (*Contd.*)	Na/NH_3, MeOH	CrO_3	I ~8 + II ~92	92
	K/NH_3, MeOH	,,	I ~9 + II ~91	
	Li/NH_3, EtOH	,,	I ~9 + II ~91	
	Na/NH_3, EtOH	,,	I ~8 + II ~92	
	K/NH_3, EtOH	,,	I ~9 + II ~91	
	Li/NH_3, *t*-BuOH	,,	I ~7 + II ~93	
	Na/NH_3, *t*-BuOH	,,	I ~8 + II ~92	
	K/NH_3, *t*-BuOH	,,	I ~7 + II ~93	

Reactant	Reduction Conditions	Quenching Agent	II (OC_2H_5) +	III +	IV +	V (OH, OC_2H_5) +	VI (OH) +	starting material I	Refs.
I (OC_2H_5)	7 eq. Li/NH_3, Et_2O, $-33°$, 1 min	NH_4Cl	15	—	—	85	—	—	215
	,, 2.5 min	,,	18	—	—	68	8	6	
	,, 4 min	,,	14	—	8	51	21	6	
	,, 9 min	,,	5	—	18	41	36	—	
	,, 10 min	,,	11	—	17	40	32	—	
	,, 11 min	,,	—	—	27	29	43	1	
	,, 15 min	,,	—	—	24	27	43	6	
	,, 20 min	,,	—	—	28	12	55	5	
	,, 26 min	,,	4	—	27	15	54	—	
	,, 30 min	,,	1	4	29	5	50	11	
	,, 31 min	,,	2	—	34	10	52	2	
	,, 40 min	,,	—	—	21	12	57	9	
	,, 41 min	,,	1	—	27	5	57	10	
	,, 60 min	,,	—	—	46	4	47	3	
	,, 1.5 min	EtOH	—	—	—	85	15	—	
	,, 2.5 min	,,	—	—	—	75	25	—	
	,, 4 min	,,	—	—	16	51	33	—	
	,, 10 min	,,	—	—	21	29	69	—	

'', 15 min	''	—	—	17	17	66	—	
'', 19 min	''	—	—	—	19	81	—	
'', 22 min	''	—	—	—	15	79	6	
'', 25 min	''	—	—	—	25	66	5	
'', 30 min	''	—	—	—	6	94	—	
'', 35 min	''	—	—	—	5	92	3	
'', 40 min	''	—	—	—	—	100	—	
7 eq. Li/NH_3, Et_2O, −78°, 1.5 min	NH_4Cl	71[c]	—	—	23	—	6	
'', 2.5 min	''	80[c]	—	—	6	—	14	
'', 5 min	''	91[c]	—	—	4	—	5	
'', 6 min	''	69[c]	—	—	20	—	11	
'', 10 min	''	83[c]	—	—	16	—	1	
''	''	77[c]	—	—	7	—	16	
''	''	58[c]	—	—	8	—	34	
'', 15 min	''	88[c]	—	—	8	—	4	
'', 18 min	''	74[c]	—	—	9	—	17	
'', 25 min	''	77[c]	—	—	16	—	7	
''	''	91[c]	—	—	4	—	5	
'', 28 min	''	77[c]	—	—	—	—	23	
'', 50 min	''	57[c]	—	—	26	—	17	
'', 57 min	''	67[c]	—	—	19	—	14	
'', 120 min	''	74[c]	—	—	26	—	—	
'', 10 min	EtOH	—	—	—	100	—	—	
''	''	—	—	—	72	—	28	
'', 15 min	''	—	—	—	78	—	22	
'', 16 min	''	—	—	—	100	—	—	
'', 21 min	''	—	—	—	100	—	—	
7 eq. Li/NH_3, Et_2O, −78° 30 min	EtOH	II	III	IV	V	VI	I	
''	''	—	—	—	100	—	—	215
''	''	20[c]	—	—	54	—	26	
'', 60 min	''	17[c]	—	—	83	—	—	

Note: References 338–543 are on pp. 253–258.

[c] This number represents the sum of the amounts of II and III obtained. Cyclohexenone (III) was assumed to arise by elimination of ethanol from II during workup or vpc analysis.

TABLE II. REDUCTIONS OF α,β-MONOCYCLIC UNSATURATED KETONES (*Continued*)

Reactant	Reduction Conditions	Quenching Agent	Product(s) (% Yield)					Refs.
C_8 (*Contd.*)	7 eq. Li/NH_3, Et_2O, $-33°$, 20 min	NH_4Cl, 22^c $-78°$	—	37	15	14	12	
	", 28 min	" 8^c	—	68	—	11	13	
	", 45 min	" 10^c	—	65	1	5	19	
	", 46 min	" 12^c	—	49	—	—	39	
C_9 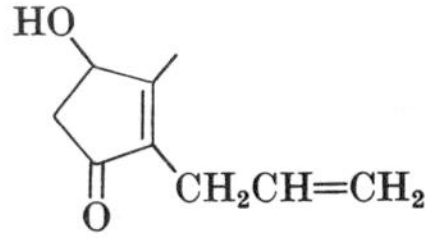 HO, O, $CH_2CH{=}CH_2$	NH_3/Li THF, EtOH (4 eq.), 45 min	NH_4Cl, H_3O^+, Ac_2O	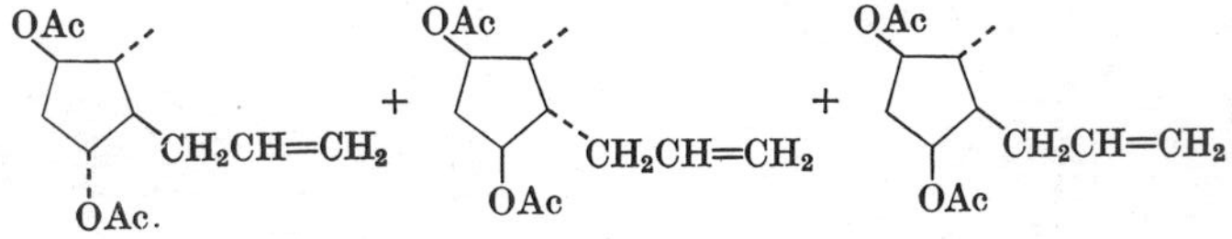 OAc, OAc, $CH_2CH{=}CH_2$ (70) + OAc, OAc, $CH_2CH{=}CH_2$ (10) + OAc, OAc, $CH_2CH{=}CH_2$ (20)					142a
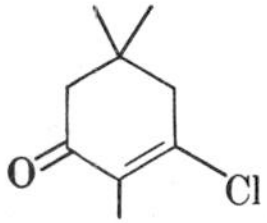 O, Cl	Li/NH_3, Et_2O, 65 min	H_2O, CrO_3	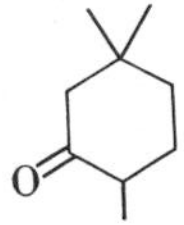 O (54)					339
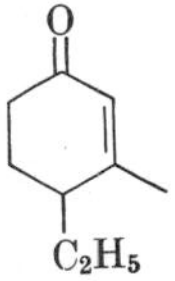 O, C_2H_5	Excess Li/NH_3, EtOH	CrO_3	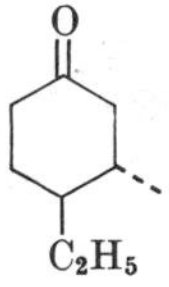 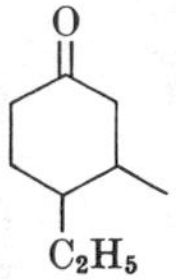 O, C_2H_5 84.3 + O, C_2H_5 15.7					109
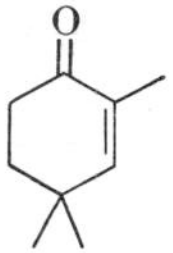 O	2.2 eq. Li/NH_3, Et_2O, 1 eq. *t*-BuOH, 30 min	NH_4Cl	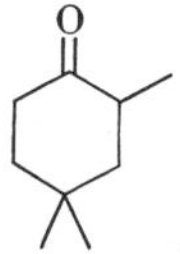 O (46)					72
	2.2 eq. Li/NH_3, Et_2O, 30 min	"	" 53 + starting material 47					147

Substrate	Conditions	Quench	Product(s) and Yield(s) (%)	Refs.
	2.2 eq. Li/NH_3, Et_2O, 1 eq. HOAc	"	" 100	
	2.2 eq. Li/NH_3, Et_2O, 1 eq. H_2O, 30 min	"	" 100	
	2.2 eq. Li/NH_3, Et_2O, 1 eq. pyrrole	"	" 100	
	2.2 eq. Li/NH_3, Et_2O, 1 eq. diphenylamine 30 min	"	" 100	
	2.2 eq. Li/NH_3, Et_2O, 1 eq triphenylmethane, 30 min	"	" 100	
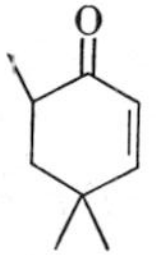 [structure]	2.2 eq. Li/NH_3, Et_2O, 1 eq t-BuOH, 30 min	"	" (80)	72
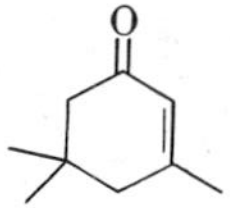 [structure]	NH_3/Li, $(CH_3)_2CDOH$, $-78°$, 13 min	H_2O, H_2CrO_4/acetone	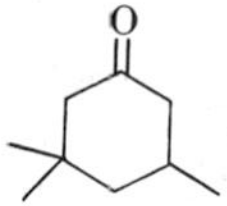 [structure] I (84) + starting material (9)	57
	Li/NH_3, 1 eq. t-BuOH, 30 min	NH_4Cl	I 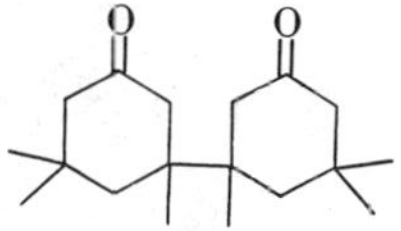 + [structure] (Low)	72

Note: References 338–543 are on pp. 253–258.

[c] This number represents the sum of the amounts of II and III obtained. Cyclohexenone (III) was assumed to arise by elimination of ethanol from II during workup or vpc analysis.

TABLE II. REDUCTIONS OF MONOCYCLIC α,β-UNSATURATED KETONES (*Contd.*)

Reactant	Reduction Conditions	Quenching Agent	Product(s) (% Yield)	Refs.
C_{10}				
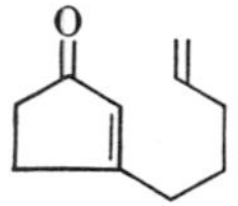	Li/NH_3, Et_2O	H_3O^+	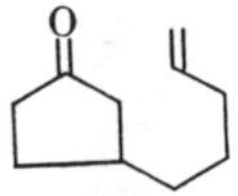 (19)	340
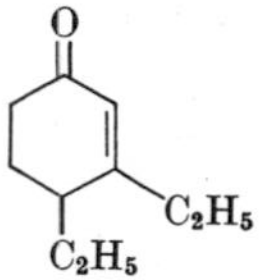 C_2H_5 C_2H_5	Excess Li/NH_3, EtOH	CrO_3	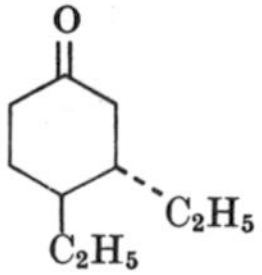 C_2H_5 C_2H_5 43.5 + C_2H_5 C_2H_5 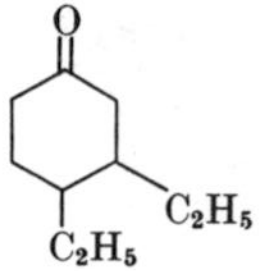 56.5	109
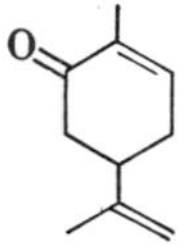	Excess Li/NH_3, Et_2O, EtOH, 3.5 hr	NH_4Cl, CrO_3	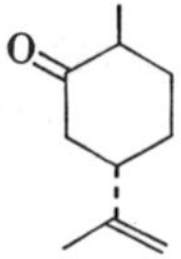 (66) I	335
	2.2 eq. Li/NH_3, Et_2O, 1 eq. *t*-BuOH, 30 min	NH_4Cl	I (61)	72
	2.2 eq. Li/NH_3, Et_2O, 30 min	"	" (50) + starting material (15)	
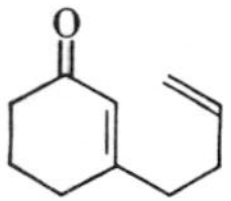	Li/NH_3, Et_2O	H_3O^+	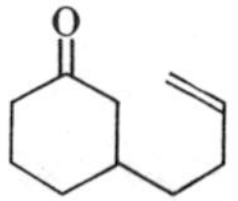 (70)	341

Substrate	Reducing Conditions	Proton Source	Product(s) and Yield(s) (%)	Refs.
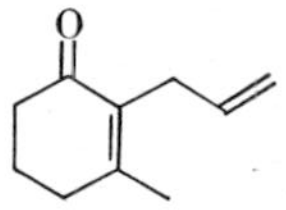	2.2 eq. Li/NH_3, Et_2O, 1 eq. H_2O, 60 min	NH_4Cl, NaOMe, MeOH	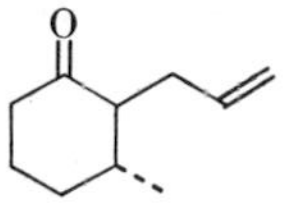 ~65 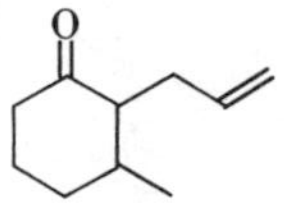 + ~35	176
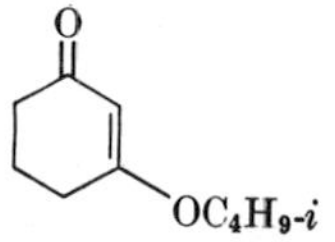 OC_4H_9-*i*	2.5 eq. Li/NH_3, Et_2O	H_2O[d]	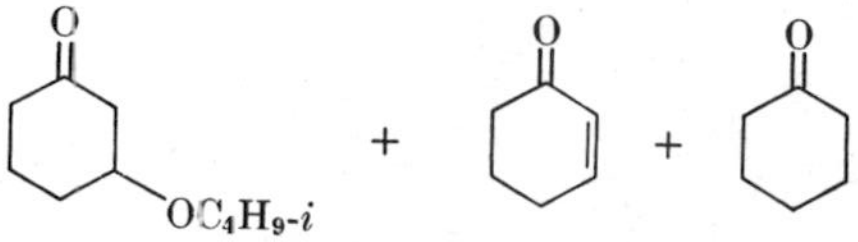 OC_4H_9-*i* + + I 95 II 4.5 III 0.5	43
	7 eq. Li/NH_3, Et_2O	,,	I 85 II 3 III 1 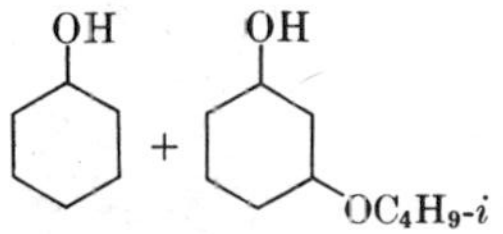 OH + OH OC_4H_9-*i* IV 0.5 V (*cis* and *trans*) 10	
	7 eq. Li/NH_3, Et_2O, 20 min	H_2O[e]	I 62 + II 5.5 + III 16.5 + IV 4 + V 12	
	7 eq. Li/NH_3, Et_2O, 6 hr	,,	I 28 + II 0.5 + III 10 + IV 40 + V (*cis* and *trans*) 21.5	
	Mg/NH_3, Et_2O, 2 hr[f]	,,	I 77 + V (*cis* and *trans*) 23	
	Mg/NH_3, Et_2O, 6 hr[f]	,,	I 62 + V (*cis* and *trans*) 38	

Note: References 338–543 are on pp. 253–258.

[d] The reaction mixture was quenched immediately after the addition of the enone by pouring it into a large excess of water.

[e] The reaction mixture was quenched after the indicated time by pouring it into a large excess of water.

[f] The Mg/NH_3 solution was prepared by electrolysis of a solution of magnesium iodide in liquid ammonia using a magnesium anode and platinum cathode with a current of 0.1 amp.

TABLE II. Reductions of Monocyclic α,β-Unsaturated Ketones (*Contd.*)

	Reactant	Reduction Conditions	Quenching Agent	Product(s) (% Yield)	Refs.
C_{11}	$COCH_3$	Li/NH_3	—	$COCH_3$	341, 342
	$COCH_3$	Li/NH_3	—	$COCH_3$	341, 342
	OH, C_3H_7-*n*, $CH_2CH{=}CH_2$, O	NH_3/Li, THF, EtOH	NH_4Cl, H_3O^+, Ac_2O	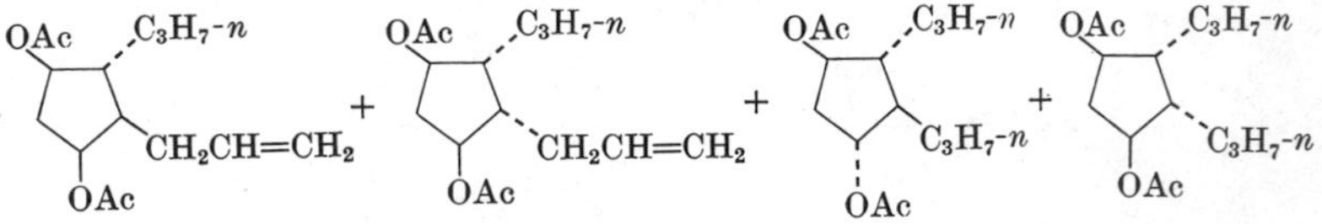 OAc, C_3H_7-*n*, $CH_2CH{=}CH_2$, OAc (33.6) + OAc, C_3H_7-*n*, $CH_2CH{=}CH_2$, OAc (30) + OAc, C_3H_7-*n*, C_3H_7-*n*, OAc (17.4) + OAc, C_3H_7-*n*, C_3H_7-*n*, OAc (12)	142a
	O, CO_2Et	Li/NH_3	CrO_3	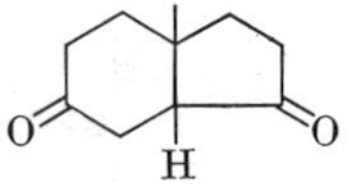 O, O, H (50)	196
		K/NH_3	"	" (60)	
	O	Li/NH_3, Et_2O	HCl	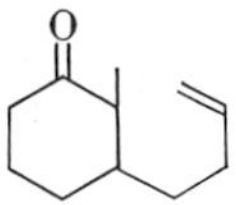 O (58)	337
	O, OH	Li/NH_3	EtOH	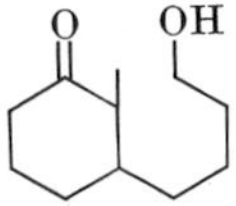 O, OH (81)	343
	O	Li/NH_3, Et_2O	H_3O^+	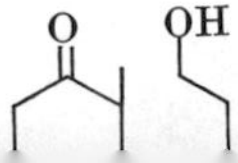 O, OH (75)	340

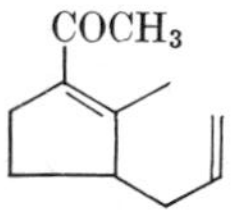

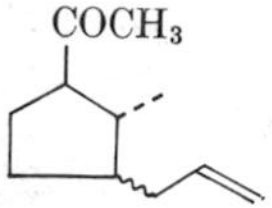

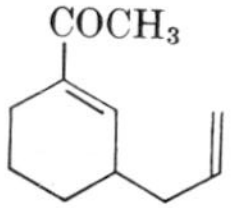

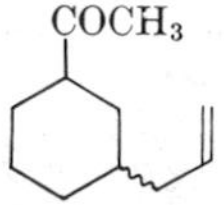

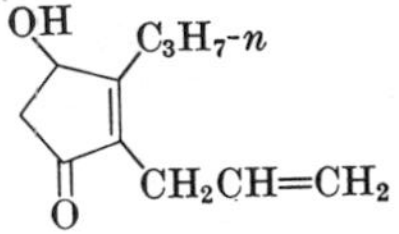

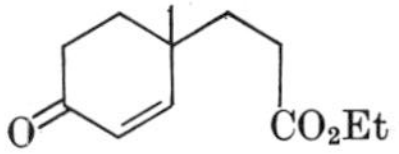

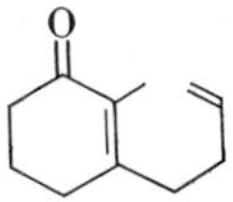

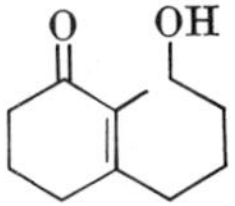

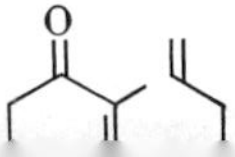

	Reactant	Conditions	Workup	Product(s) and Yield(s) (%)	Refs.
		Li/NH_3, Et_2O	H_3O^+	25 75	111
C_{12}		2.2 eq. Li/NH_3, Et_2O, 30 min	NH_4Cl	(74)	344
	$COCH_3$	Li/NH_3	—	$COCH_3$	341, 342
	CHSBu-*n*	6 eq. Li/NH_3, Et_2O, 2 eq. H_2O	NH_4Cl	(80)	170
C_{13}		Li/NH_3, Et_2O	HCl	(63)	337
	OH	Li/NH_3		OH (77)	343

Note: References 338–543 are on pp. 252–258.

TABLE II. Reductions of Monocyclic α,β-Unsaturated Ketones (*Contd.*)

Reactant	Reduction Conditions	Quenching Agent	Product(s) (% Yield)	Refs.
C_{13} (*Contd.*)				
t-Bu–cyclohexylidene=CHCOCH$_3$	Li/NH$_3$, THF, t-BuOH, 30 min	H$_2$O, H$_2$CrO$_4$	t-Bu–C$_6$H$_{10}$(CH$_2$COCH$_3$)(H) I (15) + t-Bu–C$_6$H$_{10}$(H)(CH$_2$COCH$_3$) II (73)	57
	Li/NH$_3$, THF	″	I (8) + II (64) + starting material (6)	
	NH$_3$/Li,[g] THF, t-BuOH, 30 min	″	″ (3) + ″ (31) + ″ (41)	
	Na/NH$_3$, THF t-BuOH, 30 min	″	″ (10) + ″ (84)	
3-C$_6$H$_5$-4-methylcyclohex-2-enone	Excess Li/NH$_3$, EtOH	CrO$_3$	3-C$_6$H$_5$-4-methylcyclohexanone (6) + 3-C$_6$H$_5$-4-methylcyclohexanone (94)	109
3-C$_6$H$_5$-5-methylcyclohex-2-enone	Li/NH$_3$, THF, t-BuOH	H$_2$O, H$_2$CrO$_4$	3-C$_6$H$_5$-5-methylcyclohexanone (11) + 3-C$_6$H$_5$-5-methylcyclohexanone (70)	57
	Li/NH$_3$, THF, t-BuOH, −78°	″	″ (9) ″ (75)	
C_{14} 3-CO$_2$H-4-C$_6$H$_4$OCH$_3$-p-cyclohex-2-enone	Li/NH$_3$ THF, 2 hr,	FeCl$_3$, t-BuOH, NH$_4$Cl	3-CO$_2$H-4-C$_6$H$_4$OCH$_3$-p-cyclohexanone (H, H)	112
CH$_3$C(=O)–spiro compound	Excess Li/NH$_3$, Et$_2$O, EtOH	NH$_4$Cl, CrO$_3$	CH$_3$C(=O)–spiro compound (∼70)	318

	Reactant	Conditions	Workup	Product(s) and Yield(s) (%)	Refs.
C_{15}	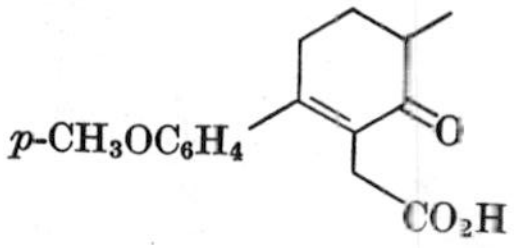 p-$CH_3OC_6H_4$, O, CO_2H	NH_3/Li, Et_2O, 7 min	H_2O, H_3O^+	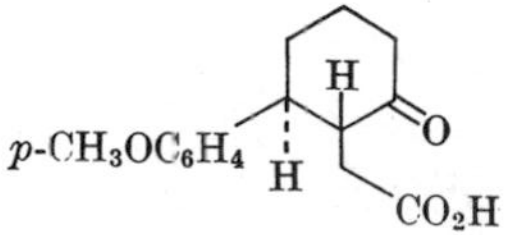 p-$CH_3OC_6H_4$, H, H, O, CO_2H (82)	345
	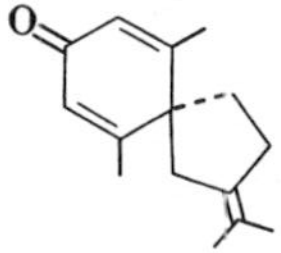 O	Li/NH_3, Et_2O, 1 eq. t-BuOH, 35 min	NH_4Cl	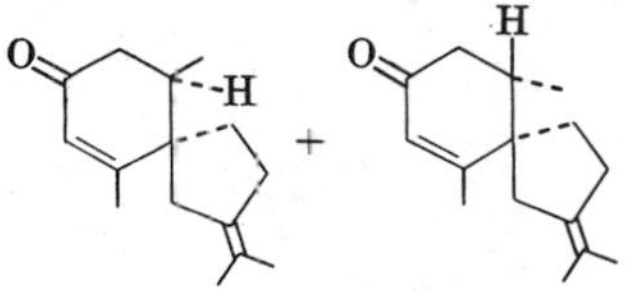 O, H (24) + O, H (56)	345a
C_{16}	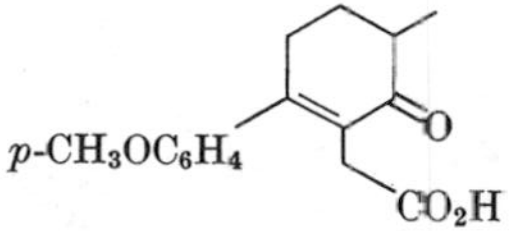 p-$CH_3OC_6H_4$, O, CO_2H	NH_3/Li, Et_2O, 7 min	H_2O, H_3O^+	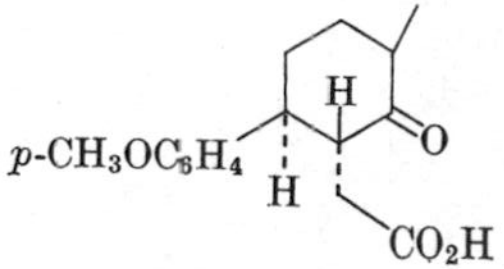 p-$CH_3OC_6H_4$, H, H, O, CO_2H (81)	345a
C_{17}	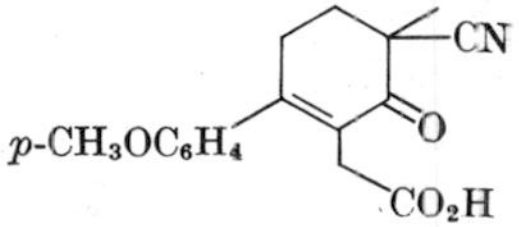 p-$CH_3OC_6H_4$, CN, O, CO_2H	Li/NH_3, Et_2O	H_2O[h]	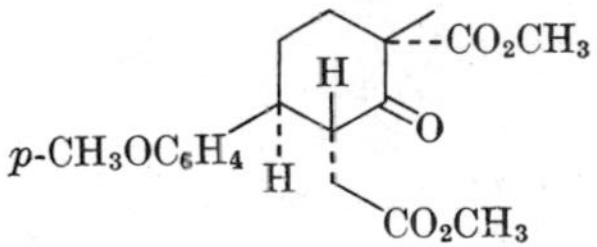 p-$CH_3OC_6H_4$, H, H, CO_2CH_3, O, CO_2CH_3 (34)	345
	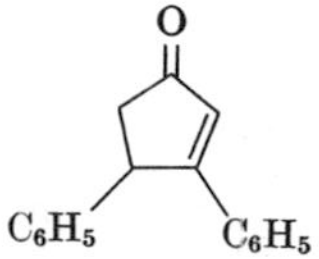 O, C_6H_5, C_6H_5	Li/NH_3	—	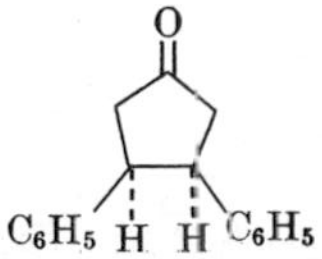 O, C_6H_5 H H C_6H_5	10

Note: References 338–543 are on pp. 253–258.

[g] A solution of lithium in liquid ammonia was added dropwise to the enone in NH_3—THF.
[h] After workup the nitrile function was hydrolyzed and the carboxyl group esterified with methanol and hydrogen chloride.

TABLE II. REDUCTIONS OF MONOCYLIC α,β-UNSATURATED KETONES (*Contd.*)

Reactant	Reduction Conditions	Quenching Agent	Product(s) (% Yield)	Refs.
C_{17} (*Contd.*)				
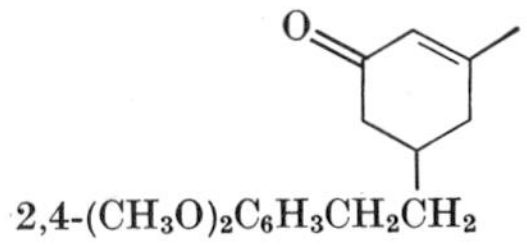 [structure] 2,4-$(CH_3O)_2C_6H_3CH_2CH_2$	Li/NH_3, EtOH	HCl-H_2O, CrO_3	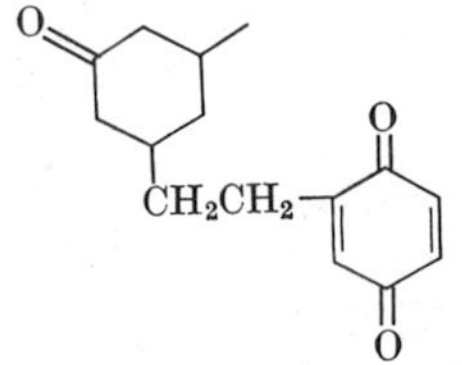 [structure] CH_2CH_2 (18)	346
C_{18} 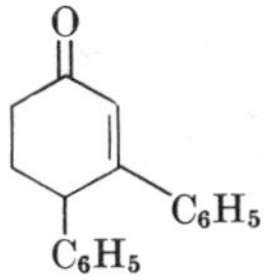 [structure] C_6H_5, C_6H_5	Excess Li/NH_3, EtOH	CrO_3	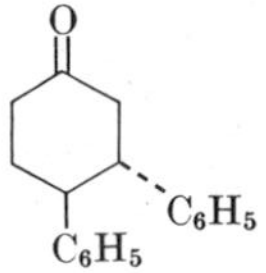 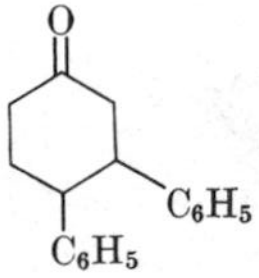 [structure] C_6H_5, C_6H_5 2 + [structure] C_6H_5, C_6H_5 (34)	109
	4 eq. Li/NH_3, Et_2O, 30 min	NH_4Cl	" (34)	347
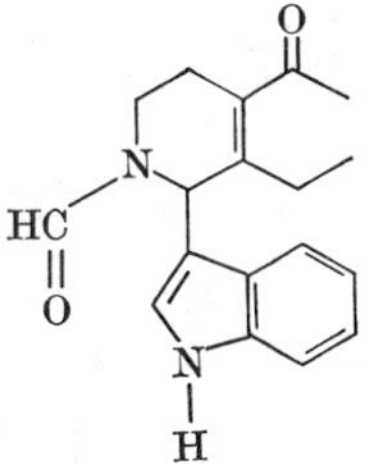 [structure]	Li/NH_3, THF, 2 min	C_6H_5Br, $CH_3C_6H_5$, H_2O	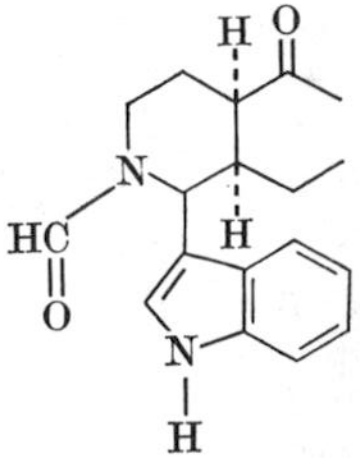 [structure] (47) 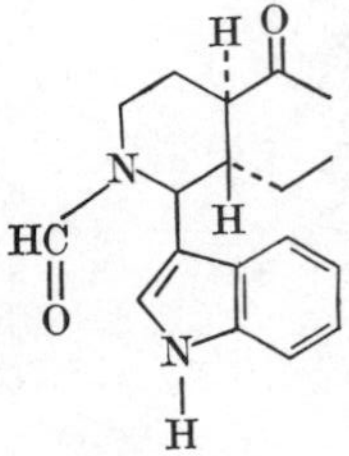 + [structure] (25)	166

Reactant	Reducing conditions	Quenching	Product(s) and Yield(s) (%)	Refs.
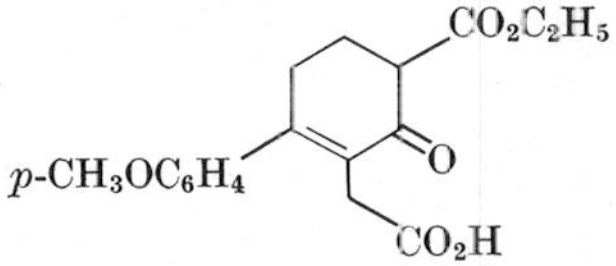 p-$CH_3OC_6H_4$ (cyclohexenone bearing $CO_2C_2H_5$, O, CH_2CO_2H)	Li/NH_3, dioxane, Et_2O	H_2O, CH_2N_2	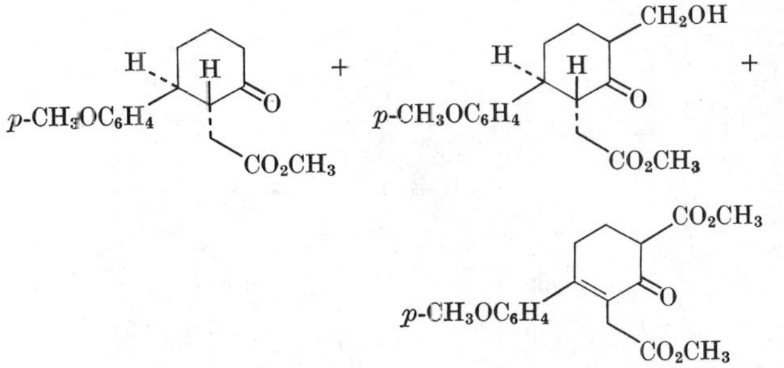 p-$CH_3OC_6H_4$ (H, H, O, CO_2CH_3) + p-$CH_3OC_6H_4$ (H, H, CH_2OH, O, CO_2CH_3) + p-$CH_3OC_6H_4$ (CO_2CH_3, O, CO_2CH_3)	345
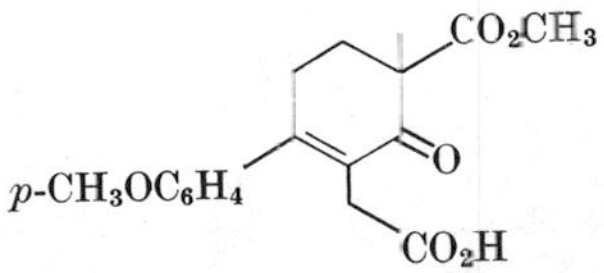 p-$CH_3OC_6H_4$ (cyclohexenone bearing CO_2CH_3, O, CH_2CO_2H)	Li/NH_3, dioxane, Et_2O	H_2O, CH_2N_2	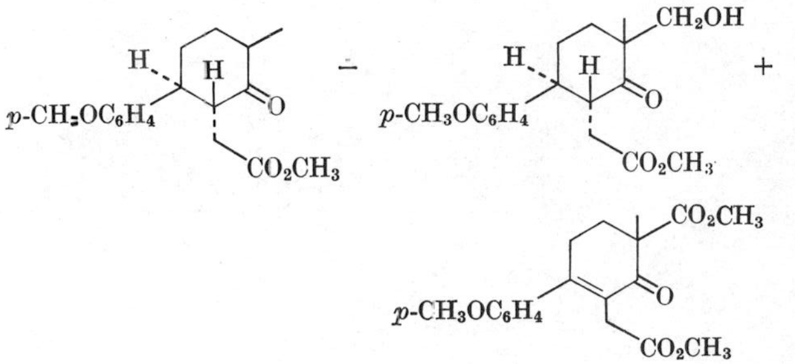 p-$CH_3OC_6H_4$ (H, H, O, CO_2CH_3) + p-$CH_3OC_6H_4$ (H, H, CH_2OH, O, CO_2CH_3) + p-$CH_3OC_6H_4$ (CO_2CH_3, O, CO_2CH_3)	
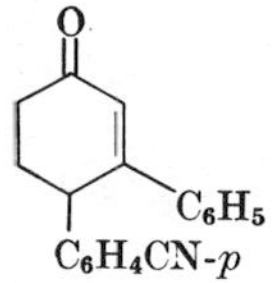 C_{19} O (cyclohexenone, C_6H_5, C_6H_4CN-p)	NH_3/3 eq. Li, Et_2O, 30 min	NH_4Cl	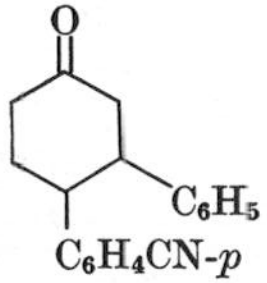 O (cyclohexanone, C_6H_5, C_6H_4CN-p) (21)	348

Note: References 338–543 are on pp. 253–258.

TABLE II. Reductions of Monocylic α,β-Unsaturated Ketones (*Contd.*)

Reactant	Reduction Conditions	Quenching Agent	Product(s) (% Yield)	Refs.
C_{19} (*Contd.*)				
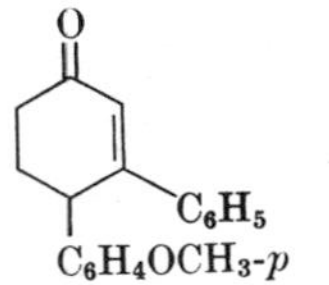 3-C_6H_5-4-($C_6H_4OCH_3$-*p*)-cyclohex-2-enone	3 eq. Li/NH_3, Et_2O, $-78°$, 1 hr	NH_4Cl	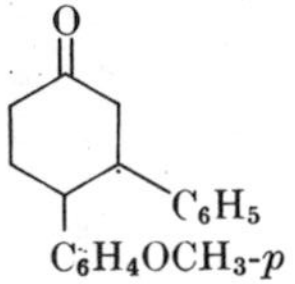 3-C_6H_5-4-($C_6H_4OCH_3$-*p*)-cyclohexanone (87)	348
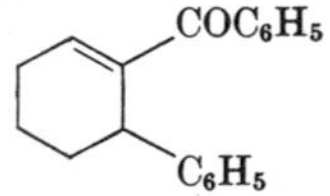 1-COC_6H_5-6-C_6H_5-cyclohexene	Li/NH_3	H_2O^i	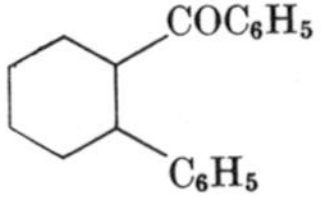 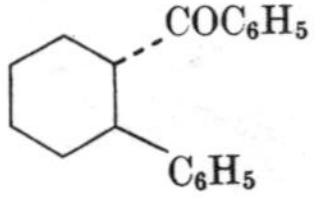 1-COC_6H_5-2-C_6H_5-cyclohexane 80 + 1-COC_6H_5-2-C_6H_5-cyclohexane (dashed COC_6H_5) 20	25, 143
	Na/NH_3	"	" 72 + " 28	
	K/NH_3	"	" 79 + " 21	
	Cs/NH_3	"	" 76 + " 24	
	Mg/NH_3 [j]	"	" 76 + " 24	
	Li/NH_3	$MeOH^i$	" 80 + " 20	143
	"	0.2 *M* NaOMe-$MeOH^i$	" 44 + " 56	
	"	0.2 *M* LiOMe-$MeOH^i$	" 68 + " 32	
	"	0.2 *M* NaOH-H_2O^i	" 71 + " 29	
	"	2 *M* NaOMe-MeOH	" 30 + " 70	
	"	2 *M* NaOH-H_2O^i	" 68 + " 32	
	Ca/NH_3	—	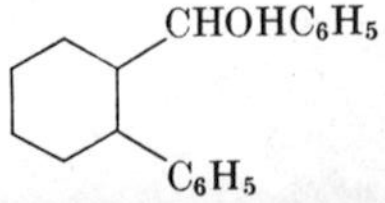 1-$CHOHC_6H_5$-2-C_6H_5-cyclohexane	42

	Substrate	Reducing Conditions	Quench	Product(s) (Yield, %)	Refs.
C_{20}	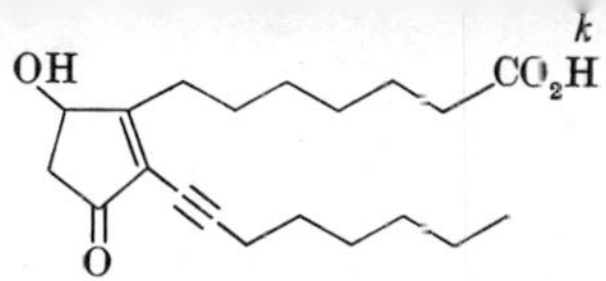 OH, CO_2H [k]	NH_3/Li, EtOH, THF	NH_4Cl	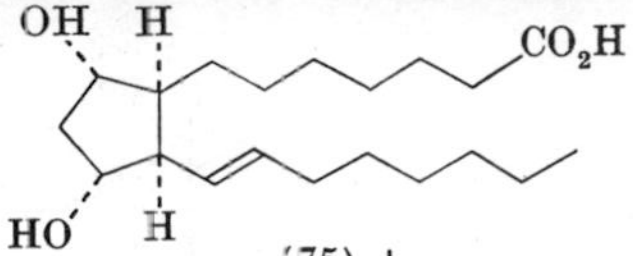 OH, H, CO_2H, HO, H (75) 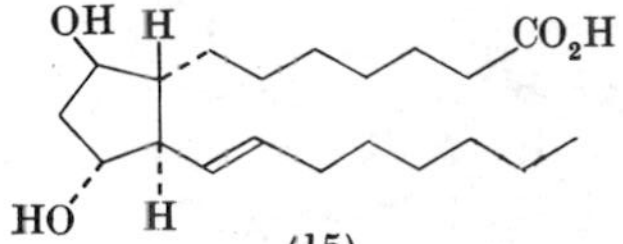 + OH, H, CO_2H, HO, H (15)	142b
C_{21}	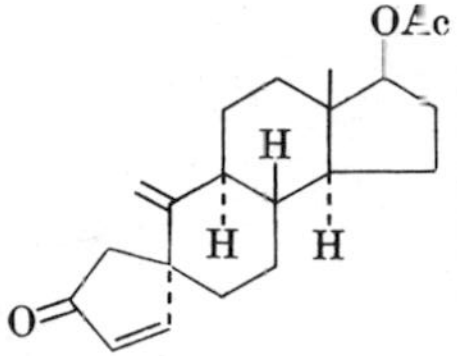 OAc, H, H, H, O	Li/NH_3, THF, 30 min	NH_4Cl	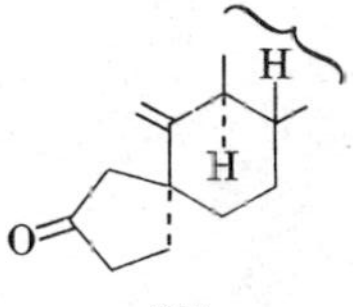 H, H, O (37)	348a
C_{24}	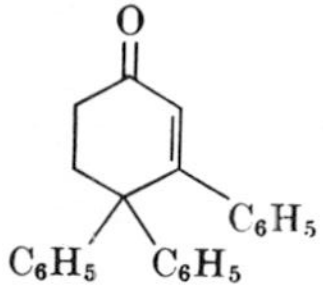 O, C_6H_5, C_6H_5, C_6H_5	NH_3/3 eq. Li, Et_2O-THF	NH_4Cl	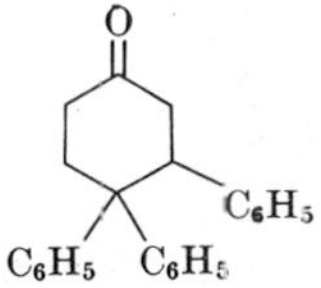 O, C_6H_5, C_6H_5, C_6H_5 (78)	347
	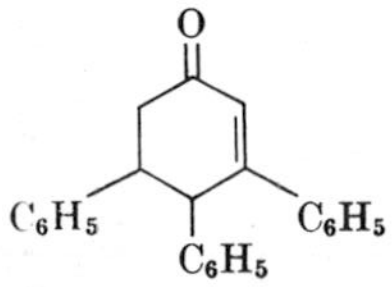 O, C_6H_5, C_6H_5, C_6H_5	NH_3/5 eq. Li, Et_2O-THF, 3 hr	"	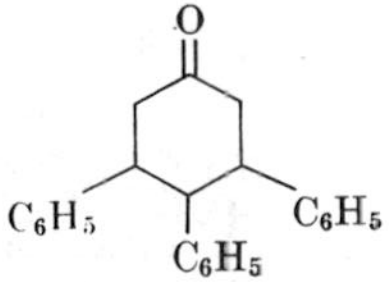 O, C_6H_5, C_6H_5, C_6H_5 (57.4)	347

Note: References 338–543 are on pp. 253–258.

[i] The metal enolate was quenched by pouring the reaction mixture into a large excess of the indicated reagent.

[j] The Mg/NH_3 solution was produced electrochemically.

[k] The enone was converted into the dianion with 2 eq. of sodium hydride in THF prior to reduction.

TABLE III. REDUCTIONS OF BICYCLIC α,β-UNSATURATED KETONES

Reactant	Reduction Conditions	Quenching Agent	Product(s) (% Yield)	Refs.
C_9				
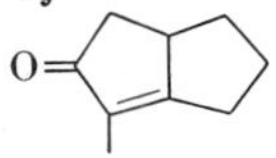 [structure]	Li/NH_3 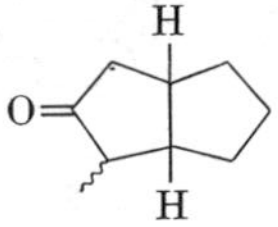	—	[structure] (—)	116
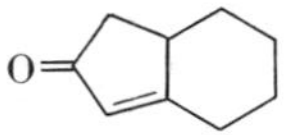 [structure]	Li/NH_3 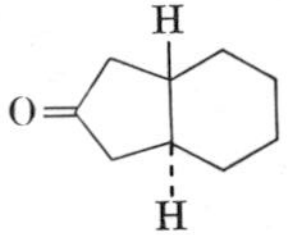 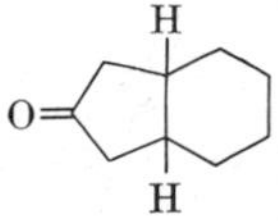	—	[structure] 99 + [structure] 1	119
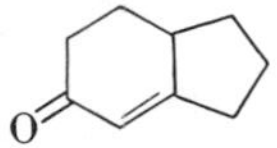 [structure]	Li/NH_3	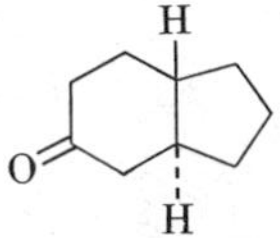 —	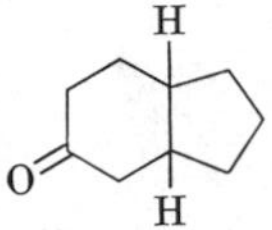 [structure] 85 + [structure] 15	119
	Li/NH_3, Et_2O, 1 hr	NH_4Cl	" 80 + " 20	14
	NH_3/Na, dioxane, −40°, 1 hr	"	" 87 + " 13	120, 121
C_{10}				
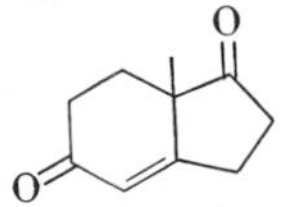 [structure]	Li/NH_3, Et_2O, 70 min	NH_4Cl	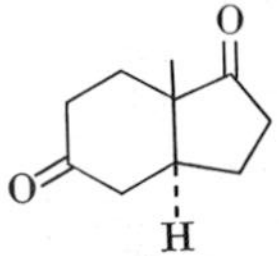 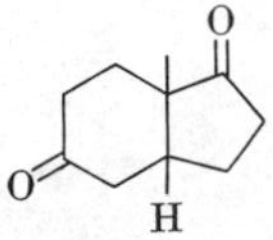 [structure] (27) + [structure] (16)	349
	Li/NH_3, THF −60°, 1 hr		" (—) + " (59)	349a
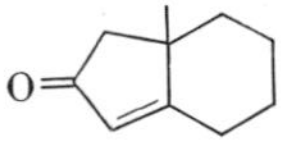 [structure]	Li/NH_3 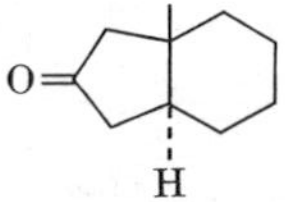 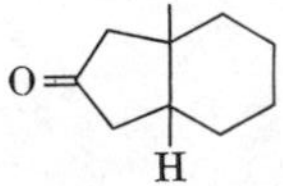	—	[structure] 30 + [structure] 70	119

Reactant	Conditions	Quench	Product(s) and Yield(s) (%)				Refs.
	Li/NH_3, Et_2O	—	"	(24) +	"	(56)	122, 269
	Li/ND_3, Et_2O 30 min	Aq. NH_4Cl	(*cis-trans* mixture)	(69)			59
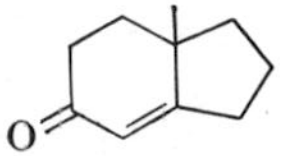	Li/NH_3	—		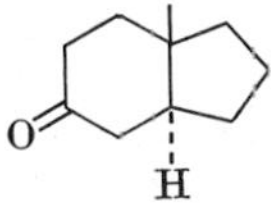 10 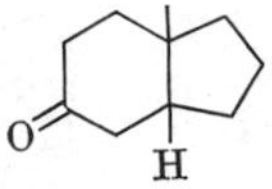 +		90	119
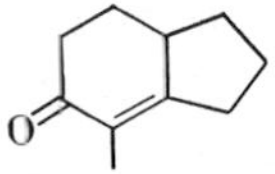	Li/NH_3, Et_2O, 60 min	NH_4Cl		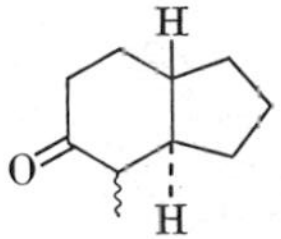 (55) 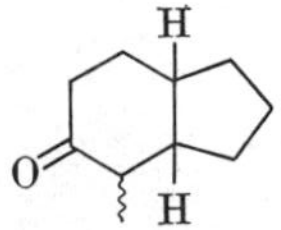 +		(18)	14
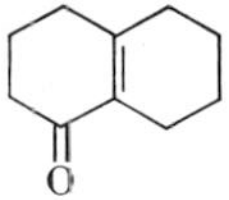	Li/NH_3, THF, 10 min	EtOH, H_2O		(–)			16
	Li/NH_3	Satd. NH_4Cl[a]		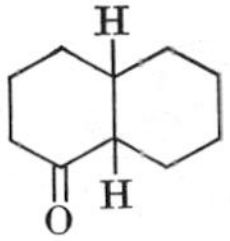 42 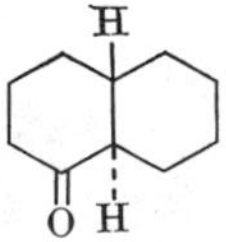 +		58	41

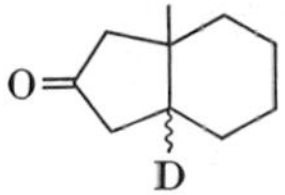

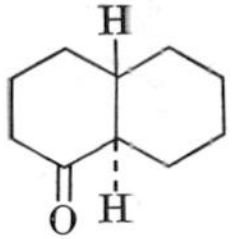

Note: References 338–543 are on pp. 253–258.

[a] A solution of the lithium enolate was added slowly to a saturated aqueous solution of ammonium chloride.

TABLE III. REDUCTIONS OF BICYCLIC α,β-UNSATURATED KEOTNES (*Continued*)

Reactant	Reduction Conditions	Quenching Agent	Product(s) (% Yield)	Refs.
C_{10} (*Contd.*)				
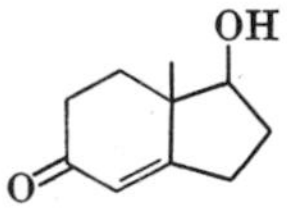 [structure]	Li/NH_3, Et_2O, 10 min	NH_4Cl	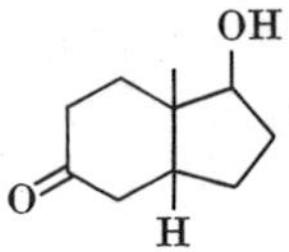 [structure] (90)	307
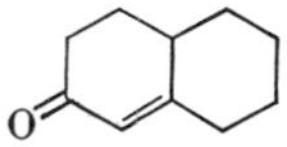 [structure]	NH_3/Li, Et_2O, 10 min	NH_4Cl (slow addn.), $H_2O - CrO_3$	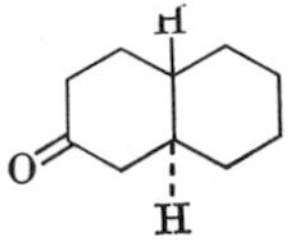 [structure] (66)[b]	350
	NH_3/Li, Et_2O-dioxane, 45 min	NH_4Cl	" (64)	351
	NH_3/Li, Et_2O, 15 min	"	" (70)	261
	Li/NH_3, Et_2O, 1 eq. *t*-BuOH	"	" (76) + starting material (1–1.5)	14
	NH_3/Na, dioxane, −40°, 1 hr	"	" 99 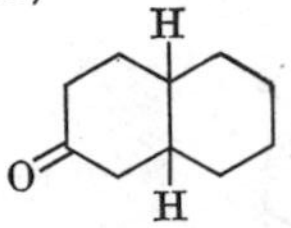 + [structure] 1	120, 121
	Li/NH_3, THF, 2 eq. *t*-BuOH, 2 min	"	" >98 + " <2	57
	Excess Na/NH_3, Et_2O-MeOH	NH_4Cl, CrO_3	" 99 + " 1	78
	Na/NH_3	—	" 95 + " 5	63
	Ba/NH_3	—	" 94 + " 6	
	NH_3/excess Li, 1 hr.	MeI,[c] —NH_3, benzene, D_2O, 0.2% KOH/MeOH	" + starting material	14
	Li/NH_3, Et_2O	NH_4Cl	" (49) + " (11)	110

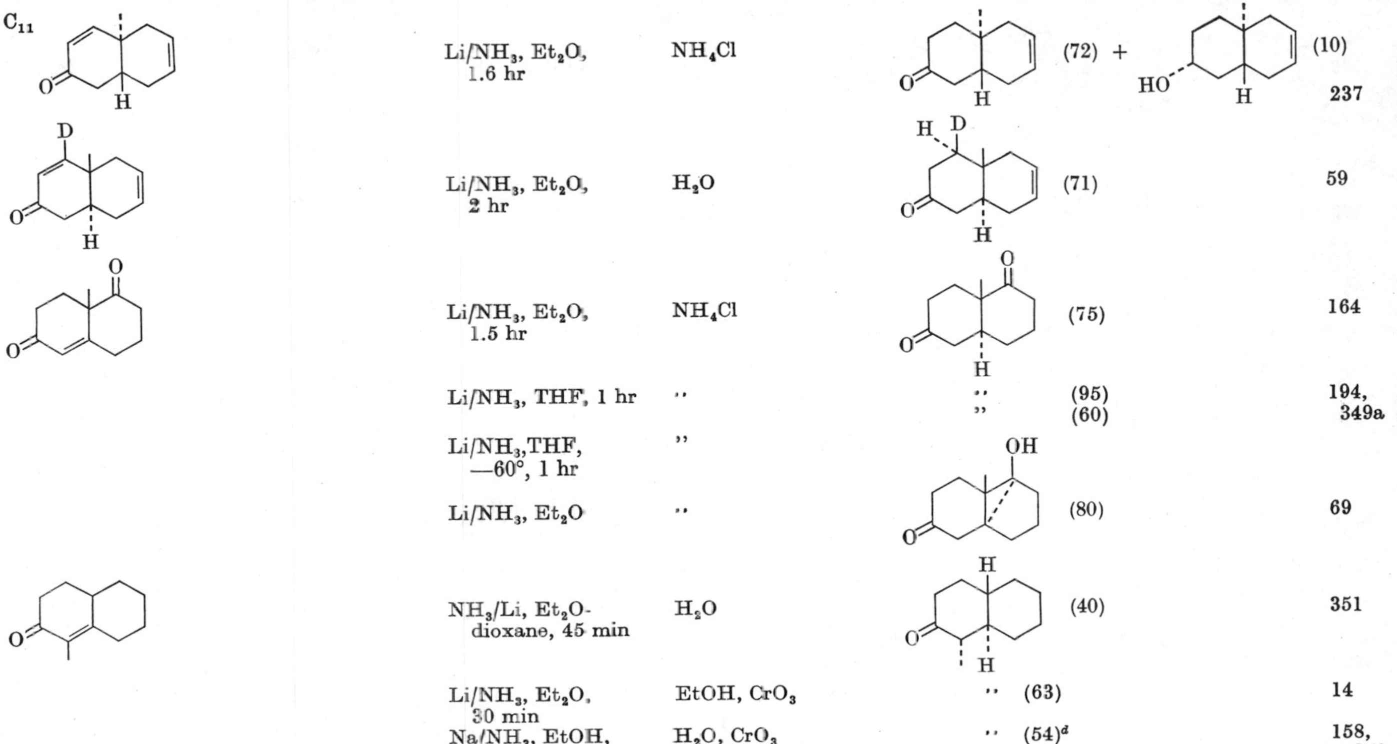

C_{11}

Substrate	Conditions	Quench	Product (%)	Refs.
Substrate	Li/NH_3, Et_2O, 1.6 hr	NH_4Cl	Product (72) + Product (10)	237
Substrate	Li/NH_3, Et_2O, 2 hr	H_2O	Product (71)	59
Substrate	Li/NH_3, Et_2O, 1.5 hr	NH_4Cl	Product (75)	164
	Li/NH_3, THF, 1 hr	"	" (95)	194,
			" (60)	349a
	Li/NH_3,THF, —60°, 1 hr	"		
	Li/NH_3, Et_2O	"	Product (80)	69
Substrate	NH_3/Li, Et_2O-dioxane, 45 min	H_2O	Product (40)	351
	Li/NH_3, Et_2O, 30 min	EtOH, CrO_3	" (63)	14
	Na/NH_3, EtOH, —78°, 4.5 hr	H_2O, CrO_3	" (54)[d]	158, 352

Note: References **338–543** are on pp. 253–258.

[b] The authors state that the product contained more than 90% of the *trans* isomer.

[c] The methyl iodide was used to destroy the excess lithium.

[d] The corresponding 2β-hydroxydecalin was isolated in 80% yield before the oxidation.

TABLE III. REDUCTIONS OF BICYCLIC α,β-UNSATURATED KETONES (*Continued*)

Reactant	Reduction Conditions	Quenching Agent	Product(s) (% Yield)	Refs.
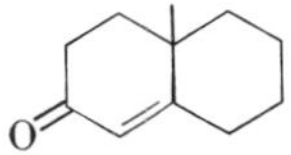	NH_3/Li, Et_2O, −40°, 30 min	NH_4Cl	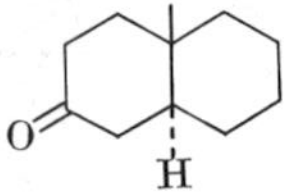 (60)[e] 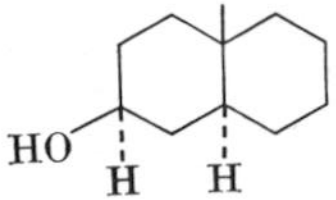 + HO H H (Trace)	353
	Li/NH_3, Et_2O, 1.5 hr	NH_4Cl, CrO_3	" (85)[f]	304
	NH_3/Li, Et_2O, 1.5 hr, Et_2O-EtOH, 4 hr	"	" (87)	354
	Li/NH_3, Et_2O-EtOH[g]	"	" 98.7 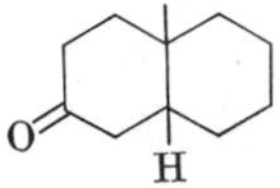 + H 1.3	78
	Li/NH_3, THF, 2 eq. *t*-BuOH, 2 min	NH_4Cl	" 98 + " <2	57
	Li/NH_3, *t*-BuOH, 2.25 hr	"	" (83)	355
	Li/NH_3, Et_2O, 1.5 hr, EtOH	H_2O	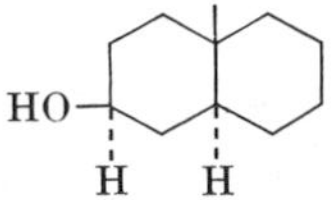 HO H H (74)	356
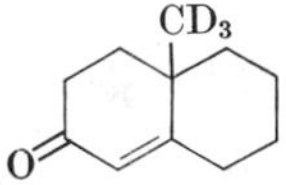 CD_3	Li/NH_3, Et_2O, 1.5 hr	MeOH, H_2O, CrO_3	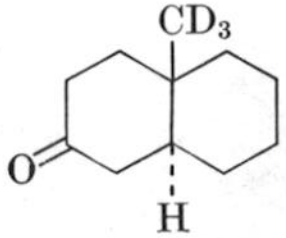 CD_3 H (59)	357

Reactant	Conditions	Work-up	Product(s) and Yield(s) (%)		Refs.
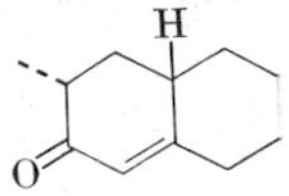 H	Li/NH_3, Et_2O, 30 min	EtOH, CrO_3	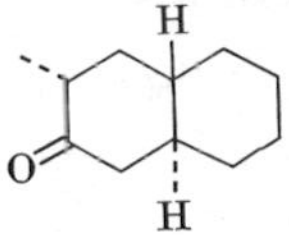 H, H (62)		14
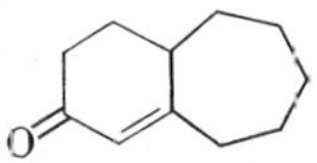	NH_3/Li, dioxane, −40°, 1 hr	NH_4Cl	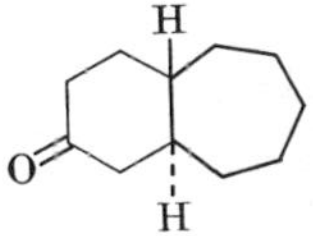 H, H 76 +	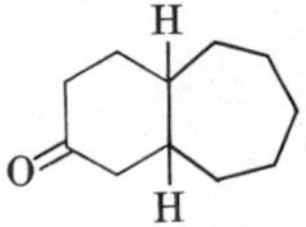 H, H 24	120, 121
	Li/NH_3, Et_2O	"	" (46) +	" (15)	110
	NH_3/Na, dioxane, −40°, 1 hr	"	" 87 +	" 13	120, 121
	NH_3/Ca, dioxane, −40°, 1 hr	"	" 59 +	" 41	
	NH_3/Ba, dioxane, −40°, 1 hr	"	" 44 +	" 56	
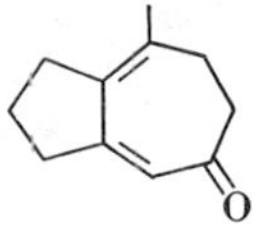	Li/NH_3	NH_4Cl	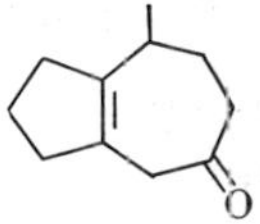 (—)		231
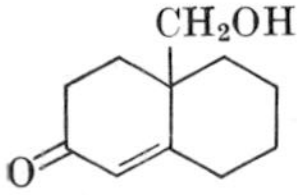 CH_2OH	Li/NH_3, Et_2O, 1 eq. t-BuOH	H_2O	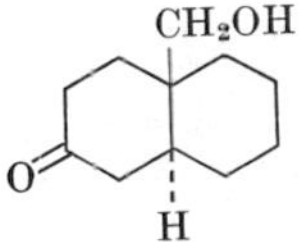 CH_2OH, H (—)		89

Note: References 338–543 are on pp. 253–258.

[e] This yield was based upon the 2,4-dinitrophenylhydrazone isolated after the product was treated with 2,4-dinitrophenylhydrazine.
[f] The authors state that the product contained more than 95% of the *trans* isomer.
[g] A very short reduction time was employed.

TABLE III. Reductions of Bicyclic α,β-Unsaturated Ketones (*Contd.*)

Reactant	Reduction Conditions	Quenching Agent	Product(s) (% Yield)	Refs.
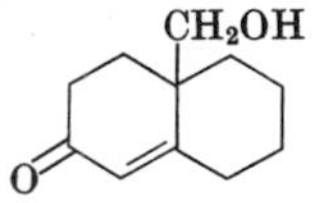 CH_2OH (*Contd.*)	Li/NH_3, Et_2O	NH_4Cl	ʺ ~90[h] + CH_2OH, H 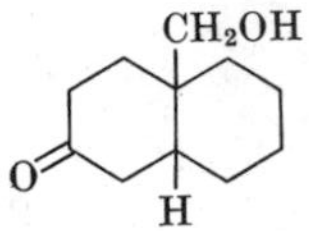 ~10[h]	89
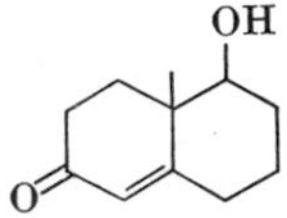 OH	Li/NH_3, THF, 1.5 hr	$NaNO_2$, H_2O	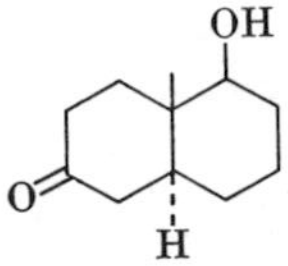 OH, H (53)	165
	Li/NH_3, THF, 15 min	EtOH	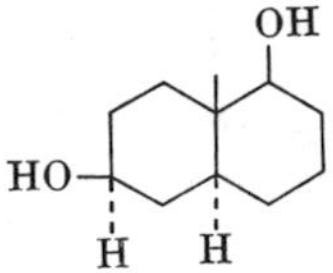 OH, HO, H, H (40)	
C_{12} 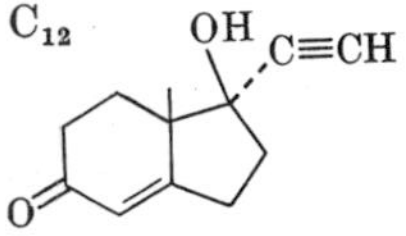 OH, C≡CH	Li/NH_3, Et_2O	NH_4Cl	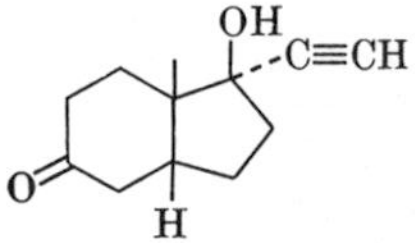 OH, C≡CH, H (—)	253
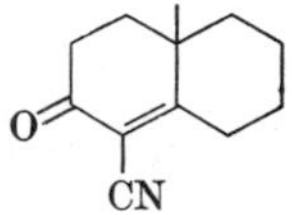 CN	Li/NH_3, Et_2O, 7 hr	NH_4Cl	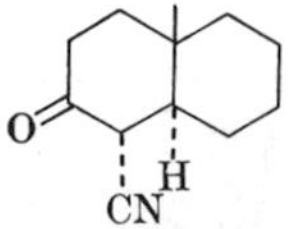 CN, H (91)	175

Substrate	Conditions	Quench	Product (Yield, %)		Refs.
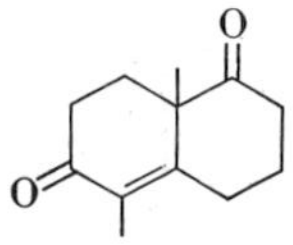	Li/NH_3, Et_2O, 30 min	NH_4Cl		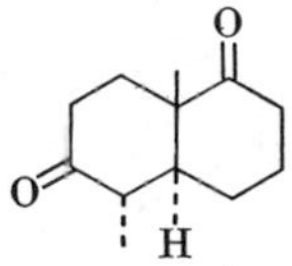 (22)	358
	NH_3/Li, THF, −70°, 15 min	EtOH, CrO_3		(—)	157
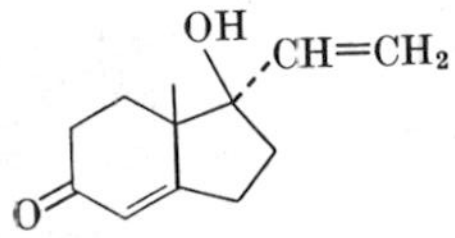 OH, CH=CH_2	Li/NH_3, Et_2O	NH_4Cl	OH, CH=CH_2	(—)	253
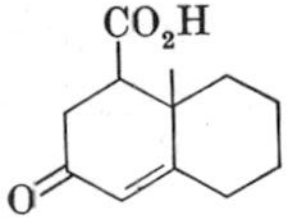 CO_2H	NH_3/Li, 1 hr	H_3O^+	CO_2H	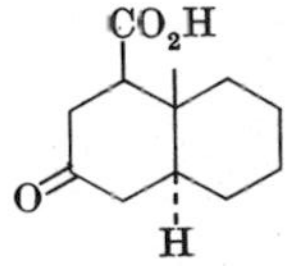 (97)	358a
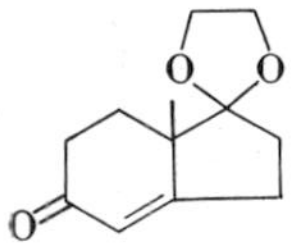	NH_3/Li, THF, −60°, 1 hr	NH_4Cl		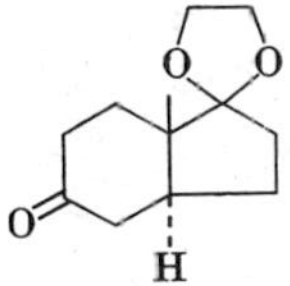 (79)	349a

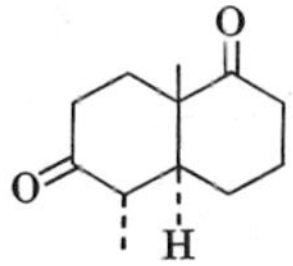

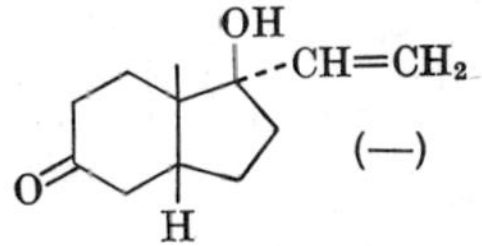

Note: References 338–543 are on pp. 253–258.

[a] In several runs conducted without a proton donor the *trans*/*cis* ratio was found to be 8/1 or larger.

TABLE III. REDUCTIONS OF BICYCLIC α,β-UNSATURATED KETONES (*Continued*)

Reactant	Reduction Conditions	Quenching Agent	Product (% Yield)	Refs.
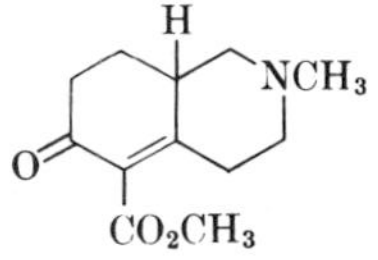 H, NCH_3, O, CO_2CH_3	Li/NH_3, Et_2O, 4 eq. *t*-BuOH, −78°	—	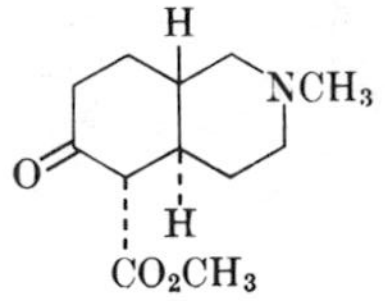 H, NCH_3, O, H, CO_2CH_3 (—)	359
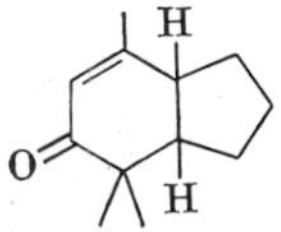 H, O, H	Li/NH_3	—	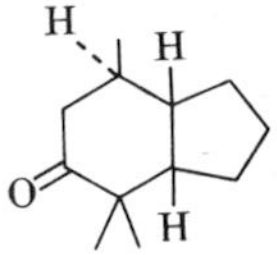 H, H, O, H (—)	360
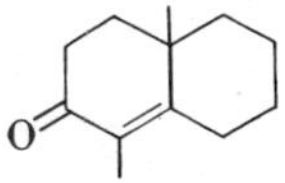 O	Excess Li/NH_3, Et_2O, 30 min	EtOH	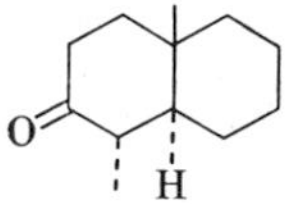 O, H (60)	14
	Na/NH_3, *t*-BuOH	CrO_3	" (—)	361
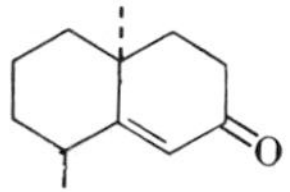 O	Li/NH_3, Et_2O, 4 hr	NH_4Cl	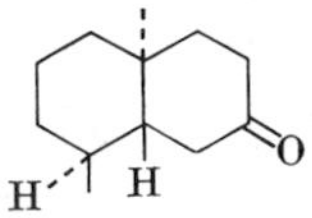 O, H, H (30)	362
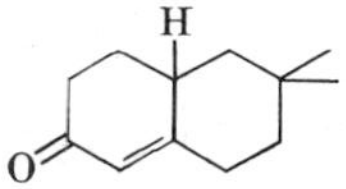 H, O	NH_3/Li, Et_2O, 3 hr	NH_4Cl, CrO_3	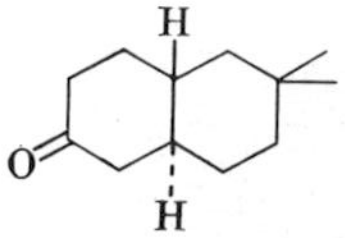 H, O, H (37)	363

Substrate	Reduction conditions	Work-up	Product(s) and Yield(s) (%)		Refs.
[structure]	Na/NH$_3$, Et$_2$O, EtOH[g]	CrO$_3$	,, 99.85 + [structure]	0.15	78
	NH$_3$/Li, Et$_2$O, 3 hr	NH$_4$Cl, CrO$_3$	[structure] (52)		363
	NH$_3$/Li, Et$_2$O, 2 hr	NH$_4$Cl	,, (80)		301
	Na/NH$_3$, Et$_2$O, EtOH[g]	CrO$_3$	,, 99.1 + [structure]	0.9	78
[structure]	Na/NH$_3$, Et$_2$O, EtOH[g]	CrO$_3$	[structure] 99.6 + [structure]	0.4	78

Note: References **338–543** are on pp. 253–258.

[g] A very short reduction time was employed.

TABLE III. REDUCTIONS OF BICYCLIC α,β-UNSATURATED KETONES (*Contd.*)

Reactant	Reduction Conditions	Quenching Agent	Product(s) (% Yield)	Refs.
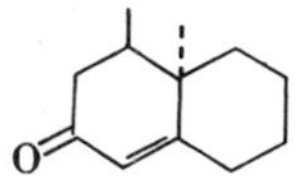	Li/NH_3, THF, 30 min	NH_4Cl, CrO_3	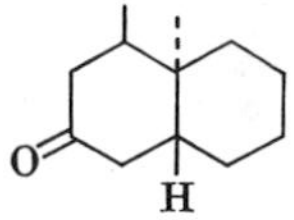 (65) + 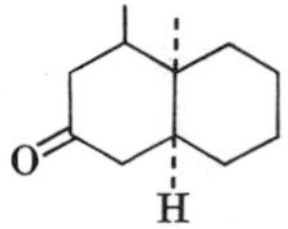 (7)	90
	Li/NH_3, Et_2O, 2.5 hr	NH_4Cl	" (80) + " (12)	91
	Li/NH_3, t-BuOH	CrO_3	"	364
	Li/NH_3, THF	—	" (79)	365
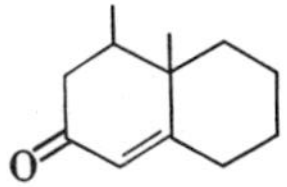	Li/NH_3, t-BuOH	CrO_3	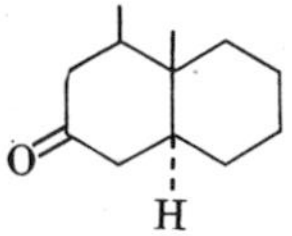 (—)	364
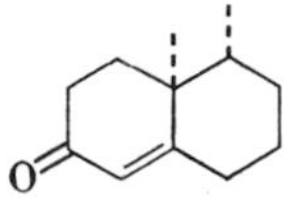	Li/NH_3, Et_2O, 1 hr EtOH, 1.5 hr	EtOH	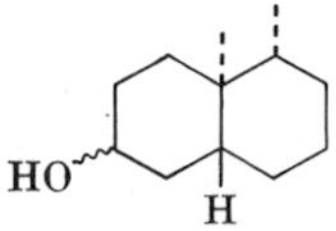 (75)	270
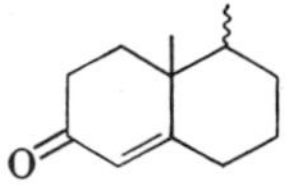	Li/NH_3	—	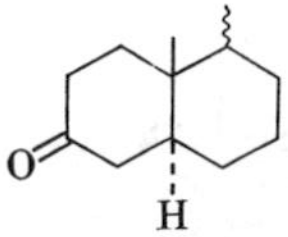 (—)	366

Substrate	Conditions	Work-up	Product (Yield, %)	Refs.
OH	Li/NH_3	—	OH (—) H	367
OH	Li/NH_3, Et_2O, 10 min	NH_4Cl	OH (89) H	358
OCH_3 + OCH_3	Li/NH_3, Et_2O, EtOH	NH_4Cl, CrO_3	OCH_3 H (57)[i] + OCH_3 H (10)[i]	10
OH CH_2CH_3	Li/NH_3, Et_2O	NH_4Cl	OH CH_2CH_3 H	253

Note: References 338–543 are on pp. 253–258.

[i] These products were isolated after Huang-Minlon reduction.

TABLE III. Reductions of Bicyclic α,β-Unsaturated Ketones (*Continued*)

Reactant	Reduction Conditions	Quenching Agent	Product(s) (% Yield)	Refs.
C_{13} C_6H_5	Li/NH_3, Et_2O	—	C_6H_5 (85) + C_6H_5 (—)	269
	Li/NH_3, Et_2O-*t*-BuOH	—	C_6H_5 (—) + C_6H_5 (—)	269
	Li/NH_3, Et_2O	—	OH (—)	193
OH, C≡CH	NH_3/Li, Et_2O	NH_4Cl	OH, C≡CH, H (50)	164
OH, CH=CH_2	Li/NH_3, Et_2O, 2 hr, −50°	$Fe(NO_3)_3$, NH_4Cl	OH, CH=CH_2, H (65)	164

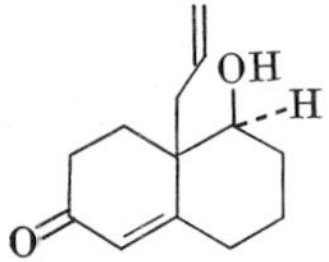 OH, H	Li/NH_3, THF	—	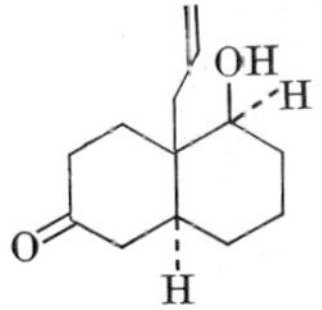 OH, H, H (—)	193
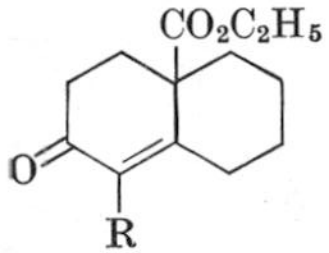 $CO_2C_2H_5$, R	NH_3/2 eq Li, Et_2O, EtOH	NH_4Cl, CrO_3	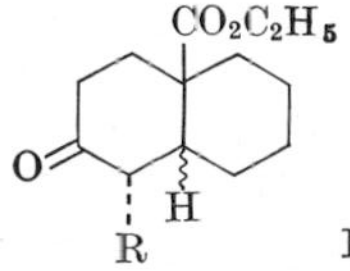 $CO_2C_2H_5$, H, R (—) R=H, *cis-trans* mixture (32)	368
	Li/NH_3, Et_2O, 20 min	NH_4Cl	" R=CH_3, *trans* isomer (80)	368a
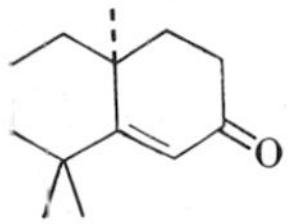	Li/NH_3, THF, 50 min	NH_4Cl	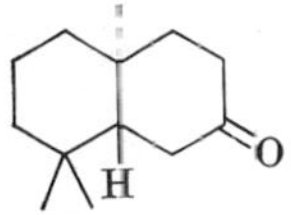 H	369
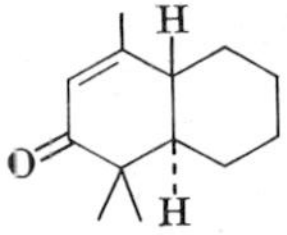 H, H	NH_3/Li, Et_2O, 70 min, EtOH slowly over 2 hr	Aq. NH_4Cl, CrO_3	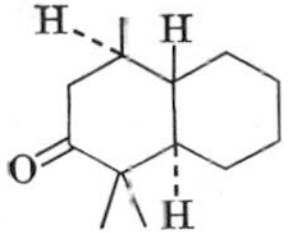 H, H, H (94)	370
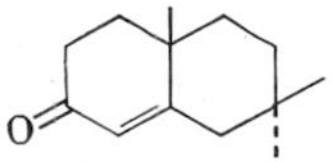	Li/NH_3, Et_2O, 30 min	NH_4Cl	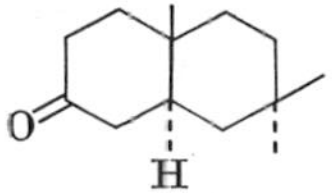 H (74)	94, 339

Note: References **338–543** are on p. 253–258.

TABLE III. REDUCTIONS OF BICYCLIC α,β-UNSATURATED KETONES (*Continued*)

Reactant	Reduction Conditions	Quenching Agent	Product(s) (% Yield)	Refs.
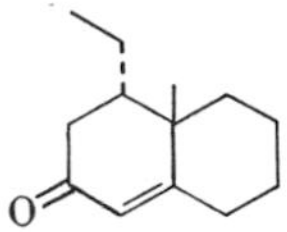	Li/NH_3, Et_2O, 2.5 hr	NH_4Cl	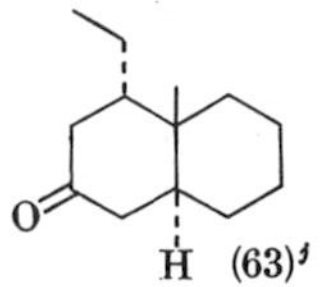 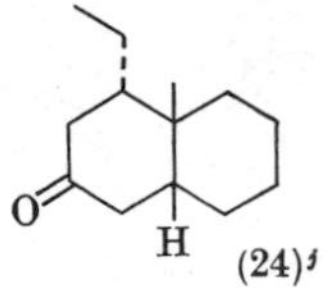 H (63)[j] + H (24)[j] + starting material (13)	91
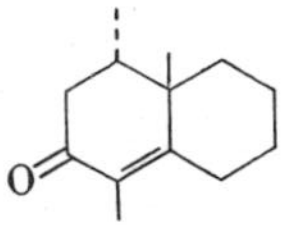	Li/NH_3, Et_2O, 2.5 hr	NH_4Cl	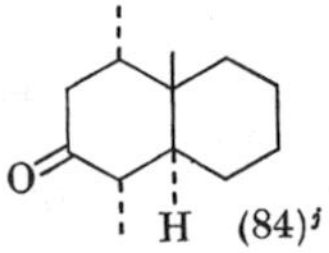 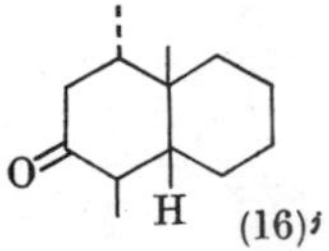 H (84)[j] + H (16)[j] + starting material (8)	91
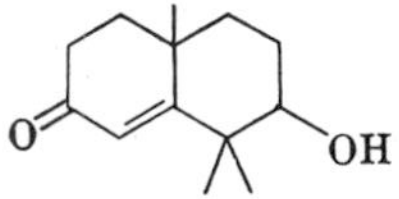 OH	Li/NH_3, Et_2O-dioxane	NH_4Cl	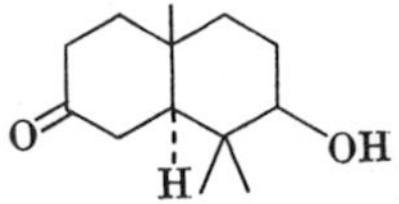 H OH (—)	371
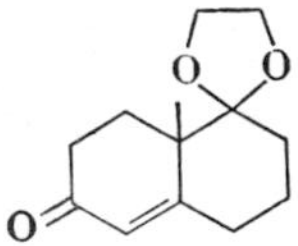 O O	Li/NH_3, Et_2O, 30 min	NH_4Cl, H_3O^+	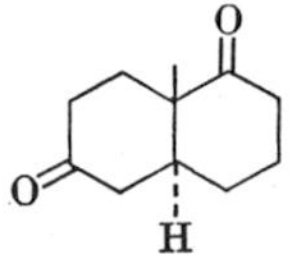 O H (~30)	195
	Li/NH_3	—	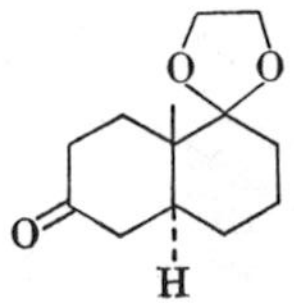 O O H (96)	372
	Li/NH_3, THF —60°, 1hr	—	" (80)	349a

Reactant	Reagents	Workup	Product(s) and Yield(s) (%)	Refs.
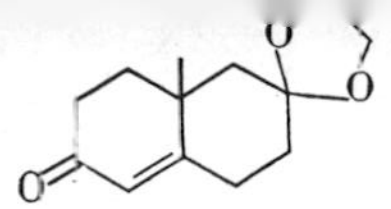	Li/NH_3, Et_2O, 1 hr	EtOH, CrO_3	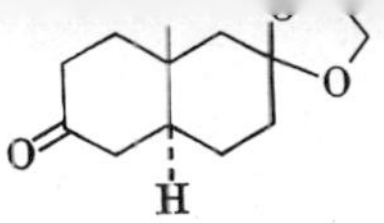 (93)	373
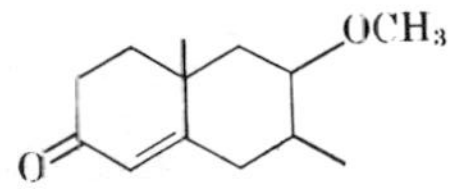 OCH_3	NH_3/Li,[k] EtOH-Et_2O	NH_4Cl	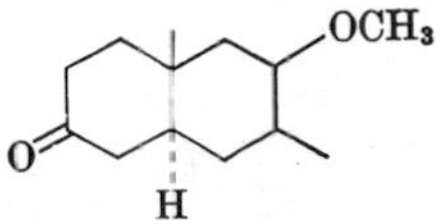 OCH_3 (77)	10
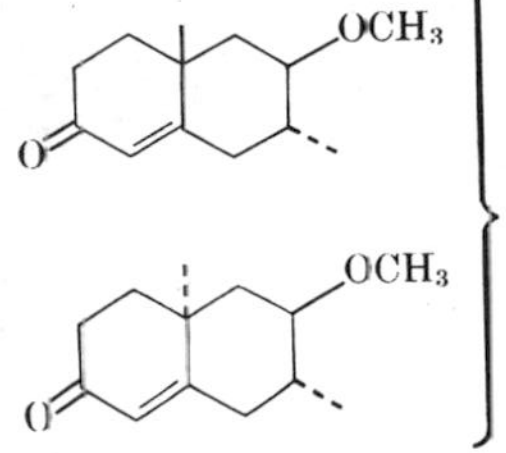 OCH_3 / OCH_3 1:1	NH_3/Li,[k] EtOH-Et_2O	NH_4Cl	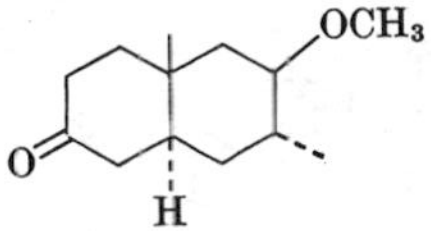 OCH_3 (35) 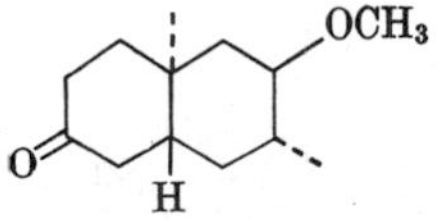 + OCH_3 (35)	10
C_{14} 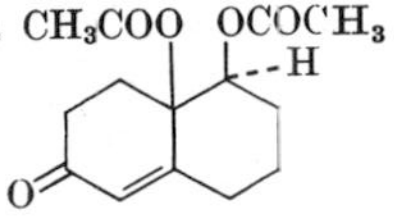 CH_3COO $OCOCH_3$ H	Excess Li/NH_3, THF, 4 min	NH_4Cl	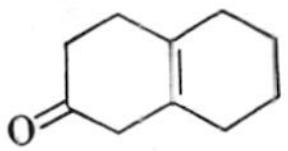 (53) 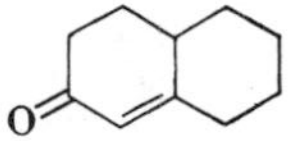 + (7)	70

Note: References 338–543 are on pp. 253–258.

[j] The yields are based upon the amount of unrecovered starting material.
[k] The lithium metal was added in pieces until the blue color persisted.

TABLE III. Reductions of Bicyclic α,β-Unsaturated Ketones (*Continued*)

Reactant	Reduction Conditions	Quenching Agent	Product(s) (% Yield)	Refs.
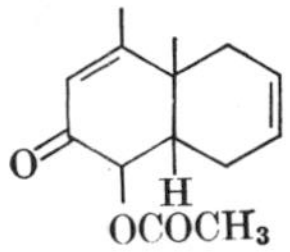 H, $OCOCH_3$	Excess Ca/NH_3, THF, 5 min, MeOH dropwise	H_2O-HCl, CrO_3	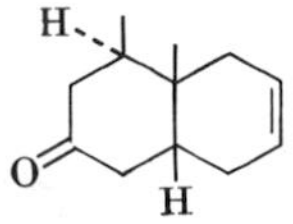 H, H (44)	265
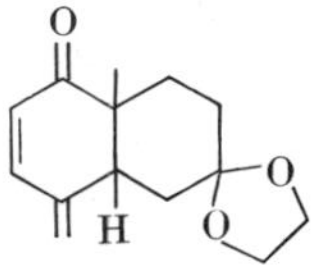 O, H, O, O	Li/NH_3, THF-EtOH	H_2O	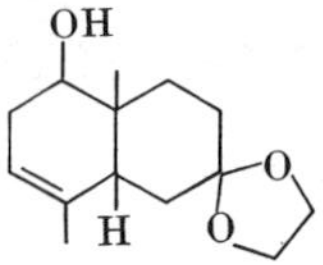 OH, H, O, O (92)	374
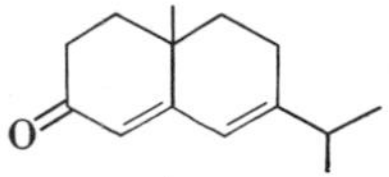	Li/NH_3, Et_2O, trace EtOH, 30 min	NH_4Cl	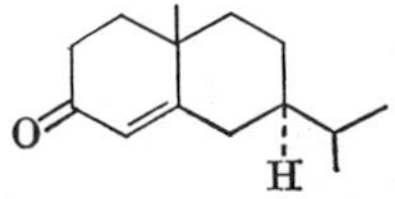 H (30)	221
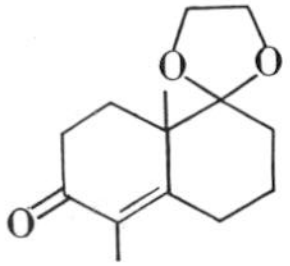 O, O	Li/NH_3, Et_2O, 30 min	$NaNO_2$, H_2O	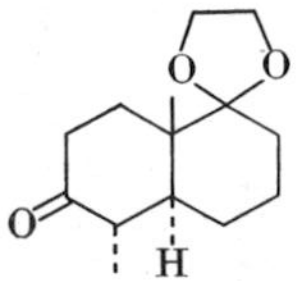 O, O, H (47)	375
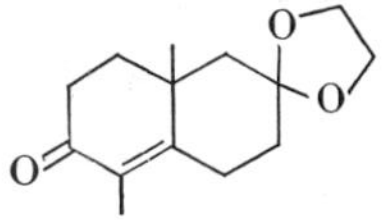 O, O	Li/NH_3, Et_2O, 1 hr	EtOH, Cr_3O	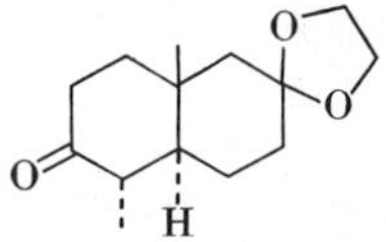 O, O, H (63)	373

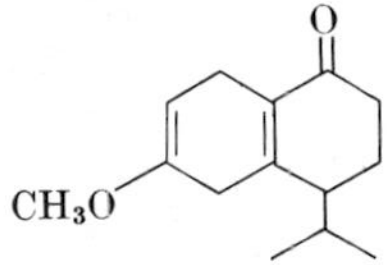	Li/NH_3, Et_2O, 20 min	H_2O, HCl, dioxane	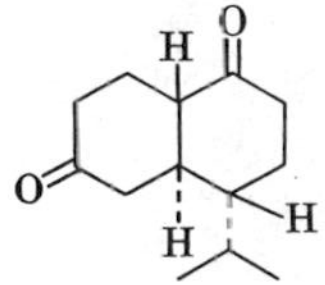 (66)	248
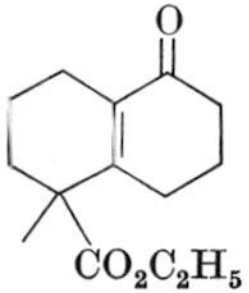	NH_3/Li, Et_2O-THF	H_2O, CrO_3, OH^-, CH_2N_2	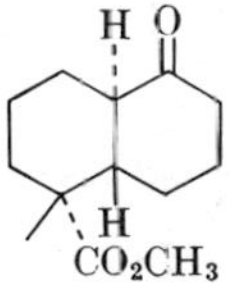 (34)	376
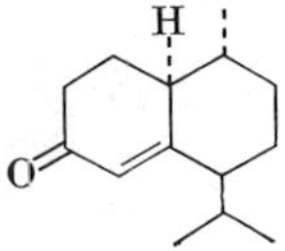	Li/NH_3	—	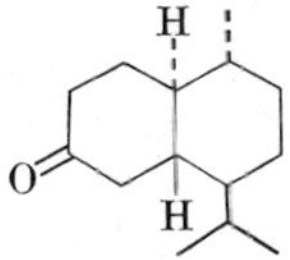 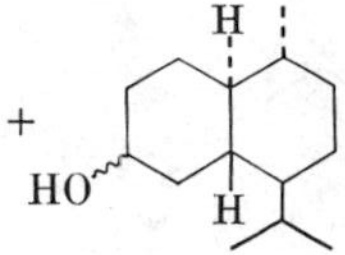 (–) + (–)	377
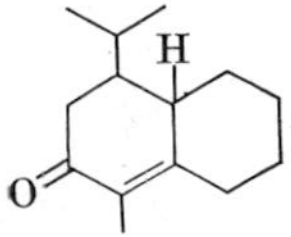	Li/NH_3, Et_2O, 2.5 hr	NH_4Cl	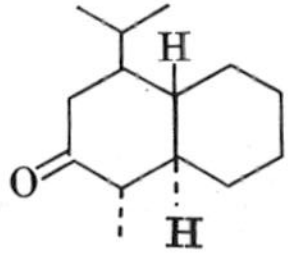 (90) 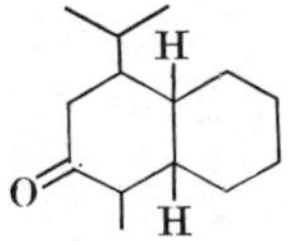 + (<1)	91

Note: References 338–543 are on pp. 253–258.

TABLE III. REDUCTIONS OF BICYCLIC α,β-UNSATURATED KETONES (*Continued*)

Reactant	Reduction Conditions	Quenching Agent	Product(s) (% Yield)	Refs.
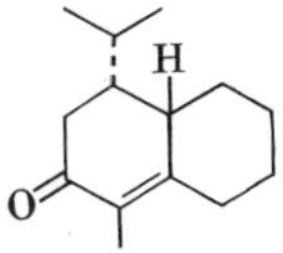	Li/NH_3, Et_2O, 2.5 hr	NH_4Cl	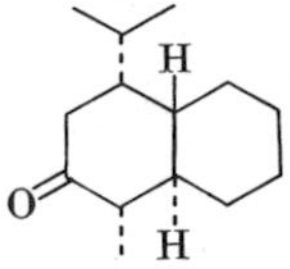 (<1) + 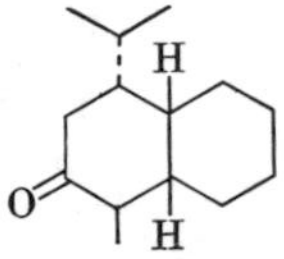 (79)	91
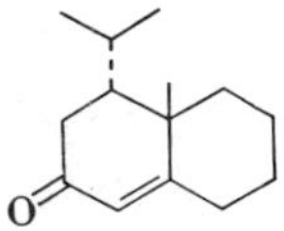	Li/NH_3, Et_2O, 2.5 hr	NH_4Cl	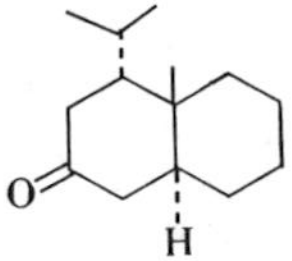 (68)[j] + 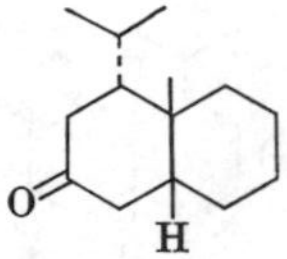 (30)[j] + starting material (14)	91
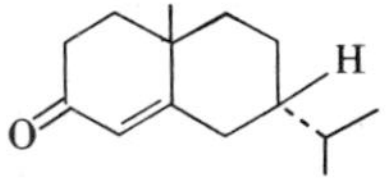	Li/NH_3, Et_2O 10 min	NH_4Cl, CrO_3	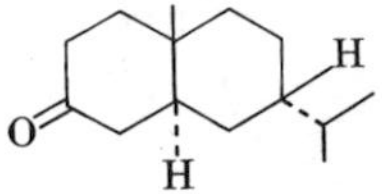 (21) + starting material (51) + 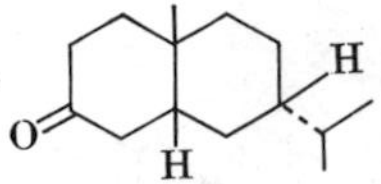 (Trace)	94
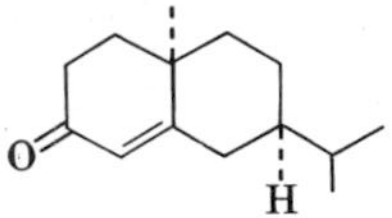	Li/NH_3, Et_2O, 1 eq. *t*-BuOH, 30 min	NH_4Cl	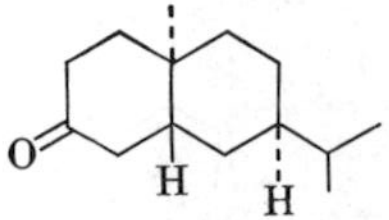 (67)	95

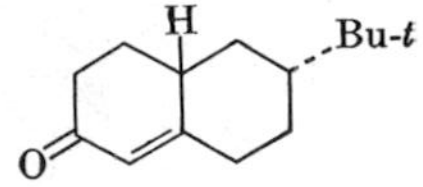	Li/NH_3	—	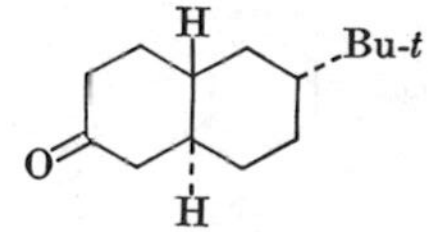 (30)	378
	NH_3/Li, Et_2O, 2.25 hr	NH_4Cl, CrO_3	" (74)	379
	Li/NH_3, Et_2O, 2 eq. t-BuOH, 2 min	"	" (78.5)	57
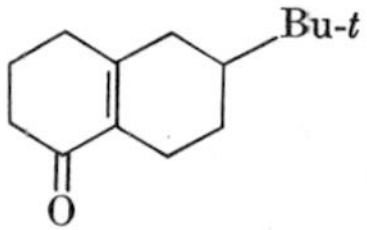	Li/NH_3, THF, 3 eq. t-BuOH, 1 hr	CrO_3	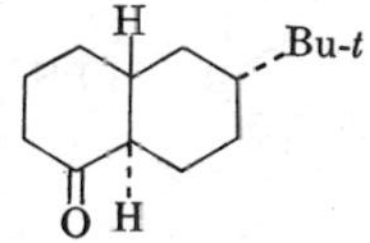 (77.5)	57
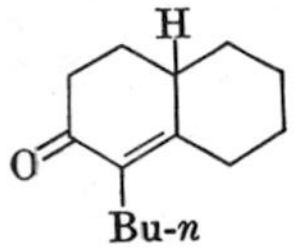	NH_3/Li, Et_2O, 1 hr	NH_4Cl	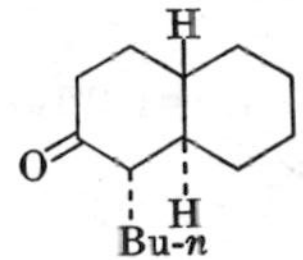 (43)	14
C_{15} 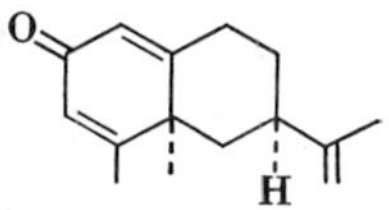	Li/NH_3, −50°, 2 min	—	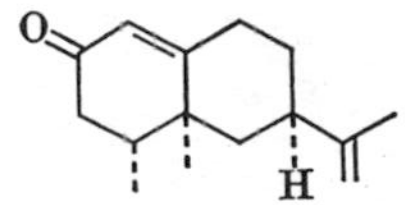 (–)	380
	Li/NH_3, Et_2O-EtOH, −70°, 2 min	—	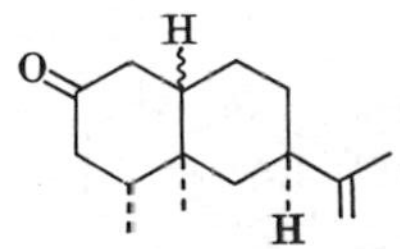 (—)	380

Note: References 338–543 are on pp. 253–258.

[j]The yields are based upon the amount of unrecovered starting material.

TABLE III. REDUCTIONS OF BICYCLIC α,β-UNSATURATED KETONES (*Continued*)

Reactant	Reduction Conditions	Quenching Agent	Product(s) (% Yield)	Refs.
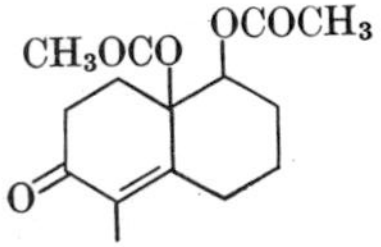 CH_3OCO, $OCOCH_3$	Li/NH_3, THF, 5 min	NH_4Cl	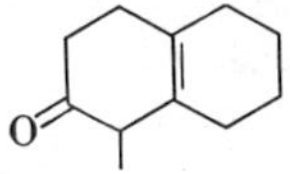 (80) 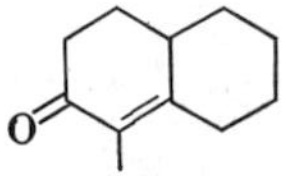 + (Trace)	70
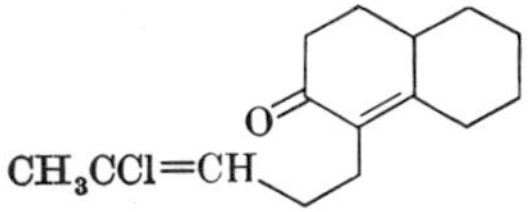 $CH_3CCl{=}CH$	NH_3/Li,[l] Et_2O, −75°	—	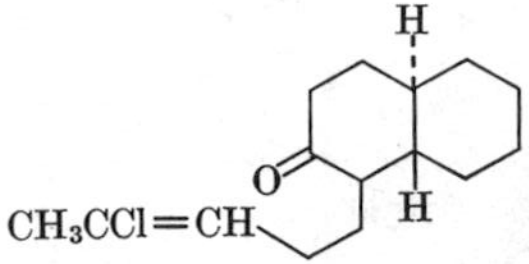 $CH_3CCl{=}CH$ H H (65)	254
	Li/NH_3, Et_2O, −33°	—	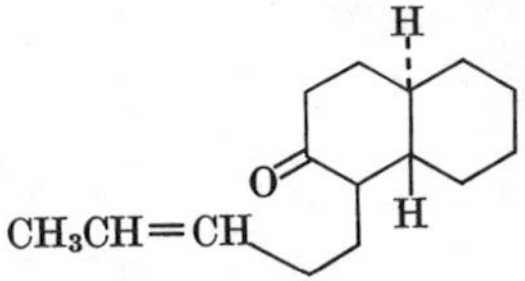 $CH_3CH{=}CH$ H H (—)	
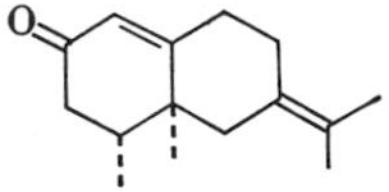	Li/NH_3, EtOH	CrO_3	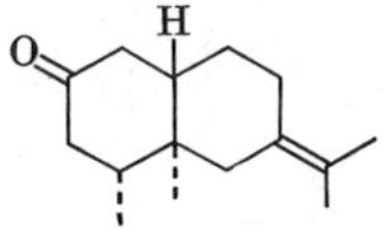 H (—)	381
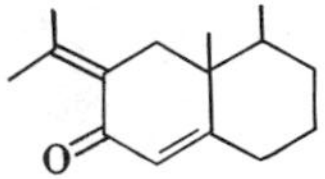	Li/NH_3, THF, 12 min	NH_4Cl	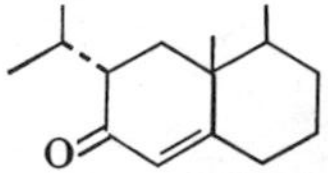 (96)	236

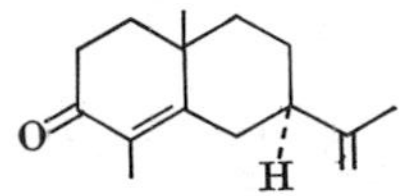	Li/NH_3, Et_2O, 40 min	NH_4Cl	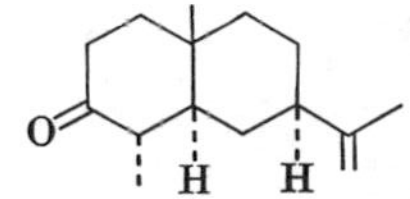 (—)	93
	Li/NH_3, Et_2O, 30 min	,,	,, (92)	148
	Li/NH_3 dioxane, 15 min	EtOH	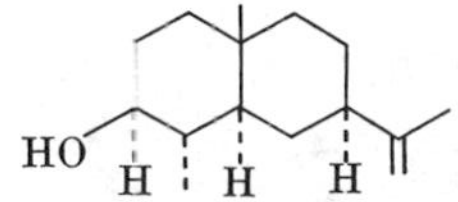 (80)	
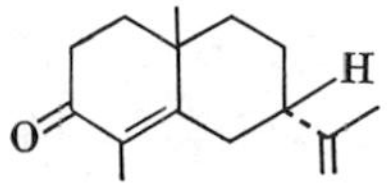	Li/NH_3, Et_2O, 30 min	NH_4Cl	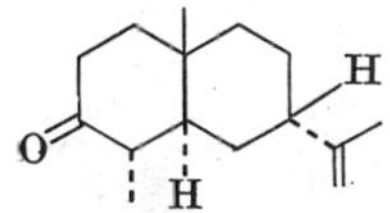 (—)	140
	Li/NH_3, Et_2O, 1 hr	,,	,, (98)[m]	139
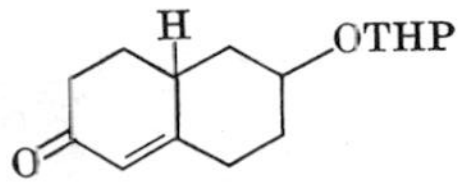	Li/NH_3, Et_2O, 16 min	NH_4Cl, 2*N* HCl	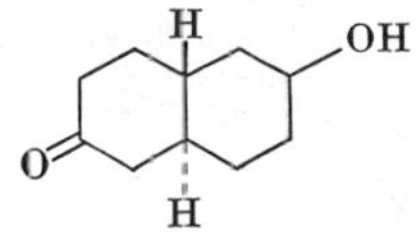 (59)	273
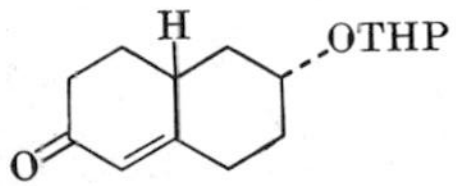	Li/NH_3, Et_2O, 16 min	NH_4Cl, 2*N* HCl	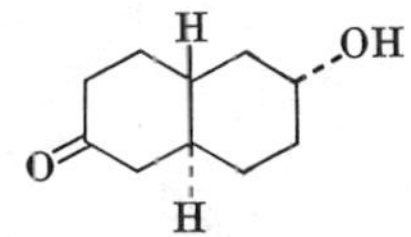 (30)	

Note: References 338–543 are on pp. 253–258.

[l] A solution of lithium in liquid ammonia was added dropwise to an ammonia-ether solution of the enone.
[m] Prolonged reduction in the presence of ethanol led to reduction of the double bond in the isopropenyl side chain.

TABLE III. Reductions of Bicyclic α,β-Unsaturated Ketones (*Continued*)

Reactant	Reduction Conditions	Quenching Agent	Product(s) (% Yield)	Refs.
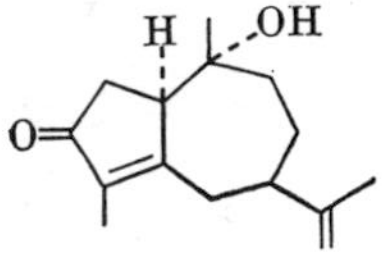 H, OH	Li/NH_3, Et_2O, 1.5 hr	MeOH, CrO_3	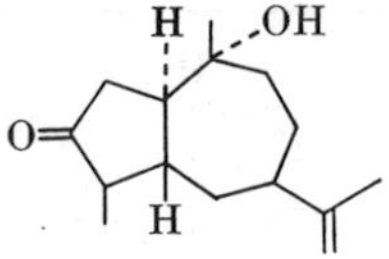 H, OH, H (90)	137
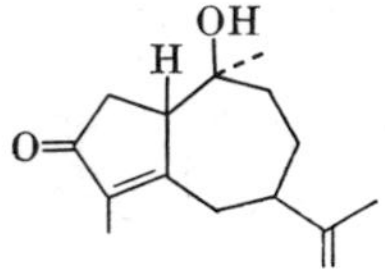 OH, H	Li/NH_3, Et_2O, 1.5 hr	MeOH, CrO_3	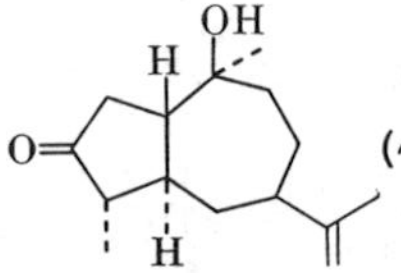 OH, H, H (~65)	137
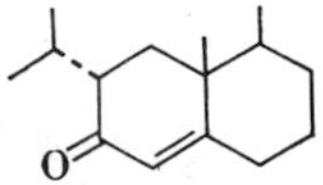	Li/NH_3, dioxane	Aq NH_4Cl, CrO_3	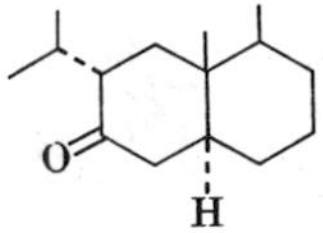 H (73)	382
	Li/NH_3, dioxane, 12 min	NH_4Cl, CrO_3	" (58)	236
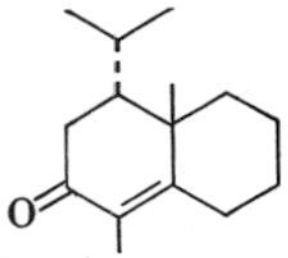	Li/NH_3, Et_2O, 2.5 hr	NH_4Cl	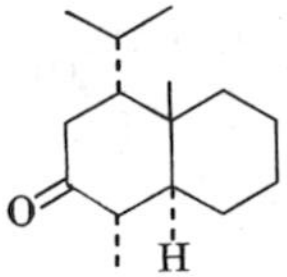 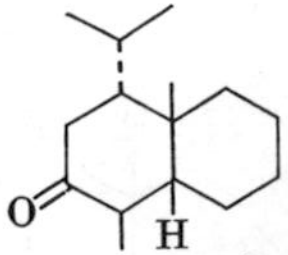 H (64)[j] + H (34)[j] + starting material (7)	91

Substrate	Reducing Conditions	Additional Reagent	Product(s) and Yield(s) (%)	Refs.
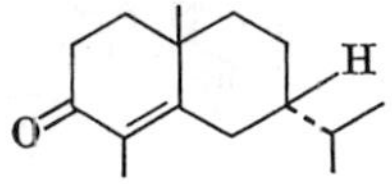	Li/NH_3, Et_2O 40 min	NH_4Cl	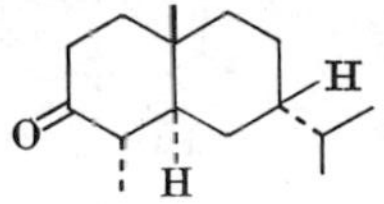 (—)	93
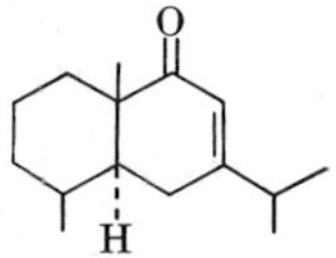	Li/NH_3	—	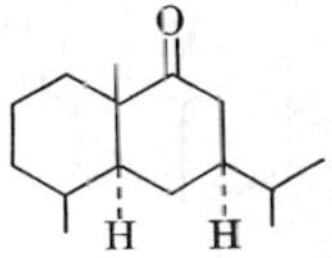 (—)	383
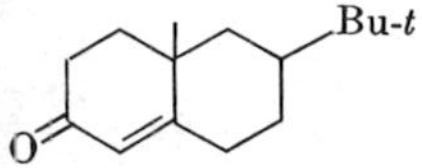 Bu-*t*	Na/NH_3, MeOH	CrO_3	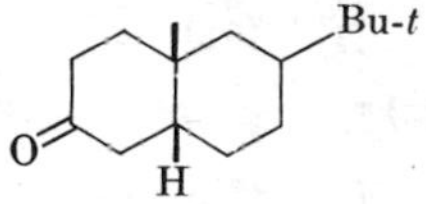 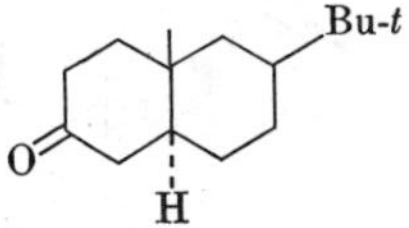 Bu-*t* >99 + Bu-*t* <1	92
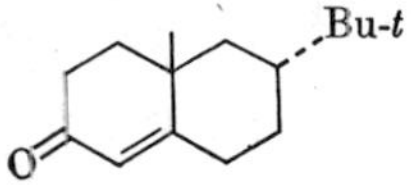 Bu-*t*	Na/NH_3, MeOH	CrO_3	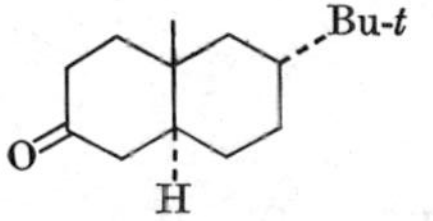 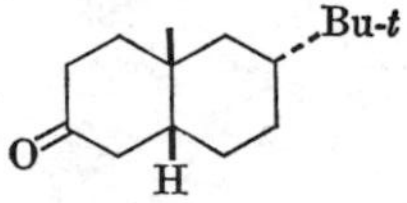 Bu-*t* (Major) + Bu-*t* (Trace)	92

Note: References 338–543 are in pp. 253–258.

[j] The yields are based upon the amount of unrecovered starting material.

TABLE III. REDUCTIONS OF BICYCLIC α,β-UNSATURATED KETONES (*Continued*)

Reactant	Reduction Conditions	Quenching Agent	Product(s) (% Yield)	Refs.
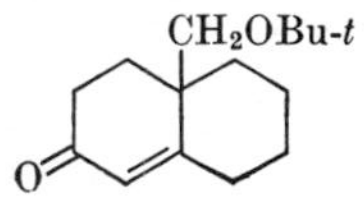 CH_2OBu-*t*	Li/NH_3, *t*-BuOH-THF	—	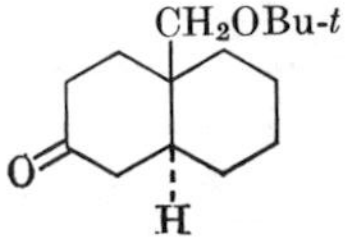 CH_2OBu-*t* H (75)	384
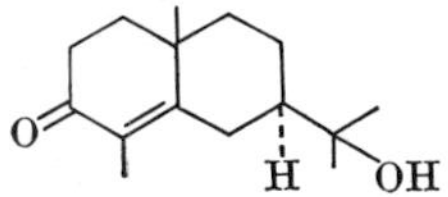 H OH	Li/NH_3, Et_2O, 40 min	NH_4Cl	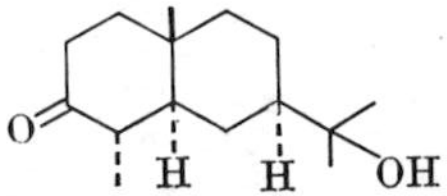 H H OH (87)	385
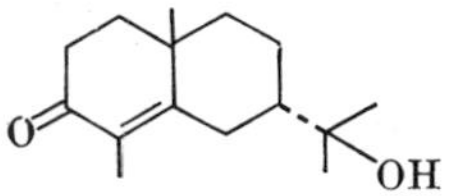 OH	Li/NH_3, 25 min	—	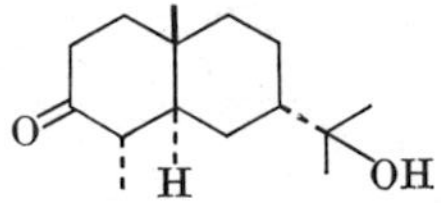 H OH I (35) 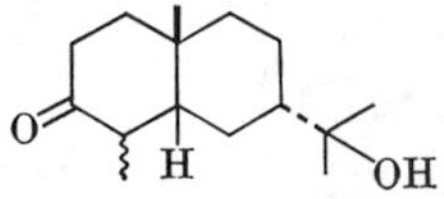 + H OH II (23)	96
	Excess $LiNH_2$, Li/NH_3	—	I (75) + II (7)	
	Excess $NaNH_2$, Li/NH_3	—	I (52) + II (24)	
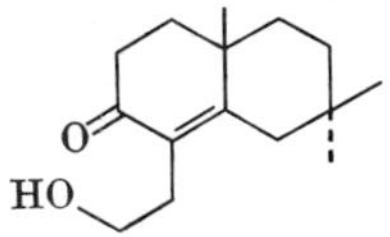 HO	NH_3/Li, Et_2O, 45 min, $EtOH^n$	H_2O	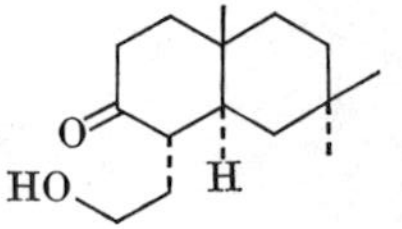 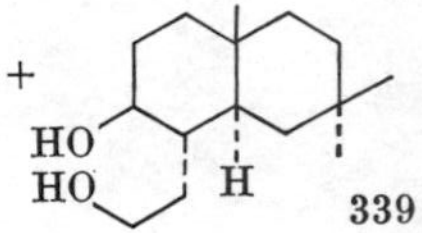 HO H (−) + HO HO H (−)	339
C_{16} 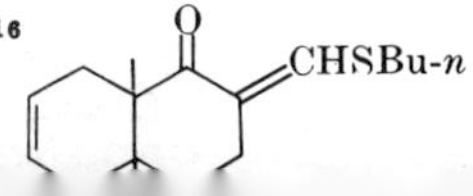 CHSBu-*n*	NH_3/Na, Et_2O-EtOH, 4 hr	H_2O	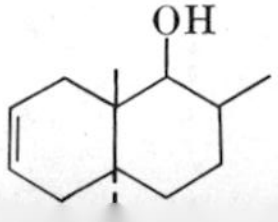 OH (34)	212

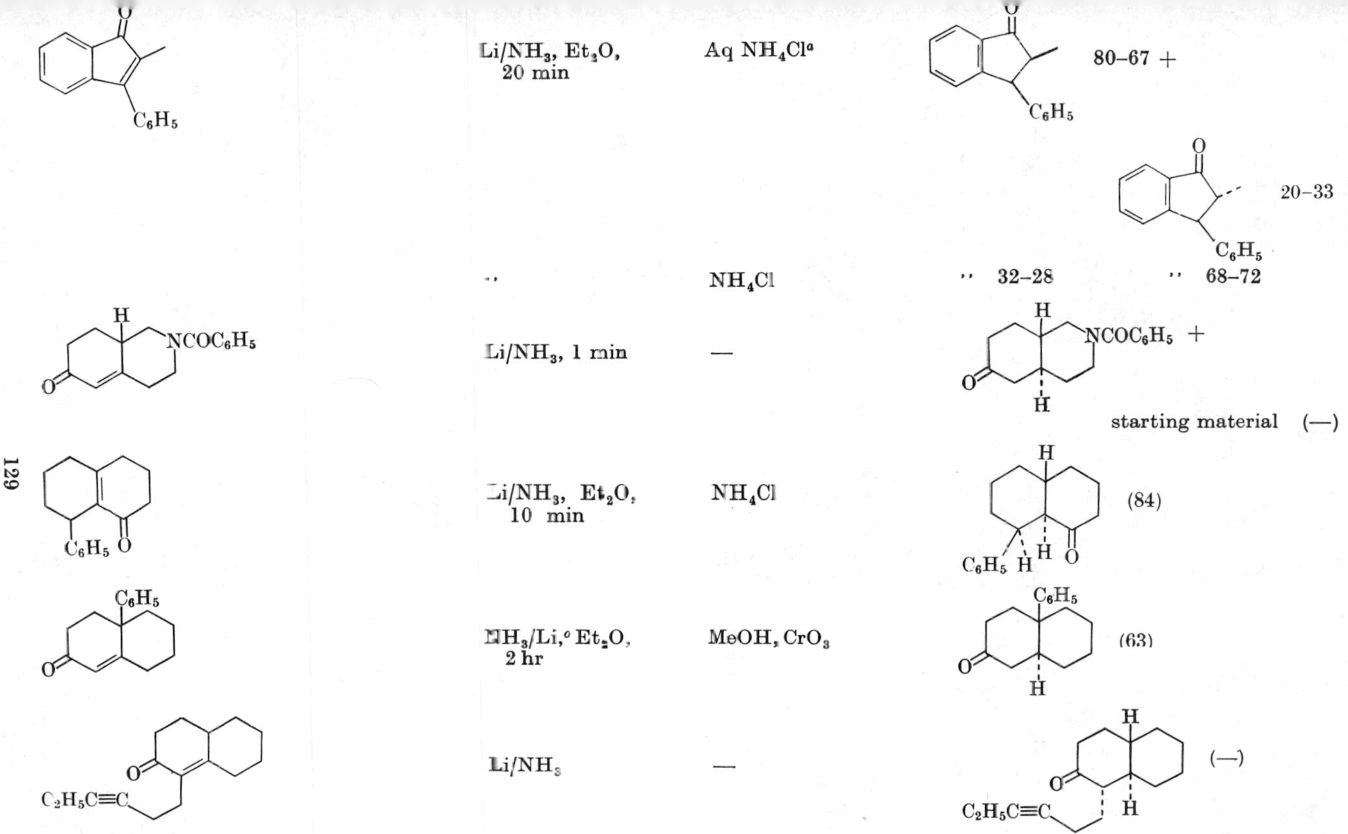

Reactant	Conditions	Workup	Product(s) and Yield(s) (%)	Refs.
C_6H_5	Li/NH_3, Et_2O, 20 min	Aq NH_4Cl[a]	C_6H_5 80–67 + C_6H_5 20–33	12
	"	NH_4Cl	" 32–28 " 68–72	
H, $NCOC_6H_5$	Li/NH_3, 1 min	—	H, $NCOC_6H_5$, H + starting material (—)	260
C_6H_5	Li/NH_3, Et_2O, 10 min	NH_4Cl	H, C_6H_5, H, H (84)	97
C_6H_5	NH_3/Li,[o] Et_2O, 2 hr	MeOH, CrO_3	C_6H_5, H (63)	386
$C_2H_5C{\equiv}C$	Li/NH_3	—	H, $C_2H_5C{\equiv}C$, H (—)	249

Note: References 338–543 are on pp. 253–258.

[a] A solution of the lithium enolate was added slowly to a saturated aqueous solution of ammonium chloride.

[n] It was reported that reduction did not occur when the addition of ethanol was omitted.

[o] A solution of lithium in liquid ammonia was added dropwise to the enone in ether.

TABLE III. Reductions of Bicyclic α,β-Unsaturated Ketones (*Continued*)

Reactant	Reduction Conditions	Quenching Agent	Product(s) (% Yield)	Refs.
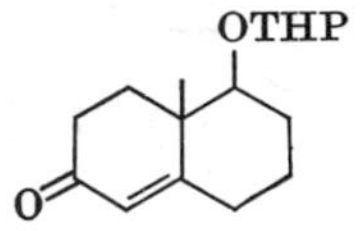 OTHP	Li/NH_3, THF	NH_4Cl	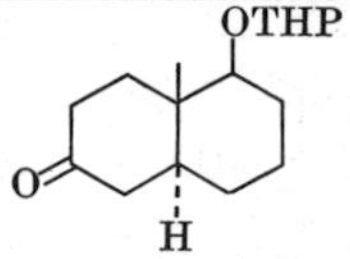 OTHP; H (78)	387
C_{17}; 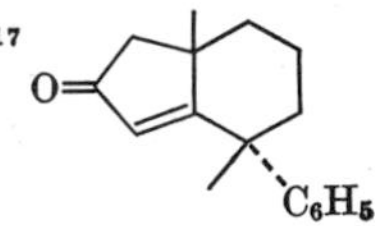 O; C_6H_5	Excess Li/NH_3, Et_2O, 1 hr	NH_4Cl, CrO_3	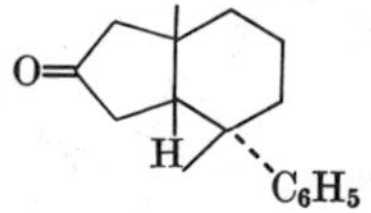 O; H; C_6H_5 (90)	129
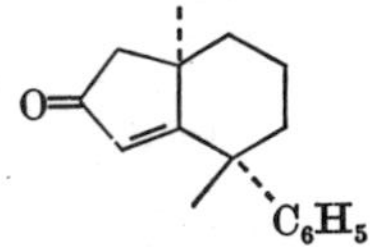 O; C_6H_5	Li/NH_3, Et_2O, 1 hr	NH_4Cl	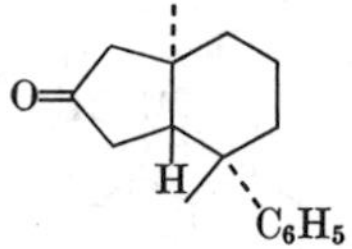 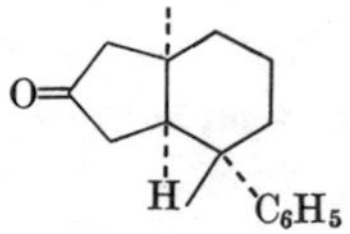 O; H; C_6H_5 (>36) + O; H; C_6H_5 (>23)	129
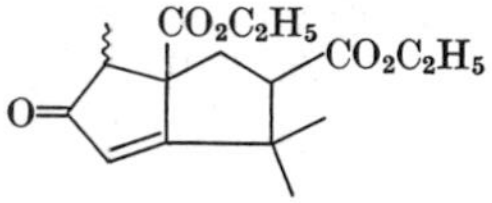 $CO_2C_2H_5$; $CO_2C_2H_5$; O	NH_3/*ca.* 2 eq. Li, THF, short time	—	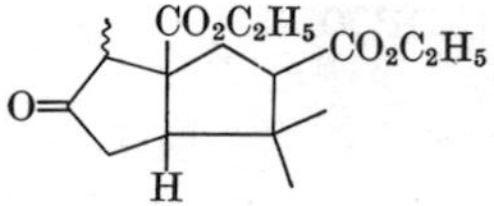 $CO_2C_2H_5$; $CO_2C_2H_5$; O; H (—)	118
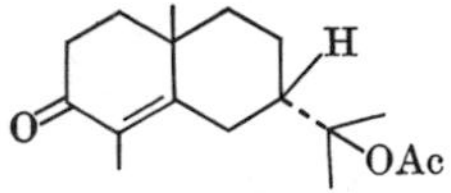 H; O; OAc	Li/NH_3	—	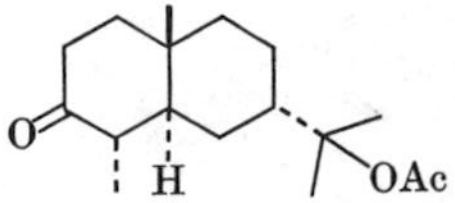 O; H; OAc (6) 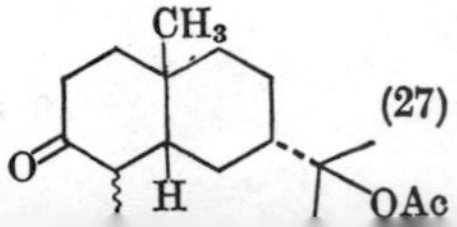 + CH_3; O; H; OAc (27)	96

Substrate	Conditions	Quench	Product(s) and Yield(s) (%)	Refs.
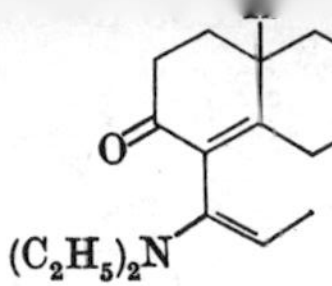 $(C_2H_5)_2N$	Li/NH_3, EtOH	H_3O^+	(—)	387a
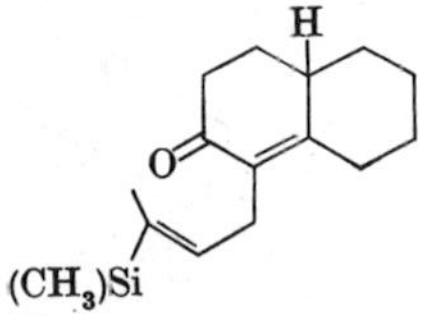 $(CH_3)Si$, H	3 eq. Li/NH_3, 1 eq. t-BuOH, THF	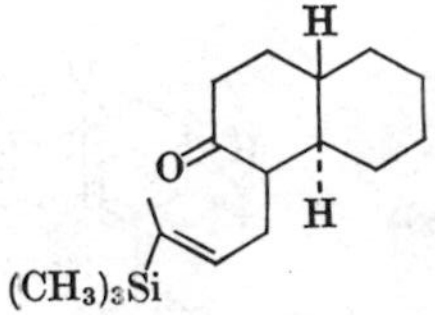	$(CH_3)_3Si$, H, H (~100)	387b
C_{18} 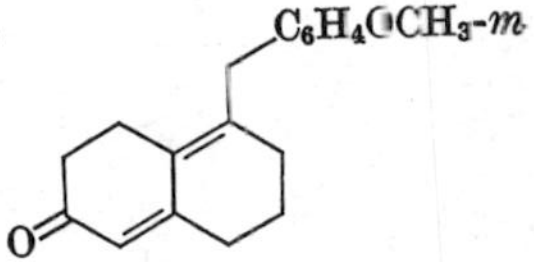 $C_6H_4OCH_3$-m	Li/NH_3, THF	H_2O, HCl-C_6H_6	$C_6H_4OCH_3$-m, H, H (—)	228
C_6H_5	Li/NH_3 Et_2O, 1 hr	NH_4Cl	H, C_6H_5 (94)	388
C_6H_5	Li/NH_3, Et_2O 1 hr	NH_4Cl	H, C_6H_5 (89)	388
CH_2OTs	Li/NH_3, Et_2O	NH_4Cl	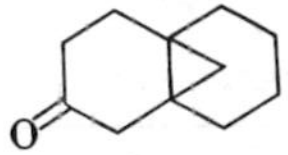 (40)	14, 68

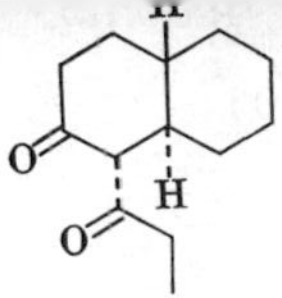

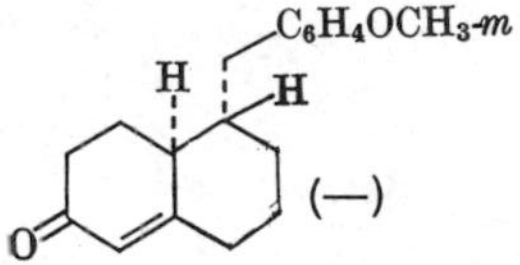

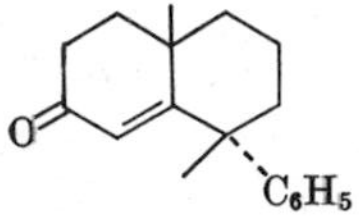

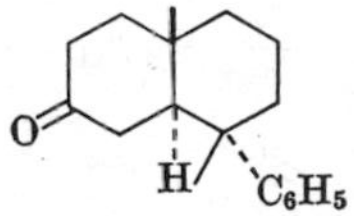

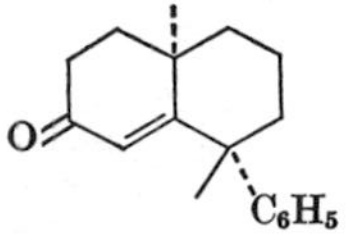

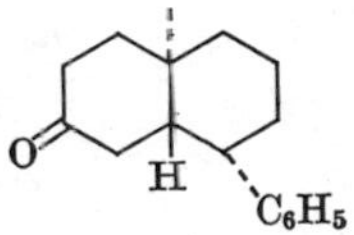

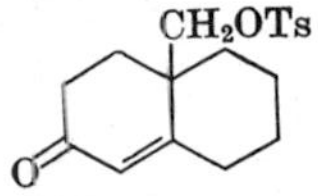

Note: References 338–513 are on pp. 253–258.

Reactant	Reduction Conditions	Quenching Agent	Product(s) (% Yield)	Refs.
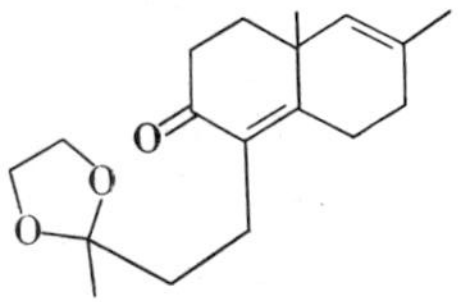	Li/NH_3, Et_2O	NH_4Cl	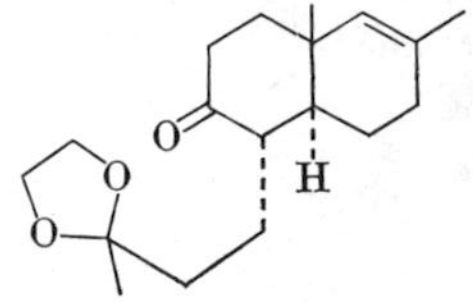 (92)	173
C_{19} 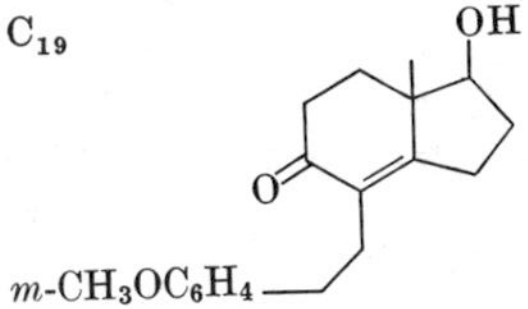 OH m-$CH_3OC_6H_4$	Li/NH_3, THF, 15 min	$NaNO_2$, CrO_3	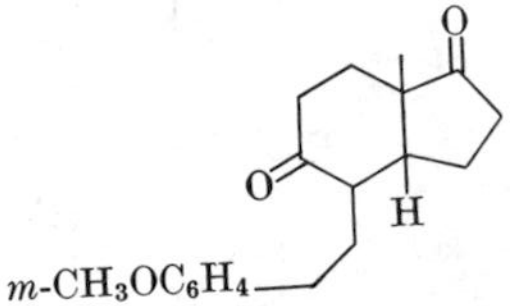 H m-$CH_3OC_6H_4$ (92) + some *trans* isomer	309
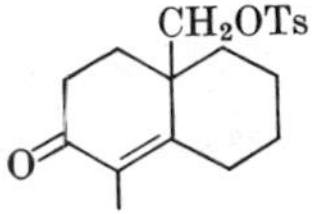 CH_2OTs	NH_3/Li, THF, 15 min	NH_4Cl	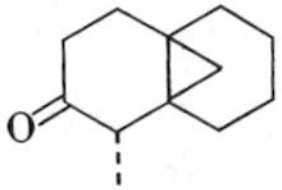 (47)	14, 68
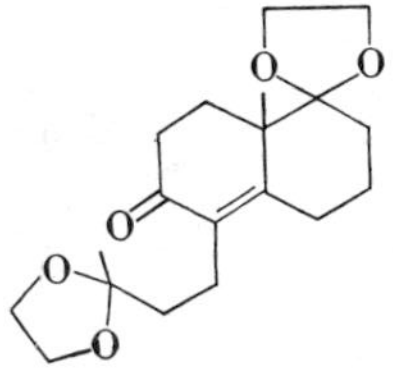	Li/NH_3, −50°	—	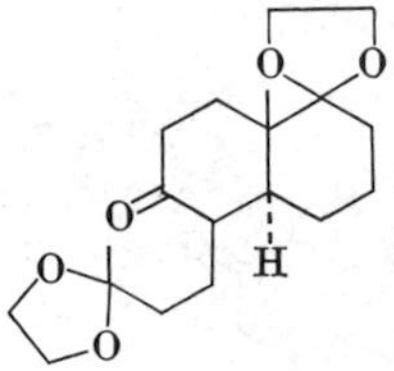 H (85)	388a
C_{20} 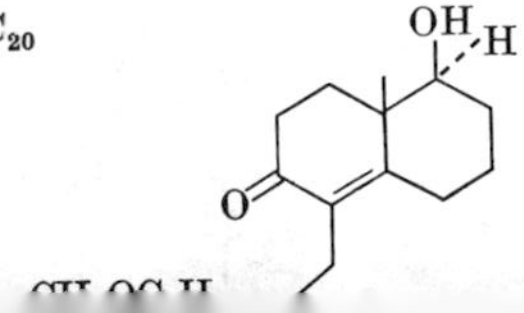 OH, H m-$CH_3OC_6H_4$	Li/NH_3, THF, 5 min	$NaNO_2$, CrO_3	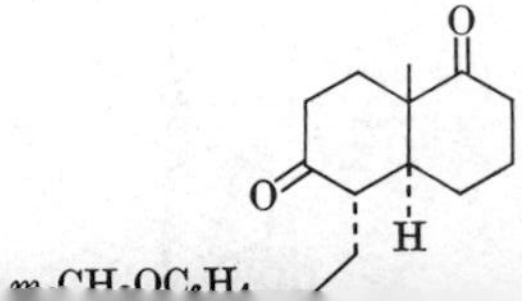 H m-$CH_3OC_6H_4$ (61)	309, 389

Substrate	Conditions	Trapping agent	Product(s) and Yield(s) (%)	Refs.
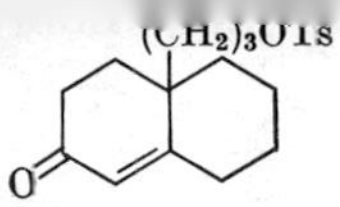 $(CH_2)_3OTs$	Li/NH_3	—	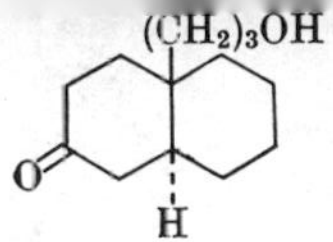 $(CH_2)_3OH$ (—)	14
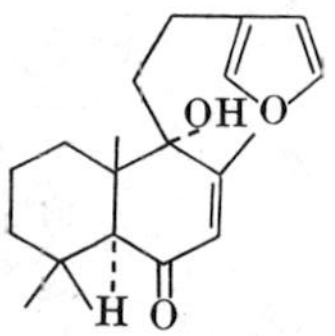 OH, O, H, O	Li/NH_3, Et_2O, 30 min	NH_4Cl	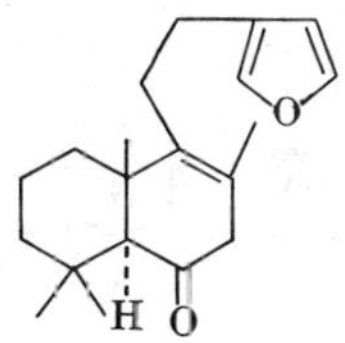 O, H, O (28)	200
C_{21} 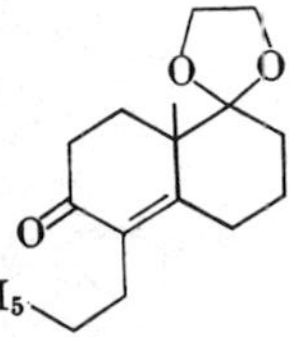 O, O, O, C_6H_5	Li/NH_3, THF, 5 min	$NaNO_2$	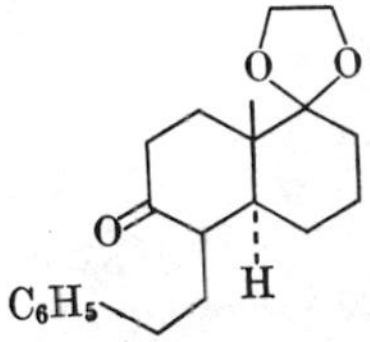 O, O, O, H, C_6H_5 (90)	309
C_{28} 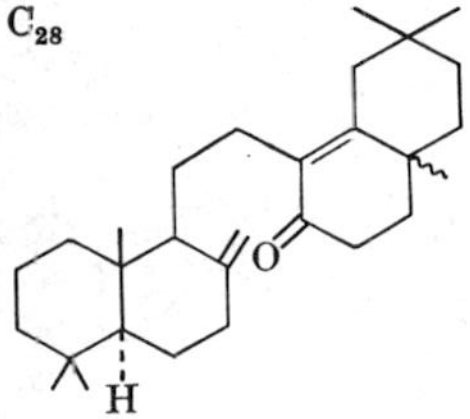 O, O, H	Li/NH_3	—	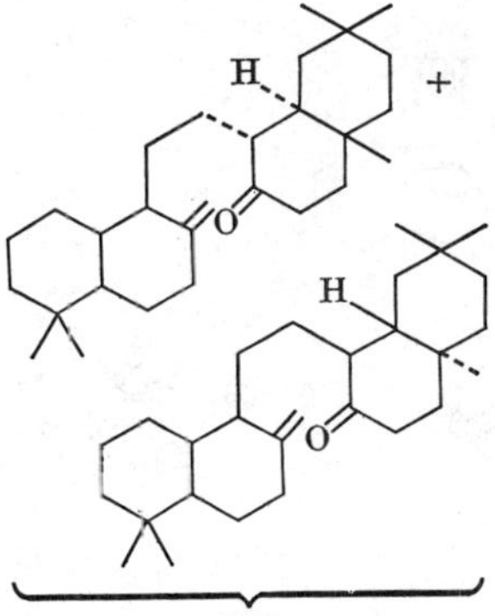 H, O, O + H, O, O (97)	390

Note: References 338–543 are on pp. 253–258.

TABLE IV. REDUCTIONS OF TRICYCLIC α,β-UNSATURATED KETONES

Reactant	Reduction Conditions	Quenching Agent	Product(s) (% Yield)	Refs.
C_{10} 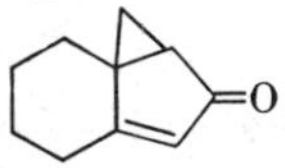	Li/NH_3	—	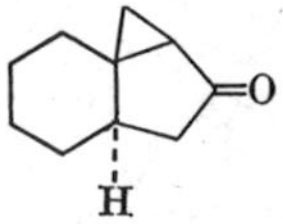 90–85 + 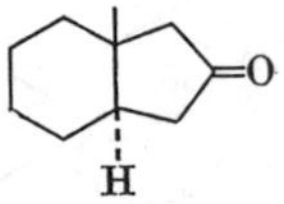 10–15	122
	Li/NH_3, *t*-BuOH	—	" 100	
C_{11} 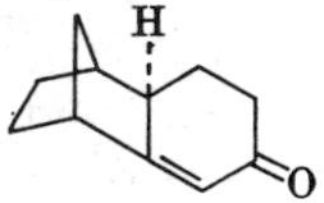	Li/NH_3, Et_2O-MeOH[a]	NH_4Cl, CrO_3	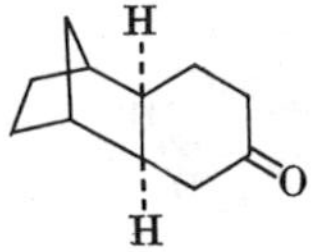 (>96) + 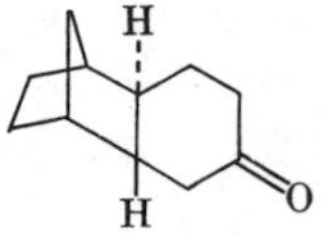 (<0.1)	80
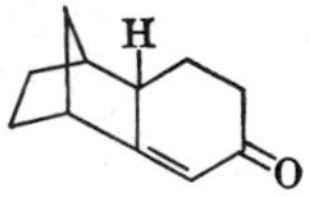	Li/NH_3, Et_2O-MeOH[a]	NH_4Cl, CrO_3	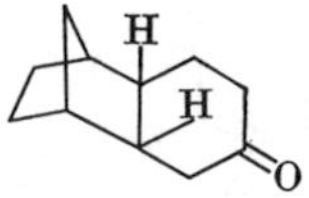 (~83) + 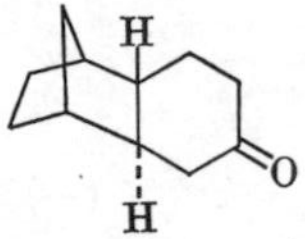 (<0.2)	80

Substrate	Reducing conditions	Work-up	Product(s) (Yield, %)	Refs.
C_{12} 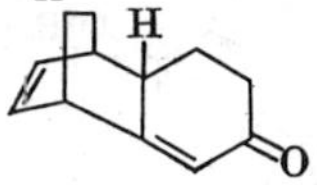	Li/NH_3, Et_2O-MeOH[a]	NH_4Cl, CrO_3	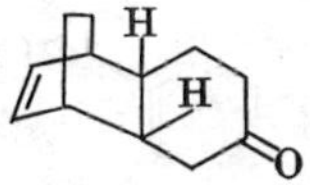 (~70) 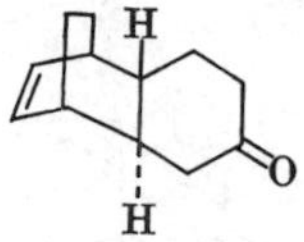 + (0.2)	80
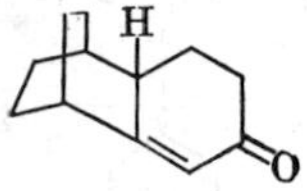	Li/NH_3, Et_2O-MeOH[a]	NH_4Cl, CrO_3	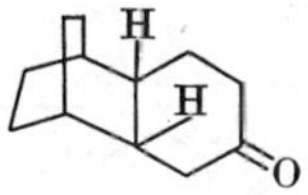 (~93) 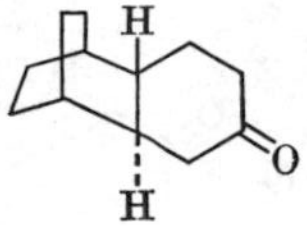 + (~1)	80
C_{13} 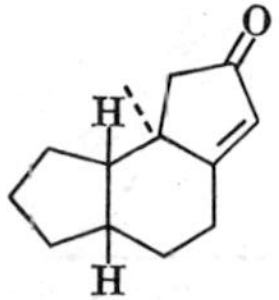	Li/NH_3	—	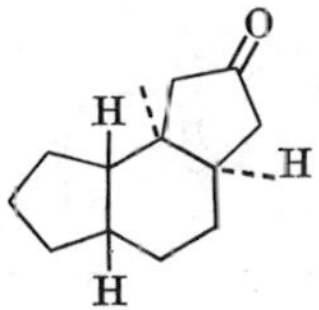 (—)	117

Note: References **338–543** are on pp. 253–258.

[a] It was stated that the reduction time employed was 1–5 minutes.

TABLE IV. REDUCTIONS OF TRICYCLIC α,β-UNSATURATED KETONES (*Continued*)

Reactant	Reduction Conditions	Quenching Agent	Product(s) (% Yield)	Refs.
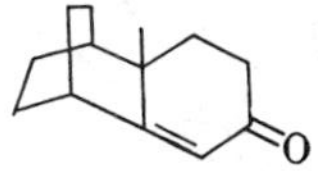 [structure]	Li/NH_3, Et_2O-MeOH[a]	NH_4Cl, CrO_3	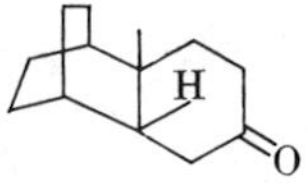 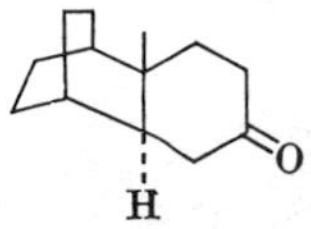 [structure] H (~98) + [structure] H (0.025)	80
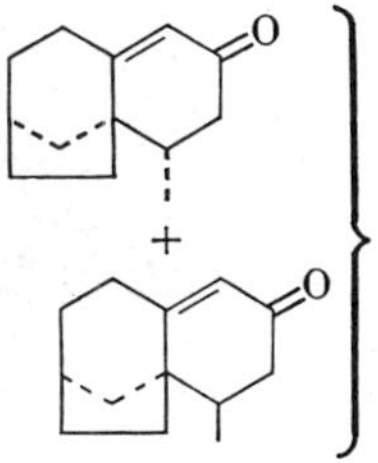 [structure] + [structure] } 1:3	Li/NH_3, Et_2O, 50 min	NH_4Cl, CrO_3	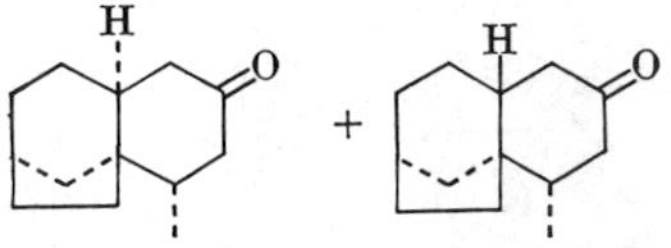 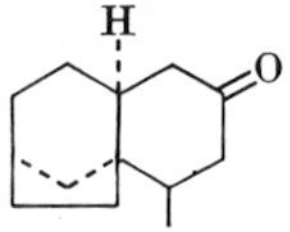 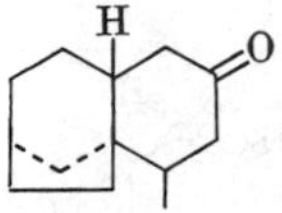 [structure] H + [structure] H; [structure] H 9 + [structure] H 91	114
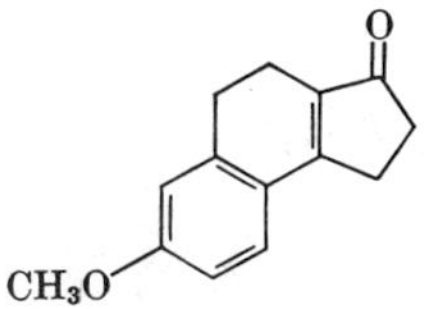 CH_3O [structure] O	Na/NH_3, Et_2O, 30 min	NH_4Cl	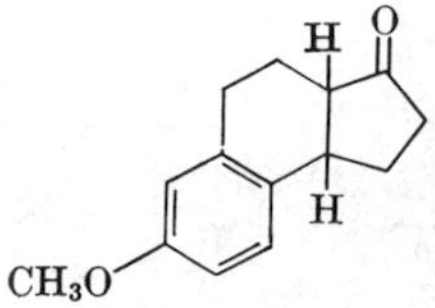 CH_3O [structure] H O H (94)	391

Substrate	Reagents	Quench	Product (Yield, %)	Refs.
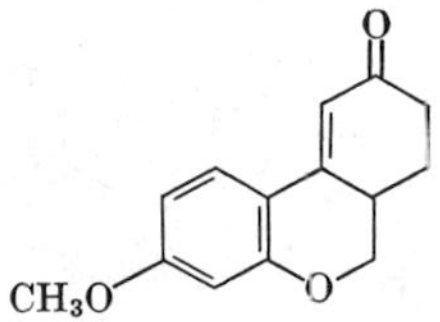 CH_3O	NH_3/Li, Et_2O	NH_4Cl	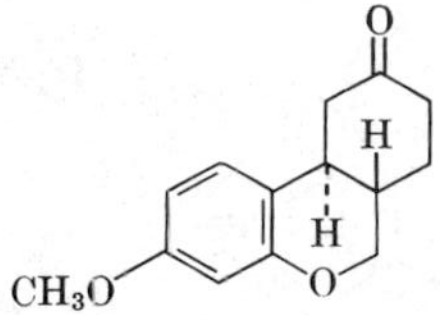 CH_3O (73)	392
	,,	NH_4Cl, portionwise	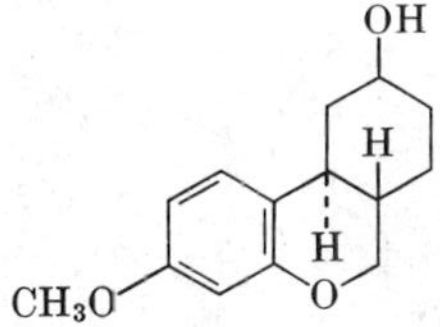 CH_3O (—)	
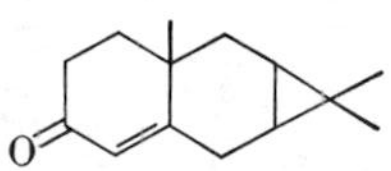	Li/NH_3, Et_2O, −45°, 1 eq. *t*-BuOH, 30 min,	NH_4Cl	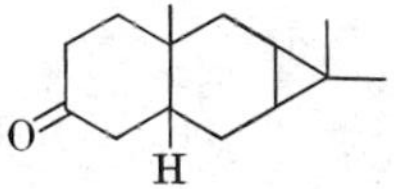 (91)	92a, 393
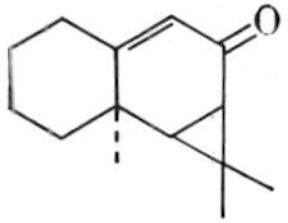	Li/NH_3, Et_2O, 1 hr	NH_4Cl	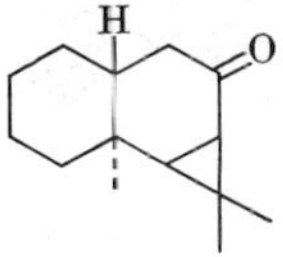 (90)	98
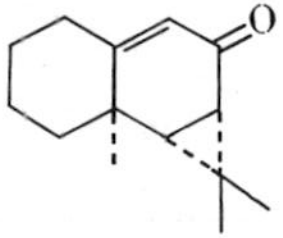	Li/NH_3, Et_2O, 1 hr	NH_4Cl	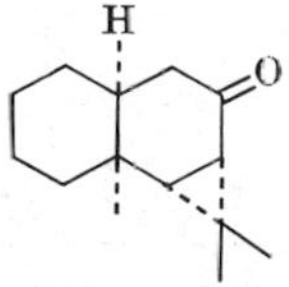 (90)	98

Note: References 338–513 are on pp. 253–2[illegible]8.

[a] It was stated that the reduction time employed was 1–5 minutes.

TABLE IV. Reductions of Tricyclic α,β-Unsaturated Ketones (*Continued*)

Reactant	Reduction Conditions	Quenching Agent	Product(s) (% Yield)	Refs.
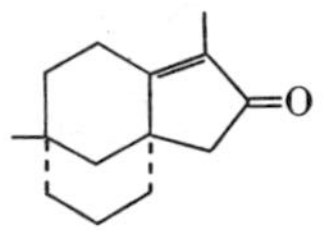	Li/NH_3, Et_2O, 2 hr	Wet Et_2O, dil HCl, CrO_3	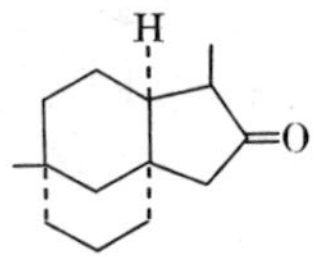 (70)	394
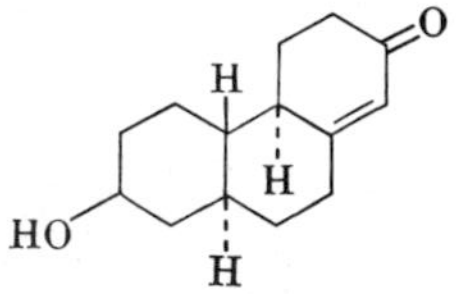 HO, H, H, H	Excess Li/NH_3 $THF\text{-}Et_2O$[b]	NH_4Cl	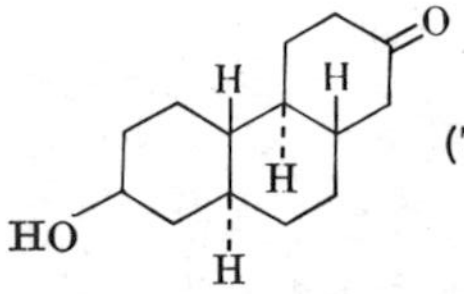 HO, H, H, H, H (73) + diol (8)	395
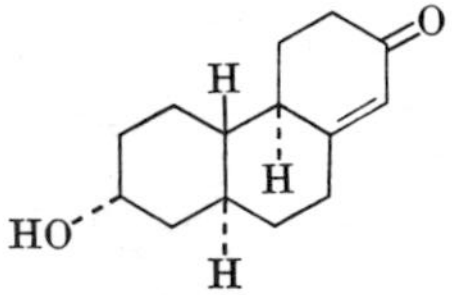 HO, H, H, H	Excess Li/NH_3, $THF\text{-}Et_2O$	NH_4Cl	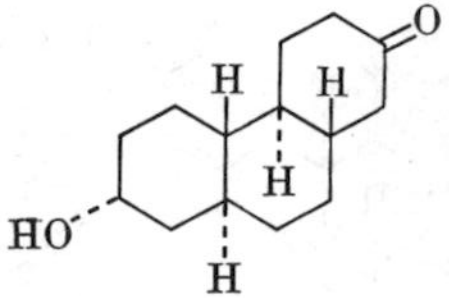 HO, H, H, H, H (50)	395
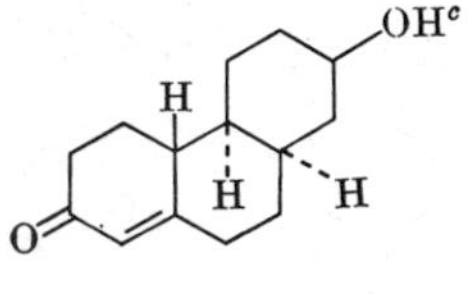 OH[c], H, H, H	NH_3/Li, THF, 1.75 hr, EtOH	$EtOH\text{-}H_2O$, Ac_2O, piperidine 2 N HCl	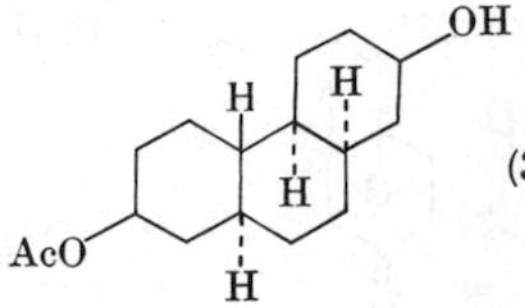 AcO, OH, H, H, H, H (32)	395

Substrate	Reduction Conditions	Proton Source	Product(s) (Yield, %)	Refs.
	NH_3/Li, THF, EtOH, 1.5 hr	EtOH-H_2O	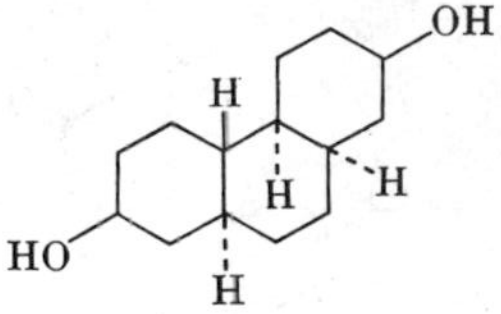 (71)	
	Li/NH_3, THF-ether[b]	NH_4Cl	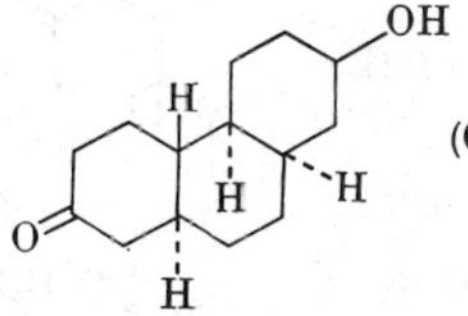 (60)	
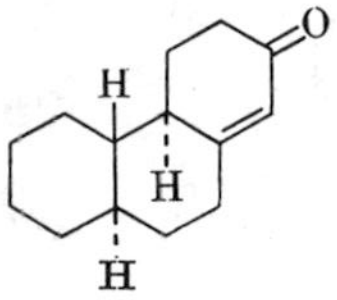	Na/NH_3	—	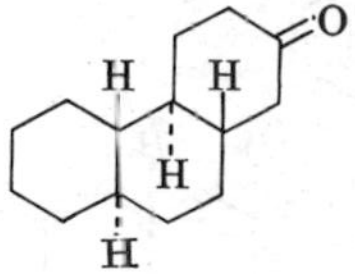 (–)	396
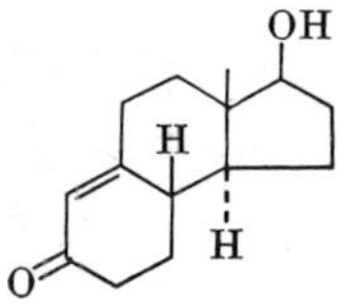	Li/NH_3	—	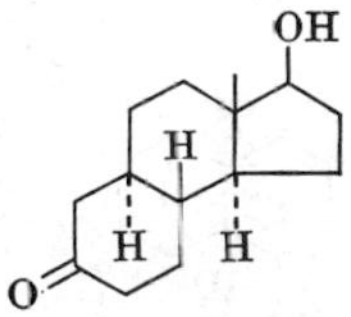 (–)	254

Note: References 338–543 are on pp. 253–258.

[b] A very short reduction time was employed.
[c] The alcohol was converted into the tetrahydropyranyl derivative before reduction.

TABLE IV. Reductions of Tricyclic α,β-Unsaturated Ketones (*Continued*)

Reactant	Reduction Conditions	Quenching Agent	Product(s) (% Yield)	Refs.
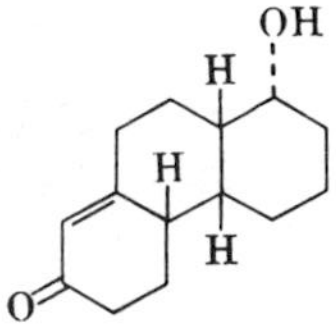	Li/NH_3, Et_2O	H_2O	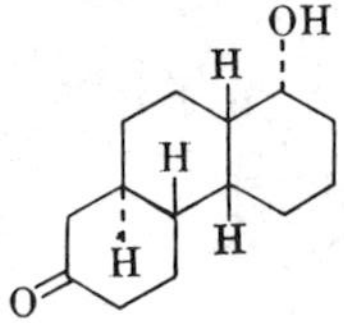 (68)	397
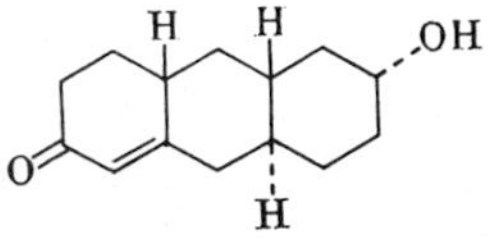	Li/NH_3,[c] Et_2O, 5 min	2 N HCl, 15 min	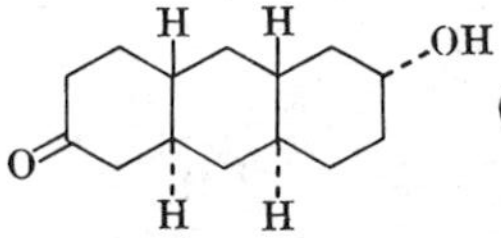 (50) + diol (Trace)	273
C_{15}				
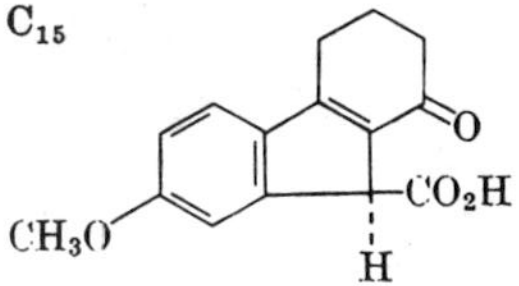	Li/NH_3	—	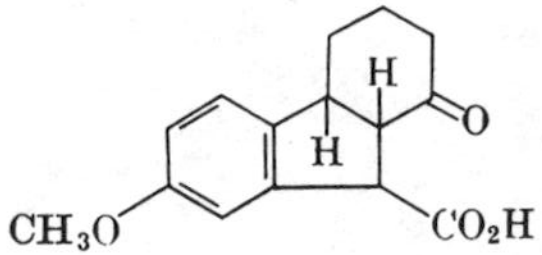 (—)	398
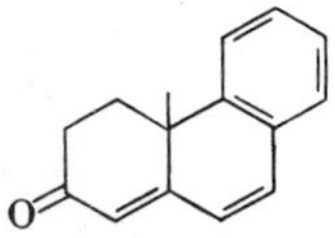	Li/NH_3, THF, 3 min	NH_4Cl, NaOMe, MeOH	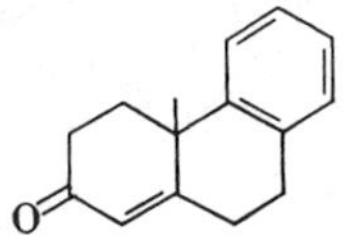 (71)	224
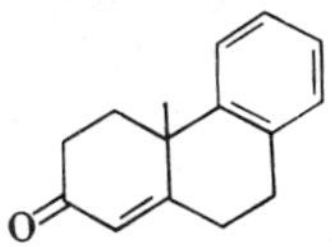	Li/NH_3, Et_2O, 30 min	NH_4Cl	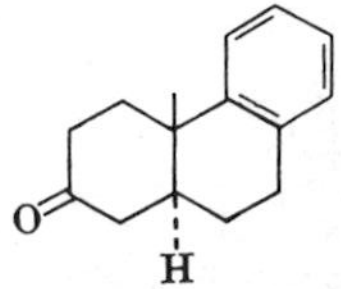 (70)	399

Substrate	Conditions	Quench	Product(s) and Yield(s) (%)	Refs.
	Li/NH_3, Et_2O, 15 min	"	" (60)	400
	Ca/NH_3, Et_2O	"	" (~60)	
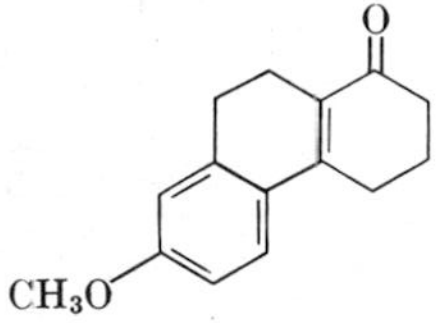 CH_3O	Li/NH_3, THF, 10 min	$EtOH\text{-}H_2O$	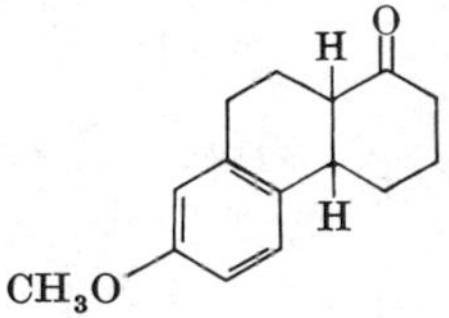 CH_3O, H, H (59)	11
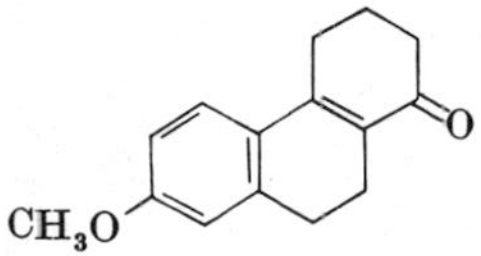 CH_3O	Li/NH_3, Et_2O, 30 min	NH_4Cl	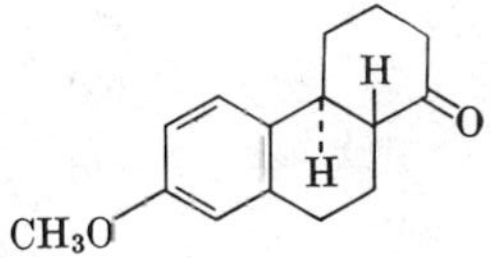 CH_3O, H, H (90)	401
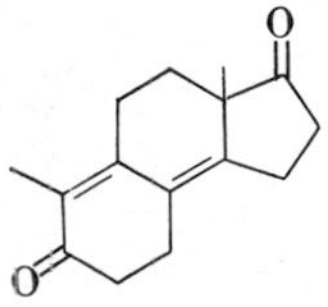	Li/NH_3, THF, MeOH, EtOH	NH_4Cl, CrO_3	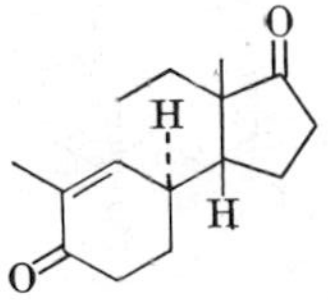 H, H (15)	402
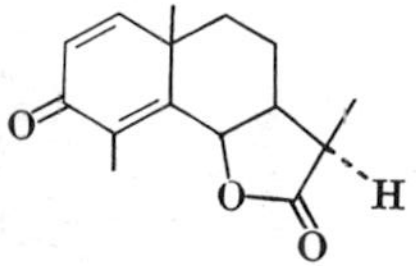 H	Li/NH_3, THF, 24 hr	Dil H_2SO_4	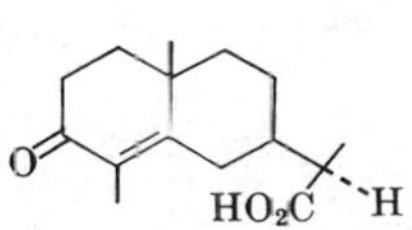 HO_2C, H (85)	203

Note: References 338–543 are on pp. 253–258.

[c] The alcohol was converted into its tetrahydropyranyl derivative before reduction.

TABLE IV. REDUCTIONS OF TRICYCLIC α,β-UNSATURATED KETONES (*Continued*)

Reactant	Reduction Conditions	Quenching Agent	Product(s) (% Yield)	Refs.
(*Contd.*)				
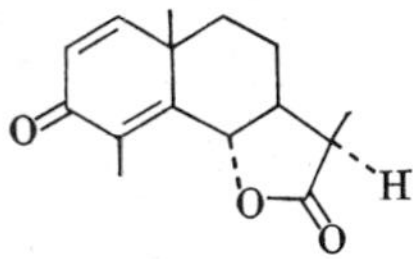 O, O, O, H	Li/NH_3, THF	H_3O^+	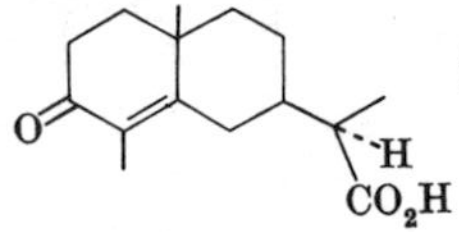 O, H, CO_2H (56)	235
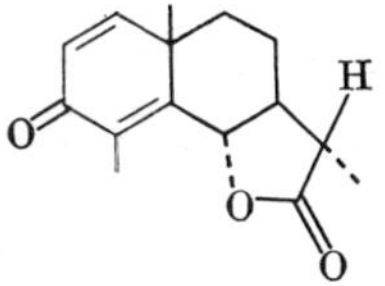 O, O, O, H	Li/NH_3, THF, 24 hr	Dil H_2SO_4	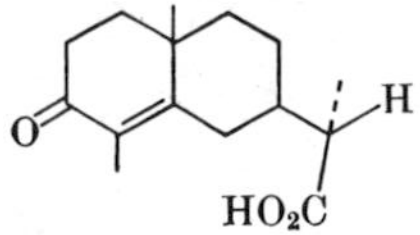 O, H, HO_2C (80)	203
	Li/NH_3, THF, 3 hr	"	" (80)	204
	NH_3/Li, THF, 30 min	NH_4Cl	" (80)	403
	"	H_3O^+, CH_2N_2	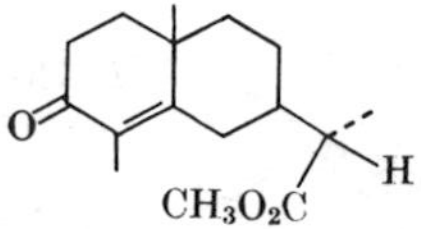 O, H, CH_3O_2C (54)	202
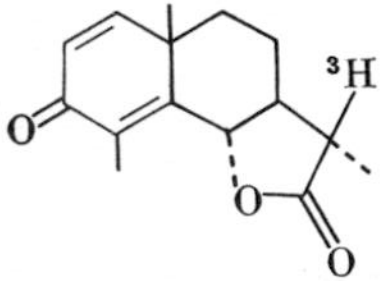 O, O, O, ^3H	Li/NH_3, THF, 30 min	H_3O^+	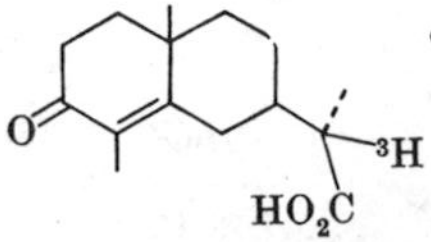 O, ^3H, HO_2C [d]	404

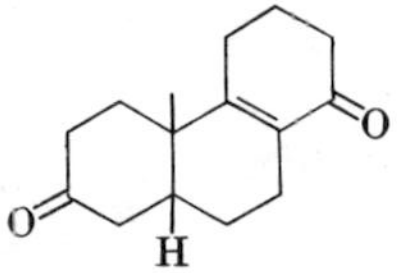	Li/NH_3, EtOH, Et_2O	CrO_3	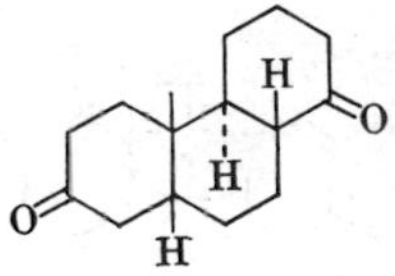 (35–40)	3
	Li/NH_3, Et_2O-dioxane	NH_4Cl	" (10–15) 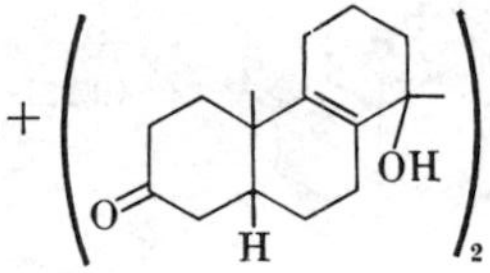 + (OH)$_2$ (25–42)	
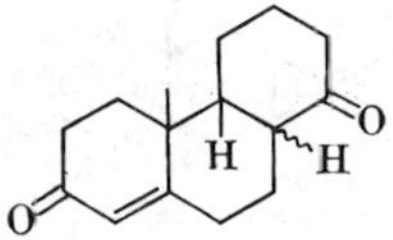	Li/NH_3, EtOH	CrO_3	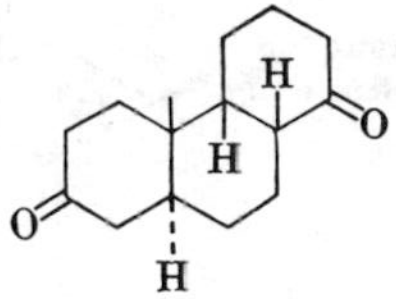 75 + 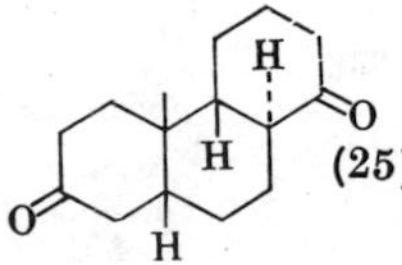 (25)	3
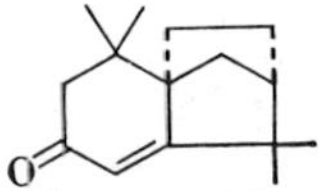	Li/NH_3	NH_4Cl	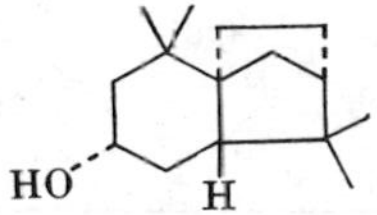 HO (—)	405

Note: References 338–543 are on pp. 253–258.

[d] It was stated that 74% of the tritium was retained in the reduction product.

TABLE IV. Reductions of Tricyclic α,β-Unsaturated Ketones (*Continued*)

Reactant	Reduction Conditions	Quenching Agent	Product(s) (% Yield)	Refs.
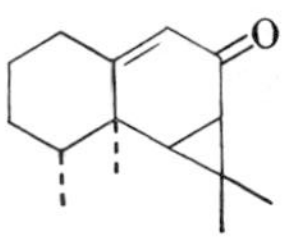	Li/NH_3, Et_2O, 45 min	NH_4Cl	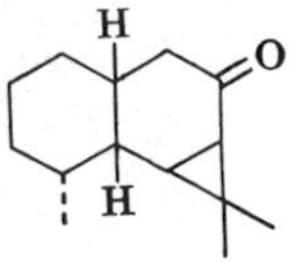 (92)	270
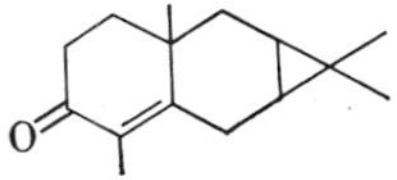	Li/NH_3, Et_2O, 1 eq. *t*-BuOH, 60 min, −45°	NH_4Cl	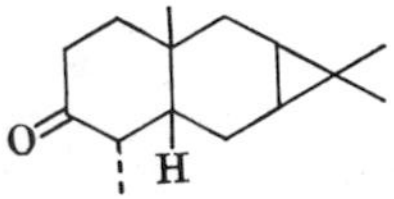 (90)	92a, 393
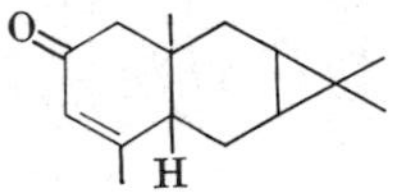	Li/NH_3, Et_2O, 1 eq. *t*-BuOH, 75 min, −45°	NH_4Cl	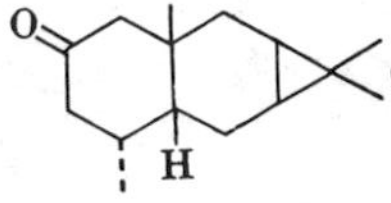 95 + starting material 5	92a, 393
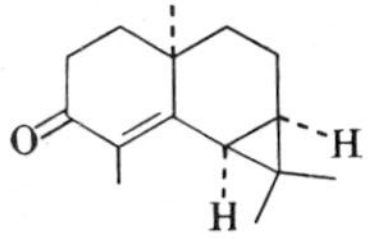	NH_3/Li, Et_2O, EtOH, 4 hr	H_2O	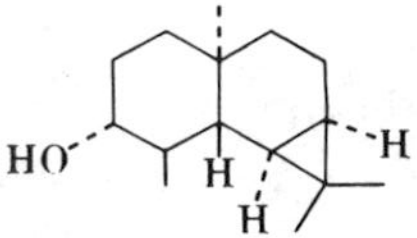 (27.5) + 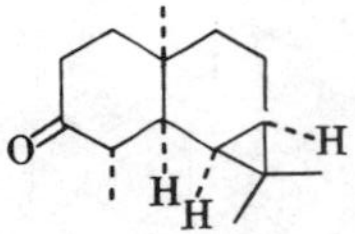 + 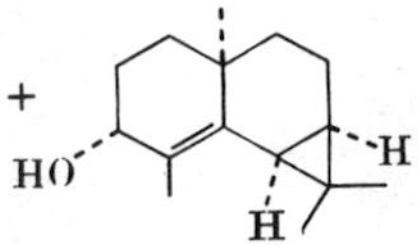 + 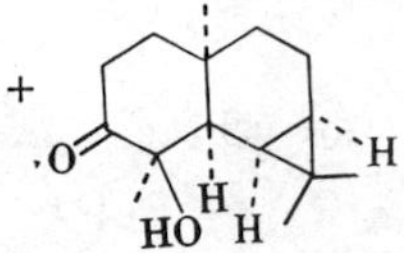	406

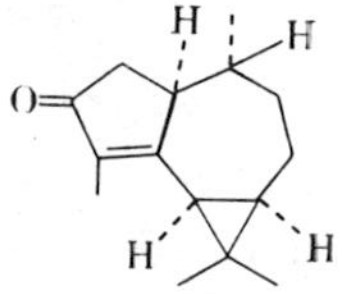	Li/NH_3, Et_2O, 50 min	NH_4Cl	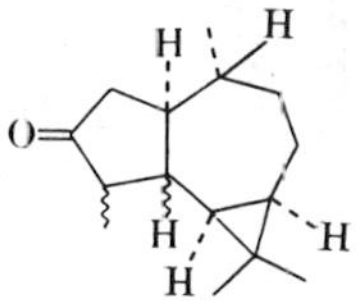 (15)	407
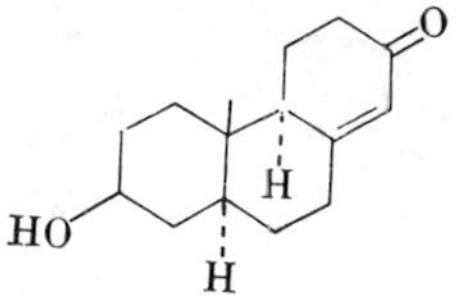	Li/NH_3; THF-Et_2O	NH_4Cl	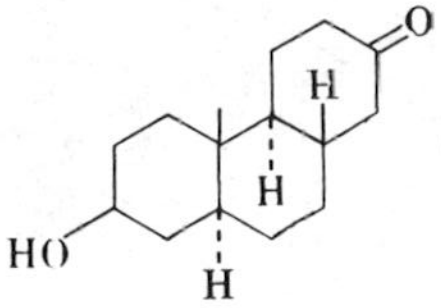 (54) 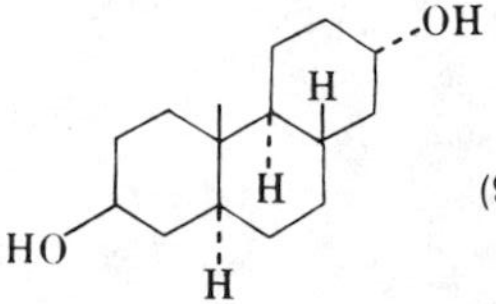 + (9)	395
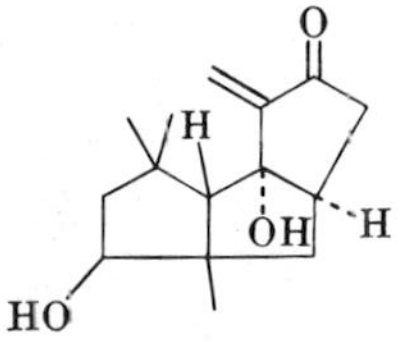	Li/NH_3	—	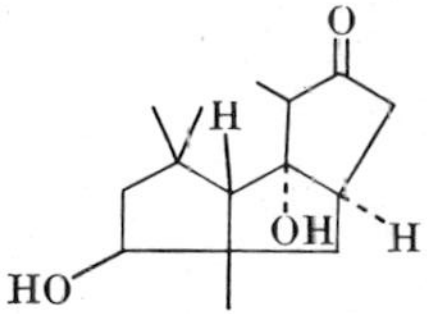 (—)	407a
C_{16} 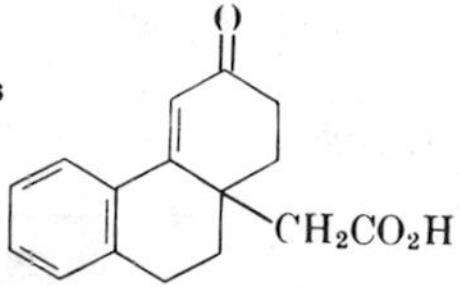 CH_2CO_2H	NH_3/Li, −60°, 30 min	NH_4Cl, CrO_3	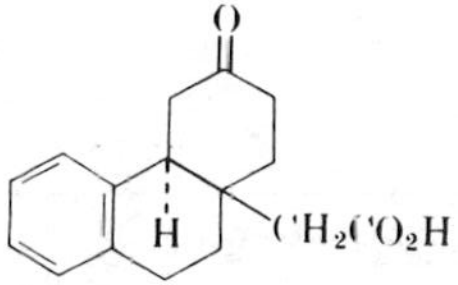 CH_2CO_2H (43)	408

Note: References 338–543 are on pp. 253–258

TABLE IV. REDUCTIONS OF TRICYCLIC α,β-UNSATURATED KETONES (*Continued*)

Reactant	Reduction Conditions	Quenching Agent	Product(s) (% Yield)	Refs.
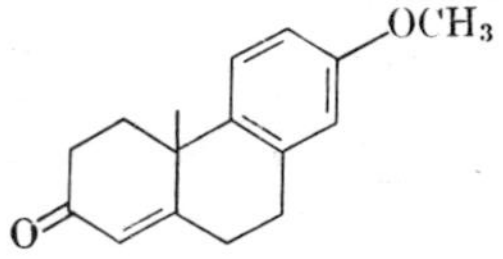 OCH_3	Li/NH_3, THF, 10 min, *t*-BuOH	NH_4Cl	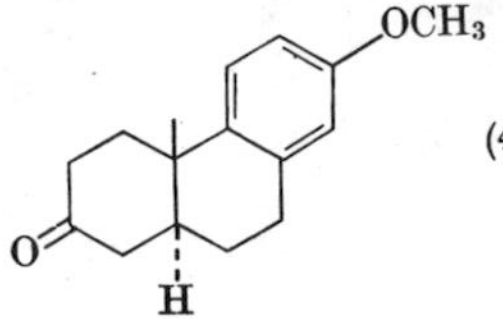 OCH_3 O H (40)	167
	Li/NH_3, THF, 30 min	"	" (59)	409
	Li/NH_3, Et_2O	"	"	400
	Li/NH_3, EtOH	—	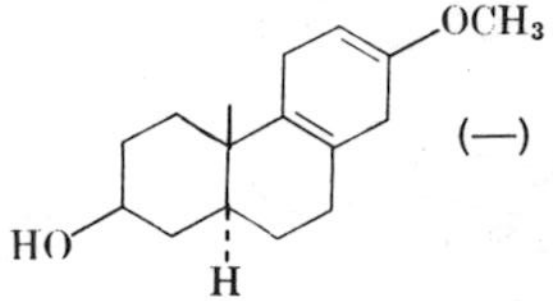 OCH_3 HO H (—)	410
	Excess Li/NH_3, THF, 25 min, *t*-BuOH, 4 hr	H_3O^+	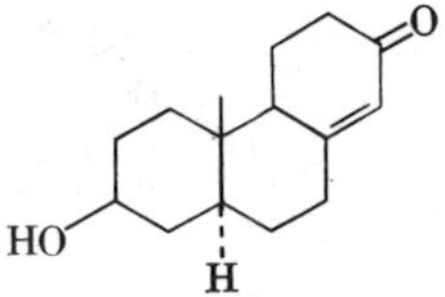 O HO H (74)	409
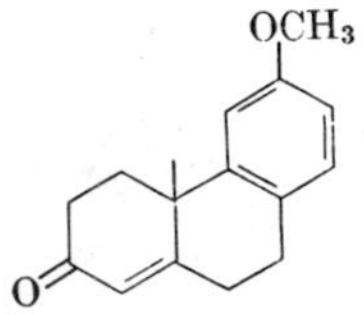 OCH_3 O	NH_3, Li, THF 10 min	$NaBrO_3$, NH_4Cl	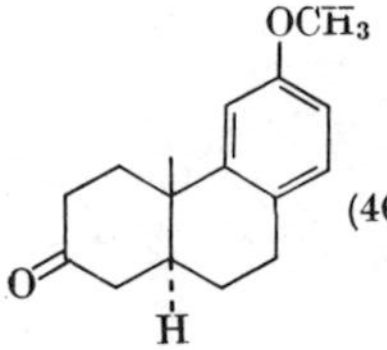 OCH_3 O H (46)	167

Substrate	Reducing conditions	Quench	Product(s) (Yield %)	Refs.
	Li/NH_3, THF, 10 min, t-BuOH	NH_4Cl	" (34)	167
	Li/NH_3, Et_2O, 5 min	NH_4Cl	" (78)	400
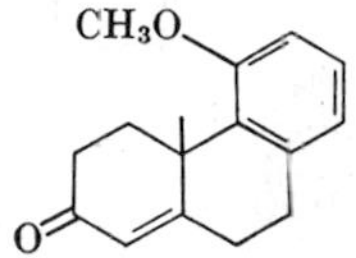	Li/NH_3	—	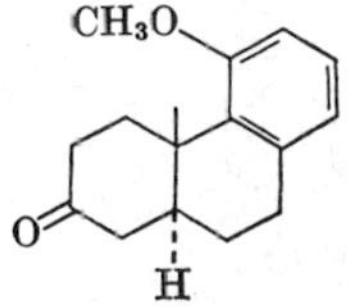 (–)	411
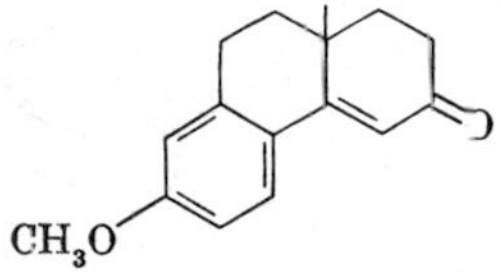	Li/NH_3	H_3O^+	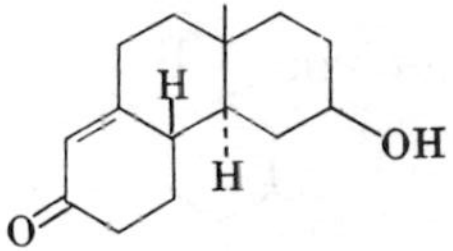 (—)	411a
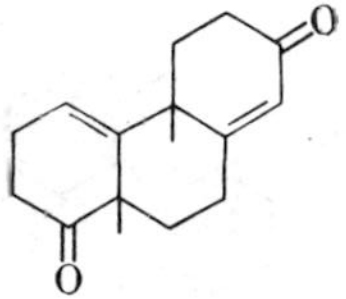	Li/NH_3, t-BuOH	CrO_3	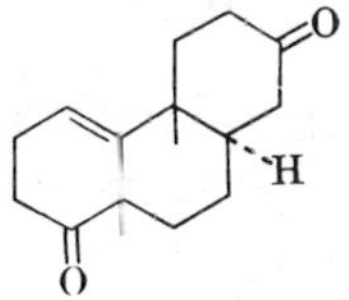 (—)	412
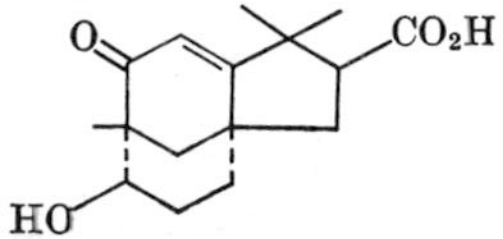	NH_3/Li, 60°, 1.5 hr	H_3O^+, CH_2N_2, CrO_3	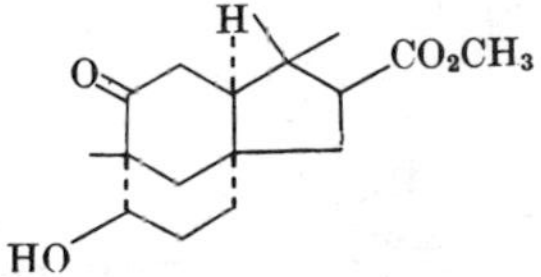 (—)	135
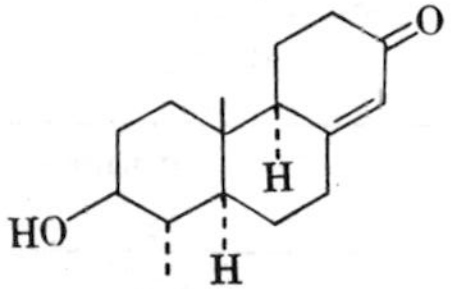	Li/NH_3, THF, 1 hr	NH_4Cl	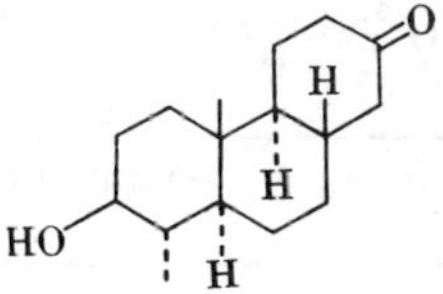 (67)	413

Note: References 338–543 are on pp. 253–258.

TABLE IV. Reductions of Tricyclic α,β-Unsaturated Ketones (*Continued*)

Reactant	Reduction Conditions	Quenching Agent	Product(s) (% Yield)	Refs.
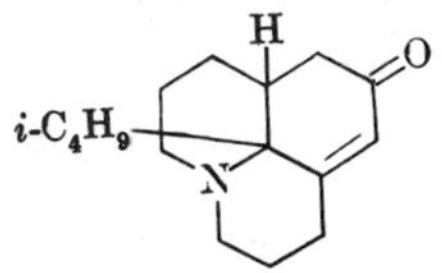 i-C_4H_9, H, O, N	Li/NH_3	—	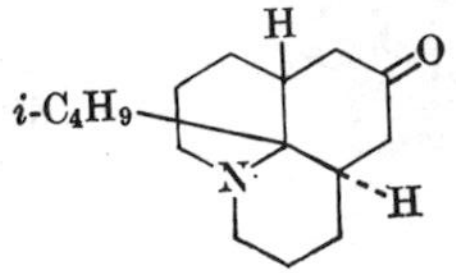 i-C_4H_9, H, O, N, H (—)	414
C_{17} 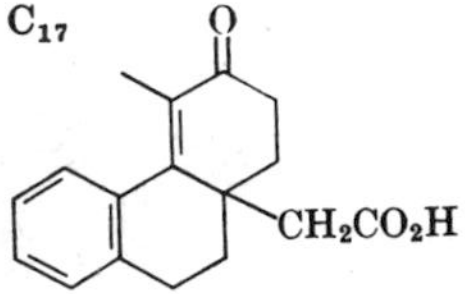 O, CH_2CO_2H	NH_3/Li, −60°, 30 min	NH_4Cl, H^+, CrO_3	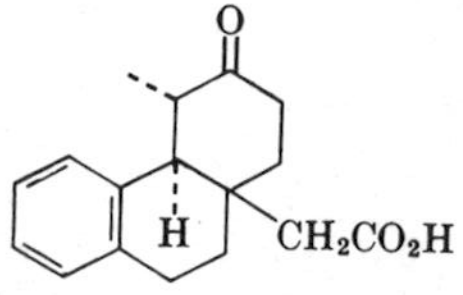 O, H, CH_2CO_2H (66)	408
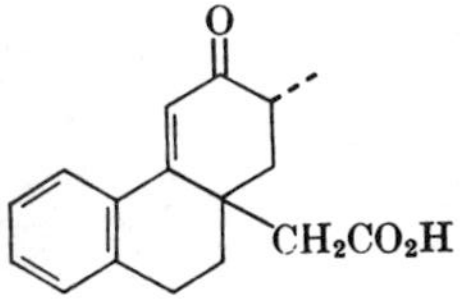 O, CH_2CO_2H	NH_3/Li, −60°, 30 min	NH_4Cl, H^+, CrO_3	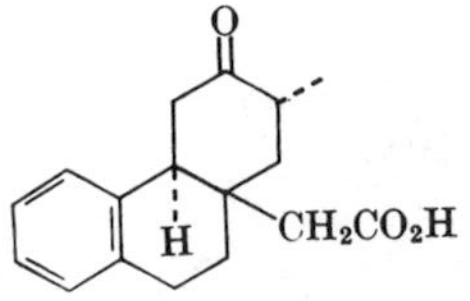 O, H, CH_2CO_2H (34)	408
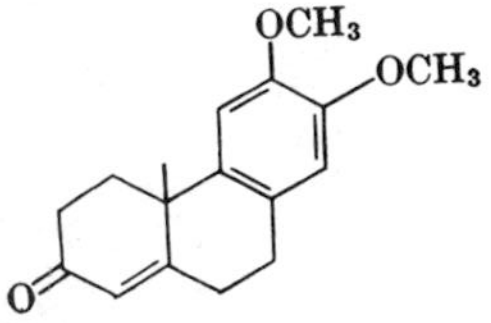 OCH_3, OCH_3, O	Li/NH_3, dioxane-Et_2O, 15 min	NH_4Cl	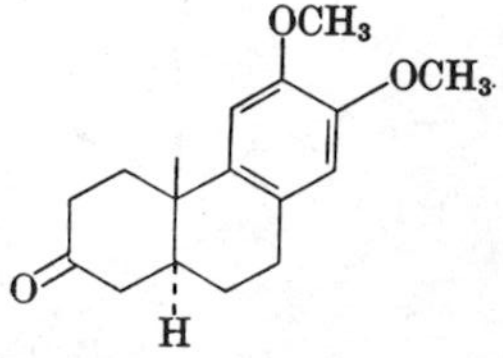 OCH_3, OCH_3, O, H (—)	400

Substrate	Conditions	Quench	Product(s) (Yield, %)	Refs.
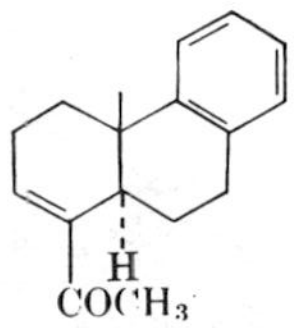 COCH$_3$	NH_3/Na, Et_2O, 15 min	NH_4Cl, CrO_3	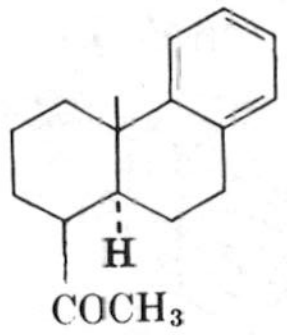 COCH$_3$ (82)	415
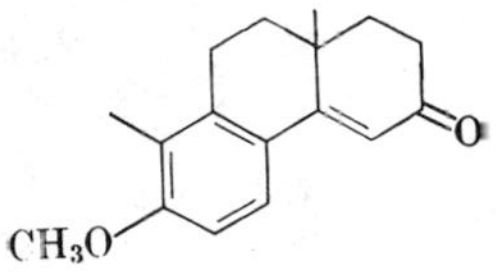 CH_3O	NH_3/Li, dioxane, Et_2O-EtOH	—	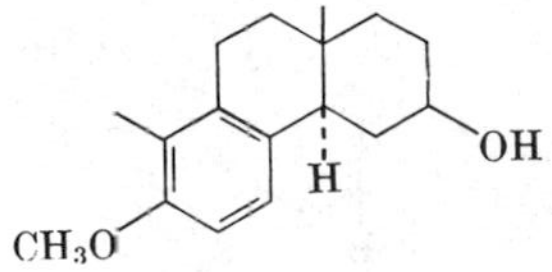 CH_3O, OH (40)	416
	Li/NH_3, $C_6H_5CH_3$-Et_2O	NH_4Cl	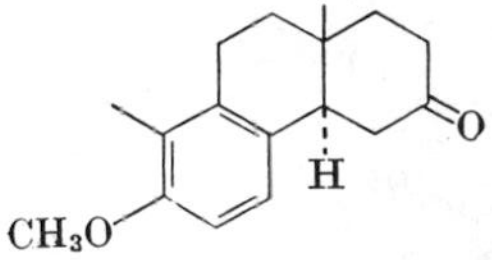 CH_3O (50) 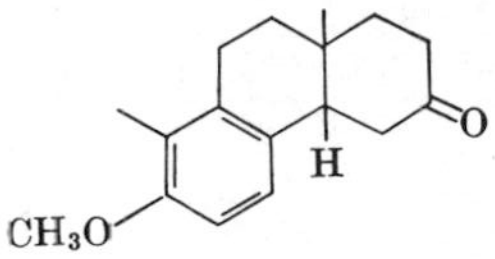 + CH_3O (Low)	
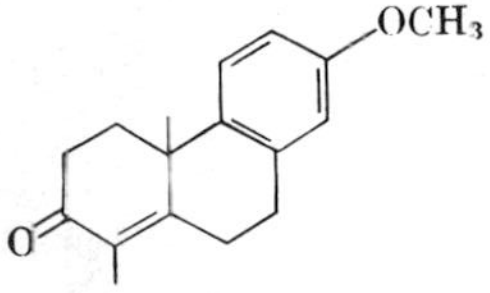 OCH_3	Li/NH_3, THF, −40°, 1 hr	NH_4Cl	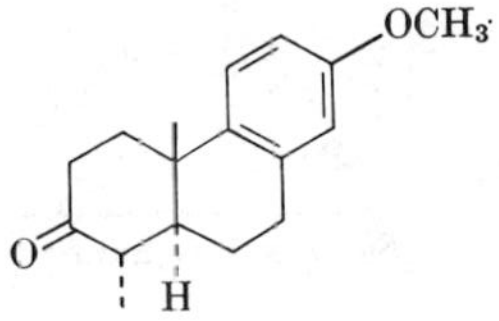 OCH_3 (78)	413

Note: References 338–543 are on pp. 253–258.

TABLE IV. Reductions of Tricyclic α,β-Unsaturated Ketones (*Continued*)

Reactant	Reduction Conditions	Quenching Agent	Products) (% Yield)	Refs.
(*Contd.*)				
OCH_3	Li/NH_3, EtOH	H_3O^+	H, H (50) +	413
			HO, H, H (37)	
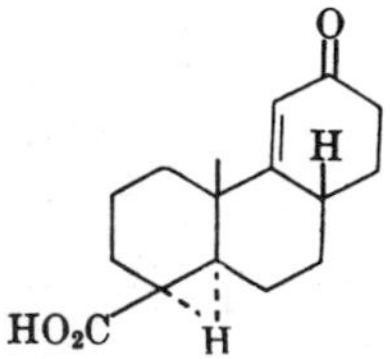 O, H, HO_2C, H	Li/NH_3, THF, 3 min	NH_4Cl, CH_2N_2, CrO_3	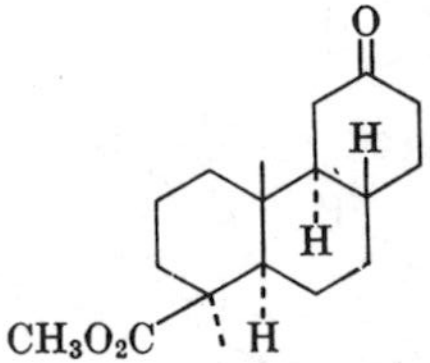 O, H, H, CH_3O_2C, H (~50)	417
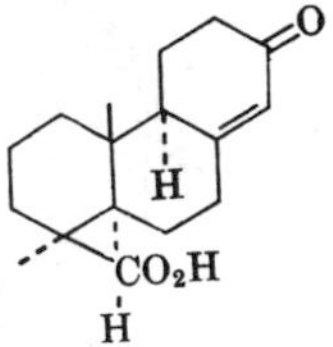 O, H, CO_2H, H	Li/NH_3, THF, 30 min	EtOH, H^+, CrO_3, CH_2N_2	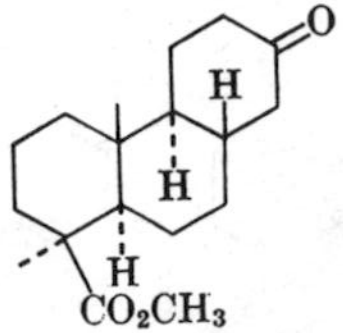 O, H, H, H, CO_2CH_3 (—)	418

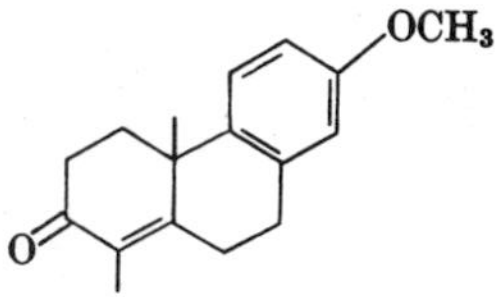

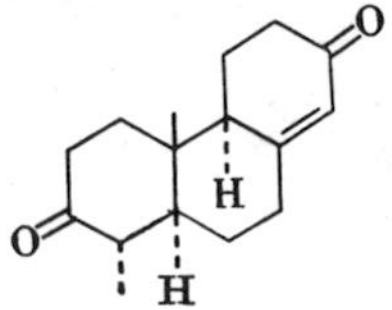

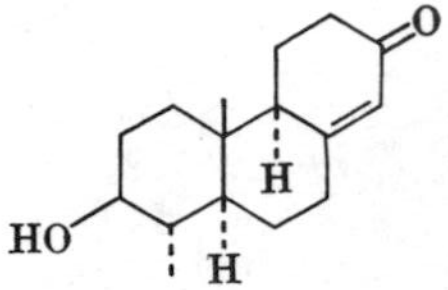

Reagents	Work-up	Product (Yield, %)	Ref.
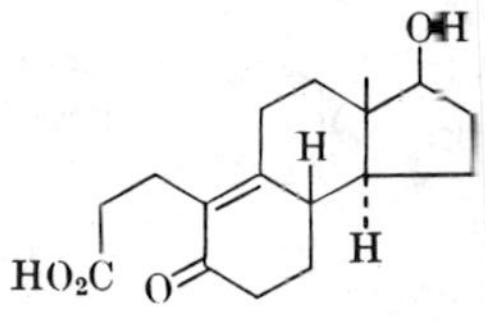 Li/NH_3, THF, 15 min	NH_4Cl, H^+	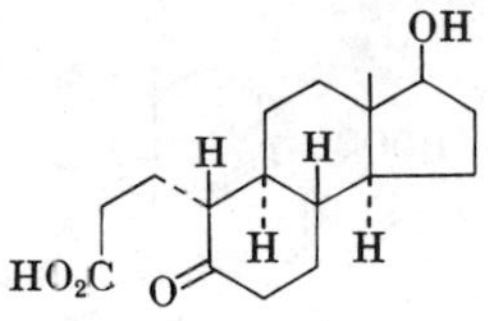 (–)	419
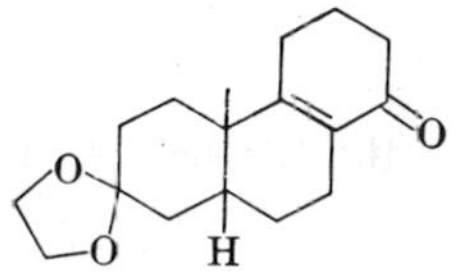 Li/NH_3, Et_2O-dioxane	NH_4Cl, H^+	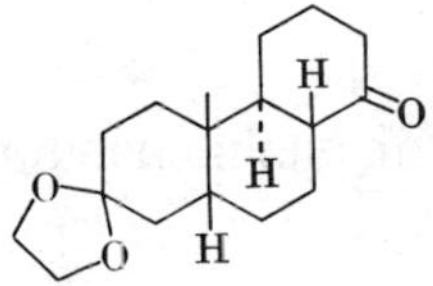 (40)	4
Li/NH_3, EtOH-Et_2O	CrO_3	" (40)	
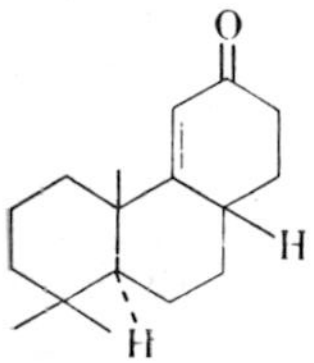 NH_3/Li, THF, –78°, 10 min	NH_4Cl	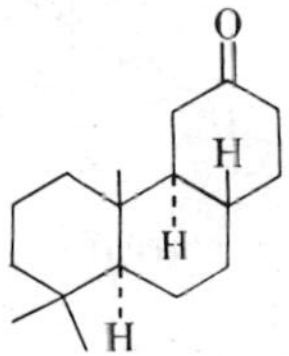 (54) + starting material (6)	420
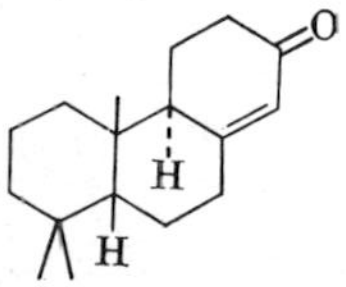 Li/NH_3, dioxane	—	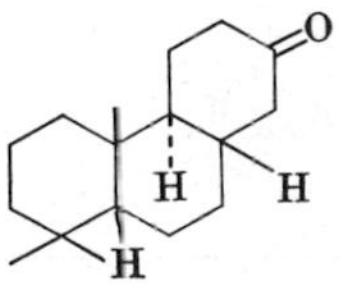 (—)	421

Note: References 338–513 are on pp. 253–258.

TABLE IV. REDUCTIONS OF TRICYCLIC α,β-UNSATURATED KETONES (*Continued*)

Reactant	Reduction Conditions	Quenching Agent	Product(s) (% Yield)	Refs.
C_{17} (*contd.*) 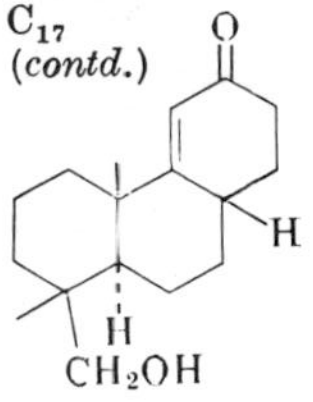 H, H, CH_2OH	NH_3/Li, *t*-BuOH THF, 3 hr	MeOH	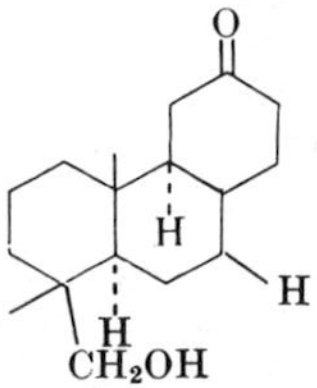 H, H, H, CH_2OH (99)	160
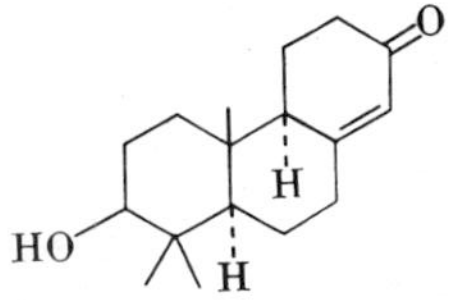 HO, H, H	Li/NH_3, THF, −50°, 3 hr	NH_4Cl	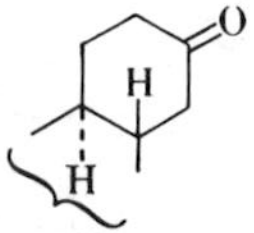 H, H (67) 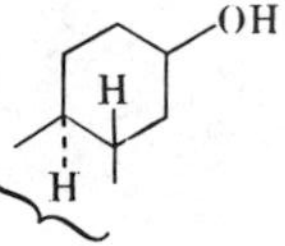 + OH, H, H (Trace)	422
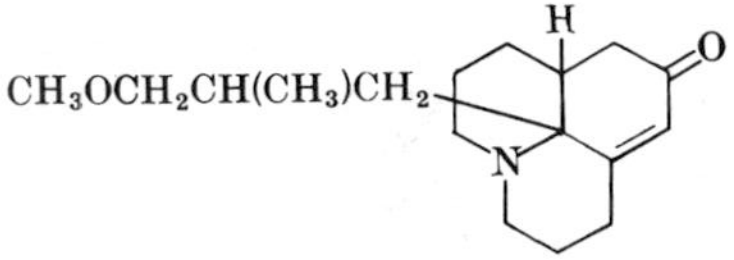 $CH_3OCH_2CH(CH_3)CH_2$, H, N	Li/NH_3	—	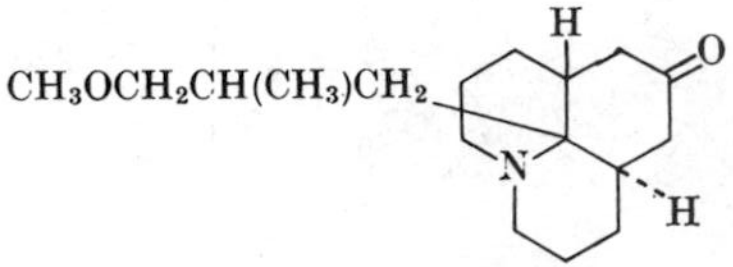 $CH_3OCH_2CH(CH_3)CH_2$, H, N, H (−)	423
C_{18} 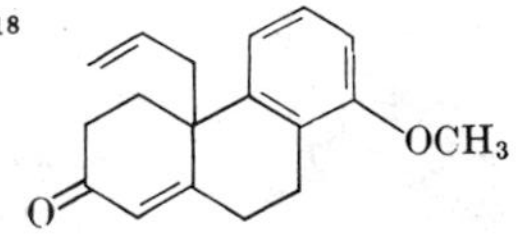 OCH_3	NH_3/Li, Et_2O, 45 min	NH_4Cl	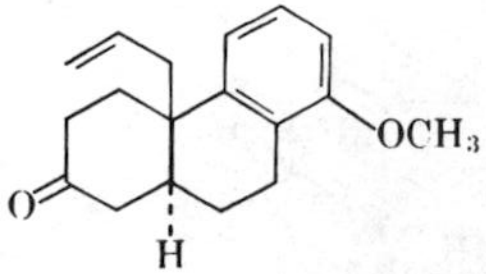 OCH_3, H (75)	424

Reactant	Conditions	Work-up	Product(s) and Yield(s) (%)	Refs.
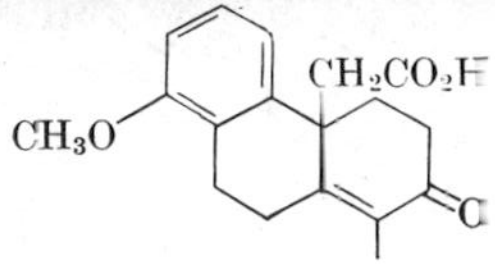 CH_3O, CH_2CO_2H	NH_3/Li, 40 min	NH_4Cl, H^+	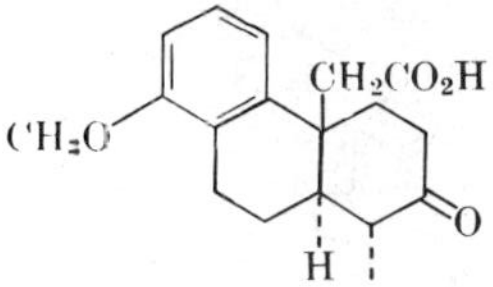 (70)	425
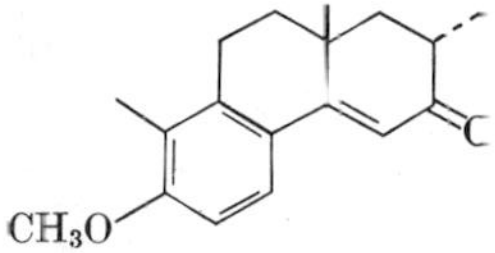 CH_3O	Li/NH_3, $C_6H_5CH_3$-Et_2O	NH_4Cl, H^+	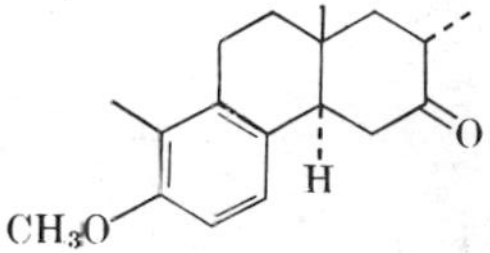 (50) 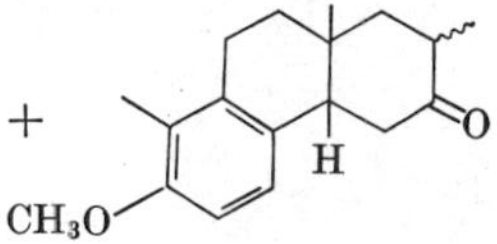 + (Low)	416
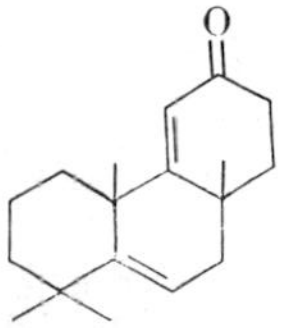	Li/NH_3	CrO_3	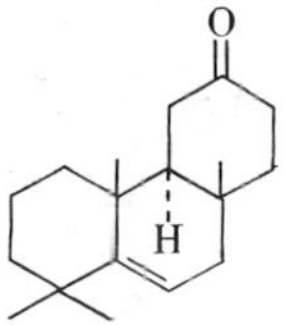 (—)	426
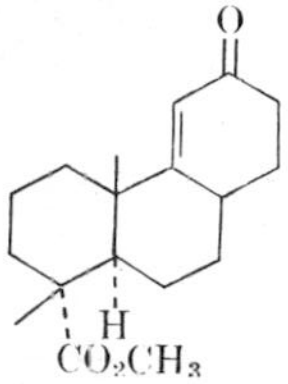	NH_3/Li,[e] Et_2O	NH_4Cl	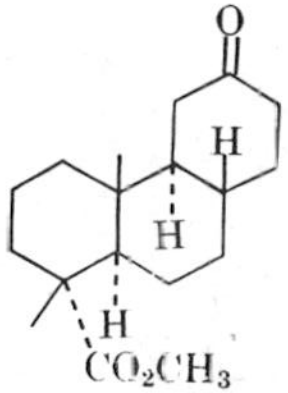 (42)	417

Note: References 338–543 are on pp. 253–258.

[e] A solution of 2.5 eq. of lithium in liquid ammonia was added dropwise to the enone in NH_3-Et_2O.

TABLE IV. REDUCTIONS OF TRICYCLIC α,β-UNSATURATED KETONES (*Continued*)

Reactant	Reduction Conditions	Quenching Agent	Product(s) (% Yield)	Refs.
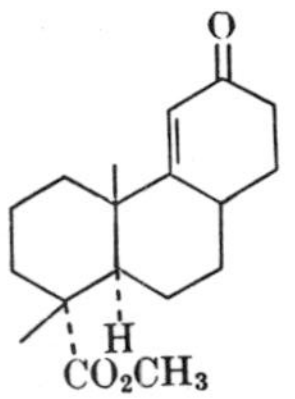 CO_2CH_3, H (*Contd.*)	Li/NH_3, EtOH, 70 min	EtOH, CrO_3	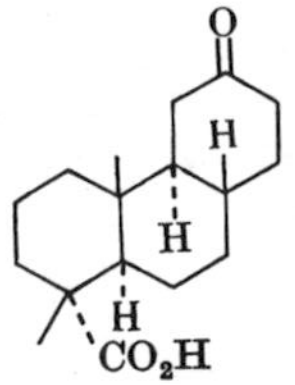 H, H, H, CO_2H (85)	417
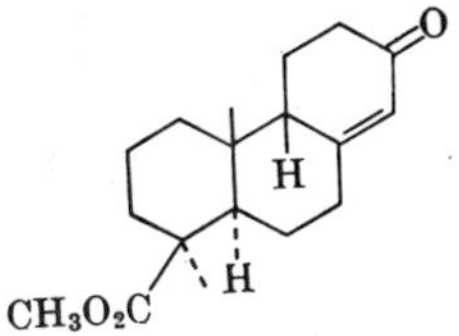 CH_3O_2C, H, H	NH_3/Li, Et_2O, 10 min	NH_4Cl, CrO_3	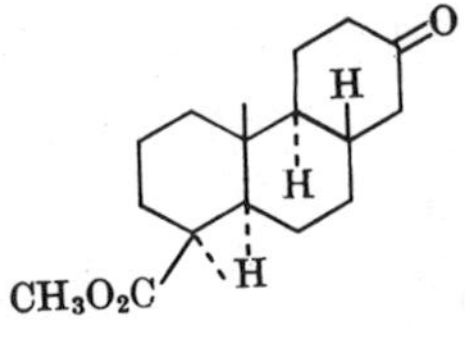 CH_3O_2C, H, H, H (75)	427
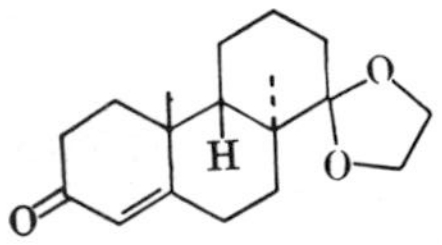 H	Li/NH_3,^{f} THF	EtOH, CrO_3	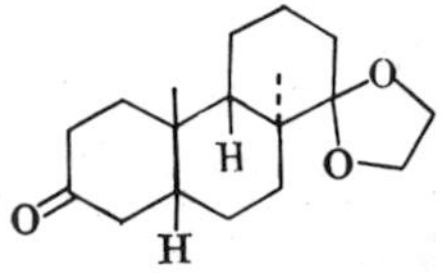 H, H (—)	79
C_{19}				
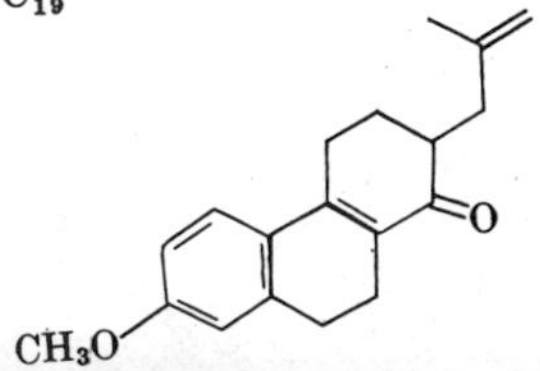 CH_3O	Li/NH_3, Et_2O, 30 min	EtOH, CrO_3	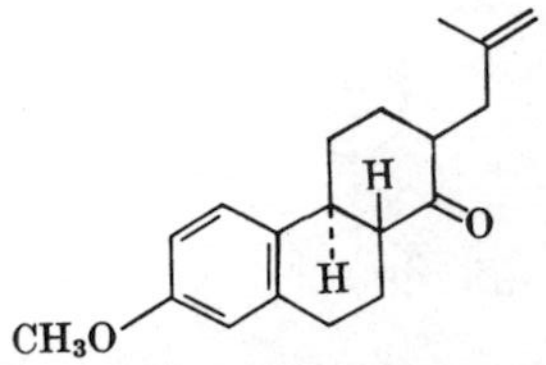 CH_3O, H, H (65)	401

Substrate	Reducing Conditions	Subsequent Treatment	Product(s) and Yield(s) (%)	Refs.
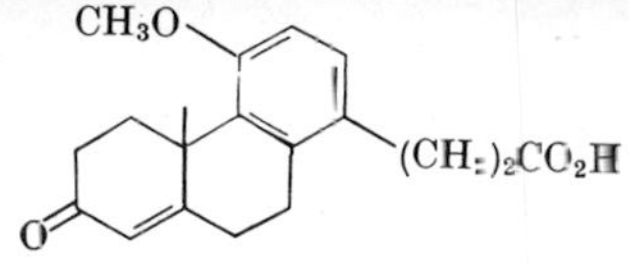 [structure]	Li/NH_3	—	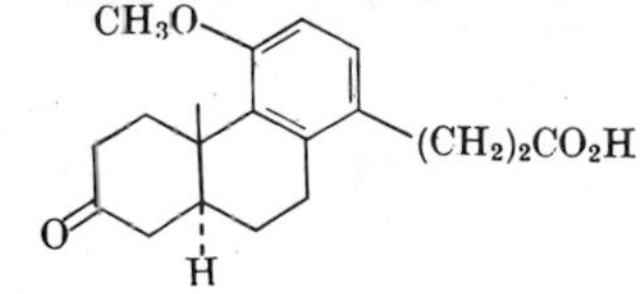 [structure] (—)	429
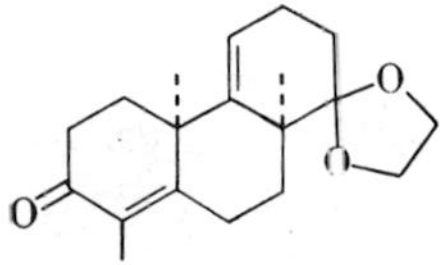 [structure]	Li/NH_3, THF	EtOH, CrO_3, H_3O^+, H_2(Pd/C), HOAc, 8 *N* H_2CrO_4	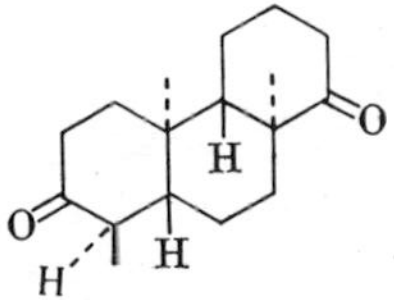 [structure] (—)	79
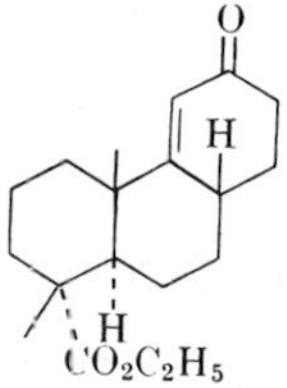 [structure]	Li/NH_3	—	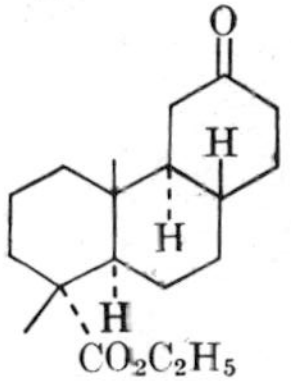 [structure] (—)	376
C_{20} [structure]	Excess Li/NH_3, Et_2O, 30 min	NH_4Cl	[structure] (14) + [structure] (68)	430

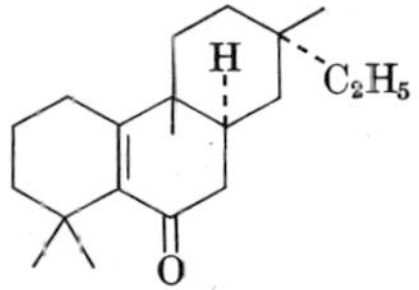

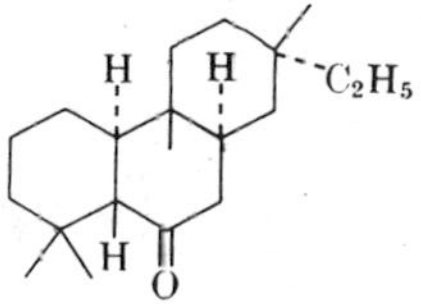

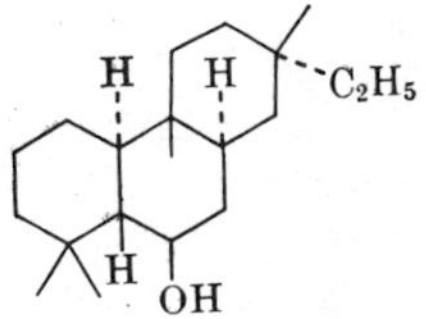

Note: References 338–543 are on pp. 253–258.

[f] Similar results were obtained with potassium.

TABLE IV. Reductions of Tricyclic α,β-Unsaturated Ketones (*Continued*)

Reactant	Reduction Conditions	Quenching Agent	Product(s) (% Yield)	Refs.
C_{21} 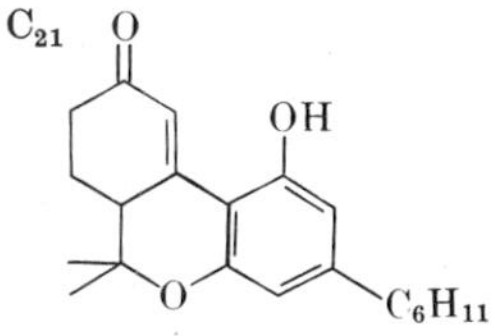 [structure: OH, C_6H_{11}]	Excess Li/NH_3, THF, −70°	NH_4Cl	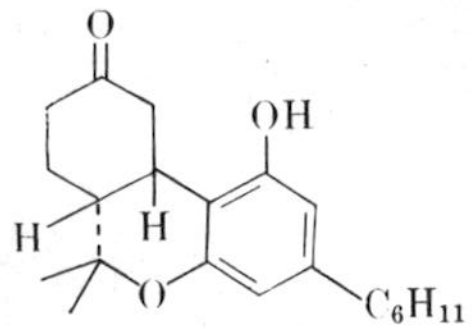 [structure: OH, H, H, C_6H_{11}] (53) 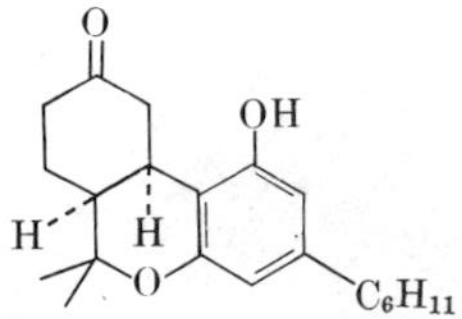 + [structure: OH, H, H, C_6H_{11}] (Trace)	431
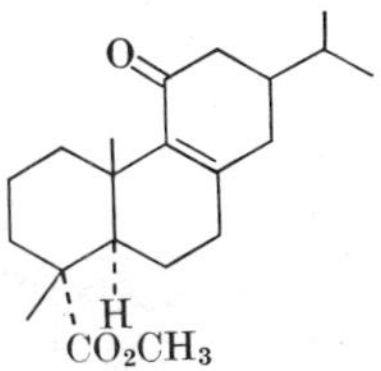 [structure: H, CO_2CH_3]	Li/NH_3, THF, 2 hr	H_2O, CrO_3, CH_2N_2	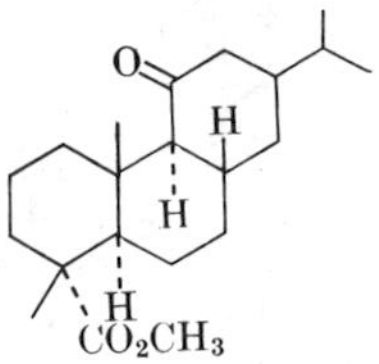 [structure: H, H, H, CO_2CH_3] (8) 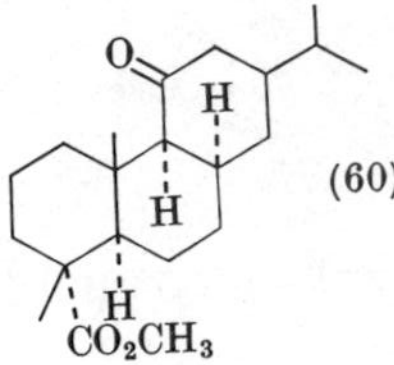 + [structure: H, H, H, CO_2CH_3] (60)	108

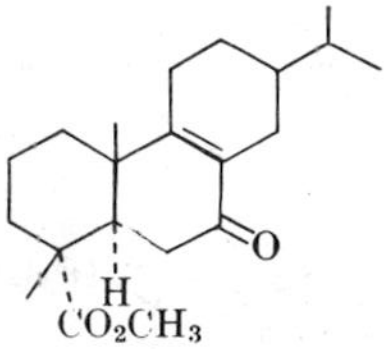

Reactant	Conditions	Quench	Product(s) and Yield(s) (%)	Refs.
H, CO_2CH_3	Li/NH_3, Et_2O, 5 min	NH_4Cl	H, H, H, CO_2CH_3 (35) + H, H, H, CH_2OH (10) + starting material (50)	
	Li/NH_3, THF 5 min	H_2O	,, (85)	108
C_{24} 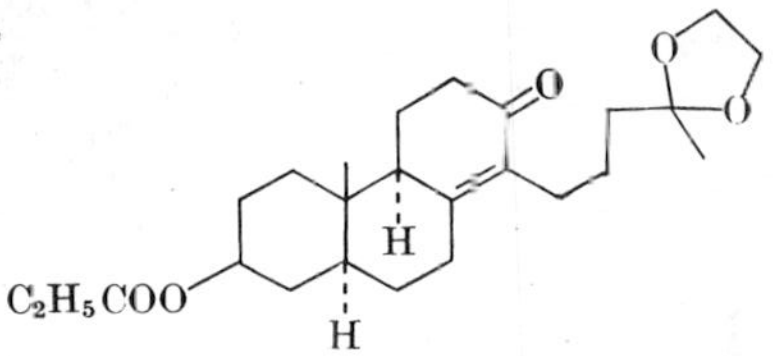 H, C_2H_5COO, H	Li/NH_3	H_3O^+	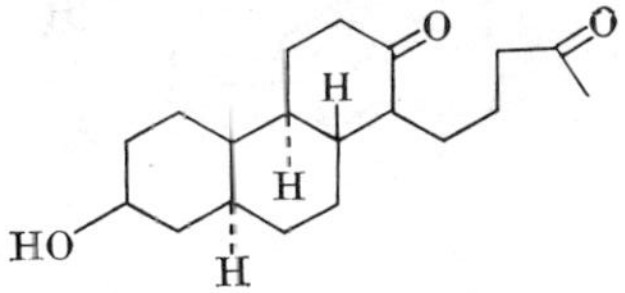 H, H, HO, H (—)	410
C_{32} 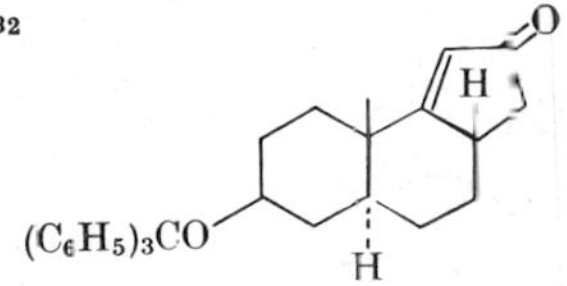 H, $(C_6H_5)_3CO$, H	Li/NH_3	—	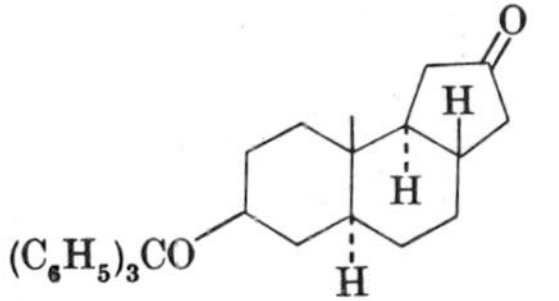 H, H, $(C_6H_5)_3CO$, H (—)	372

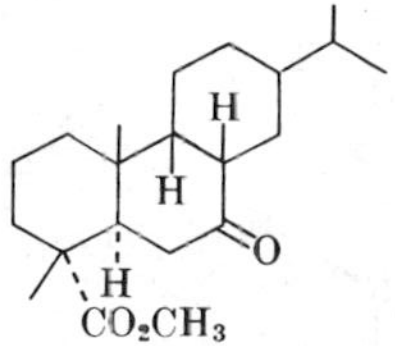

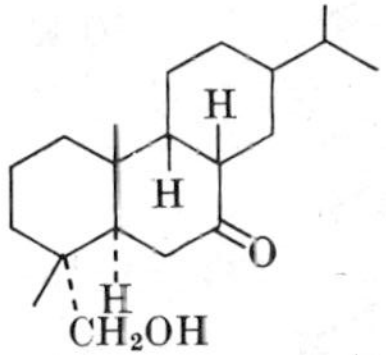

Note: References 338–543 are on pp. 253–258.

TABLE IV. REDUCTIONS OF TRICYCLIC α,β-UNSATURATED KETONES (*Continued*)

Reactant	Reduction Conditions	Quenching Agent	Product(s) (% Yield)	Refs.
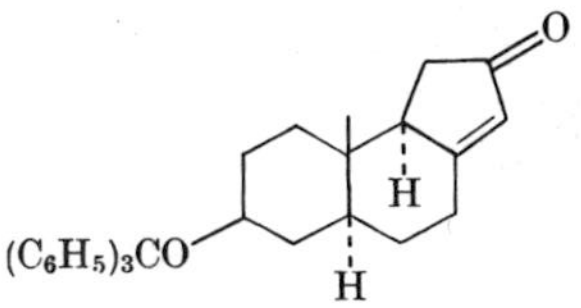 $(C_6H_5)_3CO$ [structure]	Li/NH_3	—	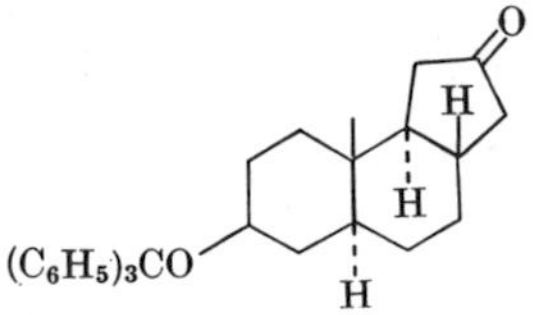 $(C_6H_5)_3CO$ [structure] (—)	372
C_{33} 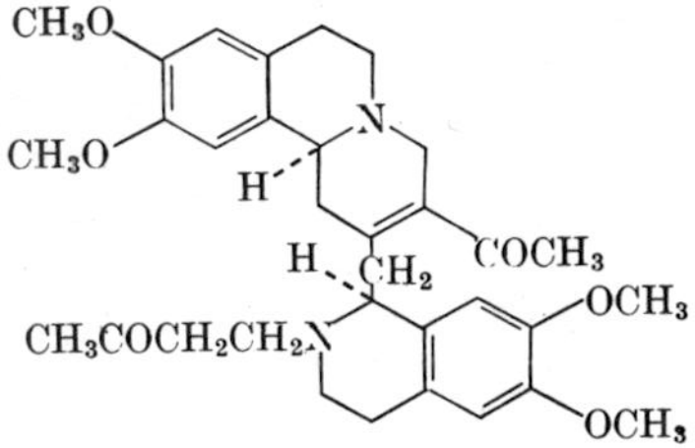 CH_3O, CH_3O, $COCH_3$, CH_2, $CH_3COCH_2CH_2N$, OCH_3, OCH_3 [structure]	Li/NH_3, THF, 30 min	Acetone, H_3O^+, heat, $HSCH_2CH_2SH$, MeOH, HCl	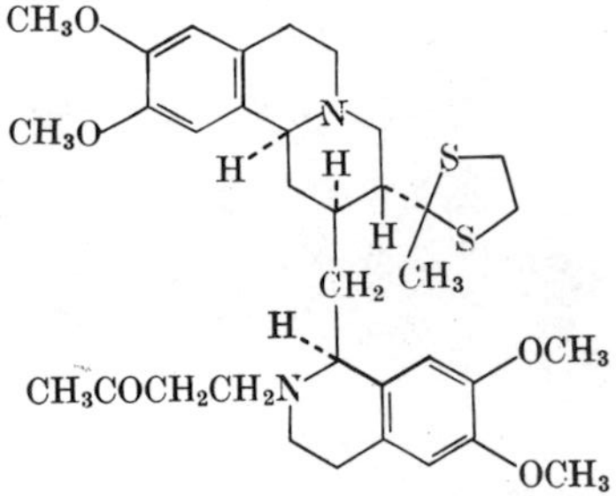 CH_3O, CH_3O, CH_2, CH_3, $CH_3COCH_2CH_2N$, OCH_3, OCH_3 [structure] (—)	432
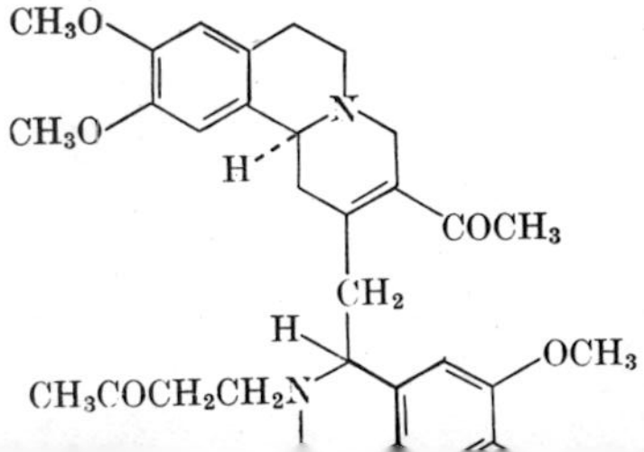 CH_3O, CH_3O, $COCH_3$, CH_2, $CH_3COCH_2CH_2N$, OCH_3 [structure]	Li/NH_3, THF 30 min	Acetone, H_3O^+, heat, $HSCH_2CH_2SH$, MeOH, HCl	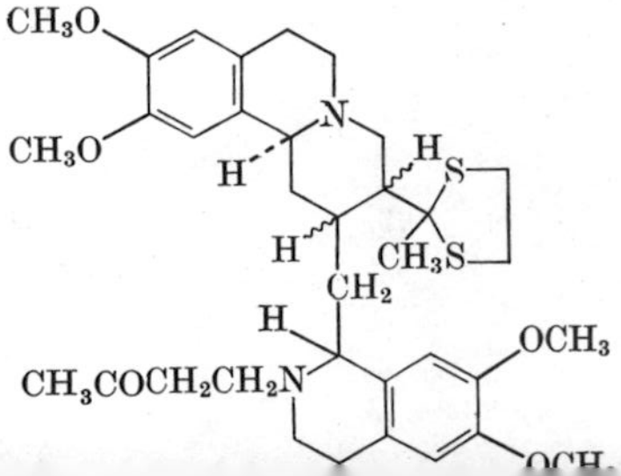 CH_3O, CH_3O, CH_3, CH_2, $CH_3COCH_2CH_2N$, OCH_3, OCH_3 [structure] (—)	432

C_{36} 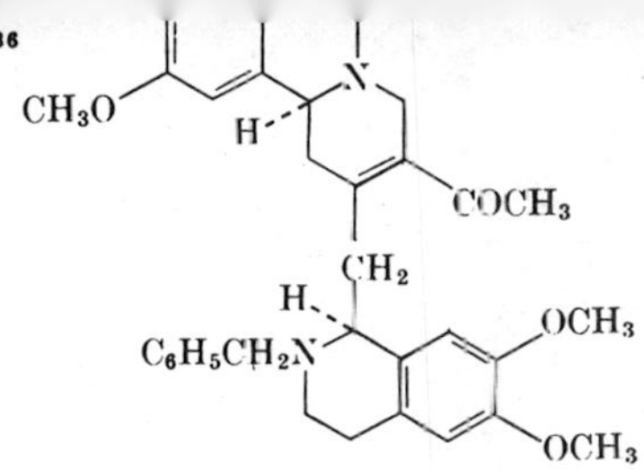	Ca/NH_3, Et_2O-C_6H_5, 10 min	Acetone, H_2O	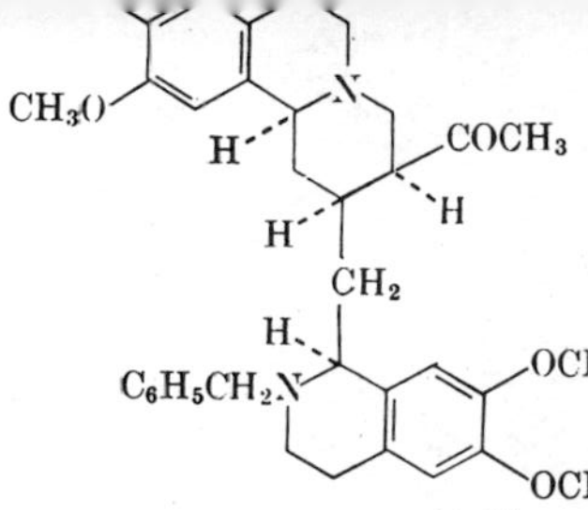 (30) + 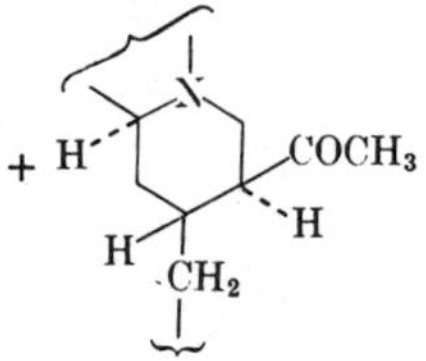 (5)	432
	Li/NH_3, THF-EtOH, 30 min	"	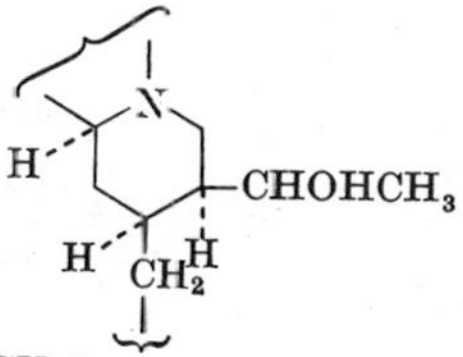 (20)	
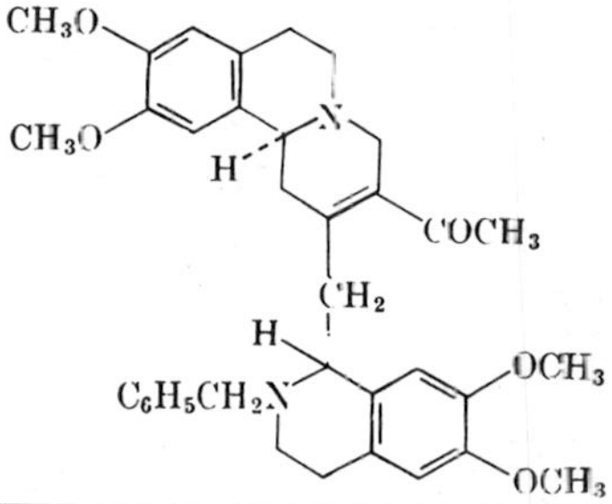	Li/NH_3, THF, 30 min	Acetone, H_3O^+, heat	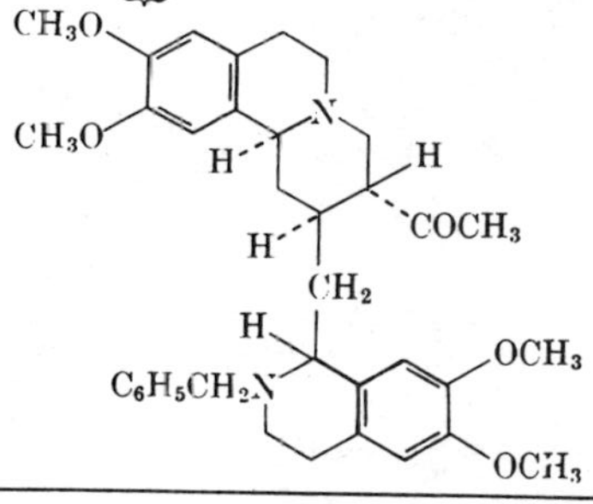 (28)	

Note: References 338–543 are on pp. 253–258

TABLE V. Reductions of Tetracyclic α,β-Unsaturated Ketones

Reactant	Reduction Conditions	Quenching Agent	Product(s) (% Yield)		Refs.
C_{15} 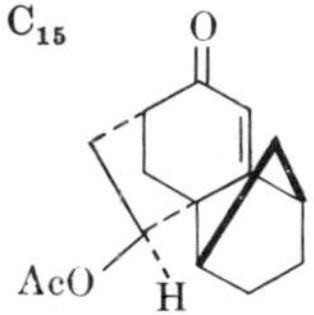	Li/NH_3	Ac_2O	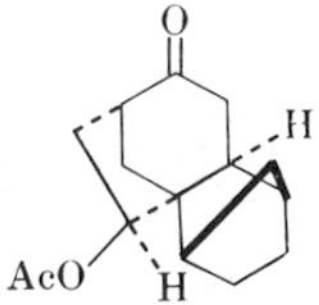	(—)	433
C_{16} 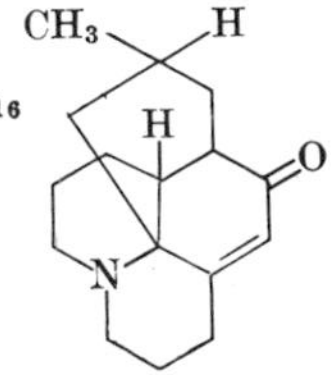	Li/NH_3	—	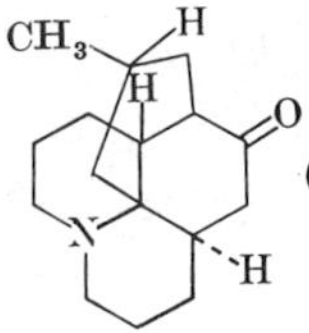 (~60)	(—)	423
C_{18} 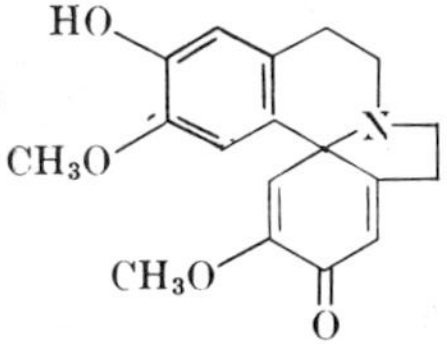	Na/NH_3	—	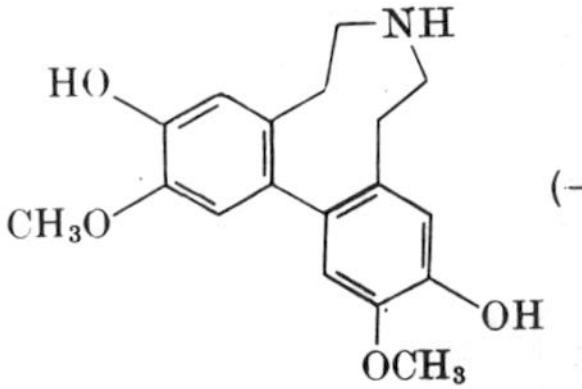	(—)	207
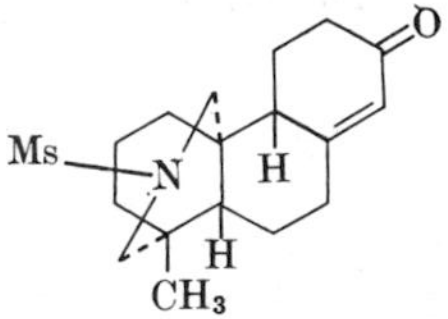	Li/NH_3, THF	NH_4Cl, MsCl/Py, CrO_3	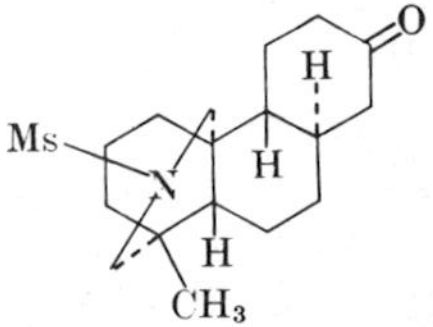 (54)		434

Compound	Reagent	Quench	Product(s) (Yield, %)	Refs.
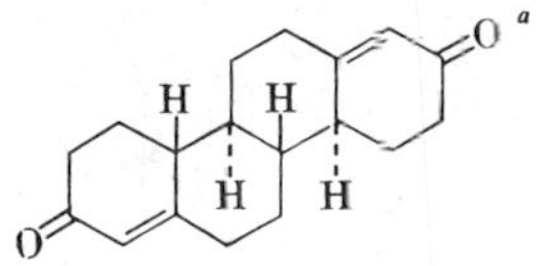 [a]	Li/NH_3	—	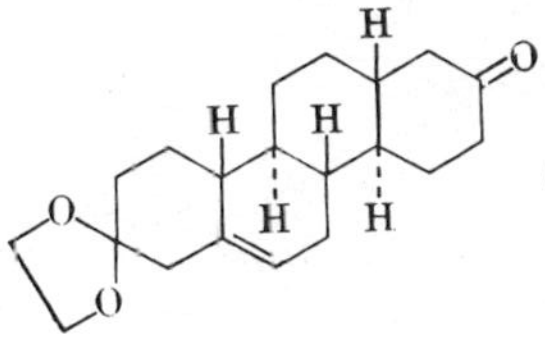 (—)	435
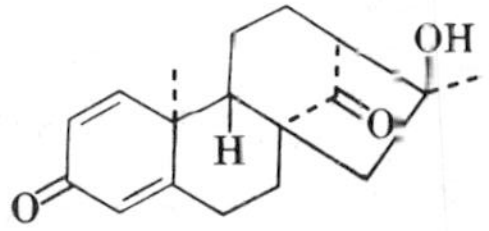	Li/NH_3	NH_4Cl	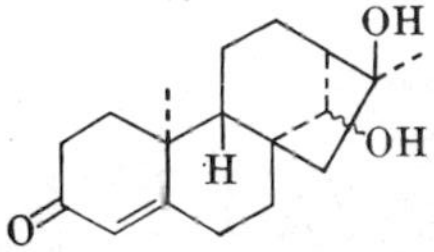 (—)	436
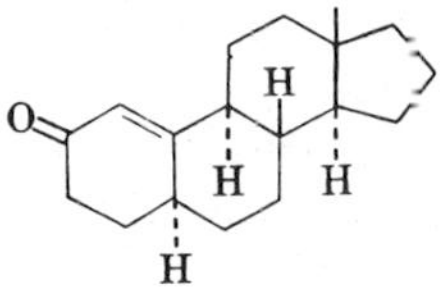	Li/NH_3, Et_2O-THF	NH_4Cl	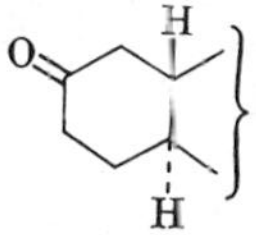 (71)	437
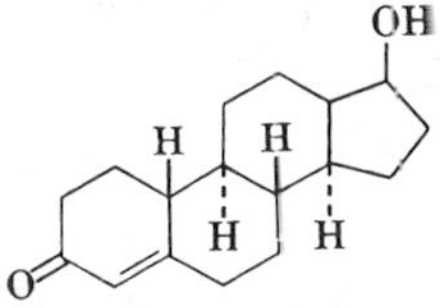	Li/NH_3, Et_2O-dioxane	NH_4Cl	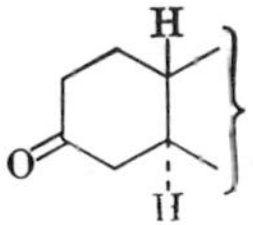 (87)	251
	Li/NH_3, Et_2O-THF	"	" + 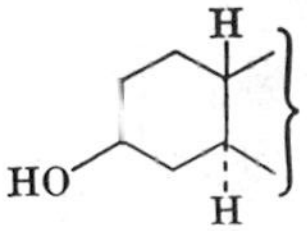 (Trace)	438
	Li/NH_3, Et_2O-dioxane, 5 min	MeOH	" (64)	251

Note: References **338–543** are on pp. 253–258.

[a] The compound was converted into its monoketal derivative before reduction.

TABLE V. REDUCTIONS OF TETRACYCLIC α,β-UNSATURATED KETONES (*Continued*)

Reactant	Reduction Conditions	Quenching Agent	Product(s) (% Yield)	Refs.
C_{18} (*Contd.*)				
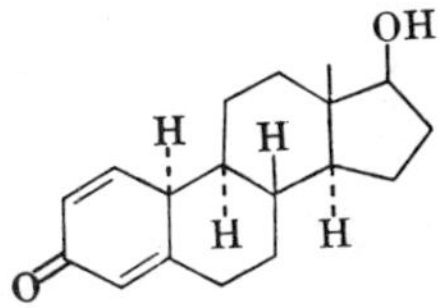	Li/NH_3, THF	NH_4Cl	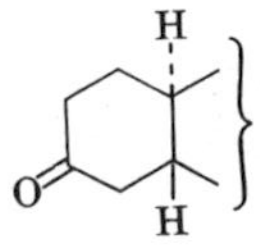 (60) 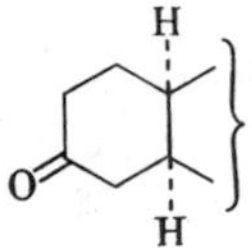 + (20)	102, 103
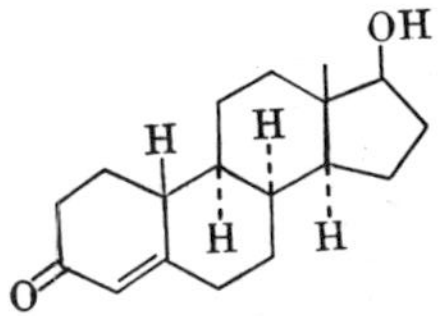	Li/NH_3, Et_2O-dioxane, 2 min, −43°	NH_4Cl	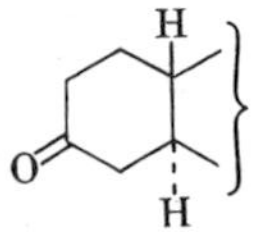 (83)	439
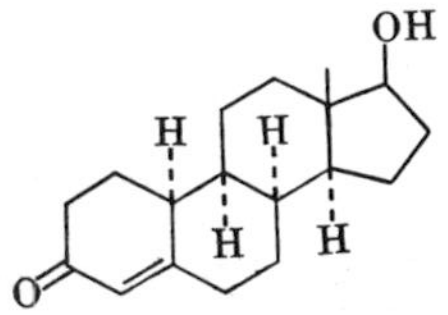	Li/NH_3, Et_2O-dioxane, 2 min, −43°	NH_4Cl	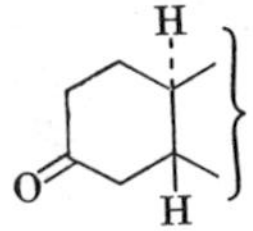 (50) 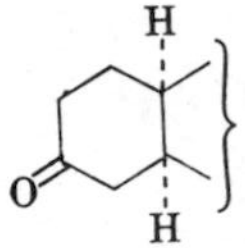 + (5–8)	439
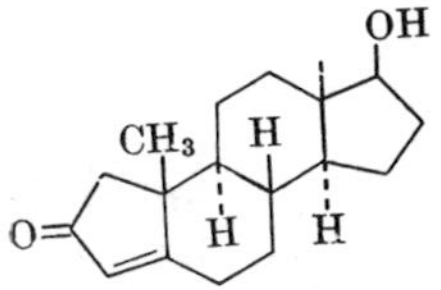	Li/NH_3, Et_2O-THF, 3 min	NH_4Cl	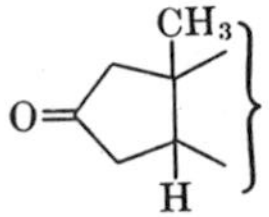 (49) 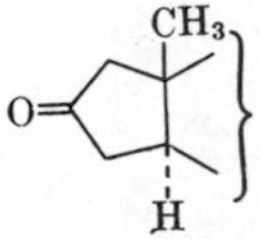 + (43)	131
	"	"	" 61 + " 39	119

Reactant	Reducing conditions	Work-up	Product(s) (Yield %)		Refs.
C_{19}					
	K/NH_3, MeOH	—		(—)	428
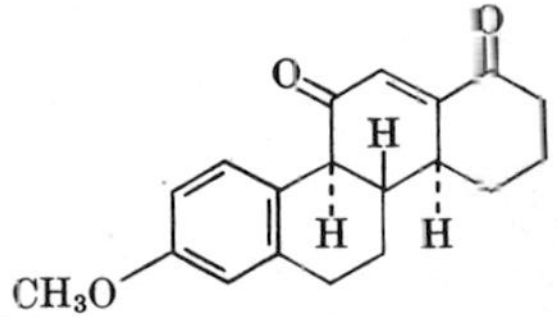	Na/NH_3, EtOH	HCl, MeOH	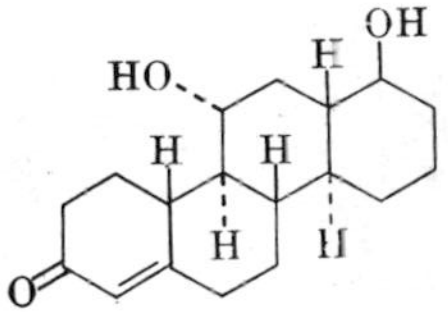	(—)	440
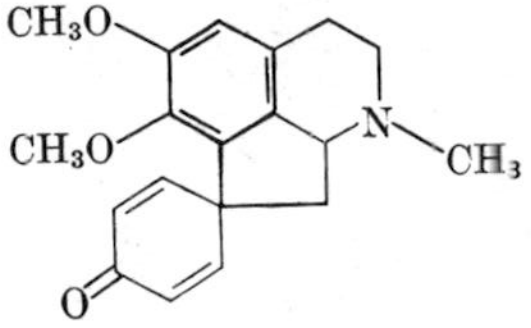	Na/NH_3	—	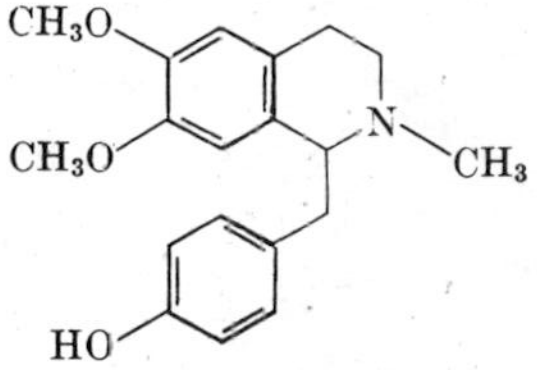	(—)	209
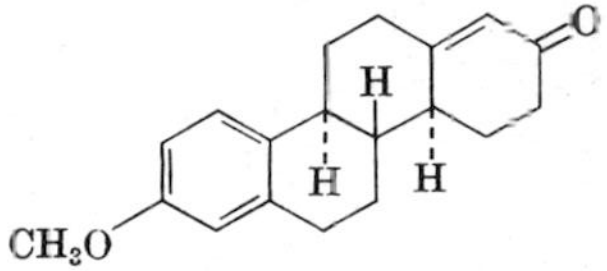	Li/NH_3	—	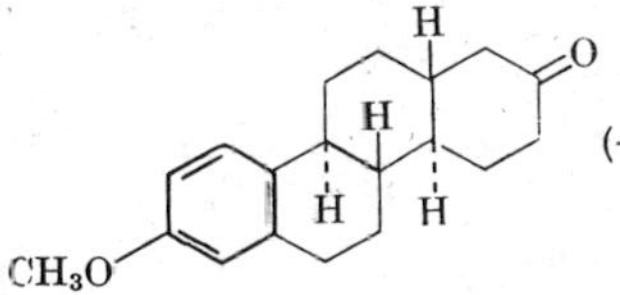	(—)	435
	Li/NH_3, Et_2O-dioxane, 5 min	NH_4Cl	"	(~70)	246, 441
	NH_3/Li, THF-*t*-BuOH, 4 hr	H_2O	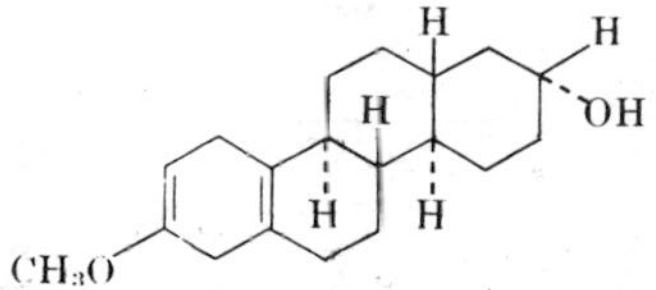	(61)	246

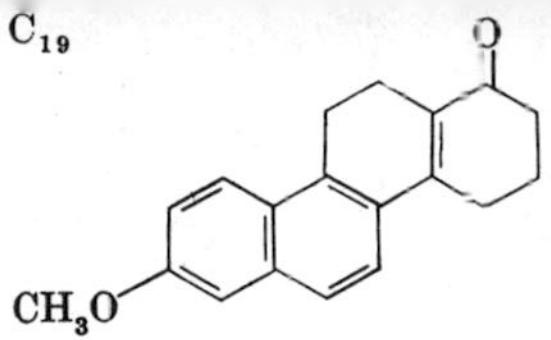

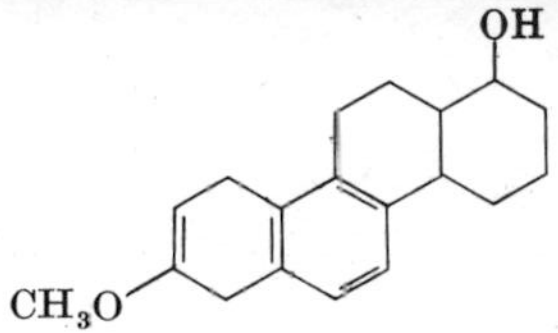

Note: References 338–543 are on pp. 253–258.

TABLE V. Reductions of Tetracyclic α,β-Unsaturated Ketones (*Continued*)

Reactant	Reduction Conditions	Quenching Agent	Product(s) (% Yield)	Refs.
C_{19} (*Contd.*)				
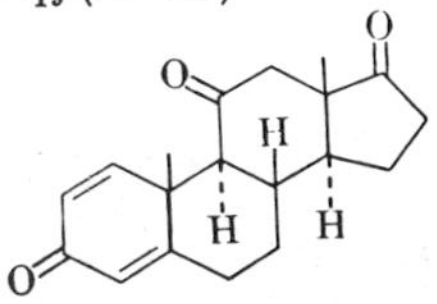	Li/NH_3	NH_4Cl	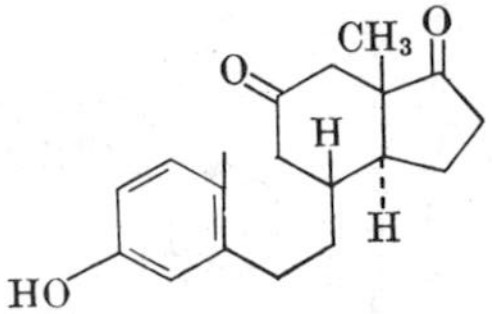 (20)	197
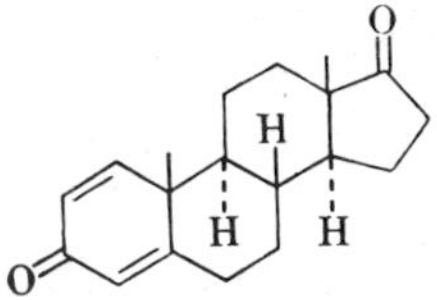	Li/NH_3, THF, −40 to −60°, 1.5 min	NH_4Cl	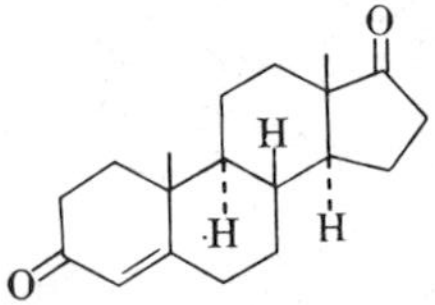 (20) + 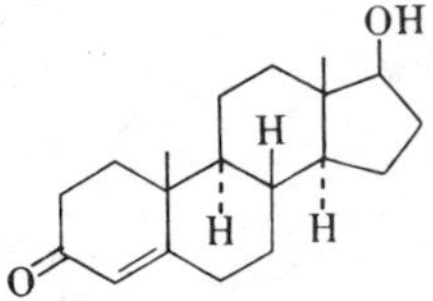 (40)	234
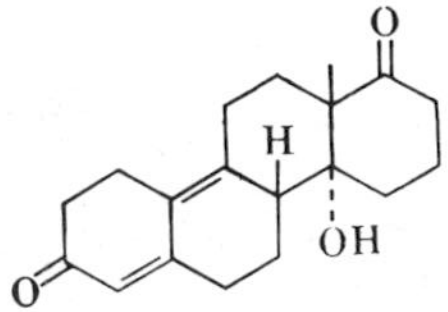	Li/NH_3, THF-dioxane, 20 min	NH_4Cl	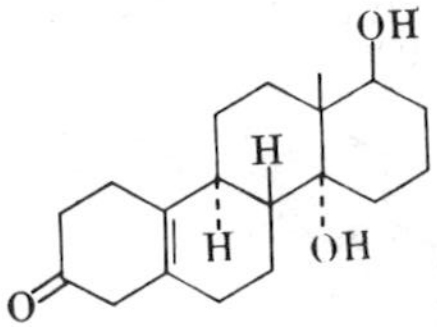 (40)	227

Substrate	Conditions	Work-up	Product(s) and Yield(s) (%)	Refs.
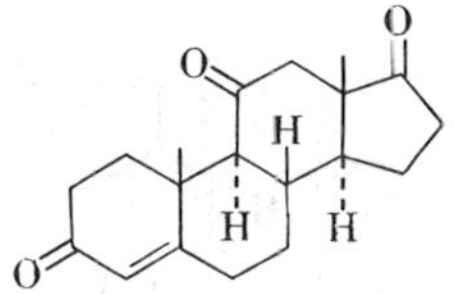	Li/NH_3, dioxane, 15 min	NH_4Cl, H_2CrO_4	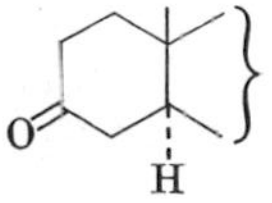 (39)	442
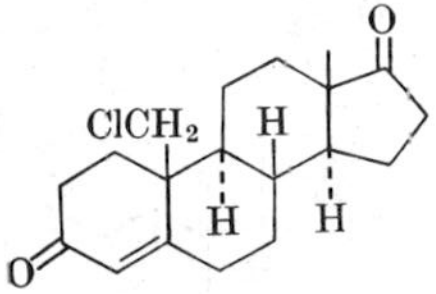	Li/NH_3, Et_2O-dioxane	NH_4Cl, CrO_3	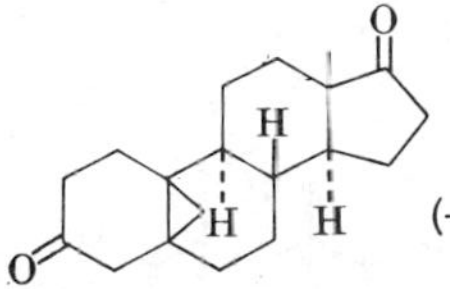 (—) + 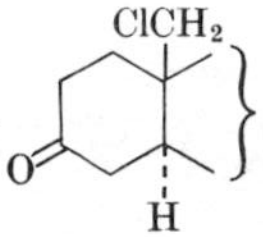 (Trace)	141
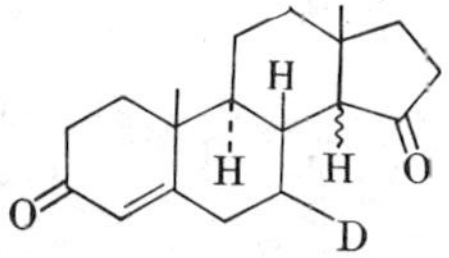	Li/NH_3, Et_2O-dioxane	NH_4Cl	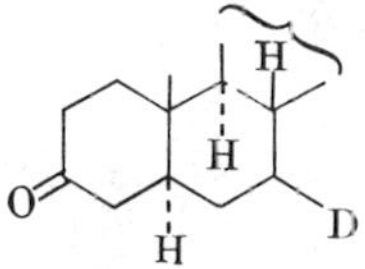 (78)	443
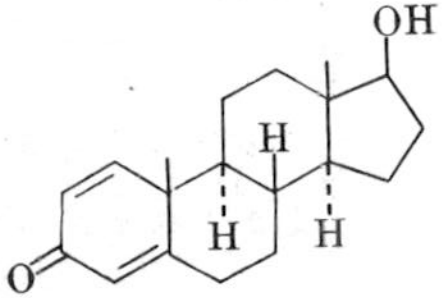	Li/NH_3, THF	NH_4Cl	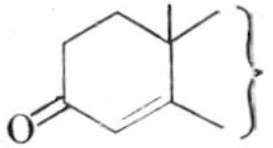 (—)	172
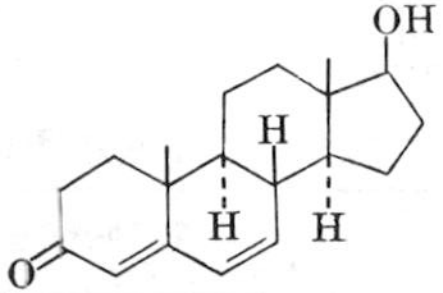	Li/NH_3, THF	NH_4Cl	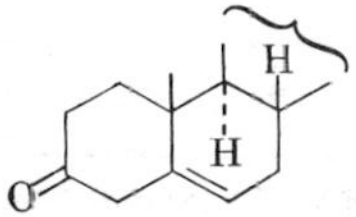 (74)	172

Note: References 338–543 are on pp. 253–258.

TABLE V. REDUCTIONS OF TETRACYCLIC α,β-UNSATURATED KETONES (*Continued*)

Reactant	Reduction Conditions	Quenching Agent	Product(s) (% Yield)	Refs.
C_{19} (*Contd.*)				
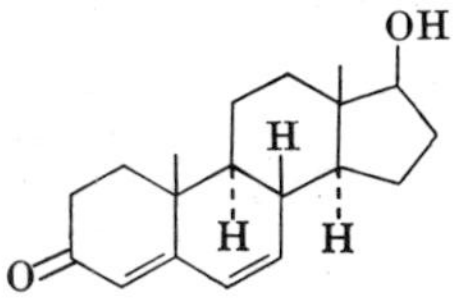 [structure]	Li/NH_3, THF	NH_4Cl, KOH	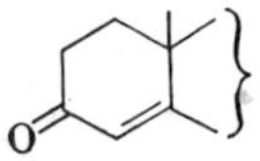 [structure] (—)	444
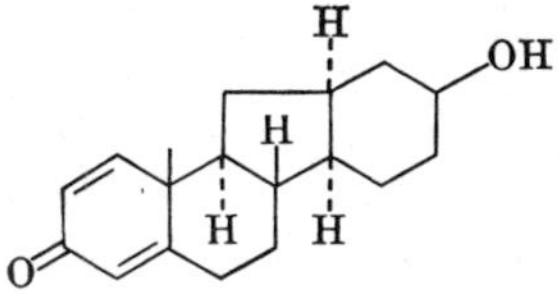 [structure]	Li/NH_3, THF, 1.6 min, −70°	NH_4Cl	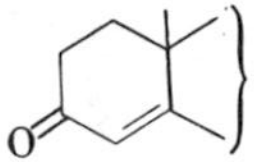 [structure] (—) + saturated ketone	233
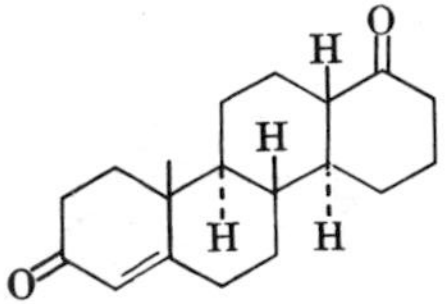 [structure]	NH_3/Li, EtOH, 20 min	H_2O, EtOH	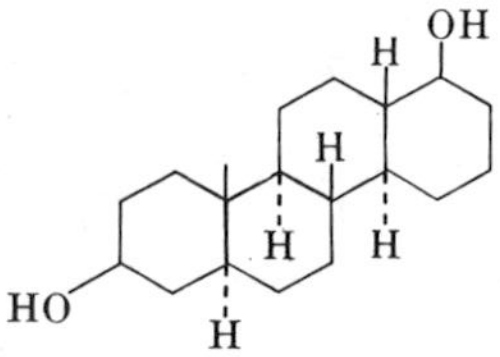 [structure] (86)	445
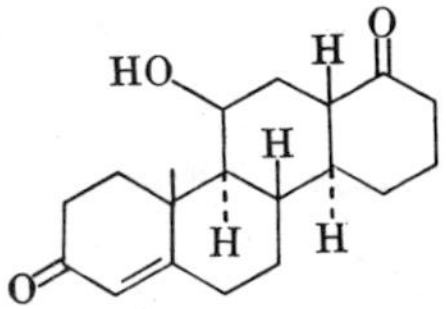 [structure]	NH_3/Li, EtOH, 30 min	H_2O, Ac_2O	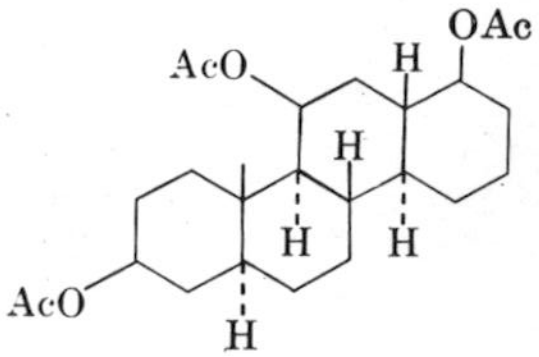 [structure] (55)	446

Substrate	Reagent	Quench	Product(s) (Yield %)		Ref.
	Li/NH_3, Et_2O-dioxane	NH_4Cl		(—)	447
	Excess Li/NH_3, Et_2O	NH_4Cl, CrO_3		(78) + starting material (~20)	448
	Li/NH_3, Et_2O, −78°	NH_4Cl, CrO_3		(70)	449
	Li/NH_3, THF	NH_4Cl, CrO_3		(—)	450
	Li/ND_3, Et_2O-dioxane, −78°, 5 min	MeOD, CrO_3		(50)	60

Note: References 338–513 are on pp. 253–258.

TABLE V. REDUCTIONS OF TETRACYCLIC α,β-UNSATURATED KETONES (*Continued*)

Reactant	Reduction Conditions	Quenching Agent	Product(s) (% Yield)	Refs.
C_{19} (*Contd.*)				
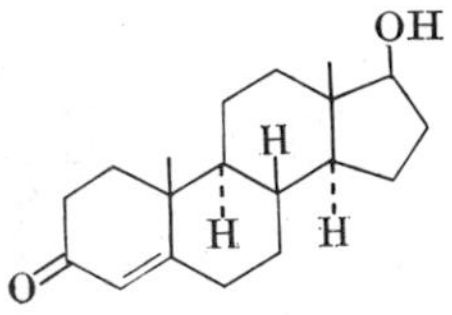	Li/NH_3, THF, 15 min	*t*-BuOH	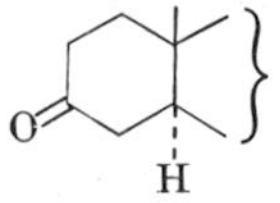 (47)	199
	Li/NH_3, Et_2O-dioxane, 40 min	NH_4Cl	" (92)	131
	"	"	" (80)	451
	"	"	" (60)	452
	Li/NH_3	—	"	453
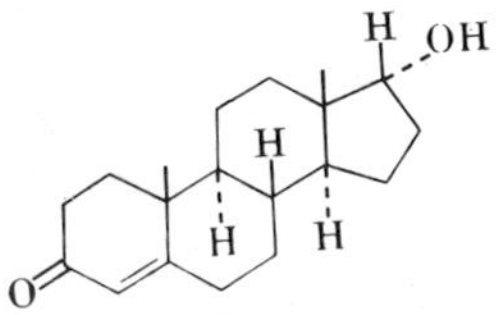	Li/NH_3, Et_2O-dioxane, 40 min	NH_4Cl	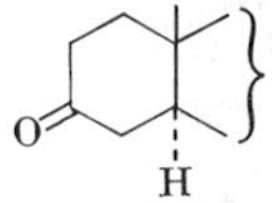 (73)	452
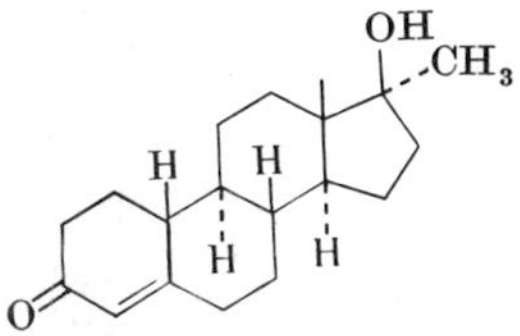	Li/NH_3, Et_2O-dioxane, 40 min	NH_4Cl	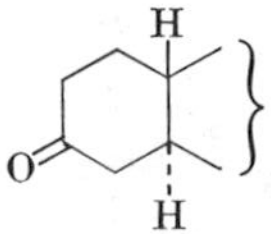 (69)	251

Substrate	Reagents	Quench	Product(s) and Yield(s) (%)	Refs.
	"	MeOH	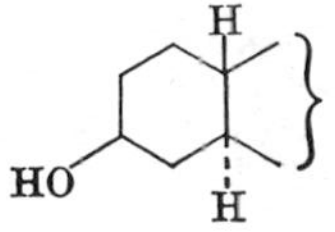 (80)	
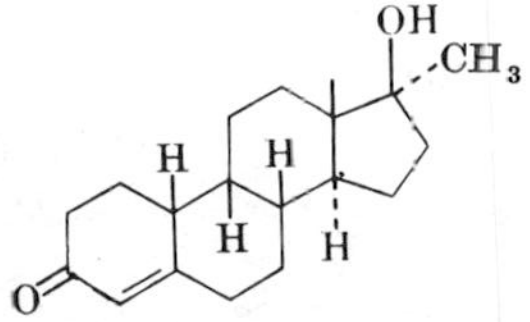	Li/NH_3, THF, 20 min	NH_4Cl	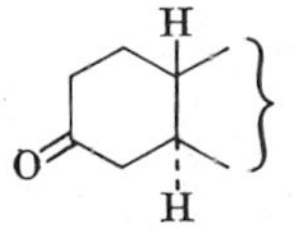 (63)	454
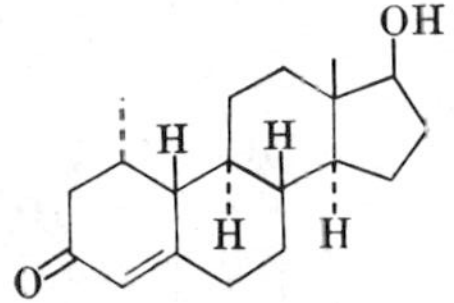	Li/NH_3, Et_2O-dioxane	NH_4Cl	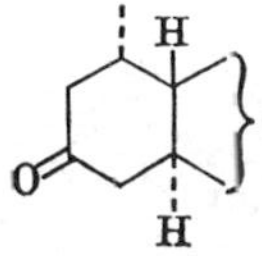 (64)	251
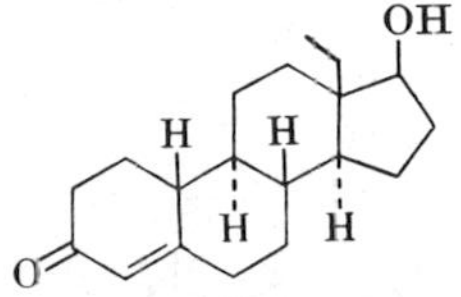	Li/NH_3, Et_2O-dioxane	NH_4Cl	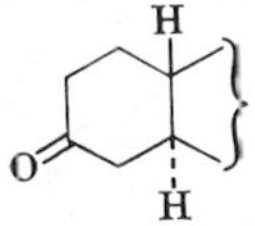 (57)	455
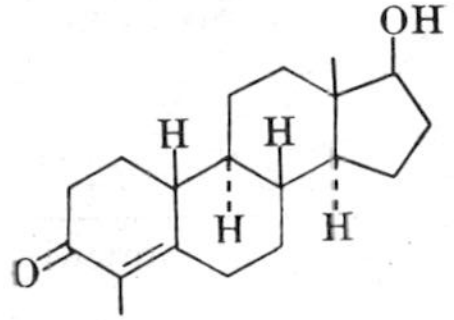	Li/NH_3, Et_2O-dioxane, 1 min	NH_4Cl	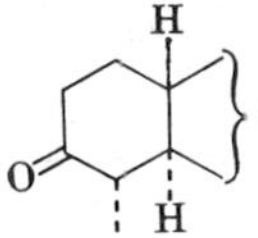 (83)	451

Note: References 338–543 are on pp. 253–258.

TABLE V. REDUCTIONS OF TETRACYCLIC α,β-UNSATURATED KETONES (*Continued*)

Reactant	Reduction Conditions	Quenching Agent	Product(s) (% Yield)	Refs.
C_{19} (*Contd.*)				
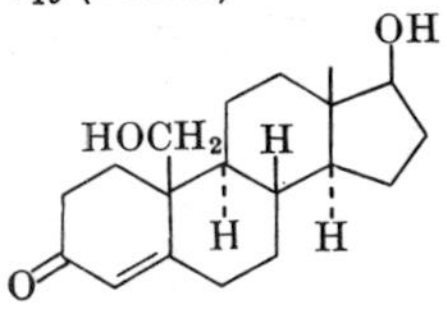 [structure: HOCH2, OH, H]	Li/NH_3, dioxane, 4 min	NH_4Cl	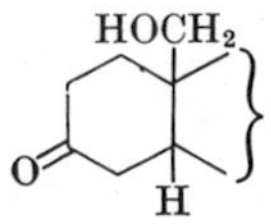 [structure: HOCH2, O, H] (53) 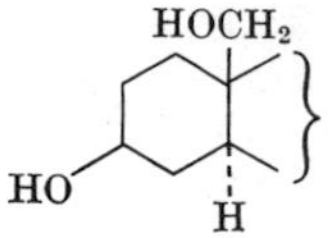 + [structure: HOCH2, HO, H] (19)	141
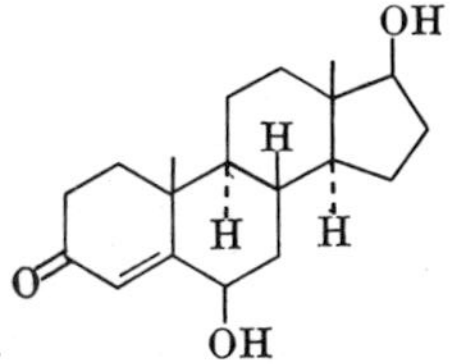 [structure: OH, H, OH]	Li/NH_3, THF, 15 min	*t*-BuOH	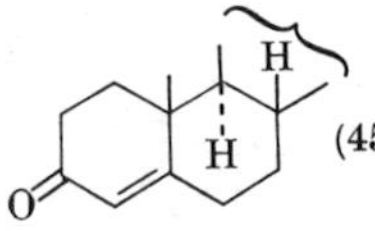 [structure: H] (45)	199
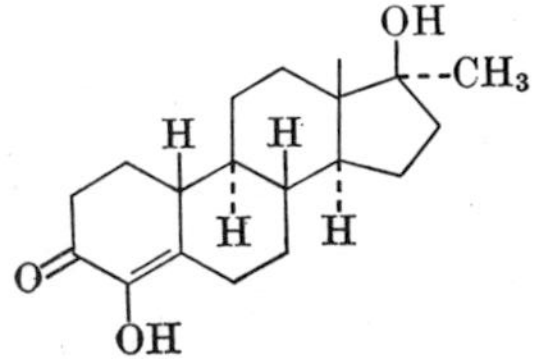 [structure: OH, --CH3, H, OH]	Li/NH_3, dioxane	NH_4Cl	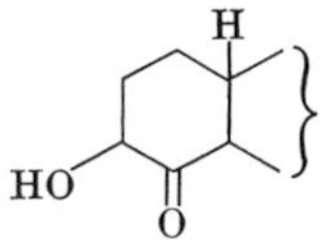 [structure: H, HO, O] (—)	456
C_{20}				
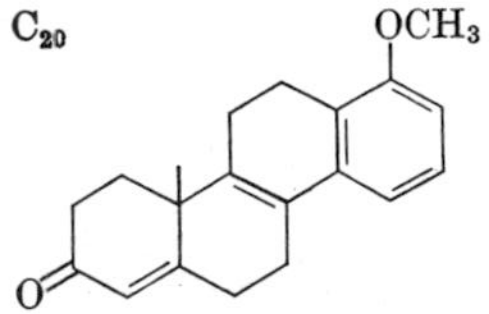 [structure: OCH3]	NH_3/excess Li,[b] EtOH-dioxane, 3 hr	EtOH, H_2O, HCl	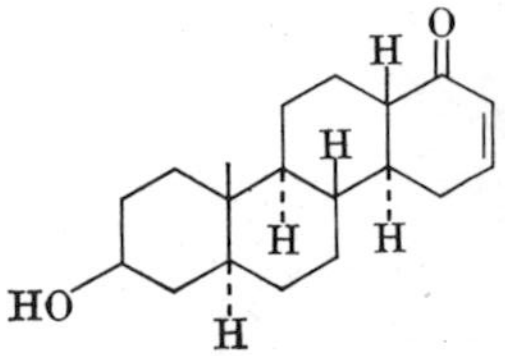 [structure: O, H, HO] (—) +	

Reactant	Conditions	Workup	Product(s) (Yield, %)	Refs.
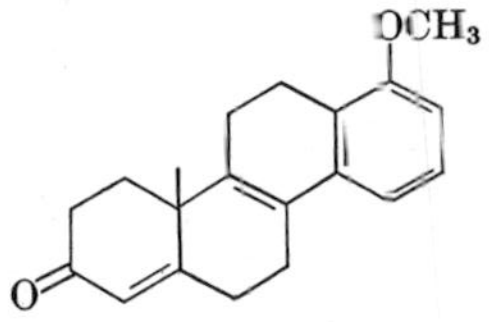 OCH_3 (steroid dienone)			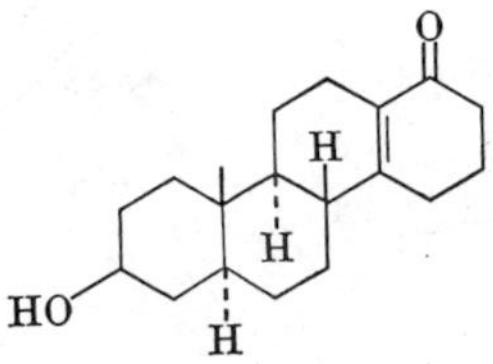 HO … H (43)	457, 458
	NH_3/excess Li, EtOH-dimethylcellosolve	H_2O, isopropenyl acetate, PTSA	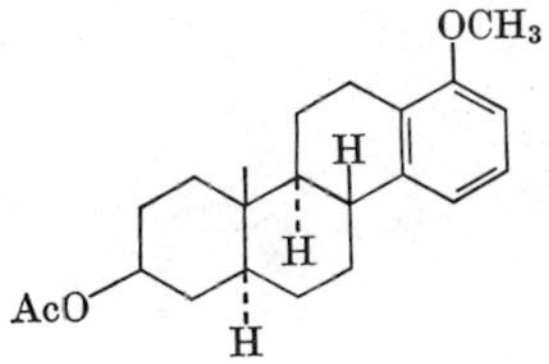 AcO … OCH_3 (37)	243
	NH_3/excess Na, EtOH-dimethylcellosolve	"	" (43)	
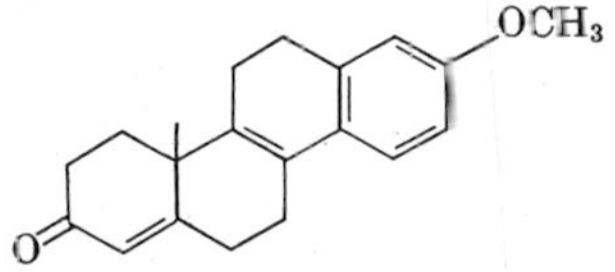 OCH_3 (steroid dienone)	Excess Li/NH_3, ether-dioxane, 2 hr, EtOH, 5 min	EtOH	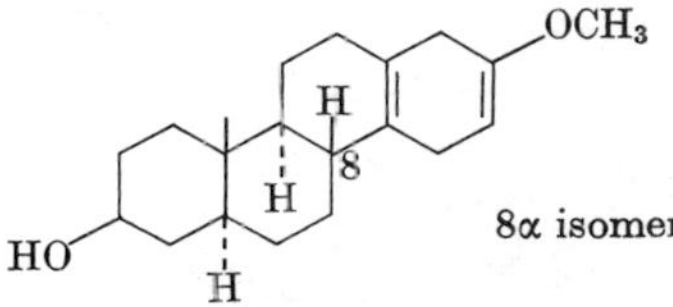 HO … OCH_3 (62) + 8α isomer (20)	242
	Limited Li/NH_3, EtOH-ether 1.5 hr	EtOH, H_3O^+	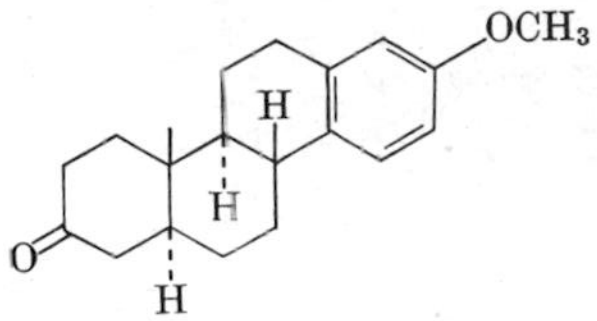 O … OCH_3 (~6) +	

Note: References 338–543 are on pp. 253–258.

[b] Sufficient lithium to develop a bronze phase was added.

TABLE V. REDUCTIONS OF TETRACYCLIC α,β-UNSATURATED KETONES (*Continued*)

Reactant	Reduction Conditions	Quenching Agent	Product(s) (% Yield)	Refs.
C_{20} (*Contd.*)				
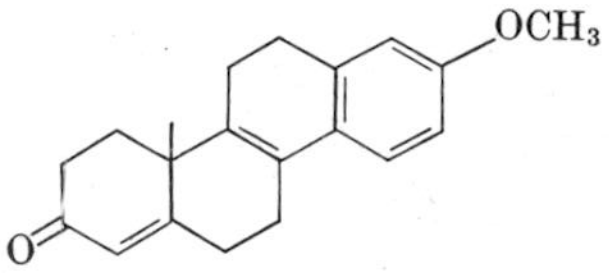 OCH$_3$	Limited Li/NH$_3$, EtOH-ether 1.5 hr	EtOH, H_3O^+	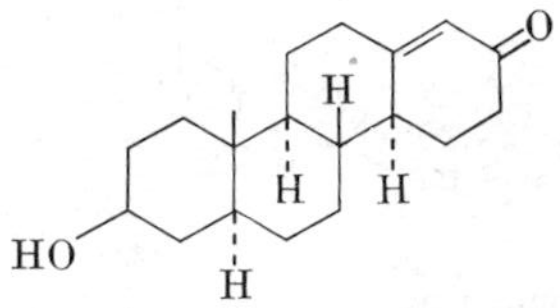 O HO H H H H (~ 14) +	
			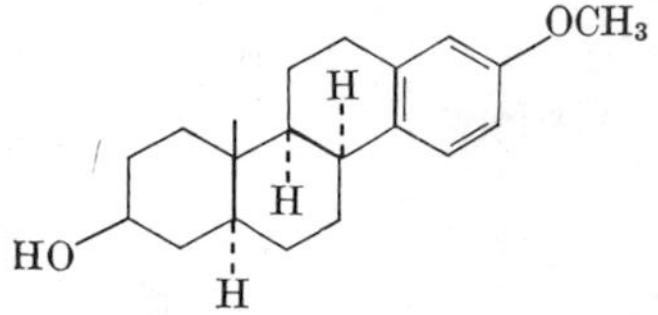 OCH$_3$ HO H H H (~ 9) +	
			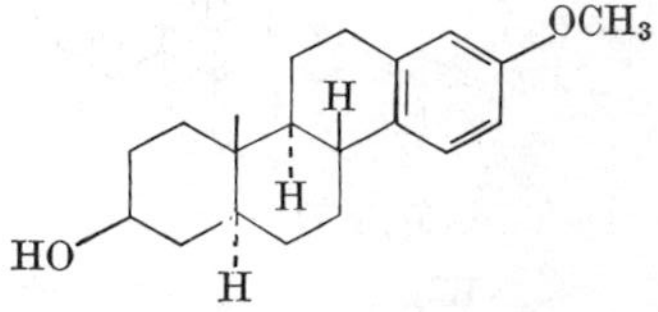 OCH$_3$ HO H H H (~ 8) +	
			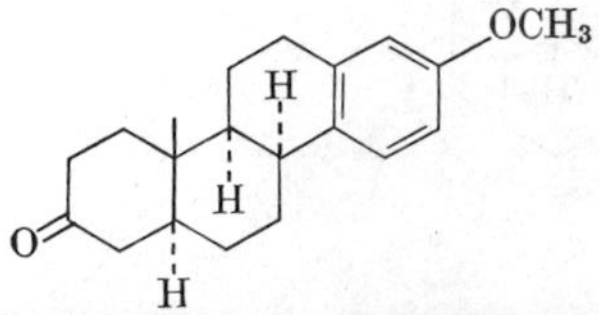 OCH$_3$ O H H H (Trace)	

Substrate	Conditions	Quench/Workup	Product(s) and Yield(s) (%)	Refs.
OCH3	2 eq. Li/NH_3, dioxane-Et_2O, −70°	NH_4Cl	OCH_3, H (53) + OH, H (6) + starting material (∼7)	242
H, O	Li/NH_3, *t*-BuOH	—	H, O, H (20)	459
OCH_3, H, H	NH_3/Excess Li, EtOH, 35 min	H_2O, isopropenyl acetate, PTSA	OCH_3, H, H, AcO, H (79)	460
H, $COCH_3$, H, H, H, CH_3O	Li/NH_3, EtOH-Et_2O	H_3O^+, CrO_3	H, $COCH_3$, H, H, H, H (−)	441

Note: References **338–543** are on pp. 253–258.

TABLE V. REDUCTIONS OF TETRACYCLIC α,β-UNSATURATED KETONES (*Continued*)

Reactant	Reduction Conditions	Quenching Agent	Product(s) (% Yield)	Refs.
C_{20} (*Contd.*)				
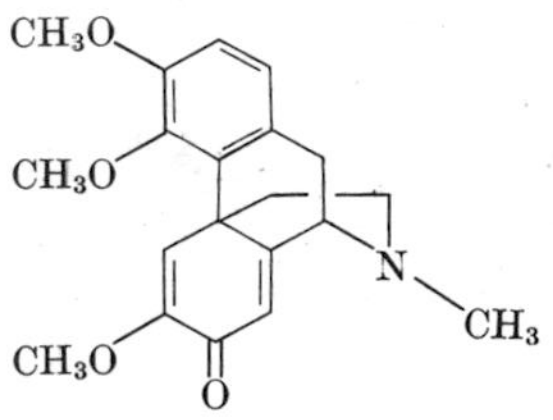 [structure: CH_3O, CH_3O, CH_3O, O, N–CH_3]	NH_3/Na, THF, 30 min	EtOH, CH_2N_2	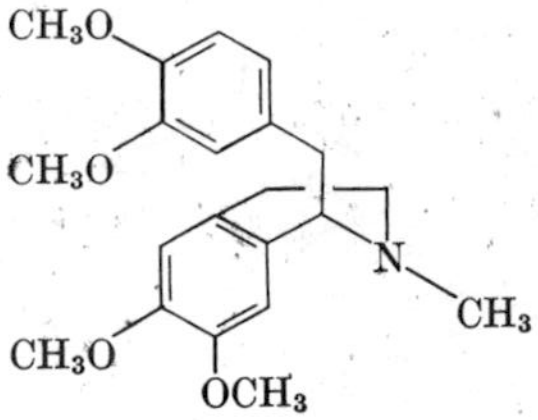 [structure: CH_3O, CH_3O, CH_3O, OCH_3, N–CH_3] (—)	206
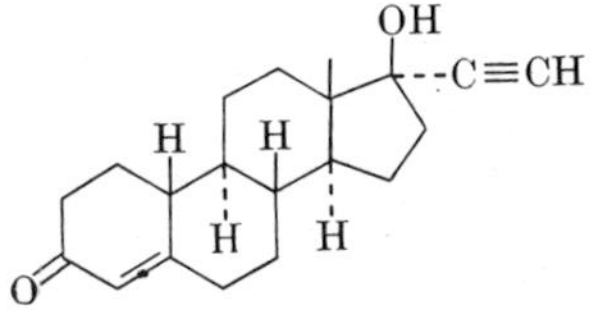 [structure: OH, C≡CH, H, H, H, H, O]	Li/NH_3, Et_2O-dioxane	NH_4Cl	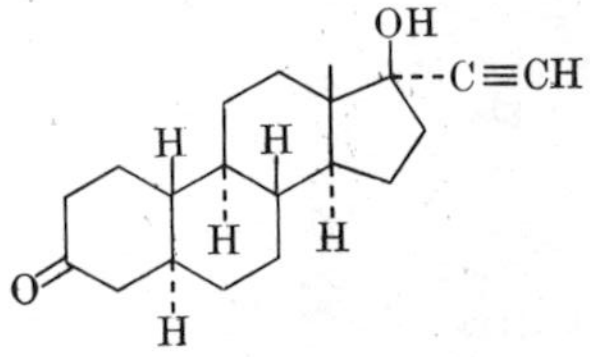 [structure: OH, C≡CH, H, H, H, H, H, O] (63)	251
	2 eq. Li/NH_3, $C_6H_5CH_3$-THF, 8 min	$BrCH_2CH_2Br$, HOAc-MeOH	" (70)	19
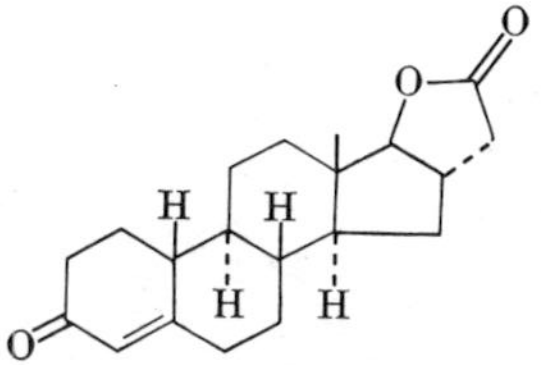 [structure: lactone O, O, H, H, H, H, O]	Li/NH_3, THF-*t*-BuOH	H_2O, CrO_3	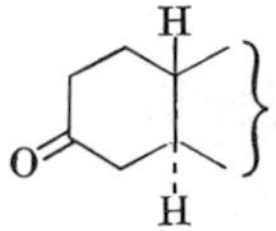 [structure: H, H, O] (—)	461

Substrate	Reagents	Workup	Product (Yield %)	Refs.
	Li/NH_3, EtOH	HCl, H_2O	(36)	136
	Li/NH_3, THF	$NaNO_2$, H_2O	(60)	462
	Li/NH_3, Et_2O-dioxane	NH_4Cl	(57)	251
	Li/NH_3, Et_2O	NH_4Cl, CrO_3	(60)	463

Note: References 338–543 are on pp. 253–258.

TABLE V. Reductions of Tetracyclic α,β-Unsaturated Ketones (*Continued*)

Reactant	Reduction Conditions	Quenching Agent	Product(s) (% Yield)	Refs.
C_{20} (*Contd.*)				
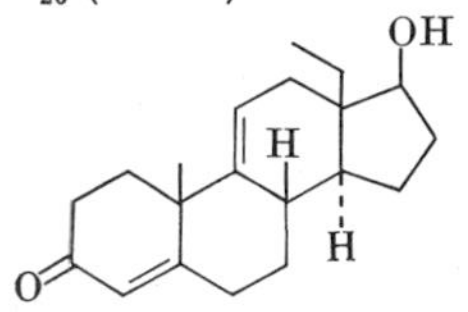	Li/NH_3, Et_2O-dioxane	NH_4Cl	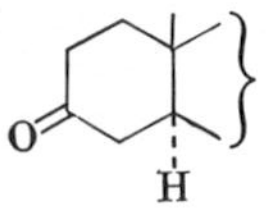 (—)	464
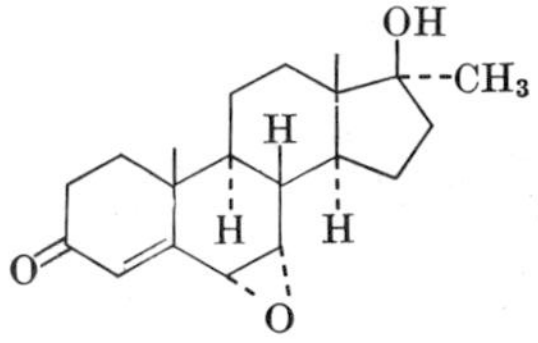	Li/NH_3, THF	—	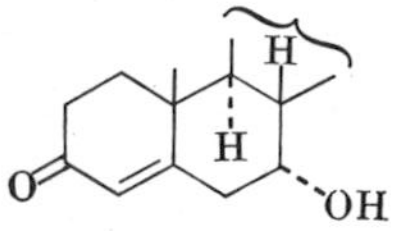 (66)	205
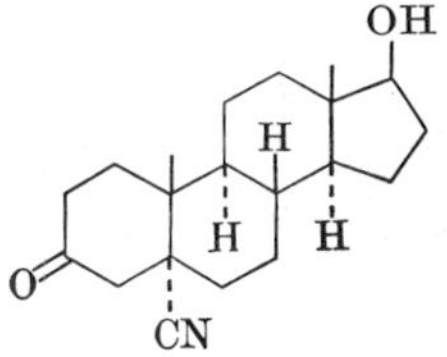	Li/NH_3, dioxane	MeOH	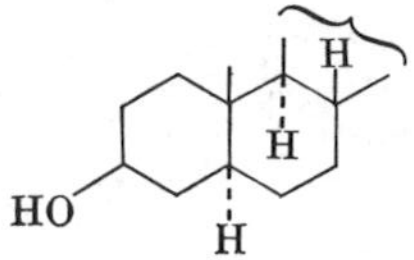 (48)	465
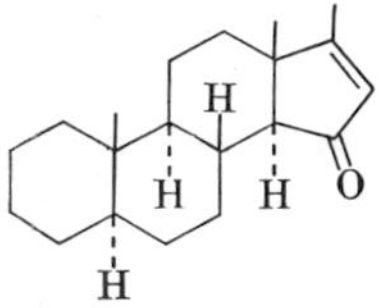	Li/NH_3, dioxane	NH_4Cl	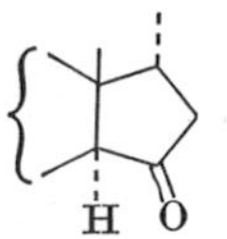 (—)	466

Substrate	Reducing Agent	Proton Source	Product(s) and Yield(s) (%)	Refs.
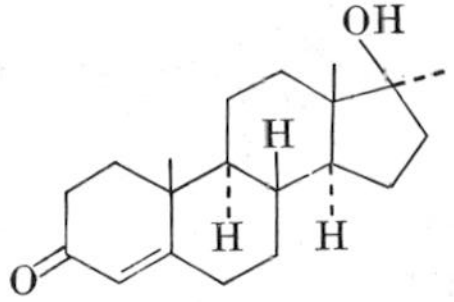 OH, H, H, H	Li/NH_3	Wolff-Kishner reduction	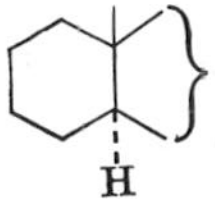 H (50)	467
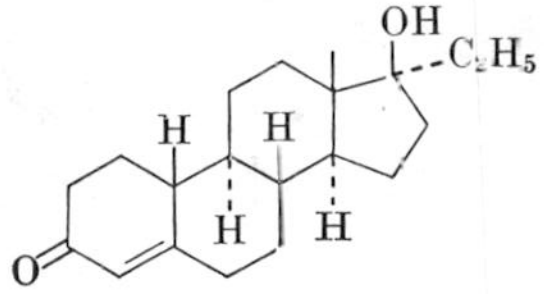 OH, C_2H_5, H, H, H, H	Li/NH_3, Et_2O-dioxane	NH_4Cl	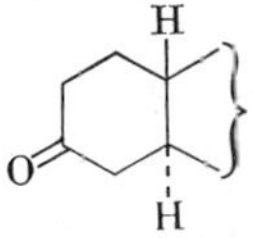 H, H, O (60)	251
	2 eq. Li/NH_3, $C_6H_5CH_3$-THF, 8 min	$BrCH_2CH_2Br$, HOAc-MeOH	" (85)	19
	2 eq. Li/NH_3, $C_6H_5CH_3$-THF, 80 min	"	" (76)	
	2 eq. Na/NH_3, $C_6H_5CH_3$-THF, 8 min	"	" (79)	
	2 eq. Na/NH_3, $C_6H_5CH_3$-THF, 80 min	"	" (77)	
	Li/NH_3, Et_2O-dioxane	MeOH	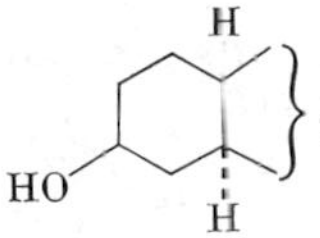 H, HO, H (100)	
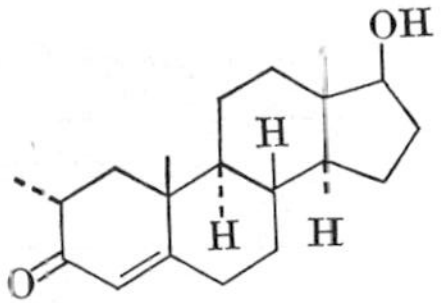 OH, H, H, H	Li/NH_3, Et_2O-dioxane, 2 min	NH_4Cl	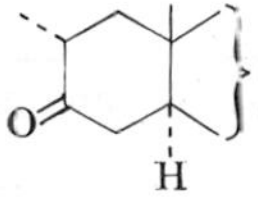 O, H (82)	468

Note: References 338–543 are on pp. 253–258.

TABLE V. Reductions of Tetracyclic α,β-Unsaturated Ketones (*Continued*)

Reactant	Reduction Conditions	Quenching Agent	Product(s) (% Yield)	Refs.
C_{20} (*Contd.*)				
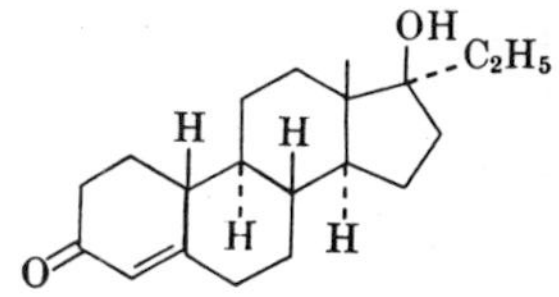 OH, C_2H_5, H, H, H, H	Li/NH_3, THF	NH_4Cl	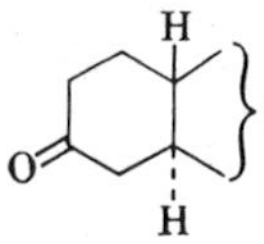 H, H (63)	463
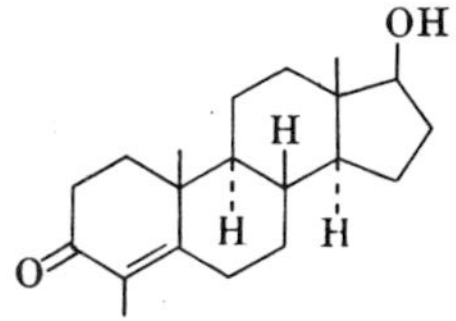 OH, H, H, H	Li/NH_3, Et_2O-dioxane, 1 min	NH_4Cl	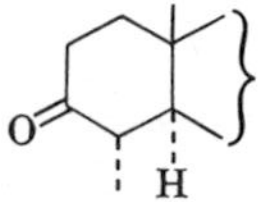 H (80)	451
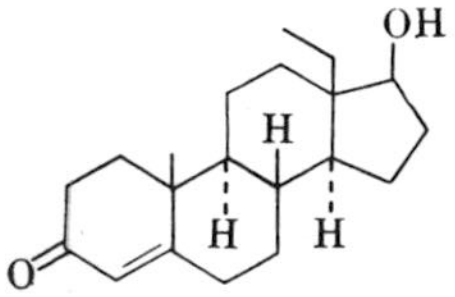 OH, H, H, H	Li/NH_3, Et_2O-dioxane	NH_4Cl	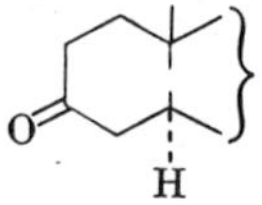 H (—)	464
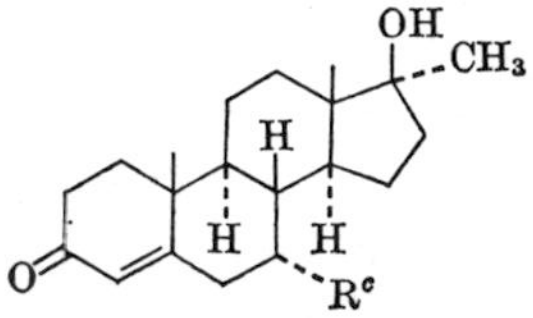 OH, CH_3, H, H, H, R^c	Li/NH_3, Et_2O-dioxane, 40 min	NH_4Cl	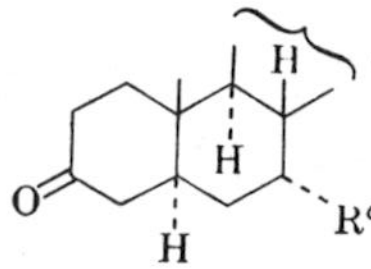 H, H, H, R^c (—)	256

R = SCH_3	"	"	R = SCH_3 (60–70)	
R = SC_2H_5	"	"	R = SC_2H_5 (60–70)	
R = SPr-*n*	"	"	R = SPr-*n* (60–70)	
R = SPr-*i*	"	"	R = SPr-*i* (60–70)	
R = SBu-*n*	"	"	R = SBu-*n* (60–70)	
R = SBu-*t*	"	"	R = SBu-*t* (60–70)	
R = $SCH_2CH{=}CH_2$	"	"	R = SH (30)	
R = $SCH_2C_6H_5$	"	"	R = SH (30)	
R = SC_6H_5	"	"	R = H (—)	
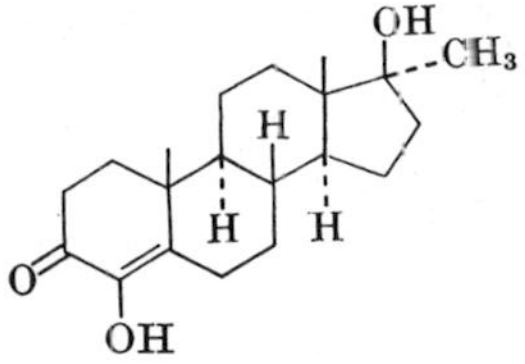	Li/NH_3, Et_2O-dioxane	"	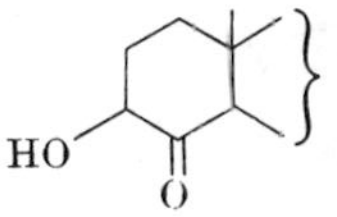 (—)	456
C_{21}	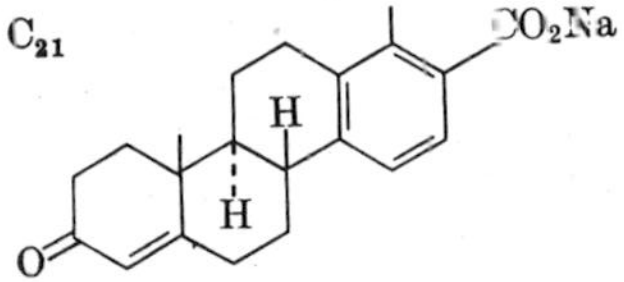 Li/NH_3	CH_2N_2	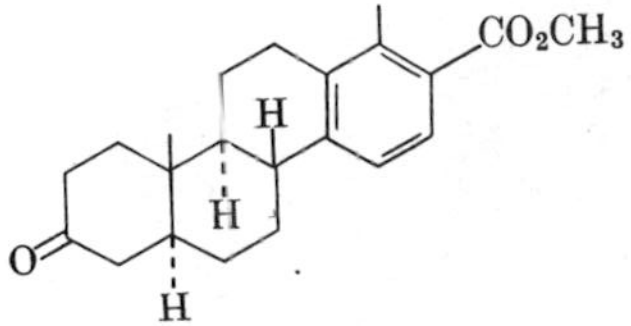 (—)	247
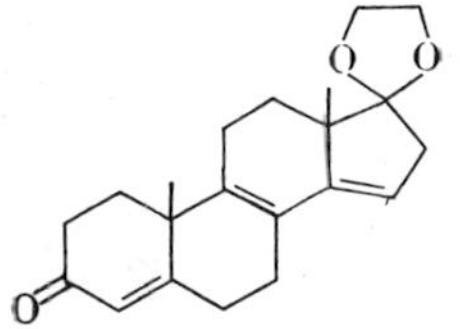	Li/NH_3, EtOH	CrO_3, H_3O^+	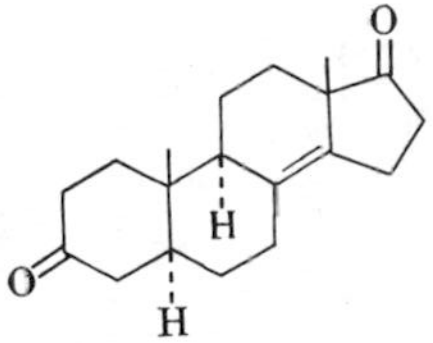 (—)	469

Note: References **338–543** are on pp. 253–258.

[c] The thioethers are not listed in order of molecular formula.

TABLE V. Reductions of Tetracyclic α,β-Unsaturated Ketones (*Continued*)

Reactant	Reduction Conditions	Quenching Agent	Product(s) (% Yield)	Refs.
C_{21} (*Contd.*)				
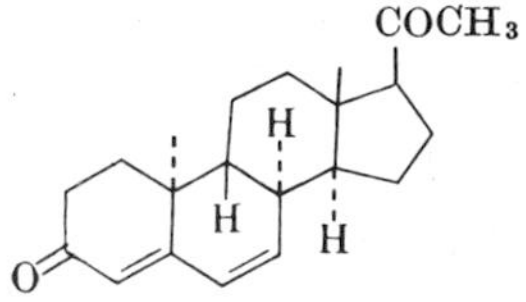 $COCH_3$	Li/NH_3, THF	H_2O	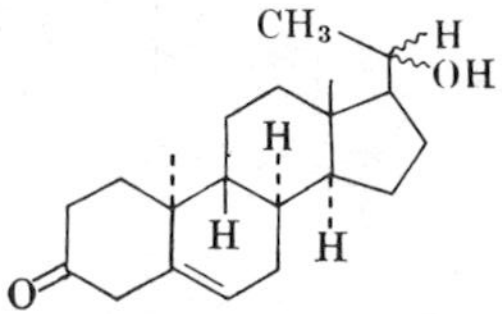 CH_3, H, OH (95)	223
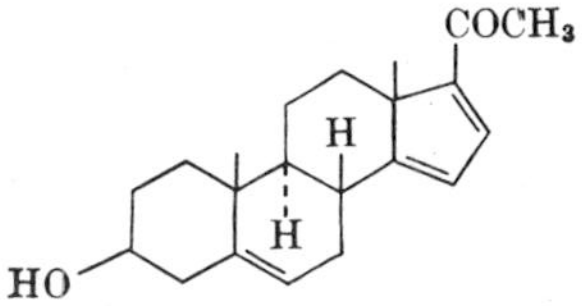 $COCH_3$, HO	Li/NH_3, Et_2O-dioxane, −45°, *n*-PrOH, 2.5 hr	H_2O	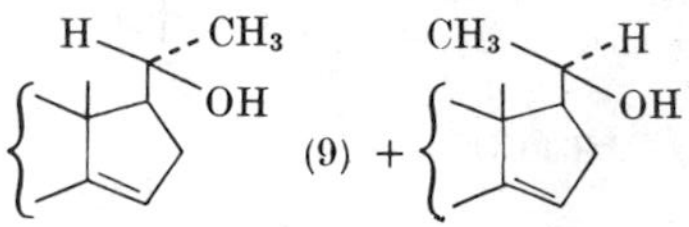 H, CH_3, OH (9) + CH_3, H, OH (22)	232
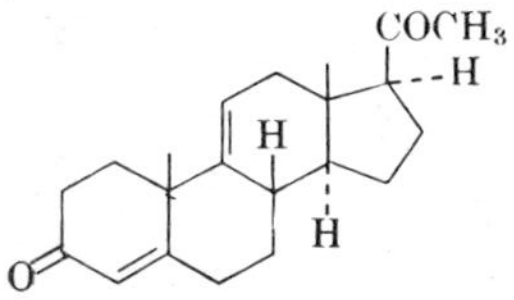 $COCH_3$, H	Li/NH_3, DME, 30 min, EtOH, 90 min	EtOH, Ac_2O/Py	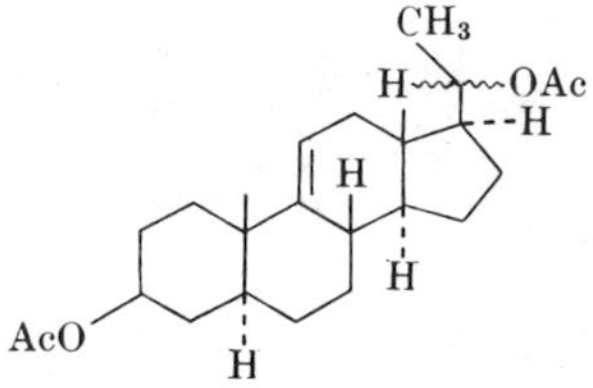 CH_3, H, OAc, AcO (—)	470
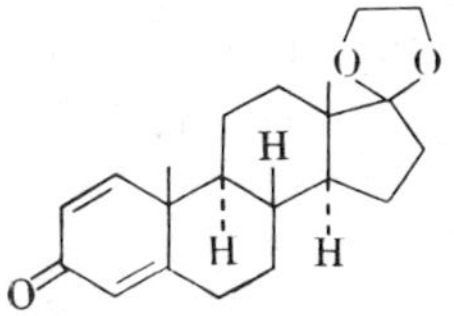 O, O	2 eq. Li/NH_3, $C_6H_5CH_3$-THF, 8 min	$BrCH_2CH_2Br$, HOAc-MeOH	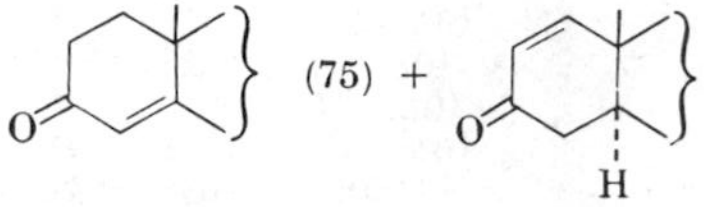 (75) + (10)	19

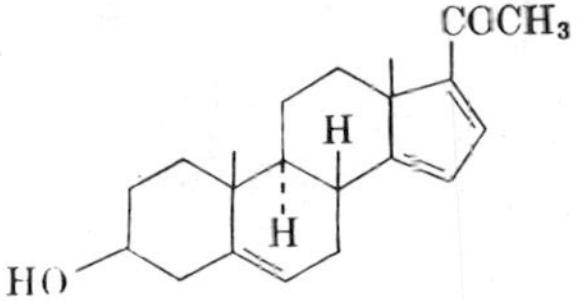 COCH$_3$, H, H, HO	Li/NH$_3$, Et$_2$O	NH$_4$Cl	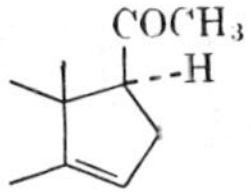 COCH$_3$, H (26) 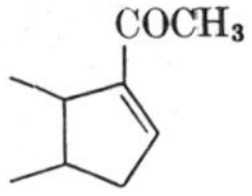 + COCH$_3$ (43)	230
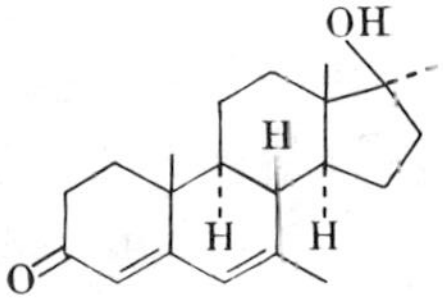 OH, H, H, H, O	Li/NH$_3$, THF, 15 min	H$_2$O	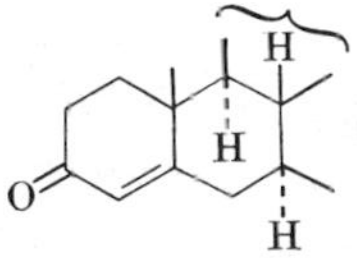 H, H, H, O (16)	218
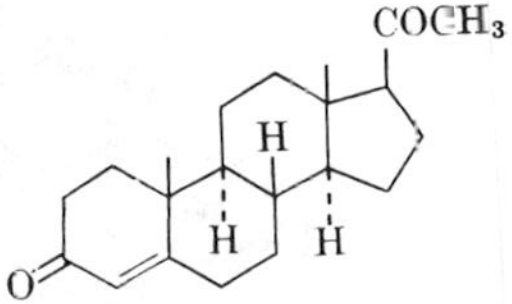 COCH$_3$, H, H, H, O	Li/NH$_3$, EtOH, 30 min, n-PrOH, Li, 30 min	H$_2$O, CrO$_3$	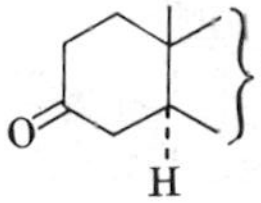 O, H (42)	232
	"	"	" (30)	471
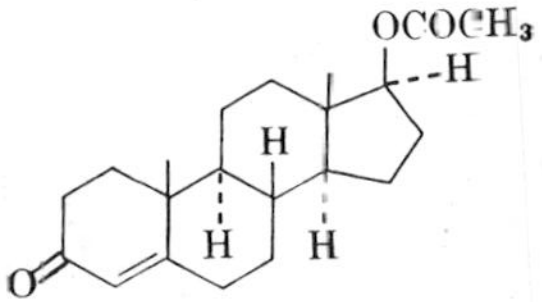 OCOCH$_3$, H, H, H, H, O	NH$_3$/Li, Et$_2$O	NH$_4$Cl	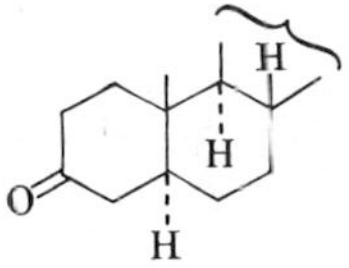 H, H, O, H (86)	472

Note: References 338–543 are on pp. 253–258.

TABLE V. REDUCTIONS OF TETRACYCLIC α,β-UNSATURATED KETONES (*Continued*)

Reactant	Reduction Conditions	Quenching Agent	Product(s) (% Yield)	Refs.
C_{21} (*Contd.*)				
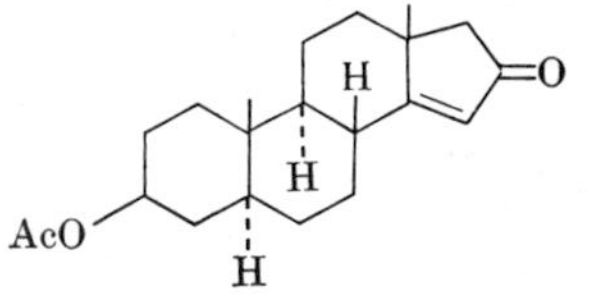 AcO ... =O	Li/NH_3	—	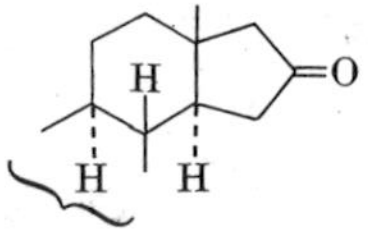 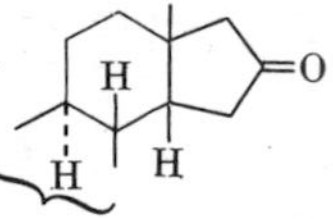 =O 10 + =O 90	119
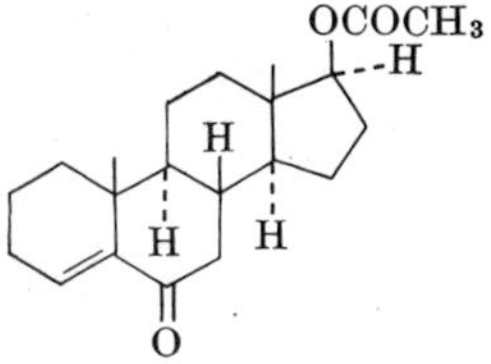 $OCOCH_3$	Li/NH_3, Et_2O	NH_4Cl, Ac_2O	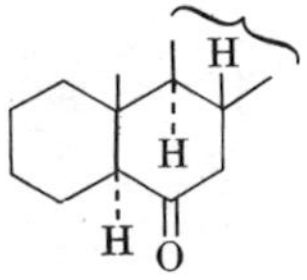 (35)	473
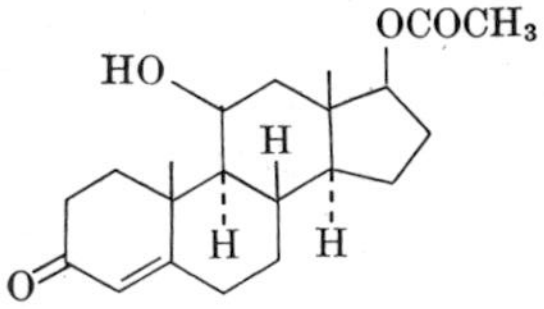 HO, $OCOCH_3$	Li/NH_3, Et_2O-dioxane, 5 min	NH_4Cl, Ac_2O/Py	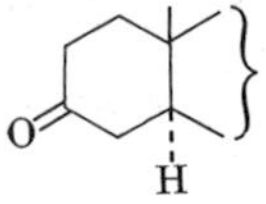 (33)	474
	Li/NH_3, Et_2O-dioxane, 1.5 min	"	" (39)	
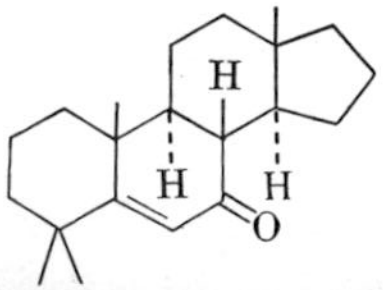	Li/NH_3, Et_2O-THF, 60 min	NH_4Cl, CrO_3	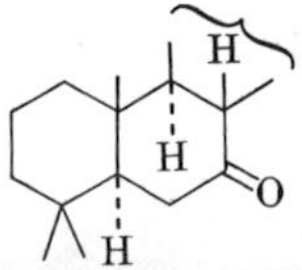 (96)	475

Substrate	Conditions	Quench	Product(s) (Yield %)	Ref.
C_{21} (*Contd.*) 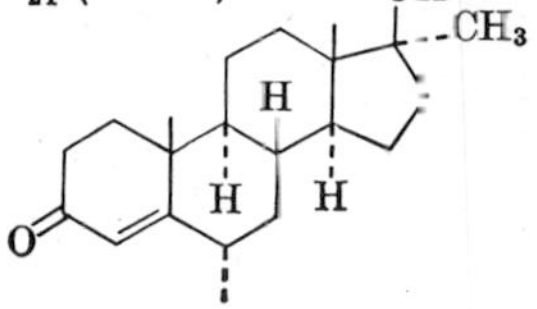 OH, CH_3, H, H, H, O	Li/NH_3, Et_2O-dioxane, 50 min	NH_4Cl, CrO_3	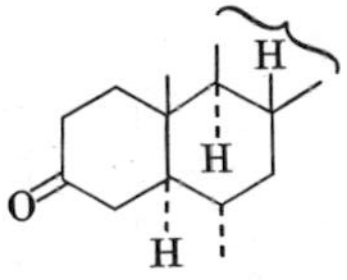 O, H, H, H (35)	476
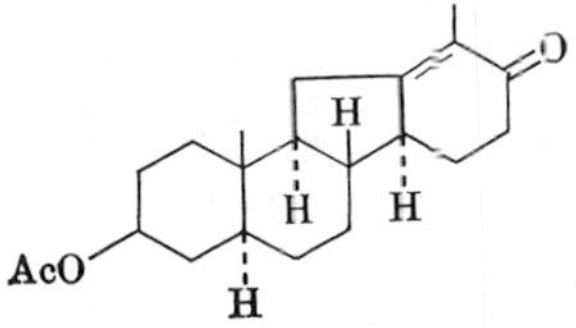 AcO, H, H, H, O	Li/NH_3, dioxane-Et_2O, 1.5 min	NH_4Cl	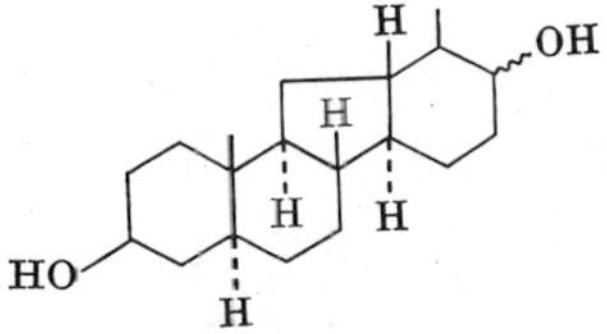 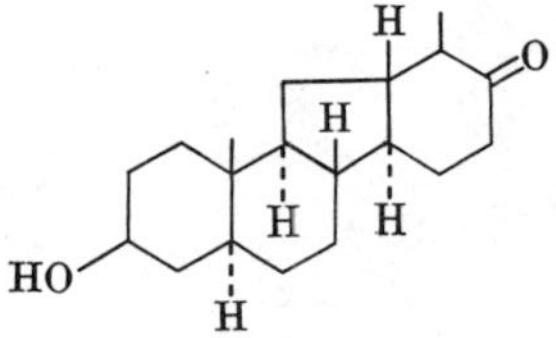 HO, H, H, H, H, OH (—) + HO, H, H, H, H, O	477
C_{22} 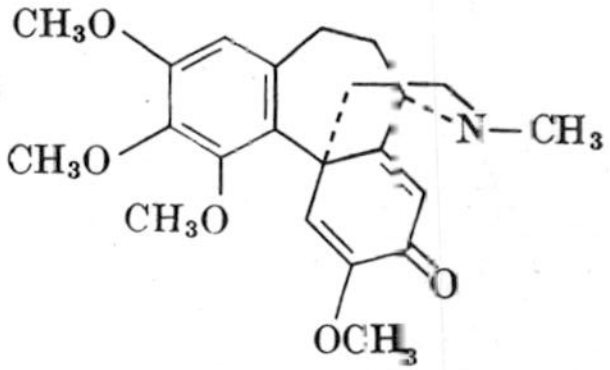 CH_3O, CH_3O, CH_3O, N—CH_3, O, OCH_3	Na/NH_3	—	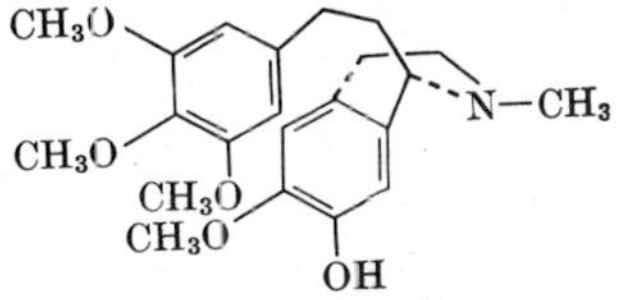 CH_3O, CH_3O, CH_3O, CH_3O, N—CH_3, OH (—)	208
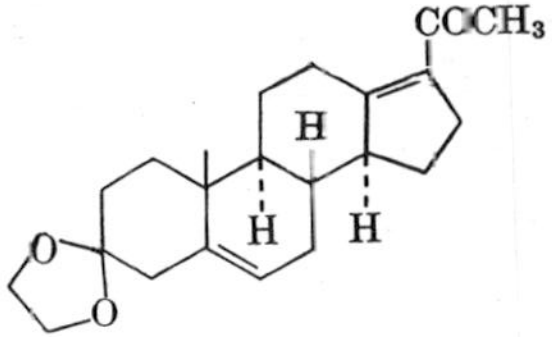 O, O, H, H, $COCH_3$	Li/NH_3, Et_2O, −53°, 130 min	H_2O	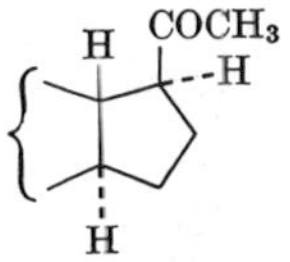 H, H, $COCH_3$, H (37)	478

Note: References 338–543 are on pp. 253–258.

TABLE V. REDUCTIONS OF TETRACYCLIC α,β-UNSATURATED KETONES (*Continued*)

Reactant	Reduction Conditions	Quenching Agent	Product(s) (% Yield)	Refs.
C_{22} (*Contd.*) 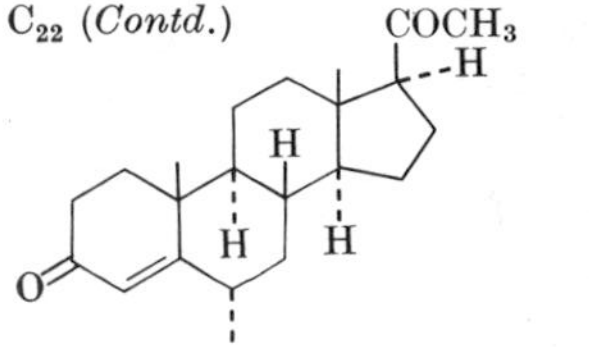 COCH$_3$, H, H, H, H	Li/NH_3, EtOH	H_2O, CrO_3	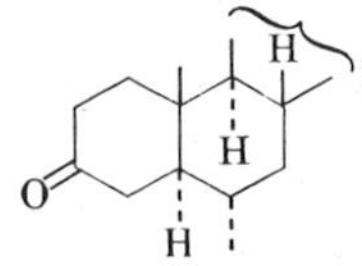 O, H, H, H (45)	479
	Li/NH_3, Et_2O, 5 min, −80°	NH_4Cl	" (36)	480
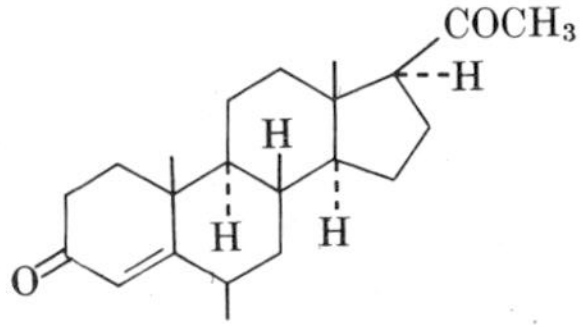 COCH$_3$, H, H, H, H	Li/NH_3, EtOH	H_2O, CrO_3	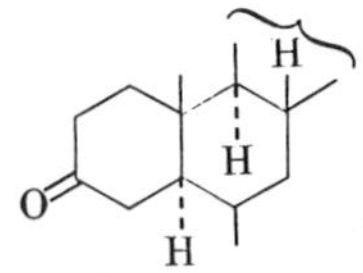 O, H, H, H (38)	479
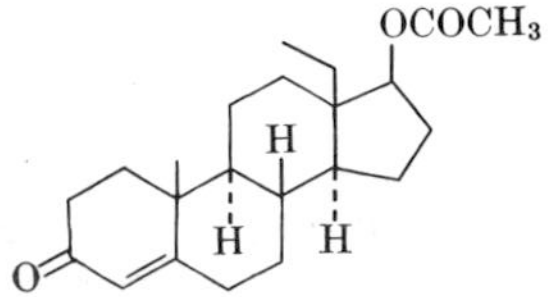 OCOCH$_3$, H, H, H	Li/NH_3, Et_2O-dioxane, 30 min, −70°	—	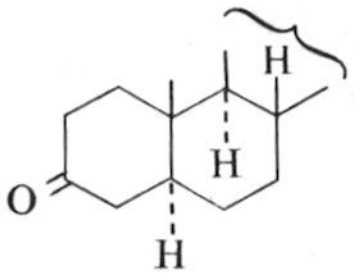 O, H, H, H (−)	481
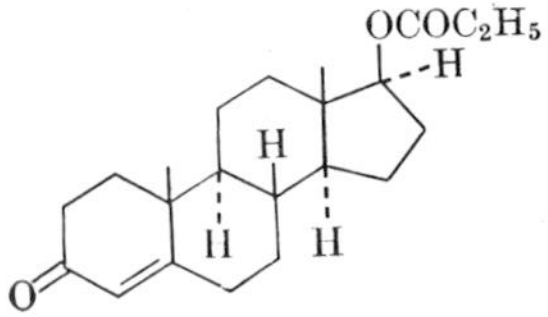 OCOC$_2$H$_5$, H, H, H, H	NH_3/Li,[d] Et_2O-dioxane, 20 min	NH_4Cl	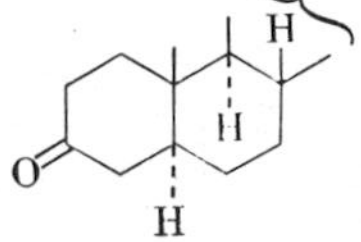 O, H, H, H (84)	472

Substrate	Conditions	Quench	Product(s) (Yield, %)	Refs.
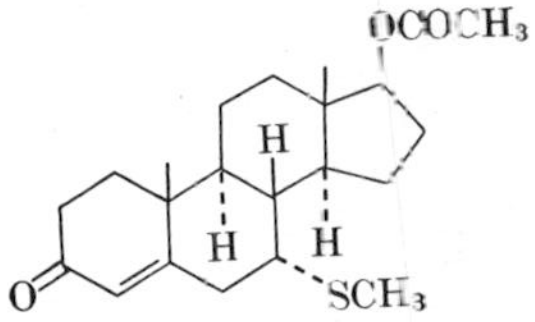	Li/NH_3, Et_2O-dioxane, 50 min	NH_4Cl, Ac_2O	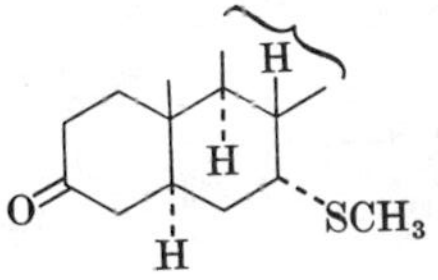 (36)	257
C_{23}	Li/NH_3	H_3O^+	(60)	197
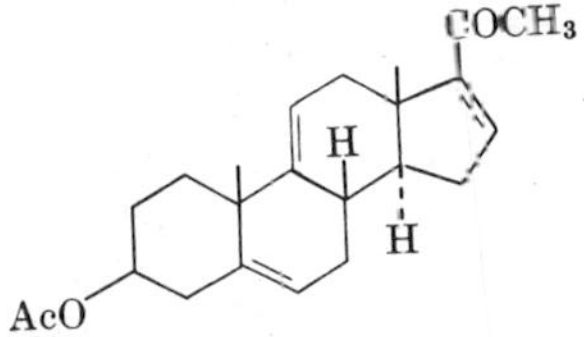	Li/NH_3, dioxane, 5 min	NH_4Cl, MeOH, KOH; Oppenhauer oxidation	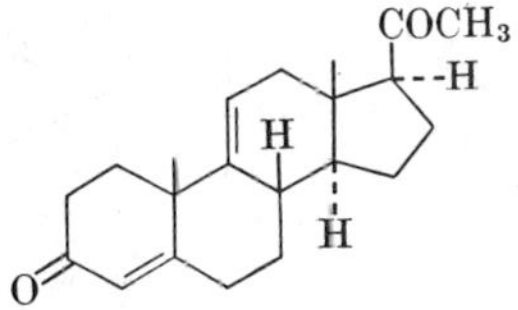 (29)	482
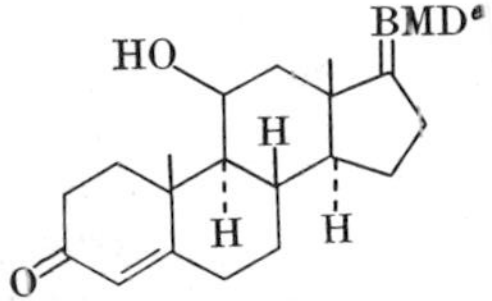	NH_3/excess Li, EtOH-THF	EtOH	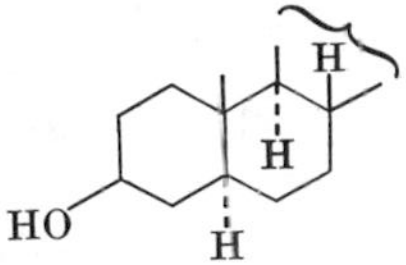 (54)	483

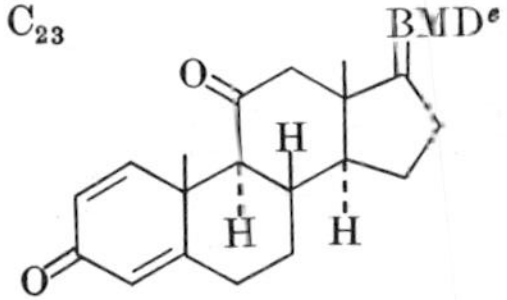

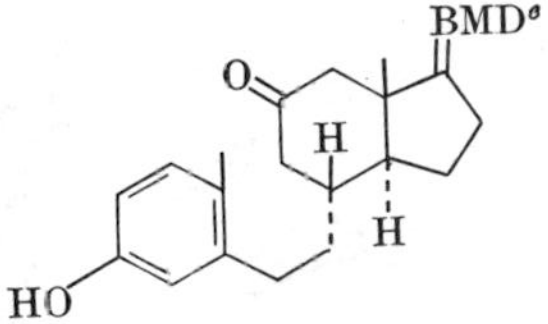

Note: References 338–543 are on pp. 253–258.

[d] A solution of lithium in liquid ammonia was added dropwise to the enone ester in ether-dioxane.

[e] BMD is the bismethylenedioxy protecting group.

TABLE V. REDUCTIONS OF TETRACYCLIC α,β-UNSATURATED KETONES (*Continued*)

Reactant	Reduction Conditions	Quenching Agent	Product(s) (% Yield)	Refs.
C_{23} (*Contd.*)				
	Li/NH_3, Et_2O-*n*-PrOH, −40°, 2 hr	H_2O, Ac_2O	(86)	484
BMD[e]	Li/NH_3, THF, −40 to −60°, 1.5 min	NH_4Cl	(52) + (—)	234
C_{24} BMD[e]	Li/NH_3, THF, 10 min	EtOH	(43) + (—)	485
	Li/NH_3, dioxane, 5 min	NH_4Cl	(—)	486

Substrate	Reduction Conditions	Proton Source	Product(s) and Yield(s) (%)	Refs.
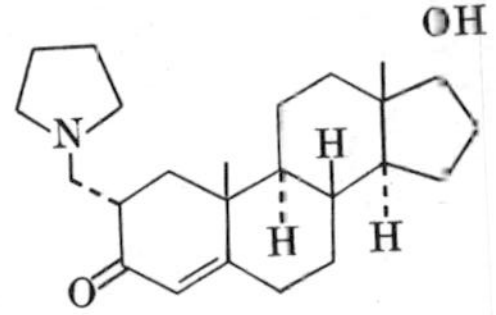 OH	Li/NH_3, Et_2O-dioxane, 40 min	NH_4Cl	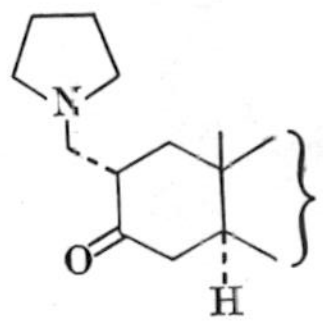 H (18)	487
C_{25} 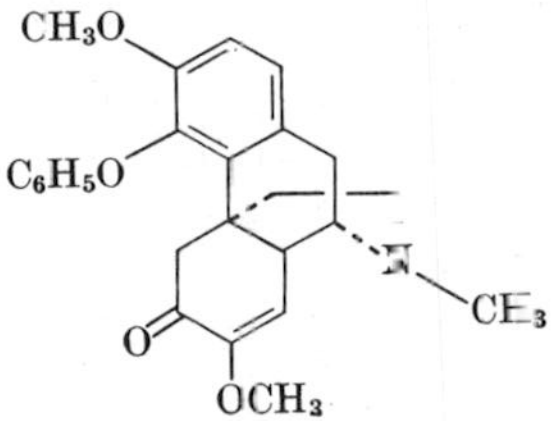 CH_3O, C_6H_5O, CH_3, OCH_3	NH_3/Na, $C_6H_5CH_3$	H_2O	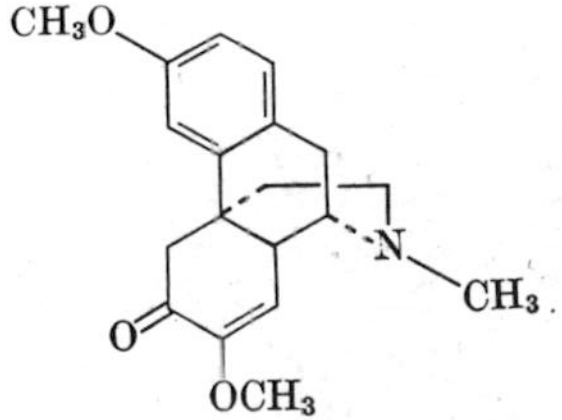 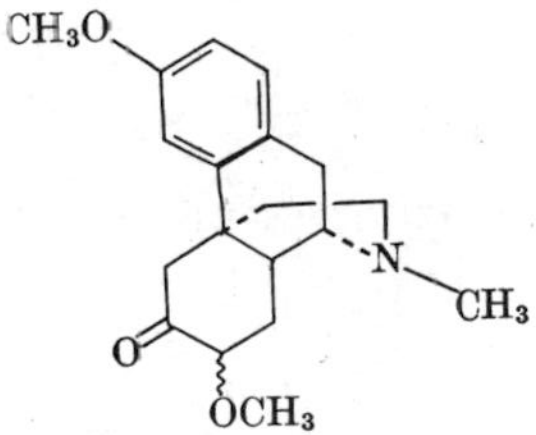 CH_3O, CH_3, OCH_3 (~27) + CH_3O, CH_3, OCH_3 (~45)	488
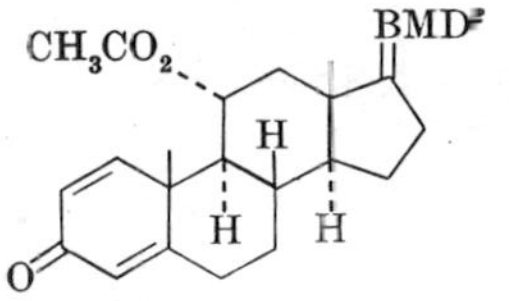 CH_3CO_2, BMD[a]	Li/NH_3, THF, 30 min	NH_4Cl	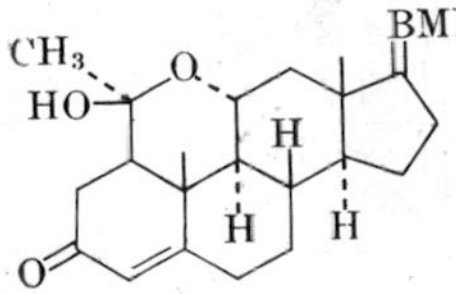 CH_3, HO, O, BMD[a] (45)	197

Note: References 338–543 are on pp. 253–258.

[a] BMD is the bismethylenedioxy protecting group.

TABLE V. REDUCTIONS OF TETRACYCLIC α,β-UNSATURATED KETONES (*Continued*)

Reactant	Reduction Conditions	Quenching Agent	Product(s) (% Yield)	Refs.
C_{25} (*Contd.*)				
[structure] + corresponding 17-ethylene ketal	Li/NH_3	NH_4Cl	[structure, OH] (43) + corresponding 17-ethylene ketal (10)	489
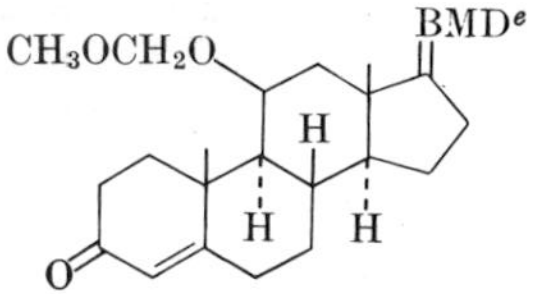 [structure, CH_3OCH_2O, BMD[e]]	NH_3/Li, EtOH-THF, 25 min	EtOH	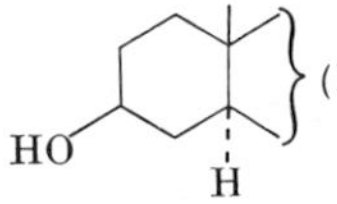 [structure, HO] (14)	483
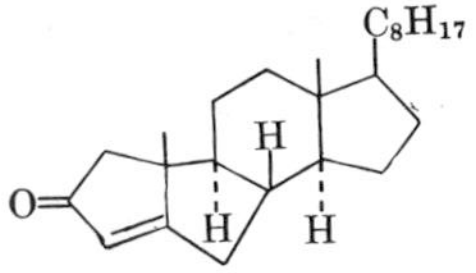 [structure, C_8H_{17}]	NH_3/Li, Et_2O		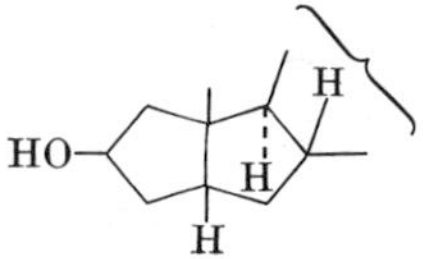 [structure, HO] (80) + [structure, HO] (20)	115

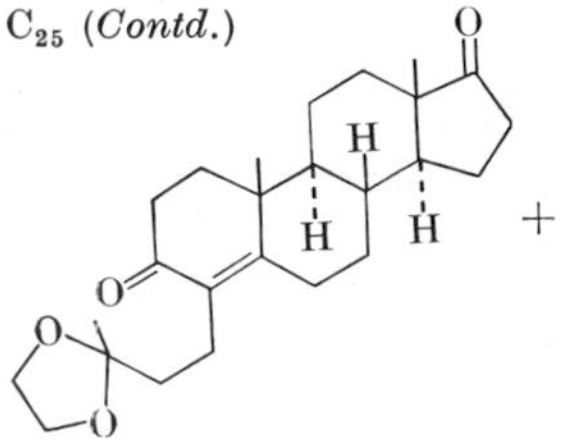

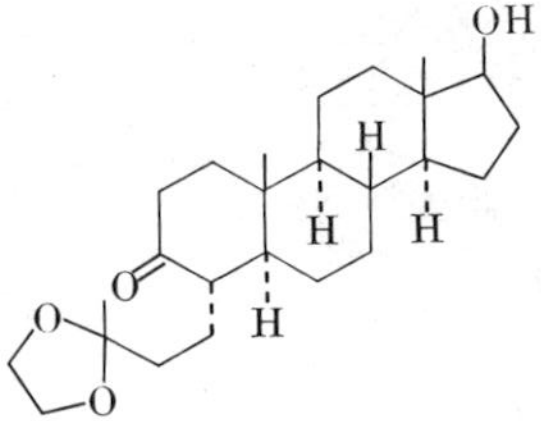

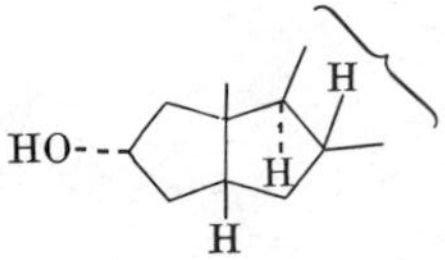

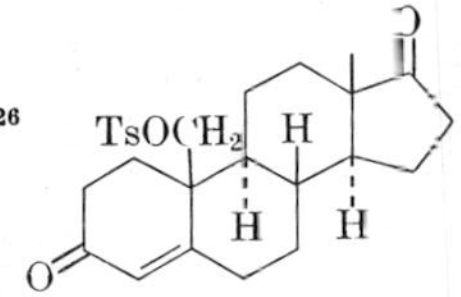 C_{26} TsOCH$_2$	Li/NH_3	—	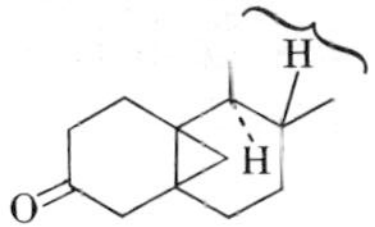 (–)	192
	Li/NH_3, Et_2O	NH_4Cl	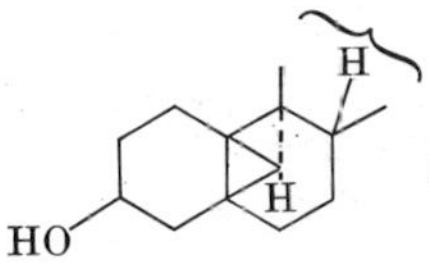 HO (14)	191
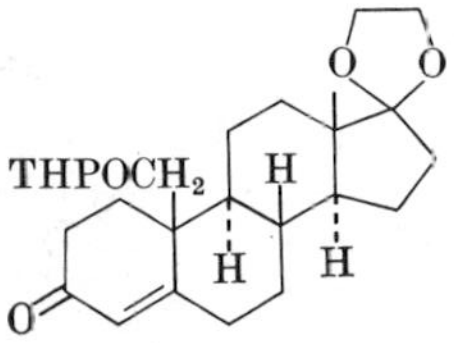 THPOCH$_2$	Li/NH_3, THF, 10 min	NH_4Cl, CrO_3	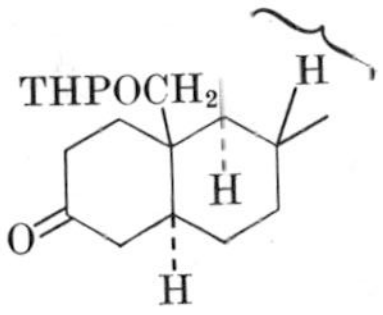 THPOCH$_2$ (80)	141
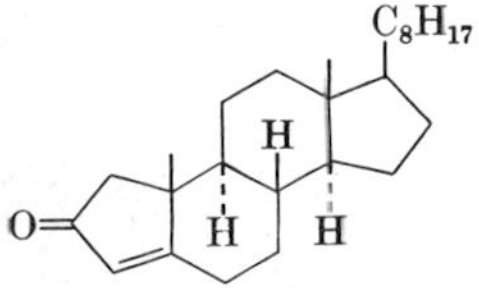 C_8H_{17}	Li/NH_3, Et_2O, 30 min	NH_4Cl	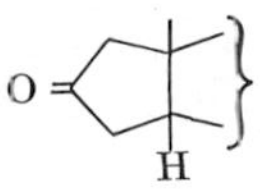 (42) 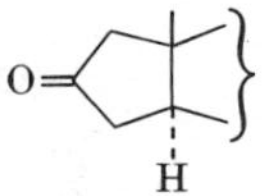 + (12)	130
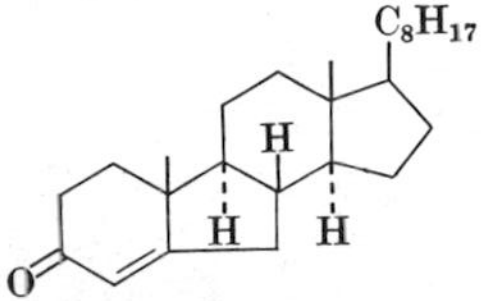 C_8H_{17}	Li/NH_3	—	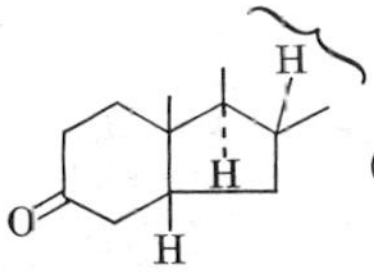 (86) 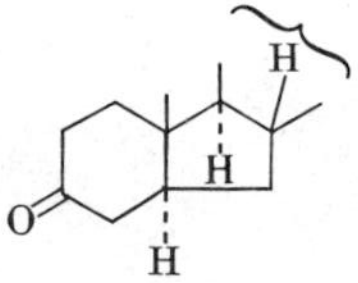 + (14)	119

Note: References 338–543 are on pp. 253–258.

[e] BMD is the bismethylenedioxy protecting group.

TABLE V. Reductions of Tetracyclic α,β-Unsaturated Ketones (*Continued*)

Reactant	Reduction Conditions	Quenching Agent	Product(s) (% Yield)	Refs.
C_{27} 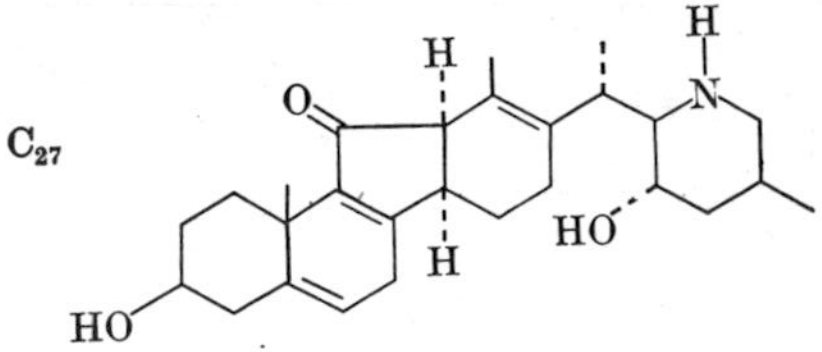	Li/NH_3, THF, 1.3 hr, −78°	NH_4Cl	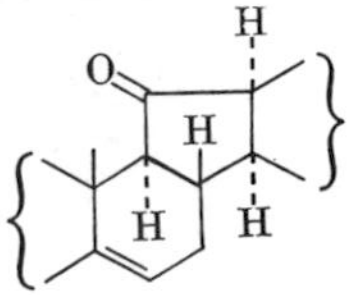 (—)	490
	NH_3/Li, THF-MeOH, −70°, 25 min	"	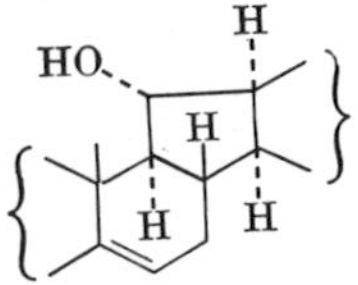 (38)	491, 492
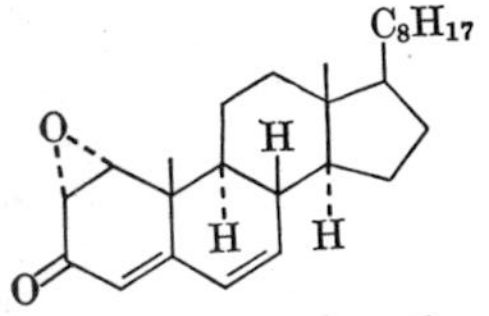	Excess Li/NH_3, THF, NH_4Cl	—	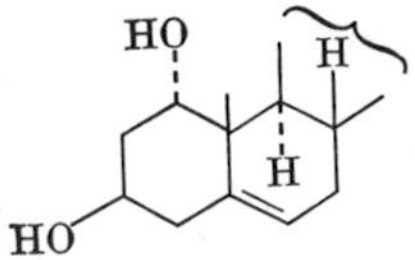 (60)	265a
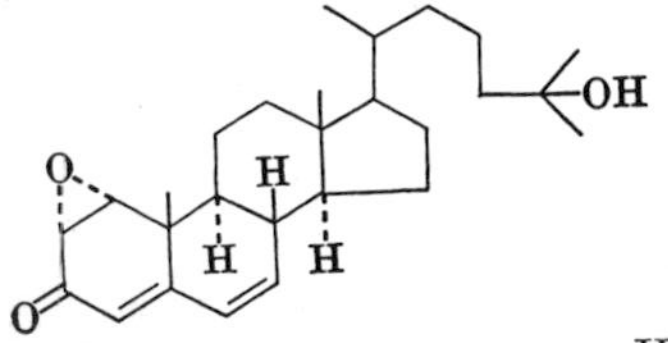	Li/NH_3, THF	NH_4Cl	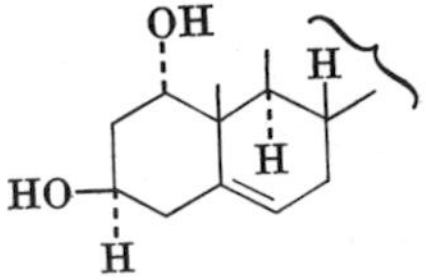 (—)	265b
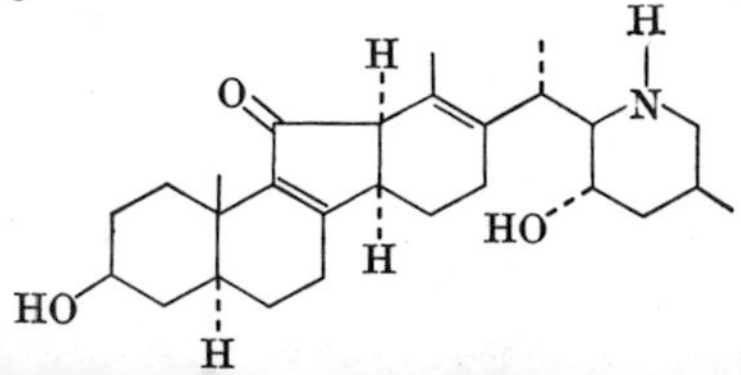	Li/NH_3, THF, 1.3 hr, −78°	NH_4Cl	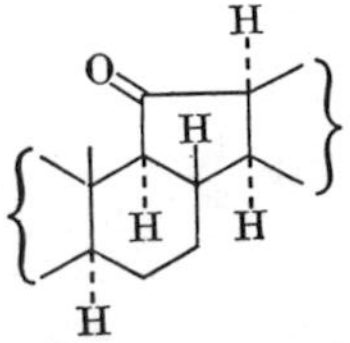 (38) + starting material (20)	490

Substrate	Conditions	Quench	Product(s) and Yield(s) (%)	Refs.
	Li/NH_3, THF, 7 min	"	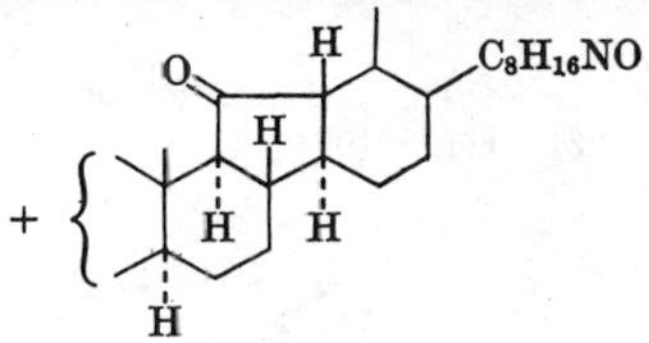 + $C_8H_{16}NO$ (—)	491
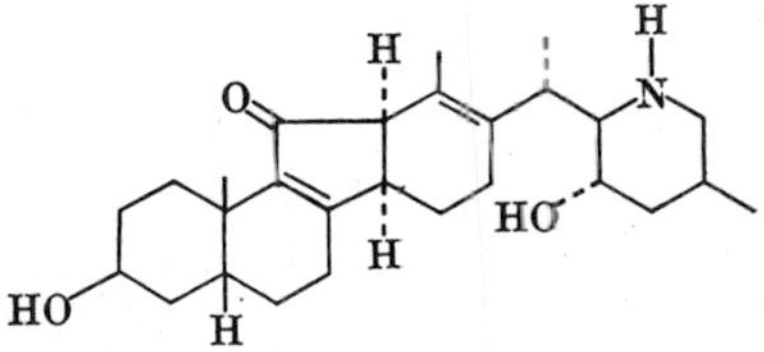 HO, HO	"	"	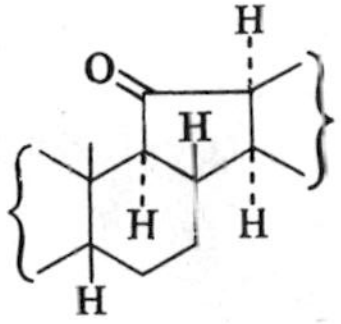 (30)	491
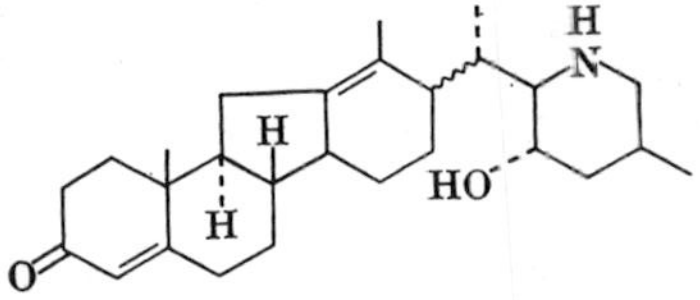 HO	Li/NH_3, THF, 7 min	NH_4Cl	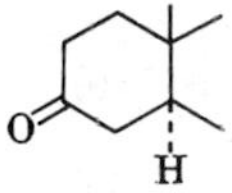 (40)	494
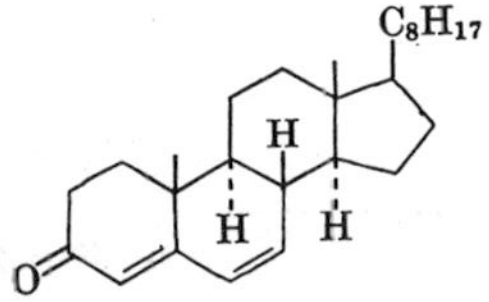 C_8H_{17}	NH_3/Li, Et_2O, 10 min	MeOH	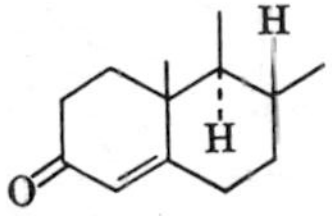 (42)	495

Note: References 338–543 are on pp. 253–258.

TABLE V. REDUCTIONS OF TETRACYCLIC α,β-UNSATURATED KETONES (*Continued*)

Reactant	Reduction Conditions	Quenching Agent	Product(s) (% Yield)	Refs.
C_{27} (*Contd.*) 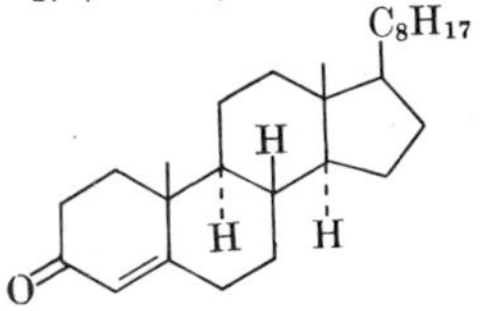 C_8H_{17}	Li/NH_3, Et_2O, 5 min	*t*-BuOH	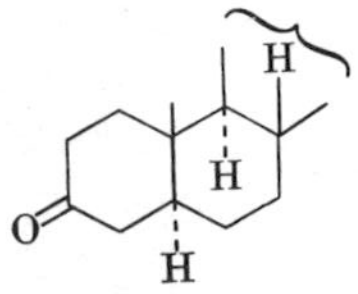 I (62)	496
	~20 eq. Li/NH_3, Et_2O, 25 min	EtOH, CrO_3	I (60)	3, 73
	~55 eq. Li/NH_3, Et_2O, 2.5 hr	EtOAc	I (0) + starting material (22) 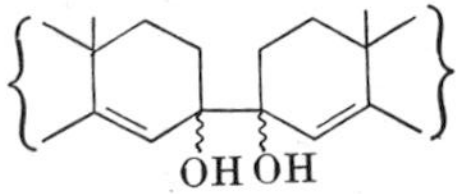 + II (28)	73
	~35 eq. Li/NH_3, DME, 8 hr	*t*-BuOH, EtOAc	I (0) + II (32) + starting material (14)	
	~35 eq. Li/NH_3, DME, ~7 hr	"	I (58) + II (6) 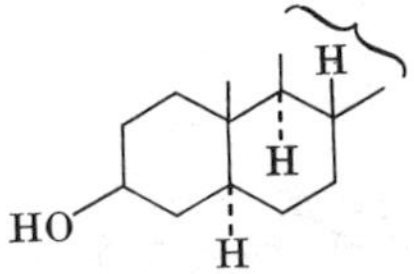 + III (6) + starting material (21)	73
	~55 eq. Li/NH_3, DME, ~6 hr	*i*-PrOH, EtOAc	I (67) + II (10) + III (23) + starting material (15)	73

	~4.5 eq. Li/NH_3, Et_2O, 7 min	*t*-BuOH	I (40) + II (25) + starting material (Trace)	
	~4.5 eq. Li/NH_3, Et_2O, 14 min	"	I (23) + II (30) + starting material (3)	
	~8 eq. Li/NH_3, Et_2O, 65 min	"	I (43) + II (17) + III (12) + starting material (1)	
	~3 eq. Li/NH_3, Et_2O, 14 min	NH_4Cl	I (51) + II (15) + starting material (7)	73
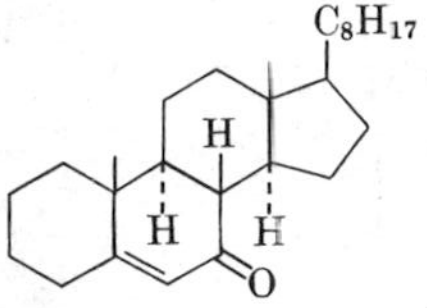	Li/NH_3, Et_2O	Acetone	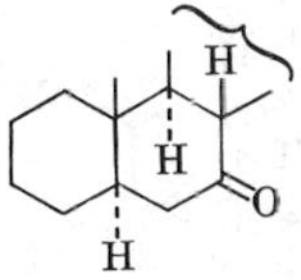 (—)	9
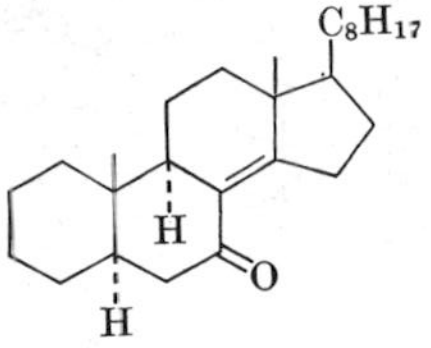	Li/NH_3, Et_2O, 2 hr	NH_4Cl, CrO_3	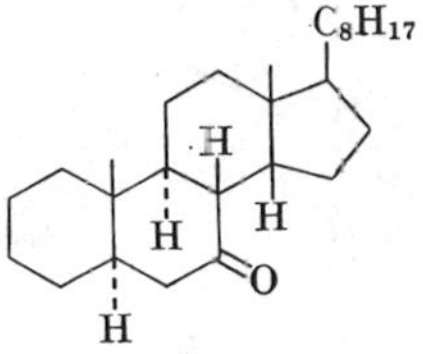 (46)	134
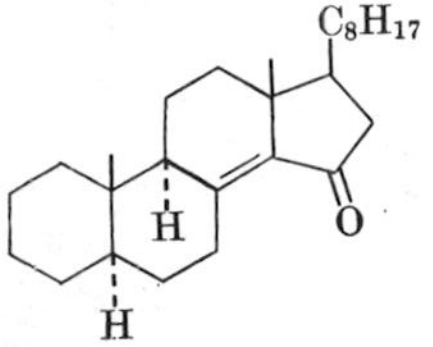	Li/NH_3, Et_2O, 1.5 hr	NH_4Cl, CrO_3	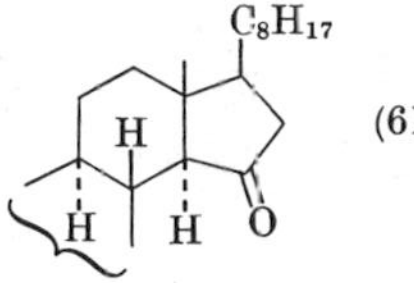 (61)	134

Note: References **338–543** are on pp. 253–258.

TABLE V. REDUCTIONS OF TETRACYCLIC α,β-UNSATURATED KETONES (*Continued*)

Reactant	Reduction Conditions	Quenching Agent	Product(s) (% Yield)	Refs.
C_{27} (*Contd.*)				
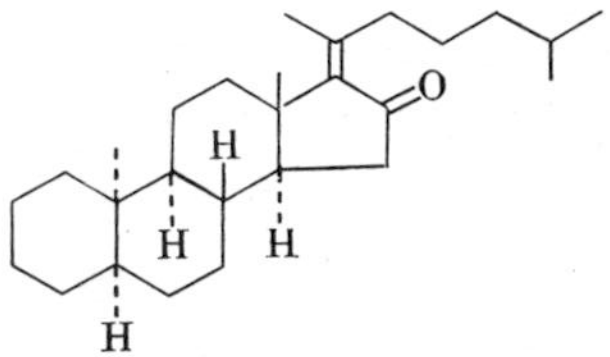	Li/NH_3, THF	Acetone	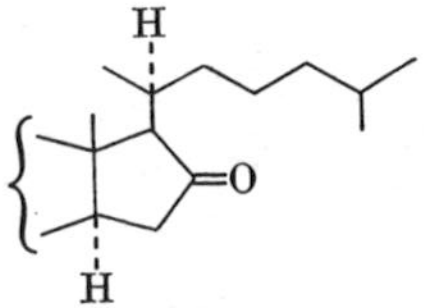 (Minor) + 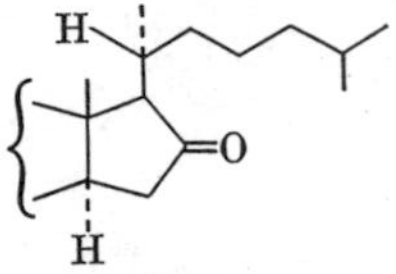 (Major)	497
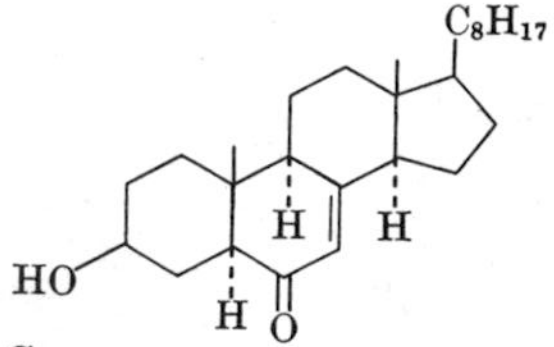 C_8H_{17}; HO; H; H; H; O	Li/ND_3, THF, 20 min	H_2O, CrO_3, dil NaOH	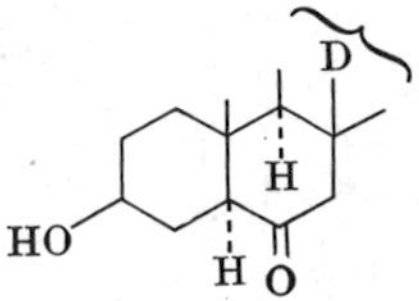 D; H; HO; H O (90)	499
C_{28}				
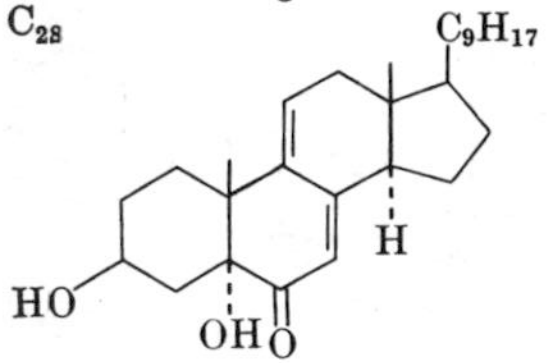 C_9H_{17}; HO; H; OH O	Excess Li/NH_3, Et_2O-dioxane, 3.5 hr	H_2O	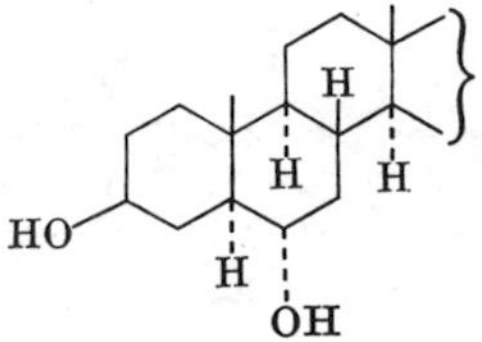 H; H; H; HO; H; OH (50)	229
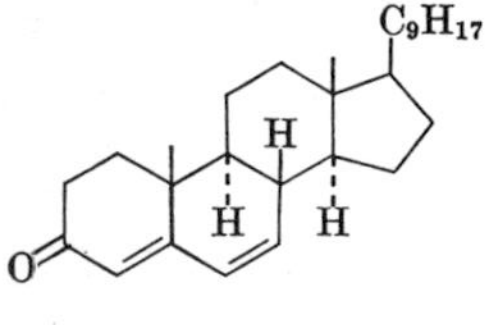 C_9H_{17}; H; H; H; O	Li/NH_3, Et_2O	EtOH	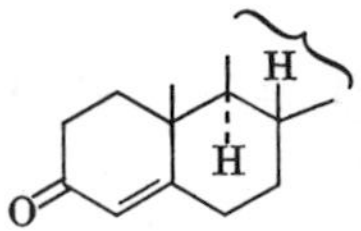 H; H; O (60)	220
	NH_3/Li, Et_2O	"	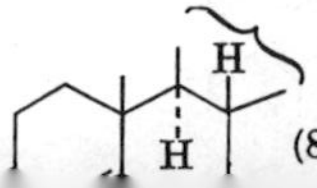 H; H (80)	219

Substrate	Reducing conditions	Work-up	Product (Yield, %)	Refs.
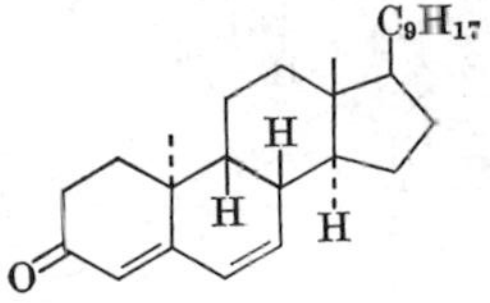 C_9H_{17} steroid	Li/NH_3, Et_2O	NH_4Cl	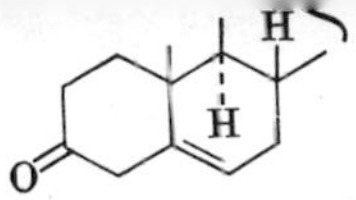 (~90)	222
	Li/NH_3, Et_2O	$EtOH$-H_2O	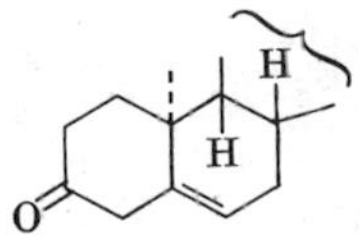 (—)	225
	NH_3/Li,[f] Et_2O	$EtOH$[g]	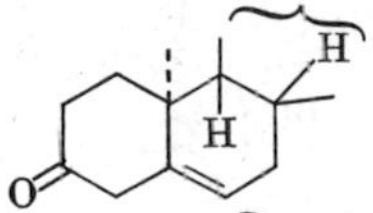 (70)	
	"	$EtOH$,[g] dil NaOH, heat	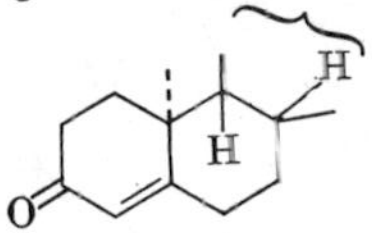 (80)	
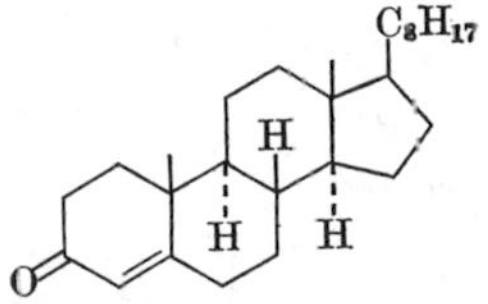 C_8H_{17} steroid	Li/NH_3, Et_2O, 5 min	t-BuOH	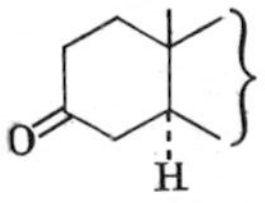 (—)	496
	"	NH_4Cl[h]	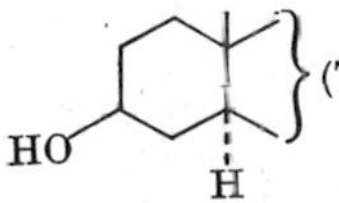 (73)	222
	Li/NH_3, Et_2O, 30 min	NH_4Cl, CrO_3	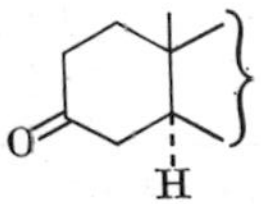 (74)	493

Note: References 338–543 are on pp. 253–258.

[f] A solution of lithium in liquid ammonia was added to the enone in ether-ammonia.

[g] Ethanol was added rapidly and the reaction mixture was worked up rapidly.

[h] The 3-keto compound was reduced with lithium aluminum hydride.

TABLE V. Reductions of Tetracyclic α,β-Unsaturated Ketones (*Continued*)

Reactant	Reduction Conditions	Quenching Agent	Product(s) (% Yield)	Refs.
C_{28} (*Contd.*)				
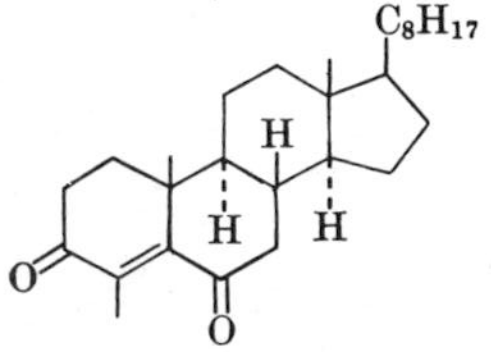 [structure: C_8H_{17}]	Li/NH_3, THF, 5 min	NH_4Cl	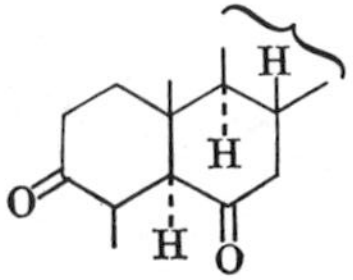 [structure] (50)	498
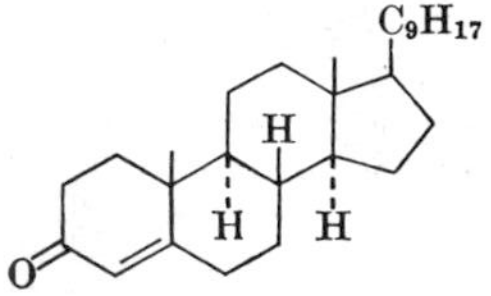 [structure: C_9H_{17}]	Li/NH_3, Et_2O, 20 min	NH_4Cl	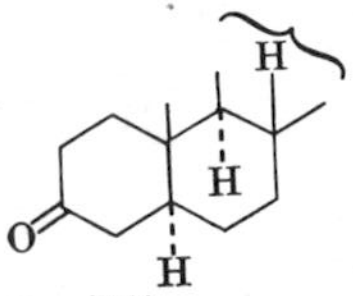 [structure] (54)	500
	"	H_3O^+	" (71)	501
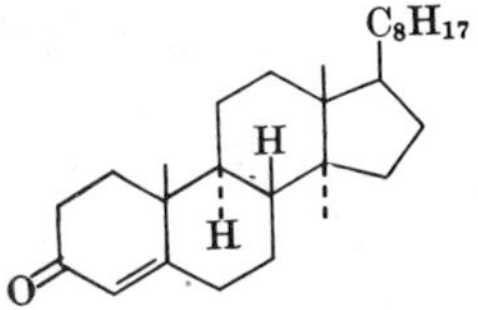 [structure: C_8H_{17}]	Li/NH_3, Et_2O, 20 min	H_3O^+	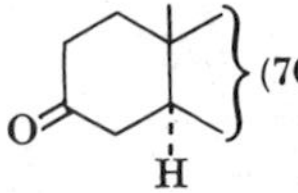 [structure] (70)	496, 502
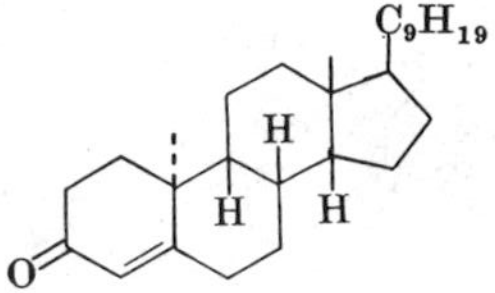 [structure: C_9H_{19}]	Li/NH_3, Et_2O EtOH	H_2O, CrO_3	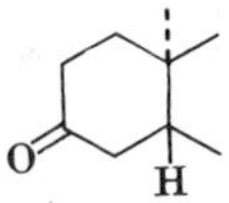 [structure] (70)	503

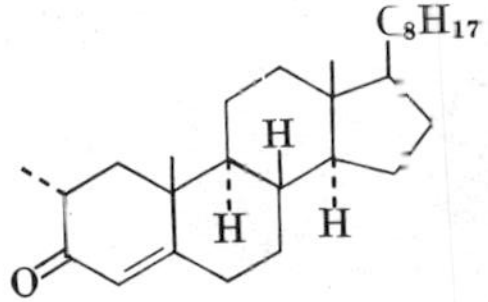 C_8H_{17}	Li/NH_3, Et_2O, 25 min	NH_4Cl	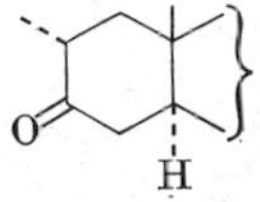 (53) 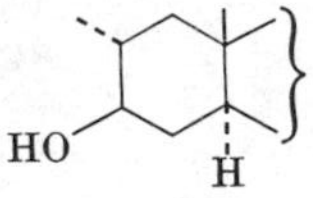 + HO (41)	501
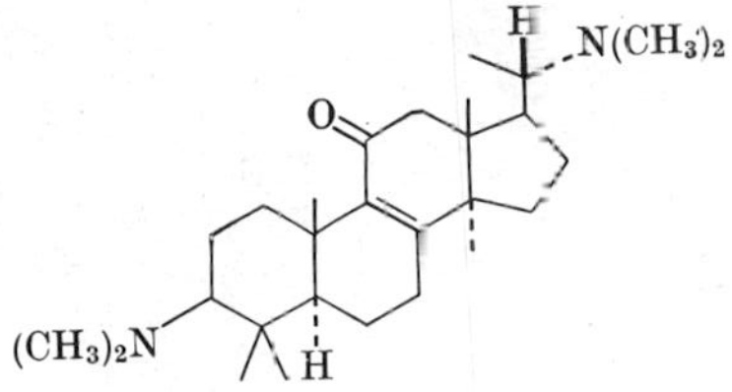 $(CH_3)_2N$, $N(CH_3)_2$	Li/NH_3, Et_2O-dioxane	MeOH, CrO_3	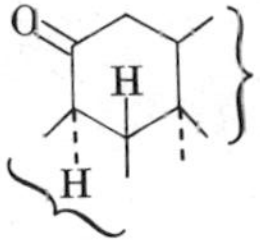 (−)	504
C_{29} 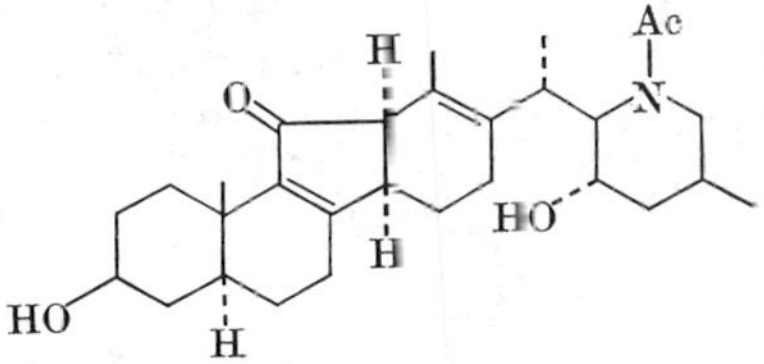 HO, Ac, HO	Li/NH_3, THF, 1 hr	NH_4Cl	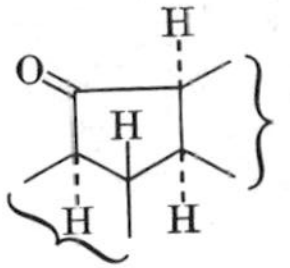 (~56)	490
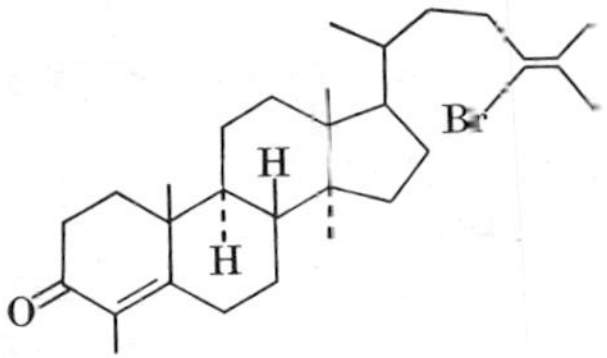 Br	Li/NH_3, Et_2O	NH_4Cl	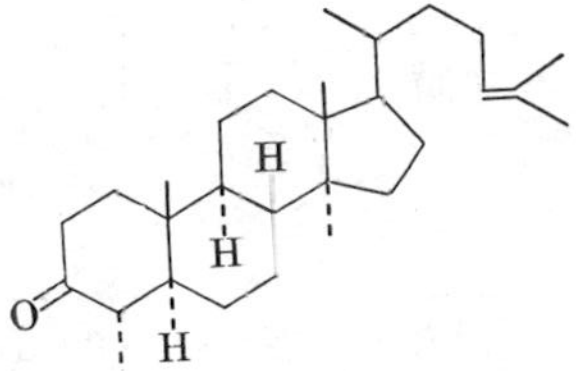 (60)	255

Note: References 338–543 are on pp. 253–258.

TABLE V. REDUCTIONS OF TETRACYCLIC α,β-UNSATURATED KETONES (*Continued*)

Reactant	Reduction Conditions	Quenching Agent	Product(s) (% Yield)	Refs.
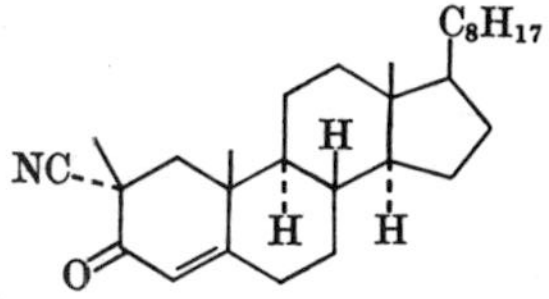 NC, C_8H_{17}, H, H, H	NH_3/Li, Et_2O, 1 hr	NH_4Cl	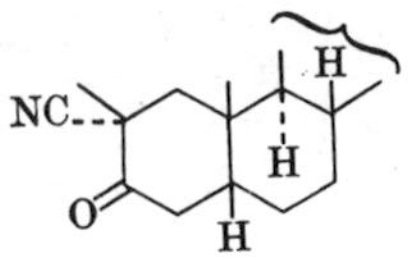 NC, H, H, H (97)	99
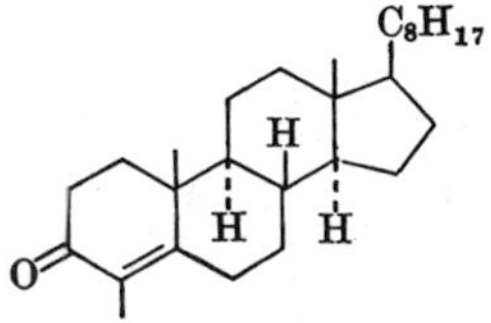 C_8H_{17}, H, H, H	Li/NH_3, Et_2O, 5 min	*t*-BuOH	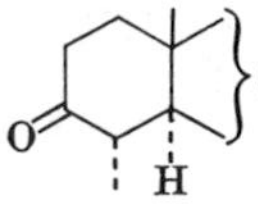 H (52)	496
	"	"	"	505
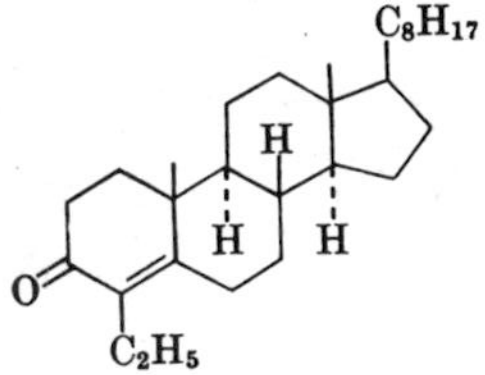 C_8H_{17}, H, H, H, C_2H_5	Li/NH_3, Et_2O, 30–45 min	NH_4Cl	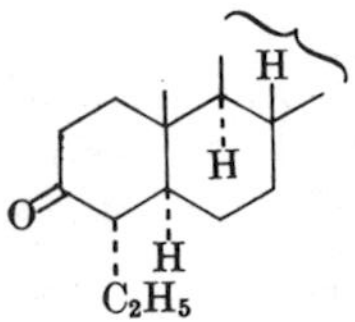 H, H, H, C_2H_5 (74)	336, 506
	Li/NH_3, dioxane, 1 hr, MeOH, 1 hr	"	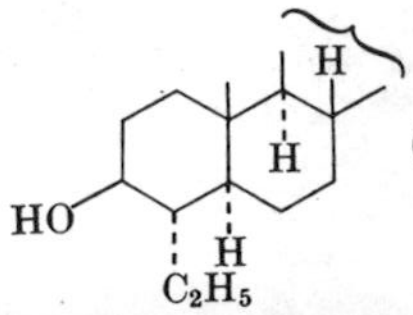 HO, H, H, H, C_2H_5 (95)	336

Substrate	Conditions	Quench	Product(s) and Yield(s) (%)	Refs.
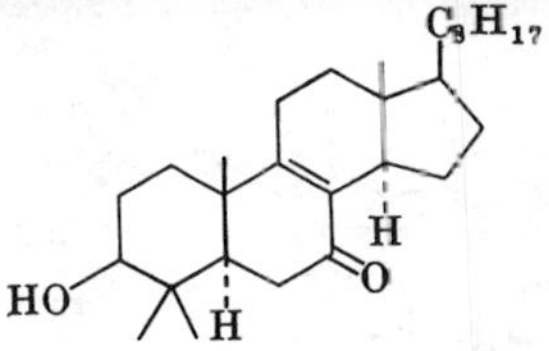 C_9H_{17}, HO, H, H, O	Li/NH_3, Et_2O, 5 min	t-BuOH	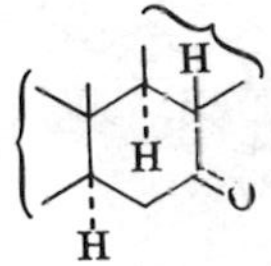 H, H, H, O (—)	507
C_{30} 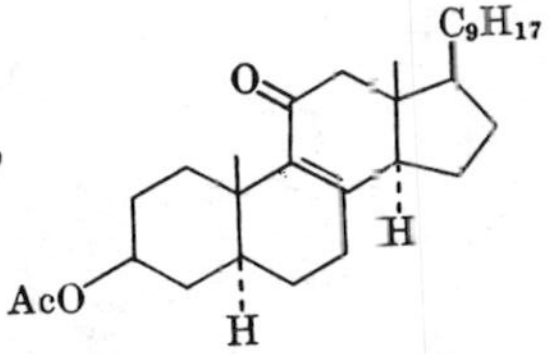 C_9H_{17}, O, AcO, H, H	Li/NH_3, Et_2O-dioxane, −45°, 2 hr	Ac_2O	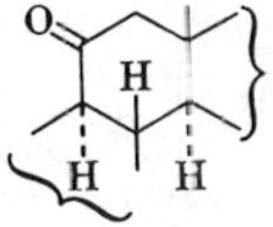 O, H, H, H (85–90)	6
	"	"	" (66)	508
	"	"	" (65)	
	Li/NH_3, Et_2O-n-PrOH, −40°, 2 hr	Ac_2O	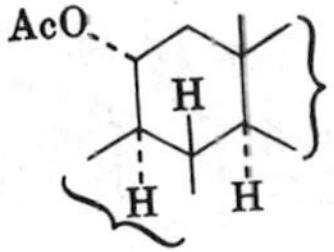 AcO, H, H, H (90)	484
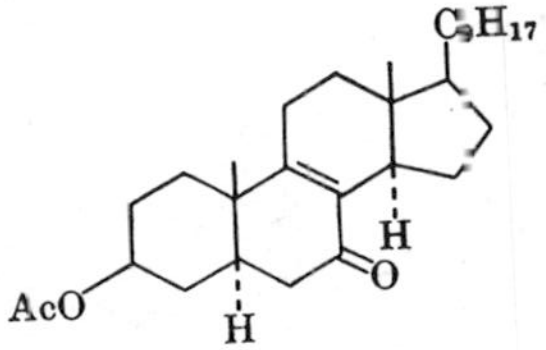 C_9H_{17}, AcO, H, H, O	Li/NH_3, Et_2O, 30 min	n-PrOH, Ac_2O	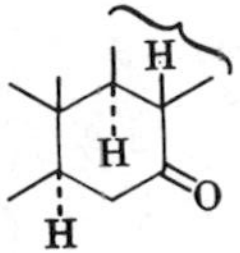 H, H, H, O (75)	509
	Li/NH_3	NH_4Cl	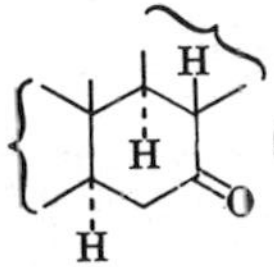 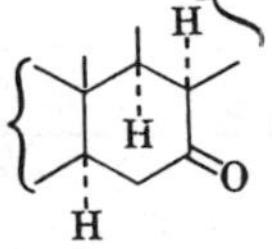 H, H, H, O (—) + H, H, H, O (—)	16

Note: References 338–543 are on pp. 253–258.

TABLE V. REDUCTIONS OF TETRACYCLIC α,β-UNSATURATED KETONES (*Continued*)

Reactant	Reduction Conditions	Quenching Agent	Product(s) (% Yield)	Refs.
C_{30} (*Contd.*)				
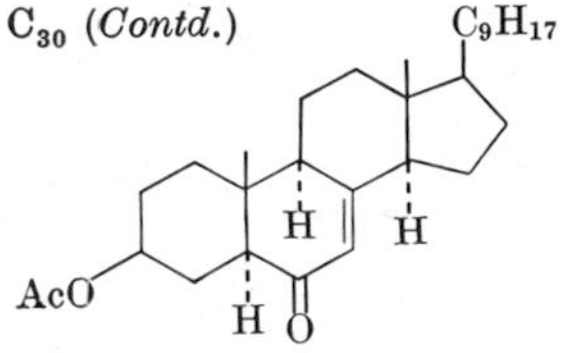 C_9H_{17}; AcO; H; H; H; O	Li/NH_3, Et_2O	NH_4Cl	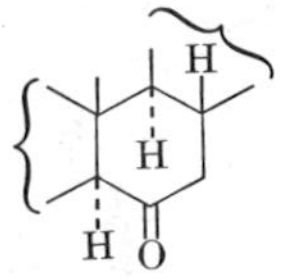 H; H; H; O (−)	9
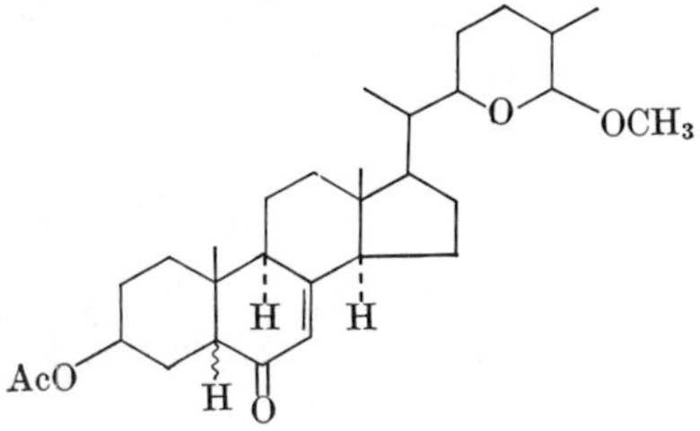 OCH_3; O; AcO; H; H; H; O	Li/NH_3, Et_2O, 20 min	NH_4Cl	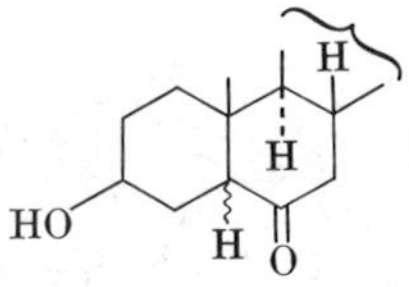 H; H; HO; H; O (28)	310
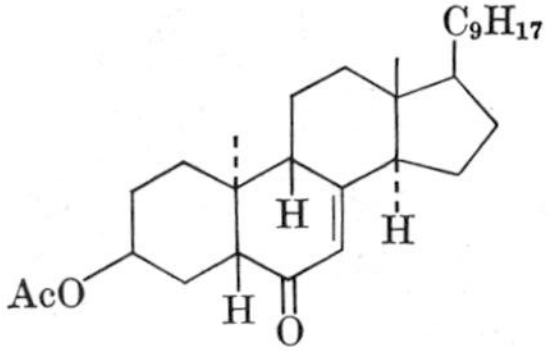 C_9H_{17}; AcO; H; H; H; O	Li/NH_3, Et_2O, 10 min	NH_4Cl, Ac_2O	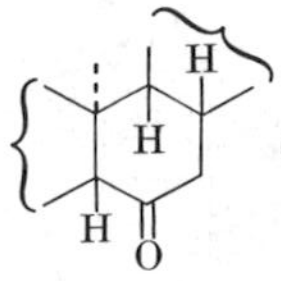 H; H; H; O (42) 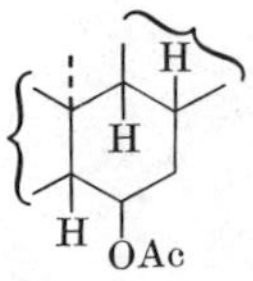 + H; H; H; OAc (23)	105
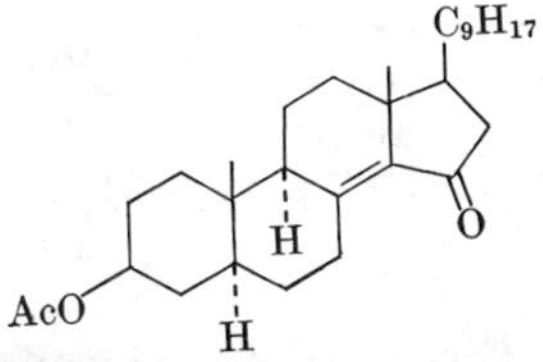 C_9H_{17}; AcO; H; H; O	Li/NH_3, Et_2O, 5 min	*t*-BuOH	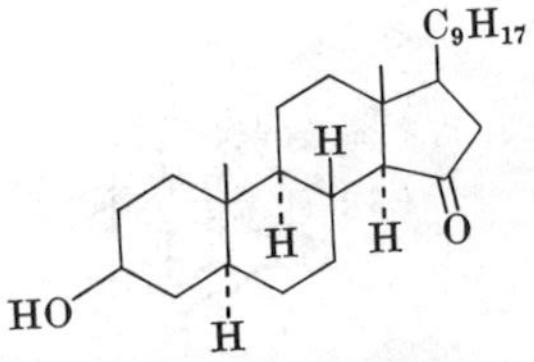 C_9H_{17}; HO; H; H; H; H; O (—)	132, 133

Reactant	Reagents	Quench	Product(s) and Yield(s) (%)	Refs.
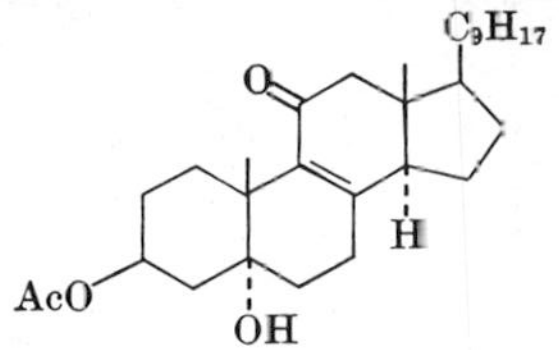 C_9H_{17}, AcO, OH, O	Li/NH_3, Et_2O, 30 min	$Fe(NO_3)_3$, Ac_2O	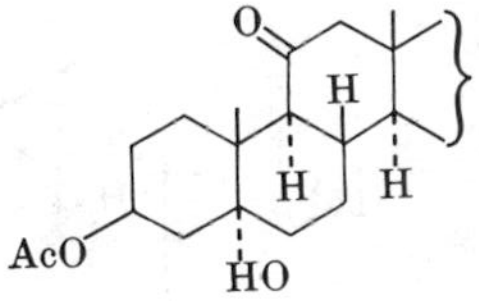 O, H, H, H, AcO, HO (55)	163
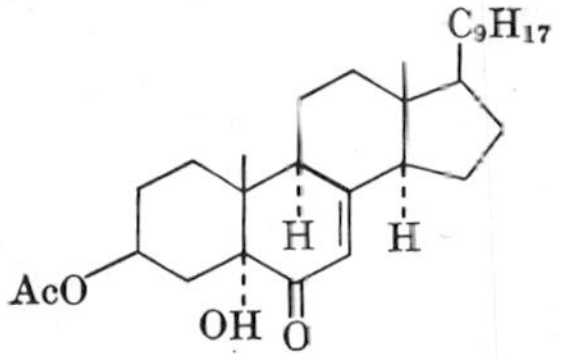 C_9H_{17}, AcO, OH, O, H, H	Li/NH_3, Et_2O	NH_4Cl, Ac_2O	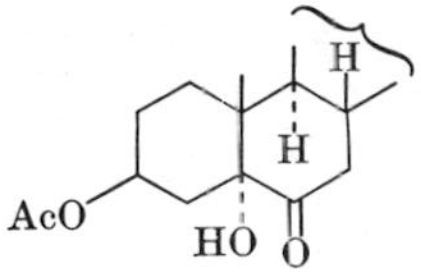 H, H, AcO, HO, O (50)	9
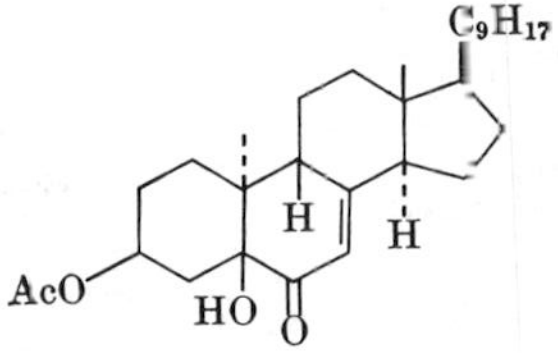 C_9H_{17}, AcO, HO, O, H, H	NH_3/Li, Et_2O, 30 min	NH_4Cl, Ac_2O	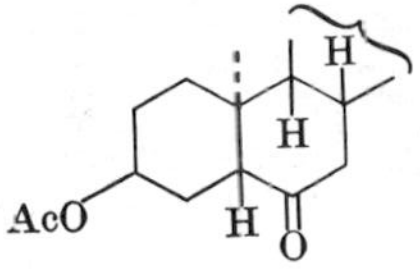 H, H, AcO, H, O (19) 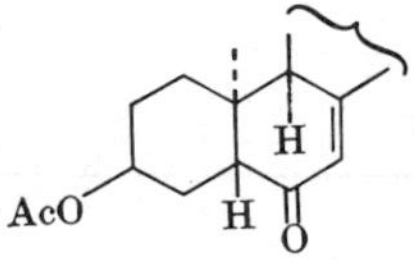 + AcO, H, H, O (27)	105

Note: References 338–543 are on pp. 253–258.

TABLE V. Reduction of Tetracyclic α,β-Unsaturated Ketones (*Continued*)

Reactant	Reduction Conditions	Quenching Agent	Product(s) (% Yield)	Refs.
C_{30} (*Contd.*)				
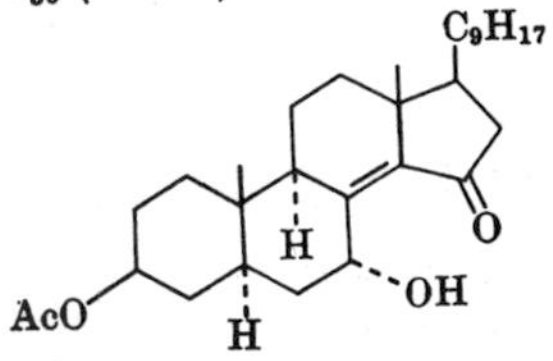 C_9H_{17}; AcO; H; OH	Li/NH_3, Et_2O, −80°, 20 min	*t*-BuOH	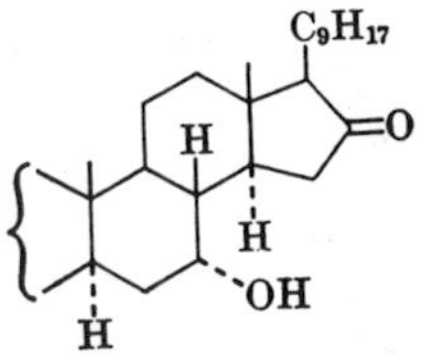 C_9H_{17}; H; OH (—)	133
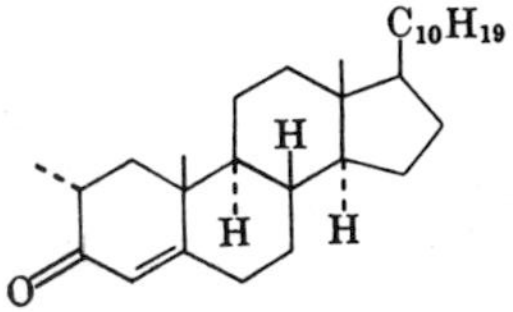 $C_{10}H_{19}$; H	Li/NH_3, Et_2O	*t*-BuOH	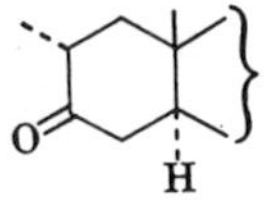 H (80)	505
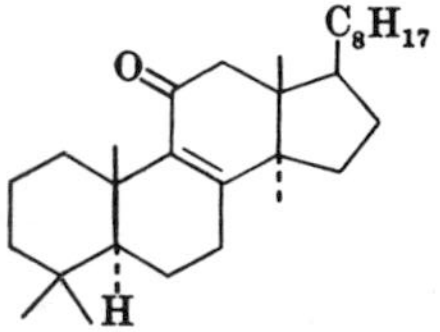 C_8H_{17}; O; H	Li/ND_3, Et_2O 30 min	NH_4Cl, CrO_3, KOH	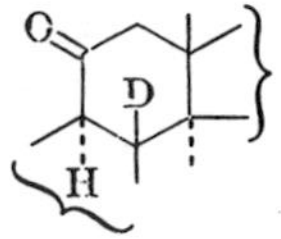 O; D; H (80)	493
	Li/NH_3, Et_2O, 30 min	NH_4Cl, CrO_3	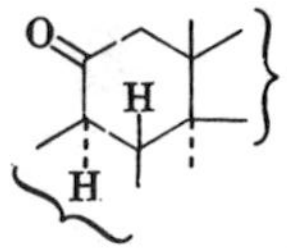 O; H (87)	

Substrate	Conditions	Workup	Product(s) and Yield(s) (%)	Refs.
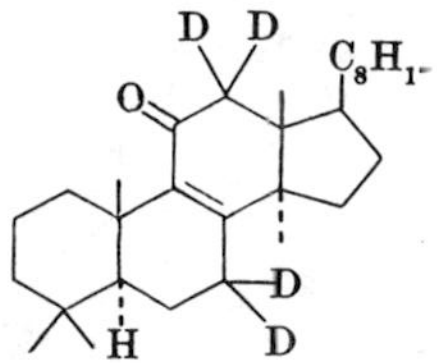	Li/NH_3, Et_2O, 30 min	NH_4Cl, KOH-MeOH	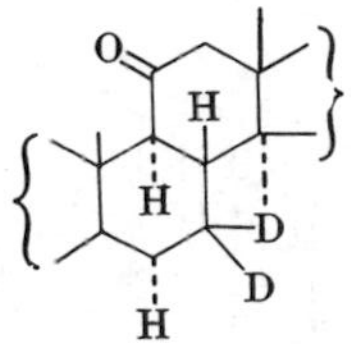 (84)	493
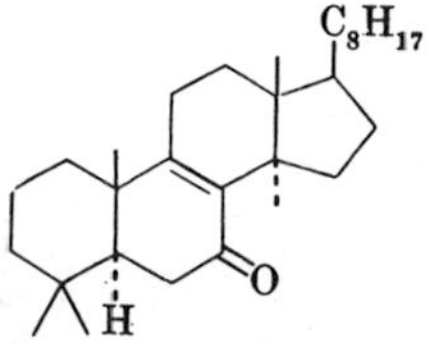	Li/ND_3, Et_2O, 30 min	NH_4Cl, CrO_3, KOH-MeOH	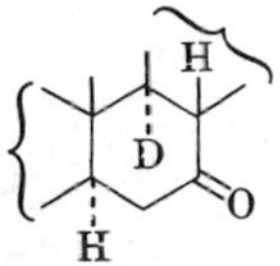 (80)	493
	Li/NH_3, Et_2O, 30 min	NH_4Cl, CrO_3	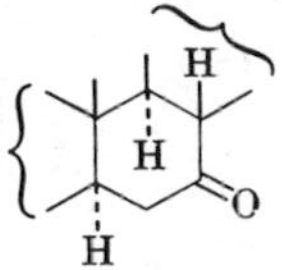 (92)	
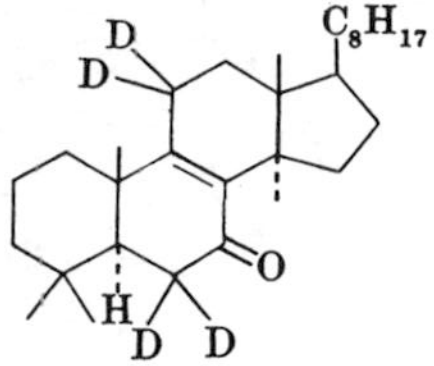	Li/NH_3, Et_2O, 30 min	NH_4Cl, KOH-MeOH	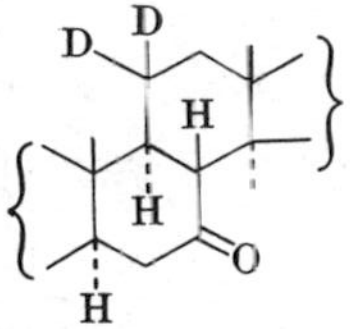 (60)	493

Note: References 338–543 are on pp. 253–258.

TABLE V. Reduction of Tetracyclic α,β-Unsaturated Ketones (*Continued*)

Reactant	Reduction Conditions	Quenching Agent	Product(s) (% Yield)	Refs.
C_{39} (*Contd.*)				
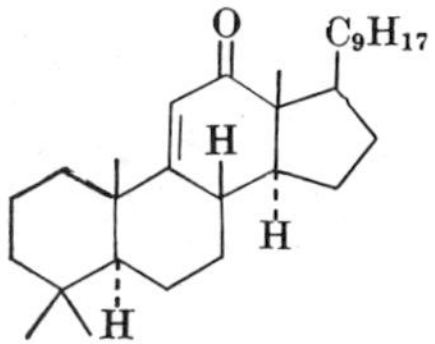 O, C_9H_{17}, H, H, H	Li/NH_3, Et_2O 30 min	NH_4Cl	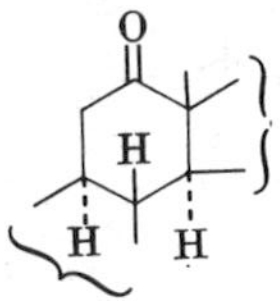 O, H, H, H (89)	493
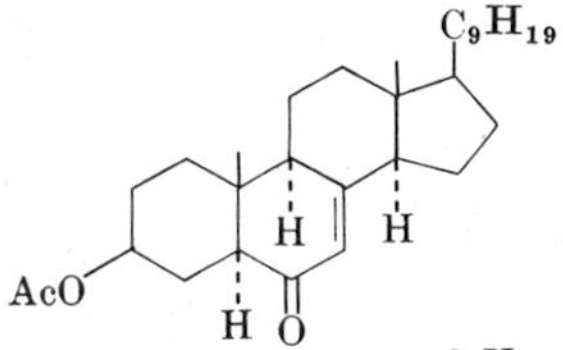 C_9H_{19}, AcO, H, H, H, O	Li/NH_3, Et_2O	NH_4Cl, Ac_2O	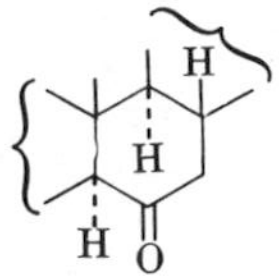 H, H, H, O (−)	9
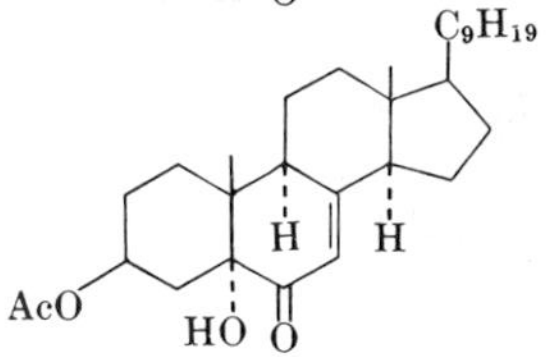 C_9H_{19}, AcO, H, H, HO, O	Li/NH_3, Et_2O	NH_4Cl, Ac_2O	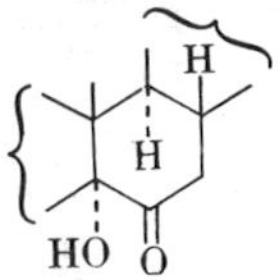 H, H, HO, O (47)	9
C_{32}				
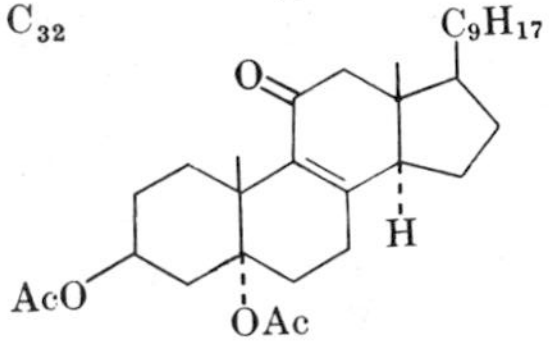 O, C_9H_{17}, H, AcO, OAc	Li/NH_3, Et_2O	NH_4Cl, Ac_2O	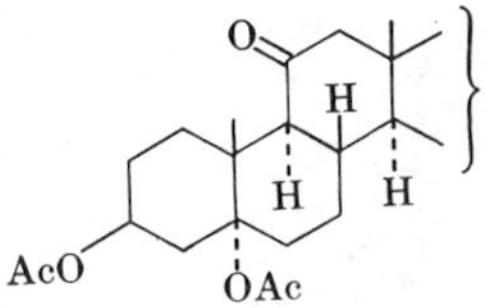 O, H, H, H, AcO, OAc (36) +	510

Substrate	Reagent	Quench	Product(s) and Yield(s) (%)	Refs.
			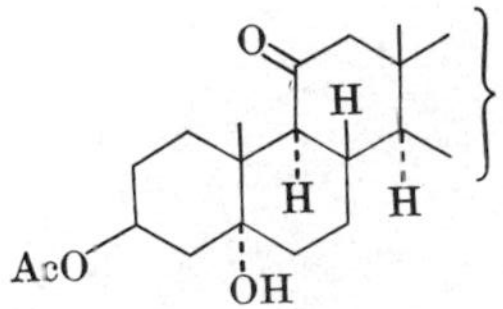 } (13) +	
			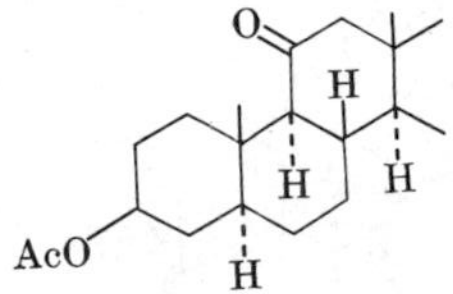 (6)	
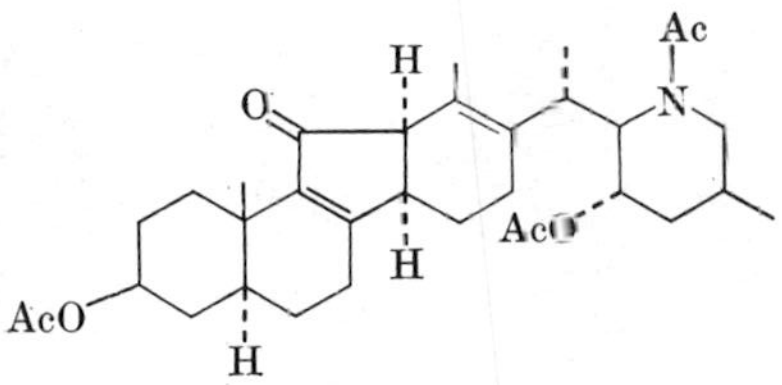	Li/NH_3, dioxane, MeOH	NH_4Cl	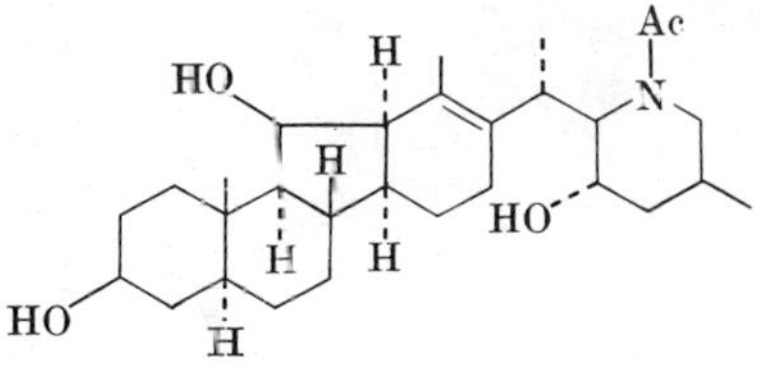 (55)	262
C_{35} 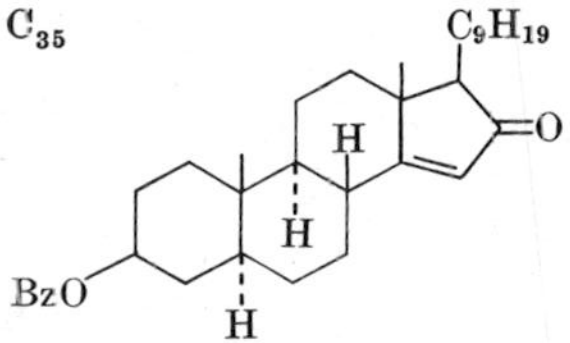	Li/NH_3	—	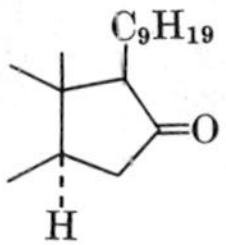 ~25 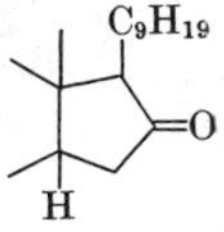 + ~75	119

Note: References **338–543** are on pp. 253–258.

TABLE VI. REDUCTIONS OF α,β-UNSATURATED KETONES CONTAINED IN SYSTEMS HAVING FIVE OR MORE FUSED RINGS

Reactant	Reduction Conditions	Quenching Agent	Product(s) (% Yield)	Refs.
C_{19} 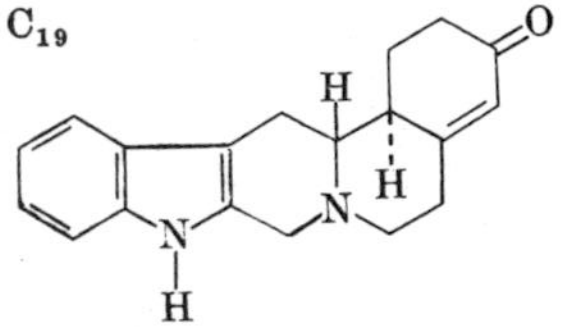	Na/NH_3	—	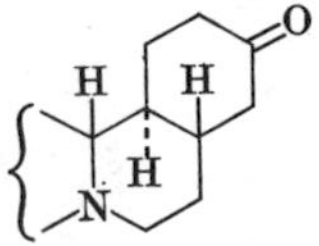 (−)	511
C_{22} 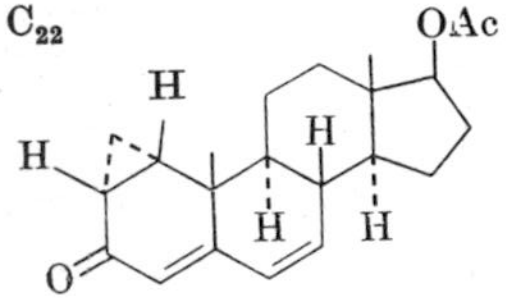	Li/NH_3, THF, −70°, 10 min	EtOH, Ac_2O	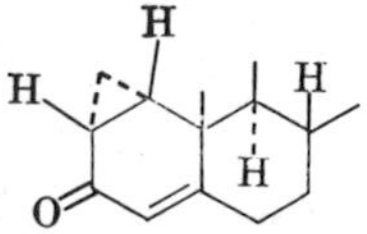 (50)	272
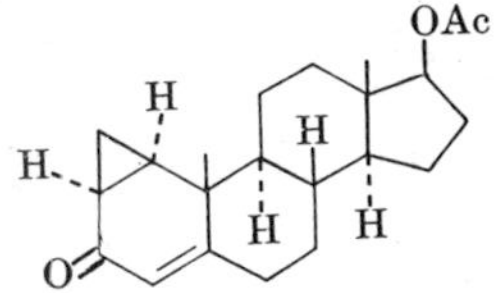	Li/NH_3, dioxane-Et_2O	NH_4Cl	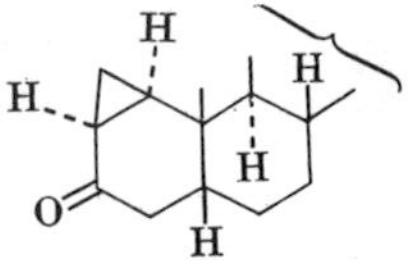 (45) + 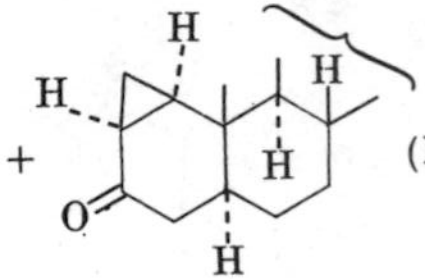 (10)	100
C_{23} 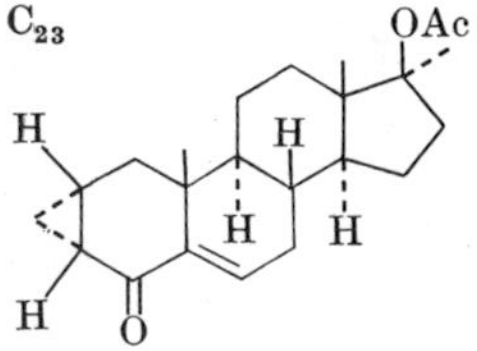	Li/NH_3, Et_2O	NH_4Cl, LAH, CrO_3	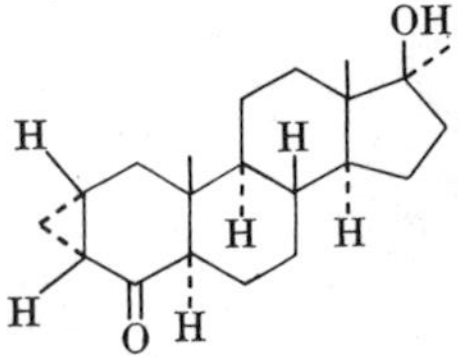 (66)	268

C_{24} 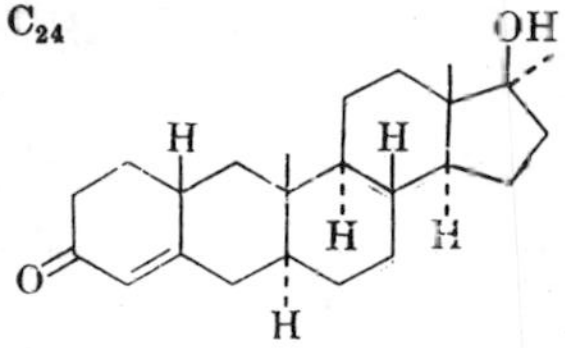	Li/NH_3, THF	NH_4Cl	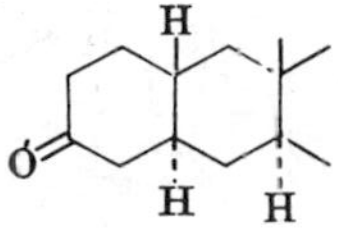 (26) + corresponding alcohol	512
C_{27} 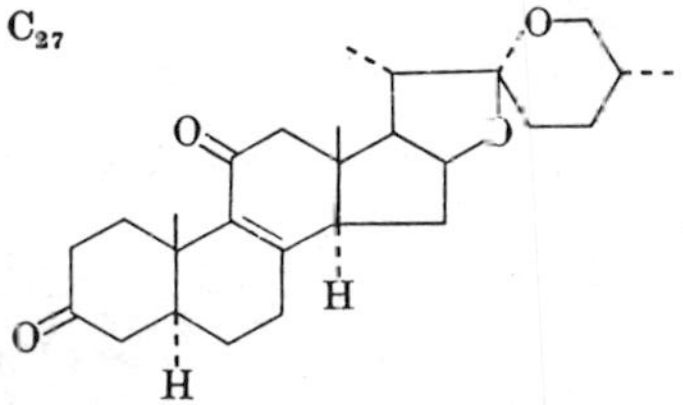	NH_3/Li, Et_2O-MeOH, 10 min	NH_4Cl	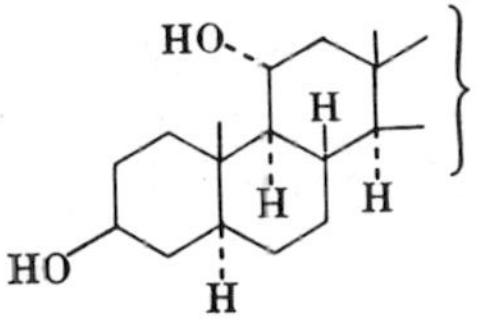 (30)	513
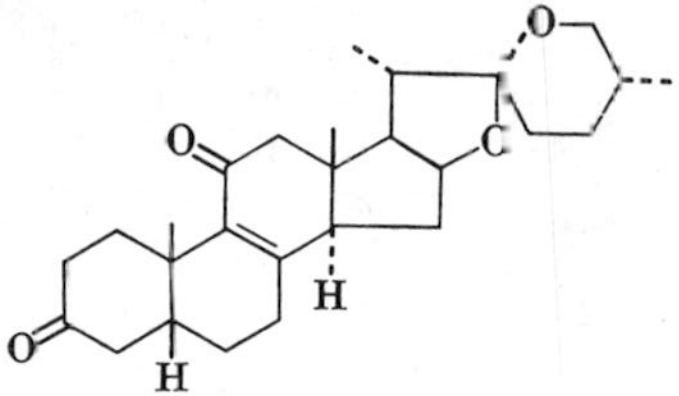	Li/NH_3, Et_2O, −40°, 10 min	NH_4Cl, CrO_3	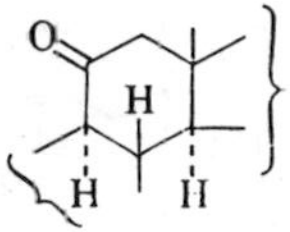 (74)	514
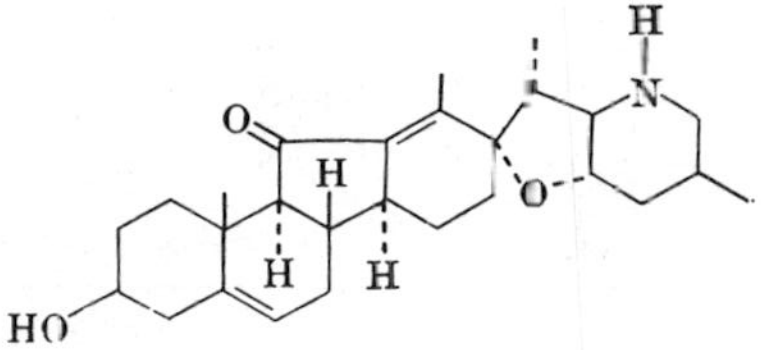	Li/NH_3, dioxane 7 min	NH_4Cl	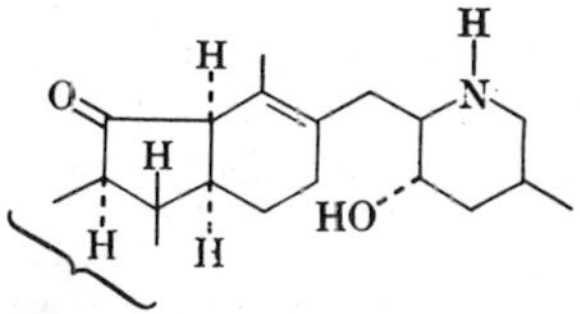 (~76)	491

Note: References 338–545 are on pp. 253–258.

TABLE VI. REDUCTIONS OF α,β-UNSATURATED KETONES CONTAINED IN SYSTEMS HAVING FIVE OR MORE FUSED RINGS (*Continued*)

Reactant	Reduction Conditions	Quenching Agent	Product(s) (% Yield)	Refs.
C_{27} (*Contd.*)				
H, N, O, H, H, H, O	Li/NH_3, dioxane	NH_4Cl	O, H, H, H (68)	516
HO, H, N, O, H, H, H, O	Li/NH_3, dioxane	NH_4Cl	O, H, H, H (50)	516
C_{28}				
C_9H_{17}, H, H, O	Li/NH_3, Et_2O	NH_4Cl	H, H, O (96)	271
C_{29}				
O, O, O, H, AcO, H	Li/NH_3	—	O, H, H, H (—)	6

Substrate	Conditions	Work-up	Product (Yield, %)	Refs.
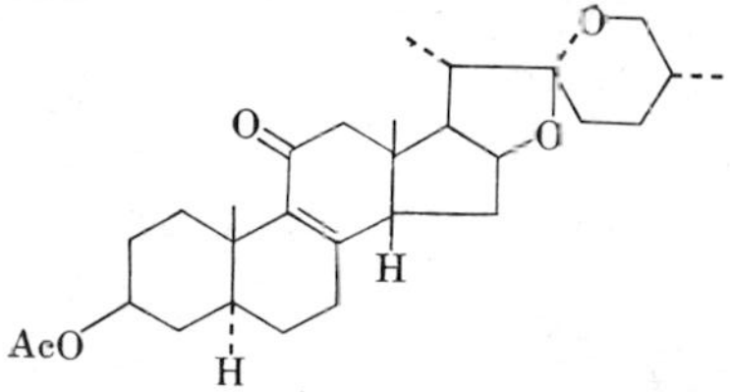 [structure]	Li/NH_3, Et_2O, −40°, 4 min	NH_4Cl, KOH, MeOH	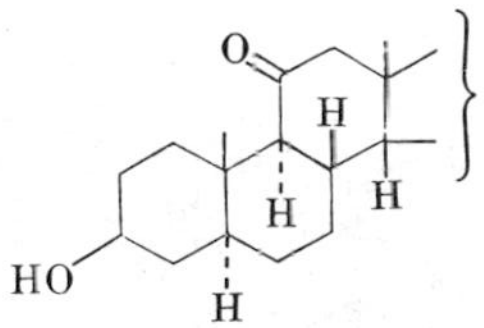 [structure] (56)	315
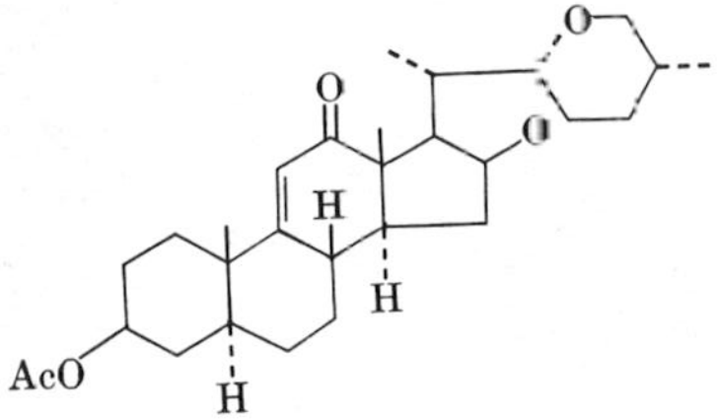 [structure]	Li/NH_3, Et_2O, 10 min	NH_4Cl, KOH, MeOH	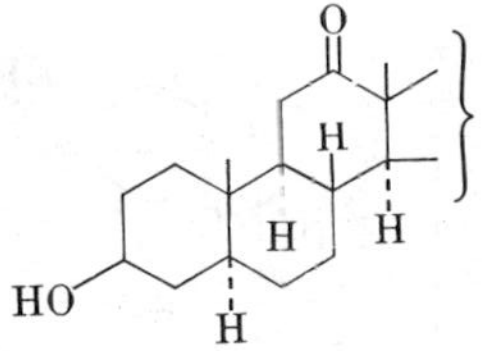 [structure] (68)	311
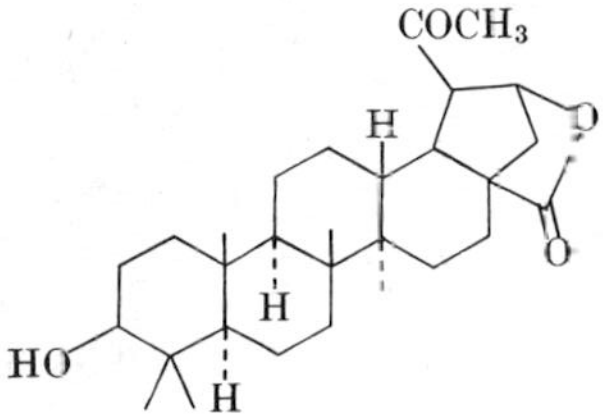 [structure]	Ca/NH_3[a]	—	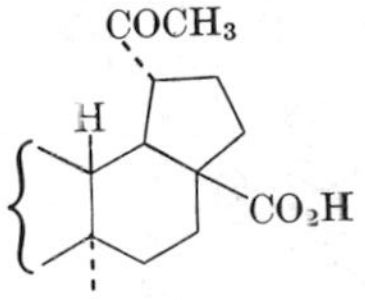 [structure] (—)	515

Note: References 338–543 are on pp. 253–253.

[a] The enone resulting from β elimination of the carboxylate grouping was considered to be an intermediate in this reduction.

TABLE VI. REDUCTIONS OF α,β-UNSATURATED KETONES CONTAINED IN SYSTEMS HAVING FIVE OR MORE FUSED RINGS (*Continued*)

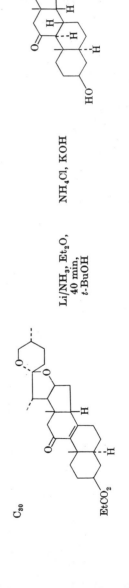

Reactant	Reduction Conditions	Quenching Agent	Product(s) (% Yield)	Refs.
C_{29} (*Contd.*)				
AcO, H, H, O, O, O	Li/NH_3[b], Et_2O-dioxane	NH_4Cl, KOH	HO, H, H, H, H, O (61)	5, 274
	Li/NH_3, MeOH, 15 min	NH_4Cl	HO, HO, H, H, H, H (58)	
C_{30}				
$EtCO_2$, H, H, O, O, O	Li/NH_3, Et_2O, 40 min, *t*-BuOH	NH_4Cl, KOH	HO, H, H, H, H, O (50)	314

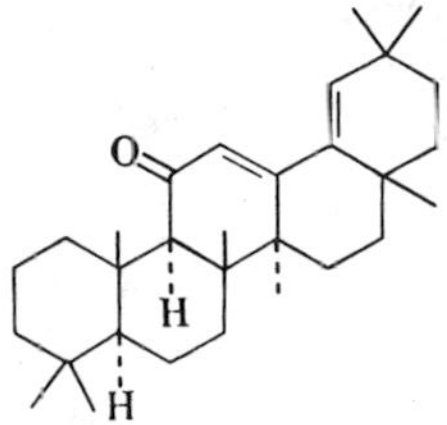	Ca/NH_3, THF, 30 min	t-BuOH	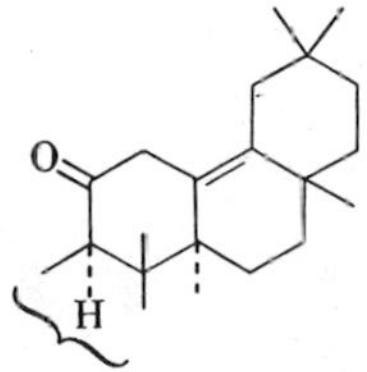 (—)	517
C_{32} 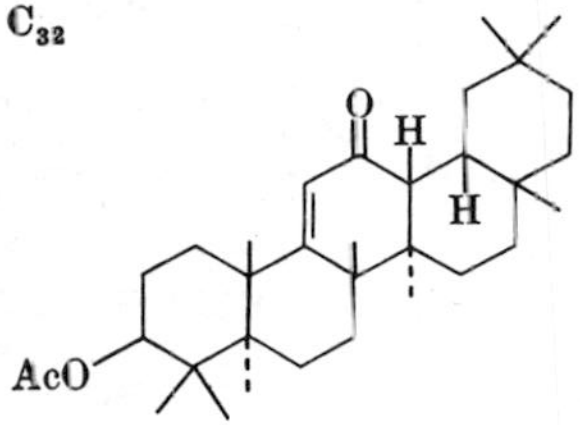	Li/NH_3, Et_2O	Acetone, Ac_2O	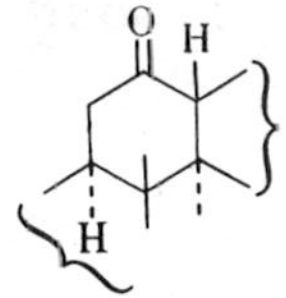 (60)	9
C_{37} 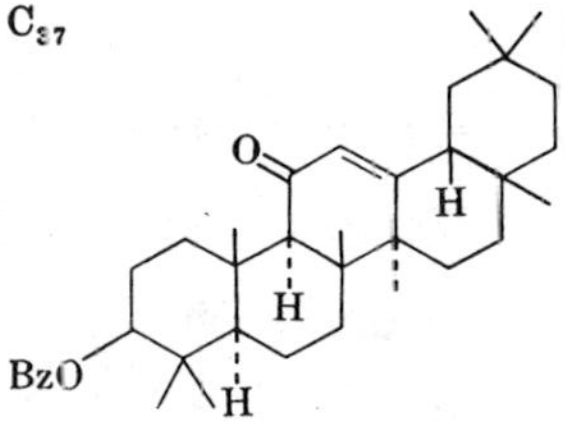	Li/NH_3, Et_2O	Acetone, KOH, Ac_2O	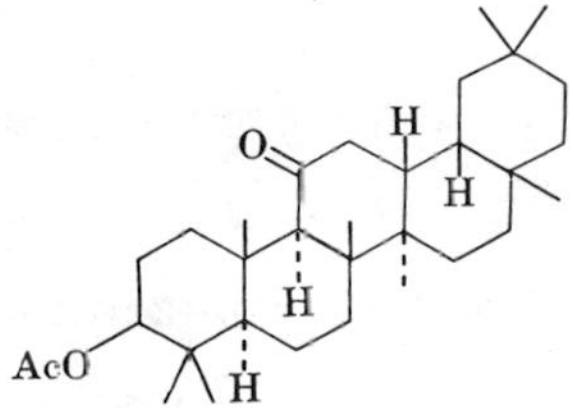 (—)	9

Note: References 338–543 are on pp. 253–258

[b] Calcium and sodium gave lower yields than lithium in this reduction.

TABLE VII. Reductions of α,β-Unsaturated Acids, Esters, and Aldehydes

Reactant	Reduction Conditions	Quenching Agent	Product(s) (% Yield)	Refs.
C_4 $CH_3CH{=}CHCO_2H$	4 eq. Li/NH_3, Et_2O, 30 min	NH_4Cl	n-$C_3H_7CO_2H$ (73)	216
C_5 $(CH_3)_2C{=}CHCO_2H$	4 eq. Li/NH_3, Et_2O, 30 min	NH_4Cl	$(CH_3)_2CHCH_2CO_2H$ (92)	216
C_6 $CH_3C(OCH_2OCH_3){=}CHCO_2H$	7 eq. Li/NH_3, Et_2O, 30 min	NH_4Cl	n-$C_3H_7CO_2H$ (78)	216
C_8 $CH_3C(OCH_2OCH_3){=}CHCO_2Et$	7 eq. Li/NH_3, Et_2O, 12 min, −33°, 6 min, −78°	NH_4Cl	n-$C_3H_7CO_2Et$ (39)[a]	211
C_9 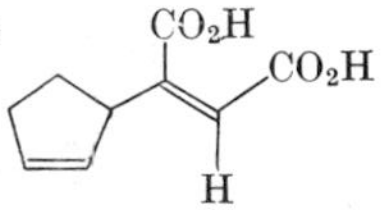 2-cyclopentenyl-C(CO_2H)=CH(CO_2H) (structure: CO_2H, CO_2H, H)	NH_3/K	i-PrOH, CH_2N_2	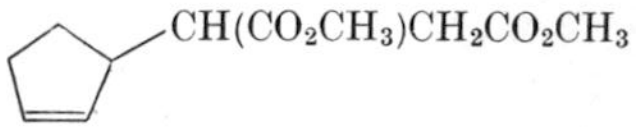 2-cyclopentenyl-$CH(CO_2CH_3)CH_2CO_2CH_3$ (20)	518
$C_6H_5CH{=}CHCO_2H$ (*trans*)	4 eq. Li/NH_3, Et_2O, 15 min, −33°, 5 min, −78°	NH_4Cl	$C_6H_5CH_2CH_2CO_2H$ (65)	211, 216
C_{10} 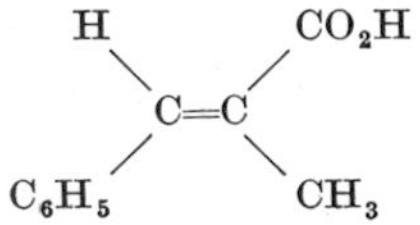 (H)(C_6H_5)C=C(CO_2H)(CH_3)	4 eq. Li/NH_3, Et_2O, 30 min	NH_4Cl	$C_6H_5CH_2CH(CH_3)CO_2H$ (95)	216
C_{11} $C_6H_5CH{=}CHCO_2Et$ (*trans*)	3 eq. Li/NH_3, Et_2O, 12 min, −33°, 6 min, −78°	NH_4Cl	$C_6H_5CH_2CH_2CO_2Et$ (18)	211
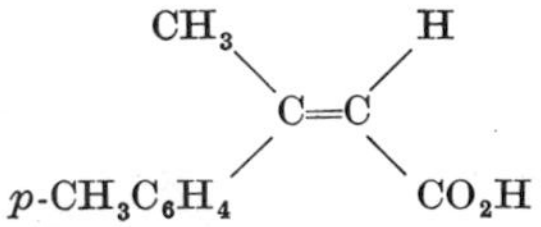 (CH_3)(p-$CH_3C_6H_4$)C=C(H)(CO_2H)	4 eq. Li/NH_3, Et_2O, 30 min	NH_4Cl	p-$CH_3C_4H_6CH(CH_3)CH_2CO_2H$ (98)	519

[a] This yield is based upon the quantity of the corresponding β-keto ester used to prepare the enol ether.

Substrate	Conditions	Quench	Product(s) (% yield)	Refs.
C_{12}				
$CH_3(CH_2)_8CH{=}CHCO_2H$	4 eq. $LiNH_3$, Et_2O, 30 min	NH_4Cl	$CH_3(CH_2)_{10}CO_2H$ (70)	216
$CH_3C(OCH_2OCH_3){=}C(Bu\text{-}n)CO_2Et$	7 eq. Li/NH_3, Et_2O, 12 min, −33°, 6 min, −78°	NH_4Cl	$C_2H_5CH(Bu\text{-}n)CO_2Et$	211
C_{13} 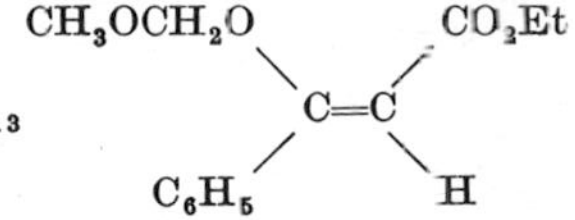 $(CH_3OCH_2O)(C_6H_5)C{=}C(CO_2Et)(H)$	7 eq. Li/NH_3, 12 min, −33°, 6 min, −78°	NH_4Cl	$C_6H_5CH_2CH_2CO_2Et$ (23)[a]	211
C_{16} 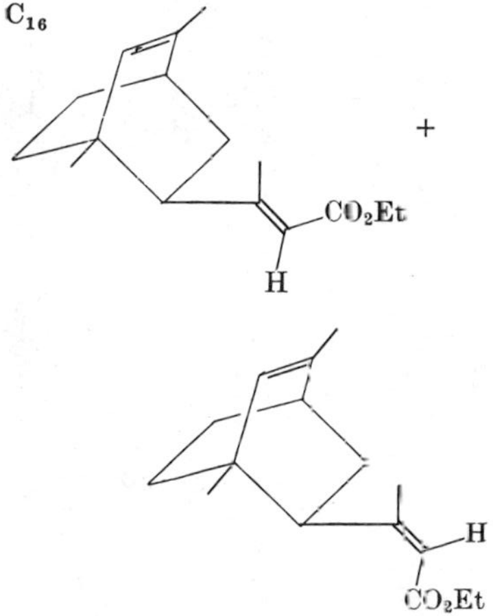 [bicyclic ester, CO_2Et / H isomer] + [bicyclic ester, H / CO_2Et isomer]	NH_3/excess Li, EtOH-Et_2O	H_2O	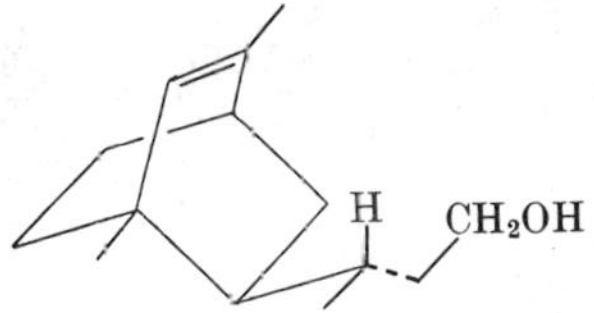 [bicyclic alcohol, H, CH_2OH] (60) 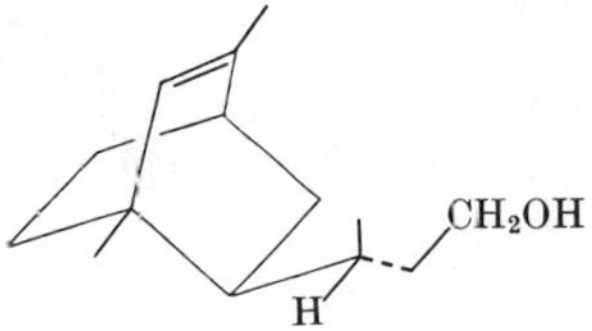 + [bicyclic alcohol, CH_2OH, H] (15)	520
C_5 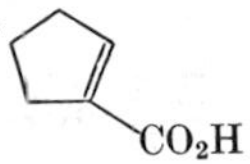 [cyclopentene-1-CO_2H]	4 eq. Li/NH_3, Et_2O, 30 min	NH_4Cl	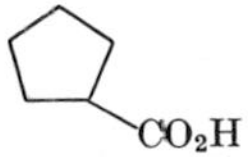 [cyclopentane-CO_2H] (82)	216

TABLE VII. Reductions of α,β-Unsaturated Acids, Esters, and Aldehydes (*Continued*)

Reactant	Reduction Conditions	Quenching Agent	Product(s) (% Yield)	Refs.
C_7 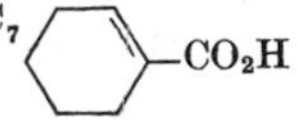 cyclohexene–CO_2H	4 eq. Li/NH_3, Et_2O, 30 min	NH_4Cl	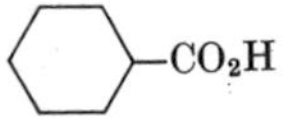 cyclohexane–CO_2H (94)	216
C_8 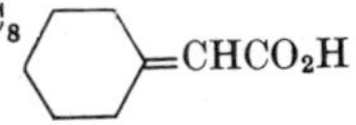 cyclohexylidene=$CHCO_2H$	4 eq. Li/NH_3, Et_2O, 30 min	NH_4Cl	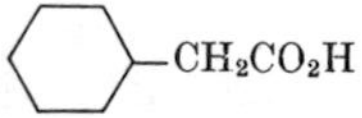 cyclohexane–CH_2CO_2H (93)	216
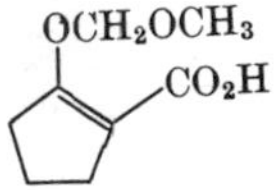 cyclopentene with OCH_2OCH_3, CO_2H	7 eq. Li/NH_3, Et_2O, 30 min	NH_4Cl	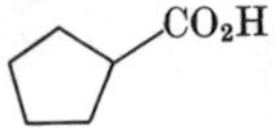 cyclopentane–CO_2H (74)	216
C_{10} 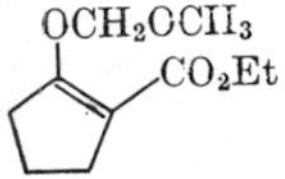 cyclopentene with OCH_2OCH_3, CO_2Et	7 eq. Li/NH_3, Et_2O, 12 min, −33°, 6 min, −78°	NH_4Cl	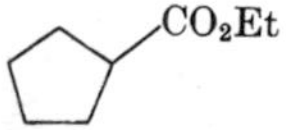 cyclopentane–CO_2Et (25)[a]	211
C_{11} 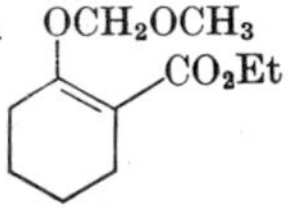 cyclohexene with OCH_2OCH_3, CO_2Et	7 eq. Li/NH_3, Et_2O, 12 min, −33°, 6 min, −78°	NH_4Cl	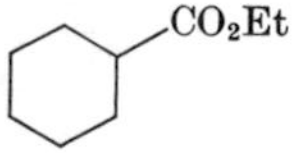 cyclohexane–CO_2Et (43)[a]	211
C_{15} 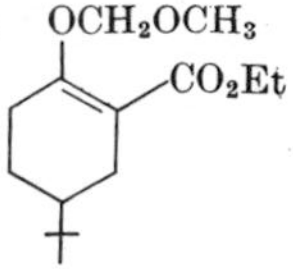 4-tert-butylcyclohexene with OCH_2OCH_3, CO_2Et	7 eq. Li/NH_3, Et_2O, 12 min, −33°, 6 min, −78°	NH_4Cl	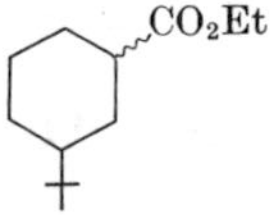 4-tert-butylcyclohexane–CO_2Et (40)[a,b]	211

Substrate	Reducing conditions	Quenching reagent	Product(s) and Yield(s) (%)	Refs.
C_{13} CHO 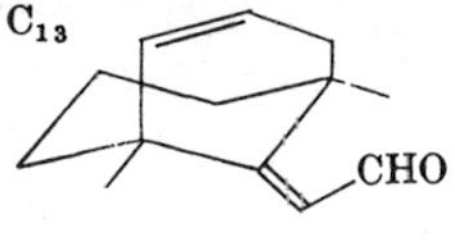	Li/NH_3, Et_2O, MeOH	CrO_3, CH_2N_2	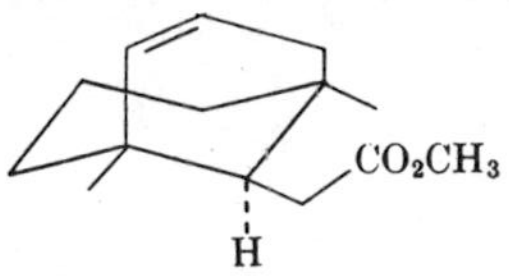 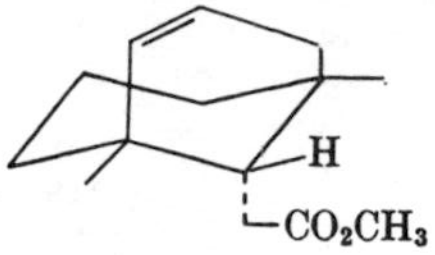 CO_2CH_3, H 35 + H, CO_2CH_3 65	521
C_{14} 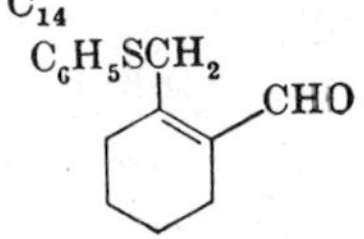 $C_6H_5SCH_2$ CHO	Li/NH_3	MeI	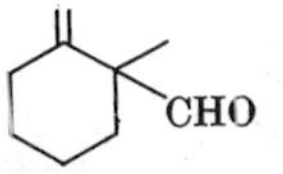 CHO (50)	521a
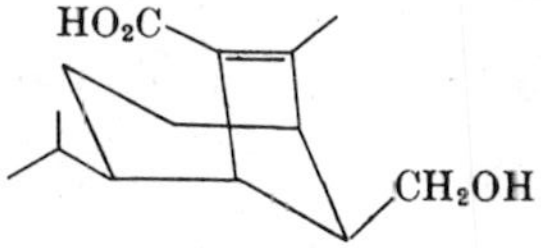 HO_2C, CH_2OH	Excess Li/NH_3, Et_2O	NH_4Cl	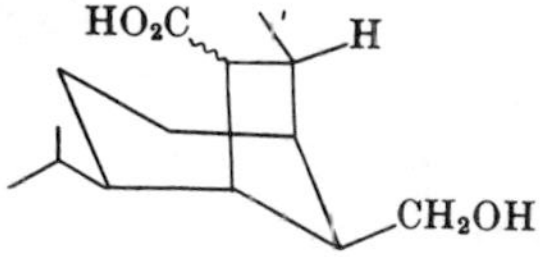 HO_2C, H, CH_2OH (70)	522
C_{15} 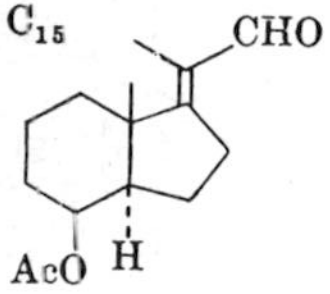 CHO, AcO, H	Li/NH_3, Et_2O	NH_4Cl, inverse addition, Ac_2O, NaOAc	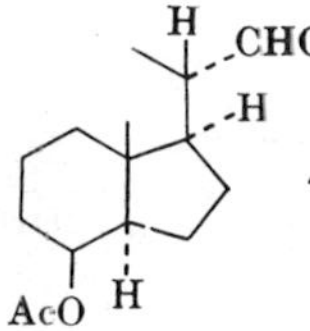 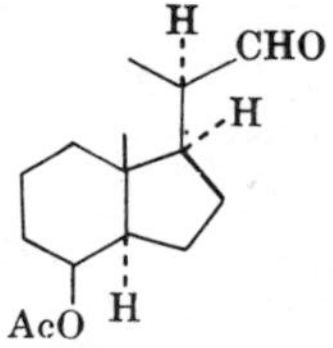 H, CHO, H, AcO, H ~40 + H, CHO, H, AcO, H ~60	523
C_{16} 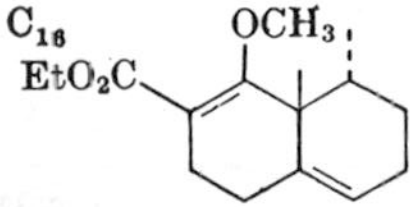 EtO_2C, OCH_3	7 eq. Li/NH_3, Et_2O, 60 min, $-33°$	NH_4Cl	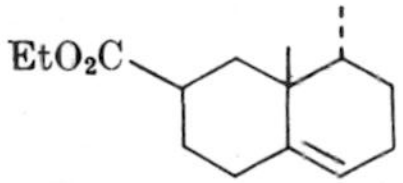 EtO_2C (67)	210

Note: References 338–543 are on pp. 253–258.

[a] This yield is based upon the quantity of the corresponding β-keto ester used to prepare the enol ether.

[b] A 70/30 mixture of *cis* and *trans* isomers was obtained.

TABLE VII. REDUCTIONS OF α,β-UNSATURATED ACIDS, ESTERS, AND ALDEHYDES (*Continued*)

Reactant	Reduction Conditions	Quenching Agent	Product(s) (% Yield)	Refs.
C_{16} (*Contd.*)				
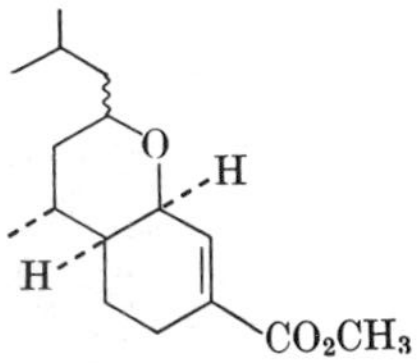 [structure] CO_2CH_3	Ca/NH_3, THF, 4 min	$C_6H_5CO_2Na$, H_2O	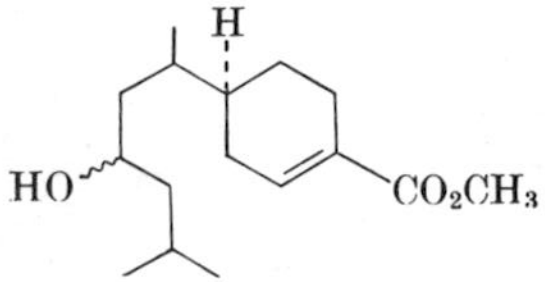 [structure] HO, CO_2CH_3 (—)	524
C_{17}				
OAc, EtO_2C [structure]	7 eq. Li/NH_3, Et_2O, 35 min, −33°, 10 min, −78°	NH_4Cl	EtO_2C [structure] (34)	210
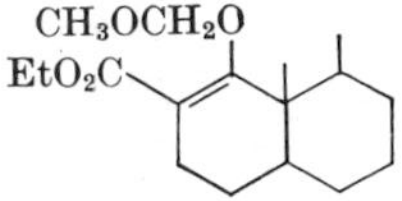 CH_3OCH_2O, EtO_2C [structure]	7 eq. Li/NH_3, 12 min, −33° Et_2O, 10 min, −78°	NH_4Cl	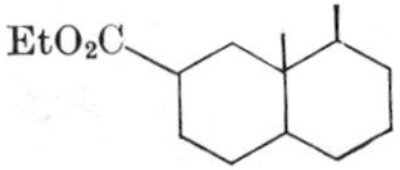 EtO_2C [structure] (60)	210
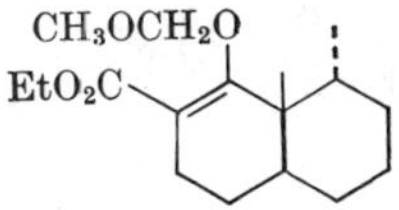 CH_3OCH_2O, EtO_2C [structure]	"	NH_4Cl	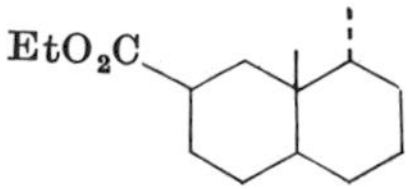 EtO_2C [structure] (61)	210
C_{18}				
HO_2C, CO_2H, $CHCO_2H$, CH_3O [structure]	NH_3/Na, Et_2O	H_3O^+	HO_2C, CO_2H, CH_2CO_2H, CH_3O [structure] (—)	525

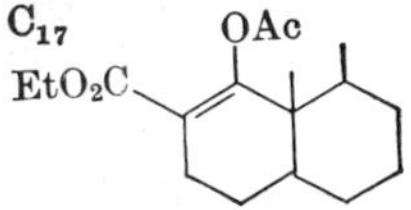

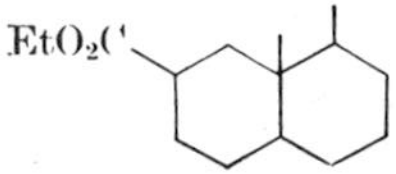

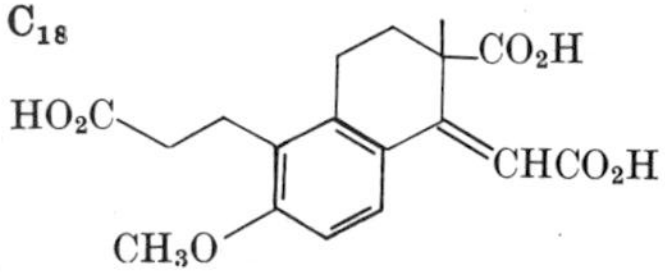

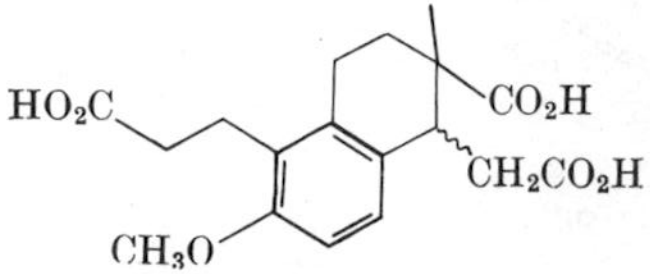

Substrate	Conditions	Quench	Product(s) and Yield(s) (%)	Refs.
C_{19} (*Contd.*) HO, H, H, H, H, CO_2H	NH_3/Li, THF 1.5 hr	EtOH	H, H, H, CH_2CO_2H	526
C_{20} OCH_3, CH_3OCH_2O, H, CO_2CH_3	Li/NH_3, DME	MeI, H_3O^+	OCH_3, H, CO_2CH_3 (80)	527
OTHP, CH_3OCH_2O, H, CO_2CH_3	Li/NH_3, DME	MeI, H_3O^+	OH, H, CO_2CH_3 (64)	528
C_{24} O, O, O, H, H, $CHCO_2H$	NH_3/Li, 2 hr, −40°	EtOAc, H_2O	HO, H, H, H, CH_2CO_2H (15) + HO, H, H, H, CH_2CO_2H (25)	7, 8

Note: References 338–543 are on pp. 253–258.

TABLE VII. Reductions of α,β-Unsaturated Acids, Esters, and Aldehydes (*Continued*)

Reactant	Reduction Conditions	Quenching Agent	Product(s) (% Yield)	Refs.
C_{24} (*Contd.*)				
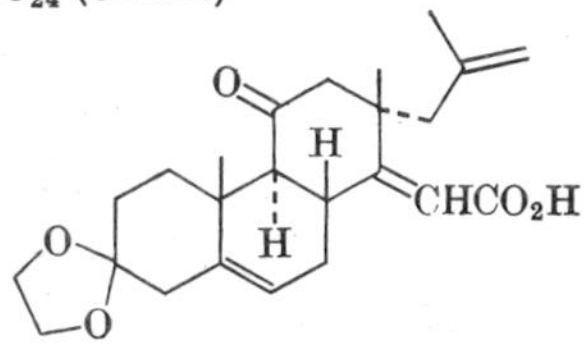 (structure: HO/O=, $CHCO_2H$)	NH_3/K, 10 min, *i*-PrOH (added dropwise)	H_2O	" (62) + " (10)	7, 8
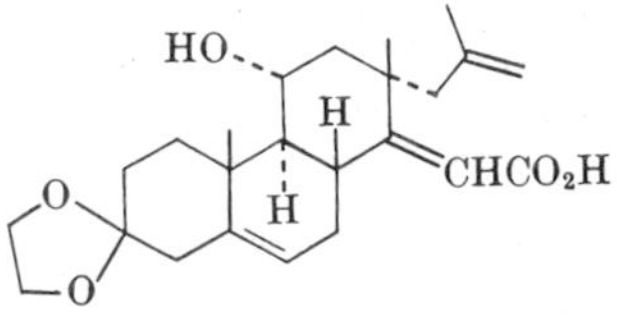 (structure: HO, $CHCO_2H$)	NH_3/Li	EtOAc, H_2O	" (30) + " (40)	7, 8
	NH_3/Li, 10 min, *i*-PrOH (added dropwise)	H_2O	" (30) + " (40)	7, 8
	NH_3/K	"	" (45) + " (—)	7, 8
	NH_3/K, *i*-PrOH (added dropwise)	"	" (80) + " (—)	7, 8
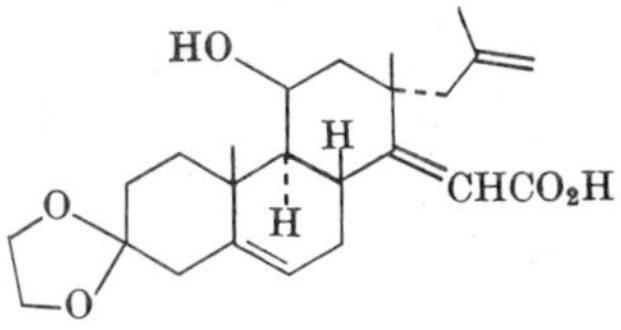 (structure: HO, $CHCO_2H$)	NH_3/Li, 2 hr, −40°	EtOAc, H_2O	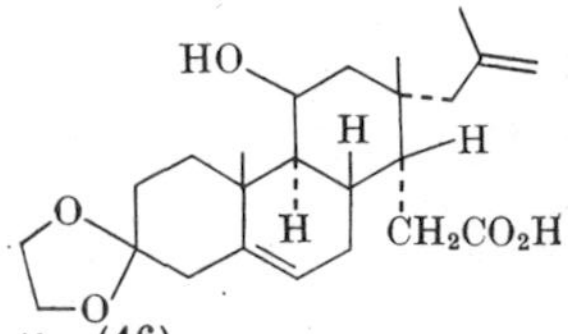 (structure: HO, H, CH_2CO_2H) (40)	7, 8
	NH_3/Na, *i*-PrOH (added dropwise)	"	" (46)	7, 8
	NH_3/K, 10 min, *i*-PrOH (added dropwise)	"	" (68)	7, 8

Substrate	Conditions	Workup	Product(s) and Yield(s) (%)	Refs.
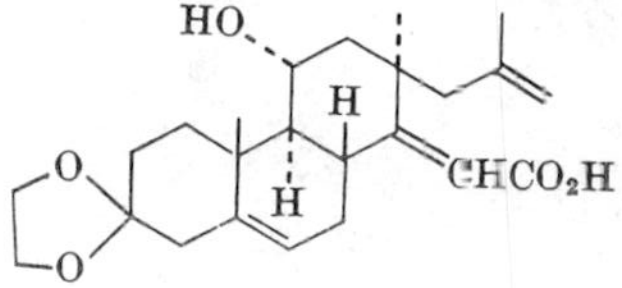 HO, H, H, O, O, $CHCO_2H$	NH_3/K, 10 min, *i*-PrOH (added dropwise)	H_2O	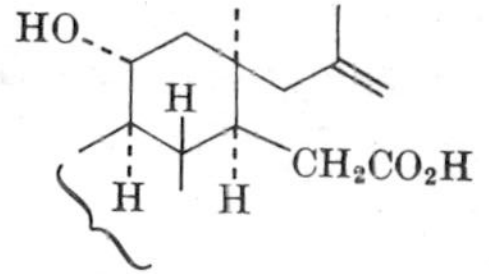 HO, H, H, H, CH_2CO_2H I (81)	7, 8
	NH_3/Li, THF, −40°, 10 min	"	I (60) 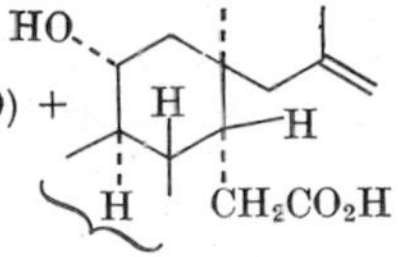 + HO, H, H, H, CH_2CO_2H (>15)	
C_{19} 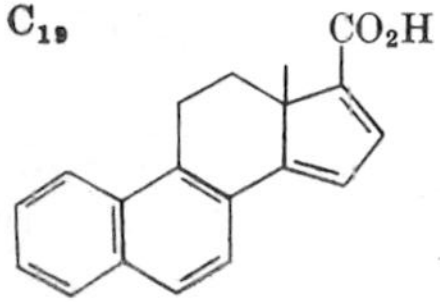 CO_2H	50 eq. Li/NH_3, dioxane, 8–11 hr	EtOAc, CH_2N_2	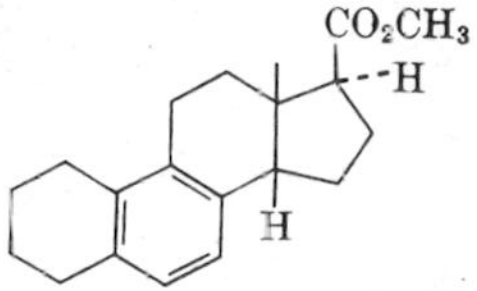 CO_2CH_3, H, H (50)	4
C_{21} 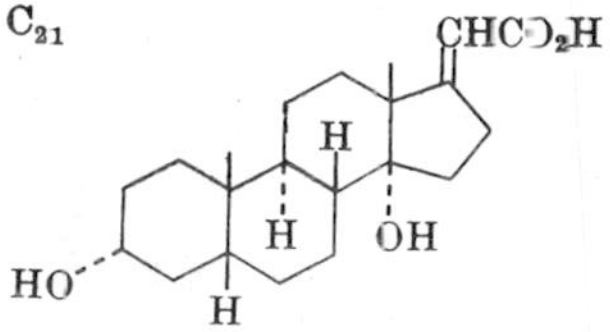 $CHCO_2H$, HO, H, H, H, OH	Excess K/NH_3, dioxane, add. −70°, 16 hr, −50°	—	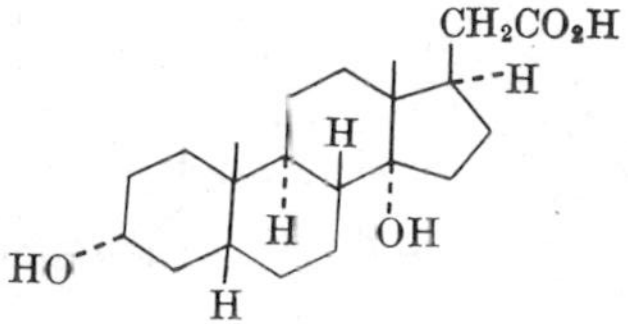 CH_2CO_2H, H, H, H, OH, HO, H (95)	529

Note: References 338–543 are on pp. 253–258.

TABLE VIII. REDUCTION-ALKYLATION AND RELATED REACTIONS OF ACYCLIC α,β-UNSATURATED KETONES

Reactant	Reduction Conditions	Quenching Agent	Product(s) (% Yield)	Refs.
C_6 $(CH_3)_2C{=}CHCOCH_3$	2 eq. Li/NH_3, Et_2O, 1 eq. t-BuOH, 30 min	1 eq. MeI	i-$C_3H_7CH(CH_3)COCH_3$ I (52) + i-$C_3H_7CH_2COCH_3$ II (21) + i-$C_3H_7CH_2CHOHCH_3$ III (8)	147
	,,	5 eq. MeI	I 34 + II 1 + III 9 + i-$C_3H_7CH(CH_3)COCH_2CH_3$ IV 33 + i-$C_3H_7CH(CH_3)COCH(CH_3)_2$ 23	
	2 eq. Li/NH_3, Et_2O, $(C_6H_5)_3COH$, 30 min	3 eq. MeI	I (75) + II (9) + III (6) + IV (8)	
C_{10} $C_2H_5COC(CH_3){=}CHSBu$-n	Li/NH_3, Et_2O, 2 eq. H_2O	$BrCH_2CH{=}CH_2$, 1 min	$C_2H_5COC(CH_3)_2CH_2CH{=}CH_2$ (82)	170
	,,	MeI, 1 min.	$C_2H_5COC(CH_3)_3$ (69)	
$C_6H_5CH{=}CHCOCH_3$ (*trans*)	Li/NH_3, Et_2O, t-BuOH, 30 min	1 eq. CH_3I	$C_6H_5CH_2CH(CH_3)COCH_3$ I (37) + $C_6H_5CH_2CH(CH_3)COCH_2CH_3$ II (11) + $C_6H_5CH_2CH_2COCH_3$ III (15) + $C_6H_5CH_2CH(CH_3)COCH(CH_3)_2$ IV (7)	147
	,,	6 eq. MeI	I 48 + II 41 + III 2 + IV 9	
	Li/NH_3, Et_2O, 1 eq. t-BuOLi, 30 min	,,	I 13 + II 20 + III 1 + IV 66	
	Li/NH_3, Et_2O, 1 eq. $(C_6H_5)_3COH$, 30 min	,,	I (50) + II (5) + III (14)	
	Li/NH_3, Et_2O, 1 eq. $(C_6H_5)_3COH$, 0.5 eq. $(C_6H_5)_3COLi$, 30 min	,,	I 58 + II 19 + III 23	

	$C_6H_5CH{=}CHCOCH_3$ (*trans*)	Li/NH_3, Et_2O, 1 eq. *t*-BuOH, 1 eq. 4,4-dimethylcyclohexanone, 30 min	5 eq. MeI	I 50 + II 26 + III 5 + IV 10	147
		Li/NH_3/*t*-BuOH, 6 eq. acetone, 30 min	14 eq. MeI	I 85 + II 1 + III 12 + IV 2	
C_{11}	$C_6H_5CH{=}C(CH_3)COCH_3$	Li/NH_3, 1 eq. *t*-BuOH, Et_2O, 30 min	1 eq. MeI	I 28 + $C_6H_5CH_2C(CH_3)_2COCH_3$ 62 + $C_6H_5CH_2C(CH_3)_2COCH_2CH_3$ 8	147
C_{15}	$C_6H_5CH{=}CHCOC_6H_5$ (*trans*)	2 eq. K/NH_3, Et_2O, 5 min	1 eq. $C_6H_5CH_2Cl$, NH_4Cl	$C_6H_5CH_2CH(C_6H_5)CH_2COC_6H_5$ (73)	50
		3 eq. K/NH_3, Et_2O, 20 min	"	" (66)	52
		2 eq. K/NH_3, Et_2O, 5 min	2 eq. $C_6H_5CH_2Cl$, NH_4Cl	$C_6H_5CH(CH_2C_6H_5)CH(CH_2C_6H_5)COC_6H_5$ (76)	50
		"	$-NH_3$, Et_2O, 2 eq. $C_6H_5CH_2Cl$	" (30)	
		"	$C_6H_5COC_6H_5$, Et_2O, 15 min, NH_4Cl	$(C_6H_5)_2C(OH)CH(C_6H_5)CH_2COC_6H_5$ (47)	
		2 eq. K/NH_3, Et_2O, 1 eq. *t*-BuOH	$-NH_3$, $C_6H_5CH_3$, $C_6H_5CH_2Cl$, reflux 4–6 hr	$(C_6H_5CH_2)_2CHCOC_6H_5$ (58)	53
		2 eq. Li/NH_3 Et_2O	MeI, $C_6H_5CH_3$	$C_6H_5CH_2CH(CH_3)COC_6H_5$ ~80 + $C_6H_5CH(CH_3)CH_2COC_6H_5$ + $C_6H_5CH_2CH_2COC_6H_5$ ~20	
		3 eq. K/NH_3, Et_2O, 20 min	$ClCH_2CH_2N(CH_3)_2$, 2–4 hr, Et_2O, NH_4Cl	$(CH_3)_2NCH_2CH_2CH(C_6H_5)CH_2COC_6H_5$ (65)	52
		"	$ClCH_2CH_2N(C_2H_5)_2$, 2–4 hr, NH_4Cl	$(C_2H_5)_2NCH_2CH_2CH(C_6H_5)CH_2COC_6H_5$ (59)	

Note: References 338–543 are on pp. 253–258.

TABLE VIII. REDUCTION-ALKYLATION AND RELATED REACTIONS OF ACYCLIC α,β-UNSATURATED KETONES (*Continued*)

Reactant	Reduction Conditions	Quenching Agent	Product(s) (% Yield)	Refs.
C_{15} $C_6H_5CH{=}CHCOC_6H_5$ (*Contd.*)	3 eq. K/NH_3, Et_2O, 20 min	$ClCH_2CH_2N$(piperidino), Et_2O, 2–4 hr, NH_4Cl	(piperidino)$NCH_2CH_2CH(C_6H_5)CH_2COC_6H_5$ (52)	52
	,,	$ClCH_2CH_2N$(morpholino), Et_2O, 2–4 hr, NH_4Cl	(morpholino)$NCH_2CH_2CH(C_6H_5)CH_2COC_6H_5$ (51)	
	,,	$Cl(CH_2)_3N(CH_3)_2$, Et_2O, 2–4 hr, NH_4Cl	$(CH_3)_2N(CH_2)_3CH(C_6H_5)CH_2COC_6H_5$ (79)	
	,,	n-PrBr, Et_2O, NH_4Cl	n-$PrCH(C_6H_5)CH_2COC_6H_5$ (63)	
	,,	$Br(CH_2)_2CH(CH_3)_2$ Et_2O, NH_4Cl	$(CH_3)_2CH(CH_2)_2CH(C_6H_5)CH_2COC_6H_5$ (65)	
	,,	$BrCH_2CH{=}CH_2$, Et_2O, NH_4Cl	$CH_2{=}CHCH_2CH(C_6H_5)CH_2COC_6H_5$ (35)	
	3 Na/NH_3, Et_2O, 20 min	,,	,, (35)	
	3 K/NH_3, Et_2O, 20 min	$I(CH_2)_4C{\equiv}CH$, NH_4Cl	$HC{\equiv}C(CH_2)_4CH(C_6H_5)CH_2COC_6H_5$ (28)	
	,,	$Br(CH_2)_4Br/Et_2O$, NH_4Cl	$C_6H_5CHCH_2COC_6H_5$ — $(CH_2)_4$ — $C_6H_5CHCH_2COC_6H_5$ (58)	
C_{16} p-$CH_3C_6H_4CH{=}CH$-COC_6H_5	2 eq. Li/NH_3, Et_2O, 20 min	NH_3, $C_6H_5CH_3$ MeI, 5 hr	p-$CH_3C_6H_4CH_2CH(CH_3)COC_6H_5$ 80 + { p-$CH_3C_6H_4CH(CH_3)CH_2COC_6H_5$ + p-$CH_3C_6H_4CH_2CH_2COC_6H_5$ } 20	53

Reactant	Conditions	Reagents	Product (Yield, %)	Refs.
	2 eq. K/NH_3, Et_2O, 1 eq. t-BuOH	NH_3, $C_6H_5CH_3$, $C_6H_5CH_2Cl$	p-$CH_3C_6H_4CH_2CH(CH_2C_6H_5)COC_6H_5$ (62)	
	3 eq. K/NH_3, Et_2O, 20 min	$ClCH_2CH_2N(CH_3)_2$, NH_4Cl	p-$CH_3C_6H_4CH(CH_2CH_2N(CH_3)_2)CH_2COC_6H_5$ (52)	52
	3 eq. K/NH_3, Et_2O, 20 min	$ClCH_2CH_2N(C_2H_5)_2$, NH_4Cl	p-$CH_3C_6H_4CH(CH_2CH_2N(C_2H_5)_2)CH_2COC_6H_5$ (57)	
	,,	$ClCH_2CH_2NC_5H_{10}$, NH_4Cl	p-$CH_3C_6H_4CH(CH_2CH_2NC_5H_{10})CH_2COC_6H_5$ (65)	
	,,	$ClCH_2CH_2NC_4H_8O$, NH_4Cl	p-$CH_3C_6H_4CH(CH_2CH_2NC_4H_8O)CH_2COC_6H_5$ (42)	
	,,	$Cl(CH_2)_3N(CH_3)_2$, NH_4Cl	p-$CH_3C_6H_4CH((CH_2)_3N(CH_3)_2)CH_2COC_6H_5$ (58)	
	,,	n-PrBr	p-$CH_3C_6H_4CH(Pr$-$n)CH_2COC_6H_5$ (59)	
	,,	$BrCH_2Bu$-i, NH_4Cl	p-$CH_3C_6H_4CH(CH_2Bu$-$i)CH_2COC_6H_5$ (57)	
$C_6H_5CH{=}CHCOC_6H_4CH_3$-$p$	2 Li/NH_3, Et_2O, 20 min	—NH_3, $C_6H_5CH_3$, MeI, heat, 5 hr	$C_6H_5CH_2CH(CH_3)COC_6H_4CH_3$-$p$ (69)	53
	2 eq. K/NH_3, Et_2O, 1 eq. t-BuOH	—NH_3, $C_6H_5CH_3$, 1 eq. $C_6H_5CH_2Cl$, NH_4Cl	$C_6H_5CH_2CH(C_6H_5)CH_2COC_6H_4CH_3$-$p$ (68)	
	3 eq. K/NH_3, Et_2O, 20 min	$ClCH_2CH_2N(CH_3)_2$, NH_4Cl	$(CH_3)_2NCH_2CH_2CH(C_6H_5)CH_2COC_6H_4CH_3$-$p$ (74)	52
	,,	$ClCH_2CH_2N(C_2H_5)_2$, NH_4Cl	$(C_2H_5)_2NCH_2CH_2CH(C_6H_5)CH_2COC_6H_4CH_3$-$p$ (78)	
	,,	$ClCH_2CH_2NC_5H_{10}$, NH_4Cl	$C_5H_{10}NCH_2CH_2CH(C_6H_5)CH_2COC_6H_4CH_3$-$p$ (56)	

Note: References 338–543 are on pp. 253–258.

TABLE VIII. Reduction-Alkylation and Related Reactions of Acyclic α,β-Unsaturated Ketones (*Continued*)

Reactant	Reduction Conditions	Quenching Agent	Product(s) (% Yield)	Refs.
$C_6H_5CH{=}CHCOC_6H_4CH_3$-*p* (*Contd.*)	3 eq. K/NH_3, Et_2O, 20 min	$ClCH_2CH_2N(CH_2CH_2)_2O$, NH_4Cl	$O(CH_2CH_2)_2NCH_2CH_2CH(C_6H_5)CH_2COC_6H_4CH_3$-*p* (56)	52
		$Cl(CH_2)_3N(CH_3)_2$, NH_4Cl	$(CH_3)_2N(CH_2)_3CH(C_6H_5)CH_2COC_6H_4CH_3$-*p* (76)	
	"	*n*-PrBr, NH_4Cl	*n*-$PrCH(C_6H_5)CH_2COC_6H_4CH_3$-*p* (77)	
	"	$BrCH_2Bu$-*i*, NH_4Cl	*i*-$C_4H_9CH_2CH(C_6H_5)CH_2COC_6H_4CH_3$-*p* (83)	
	"	1 eq. $C_6H_5CH_2Cl$, NH_4Cl	$C_6H_5CH_2CH(C_6H_5)CH_2COC_6H_4CH_3$-*p* (58)	
	"	$Br(CH_2)_4Br$, NH_4Cl	$C_6H_5CHCH_2COC_6H_4CH_3$-*p* – $(CH_2)_4$ – $C_6H_5CHCH_2COC_6H_4CH_3$-*p* (22)	
$C_6H_5CH{=}CHCOC_6H_4OCH_3$-*p*	"	$ClCH_2CH_2N(CH_3)_2$, NH_4Cl	$(CH_3)_2NCH_2CH_2CH(C_6H_5)CH_2COC_6H_4OCH_3$-*p* (80)	52
	"	$ClCH_2CH_2N(C_2H_5)_2$, NH_4Cl	$(C_2H_5)_2NCH_2CH_2CH(C_6H_5)CH_2COC_6H_4OCH_3$-*p* (76)	
		$ClCH_2CH_2NC_5H_{10}$, NH_4Cl	$C_5H_{10}NCH_2CH_2CH(C_6H_5)CH_2COC_6H_4OCH_3$-*p* (74)	
		$ClCH_2CH_2N(CH_2CH_2)_2O$, NH_4Cl	$O(CH_2CH_2)_2NCH_2CH_2CH(C_6H_5)CH_2COC_6H_4OCH_3$-*p* (72)	
	"	$Cl(CH_2)_3N(CH_3)_2$, NH_4Cl	$(CH_3)_2N(CH_2)_3CH(C_6H_5)CH_2COC_6H_4OCH_3$-*p* (81)	
	"	*n*-PrBr, NH_4Cl	*n*-$PrCH(C_6H_5)CH_2COC_6H_4OCH_3$-*p* (78)	
	"	$BrCH_2Bu$-*i*, NH_4Cl	*i*-$BuCH_2CH(C_6H_5)CH_2COC_6H_4OCH_3$-*p* (73)	
	"	1 eq. $C_6H_5CH_2Cl$, NH_4Cl	$C_6H_5CH(CH_2C_6H_5)CH_2COC_6H_4OCH_3$-*p* (69)	

		,,	$Br(CH_2)_4Br$, NH_4Cl	$C_6H_5CHCH_2COC_6H_4OCH_3$-$p$ | $(CH_2)_4$ | $C_6H_5CHCH_2COC_6H_4OCH_3$-$p$ (38)	
C_{17}	p-$CH_3C_6H_4CH{=}CH$-$COC_6H_4CH_3$-p	2 eq. Li/NH_3, Et_2O, 20 min	—NH_3, $C_6H_5CH_3$, MeI, heat, 5 hr	p-$CH_3C_6H_4CH_2CH(CH_3)COC_6H_4CH_3$-$p$ (57)	53
		2 eq. K/NH_3, Et_2O, 1 eq. t-BuOH	—NH_3, $C_6H_5CH_3$, $C_6H_5CH_2Cl$, heat	p-$CH_3C_6H_4CH_2CH(CH_2C_6H_5)COC_6H_4CH_3$-$p$ (59)	
		3 eq. K/NH_3, Et_2O, 20 min	$ClCH_2CH_2N(CH_3)_2$	$CH_2CH_2N(CH_3)_2$ | p-$CH_3C_6H_4CHCH_2COC_6H_4CH_3$-$p$ (64)	52
		,,	$ClCH_2CH_2N(C_2H_5)_2$	$CH_2CH_2N(C_2H_5)_2$ | p-$CH_3C_6H_4CHCH_2COC_6H_4CH_3$-$p$ (62)	
		,,	$ClCH_2CH_2NC_5H_{10}$ (piperidino), NH_4Cl	$CH_2CH_2NC_5H_{10}$ (piperidino) | p-$CH_3C_6H_4CHCH_2COC_6H_4CH_3$-$p$ (61)	
		,,	$ClCH_2CH_2N(CH_2CH_2)_2O$ (morpholino), NH_4Cl	$CH_2CH_2N(CH_2CH_2)_2O$ (morpholino) | p-$CH_3C_6H_5CHCH_2COC_6H_4CH_3$-$p$ (48)	
		3 eq. K/NH_3, Et_2O, 20 min	$Cl(CH_2)_3N(CH_3)_2$, NH_4Cl	$(CH_2)_3N(CH_3)_2$ | p-$CH_3C_6H_4CHCH_2COC_6H_4CH_3$-$p$ (66)	
		,,	n-PrBr, NH_4Cl	p-$CH_3C_6H_4CH(Pr$-$n)CH_2COC_6H_4CH_3$-p (72)	
		,,	$BrCH_2C_4H_9$-i, NH_4Cl	$CH_2C_4H_9$-i | p-$CH_3C_6H_4CHCH_2COC_6H_4CH_3$-$p$ (68)	

Note: References 338–543 are on pp. 253–258.

TABLE VIII. REDUCTION-ALKYLATION AND RELATED REACTIONS OF ACYCLIC α,β-UNSATURATED KETONES

Reactant	Reduction Conditions	Quenching Agent	Product(s) (% Yield)	Refs.
p-$CH_3OC_6H_4CH{=}CH$-$COC_6H_4OCH_3$-p	3 eq. K/NH_3, Et_2O, 20 min	$ClCH_2CH_2N(CH_3)_2$, NH_4Cl	p-$CH_3OC_6H_4CH(CH_2CH_2N(CH_3)_2)CH_2COC_6H_4OCH_3$-p (49)	52
	,,	$ClCH_2CH_2N(C_2H_5)_2$, NH_4Cl	p-$CH_3OC_6H_4CH(CH_2CH_2N(C_2H_5))CH_2COC_6H_4OCH_3$-p (48)	
	,,	$ClCH_2CH_2NC_5H_{10}$, NH_4Cl	p-$CH_3OC_6H_4CH(CH_2CH_2NC_5H_{10})CH_2COC_6H_4OCH_3$-p (42)	
	,,	$ClCH_2CH_2NC_4H_8O$, NH_4Cl	p-$CH_3OC_6H_4CH(CH_2CH_2NC_4H_8O)CH_2COC_6H_4OCH_3$-p (28)	
	,,	$Cl(CH_2)_3N(CH_3)_2$, NH_4Cl	p-$CH_3OC_6H_4CH((CH_2)_3N(CH_3)_2)CH_2COC_6H_4OCH_3$-p (63)	
	,,	n-PrBr, NH_4Cl	p-$CH_3OC_6H_4CH$(Pr-n)$CH_2COC_6H_4OCH_3$-p (64)	
	,,	$BrCH_2$Bu-i, NH_4Cl	p-$CH_3OC_6H_4CH(CH_2$Bu-i$)CH_2COC_6H_4OCH_3$-p (69)	
C_{21} $(C_6H_5)_2C{=}CHCOC_6H_5$	,,	$ClCH_2CH_2NC_5H_{10}$, NH_4Cl	$(C_6H_5)_2C(CH_2CH_2NC_5H_{10})CH_2COC_6H_5$	52

Note: References 338–543 are on pp. 253–258.

TABLE IX. Reduction-Alkylation and Related Reactions of Monocyclic α,β-Unsaturated Ketones

Reactant	Reduction Conditions	Quenching Agent	Product(s) (% Yield)	Refs.
C_7 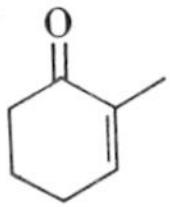	2.2 eq. Li/NH_3, Et_2O, 1 eq. *t*-BuOH, 30 min	Excess MeI	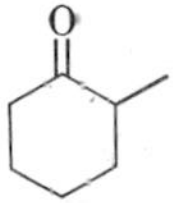 (~1) 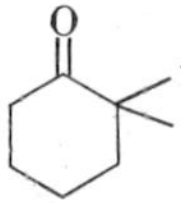 + (39) 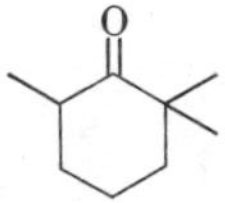 + (9)	72
	2.2 eq. Li/NH_3, Et_2O, 1 eq. H_2O, 30 min	"	" (~1) + " (60) + " (~1)	
	Li/NH_3	$ClCH_2CH{=}CClCH_3$	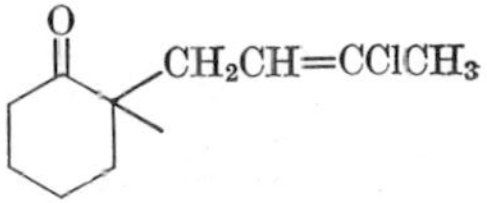 $CH_2CH{=}CClCH_3$	530
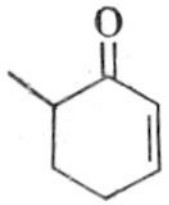	2.2 eq. Li/NH_3, Et_2O, 1 eq. *t*-BuOH, 30 min	Excess MeI	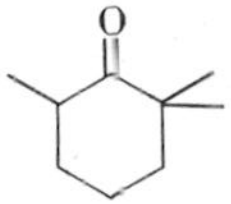 (~1) 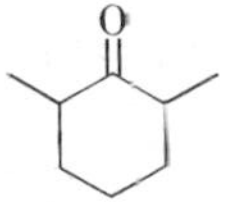 + (37) 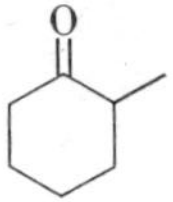 + (~5)	72
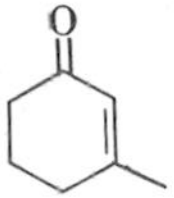	2.2 eq. Li/NH_3, Et_2O, 1 eq. *t*-BuOH, 30 min	Excess MeI	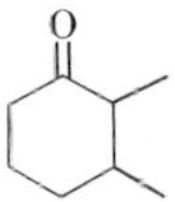 (47) 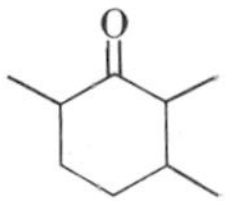 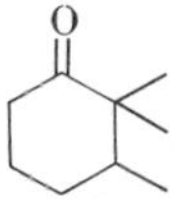 + [+] (7)	72
	2.2 eq. Li/NH_3, Et_2O, 1 eq. H_2O, 30 min	$CH_2{=}CHCH_2Br$, 6 min, NH_4Cl	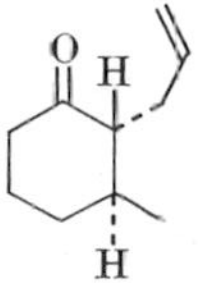 (45) 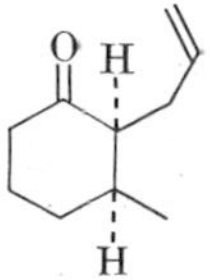 + (2)	176

TABLE IX. Reduction-Alkylation and Related Reactions of Monocyclic α,β-Unsaturated Ketones (*Continued*)

Reactant	Reduction Conditions	Quenching Agent	Product(s) (% Yield)	Refs.
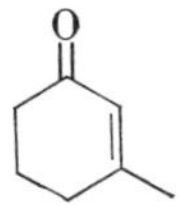	Li/NH_3	—NH_3, Et_2O, 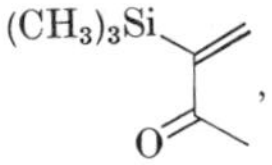 $(CH_3)_3Si$, , −20°, KOH—MeOH	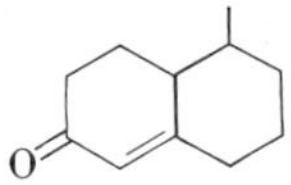 (50–60)	530a
C_8 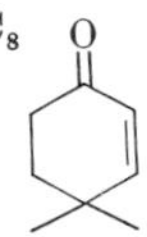	2.2 eq. Li/NH_3, Et_2O, 1 eq. *t*-BuOH, 30 min	MeI	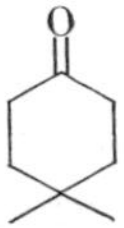 (14) 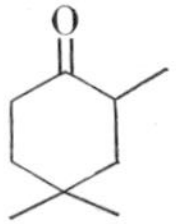 + (21)	147
	"	$C_6H_5CH_2CH_2COCH_3$ MeI	" 10 " 90	
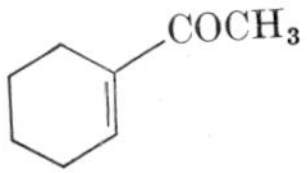 $COCH_3$	NH_3/Li, Et_2O, 20 min	MeI (slow addn.)	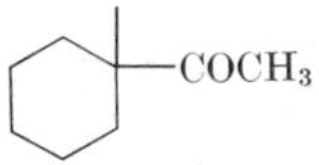 $COCH_3$ (29)	531
C_9 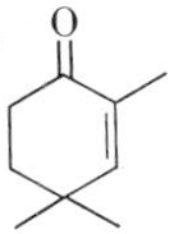	2.2 eq. Li/NH_3, Et_2O, 1 eq. *t*-BuOH, 30 min	Excess MeI	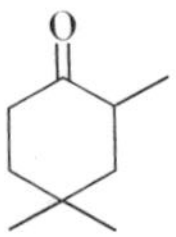 (3) 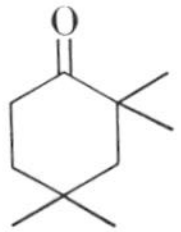 + (43)	72
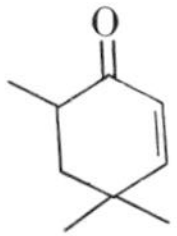	2.2 eq. Li/NH_3, 1 eq. *t*-BuOH, 30 min	Excess MeI	" (3) 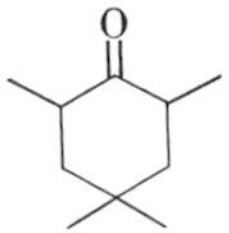 + (57)	72
C_{10} 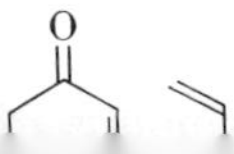	Li/NH_3, Et_2O	MeI	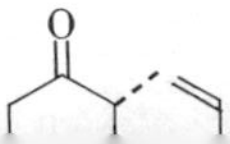 70 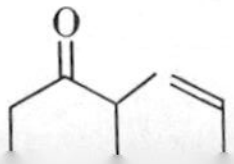 + 20 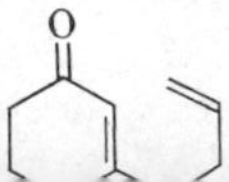 + 10	341

Reactant	Reduction Conditions	Alkylation Conditions	Product(s) and Yield(s) (%)	Refs.
2-(n-Butylthiomethylene)cyclopentanone	6 eq. Li/NH_3, Et_2O, 2 eq. H_2O, 5 min	MeI, 10 min	2,2-Dimethylcyclopentanone (51) + 2,2,5-trimethylcyclopentanone (22)	170
	6 eq. Li/NH_3, Et_2O, 2 eq. H_2O, 2 min	MeI, 2 min	" (57) + " (15)	
	6 eq. Li/NH_3, Et_2O, 2 eq. H_2O, 30 sec	MeI, 30 sec	" (70) + " (5)	
	6 eq. Li/NH_3, Et_2O, 2 eq. H_2O, 30 sec	$BrCH_2CH=CH_2$, 30 sec	2-Allyl-2-methylcyclopentanone (62) + 2,5-diallyl-2-methylcyclopentanone (15)	170
	6 eq. Li/NH_3, Et_2O, 2 eq. H_2O, 30 min	*i*-PrI, 45 hr	2-Methyl-2-Pr-*i*-cyclopentanone (40) + 2-methylcyclopentanone (15)	
C_{11} 2-(n-Butylthiomethylene)cyclohexanone (=CHSBu-*n*)	6 eq. Li/NH_3, Et_2O, 2 eq. H_2O 30 min	MeI, 30 min	2,2-Dimethylcyclohexanone (83) + 2,2,6-trimethylcyclohexanone (2)	170
	6 eq. Li/NH_3, E_2tO, 2 eq. *t*-BuOH, 30 min	"	" (61) + " (3)	
	6 eq. Li/NH_3, Et_2O, 2 eq. $(C_6H_5)_3COH$	"	" (75) + " (15)	
	6 eq. Li/NH_3, Et_2O, 2 eq. H_2O 30 min	EtI, 30 min	2-C_2H_5-2-methylcyclohexanone (75) + 2,6-di-C_2H_5-2-methylcyclohexanone (10)	
	"	*i*-PrI, THF, reflux, 50 hr	2-Pr-*i*-2-methylcyclohexanone (15–20) + 2-methylcyclohexanone (60) + 2-*i*-Pr-6-methylcyclohexanone (5–10)	

Note: References 338–543 are on pp. 253–258.

TABLE IX. REDUCTION-ALKYLATION AND RELATED REACTIONS OF MONOCYCLIC α,β-UNSATURATED KETONES (*Continued*)

Reactant	Reduction Conditions	Quenching Agent	Product(s) (% Yield)	Refs.
C_{11} (*Contd.*)				
2-(n-Butylthiomethylene)cyclohexanone ($=CHSBu$-*n*)	6 eq. Li/NH_3, Et_2O, 2 eq. H_2O, 3 min	$BrCH_2CH{=}CH_2$, 2 min	2-Allyl-2-methylcyclohexanone (85) + 2,6-diallyl-2-methylcyclohexanone (5)	170
	6 eq. Li/NH_3, Et_2O, 2 eq. H_2O, 5 min	$C_6H_5CH_2Br$, 15 min	2-Benzyl-2-methylcyclohexanone ($CH_2C_6H_5$) (82)	
	"	CO_2, 2.5 hr, CH_2N_2	2-Methyl-2-(CO_2CH_3)cyclohexanone (56) + 2-methylcyclohexanone (25)	
C_{18} 2-(HO_2C-methyl)-3-(6-CH_3O-2-naphthyl)cyclopent-2-enone	Li/NH_3	MeI	2-Methyl-2-(CH_2CO_2H)-3-(6-CH_3O-2-naphthyl)cyclopentanone (75) + stereoisomer (25)	532
C_{19} 1-(COC_6H_5)-6-(C_6H_5)cyclohexene	4 eq. Li/NH_3, Et_2O, 1.5 hr	$-NH_3$, C_6H_6, C_6H_5COCl	1-[($OCOC_6H_5$)(C_6H_5)methylene]-2-(C_6H_5)cyclohexane (68)	25
	4 eq. Na/NH_3, Et_2O, 1.5 hr	"	" (56)	
	Mg/NH_3[a]	"	" (74)	

Note: References 338–543 are on pp. 253–258.

[a] The Mg/NH_3 solution was produced electrochemically.

TABLE X. REACTION-ALKYLATION AND RELATED REACTIONS OF BICYCLIC α,β-UNSATURATED KETONES

Reactant	Reduction Conditions	Quenching Agent	Product(s) (% Yield)	Refs.
C_9 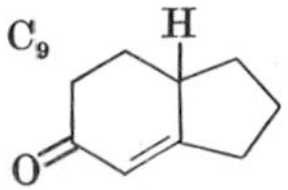	NH_3/Li, 7 hr	Excess MeI[a]	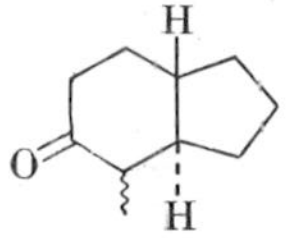 (34) + 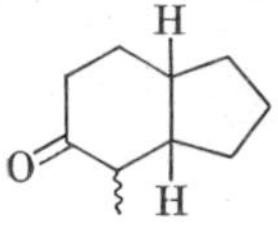 (8)	14
C_{10} 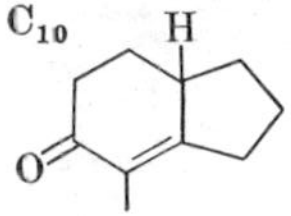	NH_3/Li, 1 hr	"	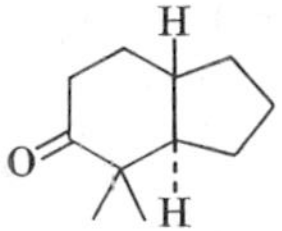 (Major) + 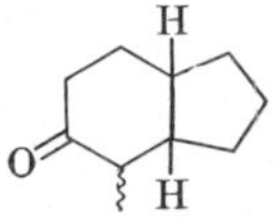 (Minor)	
	Li/NH_3, Et_2O, 35 min	$FeCl_3$, —NH_3, Et_2O, CO_2, HCl, CH_2N_2	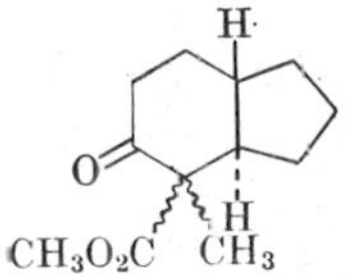 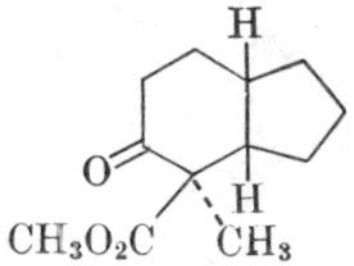 CH_3O_2C, CH_3 75 + CH_3O_2C, CH_3 25	
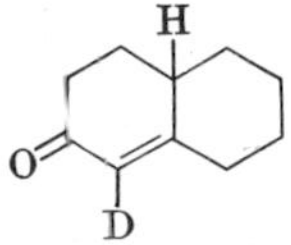 D	Li/NH_3	MeI, THF	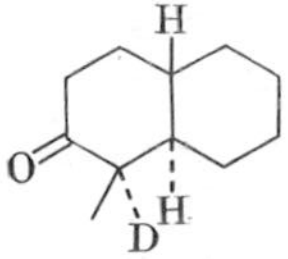 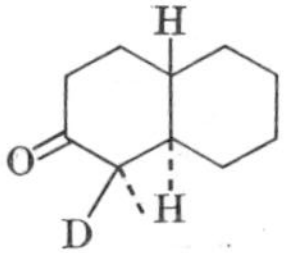 D 40 + D 60	181
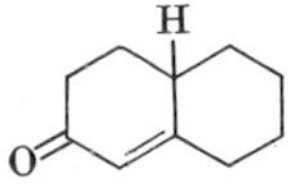	NH_3/2 eq. Li, 1 hr	Excess MeI, 30 min	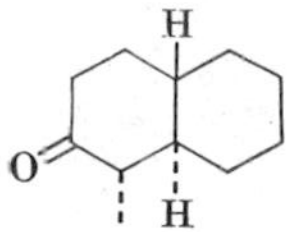 (54) + 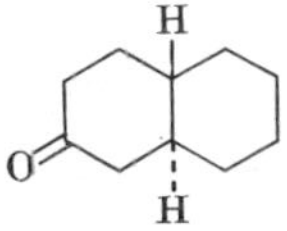 (9)	14

Note: References 338–543 are on pp. 253–258.

[a] The methyl iodide was added dropwise.

TABLE X. REACTION-ALKYLATION AND RELATED REACTIONS OF BICYCLIC α,β-UNSATURATED KETONES (*Continued*)

Reactant	Reduction Conditions	Quenching Agent	Product(s) (% Yield)	Refs.
	NH_3/2 eq. Na, 1 hr	"	Polymethylated *trans*-2-decalones	
	NH_3/2 eq. K, 1 hr	"	Polymethylated *trans*-2-decalones	
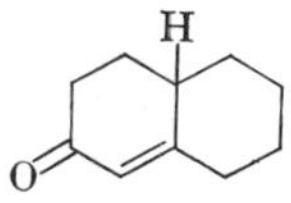	2.5 eq. Li/NH_3, 1 hr	*n*-BuI, 30 min	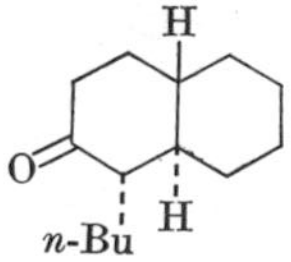 *n*-Bu + polybutylated *trans*-2-decalones	
	"	*n*-BuBr	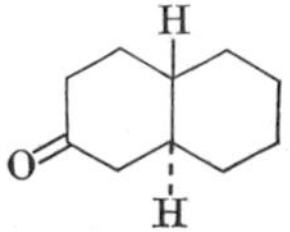 (—)	
	NH_3/2 eq. Li, 1 hr	DMSO, —NH_3, *n*-BuI	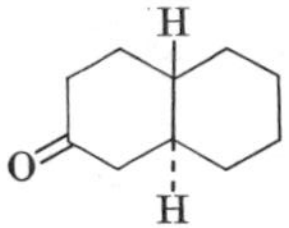 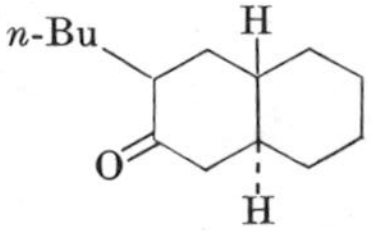 + *n*-Bu + dibutyl-*trans*-2-decalones (—)	
	NH_3/2 eq. Na, 1 hr	"	" + " + "	
	NH_3/2 eq. K, 1 hr	"	" + " + "	
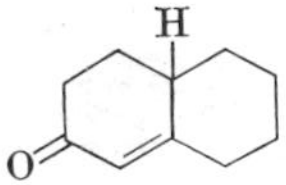	Li/NH_3, Et_2O	$ClCH_2CH_2CH{=}CClCH_3$	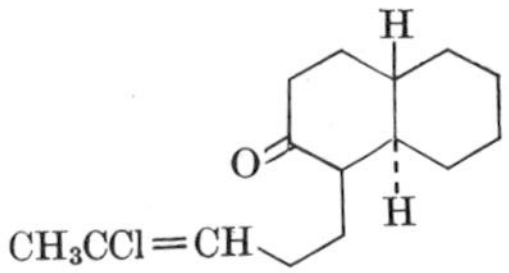 $CH_3CCl{=}CH$ (—)	254
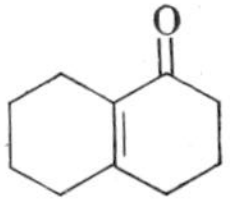	2 eq. Li/NH_3, THF, 10 min	—NH_3, THF, excess MeI	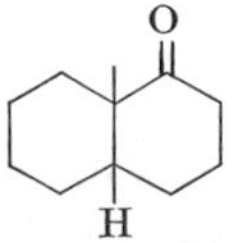 (46) + 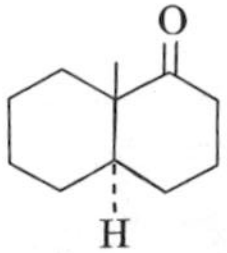 (16) + 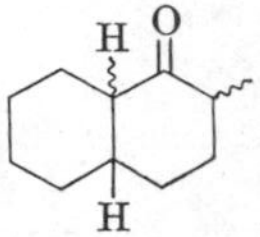 (23)	14

C_{11} 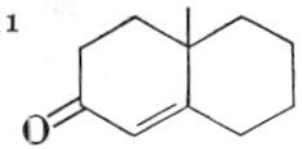	2 eq. Li/NH_3, Et_2O, 1 hr	—NH_3/Et_2O, C_6H_6, ClCN, 12 hr	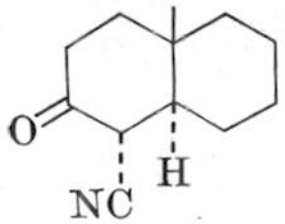 O, NC, H (11) 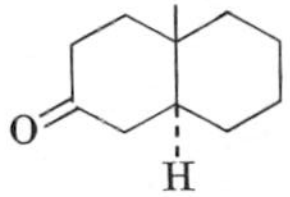 + O, H + starting material	175
	,,	$ClCH_2CH{=}CClCH_3$ 30 min	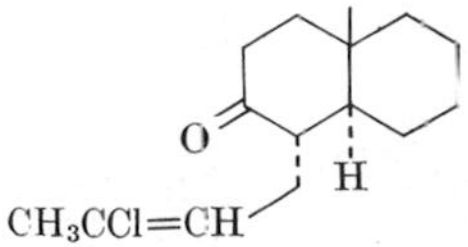 O, $CH_3CCl{=}CH$, H (40)	14
	2 eq. Li/NH_3, Et_2O, 0.5 hr	—NH_3, Et_2O, CO_2, H^+, CH_2N_2	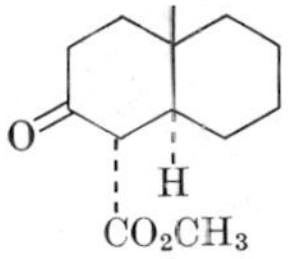 O, CO_2CH_3, H (33) 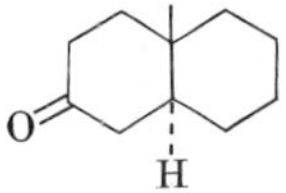 + O, H (31)	
	Li/NH_3	—NH_3, Et_2O, 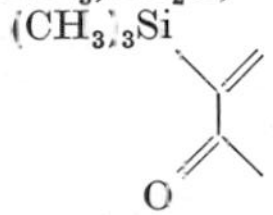 $(CH_3)_3Si$, O, −20°, KOH-MeOH	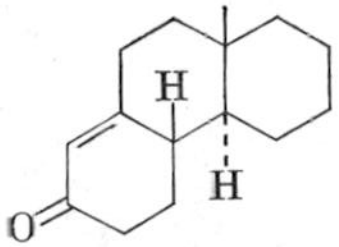 O, H, H (50–60)	530a
	3 eq. Li/NH_3, THF 0.8 eq. t-BuOH, 5 min	—NH_3, DME, 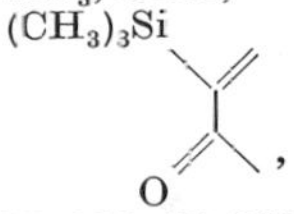 $(CH_3)_3Si$, O, NaOMe-MeOH, 3 hr	,, (—)	532b
	,,	—NH_3, THF, 1:1 $ClSi(CH_3)_3$-$(C_2H_5)_3N$, −10°	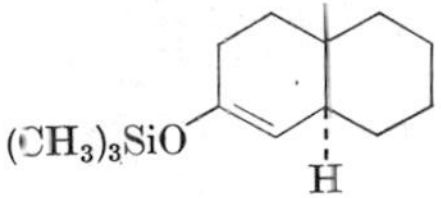 $(CH_3)_3SiO$, H (—)	532b

Note: References 338–543 are on pp. 253–258.

TABLE X. REACTION-ALKYLATION AND RELATED REACTIONS OF BICYCLIC α,β-UNSATURATED KETONE (*Continued*)

Reactant	Reduction Conditions	Quenching Agent	Product(s) (% Yield)	Refs.
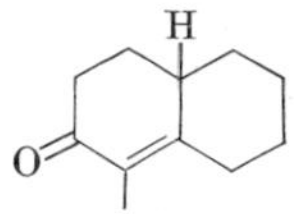	2 eq. Li/NH_3, THF, 10 min	—NH_3, THF, excess MeI	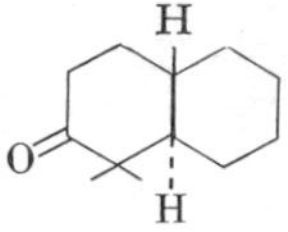 (64) 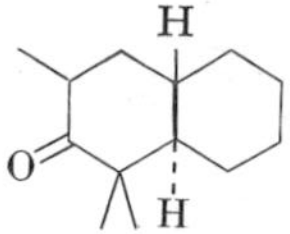 + (16)	14
	NH_3/2 eq. Li, Et_2O, 20 min	Excess MeI	" (64) + 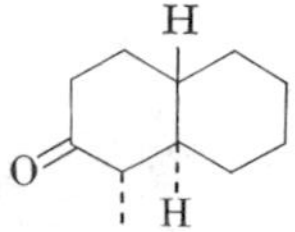 (4) + 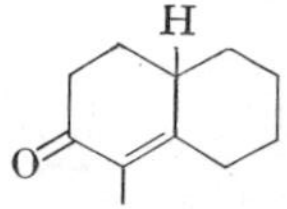 (24)	370
	Li/NH_3, Et_2O	CD_3I	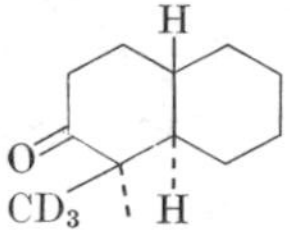 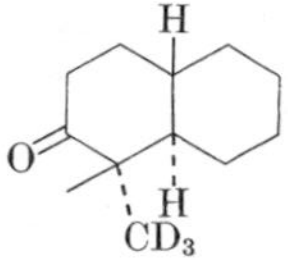 CD_3 83 + CD_3 17	179
	"	EtI, H_3O^+	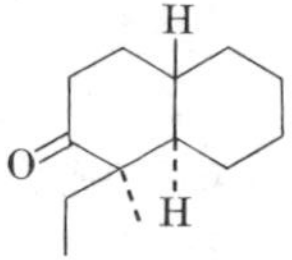 95 + 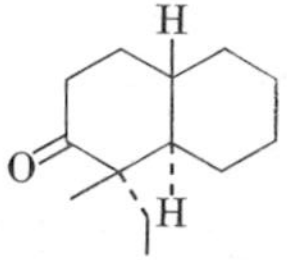 5	
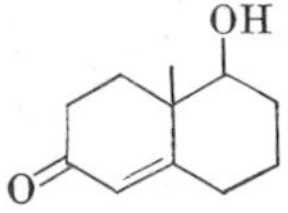 OH	Li/NH_3, Et_2O, 5 min	—NH_3, Et_2O, CO_2(gas), 2.75 hr, H^+, CH_2N_2	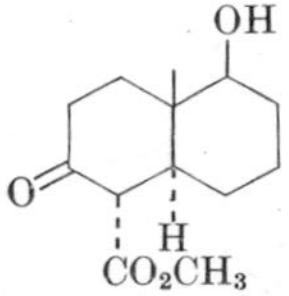 OH CO_2CH_3 (18) 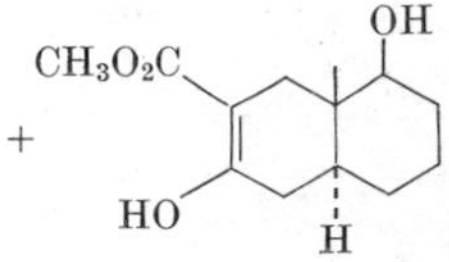 + CH_3O_2C OH HO (12)	189

Substrate	Reducing Conditions	Electrophile	Product(s) and Yield(s) (%)	Refs.
C_{12}				
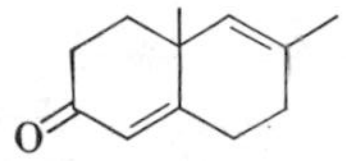	Li/NH_3, Et_2O	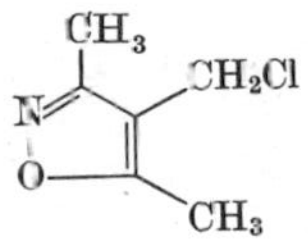 $-NH_3$, Et_2O, CH_3 / CH_2Cl / N / O / CH_3	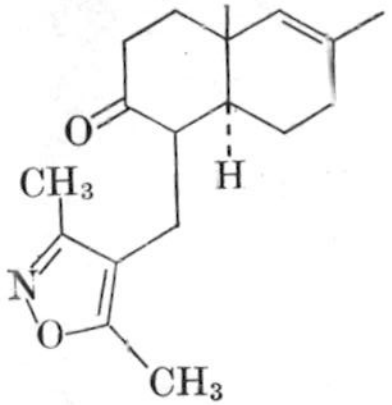 CH_3 / N / O / CH_3 / H / O (17)	173
	Li/NH_3, Et_2O	$BrCH_2CH{=}CClCH_3$, NH_4Cl, H_2O	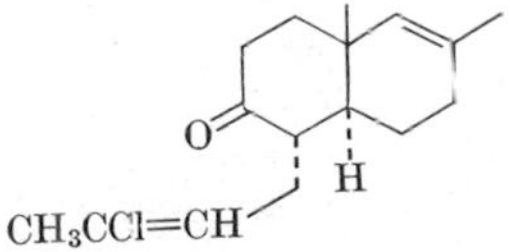 O / H / $CH_3CCl{=}CH$ (60)	173
	Li/NH_3	$-NH_3$, Et_2O, 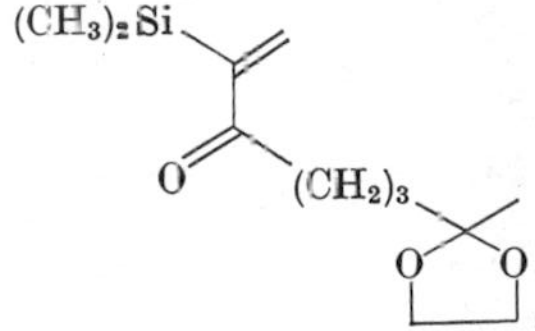 $(CH_3)_2Si$ / O / $(CH_2)_3$ / O / O, ; KOH—MeOH	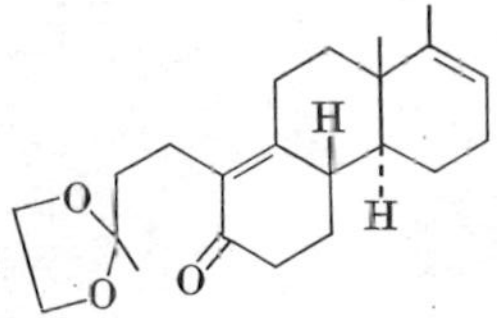 O / O / O / H / H (51)	530a
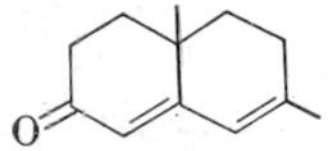	NH_3/2 eq. Na, THF, 30 min	Excess MeI	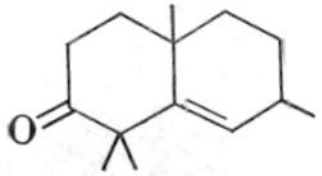 O (52)	225a
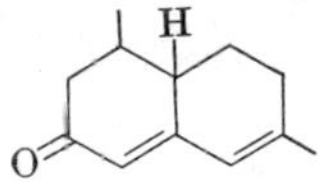 H	NH_3/2 eq. Na, THF, 30 min	Excess MeI	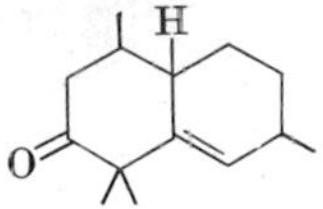 O / H (51)	225a

Note: References 338–543 are on pp. 253–258.

TABLE X. REACTION-ALKYLATION AND RELATED REACTIONS OF BICYCLIC α,β-UNSATURATED KETONE (*Continued*)

Reactant	Reduction Conditions	Quenching Agent	Product(s) (%Yield)	Refs.
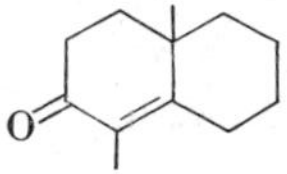	Li/NH_3, THF, 10 min	—NH_3, THF, MeI	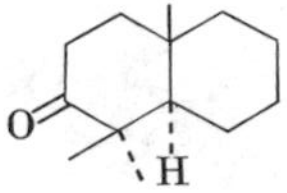 (57) + 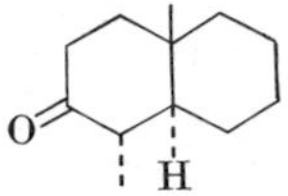 (25)	14
	"	—NH_3, THF, $C_6H_5CH_2Cl$	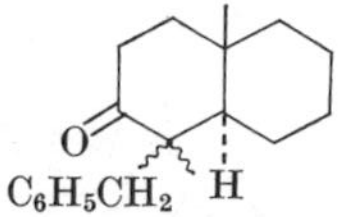 $C_6H_5CH_2$ H	
	Li/NH_3, Et_2O	CD_3I	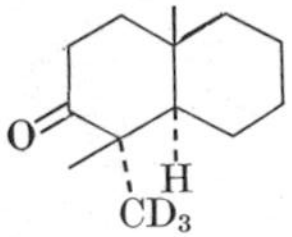 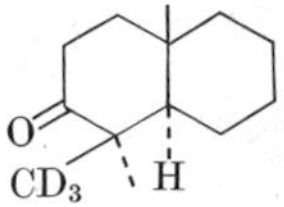 H CD_3 93 + CD_3 H 7	180
		EtI, H_5O^+	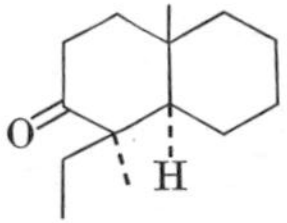 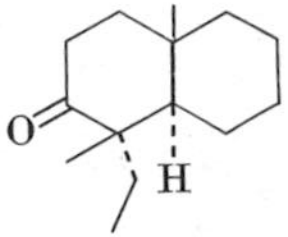 H 5 + H 95	
	Li/NH_3, THF	—NH_3/THF, C_6H_6, ClCN, 10 hr	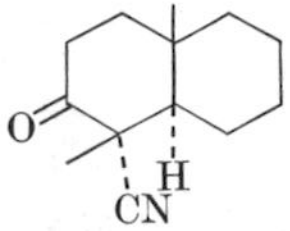 H CN (29)	175

Substrate	Reduction conditions	Electrophile	Product(s) and Yield(s) (%)	Refs.
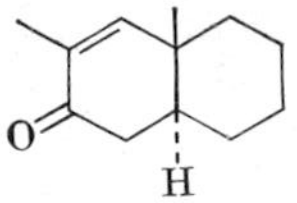	Li/NH_3, Et_2O	CD_3I	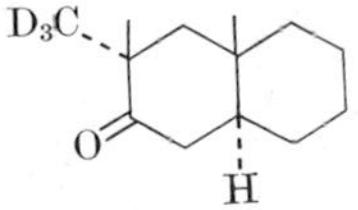 D_3C (—)	180
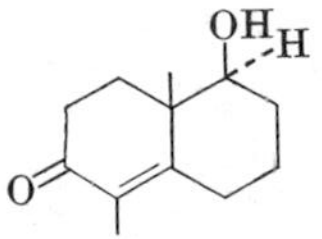	3 eq. Li/NH_3, THF, 30 min	10 eq. MeI	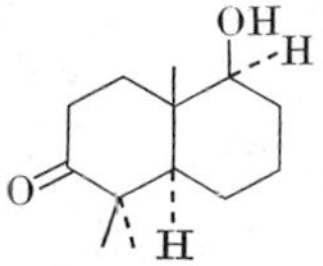 (85–90) 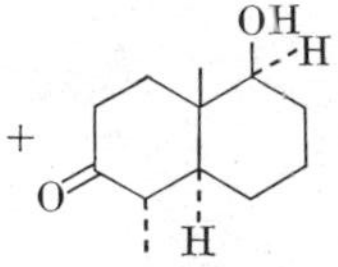 + trace 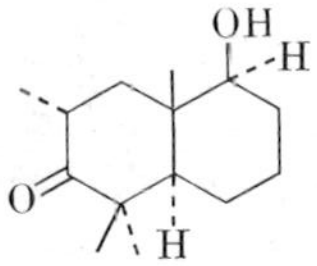 + trace	532c
C_{13} 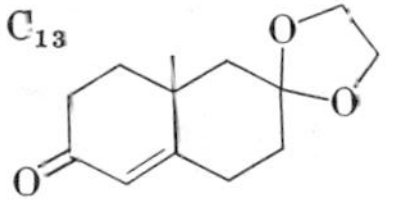	Li/NH_3, Et_2O, 1 hr	$ClCH_2CH{=}CClCH_3$, 30 min, H_3O^-	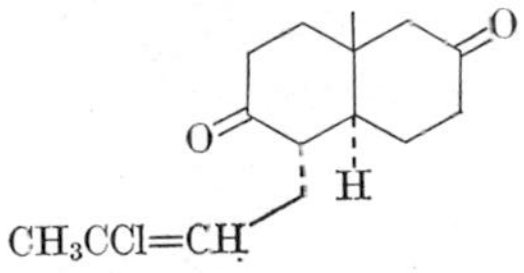 $CH_3CCl{=}CH$ (50)	14

Note: References 338–543 are on pp. 253–258.

TABLE X. REACTION-ALKYLATION AND RELATED REACTIONS OF BICYCLIC α,β-UNSATURATED KETONES (*Continued*)

Reactant	Reduction Conditions	Quenching Agent	Product(s) (% Yield)	Ref.
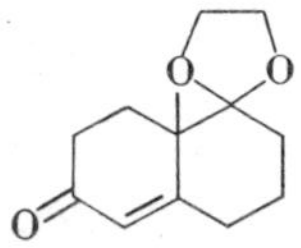	Li/NH_3, Et_2O	—NH_3, Et_2O, $(C_2H_5O)_2POCl$, Li/$C_2H_5NH_2$, *t*-BuOH	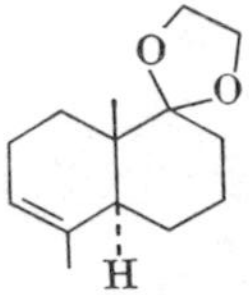 (—)	190
	3 eq. Li/NH_3, THF, 0.8 eq. *t*-BuOH, 5 min	—NH_3, THF, 1:1 $ClSi(CH_3)_3$, $(C_2H_3)_3N$, $-10°$	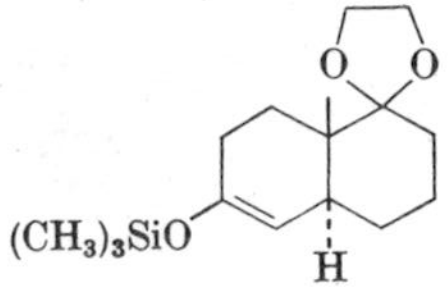 $(CH_3)_3SiO$ H (—)	532b
C_{14} 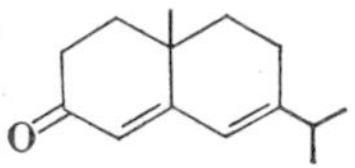	NH_3/2 eq. Na, THF, 30 min	Excess MeI	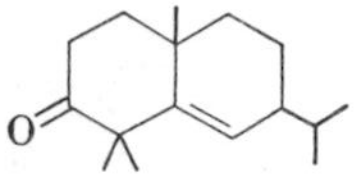 (50)	225a
C_{15} 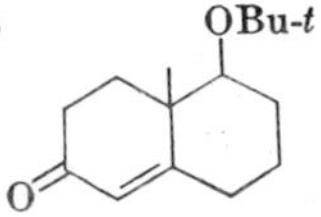 OBu-*t*	3 eq. Li/NH_3, THF, 0.8 eq. *t*-BuOH, 5 min	—NH_3, THF 1:1 $ClSi(CH_3)_3$-$(C_2H_5)_3N$, $-10°$	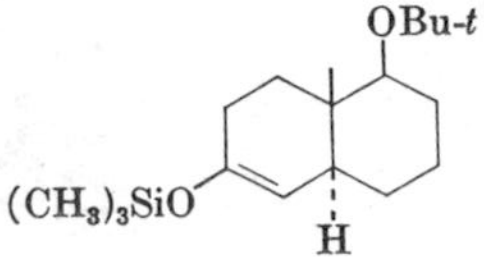 OBu-*t* $(CH_3)_3SiO$ H (88)	532b
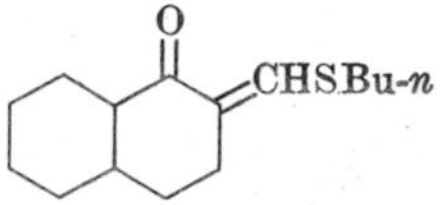 CHSBu-*n*	6 eq. Li/NH_3, Et_2O, 2 eq. H_2O, 30 min	Excess MeI, 30 min	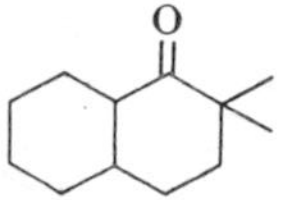 (70)	170

Substrate	Reduction conditions	Subsequent conditions	Product(s) and Yield(s) (%)	Refs.
C_{16} (decalone, =CHSBu-*n*) 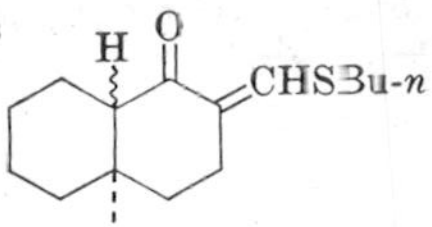	Li/NH_3, 2 eq. H_2O, Et_2O, 90 min	MeI, 30 min, NH_4Cl, NaOMe—MeOH	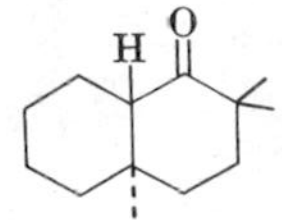 (50) 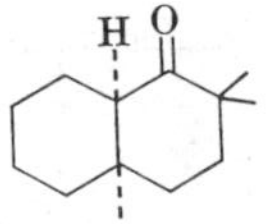 + (17)	532d
	Li/NH_3, 2 eq. H_2O, Et_2O, 15 min	MeI (to destroy excess Li), *i*-PrI to react with *n*-BuSLi, 1.5 eq. CD_3I	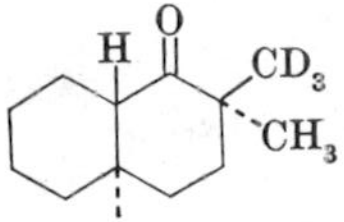 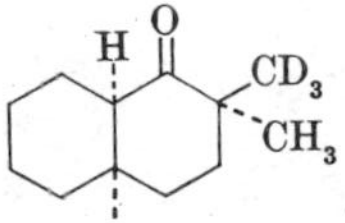 (CD_3, CH_3) + (CD_3, CH_3) (72)	
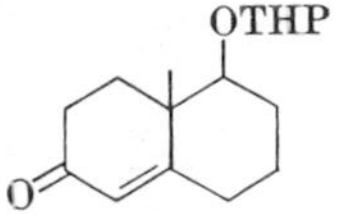 OTHP (enone)	Li/NH_3, Et_2O, 5 min	—NH_3/Et_2O, Et_2O, CO_2 (gas), 2.75 hr, H^+, CH_2N_2[b]	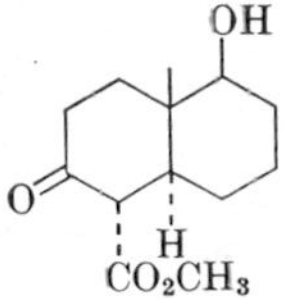 OH, H, CO_2CH_3 (30) 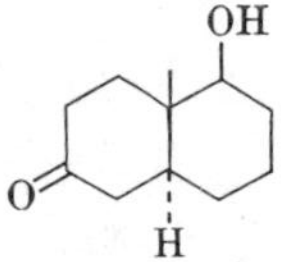 + OH, H (~10)	187
	"	—NH_3/Et_2O, Et_2O, CO_2 solid, −78°, H^+, CH_2N_2	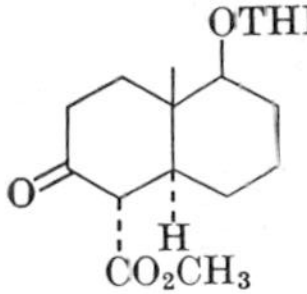 OTHP, H, CO_2CH_3 (68)	
	Li/NH_3	—NH_3, Et_2O, 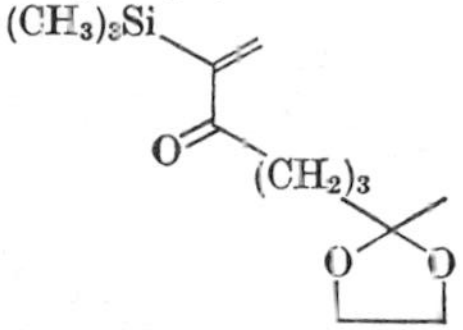 $(CH_3)_3Si$ (vinyl ketone, $(CH_2)_3$, dioxolane), KOH-MeOH	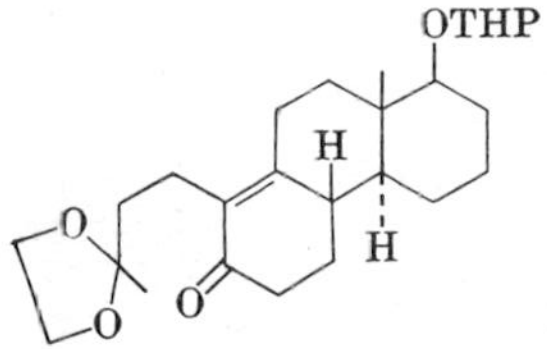 OTHP, H, H (50)	530a

Note: References 338–543 are on pp. 253–258.

TABLE X. Reaction-Alkylation and Related Reactions of Bicyclic α,β-Unsaturated Ketones (*Continued*)

Reactant	Reduction Conditions	Quenching Agent	Product(s) (% Yield)	Refs.
C_{17} 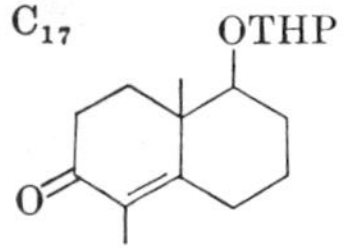 OTHP	Li/NH_3, Et_2O, 5 min	—NH_3/Et_2O, Et_2O, CO_2 (gas), H^+, CH_2N_2[b]	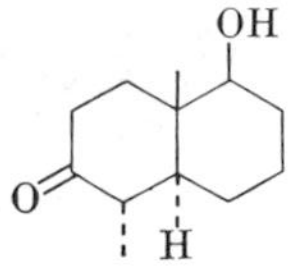 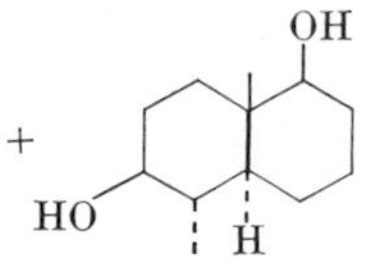 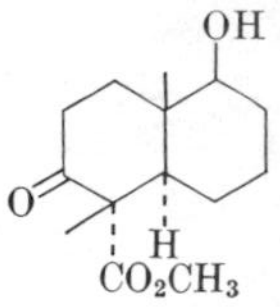 OH, O, H + OH, HO, H + OH, O, H, CO_2CH_3 I (30) II (6) III (3)	189
	"	—NH_3/Et_2O, CO_2 (solid), −78°, H^+, CH_2N_2[b]	I (34) + III (16)	
	"	—NH_3, Et_2O, excess $ClCO_2CH_3$, 3.5 hr[b]	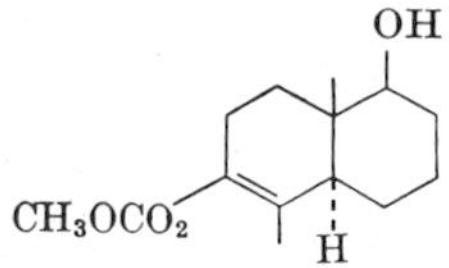 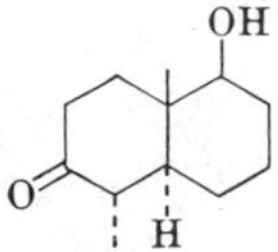 OH, CH_3OCO_2, H (—) + OH, O, H (—)	
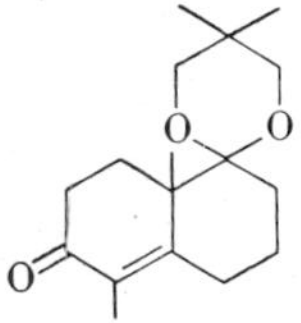 O, O, O	Li/NH_3, THF	—NH_3, THF-N,N,N′,N′-tetramethylethylene diamine, $ClPO(N(CH_3)_2)_2$	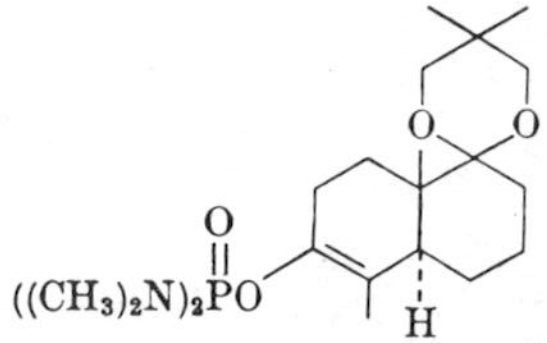 O, O, O, $((CH_3)_2N)_2PO$, H (70)	532a
C_{19} 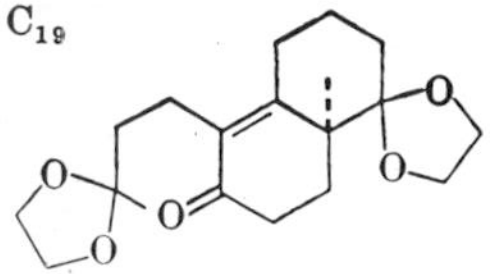 O, O, O, O, O	Li/NH_3, THF	MeI, H_3O^+	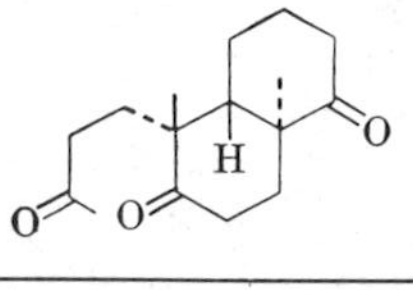 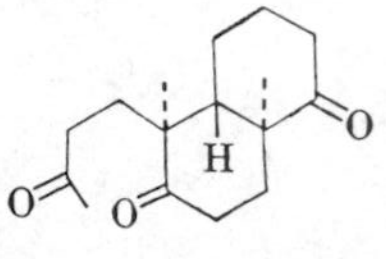 O, O, H, O 86 + O, O, H, O 14	79

Note: References 338–543 are on pp. 253–258.

[b] The tetrahydropyranyl protecting group was removed with *p*-toluenesulfonic acid in methanol.

TABLE XI. REDUCTION-ALKYLATION AND RELATED REACTIONS OF TRICYCLIC α,β-UNSATURATED KETONES

Reactant	Reduction Conditions	Quenching Agent	Product(s) (% Yield)	Refs.
C_{14} 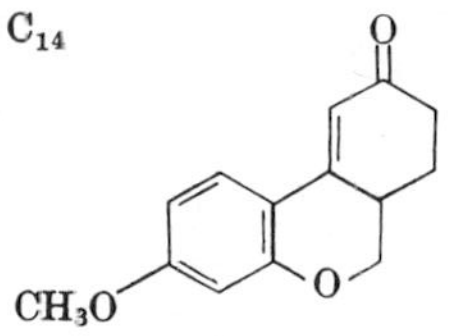 CH_3O	NH_3/Li, Et_2O, 25 min	MeI	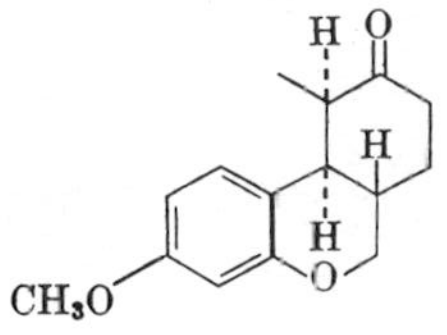 CH_3O (50)	392
C_{15} 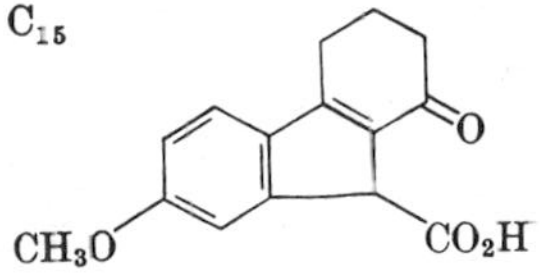 CH_3O, CO_2H	Li/NH_3	MeI	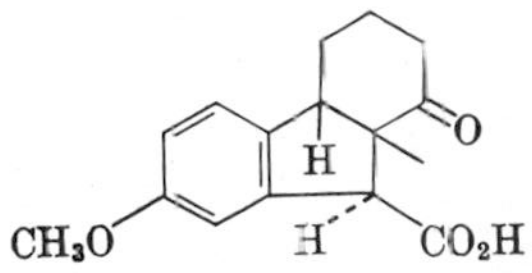 CH_3O, CO_2H	398
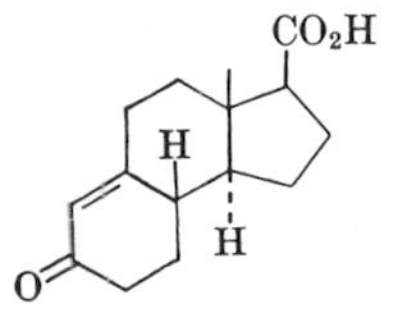 CO_2H	M/NH_3	$ClCH_2CH{=}CClCH_3$	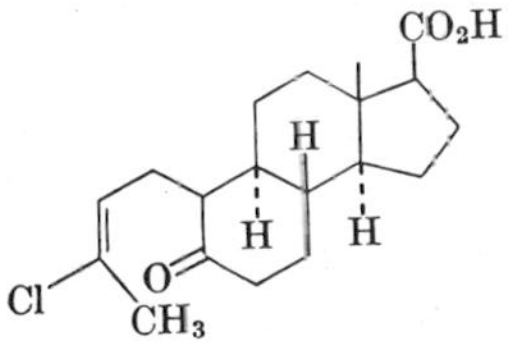 CO_2H, Cl, CH_3	533
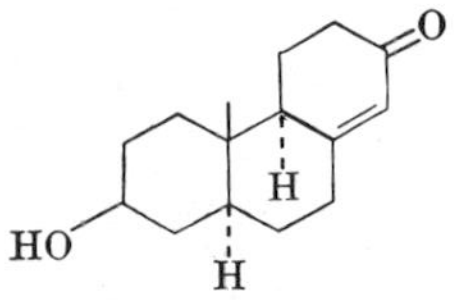 HO	2 eq. Li/NH_3, Et_2O-THF, 40 min	MeI	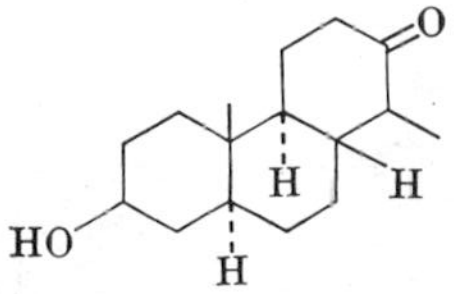 HO (34)	395

Note: References 338–543 are on pp. 253–258.

TABLE XI. REDUCTION-ALKYLATION AND RELATED REACTIONS OF TRICYCLIC α,β-UNSATURATED KETONES (*Continued*)

Reactant	Reduction Conditions	Quenching Agent	Product(s) (% Yield)	Refs.
C_{16} 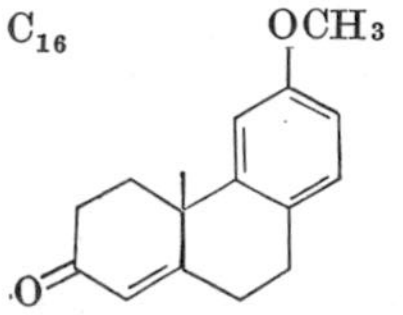 OCH₃	Li/NH_3, Et_2O	—NH_3, CO_2, H^+, CH_2N_2	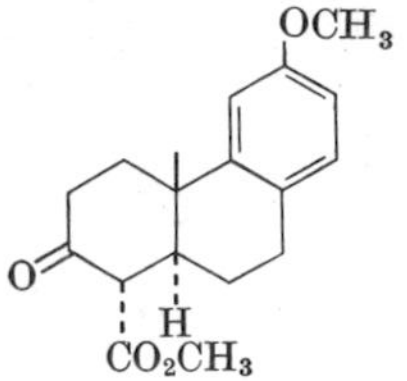 OCH₃, H, CO_2CH_3 (40)	527
C_{17} 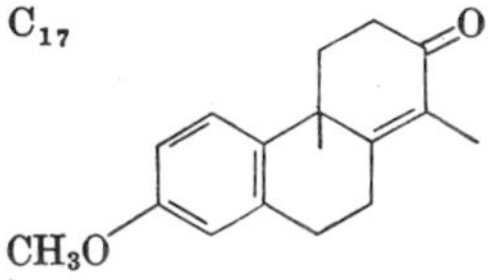 CH_3O	Li/NH_3, THF, 1 eq. H_2O, 10 min	CH_2=$CHCH_2Br$	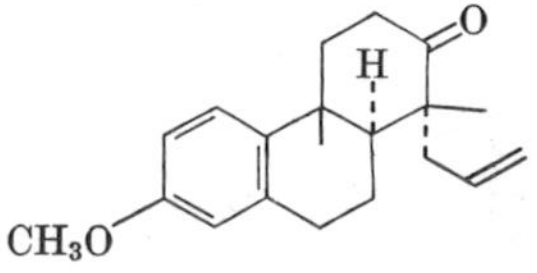 H, CH_3O (80)	534
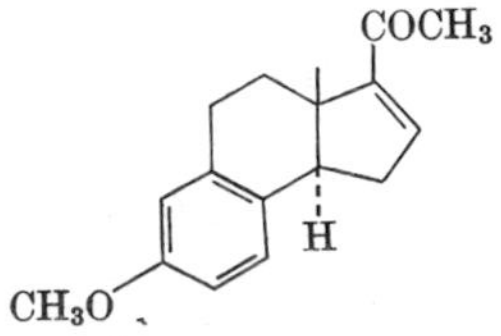 COCH₃, H, CH_3O	M/NH_3	MeI	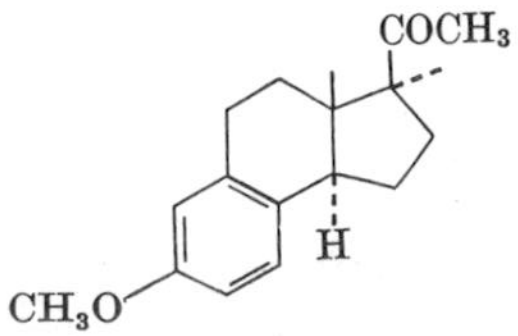 COCH₃, H, CH_3O (—)	535
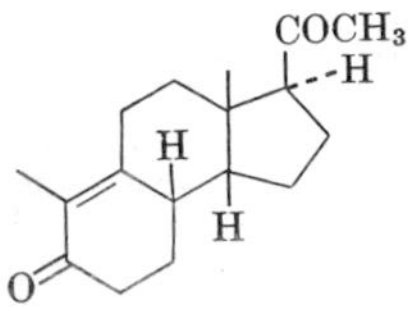 COCH₃, H, H, H	Li/NH_3	CH_3COCH=CH_2	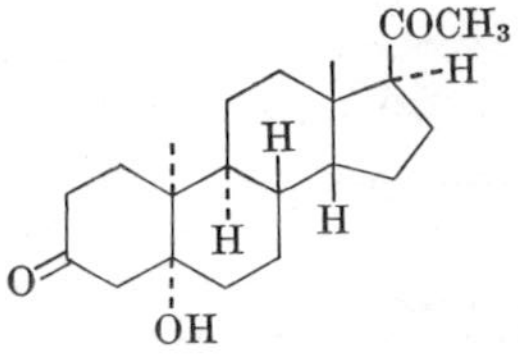 COCH₃, H, H, H, H, OH	536

Reactant	Reducing Conditions	Electrophile	Product(s) and Yield(s) (%)	Refs.
	2.5 eq. Li/NH_3, THF, 45 min	MeI, THF, (dropwise)	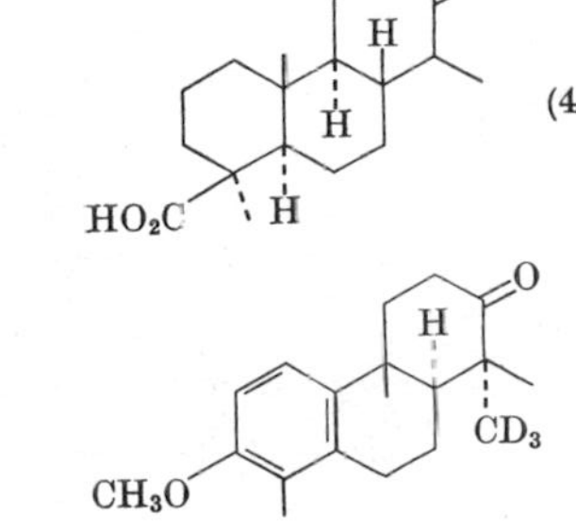 (47)	418
C_{18}	Li/NH_3, THF, 25 min	CD_3I	(—)	534
C_{19}	Li/NH_3, t-BuOH	MeI	(77)	537
C_{21}	Li/NH_3, Et_2O	MeI	(~100)	177
C_{22}	Li/NH_3	$—NH_3$, Et_2O, MeI	(67)	530a

Note: References 338–543 are on pp. 253–258.

TABLE XI. Reduction-Alkylation and Related Reactions of Tricyclic α,β-Unsaturated Ketones (*Continued*)

Reactant	Reduction Conditions	Quenching Agent	Product(s) (% Yield)	Refs.
C_{25} 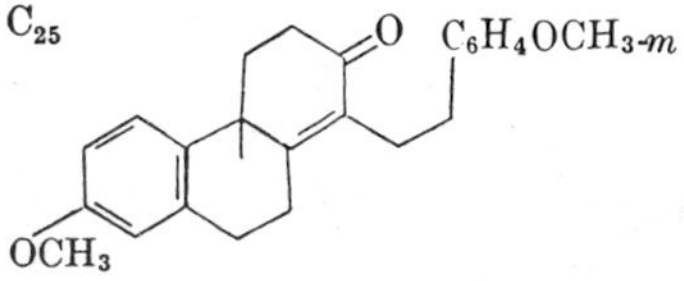 (structure; $C_6H_4OCH_3$-*m*, OCH_3)	Li/NH_3, DME, 1 eq. *t*-BuOH, 15 min	MeI, DME	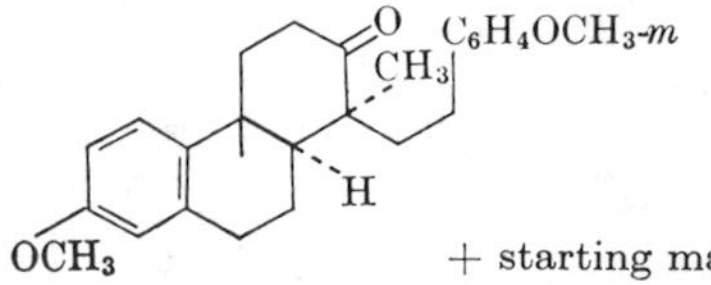 (structure; $C_6H_4OCH_3$-*m*, CH_3, H, OCH_3) (70) + starting material (15) 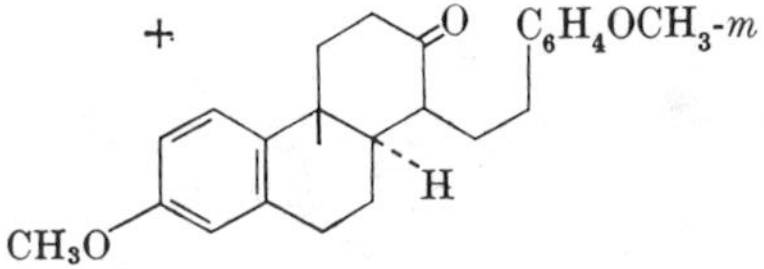 + (structure; $C_6H_4OCH_3$-*m*, H, CH_3O) (10)	538
C_{26} 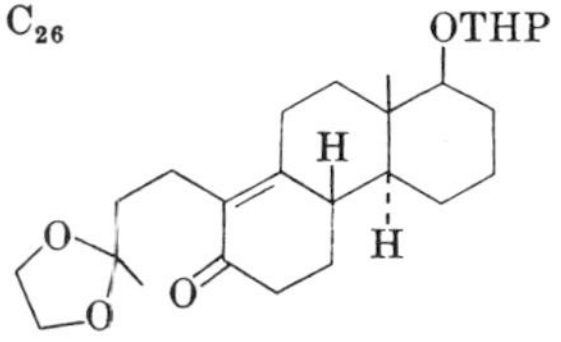 (structure; OTHP, H, H)	Li/NH_3	—NH_3, Et_2O, CH_3I	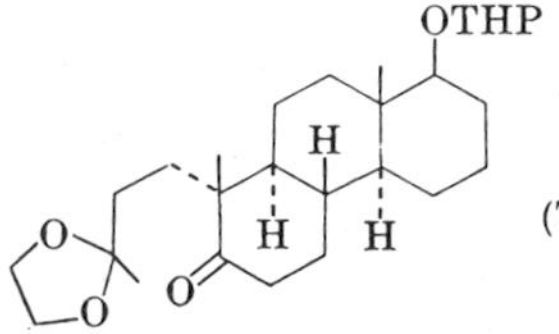 (structure; OTHP, H, H, H) (75)	530a
C_{38} 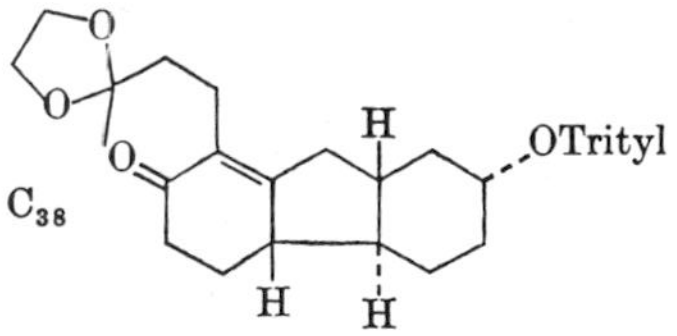 (structure; OTrityl, H, H, H)	Li/NH_3	MeI	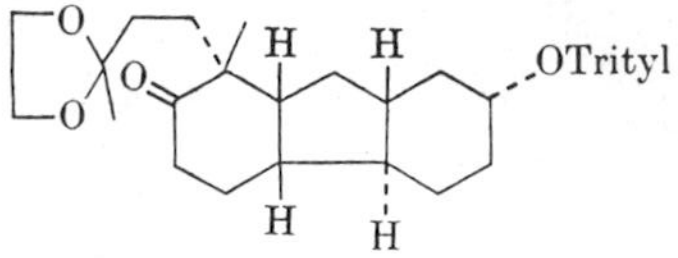 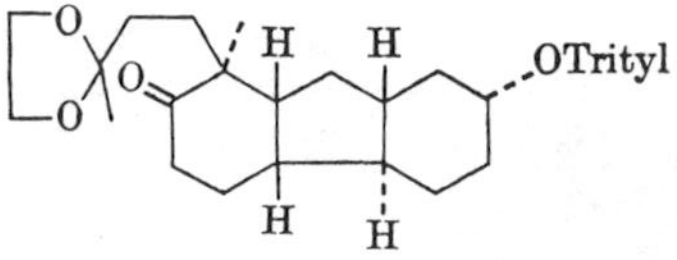 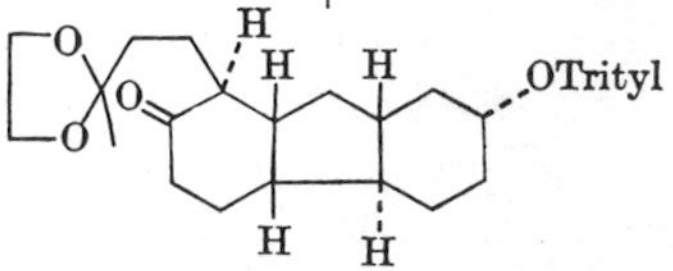 (structure; OTrityl, H, H, H, H) ~72 + (structure; OTrityl, H, H, H, H) ~19 + (structure; OTrityl, H, H, H, H, H) ~9	126

TABLE VII. REDUCTION-ALKYLATION AND RELATED REACTIONS OF α,β-UNSATURATED KETONES IN COMPOUNDS HAVING FOUR OR MORE FUSED RINGS

Reactant	Reduction Conditions	Quenching Agent	Product(s) (% Yield)	Refs.
C_{18} 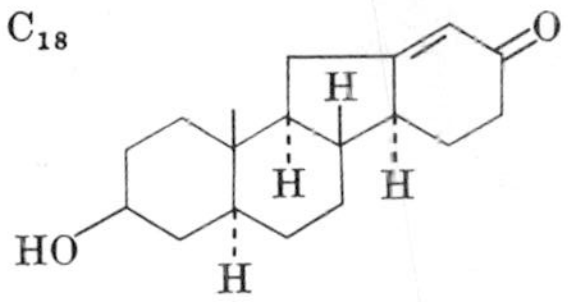	Li/NH_3, dioxane	MeI, Ac_2O	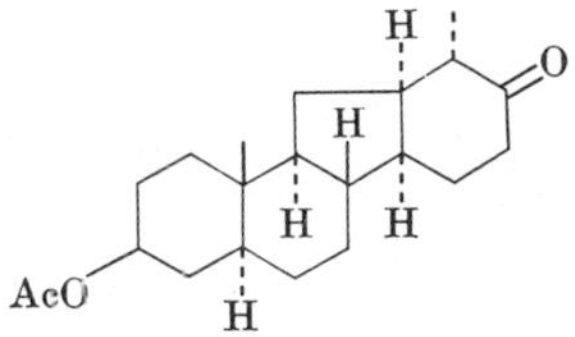 (—)	127
C_{19} 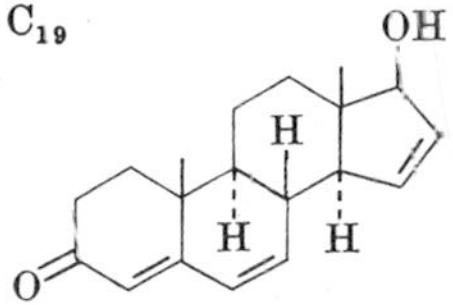	Li/NH_3, THF	MeI	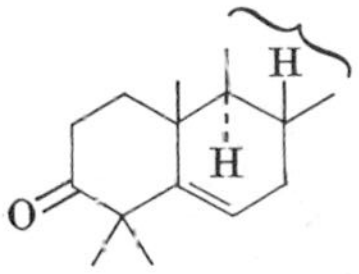 (—)	172
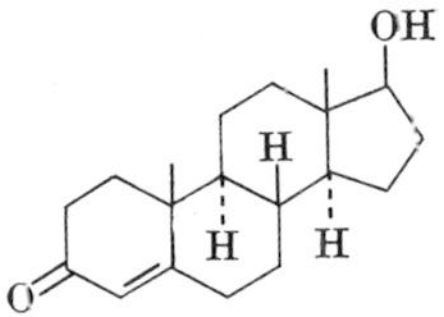	Li/NH_3, THF	MeI	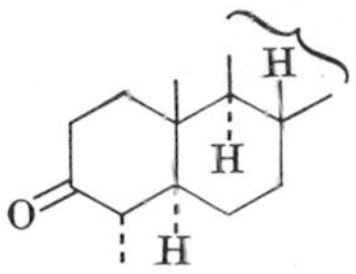 (40)	172
	Li/NH_3, THF	Amyl nitrate	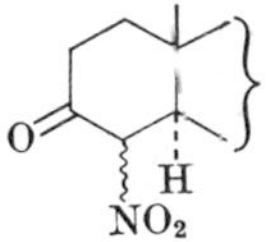 NO_2 (—)	174
C_{20} 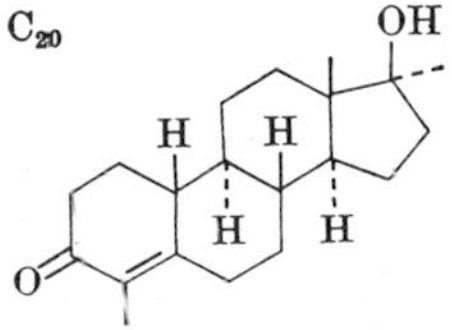	"	MeI	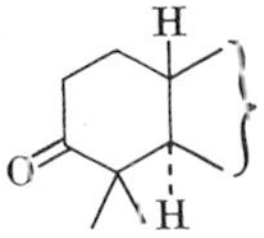 (—)	539

Note: References 338–543 are on pp. 253–258.

TABLE XII. Reduction-Alkylation and Related Reactions of α,β-Unsaturated Ketones in Compounds Having Four or More Fused Rings (*Continued*)

Reactant	Reduction Conditions	Quenching Agent	Product(s) (% Yield)	Refs.
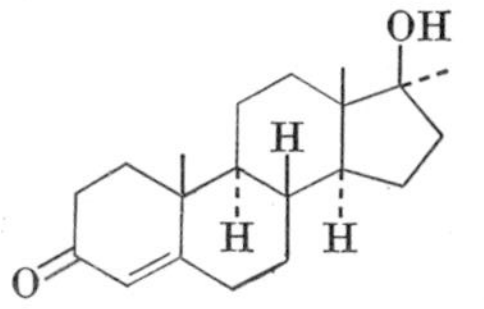	Li/NH_3, THF	MeI	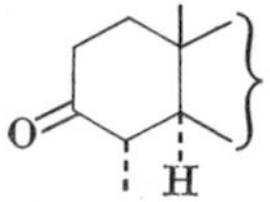 (67)	463
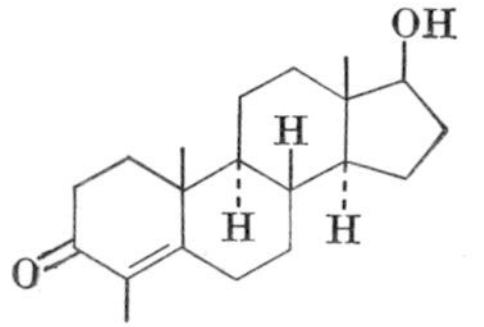	Li/NH_3, Et_2O, 5 min	MeI	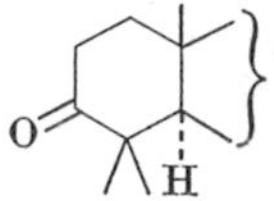 (60)	14
C_{21} 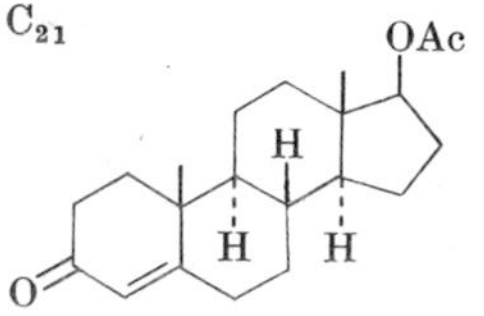	Li/NH_3, THF, 45 min	$FeCl_3$, $—NH_3$, Et_2O, CO_2, CH_2N_2, Ac_2O	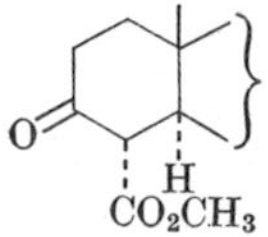 CO_2CH_3 (24)	541
C_{22} 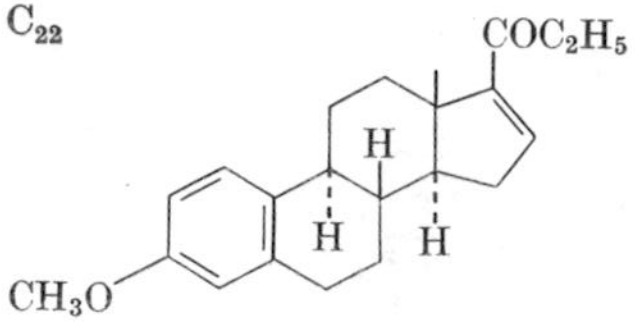	Li/NH_3, $-78°$	MeI	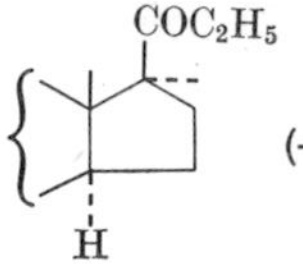 COC_2H_5 (—)	540

Substrate	Reducing Agent	Electrophile	Product (Yield, %)	Refs.
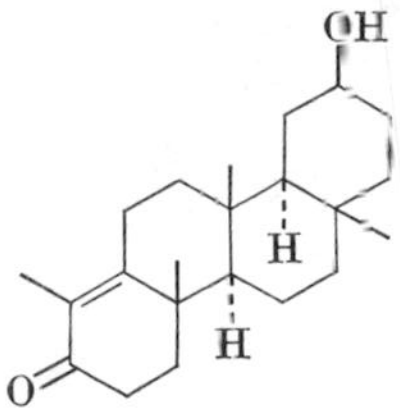 OH; H; H; O	Li/NH$_3$, Et$_2$O	CH$_2$=CHCH$_2$Br	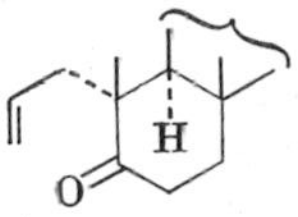 H; O (80)	178
C$_{23}$ 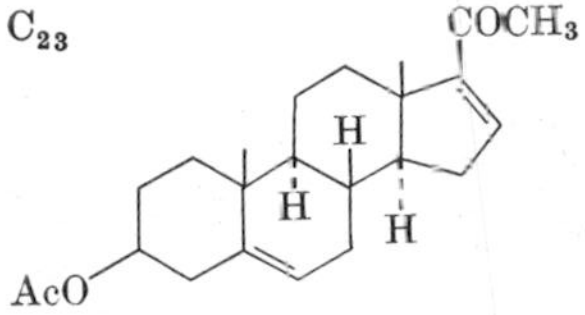 COCH$_3$; H; H; H; AcO	Li/NH$_3$, THF, −78°	MeI, Ac$_2$O	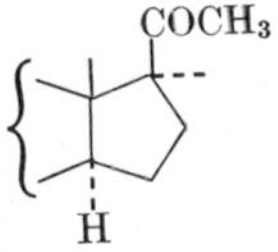 COCH$_3$; H (40)	169, 171
	,,	EtI, Ac$_2$O	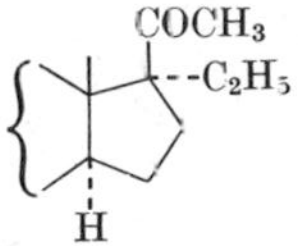 COCH$_3$; C$_2$H$_5$; H (~40)	
	,,	n-PrI, Ac$_2$O	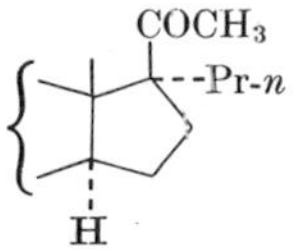 COCH$_3$; Pr-n; H (~15)	169

Note: References 338–543 are on pp. 253–258

TABLE XII. REDUCTION-ALKYLATION AND RELATED REACTIONS OF α,β-UNSATURATED KETONES IN COMPOUNDS HAVING FOUR OR MORE FUSED RINGS (*Continued*)

Reactant	Reduction Conditions	Quenching Agent	Product(s) (% Yield)	Refs.
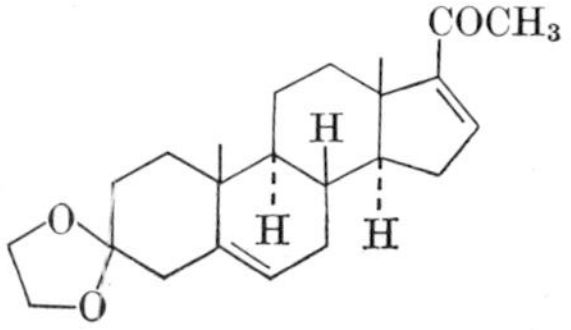 $COCH_3$ (structure)	2 eq. Li/NH_3, THF	MeI	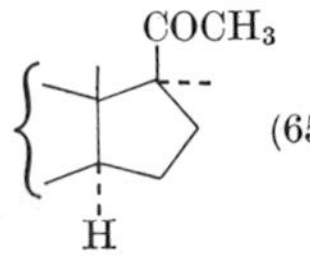 $COCH_3$ (structure) (65)	67
	Ca/NH_3, THF	"	" (21)	
	Ba/NH_3, THF	"	" (6)	
	2 eq. Li/NH_3, THF	EtI	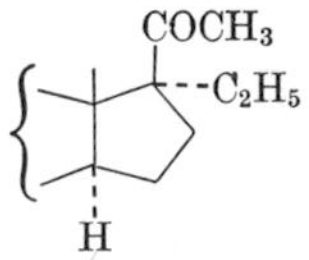 $COCH_3$, C_2H_5 (structure) (43)	
	1 eq. Li/NH_3, THF	"	" (10) + starting material (26) 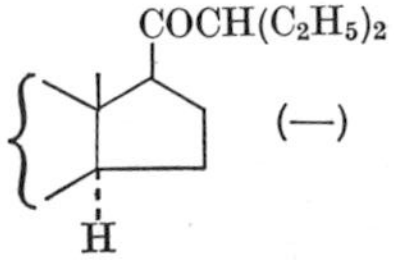 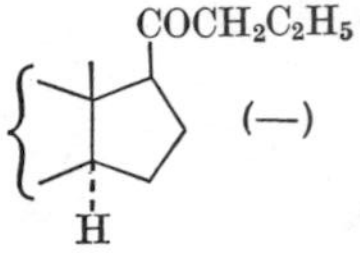 + $COCH(C_2H_5)_2$ (structure) (—) + $COCH_2C_2H_5$ (structure) (—)	
	2 eq. Li/NH_3, THF	*n*-PrI, THF	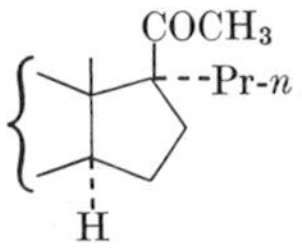 $COCH_3$, Pr-*n* (structure) (21)	

Reactant	Conditions	Product(s) and Yield(s) (%)
"	n-BuI, THF	Steroid D-ring with $COCH_3$ and ---Bu-n at C-17, H at C-14 (7)
"	n-$C_6H_{13}I$, THF	Steroid D-ring with $COCH_3$ and ---C_6H_{13}-n at C-17, H at C-14 (20)
"	n-$C_8H_{17}I$, THF	Steroid D-ring with $COCH_3$ and ---C_8H_{17}-n at C-17, H at C-14 (24)
"	$BrCH_2CH{=}CH_2$, THF	Steroid D-ring with $COCH_3$ and ---$CH_2CH{=}CH_2$ at C-17, H at C-14 (17)
"	$ClCH_2C_6H_5$, THF	Steroid D-ring with $COCH_3$ and ---$CH_2C_6H_5$ at C-17, H at C-14 (16)

Note: References 338–543 are on pp. 253–258.

TABLE XII. REDUCTION-ALKYLATION AND RELATED REACTIONS OF α,β-UNSATURATED KETONES IN COMPOUNDS HAVING FOUR OR MORE FUSED RINGS (*Continued*)

Reactant	Reduction Conditions	Quenching Agent	Product(s) (% Yield)	Ref.
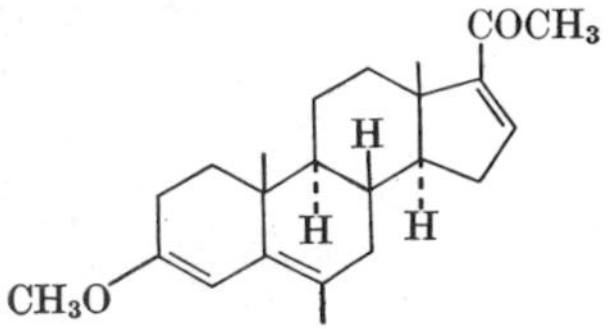 [structure: CH_3O, $COCH_3$, H, H, H]	2 eq. Li/NH_3, THF	EtI, H_3O^+	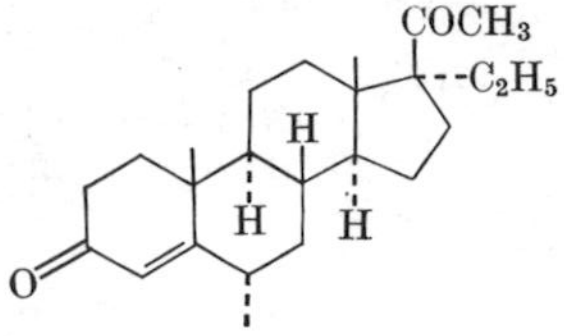 [structure: O, $COCH_3$, C_2H_5, H, H, H] (17)	67
	,,	n-PrI, H_3O^+	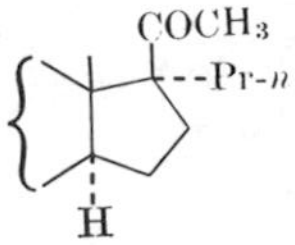 [structure: $COCH_3$, Pr-n, H] (26)	
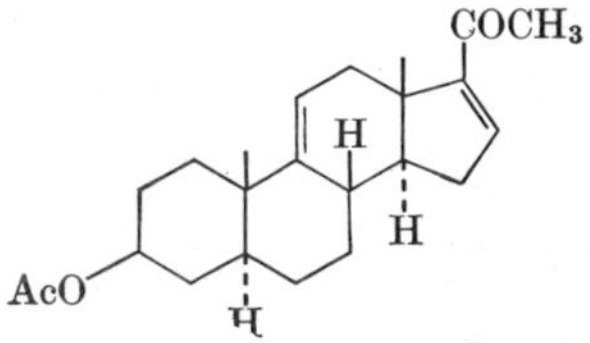 [structure: AcO, $COCH_3$, H, H, H]	Li/NH_3, THF	MeI	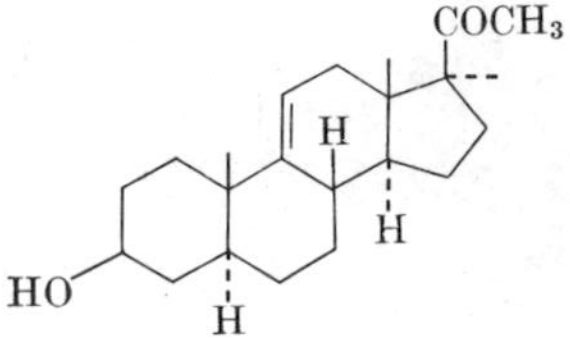 [structure: HO, $COCH_3$, H, H, H] (16)	542
	,,	EtI	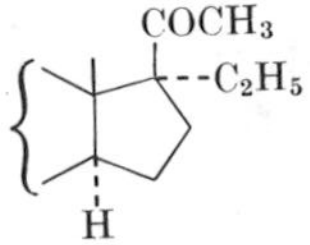 [structure: $COCH_3$, C_2H_5, H] (16)	
C_{24} 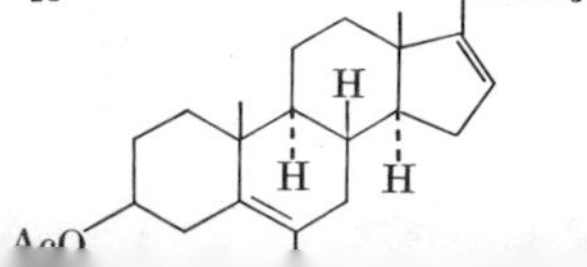 [structure: AcO, $COCH_3$, H, H, H]	Li/NH_3, THF	MeI, Ac_2O	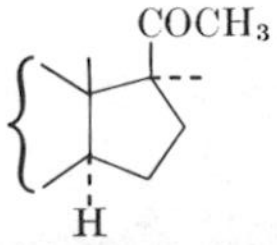 [structure: $COCH_3$, H] (40)	169, 171

Substrate	Reduction conditions	Alkylation	Product (yield %)	Refs.
	"	EtI, Ac_2O	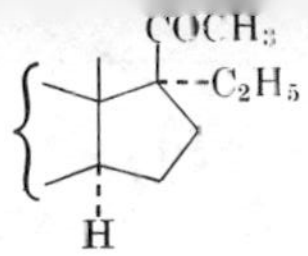 $COCH_3$, C_2H_5, H (~40)	169
AcO, $COCH_3$, H, H, H	Li/NH_3, THF	EtI, H_3O^+	O, $COCH_3$, C_2H_5, H, H, H (—)	67
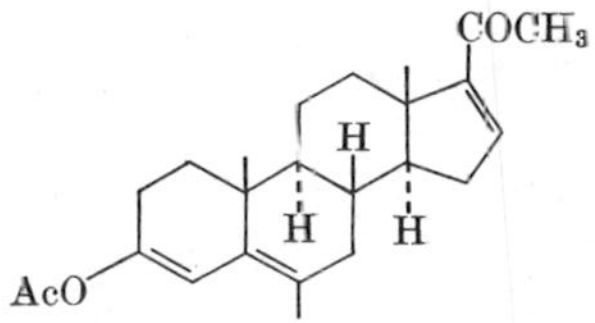 C_{27} THPO, $COCH_3$, H, H, H	Li/NH_3, THF	EtI, H_3O^+	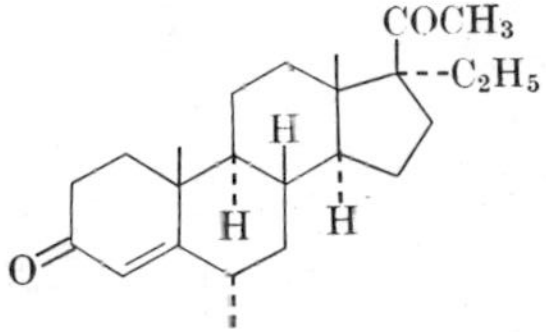 "	67
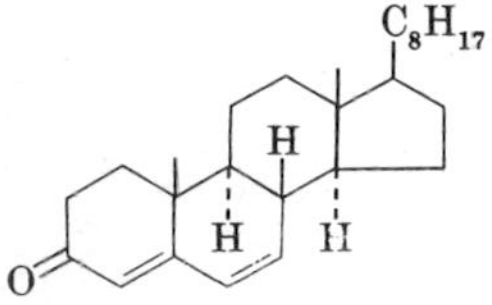 O, C_8H_{17}, H, H, H	NH_3/2 eq. Na, THF, 30 min	Excess MeI	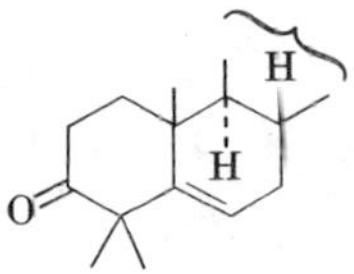 O, H, H (45)	225a
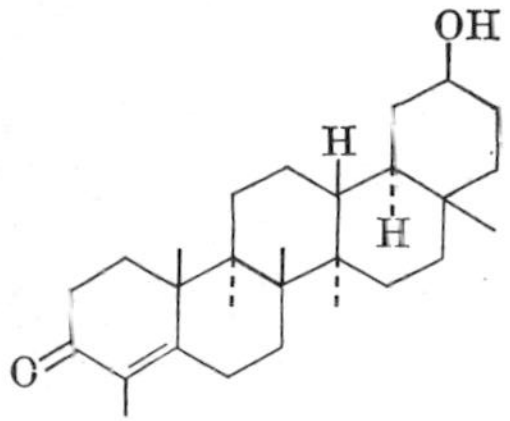 O, OH, H, H	Li/NH_3	MeI	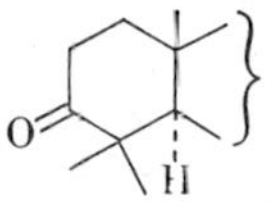 O, H (65)	178

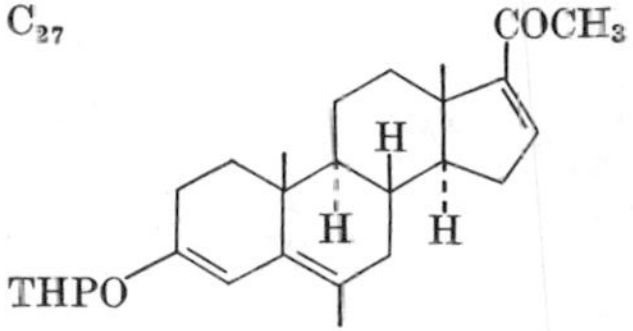

Note: References 338–543 are on pp. 253–258.

TABLE XII. REDUCTION-ALKYLATION AND RELATED REACTIONS OF α,β-UNSATURATED KETONES IN COMPOUNDS HAVING FOUR OR MORE FUSED RINGS (*Continued*)

Reactant	Reduction Conditions	Quenching Agent	Product(s) (% Yield)	Refs.
C_8H_{17}, H, H, H, O	Li/NH_3	$-NH_3$, Ac_2O, THF	AcO, H (—)	188
	,,	$-NH_3$, CO_2, THF, H^+, CH_2N_2	O, H, CO_2CH_3 (—)	543
C_8H_{17}, O, H, H	K/NH_3, Et_2O, 45 min	MeI	C_8H_{17}, H, O, H (36)	493
	,,	CD_3I	H, O, H, CD_3 (—)	
C_{28} C_8H_{17}, H, H, H, O	Li/NH_3	CD_3I	O, H, CD_3 >80 + O, D_3C, H <20	182

REFERENCES TO TABLES

[338] R. Granger, J. P. Chapat, F. Simon, J. R. Girand, and J. Crassous, *C.R. Acad. Sci., Ser. C*, **270,** 869 (1970).

[339] T. G. Halsall and D. B. Thomas, *J. Chem. Soc.*, **1956,** 2431.

[340] J.-M. Conia and G. Moinet, *Bull. Soc. Chim. Fr.*, **1969,** 500.

[341] J.-M. Conia and P. Beslin, *Bull. Soc. Chim. Fr.*, **1969,** 483.

[342] F. Rouessac, P. Beslin, and J.-M. Conia, *Tetrahedron Lett.*, **1965,** 3319.

[343] J.-M. Conia and F. Rouessac, *Tetrahedron*, **16,** 45 (1961).

[344] D. Caine and T. I. Chao, unpublished work.

[345] D. K. Banerjee and K. M. Sivanandaiah, *Tetrahedron Lett.*, No. 5, 20 (1960); *J. Indian Chem. Soc.*, **38,** 652 (1961).

[345a] D. Caine and C.-Y. Chu, *Tetrahedron Lett.*, **1974,** 703.

[346] R. Anliker, A. S. Lindsey, D. E. Nettleton, and R. B. Turner, *J. Amer. Chem. Soc.*, **79,** 220 (1957).

[347] H. E. Zimmerman and R. L. Morse, *J. Amer. Chem. Soc.*, **90,** 954 (1968).

[348] H. E. Zimmerman, R. D. Rieke, J. R. Scheffer, *J. Amer. Chem., Soc.*, **89,** 2033 (1967).

[348a] C. Ganter, E. C. Utsinger, K. Schaffner, D. Arigoni, and O. Jeger, *Helv. Chim. Acta.*, **45,** 2403 (1962).

[349] K. M. Baggaley, S. G. Brooke, J. Green, and B. T. Redman, *J. Chem. Soc., C*, **1971,** 2671.

[349a] G. Bauduin and Y. Petrasanta, *Tetrahedron*, **29,** 4225 (1973).

[350] E. E. van Tamelen and W. C. Proost, *J. Amer. Chem. Soc.*, **76,** 3632 (1954).

[351] D. K. Banerjee, S. Chatterjee, and S. P. Bhattacharya, *J. Amer. Chem. Soc.*, **77,** 408 (1955).

[352] E. E. Smissman, J. Pongman Li, and M. W. Cresse, *J. Org. Chem.*, **35,** 1352 (1970).

[353] M. Yanagita, K. Yamakawa, A. Tahara, and H. Ogura, *J. Org. Chem.*, **20,** 1767 (1955).

[354] J. A. Marshall, N. Cohen, and K. R. Arenson, *J. Org. Chem.*, **30,** 762 (1965).

[355] R. G. Carlson and N. S. Behn, *J. Org. Chem.*, **32,** 1303 (1967).

[356] B. Gaspert, T. G. Halsall, and D. Willis, *J. Chem. Soc.*, **1958,** 624.

[357] R. L. N. Harris, F. Komitsky, Jr., and C. Djerassi, *J. Amer. Chem. Soc.*, **89,** 4765 (1967).

[358] V. F. Kucherov and I. A. Gurvich, *J. Gen. Chem. USSR.*, **31,** 731 (1961).

[358a] J. E. McMurry and L. C. Blaszczak *J. Org. Chem.*, **39,** 2217 (1974).

[359] G. Stork and R. N. Guthikonka, *J. Amer. Chem. Soc.*, **94,** 5109 (1972).

[360] J. A. Marshall, N. H. Andersen, and P. C. Johnson, *J. Amer. Chem. Soc.*, **89,** 2748 (1967).

[361] N. K. Basu, U. R. Ghatak, G. Senguta, and P. C. Dutta, *Tetrahedron*, **21,** 2641 (1965).

[362] B. Maurer, M. Fracheboud, A. Grieder, and G. Ohloff, *Helv. Chim. Acta*, **55,** 2371 (1972)

[363] H. Bruderlein, N. Dufort, H. Favre, and A. J. Liston, *Can. J. Chem.*, **41,** 2908 (1963).

[364] C. J. V. Scanio and R. M. Starrat, *J. Amer. Chem. Soc.*, **93,** 1539 (1971).

[365] R. L. Hale and L. H. Zalkow, *Chem. Commun.*, **1968,** 1249.

[366] C. Berger, M. Franck-Neumann, and G. Ourisson, *Tetrahedron Lett.*, **1968,** 3451.

[367] M. Kato, H. Kosugi, and A. Yoshikoshi, *Chem. Commun.*, **1970,** 185.

[368] R. L. Kronenthal and E. I. Becker, *J. Amer. Chem. Soc.*, **79,** 1095 (1957).

[368a] F. J. McQuillin and P. L. Simpson, *J. Chem. Soc.*, **1963,** 4726.

[369] C. Enzell, *Acta Chem. Scand.*, **16,** 1553 (1962).

[370] J. A. Marshall and N. H. Andersen, *J. Org. Chem.*, **31,** 667 (1966).

[371] J. Levisalles and H. Rudler, *Bull. Soc. Chim. Fr.*, **1968,** 299.

[372] E. Brown and M. Ragault, *Tetrahedron Lett.*, **1973,** 1927.

[373] R. K. Mathur and A. S. Rao, *Tetrahedron*, **23,** 1259 (1967).

[374] E. J. Corey and D. S. Watt, *J. Amer. Chem. Soc.*, **95,** 2302 (1973).

[375] D. J. France, J. J. Hand, and M. Los, *Tetrahedron*, **25,** 4011 (1969).

[376] C. T. Mathew, G. C. Banerjee, and P. C. Dutta, *J. Org. Chem.*, **30,** 2754 (1965).

[377] T. Matsuura and A. Horinaka, *J. Chem. Soc.* **92,** *Jap.*, 1199 (1971).

[378] H. Favre and A. J. Liston, *Can. J. Chem.*, **47,** 3233 (1969).
[379] M. Maes, R. Ottinger, J. Reisse, and G. Chiurdoglu, *Tetrahedron*, **25,** 5163 (1969).
[380] G. Bozzato, M. Mario, and P. Schudel, *Ger. Pat.* 1,934,374 [*C.A.*, **72,** 100169s (1970)].
[381] J. A. Marshall and N. H. Andersen, *Tetrahedron Lett.*, **1967,** 1611.
[382] L. H. Zalkow, A. M. Shaligram, S.-E. Hu, and C. Djerassi, *Tetrahedron*, **22,** 337 (1966).
[383] S. S. Welankiwar, G. D. Joshi, S. N. Kulkarni, and S. C. Bhattacharyya, *Indian J. Chem.* **8,** 40 (1970).
[384] P. A. Grieco and K. Hiroi, *Tetrahedron Lett.*, **1973,** 1831.
[385] D. C. Humber, A. R. Pinder, and R. A. Williams, *J. Org. Chem.*, **32,** 2335 (1967).
[386] R. T. Gray and C. Djerassi, *J. Org. Chem.*, **35,** 753 (1970).
[387] R. H. Jaeger, *Tetrahedron*, **2,** 326 (1958).
[387a] J. Ficini and J. d'Angelo *C.R. Acad. Sci., Ser. C*, **276,** 803 (1973).
[387b] G. Stork and M. E. Jung, *J. Amer. Chem. Soc.*, **96,** 3682 (1974).
[388] R. E. Ireland and R. C. Kierstead, *J. Org. Chem.*, **31,** 2543 (1966).
[388a] Y. Pietrasanta and B. Pucci, *Tetrahedron Lett.*, **1974,** 1901.
[389] G. H. Hughes and H. Smith, *Proc. Chem. Soc.*, **1960,** 74.
[390] J. A. Barltrop, J. D. Littlehailes, J. P. Rushton, and N. A. J. Rogers, *Tetrahedron Lett.* **1962,** 429.
[391] N. A. Nelson, J. C. Wollensak, R. L. Foltz, J. B. Hester, Jr., J. I. Brauman, R. B. Garland, and G. H. Rasmusson, *J. Amer. Chem. Soc.*, **82,** 2569 (1960).
[392] M. Miyano and C. R. Dorn, *J. Org. Chem.*, **37,** 259 (1972).
[393] A. Taticchi and F. Fringuelli, *J. Chem. Soc., C*, **1971,** 756.
[394] P. Doyle, I. R. Maclean, R. D. H. Murray, W. Parker, and R. A. Raphael, *Proc. Chem. Soc.*, **1963,** 239; *J. Chem. Soc.*, **1965,** 1344.
[395] S. J. Daum, P. E. Shaw, and R. L. Clarke, *J. Org. Chem.*, **32,** 1427 (1967).
[396] W. S. Johnson and R. B. Kinnel, *J. Amer. Chem. Soc.*, **88,** 3861 (1966).
[397] S. K. Balasubramanian, *Tetrahedron*, **12,** 196 (1961).
[398] H. J. E. Loewenthal and H. Rosenthal, *Tetrahedron Lett.*, **1968,** 3693.
[399] E. Wenkert and T. E. Stevens, *J. Amer. Chem. Soc.*, **78,** 2318 (1956).
[400] F. H. Howell and D. A. H. Taylor, *J. Chem. Soc.*, **1958,** 1248.
[401] N. A. Nelson and R. B. Garland, *J. Amer. Chem. Soc.*, **79,** 6313 (1957).
[402] G. Saucy, W. Koch, M. Müller, and A. Fürst, *Helv. Chim. Acta*, **53,** 964 (1970).
[403] V. M. Sathe and A. S. Rao, *Indian J. Chem.*, **9,** 95 (1971).
[404] D. H. R. Barton, G. P. Moss, and J. A. Whittle, *J. Chem. Soc., C*, **1968,** 1813.
[405] D. V. Banthorpe, A. J. Curtis, and W. D. Fordham, *Tetrahedron Lett.*, **1972,** 3865.
[406] R. B. Bates, G. Büchi, T. Matsurra, and R. R. Shaffer, *J. Amer. Chem. Soc.*, **82,** 2327 (1960).
[407] R. E. Corbett and R. N. Speden, *J. Chem. Soc.*, **1958,** 3710.
[407a] M. Kaisin, Y. M. Sheikh, L. J. Durham, C. Djerassi, B. Tursch, D. Losman, and R. Karlsson, *Tetrahedron Lett.*, **1974,** 2239.
[408] H. J. E. Loewenthal and Z. Newwirth, *J. Org. Chem.*, **32,** 517 (1967).
[409] R. E. Ireland and L. M. Mander, *J. Org. Chem.*, **32,** 689 (1967).
[410] W. Nagata, T. Terasawa, and T. Aoki, *Tetrahedron Lett.*, **1963,** 865.
[411] G. Stork and M. Gregson, *J. Amer. Chem. Soc.*, **91,** 2373 (1969).
[411a] G. Stork, H. J. E. Loewenthal, and P. C. Mukharji, *J. Amer. Chem. Soc.*, **78,** 501 (1956).
[412] Y. Kitahara, A. Yoshikoshi, and S. Oida, *Tetrahedron Lett.*, **1964,** 1763.
[413] H. Hauth and D. Stauffacher, *Helv. Chim. Acta*, **55,** 1532 (1972).
[414] W. A. Ayer, W. R. Bowman, G. A. Cooke, and A. C. Soper, *Tetrahedron Lett.*, **1966,** 2021.
[415] J. A. Barltrop and A. C. Day, *Tetrahedron*, **14,** 310 (1961).
[416] M. J. T. Robinson, *Tetrahedron*, **1,** 49 (1957).
[417] T. A. Spencer, R. A. J. Smith, D. L. Storm, and R. M. Villarica, *J. Amer. Chem. Soc.*, **93,** 4856 (1971).
[418] K. Mori and M. Matsui, *Tetrahedron*, **22,** 2883 (1966).
[419] L. J. Chinn and H. L. Dryden, Jr., *J. Org. Chem.*, **26,** 3904 (1961).
[420] E. Wenkert, V. I. Stenberg, and P. Beak, *J. Amer. Chem. Soc.*, **83,** 2320 (1961).

[421] W. S. Johnson and T. K. Schaff, *Chem. Commun.*, **1969**, 611.

[422] H. Hauth and D. Stauffacher, *Helv. Chim. Acta*, **54,** 1278 (1971).

[423] W. A. Ayer, W. R. Bowman, T. C. Joseph, and P. Smith, *J. Amer. Chem. Soc.*, **90,** 1648 (1968).

[424] R. B. Turner, G. D. Diana, G. E. Fodor, K. Gebert, D. L. Simmons, A. S. Rao, O. Roos, and W. Wirth, *J. Amer. Chem. Soc.*, **88,** 1786 (1966).

[425] R. W. Gutherie, W. A. Henry, H. Immer, C. M. Wong, Z. Valenta, and K. Wiesner, *Collect. Czech. Chem. Commun.*, **31,** 602 (1966).

[426] R. V. Venkateswaran, D. Mukherjee, and P. C. Dutta, *Indian J. Chem.*, **10,** 768 (1972).

[427] J. A. Barltrop and A. C. Day, *Tetrahedron*, **22,** 3181 (1966).

[428] A. J. Birch and H. Smith, *J. Chem. Soc.*, **1951,** 1882.

[429] A. Chatterjee and B. G. Hazra, *Chem. Commun.*, **1970,** 618.

[430] R. E. Corbett and S. G. Wyllie, *J. Chem. Soc., C*, **1969,** 1747.

[431] K. E. Fahrenholtz, M. Lurie, and R. W. Kierstead, *J. Amer. Chem. Soc.*, **89,** 5934 (1967).

[432] D. E. Clark, R. F. K. Meredith, A. C. Ritchie, and T. Walker, *J. Chem. Soc.*, **1962,** 2400.

[433] K. Wiesner, A. Deljac, T. Y. R. Tsai, and M. Przybylska, *Tetrahedron Lett.*, **1970,** 1145.

[434] W. Nagata, T. Sugasawa, M. Narisada, T. Wakabayashi, and Y. Hayase, *J. Amer. Chem. Soc.*, **89,** 1483 (1967).

[435] G. Stork, H. N. Khastgir, and A. J. Solo, *J. Amer. Chem. Soc.*, **80,** 6457 (1958).

[436] M. Shiozaki, K. Mori, M. Matsui, and T. Hiraoka, *Tetrahedron Lett.*, **1972,** 657.

[437] M. Fetizon and J. C. Gramain, *Bull. Soc. Chim. Fr.*, **1968,** 3301.

[438] R. E. Counsell, *Tetrahedron*, **15,** 202 (1961).

[439] R. Bucourt, D. Hainaut, J. C. Gasc, and G. Nomine, *Bull. Soc., Chim. Fr.*, **1969,** 1920.

[440] A. J. Birch and J. A. K. Quartey, *Chem. Ind.* (London), **1953,** 489.

[441] W. F. Johns, *J. Amer. Chem. Soc.*, **80,** 6456 (1958).

[442] A. Bowers and E. Denot, *J. Amer. Chem. Soc.*, **82,** 4956 (1960).

[443] C. Djerassi, G. von Mutzenbecher, J. Fajkos, D. H. Williams, and H. Budzikiewicz, *J. Amer. Chem. Soc.*, **87,** 817 (1965).

[444] E. J. Bailey, A. Gale, G. H. Phillipps, P. T. Siddons, and G. Smith, *Chem. Commun.*, **1967,** 1253.

[445] W. S. Johnson, B. Bannister, R. Pappo, and J. E. Pike, *J. Amer. Chem. Soc.*, **78,** 6354 (1956).

[446] W. S. Johnson, R. Pappo, and W. F. Johns, *J. Amer. Chem. Soc.*, **78,** 6330 (1956).

[447] M. Fotizon, J. C. Gramain, and P. Mourgues, *Bull. Soc. Chim. Fr.*, **1969,** 1673.

[448] C. Djerassi, P. A. Hart, and C. Beard, *J. Amer. Chem. Soc.*, **86,** 85 (1964).

[449] R. H. Shapiro, D. H. Williams, H. Budzikiewicz, and C. Djerassi, *J. Amer. Chem. Soc.*, **86,** 2837 (1964).

[450] J. Gutzwiller and C. Djerassi, *Helv. Chim. Acta*, **49,** 2108 (1966).

[451] B. Berkoz, E. P. Chavez, and C. Djerassi, *J. Chem. Soc.*, **1962,** 1323.

[452] J. B. Jones and J. M. Zander, *Can. J. Chem.*, **46,** 1913 (1968).

[453] F. V. Brutcher and W. Bauer, *J. Amer. Chem. Soc.*, **84,** 2236 (1962).

[454] E. Farkas, J. M. Owen, and D. J. O'Toole, *J. Org. Chem.*, **34,** 3022 (1969).

[455] D. C. DeJongh, J. D. Hribar, P. Littleton, K. Fotherby, R. W. A. Rees, S. Shrader, T. J. Foell, and H. Smith, *Steroids*, **11,** 649 (1968).

[456] R. Sciaky, *Gazz. Chim. Ital.*, **92,** 561 (1962).

[457] W. S. Johnson, B. Bannister, and R. Pappo, *J. Amer. Chem. Soc.*, **78,** 6331 (1956).

[458] W. S. Johnson, B. Bannister, B. M. Bloom, A. D. Kemp, R. Pappo, E. R. Rogier, and J. Szmuszkovic, *J. Amer. Chem. Soc.*, **75,** 2275 (1953).

[459] W. G. Dauben, G. Ahlgren, T. J. Leitereg, W. C. Schwarzel, and M. Yoshioko, *J. Amer. Chem. Soc.*, **94,** 8593 (1972).

[460] W. S. Johnson, J. Ackerman, J. F. Eastham, and H. A. DeWalt, Jr., *J. Amer. Chem. Soc.*, **78,** 6302 (1956).

[461] J. A. Cella, E. A. Brown, and R. R. Burtner, *J. Org. Chem.*, **24,** 743 (1959).

[462] A. J. Birch, G. A. Hughes, and H. Smith, *J. Chem. Soc.*, **1958**, 4774.

[463] R. F. R. Church, A. S. Kende, and M. J. Weiss, *J. Amer. Chem., Soc.*, **87**, 2665 (1965).

[464] D. R. Herbst and H. Smith, *Steroids*, **11**, 935 (1968).

[465] A. Bowers, *J. Org. Chem.*, **26**, 2043 (1961).

[466] A. R. Van Horn and C. Djerassi, *J. Amer. Chem. Soc.*, **89**, 651 (1967).

[467] J. B. Jones and J. P. Leman, *Can. J. Chem.*, **49**, 2421 (1971).

[468] R. Villotti, H. J. Ringold, and C. Djerassi, *J. Amer. Chem. Soc.*, **82**, 5693 (1960).

[469] S. Danishefsky, P. Solomon, L. Crawley, M. Sax, S. C. Yoo, E. Abola, and J. Pletcher, *Tetrahedron Lett.*, **1972**, 961.

[470] N. S. Crossley and R. Dowell, *J. Chem. Soc., C*, **1971**, 2496.

[471] H. Hasegawa, Y. Sato, T. Tanaka, and K. Tsuda, *Chem. Pharm. Bull.* (Tokyo), **9**, 740 (1961).

[472] H. R. Nace and J. L. Pyle, *J. Org. Chem.*, **36**, 81 (1971).

[473] C. H. Robinson, O. Gnoj, and F. E. Carlon, *Tetrahedron*, **21**, 2509 (1965).

[474] A. J. Liston and M. Howarth, *J. Org. Chem.*, **32**, 1034 (1967).

[475] M. Fetizon and M. Golfier, *Bull. Soc. Chim. Fr.*, **1966**, 859.

[476] R. O. Clinton, A. S. Manson, F. W. Stonner, H. C. Neumann, R. G. Christiansen, R. L. Clarke, J. H. Ackerman, D. F. Page, J. W. Dean, W. B. Dickinson, and C. Carabateas, *J. Amer. Chem. Soc.*, **83**, 1478 (1961).

[477] H. Mitsuhashi and N. Kawahara, *Tetrahedron*, **21**, 1215 (1965).

[478] R. Anliker, M. Müller, M. Perelman, J. Wohlfahrt, and H. Heusser, *Helv. Chim. Acta*, **42**, 1071 (1959).

[479] P. F. Beal, M. A. Rebenstorf, and J. E. Pike, *J. Amer. Chem. Soc.*, **81**, 1231 (1959).

[480] A. L. Nussbaum, T. L. Popper, E. P. Oliveto, S. Freidman, and I. Wender, *J. Amer. Chem. Soc.*, **81**, 1228 (1959).

[481] Schering A.-G., *Fr. Pat.*, 2,002,566 (Cl. A61k C 07c), 1969 [*C.A.* **72**, 101003v (1970)].

[482] A. Bowers, E. Denot, M. B. Sanchez, F. Neumann, and C. Djerassi, *J. Chem. Scc.*, **1961**, 1859.

[483] D. K. Fukushima and S. Daum, *J. Org. Chem.*, **26**, 520 (1961).

[484] K. Heusler, H. Heusser, and R. Anliker, *Helv. Chim. Acta*, **36**, 652 (1953).

[485] J. H. Fried, G. E. Arth, and L. H. Sarett, *J. Amer. Chem. Soc.*, **81**, 1235 (1959).

[486] D. Burn and V. Petrow, *J. Chem. Soc.*, **1962**, 1223.

[487] G. DeStevens and A. Halamandaris, *J. Org. Chem.*, **26**, 1614 (1961).

[488] Y. K. Sawa, N. Tsuji, and S. Maeda, *Tetrahedron*, **15**, 148 (1961).

[489] J. M. H. Graves and H. J. Ringold, *Steroids Suppl.* **I**, 23 (1965).

[490] O. Wintersteiner and M. Moore, *J. Org. Chem.*, **29**, 262 (1964).

[491] T. Masamune, M. Takasugi, M. Gohda, H. Suzuki, S. Kawahara, and T. Irie, *J. Org. Chem.*, **29**, 2282 (1964).

[492] T. Masamune, M. Takasugi, H. Suzuki, S. Kawahara, M. Gohda, and T. Irie, *Bull. Chem. Soc. Jap.*, **35**, 1749 (1962).

[493] R. R. Muccino and C. Djerassi, *J. Amer. Chem. Soc.*, **96**, 556 (1974).

[494] T. Masamune, K. Kobayashi, M. Takasugi, Y. Mori, and A. Murai, *Tetrahedron*, **24**, 3461 (1968).

[495] J. C. Bloch and G. Ourisson, *Bull. Soc. Chim. Fr.*, **1964**, 3018.

[496] D. H. R. Barton, D. A. J. Ives, and B. R. Thomas, *J. Chem. Soc.*, **1954**, 903.

[497] C. Beard, J. N. Wilson, H. Budzikiewicz, and C. Djerassi, *J. Amer. Chem. Soc.*, **86**, 269 (1964).

[498] S. K. Pradhan, G. Subrahmanyam, and H. J. Ringold, *J. Org. Chem.*, **32**, 3004 (1967).

[499] L. Tökés, G. Jones, and C. Djerassi, *J. Amer. Chem. Soc.*, **90**, 5465 (1968).

[500] G. D. Meakins and O. R. Rodig, *J. Chem. Soc.*, **1956**, 4679.

[501] Y. Mazur and F. Sondheimer, *J. Amer. Chem. Soc.*, **80**, 5220 (1958).

[502] D. H. R. Barton, D. A. J. Ives, and B. R. Thomas, *Chem. Ind.* (London), **1953**, 1180.

[503] J. Castells, G. A. Fletcher, E. R. H. Jones, G. D. Meakins, and R. Swindels, *J. Chem. Soc.*, **1960**, 2627.

[504] D. Herlem-Gaulier, F. Khuong-Huu-Lainé, and M. R. Goutarel, *Bull. Soc. Chim. Fr.*, **1966**, 3478.

505 K. Tsuda and S. Nozoe, *Chem. Pharm. Bull.* (Tokyo), **7,** 232 (1959).

506 B. R. Brown, P. W. Trown, and J. M. Woodhouse, *J. Chem. Soc.*, **1961,** 2478.

507 D. H. R. Barton and B. R. Thomas, *J. Chem. Soc.*, **1953,** 1842.

508 G. D. Laubach, E. C. Schreiber, E. J. Agnello, and K. J. Brunings, *J. Amer. Chem. Soc.*, **78,** 4746 (1956).

509 G. H. Alt and D. H. R. Barton, *J. Chem. Soc.*, **1954,** 1356.

510 P. Bladon, H. B. Henbest, E. R. H. Jones, B. J. Lovell, and G. H. Woods, *J. Chem. Soc.*, **1954,** 125.

511 G. C. Morrison, R. O. Waite, and J. Shavel, Jr., *J. Heterocycl. Chem.*, **8,** 1025 (1971).

512 J. C. Bloch, P. Crabbé, F. A. Kincl, G. Ourisson, J. Perez, and J. A. Zderic, *Bull. Soc. Chim. Fr.*, **1961,** 559.

513 M. Nussim, Y. Mazur, and F. Sondheimer, *J. Org. Chem.*, **29,** 1131 (1964).

514 A. J. Lemin and C. Djerassi, *J. Amer. Chem. Soc.*, **76,** 5672 (1954).

515 M. Marx, J. Leclercq, B. Tursch, and C. Djerassi, *J. Org. Chem.*, **32,** 3150 (1967).

516 T. Masamune, N. Sato, K. Kobayashi, I. Yamazaki, and Y. Mori, *Tetrahedron*, **23,** 1591 (1967).

517 D. H. R. Barton, E. F. Lier, and J. F. McGhie, *J. Chem. Soc., C*, **1968,** 1031.

518 P. G. Gassman and K. T. Manfield, *J. Amer. Chem. Soc.*, **90,** 1517 (1968).

519 P. A. Grieco and R. S. Finkelhor, *J. Org. Chem.*, **38,** 2245 (1973).

520 K. J. Schmalzl and R. N. Mirrington, *Tetrahedron Lett.*, **1970,** 3219; R. N. Mirrington and K. J. Schmalzl, *J. Org. Chem.*, **37,** 2871 (1972).

521 J. Martin, W. Parker, and R. A. Raphael, *J. Chem. Soc., C*, **1967,** 348.

521a R. L. Sowerby and R. M. Coates, *J. Amer. Chem. Soc.*, **94,** 4758 (1972).

522 A. Sakurai and S. Tamura, *Agr. Biol. Chem.* (Tokyo), **30,** 793 (1960).

523 H. H. Inhoffen, K. Irmscher, G. Friedrich, D. Kampe, and O. Berges, *Chem. Ber.*, **92,** 1772 (1959).

524 A. J. Birch, P. L. MacDonald, and V. H. Powell, *Tetrahedron Lett.*, **1969,** 351, *J. Chem. Soc., C*, **1970,** 1469.

525 G. Subrahmanyan, *Indian J. Chem.*, **8,** 210 (1970).

526 H. Hauth and D. Stauffacher, *Helv. Chim. Acta*, **54,** 2197 (1971).

527 S. C. Welch and C. P. Hagan, *Syn. Commun.* **3,** 29 (1973).

528 S. C. Welch and C. P. Hagan, *Syn. Commun.*, **2,** 221 (1972).

529 F. Sondheimer, W. McCrae, and W. G. Salmond, *J. Amer. Chem. Soc.*, **91,** 1228 (1969).

530 J. Schreiber, Ph.D. Dissertation, Zurich, 1953; see G. Stork *et al.*, *J. Amer. Chem. Soc.*. **87,** 275 (1965), footnote 14.

530a R. K. Boeckman, Jr., *J. Amer. Chem. Soc.*, **96,** 6179 (1974).

531 J. A. Marshall and A. E. Greene, *Tetrahedron*, **1969,** 4183.

532 A. J. Birch and G. S. R. Subba Rao, *Tetrahedron Lett.*, **1967,** 2763; *Australian J. Chem.*, **23,** 547 (1970).

532a R. E. Ireland, D. C. Muchmore, and U. Hengartner, *J. Amer. Chem. Soc.*, **94,** 5098 (1972).

532b G. Stork and J. Singh, *J. Amer. Chem. Soc.*, **96,** 6181 (1974).

532c J. S. Dutcher, J. G. Macmillan, and C. H. Heathcock, *Tetrahedron Lett.*, **1974,** 929.

532d R. M. Coates and S. K. Chung, *J. Org. Chem.*, **38,** 3677 (1973).

533 D. K. Banerjee, W. S. Johnson, and K. Srinivison, unpublished work; see D. K. Banerjee, *J. Indian Chem. Soc.*, **47,** 1 (1970).

534 J. W. ApSimon, P. Baker, J. Buccini, J. W. Hooper, and S. Macaulay, *Can. J. Chem.*, **50,** 1944 (1972).

535 G. Nominé, R. Bucourt, J. Tessier, A. Pierdet, G. Costerousse, and J. Mathieu, *C.R. Acad. Sci., Ser. C*, **260,** 4545 (1965).

536 M. Uskoković, J. Iacobelli, R. Philion, and T. Williams, *J. Amer. Chem. Soc.*, **88,** 4538 (1966).

537 R. E. Ireland, S. W. Baldwin, D. J. Dawson, M. I. Dawson, J. E. Dolfini, J. Newbould, W. S. Johnson, M. Brown, R. J. Crawford, P. F. Hudrlik, G. H. Rasmussen, and K. K. Schmiegel, *J. Amer. Chem. Soc.*, **92,** 5743 (1970).

[538] R. E. Ireland, D. A. Evans, D. Glover, G. M. Rubottom, and H. Young, *J. Org. Chem.*, **34,** 3717 (1969).

[539] J. M. Midgley, W. B. Whalley, G. F. Katekar, and B. A. Lodge, *Chem. Commun.*, **1965,** 169.

[540] J. Warnant and A. Farcilli, *Ger. Pat.*, 2,107,835 [*C.A.* **75,** 151985z (1971)].

[541] A. Afonso, *J. Amer. Chem. Soc.*, **90,** 7375 (1968); *J. Org. Chem.*, **35,** 1949 (1970).

[542] R. E. Schaub and M. J. Weiss, *J. Med. Chem.*, **10,** 789 (1967).

[543] K. B. Sharpless, T. E. Snyder, T. A. Spencer, K. K. Maheshwari, G. Gahn, and R. B. Clayton, *J. Amer. Chem. Soc.*, **90,** 6874 (1968).

CHAPTER 2

THE ACYLOIN CONDENSATION

JORDAN J. BLOOMFIELD, DENNIS C. OWSLEY,
AND
JANICE M. NELKE

Corporate Research Department
Monsanto Company

CONTENTS

INTRODUCTION

The acyloin condensation usually involves the reductive dimerization of a carboxylic ester, although acid chlorides and anhydrides have been used. The reducing agent is an alkali metal and the product is an ene-diolate. Two gram-atoms of metal are required for each mole of ester with the concomitant formation of a mole of alkoxide and one-half mole of the ene-diolate. When an α,ω-diester is used, the product is a cyclic ene-diolate.

$$2\,C_2H_5CO_2C_2H_5 + 4\,Na \rightarrow C_2H_5\overset{O^-Na^+}{\overset{|}{C}}{=\!=}\overset{O^-Na^+}{\overset{|}{C}}C_2H_5 + 2\,NaOC_2H_5$$

$$C_2H_5\overset{O^-}{\overset{|}{C}}{=\!=}\overset{O^-}{\overset{|}{C}}C_2H_5 \xrightarrow{H_3O^+} C_2H_5\overset{OH}{\overset{|}{C}}H{-\!-}\overset{O}{\overset{\|}{C}}C_2H_5$$

$$(CH_2)_{10}\begin{cases}CO_2CH_3\\ CO_2CH_2\end{cases} + 4\,Na \rightarrow (CH_2)_{10}\begin{cases}C{-}O^-Na^+\\ \| \\ C{-}O^-Na^+\end{cases} + 2\,CH_3ONa$$

$$\xrightarrow{H_3O^+} (CH_2)_{10}\begin{cases}C{=}O\\ | \\ CHOH\end{cases} + 2\,CH_3OH$$

Neutralization of the reaction mixture produces an α-hydroxyketone—an acyloin.

The history of the early discovery[1,2] and later identification of acyloin derivatives[3,4] by reduction of acid chlorides was described in the *Organic Reactions* chapter, "The Acyloins,"[5] which also provides further references to the early work as well as alternative methods of synthesis. A more up-to-date review on synthesis of α-hydroxyketones, including acyloins and benzoins and their tautomeric structures, was published in 1964.[6]

A number of reviews have been published on the synthesis of acyloins: as intermediates in the manufacture of perfumes,[7] for large rings,[8–11] as routes to cyclophanes,[12,13] and for preparation of heterocyclic ketones used to study transannular interactions.[14] A thorough review of the literature concerning the acyloin condensation as a cyclization method was published in 1964.[15] Recently Rühlmann reviewed much of the work on the acyloin condensation conducted in the presence of trimethylchlorosilane.[16]

The acyloins themselves, as many of the review articles mentioned above point out, are frequently only intermediates. They can be reduced catalytically[17] or by hydride reducing agents[18] to diols. Diols can also be prepared by catalytic reduction of the bistrimethylsilyloxy derivatives of ene-diolates.[19] Olefins can subsequently be produced from the diols by

[1] A. Freund, *Ann.*, **118**, 33 (1861).

[2] J. W. Brühl, *Ber.*, **12**, 315 (1879).

[3] H. Klinger and L. Schmitz, *Ber.*, **24**, 1271 (1891).

[4] A. Basse and H. Klinger, *Ber.*, **31**, 1217 (1898).

[5] S. M. McElvain, *Org. Reactions*, **4**, 256 (1948).

[6] M. Bracke, *Mededel. Vlaamse Chem. Ver.*, **26**, 129–87 (1964) [*C.A.*, **62**, 6353b (1965)].

[7] J. A. Van Allan, *Amer. Perfum. Essential Oil Rev.*, **1949**, 33.

[8] V. Prelog, *J. Chem. Soc.*, **1950**, 420.

[9] M. Stoll, *Chimia*, **2**, 217 (1948) [*C.A.*, **43**, 2374c (1949)].

[10] K. Ziegler in Houben-Weyl, *Methoden der Organischen Chemie*, E. Müller, Ed., Vol. 4/2, G. Thieme Verlag, Stuttgart, 1955. pp. 739, 755, and H. Herlinger. *ibid.*, Vol. 7/2a, 1973, p. 642ff.

[11] L. I. Belen'kii, *Usp. Khim.*, **33**, 1265 (1964). [*Russian Chem. Rev.* (*Engl. Transl.*) **33**, 551 (1964)].

[12] D. J. Cram, *Rec. Chem. Progr.*, **20**, 71 (1959).

[13] B. H. Smith, *Bridged Aromatic Compounds*, Academic Press, New York, 1964, pp. 27–42.

[14] N. J. Leonard, *Rec. Chem. Progr.*, **17**, 243 (1956).

[15] K. T. Finley, *Chem. Rev.*, **64**, 573 (1964).

[16] K. Rühlmann, *Synthesis*, **1971**, 236.

[17] A. T. Blomquist and A. Goldstein, *Org. Syn., Coll. Vol.*, **4**, 216 (1963).

[18] A. T. Blomquist, R. E. Burge, Jr., and A. C. Sucsy, *J. Amer. Chem. Soc.*, **74** 3636 (1952).

[19] H.-M. Fischler, H.-G. Heine, and W. Hartmann, *Tetrahedron Lett.*, **1972**, 860.

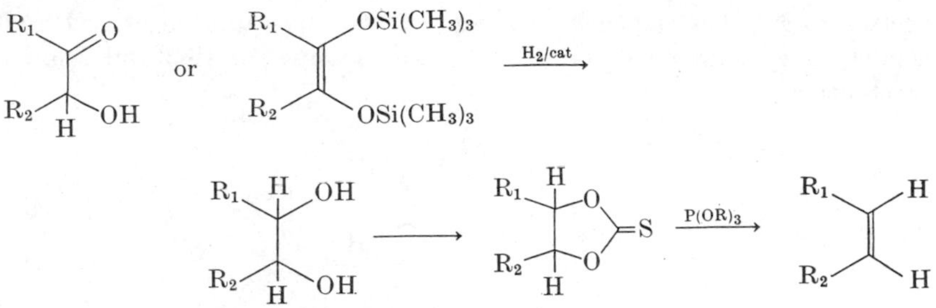

R_1, R_2 acyloin or bis(trimethylsilyl) ene-diolate $\xrightarrow{H_2/cat}$ diol → thionocarbonate $\xrightarrow{P(OR)_3}$ olefin

the Corey-Winter procedure.[20,21] This sequence has been applied to synthesis of cyclobutenes[22–24] as well as other olefins.

Oxidation of acyloins to diketones can be accomplished by a variety of reagents. Among the most familiar are cupric acetate[25] and bismuth trioxide.[26] A relatively new technique utilizing dimethyl sulfoxide-acetic anhydride may be particularly useful for acyloin oxidations.[27–29] Another route to α-diketones, with the virtue that the acyloin itself need not be isolated, involves the bromine oxidation of the bistrimethylsilyl derivative of the ene-diolate.[30–32] Some diketones can be converted into

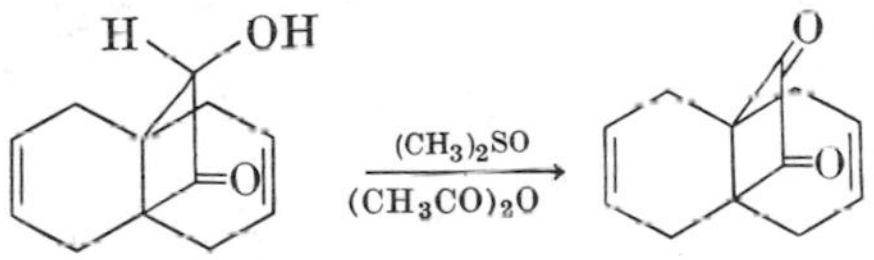

$\xrightarrow[(CH_3CO)_2O]{(CH_3)_2SO}$ (Ref. 29)

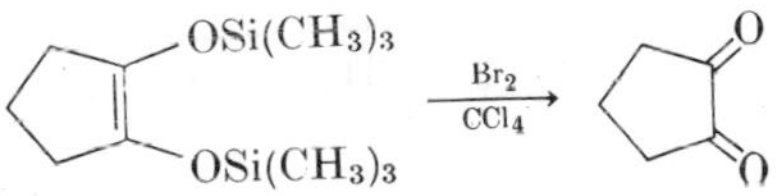

$\xrightarrow[CCl_4]{Br_2}$ (Ref. 30)

[20] E. J. Corey and R. A. E. Winter, *J. Amer. Chem. Soc.*, **85**, 2677 (1963); E. J. Corey, F. A. Carey, and R. A. E. Winter, *ibid.*, **87**, 934 (1965); E. J. Corey, *Pure Appl. Chem.*, **14**, 19 (1967).

[21] A recent modification of this procedure involves the use of iron pentacarbonyl in place of an alkyl phosphite with the advantage of lower temperatures and shorter reaction times; J. Daub, V. Trautz, and U. Erhardt, *Tetrahedron Lett.*, **1972**, 4435.

[22] J. R. S. Irelan, Ph.D. Dissertation, University of Oklahoma, 1968 [*Diss. Abstr.*, **29B**, 2808 (1969)].

[23] L. A. Paquette and J. C. Philips, *Tetrahedron Lett.*, **1967**, 4645.

[24] W. Hartmann, H.-M. Fischler, and H.-G. Heine, *Tetrahedron Lett.*, **1972**, 853.

[25] A. T. Blomquist and A. Goldstein, *Org. Syn., Coll. Vol.*, **4**, 838 (1963).

[26] W. Rigby, *J. Chem. Soc.*, **1951**, 793.

[27] J. D. Albright and L. Goldman, *J. Amer. Chem. Soc.*, **87**, 4214 (1965); **89**, 2416 (1967).

[28] M. Van Dyke and N. D. Pritchard, *J. Org. Chem.*, **32**, 3204 (1967).

[29] J. J. Bloomfield, J. R. S. Irelan, and A. P. Marchand, *Tetrahedron Lett.*, **1968**, 5647.

[30] J. Strating, S. Reiffers, and H. Wynberg, *Synthesis*, **1971**, 211.

[31] J. Strating, S. Reiffers, and H. Wynberg, *Synthesis*, **1971**, 209.

[32] H. Wynberg, J. Strating, and S. Reiffers, *Rec. Trav. Chim. Pays-Bas*, **89**, 982 (1970).

acetylenes.[33,34] The sequence starting with the unsaturated ester **1** and leading to cyclodecene-6-yne provides an example of this and related reactions.[35]

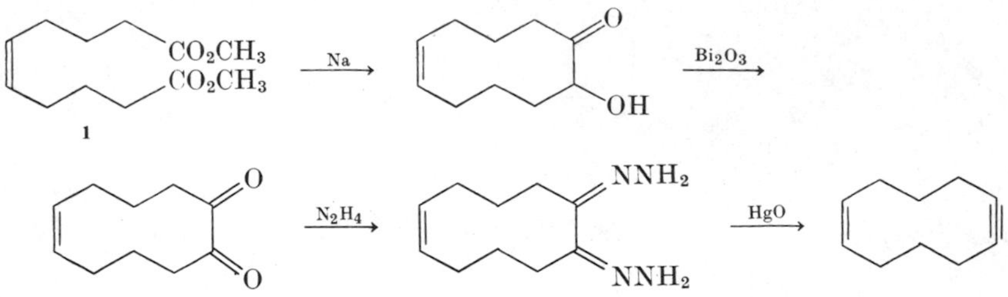

Acyloins can also be reduced to ketones by various modifications of the Clemmensen technique. Under mild conditions the ketone will predominate. However, vigorous conditions will lead to complete reduction. The accompanying examples give some idea of results that can be expected.

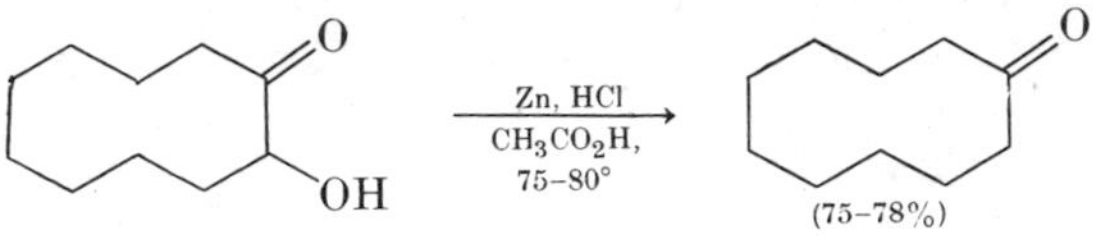

(Ref. 36)

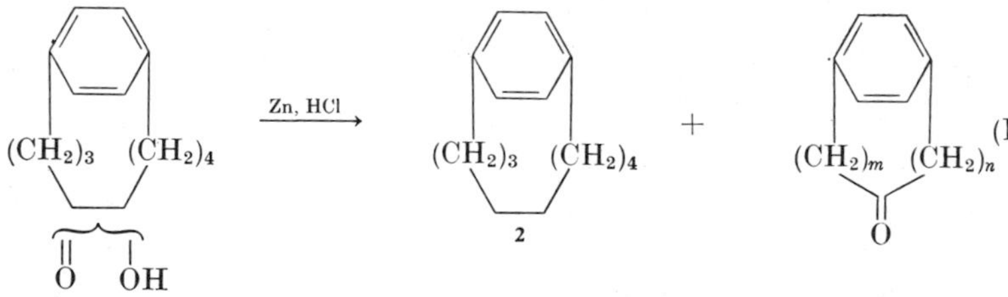

(75–78%)

(Ref. 37)

Granular Zn, 100°, **2**:**3** = 2
Zn foil, 75–80°, **2**:**3** = 0.5

$m = n = 4$ or
$m = 3, n = 5$

Another useful technique for reduction of acyloins to ketones involves preparation of the acetate followed by treatment with calcium in liquid ammonia. This procedure has been of particular value in steroid work.[38,39]

[33] V. Prelog, K. Schenker, and H. H. Günthard, *Helv. Chim. Acta*, **35**, 1598 (1952).

[34] A. T. Blomquist, R. E. Burge, Jr., L. H. Liu, J. C. Bohrer, A. C. Sucsy, and J. Kleis, *J. Amer. Chem. Soc.*, **73**, 5510 (1951).

[35] D. J. Cram and N. L. Allinger, *J. Amer. Chem. Soc.*, **78**, 2518 (1956).

[36] A. C. Cope, J. W. Barthel, and R. D. Smith, *Org. Syn., Coll. Vol.*, **4**, 218 (1963).

[37] D. J. Cram and M. F. Antar, *J. Amer. Chem. Soc.*, **80**, 3109 (1958).

[38] J. H. Chapman, J. Elks, G. H. Phillipps, and L. J. Wyman, *J. Chem. Soc.*, **1956**, 4344.

[39] J. S. Mills, H. J. Ringold, and C. Djerassi, *J. Amer. Chem. Soc.*, **80**, 6118 (1958).

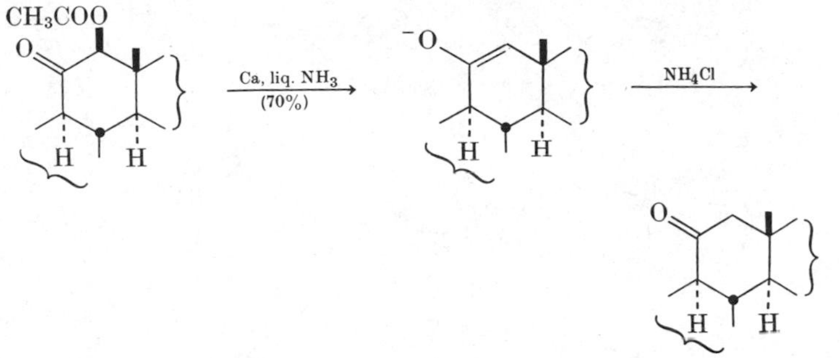

$$\xrightarrow[\text{(70\%)}]{\text{Ca, liq. NH}_3} \xrightarrow{NH_4Cl}$$

(Ref. 38)

Iron carbonyl has also been suggested as a useful reagent for this transformation.[40] A related process involves the reduction of the trimethylsilyl derivative of an acyloin with sodium and trimethylchlorosilane in ether.[16] The intermediate silylated enolate is readily hydrolyzed to the free ketone, or it can be utilized directly for subsequent reactions.

$$\text{RCH(OSi(CH}_3)_3)\text{C(=O)R}' + 2\,\text{Na} + 2\,\text{ClSi(CH}_3)_3 \xrightarrow{(C_2H_5)_2O} \text{RCH{=}C(OSi(CH}_3)_3)\text{R}' + 2\,\text{NaCl} + [(\text{CH}_3)_3\text{Si}]_2\text{O}$$

The air oxidation of acyloins in basic dimethyl sulfoxide solutions produces semidiones, relatively stable radical anions of the acyloin enediol. These radicals have been extensively investigated in recent years.[41]

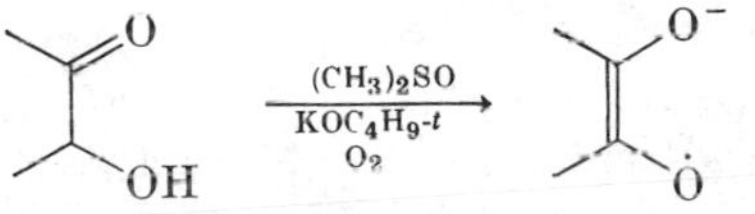

$$\xrightarrow[\substack{KOC_4H_9\text{-}t \\ O_2}]{(CH_3)_2SO}$$

Two other potentially useful reactions of acyloins have been described. In the first, the reaction of acyloins with phosgene produces vinylene carbonate derivatives which under photosensitized irradiation add olefins to yield saturated carbonates.[42] The accompanying example gives some idea of the scope of this reaction.

The second reaction involves the addition of methylene to the bis-trimethylsilyl derivative of the ene-diolate.[43] This reaction makes protected cyclopropanediols of a wide range of structure available for reactivity studies as well as for intermediates.

[40] S. J. Nelson, G. Detre, and M. Tanabe, *Tetrahedron Lett.*, **1973,** 447.

[41] G. A. Russell, P. R. Whittle, R. G. Keske, G. Holland, and C. Aubuchon, *J. Amer. Chem. Soc.*, **94,** 1693 (1972), and references cited therein.

[42] H.-M. Fischler, H.-G. Heine, and W. Hartmann, *Tetrahedron Lett.*, **1972,** 1701.

[43] M. Audibrand, R. LeGoaller, and P. Arnaud, *C. R. Acad. Sci., Ser. C*, **268,** 2322 (1969).

O OH $+\ COCl_2$ $\xrightarrow[C_6H_5N(CH_3)_2]{CH_2Cl_2}$ O O O

$\xrightarrow[(CH_3)_2CO,\ h\nu]{C_2H_4}$ O O O $\xrightarrow[CH_3OH]{KOH}$ OH OH (Ref. 42)

$\xrightarrow[H_2O]{IO_4^-}$ O O $\longrightarrow$ O

$OSi(CH_3)_3$ $OSi(CH_3)_3$ $+\ CH_2I_2$ $\xrightarrow{Zn-Cu}$ $OSi(CH_3)_3$ $OSi(CH_3)_3$ (Ref. 43)

This chapter presents a complete picture of both the linear and the cyclic acyloin condensation, with particular emphasis on developments since about 1960. The coverage of the literature through most of 1974 is comprehensive, with complete tables. The discussion is closely limited to the acyloin condensation and its modifications. Alternative methods of synthesis have been described by Bracke.[6]

MECHANISM

The generally accepted mechanistic schemes for the acyloin condensation involve production of the dianion **6** either (a) by coupling of two initially formed radical anions **4**[5,15] or (b) by a two-electron reduction of an ester to a dianion **5** followed by its addition to a second molecule of ester.[44,45] The diketone **7** has been assumed to be an intermediate produced by loss of alkoxide from **6**. Its subsequent two-electron reduction leads to the acyloin ene-diolate **8**. Neutralization of **8** produces the free acyloin **9**.*

In alkylation experiments several acyloin reaction mixtures in liquid ammonia produced ketone **(10)** derived from the starting esters.[45] This result caused the speculation that the ene-diolate **8** was in equilibrium with an acyl anion **11**. Despite the fact that other authors obtained **12**,

* For simplicity in presentation we have chosen to use e^- as the indicated reagent in the mechanistic schemata. The metal employed certainly has some effect on the course of the reduction, partly through differences in ease of electron release to the ester and partly through differences in stability of intermediate salts and their degree of association. However, at the present there is no way to measure these effects and no experimental data are available.

[44] F. F. Blicke, *J. Amer. Chem. Soc.*, **47**, 229 (1925).

[45] M. S. Kharasch, E. Sternfeld, and F. R. Mayo, *J. Org. Chem.*, **5**, 362 (1940).

$$2\,RCO_2R' \xrightarrow{2e^-} 2\,R\!-\!\overset{O^-}{\underset{\bullet}{\overset{|}{C}}}\!-\!OR' \;(\mathbf{4})\;\text{ or }\;\left[R\overset{O^-}{\underset{-}{\overset{|}{C}}}\!-\!OR' + RCO_2R'\right] (\mathbf{5}) \longrightarrow R\overset{O^-}{\underset{OR'}{C}}\!-\!\!-\!\overset{O^-}{\underset{OR'}{C}}R \;(\mathbf{6})$$

$$\mathbf{6} \xrightarrow{-2R'O^-} R\overset{O}{\overset{\|}{C}}\!-\!\overset{O}{\overset{\|}{C}}R \;(\mathbf{7}) \xrightarrow{2e^-} R\overset{O^-}{\overset{|}{C}}\!=\!\overset{O^-}{\overset{|}{C}}R \;(\mathbf{8}) \xrightarrow{H_3O^+} R\overset{OH}{\overset{|}{C}H}\!-\!\overset{O}{\overset{\|}{C}}R \;(\mathbf{9})$$

the normal result of simple enolate alkylation,[46–48] the acyl anion (or an alkoxide adduct of it) interpretation has been uncritically accepted by

$$R\overset{O^-}{\overset{|}{C}}\!=\!\overset{O^-}{\overset{|}{C}}R \;(\mathbf{8}) \rightleftharpoons 2\,RCO^- \;(\mathbf{11}) \xrightarrow{2\,R'Br} 2\,R\overset{O}{\overset{\|}{C}}R' \;(\mathbf{10}) + 2\,Br^- \qquad \text{(Ref. 45)}$$

authors of reference works.[49] Other experimental work suggests that compounds with structure **12** may have decomposed under the drastic workup conditions used, which included base extractions in the air and atmospheric pressure distillation.

$$R\overset{O^-}{\overset{|}{C}}\!=\!\overset{O^-}{\overset{|}{C}}R \;(\mathbf{8}) \xrightarrow{R'Br} R\!-\!\overset{OH}{\underset{R'}{C}}\!-\!\overset{O}{\overset{\|}{C}}R \;(\mathbf{12})$$

$$\mathbf{12} \xrightarrow{OH^-} RR'CHOH + RCO_2H + R_2CHOH + R'CO_2H \qquad \text{(Ref. 50)}$$

$(R = C_6H_5;\ R' = H,\ CH_3,\ C_6H_5CH_2,\ o,m,p\text{-}CH_3C_6H_4,\ C_6H_5)$

That the initial step in the reaction involves the addition of one electron to the ester to form **4** does not seem arguable. What subsequently occurs may still be open to considerable debate.

Evidence for the initial radical nature of the reaction was obtained in reductions of esters in which the intermediate radical anions could decompose to carbon monoxide, alkoxide, and a resonance-stabilized

[46] J. C. Speck, Jr., and R. W. Bost, *J. Org. Chem.*, **11**, 788 (1946).

[47] (a) J. H. Van de Sande and K. R. Kopecky, *Can. J. Chem.*, **47**, 163 (1969); (b) J. Colonge and P. Brison, *Bull. Soc. Chim. Fr.*, **1962**, 175.

[48] F. Chen, R. E. Robertson, and C. Ainsworth, *J. Chem. Eng. Data*, **16**, 121 (1971).

[49] *Cf.* (a) H. Smith, *Chemistry in Nonaqueous Ionizing Solvents*, Vol. I, Part II, "Organic Reactions in Liquid Ammonia," Interscience, New York, 1963, p. 174; (b) C. Walling, *Free Radicals in Solution*, Wiley, New York, 1957, p. 585; (c) A. J. Birch, *Quart. Rev.*, **4**, 69 (1950).

[50] D. B. Sharp and E. L. Miller, *J. Amer. Chem. Soc.*, **74**, 5643 (1952).

tertiary radical.[51,52] Reduction of the esters $RCO_2C_2H_5$ [R= C_6H_5-cyclopentyl, C_6H_5-cyclohexyl, and $(CH_3)_2C{=}C(CH_3)C(CH_3)_2$ (with the attachment at the last C)] produced the acyloin RCHOHCOR, the hydrocarbons RR and RH, and the ketone RCOR. The sequence of reactions suggested to account for these results was:

$$RCO_2C_2H_5 \xrightarrow{e^-} R\dot{C}(O^-)OC_2H_5 \ (\mathbf{13}) \xrightarrow{-C_2H_5O^-} R\dot{C}O \ (\mathbf{14}) \xrightarrow{-CO} R\cdot \ (\mathbf{15})$$

The radical **15** was presumed to dimerize to form RR, react with the radical **14** to produce the ketone, or abstract a hydrogen atom to form RH. The acyloin could be formed by dimerization of **13** or **14** followed by reduction.

Recently the reduction of phenylacetic esters in the presence of trimethylchlorosilane, which produced a similar variety of products, was also explained in terms of free-radical intermediates.[16]

$$C_6H_5CH_2CO_2R' \xrightarrow{e^-} C_6H_5CH_2\dot{C}(O^-)OR' \ (\mathbf{16}) \longrightarrow C_6H_5CH_2C(O^-)(OR')\text{—}C(O^-)(OR')CH_2C_6H_5 \ (\mathbf{17})$$

$$\mathbf{16} \xrightarrow{-R'O^-} C_6H_5CH_2CO\cdot \ (\mathbf{18}) \xrightarrow{-CO} C_6H_5CH_2\cdot \ (\mathbf{19}) \xrightarrow[(CH_3)_3SiCl]{e^-} C_6H_5CH_2Si(CH_3)_3 \ (\mathbf{20})$$

$$\mathbf{17} \xrightarrow[(CH_3)_3SiCl]{-2\,R'O^-,\ +2\,e^-} C_6H_5CH_2C(OSi(CH_3)_3){=}C(OSi(CH_3)_3)CH_2C_6H_5 \ (\mathbf{21})$$

$$\mathbf{18} + \mathbf{19} \rightarrow (C_6H_5CH_2)_2CO \ (\mathbf{22})$$

Bibenzyl, which is usually found in reactions involving benzyl radicals, surprisingly is absent from the reaction mixture. This suggests that the

$$C_6H_5CH{=}C(OSi(CH_3)_3)(OR') \ (\mathbf{23})$$

[51] E. Van Heyningen, *J. Amer. Chem. Soc.*, **74**, 4861 (1952).
[52] E. Van Heyningen, *J. Amer. Chem. Soc.*, **77**, 4016 (1955).

benzyl radical might not be present at all. This point is discussed in detail later. The silylated derivative **23** of the original ester was also isolated. It was suggested that this product was formed by hydrogen atom abstraction from **16** to give the enolate corresponding to **23** which was rapidly silylated.

The results of all the experiments described so far can be, and have been, explained by the mechanisms presented above. However, there are instances of reductions in liquid ammonia[45,53–56] and aromatic hydrocarbon solvents[57–59],* where acids, alcohols, and other anomalous products are found. The mechanism for any reaction should be able to account for all reported results, but the accepted mechanism for the acyloin condensation described above does not accomplish this objective. Furthermore, the mechanism postulates an intermediate α-diketone **7**, which has been shown not to be an intermediate.[16,60,61] In fact, reduction of enolizable α-diketones gives the acyloin as only a minor product,[60] although nonenolizable α-diketones are reduced normally.[62]

A mechanism which does not involve ketones (either mono- or di-) as intermediates and which can explain the formation of most reaction products has recently been developed.[63] This mechanism differs from those above in that the key step following the initial one-electron addition is attack of the radical anion **4** on an unreduced ester molecule to form the oxybridged dimer **24**. Reduction of **24** before or after loss of alkoxide would give **25** or **26**. Either **25** or **26** can cyclize to **27**, which should undergo rapid reduction to **28**. Loss of alkoxide from **28** gives the semidione **29**. (Semidiones have been observed in acyloin condensations.[41]) Reduction of the semidione completes the sequence. Although **25** is a dianion, it does not have the disadvantage of adjacent negative charges required in the 1,2-dianion proposal.[44,45] The steps involving conversion of **26** into **27** parallel very closely, in reverse, the pathway suggested for cleavage

* References 53–59 are meant to be typical. Search of the tables will provide further examples.

[53] E. Wenkert and B. G. Jackson, *J. Amer. Chem. Soc.*, **80,** 217 (1958).

[54] J. J. Bloomfield and J. R. S. Irelan, *J. Org. Chem.*, **31,** 2017 (1966).

[55] T. Okubo and S. Tsutsumi, *Technol. Rep. Osaka Univ.*, **12,** 457 (1962) [*C.A.*, **59,** 7422g (1963)].

[56] F. Chen and C. Ainsworth, *J. Amer. Chem. Soc.*, **94,** 4037 (1972).

[57] A. E. Kober and T. L. Westman, *J. Org. Chem.*, **35,** 4161 (1970).

[58] D. Machtinger, *Bull. Soc. Chim. Fr.*, **1961,** 1341.

[59] D. Machtinger, *J. Rech. Centre Natl. Rech. Sci. Lab. Bellvue* (Paris), **60,** 231 (1962).

[60] K. Rühlmann, B. Fichte, T. Kiriakidis, C. Michael, G. Michael, and E. Gründemann, *J. Organometal. Chem.*, **34,** 41 (1972).

[61] R. E. Robertson, M.S. Dissertation, Colorado State University, 1971.

[62] T. Murakawa, K. Fujii, S. Murai, and S. Tsutsumi, *Bull. Chem. Soc. Jap.*, **45,** 2520 (1972).

[63] J. J. Bloomfield, D. C. Owsley, C. Ainsworth, and R.E. Robertson, *J. Org. Chem.*, **40,** 393 (1975).

$$RCO_2R' \xrightarrow{e^-} R\dot{C}(O^-)OR' \xrightarrow{RCO_2R'} R\dot{C}(OR')\text{—O—}C(OR')(O^-)R \xrightarrow{e^-} R\underline{C}(OR')\text{—O—}C(OR')(O^-)R$$

4 24 25

$$24 \xrightarrow{-R'O^-} R\dot{C}(OR')\text{—O—}C(=O)R \xrightarrow{e^-} 26 \rightleftarrows 27$$

$$25 \xrightarrow{-R'O^-} 26; \quad 25 \xrightarrow{-R'O^-} 27$$

26 27

$$27 \xrightarrow{e^-} R\dot{C}(OR')(O^-)\text{—}C(O^-)R \xrightarrow{-R'O^-} R\overset{O^-}{C}{=}\overset{\dot{O}}{C}R \xrightarrow{e^-} R\overset{O^-}{C}{=}\overset{O^-}{C}R$$

28 29 8

of benzils by cyanide ion[64] or by methyl sulfinyl carbanion.[65–68,*] The anion **30** is comparable to **26**.

$$C_6H_5COCOC_6H_5 + CN^- \longrightarrow C_6H_5C(CN)(O^-)\text{—}C(=O)C_6H_5 \longrightarrow C_6H_5\dot{C}(CN)\text{—}C(O^-)C_6H_5\ (\text{O-bridged}) \longrightarrow$$

$$C_6H_5\underline{C}(CN)\text{—O—}C(=O)C_6H_5 \xrightarrow{+ROH} \xrightarrow{-HCN} C_6H_5CO_2R + C_6H_5CHO$$

30

Examination of this mechanism shows that it can account for the reduction products of phenylacetic esters[16] in a way that explains the absence of bibenzyl but without requiring free radicals other than as very transient radical anion intermediates.

Thus intermediate **31** can fragment, producing the resonance-stabilized

* In fact the entire concept of the oxybridged anion as an intermediate is hoary with age. The general concept was suggested by Favorsky[68] as providing possible intermediates for the Cannizzaro reaction, benzilic acid rearrangement, and acyloin rearrangement as well as the reaction that now carries his name.

[64] H. Kwart and M. M. Baevsky, *J. Amer. Chem. Soc.*, **80,** 580 (1958).

[65] J. C. Trisler, J. K. Doty, and J. M. Robinson, *J. Org. Chem.*, **34,** 3421 (1969).

[66] An intermediate very similar to **27** was proposed to account for the products of peracid oxidation of β-diketones; H. O. House and W. F. Gannon, *J. Org. Chem.*, **23,** 879 (1958).

[67] The hydrate of 1,2-cyclohexanedione is formulated as dihydroxyepoxycyclohexane, again very similar to **27**; L. De Borger, M. Anteunis, H. Lammens, and M. Verzele, *Bull. Soc. Chim. Belg.*, **73,** 73 (1964).

[68] Al. Favorsky, *Bull. Soc. Chim. Fr.*, [4], **43,** 551 (1928).

$$C_6H_5CH_2C(OR')(-O-)C(=O)CH_2C_6H_5 \longrightarrow C_6H_5CH_2CO_2R' + CO + C_6H_5CH_2^-$$

31

benzyl anion and starting ester. In the presence of trimethylchlorosilane the anion can be trapped to produce benzyltrimethylsilane **(20)**, or it can attack the starting ester (perhaps in a "cage" reaction). Attack at the carbonyl group would give dibenzyl ketone **(22)**, whereas attack at the benzylic hydrogen would produce toluene and the enolate of the starting ester which is trapped by the silylating reagent to give **23**. No bibenzyl is predicted by this mechanism.

A similar fragmentation pathway in the reactions involving tertiary benzylic or allylic compounds[51,52] also produces resonance-stabilized anions which account for all the products except the dimer hydrocarbons.

If R is very crowded sterically, it may be difficult for the bonding distance to be readily achieved in the transition from **25** or **26** to **27**. One possible result is the elimination of carbon monoxide shown above. Another is that, if a proton source is available (*e.g.*, ammonia or even another molecule of ester), **25** can fragment to an aldehyde and the starting ester. Further reduction of the aldehyde would ultimately produce only alcohol if sufficient reducing agent were present. This is, in fact, the result

$$\underset{\mathbf{25}}{R\underset{-}{C}(OR')-O-CR(OR')(O^-)} \xrightarrow{XH} \underset{\mathbf{32}}{RC(OR')(H)-O-CR(OR')(O^-)} + X^- \longrightarrow RCHO + RCO_2R' + R'O^-$$

obtained in the reduction of methyl mesitoate with lithium in liquid ammonia.[53]

In appropriate cases (nonenolizable aldehydes) the aldehyde could undergo a Cannizzaro reaction to produce an alcohol and an acid.[45,53,58,59] In ammonia, **26** would lead to amide, alcohol, and even aldehyde if insufficient reducing agent were present.[45,54,56]

The fragmentation of 1,2 diesters is accounted for if the initially formed

$$\mathbf{26} \xrightarrow{NH_3} RC(OR')(H)-O-C(=O)R \;(:NH_3) \longrightarrow [RCHO] + RCONH_2$$

$$[RCHO] \xrightarrow{2\,H^+,\ 2e^-} RCH_2OH$$

radical anion fragments or an intermediate like **25** fragments. Reduction in liquid ammonia or exceptional strain favors this reaction.[69,70]

In liquid ammonia the reaction course may be altered with a preference for two-electron reduction to a biradical dianion followed by cleavage to the enolate.[70] Alternatively, the lower temperature may simply slow

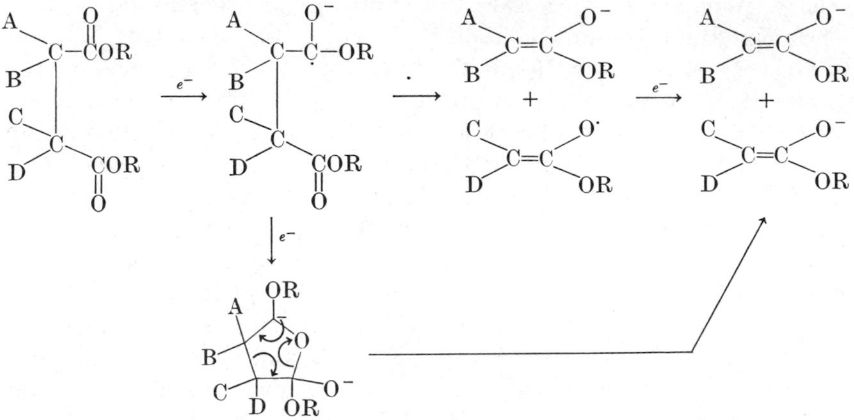

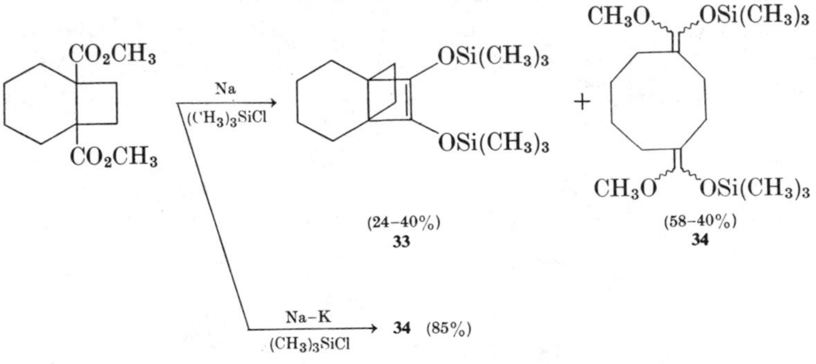

(24–40%) **33**

(58–40%) **34**

34 (85%)

(Ref. 69)

(88%)

(54%)

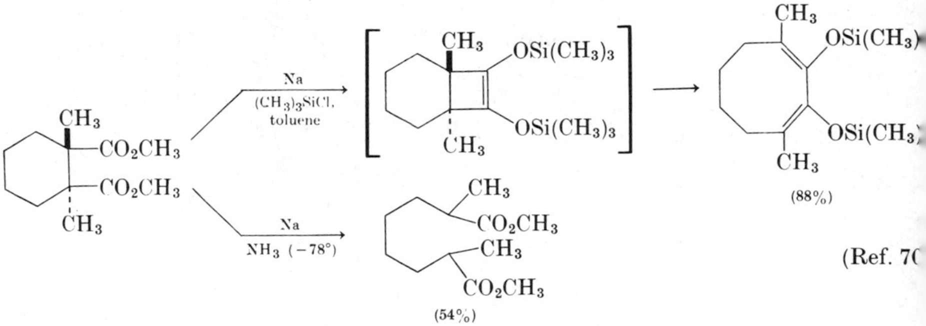

(Ref. 70)

[69] J. J. Bloomfield, R. A. Martin, and J. M. Nelke, *Chem. Commun.*, **1972**, 96.

[70] P. G. Gassman and X. Creary, *Chem. Commun.*, **1972**, 1214.

the rate of elimination of alkoxide and subsequent reduction relative to the cleavage process.

Simultaneous reduction of a mixture of two different long-chain diesters gives an unusual result that is difficult to explain. When dimethyl nonandioate is reduced alone, the yield of acyloin is 16%. However, in the presence of an equimolar amount of another long-chain diester a yield of 52–62% is obtained.[71] One possible explanation for this phenomenon is suggested by examining the data collected in Table A (p. 274).

For rings in the C_9 to C_{12} range the Dieckmann cyclization is worthless for monomer; cyclic dimer is produced, but the yield is not good.

In the normal acyloin condensation the yield for all ring sizes varies from fair to excellent. Prelog suggested that the two electrophilic ends of a diester are attracted to the metal surface and then slide across that surface toward each other (buffeted by intermolecular collisions) until finally the bond is formed.[8,72] A clue to what may be happening comes from comparison of the two cases in Table A where the reduction was carried out in the presence of trimethylchlorosilane under normal and under high-dilution conditions. Under the normal conditions the yield of dimer is considerable. If the silane could trap an intermediate *bimolecular* product, then the large-ring dimer would subsequently be produced with little difficulty. The key here seems to be the crowding in the transition state for formation of medium-size rings. (The final enolate in the Dieckmann reaction and the ene-diolate in the acyloin condensation have similar steric requirements, but they are effectively removed from the equilibria involving the earlier stages of each reaction.)

High-dilution conditions force the reaction to be monomolecular, and the yield of monomer goes up although the reaction rates are low. This is shown in Finley's paper where the relative rates are C_9, 16; C_{10}, 6; C_{11}, 5.5; C_{12}, 2; and C_{14}, 1.[71]

The details of this reaction are likely to be quite complex, and clearly much more work is needed before these results can be well understood.

The mechanism involving the oxybridged intermediates **26** and **27** (p. 270) seems to rationalize virtually all reductions of esters. A most difficult question arises in attempting to explain the formation of acids. One possible route involves the fragmentation of **26** to an alkoxycarbene and a carboxylate anion. If the carbene has a hydrogen atom on the adjacent carbon, a fairly rapid shift should occur to produce an enol ether. During the typical acid hydrolysis workup free aldehyde and acid would be formed. Typical carbene insertion or addition products have never been reported. The observation that certain highly hindered (nonenolizabile)

[71] K. T. Finley and N. A. Sasaki, *J. Amer. Chem. Soc.*, **88**, 4267 (1966).

[72] V. Prelog, L. Frenkiel, M. Kobelt, and P. Barman, *Helv. Chim. Acta*, **30**, 1741 (1947).

TABLE A. Comparison of Yields (%) for Cyclization of $RO_2C(CH_2)_nCO_2R$ under Various Conditions

N = Final Ring Size	Dieckmann[a,b]	Acyloin Ref. 71[c]	Ref. 71[d]	Normal[e]	$(CH_3)_3SiCl$[f]	$(CH_3)_3SiCl$[g]
8	15 (dimer = 11)	—	—	37–56	—	72–85
9	0 (dimer = 28)	5	5 (16)	97	22 (dimer = 62)	68
10	0 (dimer = 12)	52	69 (52)	24–67	53 (dimer = 20) 22 (dimer = 73)	58–69
11	0.5 (dimer = 23)	60	62 (69)	61–71	—	48
12	0.5 (dimer = 16)	57	84 (72)	64–89	—	68
13	24 (dimer = 19)	62	52 (64)	68–82	—	84
14	32 (dimer = 2)	58	87 (84)	47–82	—	67

[a] Data are taken from Table II in J. P. Schaefer and J. J. Bloomfield, *Org. Reactions*, **15**, 47–121 (1967).

[b] Note that for the Dieckmann reaction the value of n to obtain a ring of a given size must be one larger than in the acyloin condensation.

[c] The yield of C_9 cycle is shown when 0.01 mol of C_9 diester was cyclized with 0.01 mol of the diester for which $n = N - 2$.

[d] The yield of C_N cycle is shown when 0.01 mol of C_N diester was cyclized with 0.01 mol of C_9 diester. Numbers in parentheses show the yield when no added C_9 diester was present.

[e] The data are from Table II; no special reaction conditions.

[f] The data are from Table IV; the ester was added fairly rapidly.

[g] The data are from Table IV; the ester was slowly added via high-dilution cycle.

$$R^2R^3CH\overset{OR'}{\overset{|}{C}}-O-\underset{\underset{O}{\|}}{C}R \longrightarrow \left[R^2R^3CH\overset{OR'}{\overset{|}{C}}:\right] + RCO_2^- \xrightarrow{H_3O^+} RCO_2H + RCHO$$

26

$$\longrightarrow \left[R^2R^3C{=}C\begin{matrix}OR' \\ H\end{matrix}\right]$$

esters are about 25% reduced to alcohol and 75% converted to free acid by lithium in liquid ammonia[53] does not seem to be explained as readily.

Another possible source of both aldehyde and acid is the extraction by **26** of a proton from the solvent (or starting ester) followed by attack at the acetal carbon atom by alkoxide and loss of carboxylate. In liquid ammonia this would provide the aldehyde as an acetal protected from attack by ammonia. Hydrolysis would again produce acid and aldehyde.

Another route to carboxylic acids is shown in the reaction below. In this case the sodium suspension was prepared with a Hershberg stirrer which

$$\mathbf{26} + XH \longrightarrow R'O^- \; R-\overset{OR'}{\overset{|}{\underset{H}{\underset{|}{C}}}}-O-\underset{\underset{O}{\|}}{C}R \longrightarrow RCH(OR')_2 + RCO_2^- \xrightarrow{H_3O^+} RCHO + RCO_2H$$

produces a fairly large particle size, typical of that obtained in Dieckmann condensations.[73] With no alpha hydrogen atoms available, the aromatic esters lose the alkyl group as a radical which can dimerize or abstract a hydrogen atom from the solvent. Disproportionation should also occur but was not reported. Bibenzyl was not reported from methyl or ethyl benzoate where the yield of benzoic acid was 72%, but the nonacidic fraction contained more than 25 substances. No benzoin was isolated.

$$C_6H_5COR \xrightarrow[\text{toluene}]{Na} C_6H_5\overset{O^-}{\overset{|}{C}}{\cdot}-O-R \longrightarrow C_6H_5\overset{O^-}{\overset{|}{C}}{=}O + R\cdot$$

$$2R\cdot \longrightarrow R_2 \qquad R\cdot + C_6H_5CH_3 \longrightarrow RH + C_6H_5CH_2\cdot$$

$R = CH_3, C_2H_5$

R·	R_2
$CH_2{=}CHCH_2\cdot$	34%
$C_6H_5CH_2\cdot$	38%
$C_6H_5CH(CH_3)\cdot$	30%
$C_6H_5CH{=}CHCH_2\cdot$	39%
$(C_6H_5)_2CH\cdot$	48%

[73] H. Stetter and K.-A. Lehmann, *Ann.*, **1973**, 499.

There is a group of reactions, collected in Table VIE, that cannot be explained by any mechanism. These are reactions of esters that have failed to be reduced at all!

SCOPE AND LIMITATIONS

Condensations of Monocarboxyl Derivatives

The very first uses of the acyloin condensation involved simple aliphatic esters in the preparation of symmetrical compounds. This area was covered in the earlier *Organic Reactions* review.[5] Sometimes the yield is very good, but often it is poor. The results for esters and for acid chlorides are assembled in Tables IA and IB, respectively.

The acyloin condensation has been used very little to synthesize unsymmetrical derivatives by simultaneous reduction of two different esters (Table IC), and with modest success.[74,75] Yields are low and the mixtures are difficult to separate.

$$C_2H_5CO_2C_2H_5 + C_6H_5CO_2C_2H_5 \xrightarrow[C_6H_6,\ \text{heat}]{Na} C_2H_5CH(OH)COC_6H_5 \ (16\text{–}23\%) \qquad \text{(Ref. 74)}$$

$$n\text{-}C_3H_7CO_2C_2H_5 + n\text{-}C_{10}H_{21}CO_2C_2H_5 \xrightarrow{Na,\ \text{xylene}} \left.\begin{matrix} n\text{-}C_3H_7C \\ | \\ n\text{-}C_{10}H_{21}C \end{matrix}\right\} \begin{matrix} H,OH, \\ \\ =O \end{matrix} \ (41\%) \qquad \text{(Ref. 75)}$$

When acyloin condensations are carried out in the presence of trimethylchlorosilane (TMCS), a new dimension is added to the reaction. Since the discovery of the effects of this new reagent[76,16] (higher yields, easier isolation and storage, freedom from side reactions), the addition of TMCS has become almost routine as *the method of choice in all acyloin condensations.*

The effect of TMCS was discovered in an attempt to prepare alkyl silyl ketones by simultaneous alkali metal reduction of TMCS and an acyl halide.[76] The products were silylated ene-diolates. Better yields of purer products were obtained when the acid chloride was replaced by an ester.[76] These acyclic acyloin derivatives are produced as *cis-trans* mixtures. They are readily stored or they can be hydrolyzed to the free acyloins.

The reduction of monoesters in the presence of TMCS followed by treatment of the *bis*-trimethylsilylated enediol with (a) methyl lithium, (b) an alkyl halide, (c) sodium borohydride and (d) lead tetraacetate provides a new route to ketones in very good overall yield.[76a] When the

[74] J. W. Lynn and J. English, Jr., *J. Amer. Chem. Soc.*, **73,** 4284 (1951).
[75] D. E. Ames, G. Hall, and B. T. Warren, *J. Chem. Soc., C*, **1968,** 2617.
[76] K. Rühlmann and S. Poredda, *J. Prakt. Chem.*, **12** [4], 18 (1960).

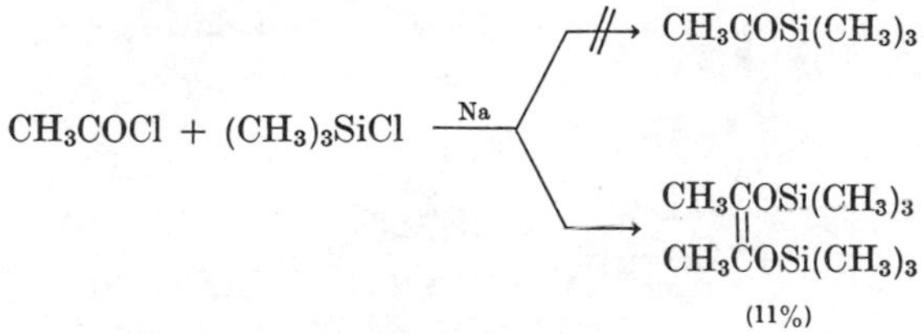

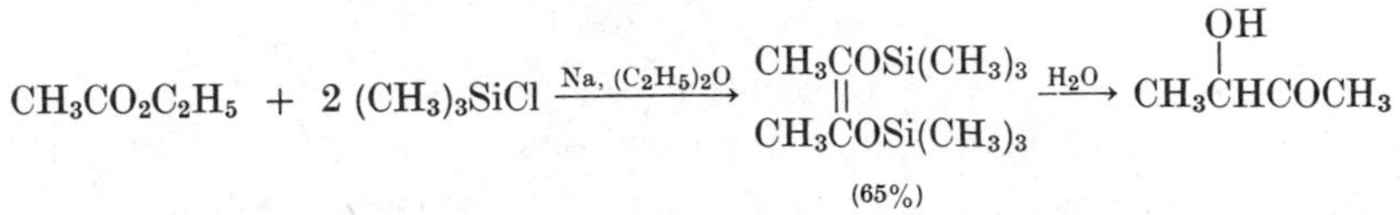

$$CH_3COCl + (CH_3)_3SiCl \xrightarrow{Na} \begin{cases} \not\rightarrow CH_3COSi(CH_3)_3 \\ \rightarrow \begin{array}{c} CH_3COSi(CH_3)_3 \\ \| \\ CH_3COSi(CH_3)_3 \end{array} \; (11\%) \end{cases}$$

$$CH_3CO_2C_2H_5 + 2\,(CH_3)_3SiCl \xrightarrow{Na,\ (C_2H_5)_2O} \begin{array}{c} CH_3COSi(CH_3)_3 \\ \| \\ CH_3COSi(CH_3)_3 \end{array} (65\%) \xrightarrow{H_2O} CH_3CH(OH)COCH_3$$

procedure is applied to methyl levulinate ethylene ketal and the oxidation product is hydrolyzed with acid, 1,4-diketones are produced. In the example, dihydrojasmone is produced in 49% overall yield from the levulinic ester.[76b]

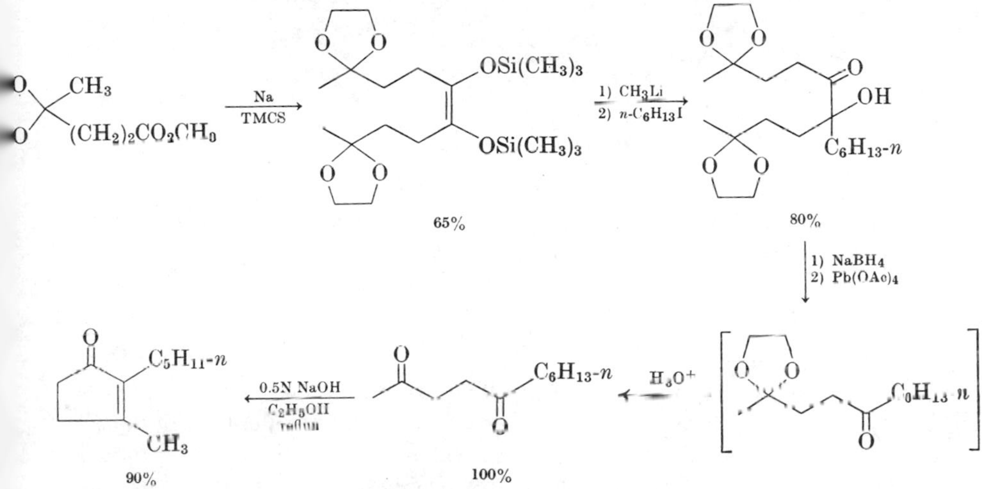

Na, TMCS → 65%; 1) CH_3Li, 2) n-$C_6H_{13}I$ → 80%; 1) $NaBH_4$, 2) $Pb(OAc)_4$ → H_3O^+ → 100%; 0.5N NaOH, C_2H_5OH, reflux → 90%

The results for the reduction of esters and of acid chlorides in the presence of TMCS to produce linear acyloin derivatives are compiled in Tables IIIA and IIIB. Strangely, this technique has not been applied to any mixed acyloins such as described above.[74,75] The ease of handling of the bis-silyloxy ethers suggests that separation by distillation should be easy. Esters with halo or hetero atoms in the α- and β-positions have been examined.[16]

Anhydrides and lactones have also been reduced, but the yields are poor (see Table IIIB).

[76a] T. Wakamatsu, K. Akasaka, and Y. Ban, *Tetrahedron Lett.*, **1974**, 3879.
[76b] T. Wakamatsu, K. Akasaka, and Y. Ban, *Tetrahedron Lett.*, **1974**, 3883.

$$(CH_3CO)_2O + Na + TMCS \xrightarrow[\text{reflux}]{\text{Xylene}} \begin{matrix} CH_3COSi(CH_3)_3 \\ \| \\ CH_3COSi(CH_3)_3 \end{matrix} \quad \text{(Ref. 77)}$$

(20%)

$$\text{(butyrolactone)} + Na + TMCS \xrightarrow[\text{reflux}]{\text{Xylene}} \begin{matrix} (CH_3)_3SiOC(CH_2)_3OSi(CH_3)_3 \\ \| \\ (CH_3)_3SiOC(CH_2)_3OSi(CH_3)_3 \end{matrix} \quad \text{(Ref. 77)}$$

(35%)

The reduction of aryl carboxylates is a special case. The yields are generally very poor.[44] In liquid ammonia the yield can be as high as 50%.[45] However, by using TMCS the yield can be considerably improved. The use of the trimethylsilyl ester appears to offer an advantage as well, increasing the yield of the *p*-toluoin derivative by a factor of two.[16]

$$p\text{-}XC_6H_4CO_2R \xrightarrow{\text{Na, TMCS}} \begin{matrix} p\text{-}XC_6H_4C{=}CC_6H_4X\text{-}p \\ | \quad\quad | \\ (CH_3)_3SiO \quad OSi(CH_3)_3 \end{matrix}$$

a, $X = H$, $R = C_2H_5$	39%
b, $X = H$, $R = Si(CH_3)_3$	41%
c, $X = CH_3$, $R = C_2H_5$	41%
d, $X = CH_3$, $R = Si(CH_3)_3$	86%

Reduction of Dicarboxylic Acid Derivatives

Since Finley's exhaustive 1964 review of the cyclic acyloin condensation,[15] much new work has been done in which trimethylchlorosilane has been used to trap the ene-diolate and other alkoxides.

Oxalates and Malonates

The reduction of diethyl oxalate is reported to produce a substituted tetrahydroxyethylene derivative.[78] The same paper describes the potentially useful dealkoxycarbonylation of disubstituted malonic ester derivatives,[78] which was first described in 1936.[79] In the presence of TMCS the ester enolate is silylated directly as it is produced, and the yield of product is much higher than in the absence of TMCS (Table VIA). Addition of an alkyl halide upon completion of the reduction should make it possible to prepare trisubstituted acetic esters from malonic esters in essentially one step when TMCS is omitted, as shown in Eq. 1.*

An entirely different result is obtained when the reduction of dimethyl dimethylmalonate is conducted in liquid ammonia followed by silylation

* This is an unknown reaction at the time of writing.

[77] F. M. F. Chen, Ph.D. Dissertation, Colorado State University, 1969 [*Diss. Abstr.*, **31B**, 1150 (1970)].

[78] Y. N. Kuo, F. Chen, C. Ainsworth, and J. J. Bloomfield, *Chem. Commun.*, **1971**, 136.

[79] F. Krollpfeiffer and A. Rosenberg, *Ber.*, **69**, 465 (1936).

$$\begin{matrix} CO_2C_2H_5 \\ | \\ CO_2C_2H_5 \end{matrix} \xrightarrow[(C_2H_5)_2O,\ TMCS]{Na\text{–}K} (C_2H_5O)((CH_3)_3SiO)C{=}C(OC_2H_5)(OSi(CH_3)_3) \quad (64\%)$$

(Ref. 78)

$$(CH_3)_2C(CO_2C_2H_5)_2 \xrightarrow{Na} \left[(CH_3)_2C{=}C(O)(OC_2H_5)\right]^- \xrightarrow{TMCS} (CH_3)_2C{=}C(OSi(CH_3)_3)(OC_2H_5) \quad (86\%)$$

(Ref. 78)

$$\xrightarrow{H_3O^+} (CH_3)_2CHCO_2C_2H_5 \quad (36\%)$$

(Ref. 79)

$$C_6H_5C(C_2H_5)(CO_2C_2H_5)_2 \xrightarrow{Na} \left[C_6H_5C(C_2H_5){=}C(O)(OC_2H_5)\right]^- \xrightarrow{RBr} C_6H_5C(C_2H_5)(R)CO_2C_2H_5 \qquad \text{(Eq. 1)}$$

of the ammonia-free reaction mixture.[56,63] Although reduction to a cyclopropane derivative was achieved, the product was not the expected cyclopropenediol derivative **35**, but rather the cyclopropane derivative **36**. In addition, ketene acetal **37**, diol **38**, amide **39**, and amide-alcohol **40** (as the silylated derivatives) were obtained in amounts varying with the reaction conditions (*cf.* Table VI). The formation of these products has been rationalized using the mechanism involving oxybridged anions (p. 270).[63]

$$(CH_3)_2C(CO_2CH_3)_2 \xrightarrow[NH_3(l),\ -34^\circ]{Na} (CH_3)_2\text{C}<\text{[CH(H)(OSi(CH}_3)_3)]_2 \ \textbf{36}\ (25\%) \quad \left[\text{not } (CH_3)_2\text{C}<\text{[C(OSi(CH}_3)_3)]_2 \ \textbf{35}\right]$$

(Ref. 56)

$$+ (CH_3)_2C{=}C(OSi(CH_3)_3)(OCH_3) \ \textbf{37}\ (6\%) \quad + (CH_3)_2C[CH_2OSi(CH_3)_3]_2 \ \textbf{38}\ (3\%)$$

$$+ (CH_3)_2CHCONHSi(CH_3)_3 \ \textbf{39}\ (25\%) \quad + (CH_3)_2C(CH_2OSi(CH_3)_3)(CONHSi(CH_3)_3) \ \textbf{40}\ (25\%)$$

There is a particularly interesting case in which a malonic ester did not fragment or form a cyclopropane derivative. In an elegant synthesis of corannulene, the final ring was formed by cyclization of the triester **41**.[80] Under carefully controlled conditions the product was **42** but, if excess metal was used at a higher temperature, the major product **43** had all the ester groups reduced.

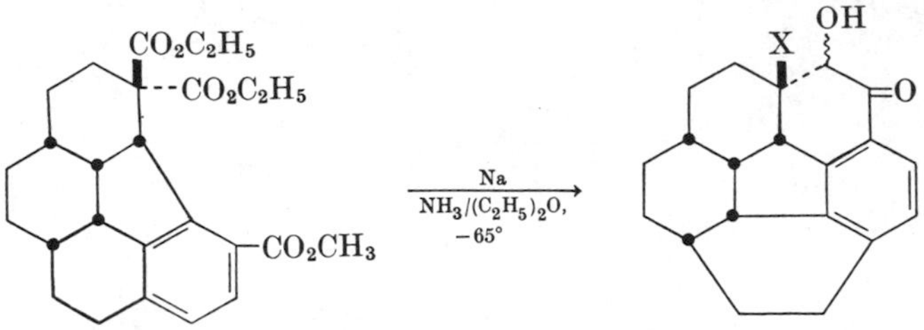

$CO_2C_2H_5$ $CO_2C_2H_5$ CO_2CH_3

Na / $NH_3/(C_2H_5)_2O$, −65°

OH X O

41

42, X = $CO_2C_2H_5$ (48%)
43, X = CH_2OH

Reduction of diethyl 1,1-cyclopropanedicarboxylate in liquid ammonia produced diethyl ethylmalonate, the ring reduction product.[77]

Succinates

Until 1966 there was only one example of a cyclobutane derivative **(44)** obtained by an acyloin cyclization. Then cyclization of the bicyclic succinate derivative **45**, which has no enolizable hydrogen, was shown to proceed nicely under appropriate reducing conditions.[54]

The introduction of trimethylchlorosilane has had a particularly dramatic effect in the cyclization of succinates. As the previous examples show, only the nonenolizable ester **45** gave a good yield of acyloin in the

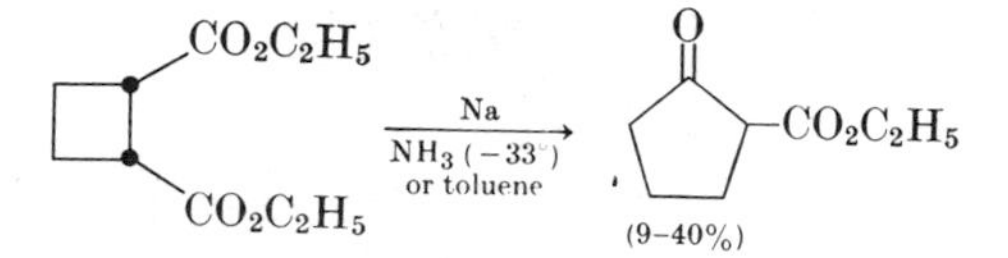

$CO_2C_2H_5$ $CO_2C_2H_5$ Na / Xylene → O OH

44 (12%) (Ref. 81)

$CO_2C_2H_5$ $CO_2C_2H_5$ Na / NH_3 (−33°) or toluene → O $CO_2C_2H_5$ (9–40%) [not O OH] (Ref. 82)

[80] W. E. Barth and R. G. Lawton, *J. Amer. Chem. Soc.*, **93**, 1730 (1971).

[81] A. C. Cope and E. C. Herrick, *J. Amer. Chem. Soc.*, **72**, 983 (1950).

[82] J. J. Bloomfield, R. G. Todd, and L. T. Takahashi, *J. Org. Chem.*, **28**, 1474 (1963), and references therein.

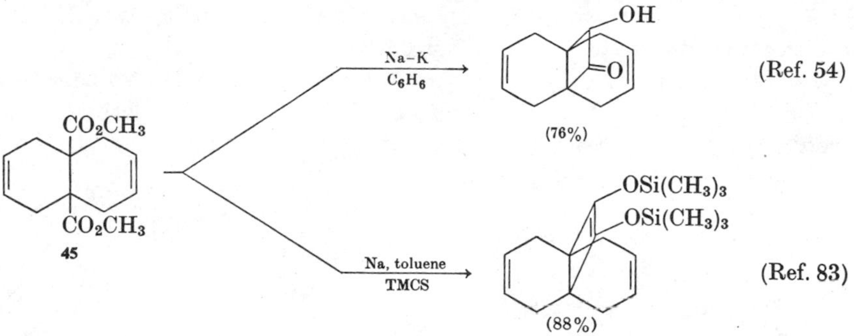

(Ref. 54)

(Ref. 83)

absence of TMCS. But, even in this instance, sodium in toluene was ineffective without TMCS.[54,83] A similar marked effect of TMCS is noted in the cyclization of dimethyl tetramethylsuccinate.[84]

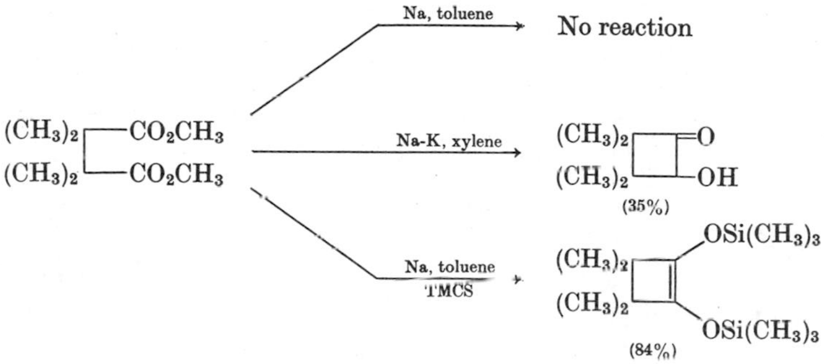

With enolizable succinates the improvement is even more marked. Diethyl succinate itself cannot be cyclized at all in the absence of TMCS, but in the presence of this trapping agent the yield can be as high as 90%.[83,85,86] Application of the TMCS technique to the cyclization of *cis* diethyl 1,2-cyclohexanedicarboxylate produces the bicyclo[4.2.0]octane derivative **46** in 90% yield.[83] Methanolysis of freshly distilled bicyclooctane **46** in anhydrous methanol under nitrogen gives the acyloin **44**

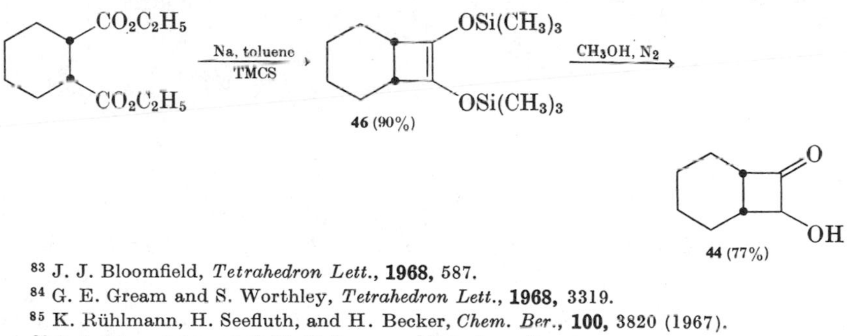

[83] J. J. Bloomfield, *Tetrahedron Lett.*, **1968,** 587.

[84] G. E. Gream and S. Worthley, *Tetrahedron Lett.*, **1968,** 3319.

[85] K. Rühlmann, H. Seefluth, and H. Becker, *Chem. Ber.*, **100,** 3820 (1967).

[86] J. J. Bloomfield and J. M. Nelke, unpublished results.

in 77% yield.[86] The overall yield is nearly six times better than that in the direct procedure previously reported.[81]

The cyclization of *trans*-cyclohexane-1,2-dicarboxylates is also possible. In this instance the product, a *trans*-bicyclo[4.2.0]oct-7-ene derivative is readily converted, via a conrotary electrocyclic ring opening, into a 1,3-cyclooctadiene derivative.[70,83,86]

CH_3, CO_2CH_3, CO_2CH_3, H $\xrightarrow[(C_2H_5)_2O,\ 0-5^\circ]{Na-K}$ CH_3, $OSi(CH_3)_3$, $OSi(CH_3)_3$, H **47** (86%) $\xrightarrow{90-100^\circ}$ CH_3, $OSi(CH_3)_3$, $OSi(CH_3)_3$ (100%)

The rearrangement is quite general.[86] It works for nearly all cyclobutenes except some of the smaller *cis*-bicyclic systems [(3.2.0) and (4.2.0)], which appear to be stable to at least 350°.[86] The gas-phase kinetic parameters for the rearrangement of 1,2-bis(trimethylsilyloxy)cyclobutene have been determined.[87]

The potential of the rearrangement for preparation of expanded ring systems was realized very quickly. Even *cis*-bicyclo[*n*.2.0] systems undergo ring opening when *n* is sufficiently large. Thus the cyclization of a *cis*-*trans* mixture of the medium-ring 1,2-diesters leads via the bicyclic systems **48** to $(n + 2)$ cyclic diketones **49** in 71–74% overall yield.[88]

$(CH_2)_{n-2}$, CO_2CH_3, CO_2CH_3 (n = 11, 12, 13) $\xrightarrow[TMCS]{Na,\ xylene}$ [$(CH_2)_{n-2}$, $OSi(CH_3)_3$, $OSi(CH_3)_3$] **48** $\xrightarrow{140^\circ}$ [$(CH_2)_{n-2}$, $OSi(CH_3)_3$, $OSi(CH_3)_3$] $\xrightarrow{H_3O^+}$ $(CH_2)_n$, O, O **49**

The rearrangement does not always proceed as nicely as the examples suggest. For instance, sodium-potassium alloy converts *trans*-dimethylcyclohexene-4,5-dicarboxylate to the bicyclic system in 65% yield at

[87] J. J. Bloomfield, H. M. Frey, and J. Metcalfe, *Int. J. Chem. Kinetics*, **3**, 85 (1971).

[88] T. Mori, T. Nakahara, and H. Nozaki, *Can. J. Chem.*, **47**, 3266 (1969).

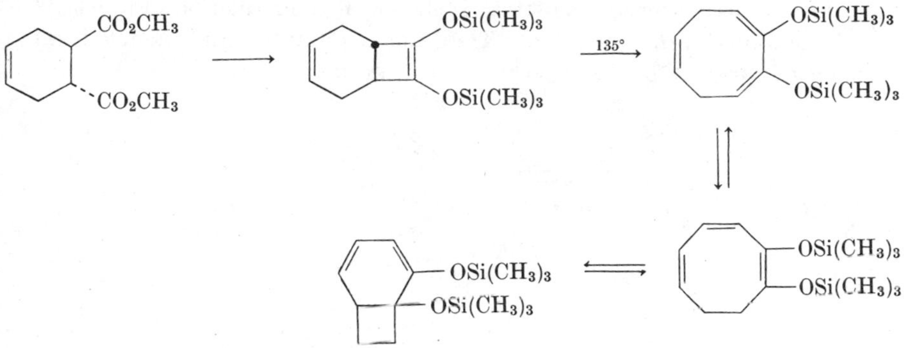

0–5° in ether, but the ring opening is not so simple.[86] The temperature required for the cyclobutene to butadiene rearrangement is so much higher when the methyl group at C_1 in **47** is gone that 1,5 hydrogen shifts become important and the whole range of equilibria shown ensues. The 1-methyl homolog **47** undergoes a similar series of rearrangements above 135°.

Three other interesting examples of cyclobutane ring formation are shown in the accompanying reactions. The second **(50 → 51)** is the only cyclization of an enolizable succinate in good yield in the absence of TMCS.

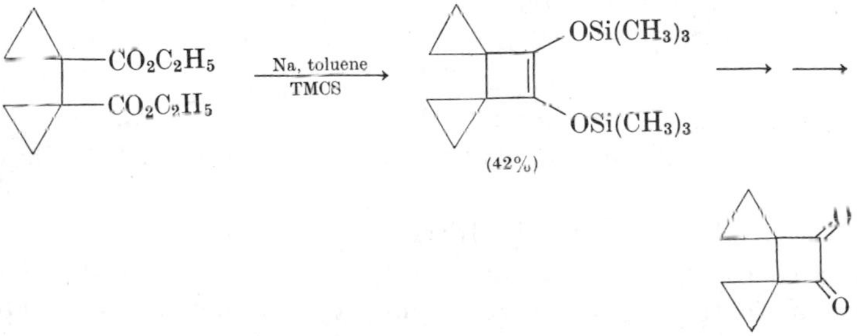

(42%)

(Ref. 89)

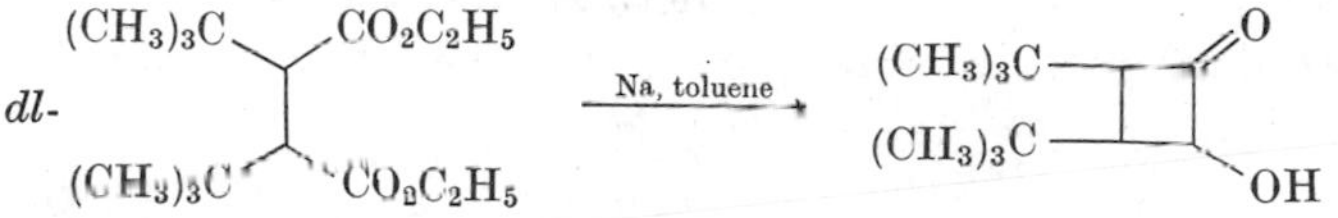

dl-

50

51 (90%)

(Ref. 90)

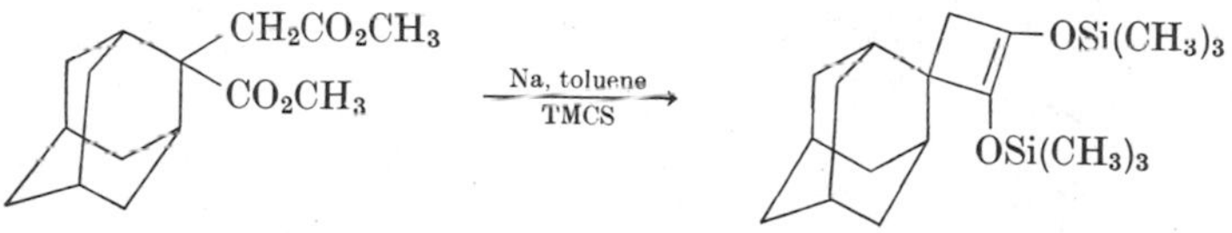

(75%)

(Ref. 32)

[89] J. M. Conia and J. M. Denis, *Tetrahedron Lett.*, **1969**, 3545.

[90] Ae. deGroot, D. Oudman, and H. Wynberg, *Tetrahedron Lett.*, **1969**, 1529.

An unusual effect of ring strain is shown in the reduction of dimethyl adamantane-1,2-dicarboxylate. This ester produces the double-bond isomer **52** instead of the expected compound **53**.[91]

CO_2CH_3 CO_2CH_3 —Na, toluene / TMCS→ $OSi(CH_3)_3$ $OSi(CH_3)_3$

52 (50–60%)

[not $OSi(CH_3)_3$ $OSi(CH_3)_3$ **53**]

Some succinates can cyclize or cleave, depending on the reaction conditions. Two examples have already been given (p. 272). Table VIB provides a list of esters that undergo reduction at the 1,2 bond.

Glutarates and Adipates

The cyclization of glutaric esters and adipic esters has also been enhanced by the introduction of TMCS. The best yield of cyclic product from a glutaric ester in the absence of TMCS is 16%;[92] with TMCS added the yield is 91% or greater.[93] Similarly the cyclization of an adipic ester proceeded in 57% yield without TMCS[92] and in 89–90% yield with it added.[86,93]

The following additional examples from Tables IIA (no TMCS) and IV (TMCS added) show the scope of the reaction for five- and six-membered ring syntheses. In the first example, acyloin cyclization of a glutarate proved a convenient route to an estradiol derivative labeled at C_{16}.[94]

The use of a sodium-potassium alloy permitted the reduction of spiro

CO_2CH_3 $CH_2{}^{14}CO_2CH_3$ CH_3O —1. Na/NH_3, $(C_2H_5)_2O$ / 2. $(CH_3CO)_2O$→ $OCOCH_3$ * $=O$

(77%)

[91] A. H. Alberts, H. Wynberg, and J. Strating, *Tetrahedron Lett.*, **1973**, 543.

[92] J. C. Sheehan, R. C. O'Neill, and M. A. White, *J. Amer. Chem. Soc.*, **72**, 3376 (1950).

[93] U. Schräpler and K. Rühlmann, *Chem. Ber.*, **97**, 1383 (1964).

[94] M. Levitz, *J. Amer. Chem. Soc.*, **75**, 5352 (1953).

Na–K/xylene, reflux

54

55 (26%)

56 (R = H, 24%)
(R = $CO_2C_2H_5$, 5%)

ester **54** to the acyloin **55**, but it was accompanied by a considerable amount of Dieckmann condensation products **56**.[95] The reduction failed with sodium metal alone.

A twistene synthesis starts by reduction of *endo*-dimethyl bicyclo[2.2.2]-octane-2,5-dicarboxylate.[96]

1. Na/NH_3, $(C_2H_5)_2O$
2. H_2/PtO_2

(49%)

The [4.4]spirononane system is prepared in much better yield when TMCS is added.

Na, xylene
R = C_2H_5

(18%) (Ref. 97)

R = CH_3
Na, toluene, TMCS

(87%) (Ref. 48)

Pimelates and Longer-Chain Esters

The synthesis of large rings is the area in which the acyloin condensation has traditionally found greatest utility, although many of the foregoing examples show that the reaction has become important in the synthesis of small and normal ring systems.

[95] K. R. Varma, M. L. Maheshwari, and S. C. Bhattacharyya, *Tetrahedron*, **21,** 115 (1965).
[96] M. Tichý and J. Sicher, *Tetrahedron Lett.*, **1969,** 4609.
[97] G. A. R. Kon, *J. Chem. Soc.*, **121,** 513 (1922).

Finley's review covered all cyclic cases reported up to about 1964.[15] The most important area outside of the perfume type of large-ring ketones was the synthesis of bridged aromatics, a subject reviewed by Smith.[13] The multitude of examples found in Table IIA are for reactions run in the absence of TMCS. Much less work on these larger-ring systems has been done with added TMCS (see, however, the discussion on pp. 273–274), Table IVA. Examples of some of the more recent work follow with some comparisons of the effect of TMCS on the course of the reaction.

$(CH_2)_5(CO_2C_2H_5)_2$ —Na, xylene→ cyclohexanone (13%) + 2-hydroxycycloheptanone (15%) (Ref. 98)

$(CH_2)_5(CO_2C_2H_5)_2$ —Na, toluene/TMCS→ 1,2-bis($OSi(CH_3)_3$)cycloheptene (75%) (Ref. 99)

$(CH_2)_6(CO_2CH_3)_2$ —Na/xylene→ 2-hydroxycyclooctanone (37–56%) (Refs. 100–104)

$(CH_2)_6(CO_2CH_3)_2$ —Na, toluene/TMCS, High-dilution cycle→ 1,2-bis($OSi(CH_3)_3$)cyclooctene (85%) (Ref. 86)

1,2-bis($(CH_2)_2CO_2CH_3$)cyclopropane —Na, xylene→ bicyclic hydroxy ketone (O, OH) (76%) (Ref. 105)

[98] P. D. Gardner, G. R. Haynes, and R. L. Brandon, *J. Org. Chem.*, **22**, 1206 (1957).
[99] J. J. Bloomfield, *Tetrahedron Lett.*, **1968**, 591.
[100] A. C. Cope, S. W. Fenton, and C. F. Spencer, *J. Amer. Chem. Soc.*, **74**, 5884 (1952).
[101] A. T. Blomquist and L. H. Liu, *J. Amer. Chem. Soc.*, **75**, 2153 (1953).
[102] V. L. Hansley, U.S. Pat. 2,228,268. (1941) [*C.A.*, **35**, 2534[2] (1941)].
[103] M. Rosenblum, V. Nayak, S. K. Das Gupta, and A. Longroy, *J. Amer. Chem. Soc.*, **85**, 3874 (1963).
[104] C. D. Hurd and W. H. Saunders, Jr., *J. Amer. Chem. Soc.*, **74**, 5324 (1952).
[105] P. G. Gassman, J. Seter, and F. J. Williams, *J. Amer. Chem. Soc.*, **93**, 1673 (1971).

$CH_2CO_2C_2H_5$ $CH_2CO_2C_2H_5$ —Na, toluene/TMCS, 60°→ [product with $OSi(CH_3)_3$, $OSi(CH_3)_3$] (40%) + [product with $OSi(CH_3)_3$, O] (5%) (Ref. 106)

The following example shows the utility of the acyloin condensation in the synthesis of the unique "in, in" and "out, in" hydrocarbons **57** and **58** from the *cis* and *trans* starting esters, respectively.[107]

$CH_3O_2C(CH_2)_3$ $(CH_2)_8$ $(CH_2)_3CO_2C_2H_5$ C C H $(CH_2)_8$ H —1. Na, xylene 2. Clemmensen→

$(CH_2)_8$ C—H H—C $(CH_2)_8$ $(CH_2)_8$ + H—C $(CH_2)_8$ H—C $(CH_2)_8$ $(CH_2)_8$

57 **58**

The TMCS-modified acyloin condensation was also used recently to prepare a mixture of macrocyclic hydrocarbons having 14–42 carbons in the ring.[108] The mixture was used to study the effect of ring size on threading reactions. An earlier study of threading using the acyloin condensation of diethyl tetratriacontanedioate in a solvent containing deuterated cyclotetratriacontane produced a very small amount of the desired catenane.[109]

Many acyloin cyclizations have been carried out with double bonds in the molecule to be cyclized. In all cases except one[217b] no products resulting from transannular interaction of intermediate radicals or radical anions with the double bond have been detected. The one example of trans-

[106] Personal communication from Prof. D. G. Farnum.

[107] C. H. Park and H. E. Simmons, *J. Amer. Chem. Soc.*, **94**, 7184 (1972), and personal communication from Dr. C. H. Park.

[108] I. T. Harrison, *Chem. Commun.*, **1972**, 231.

[109] E. Wasserman, *J. Amer. Chem. Soc.*, **82**, 4433 (1960).

$CH_2{=}C[(CH_2)_3CO_2CH_3]_2$ →(Na) 50% + 2% + 2% + 10% (CH_3, O, OH)

→(Na, TMCS) $(CH_3)_3SiO$, $OSi(CH_3)_3$, CH_3 — good yield

$CH_2{=}C[(CH_2)_2CO_2C_2H_5]_2$ →(Na) 27% (CH_2) + 12% ($CO_2C_2H_5$, CH_2) + 12% (O, OH, CH_2)

→(Na, TMCS) $(CH_3)_3SiO$, $OSi(CH_3)_3$, CH_2 → 12%

$CH_3OCH{=}C[(CH_2)_3CO_2C_2H_5]_2$ →(Na) O, OH, CH_2OCH_3 (major) + { CH_2OCH_3 + CH_2OCH_3 } (minor)

$(CH_2)_2CO_2C_2H_5$, $=CH_2$, $(CH_2)_2CO_2C_2H_5$ → CH_3, O, OH (30%) + [O, CH_2, OH]

annular reaction described in the literature has recently been reinvestigated.[217a] The reaction of dimethyl 5-methylenenonane dicarboxylate does not give 1-hydroxybicyclo[4.3.1]decane-2-one as was originally reported.[217b] The correct structure is instead 6-methyl-1-hydroxybicyclo[4.3.0]nonane-2-one. This discovery has prompted additional studies of other systems. Some of the early results of this work are shown on page 288.[118]

Reduction of Heterocyclic Esters

Cyclic compounds containing nitrogen, oxygen, sulfur, silicon, or germanium in the ring have been prepared by acyloin condensation (see Tables IIB and IVC, D, and E). The ferrocene system has also been built into ring systems by the acyloin condensation (Table IIC).

The nitrogen compounds are all tertiary amines. Ring sizes from six to twenty-three have been prepared. The yields of the two reported six-membered rings are poor.[110] In the larger ring sizes, nine or more atoms, the yields are usually quite good, as the accompanying examples show.[111–116]

$CH_2CO_2C_2H_5$ / $NCH_2CO_2C_2H_5$ (piperidine) $\xrightarrow[\text{reflux}]{\text{Na, toluene}}$ (quinolizidine) =O / –OH (3%) (Ref. 110)

$i\text{-}C_4H_9N(CH_2CH_2CH_2CO_2C_2H_5)_2$ $\xrightarrow[\text{reflux}]{\text{Na, xylene}}$ O, HO, $NC_4H_9\text{-}i$ (48%) (Ref. 114)

$C_2H_5N[(CH_2)_6CO_2C_2H_5]_2$ $\xrightarrow[\text{reflux}]{\text{Na, xylene}}$ C_2H_5N =O, OH (88%) (Ref. 112)

$CH_3N(CH_2CH_2CO_2C_2H_5)_2$ $\xrightarrow[\text{ClSi(CH}_3)_3\text{, reflux}]{\text{Na, toluene}}$ CH_3N $OSi(CH_3)_3$ / $OSi(CH_3)_3$ (67%) (Ref. 85)

[110] K. Winterfeld and K. Nonn, *Pharmazie*, **20**, 337 (1965) [*C.A.*, **63**, 8314c (1965)].
[111] N. J. Leonard, R. C. Fox, M. Ōki, and S. Chiavarelli, *J. Amer. Chem. Soc.*, **76**, 630 (1954).
[112] N. J. Leonard, R. C. Fox, and M. Ōki, *J. Amer. Chem. Soc.*, **76**, 5708 (1954).
[113] N. J. Leonard and M. Ōki, *J. Amer. Chem. Soc.*, **77**, 6245 (1955).
[114] N. J. Leonard, M. Ōki, J. Brader, and H. Boaz, *J. Amer. Chem. Soc.*, **77**, 6237 (1955).
[115] N. J. Leonard and M. Ōki, *J. Amer. Chem. Soc.*, **77**, 6241 (1955).
[116] N. J. Leonard and M. Ōki, *J. Amer. Chem. Soc.*, **76**, 3463 (1954).

There are no examples of oxygen or sulfur heterocycles in smaller than a seven-membered ring. The yields are generally good. Thiophene derivatives are best cyclized using a sodium-potassium alloy.[117]

$(CH_2)_4CO_2CH_3$ / S / $(CH_2)_4CO_2CH_3$ $\xrightarrow[55-60°]{Na-K/xylene, (C_2H_5)_2O}$ CH_3, CH_3, S, O, OH (40%) (Ref. 117)

$O[CH_2C(CH_3)_2CO_2C_2H_5]_2$ $\xrightarrow[reflux]{Na, toluene}$ CH_3 CH_3, O, O, OH, CH_3 CH_3 (80%) (Ref. 118)

$(C_2H_5)_2Ge[(CH_2)_4CO_2C_2H_5]_2$ $\xrightarrow[reflux]{Na, xylene}$ $(C_2H_5)_2Ge$, O, OH (60%) (Ref. 119)

$(C_6H_5)_2Si[(CH_2)_2CO_2CH_3]_2$ $\xrightarrow{Na, ClSi(CH_3)_3}$ $(C_6H_5)_2Si$, O, OH (60%) (Ref. 120)

Fe, $(CH_2)_3CO_2CH_3$, $(CH_2)_3CO_2CH_3$ $\xrightarrow[reflux]{Na, xylene}$ Fe, $(CH_2)_3C{=}O$, $(CH_2)_3C{-}OH$ (58%) (Ref. 121)

117 Ya. L. Gol'dfarb, S. Z. Taits, and L. I. Belen'kii, *Tetrahedron*, **19**, 1851 (1963).

118 P. Johnson, personal communication.

119 P. Mazerolles and A. Faucher, *Bull. Soc. Chim. Fr.*, **1967**, 2134.

120 K. E. Koenig, R. A. Felix, and W. E. Weber, *J. Org. Chem.*, **39**, 1539 (1974).

121 K. Schlögl and H. Seiler, *Monatsh. Chem.*, **91**, 79 (1960).

Acyloin Condensations That Fail

(See Table VI.) In the course of the preceding discussion, several examples of the failure of the acyloin condensation have been noted. This section systematically describes the kinds of esters which can be expected to fail to reduce in the normal way and the products that will be produced.

A. Disubstituted malonic esters reduced in hydrocarbon or ether solvent by sodium or sodium-potassium alloy lose carbon monoxide and produce the enolate of a disubstituted acetic ester.[78,79] A more complicated course may ensue in liquid ammonia, but only one example is known so far.[56]

B. Succinic esters that are strained or that will produce an exceptionally strained product can be expected to cleave, especially if the reduction is carried out in liquid ammonia at −78°.[70] If the reducing agent is powerful enough, *e.g.*, sodium-potassium alloy, cleavage may also occur.[69] A particularly interesting example utilized in corrin synthesis involved the reduction of the cyclic 1,2-dimalonate **59**; this on reduction with sodium in liquid ammonia produced the stable bismalonate dianion **60**.[122] This method may be very useful in the synthesis of linear diesters and tetraesters difficult to prepare by any other route. The yields of cleavage products are usually quite good.[70]

$$\mathbf{59}\ (CO_2CH_3)_2,\ (CO_2CH_3)_2 \xrightarrow[-70\ \text{to}\ -30^\circ]{Na/NH_3,\ THF} \mathbf{60}\ \bar{C}(CO_2CH_3)_2,\ \bar{C}(CO_2CH_3)_2$$

C. Some α,β-unsaturated esters undergo conjugate reductive coupling tail-to-tail, and the resulting anion simply undergoes Dieckmann condensation.[123-126] The yields are fair to poor. It would be interesting to see if the reaction would take the same course if TMCS were present during the reduction.

$$2\,C_6H_5CH{=}CHCO_2C_2H_5 \xrightarrow{Na,\ (C_2H_5)_2O} \begin{array}{l} C_6H_5CH-CH_2 \\ \quad | \qquad\quad \searrow CO \\ C_6H_5CH-CHCO_2C_2H_5 \end{array}$$

(Refs. 123, 124)

[122] E. Bertele, H. Boos, J. D. Dunitz, F. Elsinger, A. Eschenmoser, I. Felner, H. P. Gribi, H. Gschwend, E. F. Meyer, M. Pesaro, and R. Scheffold, *Angew. Chem.*, **76**, 281 (1964).

[123] E. L. Totton, G. R. Kilpatrick, N. Horton, and S. A. Blakeney, *J. Org. Chem.*, **30**, 1647 (1965).

[124] E. L. Totton, R. C. Freeman, H. Powell, and T. L. Yarboro, *J. Org. Chem.*, **26**, 343 (1961).

[125] K. Bernhauer and R. Hoffmann, *J. Prakt. Chem.*, [2], **149**, 317 (1937).

[126] H. A. Weidlich, *Ber.*, **71**, 1601 (1938).

D. Some diesters, particularly adipates and pimelates, undergo Dieckmann condensation partially or wholly under acyloin conditions. Whenever the molecular structure is just right (*e.g.*, short-chain esters), enolization of an ester can compete well with reduction. Claisen or Dieckmann condensations then ensue. Two examples have been provided (pp. 285–286). TMCS can prevent these base-catalyzed side reactions by trapping the alkoxide and the acyloin dianion (p. 299).

E. Reduction of esters in liquid ammonia frequently gives a variety of products in addition to acyloin. The ammonia acts as a proton source and, unless the acyloin formation is kinetically much faster than proton transfer, products ranging through aldehydes, alcohols, amides, and free acids, as well as Claisen-type products, can be expected.

Some additional reactions that gave reduction to nonacyloins or to mixtures of products are shown below.

1. High concentration of ester as in the following two examples.

$CH_2(CH_2CO_2CH_3)_2$ $\xrightarrow{\text{Na, xylene}}$ [OH, OH, O] (42%) (Ref. 127)

$(CH_2)_8(CO_2C_2H_5)_2$ $\xrightarrow{\text{Na, xylene}}$ [O, $(CH_2)_8$, $(CH_2)_8$, OH, O] (15%) +

[OH, O, $(CH_2)_8$, $(CH_2)_8$, O, OH] (0.5%) + { [O, O] + [=O] } (14%) + $(CH_2)_8(CH_2OH)_2$ (3%) +

$(CH_2)_8(CO_2H)_2$ (17%) + $O{=}C[(CH_2)_8CHOH(CH_2)_8CO_2H]_2$ (49%) (Ref. 58)

[127] M. Cordon. J. D. Knight, and D. J. Cram, *J. Amer. Chem. Soc.*, **76,** 1643 (1954).

2. Lithium in ammonia, which leads to alcohols and/or acids and has never been reported to reduce esters to acyloins.

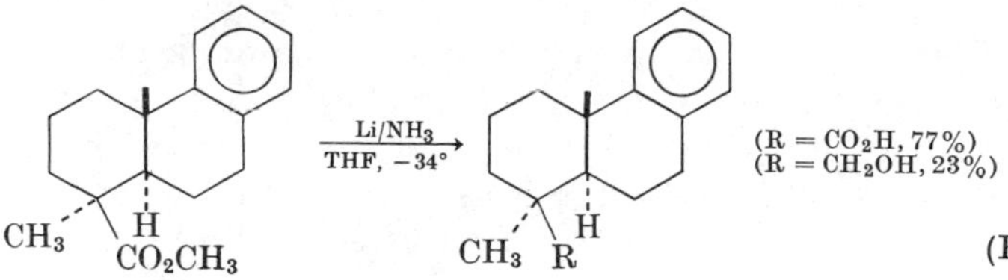

CH_3 H CO_2CH_3 $\xrightarrow[\text{THF, }-34^\circ]{\text{Li/NH}_3}$ CH_3 R H (R = CO_2H, 77%) (R = CH_2OH, 23%)

(Ref. 53)

$CO_2C_2H_5$ H $\xrightarrow[\text{THF, }-34^\circ]{\text{Li/NH}_3}$ R H (R = CO_2H, 40%) (R = CH_2OH, 30%)

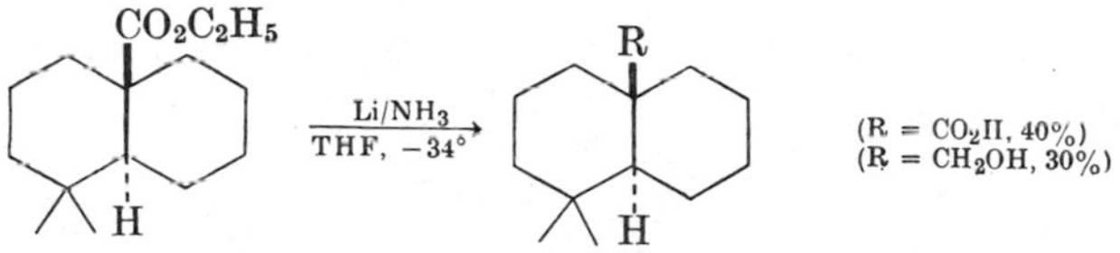

(Ref. 128)

3. Sodium in liquid ammonia, which, as has been pointed out (p. 269ff), leads in many instances to a great variety of products. Simple esters or short-chain diesters are particularly likely to give mixtures. Table VIE gives the examples where no acyloin is produced. Reactions in which the acyloin is at least one of the products are scattered throughout the tables.

Arbitrarily, the reduction of oxalates is included in Table VIE. The disilyloxydialkoxyethylenes produced here do not hydrolyze to α-hydroxyketones.

Table VIE does not include the nonacyloin reduction-fragmentation products already listed in Tables VIA and B.

F. Table VIF lists compounds which are not reduced under a variety of acyloin conditions. Lithium metal and sodium amalgam appear to be ineffective in reductions that work with sodium in the same solvents.[86] It is likely that conditions can be found for reduction of the recalcitrant compounds. It does not seem reasonable that the metal would not transfer electrons to at least one of the ester groups which, if it did not react with the other ester group in the molecule, would find a partner elsewhere in the solution.

G. Many esters are reduced, but the products are either polymers or uncharacterized mixtures. It is possible that TMCS might make isolation of products simpler. However, in the reduction of dimethyl 1,2-cyclobutanedicarboxylate the mixture of silylated products was extremely complex, and no serious attempt was made to separate the five or six major components from the multitude.[86] Hydrolysis of the silylated reaction mixture did produce three readily identifiable products: 2-carbomethyoxycyclopentanone, dimethyl adipate, and adipic acid.

[128] W. L. Meyer and A. S. Lovinson, *J. Org. Chem.*, **28**, 2184 (1963).

Condensation between Esters and Ketones

(See Table V.) Although the co-reduction of an ester and a ketone is not strictly an acyloin condensation, the product is a nonsymmetrical α-hydroxyketone. For completeness, examples of this reaction have been included. These reductions are especially interesting because the first, which involves the use of sodium-naphthalene radical anion,[129,130] was reported not to work in the acyloin condensation,[131] although very recently it has been used to reduce ethyl benzoate to benzoin when the normal conditions gave no benzoin at all.[73] The second method is electrochemical reduction,[132] which has not yet been used for reductions of esters by themselves.

The examples below give an idea of the scope of this reaction. Yields are not given in most of the work. In many of the cyclizations so many products are obtained that it is no surprise that little use has been made of this procedure.

CO_2CH_3 ; $Na-NH_3, (C_2H_5)_2O$, high dilution → (21%) + others ; Na-naphthalene, THF → HO, O (9%) (Ref. 130)

$$(CH_3CO)_2O + C_6H_5COCH_3 \xrightarrow{e^-(Hg)/CH_3CN/25^\circ} C_6H_5-C(OCOCH_3)(CH_3)-COCH_3 \quad \text{(Ref. 132)}$$

[129] C. D. Gutsche and I. Y. C. Tao, *J. Org. Chem.*, **28**, 883 (1963).
[130] C. D. Gutsche, I. Y. C. Tao, and J. Kozma, *J. Org. Chem.*, **32**, 1782 (1967).
[131] H. Gusten and L. Horner, *Angew. Chem., Int. Ed. Engl.*, **1**, 455 (1962).
[132] T. J. Curphey, personal communication.

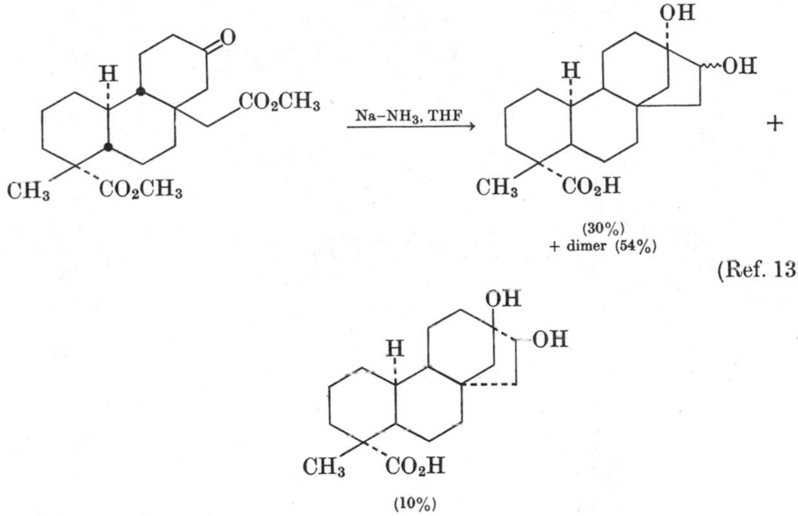

$\xrightarrow{\text{Na-NH}_3\text{, THF}}$

(30%)
+ dimer (54%)

(Ref. 133)

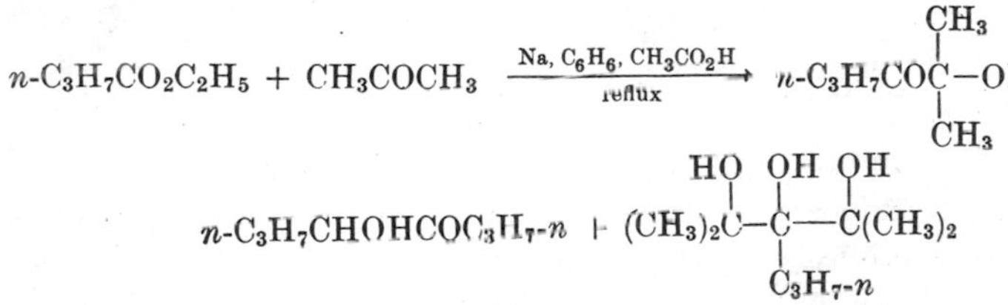

(10%)

$$n\text{-}C_3H_7CO_2C_2H_5 + CH_3COCH_3 \xrightarrow[\text{reflux}]{\text{Na, } C_6H_6\text{, } CH_3CO_2H} n\text{-}C_3H_7COC(CH_3)_2\text{-}OH +$$

$$n\text{-}C_3H_7CHOHCOC_3H_7\text{-}n + (CH_3)_2C(OH)\text{-}C(OH)(C_3H_7\text{-}n)\text{-}C(OH)(CH_3)_2$$ (Refs. 134, 135)

Semidiones

(See Table VII.) Semidiones are the unstable (but relatively long-lived) radical anions derivable from α-hydroxyketones, usually in basic solution in dimethyl sulfoxide. Table VII does not attempt to list all the semidiones ever examined, but only those prepared (a) by direct reduction of a diester with a sodium-potassium alloy in dimethoxyethane or (b) by reduction of the diester in the presence of TMCS followed by treatment with potassium *t*-butoxide in dimethyl sulfoxide (DMSO) to give the radical anion. In none of the examples tabulated were any physical or yield data reported on the acyloins or their TMCS derivatives.[41,136–139,*]

* Because of the lack of published supporting data we are unable to judge how useful the acyloin condensation actually is in these cases. This is a shame, for some of the systems prepared have quite unusual structures.

133 I. F. Cook and J. R. Knox, *Tetrahedron Lett.*, **1970**, 4091.

134 J. Kapron and J. Wiemann, *Bull. Soc. Chim. Fr.*, [5], **12**, 945 (1945).

135 J. Kapron, *C. R. Acad. Sci.*, **223**, 421 (1946).

136 G. A. Russell and P. R. Whittle, *J. Amer. Chem. Soc.*, **89**, 6781 (1967).

137 G. A. Russell, J. J. McDonnell, P. R. Whittle, R. S. Givens, and R. G. Keske, *J. Amer. Chem. Soc.*, **93**, 1452 (1971).

138 G. A. Russell, P. R. Whittle, and R. G. Keske, *J. Amer. Chem. Soc.*, **93**, 1467 (1971).

139 G. A. Russell and G. W. Holland, *J. Amer. Chem. Soc.*, **91**, 3968 (1969).

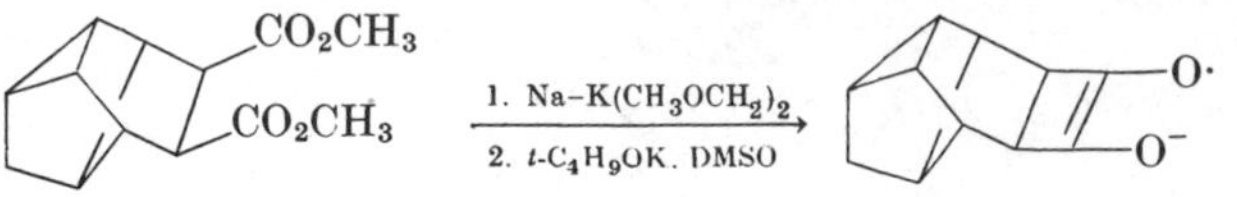

1. Na–K$(CH_3OCH_2)_2$, 2. t-C_4H_9OK, DMSO (Ref. 41)

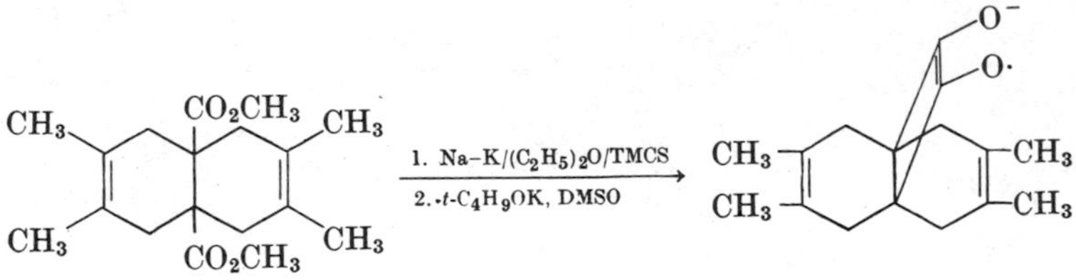

1. Na–K/$(C_2H_5)_2O$/TMCS, 2. t-C_4H_9OK, DMSO (Ref. 41)

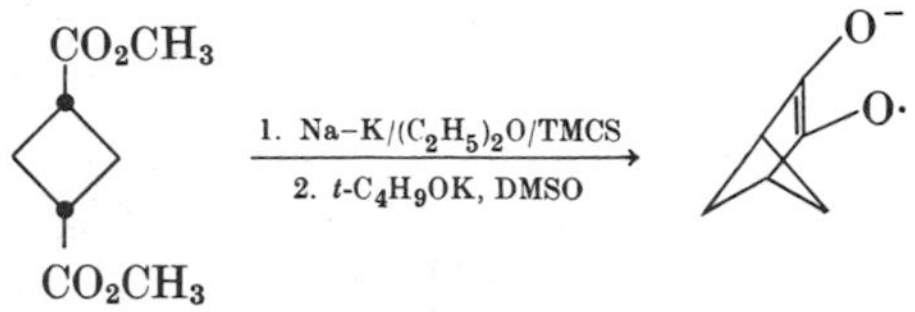

1. Na–K/$(C_2H_5)_2O$/TMCS, 2. t-C_4H_9OK, DMSO (Ref. 138)

OTHER ROUTES TO ACYLOINS

There are a large number of other routes to α-hydroxyketones. The well-known benzoin condensation for converting aromatic aldehydes into diaryl α-hydroxyketones is one example.[140] Both McElvain[5] and, more recently, Bracke[6] have reviewed the alternative procedures. The interested reader is directed to these sources. However, some of the procedures are sufficiently interesting and useful in certain instances to be included in the accompanying examples.

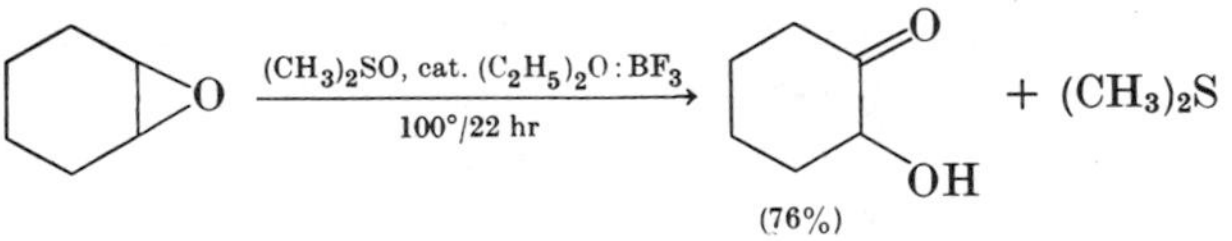

$(CH_3)_2SO$, cat. $(C_2H_5)_2O:BF_3$, 100°/22 hr → (76%) + $(CH_3)_2S$ (Ref. 141)

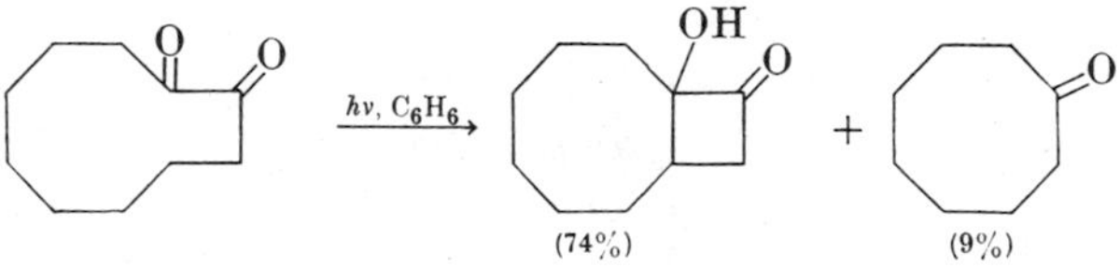

$h\nu$, C_6H_6 → (74%) + (9%) (Ref. 142)

+ 17% starting material

[140] W. S. Ide and J. S. Buck, *Org. Reactions*, **4,** 269 (1948).

[141] T. Cohen and T. Tsuji, *J. Org. Chem.*, **26,** 1681 (1961).

[142] W. H. Urry, D. J. Trecker, and D. A. Winey, *Tetrahedron Lett.*, **1962,** 609. For earlier work see W. H. Urry and D. J. Trecker, *J. Amer. Chem. Soc.*, **84,** 118 (1962).

$$\text{cyclohexanone} \xrightarrow{Tl(NO_3)_3,\ HOAc,\ \text{room temp}} \text{2-hydroxycyclohexanone } (84\%)$$

$$\text{cyclohexanone} \xrightarrow{Tl(NO_3)_3,\ HOAc,\ 40^\circ} \text{cyclopentanecarboxylic acid } (84\%)$$

(Ref. 143)

The oxidation of cyclohexanone with thallium(III) nitrate is noteworthy because of the great dependence of the product obtained on reaction conditions.[143] Oxidation of acetylenes also produces α-hydroxyketones,[144] as can oxidation of β-keto esters.[145] Another oxidative route

$$CH_3COCH(C_5H_{11}\text{-}n)CO_2C(CH_3)_3 \xrightarrow[\text{3. } p\text{-}CH_3C_6H_4SO_3H,\ 160^\circ \quad \text{4. } OH^-]{\text{1. NaH} \quad \text{2. } (C_6H_5CO_2)_2,\ 0^\circ} CH_3COCH(OH)C_5H_{11}\text{-n} \quad (42\%)$$

recently reported involves the oxidation of trimethylsilyl enol ethers with *m*-chloroperbenzoic acid. Good yields of the acyloins are reported on hydrolysis of the intermediate keto-silylethers.[145a] Still another oxidative route to acyloins involves the bromine oxidation of cyclic tin derivatives of *cis*- or *trans*-1,2-diols as shown below.[145b]

$$(CH_2)_n\langle CHOH,\ CHOH\rangle + (C_4H_9)_2SnO \xrightarrow[-H_2O]{C_6H_6,\ \text{reflux}} (CH_2)_n\langle CHO,\ CHO\rangle Sn(C_4H_9)_2$$

n=2–5

$$\xrightarrow{CH_2Cl_2,\ Br_2} (CH_2)_n\langle CHOH,\ C{=}O\rangle + (C_4H_9)_2SnBr_2$$

An unusual acyloin is formed in the acid-catalyzed decarboxylation of ethyl benzothiazole-2-glyoxalate.[146] The generality of this reaction has not been established.

[143] A. McKillop, J. D. Hunt, and E. C. Taylor, *J. Org. Chem.*, **37**, 3381 (1972).

[144] A. McKillop, O. H. Oldenziel, B. P. Swann, E. C. Taylor, and R. L. Robey, *J. Amer. Chem. Soc.*, **93**, 7331 (1971).

[145] S.-O. Lawesson and S. Grönwall, *Acta Chem. Scand.*, **14**, 1445 (1960).

[145a] G. M. Rubottom, M. A. Vazquez, and D. R. Pelegrina, *Tetrahedron Lett.*, **1974**, 4319.

[145b] M. S. David, *C. R. Acad. Sci.*, [**C**], 1051 (1974).

[146] P. Baudet and Cl. Otten, *Chem. Ind.* (London), **1968**, 485.

$$\text{2-(COCO}_2\text{C}_2\text{H}_5\text{)benzothiazole} \xrightarrow[\text{10\% Suspension, 24 hr}]{6N\ \text{HCl}} \text{(benzothiazol-2-yl)C(OH)=C(OH)(benzothiazol-2-yl)}\ (73\%) + \text{benzothiazole-2-CO}_2\text{H}\ (27\%)$$

The reaction of organolithium reagents with iron or nickel carbonyls has been reported to lead to ketones or acyloins. The yields and products depend somewhat on the reaction conditions and on the particular lithium reagent chosen.[147–149]

$$p\text{-CH}_3\text{C}_6\text{H}_4\text{Li} \xrightarrow[\text{2. HCl/C}_2\text{H}_5\text{OH, }-70^\circ]{\text{1. Ni(CO)}_4\text{, }-70^\circ} p\text{-CH}_3\text{C}_6\text{H}_4\text{COCH(OH)C}_6\text{H}_4\text{CH}_3\text{-}p\ (71\%)$$

$$p\text{-CH}_3\text{C}_6\text{H}_4\text{Li} \xrightarrow[\text{2. No hydrolysis}]{\text{1. Ni(CO)}_4\text{, }-70^\circ} (p\text{-CH}_3\text{C}_6\text{H}_4)_2\text{CO}\ (29\%)$$

(Ref. 149)

Thiazolium salts have been used as catalysts for the conversion of aliphatic[150] and aromatic[150,151] aldehydes to acyloins and benzoins in good to excellent yields. The choice and amount of thiazolium salt is critical. The micelle-forming N-laurylthiazolium bromide converts *n*-butyraldehyde into butyroin in 20% yield when the substrate to catalyst ratio is 28 but in 76% yield when the ratio is 3.5. Under the same conditions, N-butylthiazolium bromide gives at best a trace of product. The reactions with hexanal, octanal, benzaldehyde, and furfural are heterogeneous with yields of 67%, 63%, 95%, and 80%, respectively, with a substrate to N-laurylthiazolium bromide ratio of 14.[150] These new data make the conversion of aldehydes to acyloins an important reaction which should receive more attention in the future. Thiazolium salts have also been used for the

[147] M. Ryang, Y. Sawa, H. Masada, and S. Tsutsumi, *J. Chem. Soc. Jap., Ind. Chem. Sect.*, **66**, 1086 (1963).

[148] M. Ryang, I. Rhee, and S. Tsutsumi, *Bull. Chem. Soc. Jap.*, **37**, 341 (1964).

[149] S. K. Myeong, Y. Sawa, M. Ryang, and S. Tsutsumi, *Bull. Chem. Soc. Jap.*, **38**, 330 (1965).

[150] W. Tagaki and H. Hara, *Chem. Commun.*, **1973**, 891, and references cited therein.

[151] A. F. Babicheva, O. M. Polumbrik, and A. A. Yasnikov, *Reakts. Sposobnost. Org. Soedin.*, **5** [3], 802 (1968) [*Org. Reactivity, USSR, Engl. Transl.* **1968**, 332.]

reductive decarboxylation of pyruvic acid to acetoin.[152] The generality of this reaction is unknown.

EXPERIMENTAL CONSIDERATIONS

The effect of a number of factors relating to the ease of cyclization of diesters was considered in detail in Finley's review.[15] They will not be discussed here. The effect of reaction conditions, however, requires further attention. At least five factors are of potential significance: addition of TMCS; structure of the ester group; solvent; the metal; and workup procedure.

Addition of Trimethylchlorosilane

The remarkable improvement in yield when TMCS is present has been repeatedly noted. Other silylating reagents would very likely be just as efficient for scavenging alkoxides and thus preventing Claisen condensations, but they would not be as cheap. The routine use of a base scavenger in the acyloin condensation now seems to be the best way to run the reaction. When reactions are conducted in liquid ammonia, the workup is greatly simplified if TMCS is added after the solvent has evaporated and before the reaction mixture is exposed to air. This procedure was used successfully in the reduction of tetramethyl methanetetraacetate (**61**) to the [4,4]-spirononane **62**.[153]

$$C(CH_2CO_2CH_3)_4 \xrightarrow[2.\ TMCS]{1.\ Na,\ NH_3} \text{62 (88\%)}$$

61 $\qquad$ $(CH_3)_3SiO$, $(CH_3)_3SiO$, $OSi(CH_3)_3$, $OSi(CH_3)_3$ $\qquad$ **62** (88%)

Additional advantages are: (1) the silylated ene-diol is relatively stable to storage whereas many free acyloins rapidly form dimers;* (2) the silylated compound is subject to a variety of other reactions without the

* The dimers of acyloins have long been considered to be dioxane derivatives. A review in 1950 discussed the evidence for the structures of acyloins, their dimers, and the ethers derived from both of them.[154] Recently, evidence has been presented that at least one α-hydroxyketone dimer, that from 2-hydroxy-2-methylcyclobutanone, is actually the dioxolane,

OH HO CH3 / O / O / CH3

.[155] The observation may in fact be general for all acyloin dimers, or the two forms may be in equilibrium with each other and with monomer.

[152] J. E. Downes and P. Sykes, *Chem. Ind.* (London), **1957,** 1095.

[153] C. Ainsworth and F. Chen, *J. Org. Chem.*, **35,** 1272 (1970).

[154] R. Jacquier, *Bull. Soc. Chim. Fr.*, **1950,** D83.

[155] J. C. Duggan, W. H. Urry, and J. Schaefer, *Tetrahedron Lett.*, **1971.** 4197.

need to isolate the free acyloin (p. 263, 266); and (3) the workup is much simpler, requiring only filtration, evaporation of organic solvent, and distillations of the product.

There is one potential disadvantage to the use of TMCS. Unless the ester is added cautiously to the metal, a dangerous excess of ester can build up which subsequently may lead to an explosively exothermic reaction. This point is considered in detail later on p. 306.

Structure of the Ester Group

No systematic effort to examine the effect of the structure of the ester group has been reported. In most instances where different esters have been used, the effect seems to be minimal. But there are exceptions, the most

CH_3O ... CO_2CH_3 $\xrightarrow{\text{Na, } NH_3/(C_2H_5)_2O}$ ... OH, OH

63 (9%)

notable of which are found in those cases where an internal lactone-ester was cyclized. In the synthesis of colchicine, the lactone **63** was found to cyclize, albeit in low yield, while a diester of **63** was dismissed as a possible intermediate because of the expected problems from a free hydroxyl group.[156]

Other examples of lactone-ester reduction are reported for some diterpene derivatives;[157–159] unfortunately, yields are not given.

CO_2CH_3 $\xrightarrow{\text{Na, } NH_3(C_2H_5)_2O}$ OH, OH

(Ref. 157)

[156] (a) E. E. van Tamelen, T. A. Spencer, Jr., D. S. Allen, Jr., and R. L. Orvis, *J. Amer. Chem. Soc.*, **81,** 6341 (1959); (b) *Tetrahedron*, **14,** 8 (1961).

[157] E. Fujita, T. Fujita, H. Katayama, and S. Kunishima, *Chem. Commun.*, **1967,** 258.

[158] E. Fujita, T. Fujita, K. Fuji, and N. Ito, *Tetrahedron*, **22,** 3423 (1966).

[159] E. Fujita, T. Fujita, Y. Nagao, H. Katayama, and M. Shibuya, *Tetrahedron Lett.*, **1969,** 2573.

One very interesting effect of changing ester groups occurs in the reduction of phenylacetic esters.[16] The reaction was discussed earlier (pp. 268, 269, 270, and 271). The product distribution is dependent on whether an ethyl or a trimethylsilyl ester is used. For example, in ether the four products, bis(silyl) acyloin **(21)**, dibenzyl ketone **(22)**, enol **(23)**, and benzyltrimethylsilane **(20)** are obtained in 40%, 0%, 11%, and 26% yield from the ethyl ester and in 21%, 15%, 15%, and 17% yield from the trimethylsilyl ester. The more than two-fold increase in yield on changing the ester group from ethyl to trimethylsilyl in the reduction of *p*-toluic esters in the presence of TMCS[16] was noted on p. 278. But, when the metal is in a coarse suspension, without TMCS present, ethyl benzoate gives benzoic acid and ethane while the trimethylsilyl ester gives benzoin in 12% yield.[73] These results may be atypical. In most instances the effect is mainly kinetic and, if sufficient time is allowed for reduction to proceed to completion, the yields are unaffected by the ester group.[60,160]

In many kinds of reactions involving esters it is possible to substitute other carboxylic acid derivatives with good success. However, in the acyloin condensation only esters seem to work well. Nitriles take an entirely different path. In 1956 the reaction of acetonitrile with sodium in the presence of TMCS was reported to produce several products.[161] More recently it has been shown that alkali metal solutions convert nitriles to

$$CH_3CN \xrightarrow[(C_2H_5)_2O]{Na,\ TMCS} \underset{(6\%)}{(CH_3)_4Si} + \underset{(27\%)}{(CH_3)_3SiCN} + \underset{(2\%)}{(CH_3)_3SiCH_2CN}$$

$$+ \underset{(7\%)}{[(CH_3)_3Si]_2CHCN} + \underset{(25\%)}{(CH_3)_3SiCH{=}C{=}NSi(CH_3)_3}$$

hydrocarbons and sodium cyanide. Succinonitrile derivatives are converted to olefins.[162] Similar reductions have been reported with sodium-naphthalene radical anion,[163] with potassium in hexamethylphosphoramide and alcohol,[164] and with sodium plus ferric acetylacetonate.[165] In none of these instances was there any indication of the formation of a new carbon-carbon bond.

Solvent

The reduction of phenylacetic esters is also useful to consider in examining the effect of solvent. At reflux in tetrahydrofuran the sole product

[160] U. Schräpler and K. Rühlmann, *Chem. Ber.*, **96**, 2780 (1963).

[161] M. Prober, *J. Amer. Chem. Soc.*, **78**, 2274 (1956).

[162] P. G. Arapakos, *J. Amer. Chem. Soc.*, **89**, 6794 (1967); P. G. Arapakos and M. K. Scott, *Tetrahedron Lett.*, **1968**, 1975; P. G. Arapakos, M. K. Scott, and F. E. Huber, Jr., *J. Amer. Chem. Soc.*, **91**, 2059 (1969).

[163] S. Bank and S. P. Thomas, *Tetrahedron Lett.*, **1973**, 305.

[164] T. Cuvigny, M. Larcheveque, and H. Normant, *C. R. Acad. Sci., Ser. C*, **1972**, 797.

[165] E. E. van Tamelen, H. Rudler, and C. Bjorkland, *J. Amer. Chem. Soc.*, **93**, 7113 (1971).

from the silyl ester is benzyltrimethylsilane (**20**), while in toluene at 36° the same ester gives 47% of the acyloin derivative **21** and only 9% of **20**. In toluene the effect of temperature is also marked, with the yield of **21** falling to 15% and of **20** rising to 21% at 110° (p. 268). Minor amounts (0–5%) of the other two compounds are also formed.[16]

There are really only two significantly different solvent systems in use. The first, the more common one, is an aromatic solvent such as toluene or xylene in which heterogeneous reduction is carried out at reflux. It is especially useful for acyloins that are cyclic or of high molecular weight.[102,166] Ether or benzene has often been used with short-chain esters, but the yields are less satisfactory for the longer-chain esters. The higher-boiling solvent is apparently necessary in these cases to keep the metal molten and to help break up the cake that may have a tendency to form around the individual metal particles. Dioxane, tetrahydrofuran, and 1,2-dimethoxyethane have been used, but these solvents do not seem to offer any special advantage.

The other important solvent system is liquid ammonia-ether, which is used for *homogeneous* reduction. The use of this solvent first became important, and later widely employed, with the recognition that, with diesters, cyclizations are easily conducted to give good yields of acyloins. Although the use of liquid ammonia had been explored earlier with less than completely satisfactory results,[45,167] a re-examination showed that the reaction was very good for syntheses of steroid A, C, and D rings.[168,169]

There are instances where one solvent system is clearly superior. For example, in the colchicine synthesis (p. 300), reduction in xylene gave no acyloin product.[156] In the reduction of tetramethyl methanetetraacetate (**61**) in xylene, the yield of acyloin derivative **62** was only 20% compared to 88% in ammonia.[153]

The reverse situation also applies. Reduction of dimethyl bicyclo[4.4.0]-deca-3,8-diene-1,6-dicarboxylate (*cf.* p. 281) in liquid ammonia produces a variety of reduction products of which the acyloin is only a minor component, but reduction in an aromatic solvent with or without TMCS present permits ready isolation of the acyloin in good yield.[54,83]

Similarly the reduction of dialkyl malonates takes a different course in hydrocarbon solvent than in liquid ammonia (pp. 278, 279).[56,78,79]

On the basis of the bulk of the experimental work examined here, it

[166] V. L. Hansley, *J. Amer. Chem. Soc.*, **57**, 2303 (1935).

[167] E. Chablay, *Ann. Chim. (Paris)*, [9], **8**, 205 (1917).

[168] J. C. Sheehan and W. F. Erman, *J. Amer. Chem. Soc.*, **79**, 6050 (1957).

[169] J. C. Sheehan, R. C. Coderre, L. A. Cohen, and R. C. O'Neill, *J. Amer. Chem. Soc.*, **74**, 6155 (1952).

would appear that the first choice of reaction conditions would be heterogeneous (an aromatic solvent or ether) with TMCS present. If the yield is poor, ammonia should be investigated.

No solvents giving homogeneous reaction conditions have been studied except liquid ammonia. In an aprotic solvent, perhaps different results would obtain. Examination of sodium-hexamethylphosphoramide[170] should be made.

The Metal

The most commonly used metal is sodium. Potassium is very infrequently used, but sodium-potassium alloys have proved very advantageous. Alloys can be liquid at −10°. This characteristic permits use of lower reaction temperatures, which may be necessary to avoid side reactions. For example, the reduction of *trans*-dimethyl 1-methylcyclohex-4-ene-1,2-dicarboxylate by sodium in refluxing toluene produces a cyclooctatriene, whereas reduction with the mixed metal alloy at 0–5° in ether gives the *trans*-bicyclo[4.2.0] system (p. 282)[83]. One of the most useful applications is in the reduction of long-chain thiophene diesters[117,171] (p. 290). The alloy appears to be a more powerful reducing agent than either metal alone. Thus the reduction of dimethyl bicyclo-[4.2.0]octane-1,6-dicarboxylate produces the acyloin derivative **33** and the 1,2-bond fragmentation product **34** with sodium, but only the fragmentation product when the alloy is used (p. 272).[69]

In the reduction of dialkyl cyclohexane-1,1-diacetates there appears to be a dependence on whether sodium[97] or potassium[172] is used, with the yield varying from a trace[97] to 61%.[172] The 20% yield of diketone reported in the sodium reduction suggests that the poor yield of acyloin in this case may really represent poor technique in the reduction and workup.[97] (No effort was made to keep air away from the basic reaction mixture.)

CO_2R
CO_2R
K, xylene
$R = CH_3$
O
OH
(61%)
Na, xylene
$R = C_2H_5$
O
O
+
O
OH
(20%)
(trace)

170 G. Fraenkel, S. H. Ellis, and D. T. Dix, *J. Amer. Chem. Soc.*, **87**, 1406 (1965).

171 S. Z. Taits and Ya. L. Gol'dfarb, *Izv. Akad. Nauk, SSSR, Ser. Khim.*, **No. 7**, 1289 (1963) (*Bull. Acad. Sci. USSR, Div. Chem. Sci., Engl. Transl.*, **1963**, 1173.)

172 B. Eistert, G. Bock, E. Kosch, and F. Spalink, *Chem. Ber.*, **93**, 1451 (1960).

The use of a source of reducing agent other than a metal was mentioned in the discussion of keto ester reduction (p. 295). There seem to be no examples of the bimolecular reduction of esters at an electrode. The use of the radical anion prepared from naphthalene and sodium in tetrahydrofuran has been given only a limited trial, with mixed results.[73,131] In the positive work, ethyl benzoate was successfully reduced to benzoin in 38% yield.[73] This result compares favorably with reduction by sodium in toluene in the presence of TMCS, which gave a 39% yield (p. 278).[16] By contrast, reduction with a suspension (not a dispersion) of sodium in toluene leads to alkyl group cleavage (cf. p. 275), producing benzoate and ethane.[73]

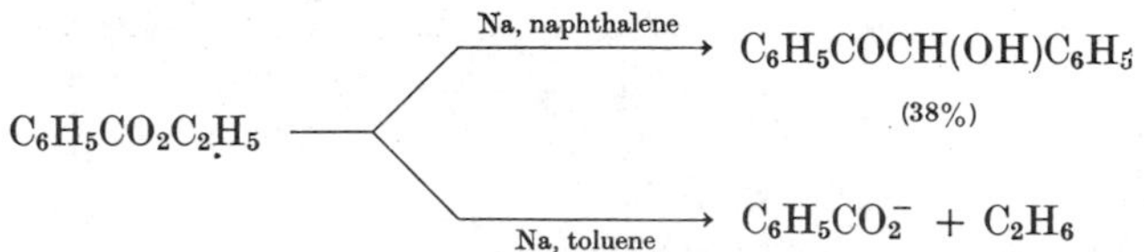

These experiments may especially point out the requirement of high surface area for the metal. Apparently quite different results can be obtained with less dispersed metal. However, careful experiments to check this point have not been made.

The use of magnesium and magnesium iodide in the reduction of aryl esters, acids, and acid chlorides produces benzoins in 30–46% yields.[173] There is no evidence that this technique would work on aliphatic acid derivatives.

A small excess of metal is usually required in reactions conducted in aromatic solvents. In liquid ammonia an exactly equivalent amount is ordinarily used. Apparently authors have universally discovered the need for an excess of metal but have never determined why. Recently it was discovered that the aromatic solvents are more or less reduced to dianions, depending on the reaction conditions. Usually, on workup, the dihydroaromatic compound would be thrown out with the solvent. In reductions in the presence of TMCS, 1,4-bis-(trimethylsilyl)-1,4-dihydrobenzene was isolated.[86,*]

* Reference 86 represents the first observation of reduction of an aromatic substrate (benzene) during an acyloin condensation. It is not the first observation of the reduction of aromatics in the presence of TMCS. The reduction of benzene and other aromatics by alkali metals was first reported by Weyenberg and Torporcer. D. R. Weyenberg and L. H. Torporcer, *J. Amer. Chem. Soc.*, **84,** 2843 (1962); *J. Org. Chem.*, **30,** 943 (1965).

[173] (a) M. Gomberg and W. E. Bachmann, *J. Amer. Chem. Soc.*, **50,** 2762 (1928); (b) R. C. Fuson, C. H. McKeever, and J. Corse, *ibid.*, **62,** 600 (1940); (c) R. C. Fuson, S. L. Scott, E. C. Horning, and C. H. McKeever, *ibid.*, **62,** 2091 (1940); (d) R. C. Fuson and E. C. Horning, *ibid.*, **62,** 2962 (1940).

The Workup Procedure

The acyloin condensation must always be conducted with the complete exclusion of oxygen. The extreme sensitivity of the ene-diolate to oxidation is widely appreciated.[5] Nevertheless, many authors have failed to recognize that base and air must not come in contact with the acyloin, even during the workup. The examples in which diketone is reported as a product and air was excluded during reduction are almost certainly produced *after* the reduction is completed. This may in part be traceable to an early *Organic Syntheses* procedure for the preparation of butyroin which calls for washing with 20% sodium carbonate solution and drying over potassium carbonate.[174] In examples described in that procedure, this appears to be harmless. However, it cannot be too strongly emphasized that oxygen must be excluded at all stages of the synthesis in which the acyloin is exposed to basic conditions.

Reactions in liquid ammonia should not be worked up until all the ammonia is allowed to evaporate. Trimethylchlorosilane can then be added or the reaction mixture can be acidified and worked up in the usual way.

The procedure for the hydrolysis of trimethylsilyl derivatives is critical. In many instances the hydrolysis can be carried out by mixing the derivative with aqueous acid. However, this procedure is not general and can lead to results varying from simply low yields to entirely different products. For example, the methanolysis of the tricyclic silyl ether **33** in methanol containing only a trace of acid leads to ring opening, but in oxygen-free, dry methanol the acyloin is obtained in high yield.[69] The latter method is general and trouble free, and there seems to be no reason to use any other procedure. The method is described on p. 310.

O OH OCH_3 (49–60%) ← MeOH, H^+ — $OSi(CH_3)_3$ $OSi(CH_3)_3$ **33** — CH_3OH, N_2 → O OH (97%)

One precaution concerning workup of TMCS reactions should be noted. The reaction mixture is simply filtered, the solvent stripped, and the residue distilled. However, if an excess of metal is used, and especially if it is sodium-potassium alloy, the solid residues from the filtration are likely to be *PYROPHORIC*. For this reason it is best, if at all possible, to filter and wash the filtrate in a nitrogen-atmosphere dry box. The solid residue must be carefully disposed of.

[174] J. M. Snell and S. M. McElvain, *Org. Syn., Coll. Vol.*, **2**, 114 (1943).

EXPERIMENTAL PROCEDURES

The usual precautions for conducting reactions under anhydrous, oxygen-free conditions should be employed.[175] Reagents should usually be freshly distilled

Note: It is particularly important that the trimethylchlorosilane be distilled, preferably from calcium hydride, under nitrogen. In at least one laboratory the use of this reagent without prior purification has been followed by explosions and injury to at least two people.[176] Trimethylchlorosilane may contain some dimethyldichlorosilane as an impurity. The dihalosilane hydrolyzes more readily than does TMCS. Cautious treatment with a small amount of water, followed by distillation from calcium hydride, under nitrogen, removes this impurity. A further cautionary note concerning TMCS reactions is also necessary. The explosions occurred in reactions run on larger than a 0.1-mol scale, using undistilled TMCS and following a published procedure. This procedure requires mixing all the reagents at 20–30° and then gradually warming the mixture.[16] When this procedure was applied to diethyl glutarate on a large scale, the reaction became uncontrollably exothermic at about 50°.[176] It is our practice, and we recommend, that the ester and TMCS be added together, dropwise, at a rate sufficient to maintain the exothermic reaction. It is usually unsafe with many esters, especially shorter-chain diesters and most aliphatic esters, to have a large amount of unreacted ester in the reaction mixture at any time.

In the heterogeneous reactions the metal must be finely dispersed. For this purpose a "Stir-O-Vac" stirrer[177] or a Vibromixer[177] or similar stirrer is needed.

Procedures without Trimethylchlorosilane

Butyroin (A Linear Acyloin). The preparation of butyroin from ethyl butyrate in 65–70% yield is described in *Organic Syntheses*.[174] The precaution noted on p. 305 against allowing oxygen and acyloin to come in contact in the presence of base should be carefully observed if this procedure is followed with a new compound.

[175] The general conditions described for carrying out the Dieckmann condensation are applicable; see J. P. Schaefer and J. J. Bloomfield, *Org. Reactions*, **15**, 40 (1967).

[176] Personal communication from Prof. P. E. Eaton.

[177] "Stir-O-Vac," Lab-Line, Cat. No. 1280, Lab-Line Instruments, Inc., Melrose Park, Illinois; Vibromixer, available in USA from Chemapec Inc., Hoboken, N.J.

Sebacoin (A Cyclic Acyloin). The preparation of sebacoin from dimethyl sebacate in 63–66% yield is also described in *Organic Synthesis*.[178] The method is general.

Cyclic Amino Acyloins (General Method for High-Dilution Reactions).[112] The reaction mixture was contained in a 1-l Morton flask with a Morton high-speed stirrer.[179] Inlets for purified nitrogen were provided through the stirrer cylinder and through one of the side necks. A high-dilution apparatus[82,180] fitted with a 500-ml addition funnel and a reflux condenser with stopcock attached was placed in the other side neck. A bubble tube filled with xylene and connected to the condenser top allowed monitoring of the nitrogen flow through the apparatus. The nitrogen flow was maintained constantly prior to the reaction and until the flask and contents were removed. Xylene or toluene was distilled from sodium into the reaction flask until it was three-fourths full. Then one-third of the solvent was redistilled from the Morton flask and the condensate was removed through the stopcock at the bottom of the reflux condenser. This step served to dry the apparatus thoroughly.

To the cooled solvent was added 94–95 g, (4.09–4.13 g-at.), about 2–3% excess, of freshly cut sodium. The solvent was heated to vigorous reflux with stirring at 7500–9000 rpm. One mole of the desired diester $RN[(CH_2)_nCO_2C_2H_5]_2$ ($R = CH_3$, C_2H_5, $n = 3$–8, 10) in 500 ml of anhydrous solvent was added via the addition funnel during 4–6 hours. When the addition was complete, the refluxing and stirring were continued for another 30 minutes. The reaction mixture was allowed to cool slowly under increased nitrogen flow. Finally the flask was cooled in an ice bath, and acetic acid was added cautiously to the moderately stirred mixture until it became slightly acidic. Water (200 ml) was added to dissolve the sodium acetate, and potassium carbonate was added to saturation. The layers were separated, and the aqueous layer was extracted with ether. The combined organic portions were dried, the solvent removed under reduced pressure, and the residue distilled through a modified Holzman column. Yields of 64–88% were obtained (see Table II).

14,16-Dimethyl-[12]-β-cyclothien-6-ol-7-one (Cyclization of 3,4-Bis-(5-methoxycarbonylpentyl)-2,5-dimethylthiophene; Use of Sodium-Potassium Alloy).[171] A reactor provided with a high-speed stirrer (9000 rpm), addition funnel, reflux condenser, and pressurized nitrogen inlet was charged with 1.3 l of xylene, a part of which was distilled to remove water from the walls. With a temperature of about 130°,

[178] N. Allinger, *Org. Syn., Coll. Vol.*, **4**, 840 (1963).
[179] A. A. Morton, B. Darling, and J. Davidson, *Ind. Eng. Chem., Anal. Ed.*, **14**, 734 (1942).
[180] N. J. Leonard and R. C. Sentz, *J. Amer. Chem. Soc.*, **74**, 1704 (1952).

3.8 g of carefully cleaned sodium and 13 g of potassium were added. The stirrer was turned on and gradually the speed was increased to 8000–8500 rpm and maintained for 1.5 hours. Heat was discontinued and, when the temperature reached 60–65°, 15 g of 3,4-bis-(5-methoxycarbonylpentyl)-2,5-dimethylthiophene in 800 ml of xylene was added slowly over 30 hours. Heating was continued for another 5 hours with stirring. The mixture was cooled to −5 to −10° and then 60 ml of methanol, 250 ml of 10% sulfuric acid, and 250 ml of water were added successively. Stirring was continued for 30 minutes. The mixture was then warmed to room temperature and stirred for another hour. Nitrogen flow was maintained continuously.

The organic layer was separated, washed with water, sodium bicarbonate solution, and water and then dried over magnesium sulfate. The xylene was removed in a stream of nitrogen and finally under vacuum. The viscous oil, 11.5 g, was treated with twice its volume of ether and cooled to −60°. Colorless crystals of 14,16-dimethyl-[12]-β-cyclothien-6-ol-7-one separated and were crystallized twice from methanol; wt 4.0 g (32%); mp 105.5–107°. The noncrystalline residue from the ether was treated with toluenesulfonyl chloride in pyridine to give 7.6 g (40%) of the tosyl derivative; mp 126–128°.

16-Keto-17β-estradiol-3-methyl Ether (Use of Sodium in Liquid Ammonia).[*,181] To a 1-l, three-necked flask fitted with a solid carbon dioxide condenser, stirrer, addition funnel, and nitrogen system[182] were added 200 ml of dry ether and 300 ml of anhydrous liquid ammonia. In this liquid 0.80 g (0.0348 g-at.) of freshly cut sodium was dissolved. The apparatus was swept thoroughly with prepurified† nitrogen and all subsequent operations up to the extractions were carried out under a slow stream of nitrogen.

A solution of 1.82 g (0.005 mol) of dimethyl marrianolate methyl ether in 180 ml of dry ether was added during 1.5 hours with efficient stirring. Stirring was continued as the flask was allowed to come slowly to room temperature. After 4 hours, only a trace of ammonia could be detected in the exit gases. To the white suspension of excess sodium and sodium

* This is the reaction described on p. 284 for preparing a labeled steroid, but the ester used in this procedure is not labeled.

† Prepurified nitrogen is a commercial grade of nitrogen, widely available in cylinders; it is dry and has a very low oxygen content. Many authors use such nitrogen to blanket the reaction while others prefer to deoxygenate the nitrogen themselves. In the authors' laboratory a quartz tube filled with copper wire and heated to 500–550° is used.[182]

[181] J. C. Sheehan, R. A. Coderre, and P. A. Cruickshank, *J. Amer. Chem. Soc.*, **75**, 6231 (1953).

[182] For a description see K. B. Wiberg, *Laboratory Technique in Organic Chemistry*, McGraw-Hill, New York, 1960, p. 218–223.

enolate of the acyloin in ether was added 2 ml of methanol in 100 ml of ether (to destroy excess sodium), and the mixture was acidified with 50 ml of 5% hydrochloric acid. After partition and separation, the aqueous layer was extracted with an ether-methylene chloride mixture, and the combined organic solution was washed with dilute aqueous sodium bicarbonate and water. Removal of the solvents under reduced pressure afforded 1.44 g (96%) of colorless crystalline product, mp 163–166°.

Procedures with Trimethylchlorosilane

1,2-Bis-(trimethylsilyloxy)cyclobutene.[183] *Method A. With Sodium.* A 1-l, three-necked, creased flask was fitted with a stirrer capable of forming a fine dispersion of molten sodium,[177] a reflux condenser, and a Hershberg addition funnel. It was maintained under an oxygen-free, nitrogen atmosphere. The flask was charged with 250–300 ml of dry toluene* and 9.6–9.8 g (0.42–0.43 g-at.) of freshly cut sodium. The toluene was brought to gentle reflux; the stirrer was operated at full speed until the sodium was fully dispersed. The stirrer speed was reduced and a mixture of 17.4 g (0.100 mol) of diethyl succinate and 45–50 g (0.41–0.46 mol) of trimethylchlorosilane in 125 ml of solvent was added over 1–3 hours. The reaction was exothermic, and a dark-purple precipitate appeared within a few minutes. The solvent was maintained at reflux during and after the addition. After 5 hours of additional stirring, the contents of the flask were cooled and filtered through a 75-mm coarse sintered disk funnel in a nitrogen dry box. (The use of a dry box was necessary to avoid hydrolysis in the air and, more important, to avoid spontaneous fire from the very likely *pyrophoric* precipitate.) The precipitate was washed several times with anhydrous ether or petroleum ether

The colorless to pale-yellow filtrate was transferred to a distilling flask and distilled under reduced pressure. After a small fore-run (0.5–1.0 g), the cyclobutene was obtained at bp 82–86° (10 mm); 18.0 g (78%); n_D^{25} 1.4331.

Method B. With Sodium-Potassium Alloy.[183] In the apparatus described above were placed 4.8–5.0 g (0.209–0.218 g-at.) of clean sodium and 8.0–8.2 g (0.205–0.210 g-at.) of clean potassium. The flask was heated with a heat gun to form the low-melting alloy; then 300–350 ml of anhydrous ether was added from a freshly opened can. The stirrer was

* Toluene, xylene, and methyl cyclohexane have all been used in similar reductions with equally good results.

[183] This procedure is taken from one submitted to *Organic Syntheses* by J. J. Bloomfield and J. M. Nelke.

operated at full speed until the alloy was dispersed, and then at a slower speed for the remainder of the reaction.

The ester (0.100 mol) and trimethylchlorosilane (0.41–0.46 mol) were then added in 125 ml of anhydrous ether at a rate sufficient to keep the reaction under control. The purple mixture was stirred for another 4–6 hours and filtered and washed as above in a nitrogen dry box. (*Caution*! The residues are *pyrophoric*.) The cyclobutene was distilled as above; fore-run 0.5–2 g; product, bp 82–86° (10 mm) 17.9–21.5 g (78–93%); n_D^{25} 1.4323–1.4330.

2,3,7,8-Tetrakis(trimethylsilyloxy)spiro[4.4]nona-2,7-diene (Reduction in Liquid Ammonia Followed by Treatment with Trimethylchlorosilane.)[153] (See p. 299.) Commercial anhydrous ammonia (75 ml) was distilled through a drying tube filled with lump barium oxide into a three-necked flask fitted with a mechanical stirrer and a Dry Ice-acetone cooled condenser. Sodium (1.84 g, 0.08 g–at.) cut in small pieces was added to the ammonia. After the system was flushed with helium, a solution of 3 g (0.01 mol) of tetramethyl methanetetraacetate in 100 ml anhydrous ether was added over 2 hours. The blue color disappeared at the end of the addition. The Dry Ice was removed from the condenser and the ammonia was evaporated using a cold-water bath. The last traces of ammonia were flushed out in a stream of helium. To the ice-bath cooled reaction mixture was added 16 ml of trimethylchlorosilane in 100 ml of dry ether. The mixture was stirred slowly for 0.5 hour and filtered. Distillation of the filtrate gave 4.2 g (90%), bp 126° (0.25 mm).

Succinoin (Preparation from the Trimethylsilyl Derivative.)[183] Methanol (450 ml; Mallinckrodt reagent grade) was placed in a 1-l, three-necked flask fitted with a magnetic stirring bar, a sintered-disk gas inlet tube, a dropping funnel, and a reflux condenser. Dry oxygen-free nitrogen was vigorously bubbled through the methanol for about 1 hour. The sintered gas inlet was replaced by a standard inlet; then 23 g (0.10 mole) of *freshly distilled* 1,2-bis-(trimethylsilyloxy)cyclobutene (p. 309) was transferred under nitrogen to the addition funnel and added dropwise to the stirred methanol. Stirring under a reduced nitrogen flow was continued for 24–30 hours.

The methanol and trimethylmethoxysilane were removed under reduced pressure. The residual succinoin was distilled through a short-path still; bp 52–57° (0.1 mm); 6.1–7.4 g (71–86%); n_D^{25} 1.4613–1.4685. Succinoin soon changed to succinic acid on exposure to air. It formed a solid dimer on storage below −10°.

This is the most general, the easiest, and the cleanest procedure for hydrolysis of silyl acyloin derivatives. It works well even for difficult compounds, such as the one described here. The rate can be increased by reflux, but the reaction should be followed by gas chromatography. Prolonged reflux can produce unknown side products.[86]

TABULAR SURVEY*

An attempt has been made to include all the acyloin condensations reported through November, 1974. The reactions are arranged in each table according to the total number of atoms in the starting carboxylic *acid*. If more than one ester group was used, they are listed also in order of increasing number of atoms: methyl, ethyl, etc.

The cyclic acyloins are tabulated according to ring size. When more than one ring size (*e.g.*, dimer) is formed in a reaction, the fact is noted and cross-referenced. When there is more than one reference, the first reference is the one that gave the best recorded yield.

The following abbreviations are used: TMCS, trimethylchlorosilane; THF, tetrahydrofuran; ether, diethyl ether.

* We are grateful to the many chemists who generously provided us with unpublished information.

TABLE I. STRAIGHT CHAIN ACYLOINS FROM MONOESTERS OR ACID CHLORIDES

A. Symmetrical Acyloin Condensations with Monoesters

	Starting Material	Metal/Solvent/Temp (°C)	Product(s) (% Yield)	Refs.
C_2	$CH_3CO_2C_2H_5$	Na/ether/0°	$CH_3COCHOHCH_3$ ("Low")	333, 246
		Na/ether/reflux	" (23), $CH_3COCOCH_3$ (7)	253, 5
		Na-NH_3	" (25), $CH_3C(NH_2){=}CHCO_2C_2H_5$ (8), CH_3CO_2H (25)	45
		1. Na/ether/0° 2. CH_3COCl	" (—), $CH_3COCH(OCOCH_3)CH_3$ (—), $CH_3(CH_3OCO)C{=}C(OCOCH_3)CH_3$ (—)	246
		1. Na/ether/0° 2. CH_3COCl	$CH_3COCH(OCOCH_3)CH_3$ (Major) $CH_3(CH_3OCO)C{=}C(OCOCH_3)CH_3$ (Minor)	332
		1. Na-NH_3 2. CH_3COCl/C_6H_6	" (83)	45
	$(C_2H_5)_2NCOCO_2C_2H_5$	Na/C_6H_6/reflux	$(C_2H_5)_2NCOCOCOCON(C_2H_5)_2$ (31)	74
	$(C_2H_5O)_2CHCO_2C_2H_5$	K/ether/0°	$(C_2H_5O)_2CHCOCHOHCH(OC_2H_5)_2$ (—)	228
C_3	$C_2H_5CO_2C_2H_5$	Na/ether/reflux	$C_2H_5COCHOHC_2H_5$ (50–65)	248, 174
		Na/ether or toluene/reflux	" (~50)	188
		Na-NH_3	" (23), $C_2H_5CO_2H$ (31), n-C_3H_7OH (26), $C_2H_5CONH_2$ (9)	45
		Na/ether/0°	" (—)	331
		Na/ether/reflux	" (52), $C_2H_5COCOC_2H_5$) (9)	253, 5
		Na/C_6H_6/reflux	" (30), " (7)	253, 5
		1. Na/ether/reflux 2. C_2H_5I	$C_2H_5COC(C_2H_5)(OH)C_2H_5$ (59)	46
C_4	n-$C_3H_7CO_2CH_3$	Na/xylene/110°	n-$C_3H_7COCHOHC_3H_7$-n (80–90)	166
	n-$C_3H_7CO_2C_2H_5$	Na/ether or C_6H_6/0°	n-$C_3H_7COCHOHC_3H_7$-n (80)	331, 333, 329
		Na/ether or toluene/reflux	" (~50)	188, 246
		K/ether/0°	" (—)	190
		Na/ether/reflux	" (72), n-$C_3H_7COCOC_3H_7$ (7)	253, 5
		Na/C_6H_6/reflux	" (61), " (7)	253, 5

		Na/ether/reflux	" (65–70)	78
		1. Na/ether/reflux 2. C_2H_5I	n-$C_3H_7COC(C_2H_5)(OH)C_3H_7$-$n$ (55)	46
	i-$C_3H_7CO_2C_2H_5$	Na/ether/0°	i-$C_3H_7COCHOHC_3H_7$-i (75)	330
		Na/ether/reflux	" (70–75)	174
		Na/ether or toluene/reflux	" (∼50)	188
		Na-NH_3/C_6H_6	" (13), i-$C_3H_7CH_2OH$ (13), i-C_3H_7CHO (26)	45
		1. Na-NH_3 2. C_2H_5Br	" (25), i-$C_3H_7COC_2H_5$ (29)	45
		K/ether/reflux	" (—)	190
		Na/ether/reflux	" (75), i-$C_3H_7COCOC_3H_7$-i (4)	253, 5
		Na/C_6H_6/reflux	" (68), " (8)	253, 5
		1. Na/ether/reflux 2. C_2H_5I	i-$C_3H_7COC(C_2H_5)(OH)C_3H_7$-$i$ (84)	46
	$(C_2H_5)_2NCO(CH_2)_2CO_2C_2H_5$	Na/C_6H_6/reflux	$[(C_2H_5)_2NCO(CH_2)_2CO]_2$ (16)	74
C_5	n-$C_4H_9CO_2C_2H_5$	Na/ether or toluene/reflux	n-$C_4H_9COCHOHC_4H_9$-n (∼50)	188
		1. Na/toluene/reflux 2. n-C_4H_9Br	n-$C_4H_9COCOH(C_4H_9$-$n)_2$ (48)	46
	sec-$C_4H_9CO_2C_2H_5$	Na/ether or toluene/reflux	sec-$C_4H_9COCHOHC_4H_9$-sec (∼50)	188
	t-$C_4H_9CO_2C_2H_5$	Na/ether/0°	t-$C_4H_9COCHOHC_4H_9$-t (80)	330
		Na/ether/reflux	" (52–60)	174, 287
		Na/NH_3	t-$C_4H_9COCHOHC_4H_9$-t (29), t-C_4H_9CHO (35), t-$C_4H_9CH_2OH$ (14), t-$C_4H_9CONH_2$ (10), t-$C_4H_9CO_2H$ (5)	45
		1. Na-NH_3 2. C_2H_5Br	t-$C_4H_9COCHOHC_4H_9$-t (24), t-$C_4H_9COC_2H_5$ (36)	45
		Na/ether/reflux	" (62), t-$C_4H_9COCOC_4H_9$-t (32)	253, 5
		Na/C_6H_6/reflux	" (63), " (15)	253, 5
C_6	n-$C_5H_{11}CO_2CH_3$	Na/xylene/110°	n-$C_5H_{11}COCHOHC_5H_{11}$-n (80–90)	166

Note: References 184–356 are on pp. 400–403.

TABLE I. STRAIGHT CHAIN ACYLOINS FROM MONOESTERS OR ACID CHLORIDES (*Continued*)

A. Symmetrical Acyloin Condensations with Monoesters (*Continued*)

	Starting Material	Metal/Solvent/Temp (°C)	Product(s) (% Yield)	Refs.
C_6	n-$C_5H_{11}CO_2C_2H_5$	Na/ether/0°	,, (50)	329, 330
(*Contd.*)	sec-$C_4H_9CH_2CO_2C_2H_5$	Na/ether or toluene/reflux	sec-$C_4H_9CH_2COCHOHCH_2C_4H_9$-$sec$ (~50)	188
	t-$C_5H_{11}CO_2C_2H_5$	Na/ether or toluene/reflux	t-$C_5H_{11}COCHOHC_5H_{11}$-t (~50)	188
C_7	$C_6H_5CO_2C_2H_5$	Na/ether/0°	$C_6H_5COCHOHC_6H_5$ (14)	227
		Na-NH_3/C_6H_6	,, (14), C_6H_5CHO (47)	45
		Na-NH_3	,, (51)[a]	45
		Na-$C_{10}H_8$/THF	,, (39)	73
		Na-NH_3 (1 g-at.)	$(C_6H_5CO)_2$ (30), $(C_6H_5)_2C(OH)CO_2H$ (28), $C_6H_5CO_2H$ (25)	45
		1. Na-NH_3 2. C_2H_5Br	$C_6H_5COCHOHC_6H_5$ (27), $C_6H_5COC_2H_5$ (34)	45
		1. Na-NH_3 2. n-C_4H_9Br	,, (20), $C_6H_5COC_4H_9$-n (30)	45
		1. Na-NH_3 2. $C_6H_5CH_2Cl$	,, (15), $C_6H_5COCH_2C_6H_5$ (5), $(C_6H_5CH_2)_2$ (8), C_6H_5CHO (30), $C_6H_5CO_2H$ (15), $C_6H_5CO[(C_6H_5)CH]_3COC_6H_5$ (10)	45
	$C_6H_5CO_2Si(CH_3)_3$	Na/toluene/reflux	$C_6H_5COCHOHC_6H_5$ (12)	73
	$C_6H_5CO_2C_6H_5$	Na (1 g-at.)/ether	$C_6H_5COCHOHC_6H_5$ (—)	44
	$cyclo$-$C_6H_{11}CO_2C_2H_5$	Na/ether	$cyclo$-$C_6H_{11}COCHOHC_6H_{11}$-$cyclo$ (63)	351,352
	n-$C_6H_{13}CO_2CH_3$	Na/xylene/120–125°	n-$C_6H_{13}COCHOHC_6H_{13}$-n (75)	187, 192
	n-$C_6H_{13}CO_2C_2H_5$	Na/ether or toluene/reflux	n-$C_6H_{13}COCHOHC_6H_{13}$-n (~50)	188
	n-$C_4H_9CH(CH_3)CO_2C_2H_5$	Na/ether or toluene/reflux	n-$C_4H_9CH(CH_3)COCHOHCH(CH_3)C_4H_9$-$n$ (~50)	188
	2-(C_4H_3S)-$(CH_2)_2CO_2C_2H_5$	Na/xylene/reflux	2-(C_4H_3S)-$(CH_2)_2COCHOH(CH_2)_2$-(C_4H_3S)-2 (38)	202
	O, O —$CO_2C_2H_5$	Na/ether/reflux	O, O —COCHOH— O, O	355

C_8	$C_6H_5CH_2CO_2C_2H_5$	Na	$C_6H_5CH_2COCHOHCH_2C_6H_5$ (—), $C_6H_5CH_2COCOCH_2C_6H_5$, $C_6H_5CH_2COCH_2C_6H_5$ (—)	191
	cyclo-$C_7H_{13}CO_2CH_3$	Na/ether/reflux	*cyclo*-$C_7H_{13}COCHOHC_7H_{13}$-*cyclo* (65)	353
	n-$C_7H_{15}CO_2CH_3$	Na/xylene/110°	*n*-$C_7H_{15}COCHOHC_7H_{15}$-*n* (80–90)	166
		Na/xylene/120–125°	" (53)	192
	p-$CH_3OC_6H_4CH_2CO_2C_2H_5$	Na	*p*-$CH_3OC_6H_4CH_2COCHOHCH_2C_6H_4OCH_3$-*p* (—)	191
	p-$ClC_6H_4CH_2CO_2C_2H_5$	Na	$C_6H_5CH_2CHOHCOCH_2C_6H_5$ (—), $C_6H_5CH_2COCOCH_2C_6H_5$ (—), $C_6H_5CH_2COCH_2C_6H_5$ (—)	191
C_9	$C_6H_5(CH_2)_2CO_2C_2H_5$	Na/xylene/reflux	$C_6H_5(CH_2)_2COCO(CH_2)_2C_6H_5$ (61)[b]	124
		Na/ether/reflux	$C_6H_5(CH_2)_2COCHOH(CH_2)_2C_6H_5$ (—)	190
	cyclo-$C_8H_{15}CO_2C_2H_5$	Na/ether/reflux	*cyclo*-$C_8H_{15}COCHOHC_8H_{15}$-*cyclo* (—)	354
	$(CH_3)_2C{=}C(CH_3)C(CH_3)_2CO_2C_2H_5$	Na/toluene/reflux	$(CH_3)_2C{=}C(CH_3)C(CH_3)_2COCHOHC(CH_3)_2$-$C(CH_3){=}C(CH_3)_2$ (—), $[(CH_3)_2C{=}C(CH_3)C(CH_3)_2]_2CO$ (—), $(CH_3)_2C{=}C(CH_3)CH(CH_3)_2$ (—), $[(CH_3)_2C{=}C(CH_3)C(CH_3)_2{-}]_2$ (—)	52
	n-$C_8H_{17}CO_2CH_3$	Na/xylene/110°	*n*-$C_8H_{17}COCHOHC_8H_{17}$-*n* (80–90)	166
		Na/xylene/120–125°	" (84)	192
	n-$C_8H_{17}CO_2C_4H_9$-*n*	Na/xylene/reflux	*n*-$C_8H_{17}COCHOHC_8H_{17}$-*n* (70–90)	356
C_{10}	$C_6H_5CH(CH_3)CH_2CO_2C_2H_5$	K/ether/reflux	$C_6H_5CH(CH_3)CH_2COCHOHCH_2CH(CH_3)C_6H_5$ (—)	190
	$C_6H_5C(CH_3)_2CO_2C_2H_5$	Na/xylene/reflux	$C_6H_5C(CH_3)_2COCHOHC(CH_3)_2C_6H_5$ (—), $[C_6H_5C(CH_3)_2{-}]_2$ (—), $C_6H_5CH(CH_3)_2$ (—)	51
	cyclo-$C_9H_{17}CO_2C_2H_5$	Na/ether/reflux	*cyclo*-$C_9H_{17}COCHOHC_9H_{17}$-*cyclo* (—)	354
	n-$C_9H_{19}CO_2CH_3$	Na/xylene/120–125°	*n*-$C_9H_{19}COCHOHC_9H_{19}$-*n* (88)	192
		Na/xylene/110°	" (80–90)	166
	p-$CH_3OC_6H_4C(CH_3)_2CO_2C_2H_5$	Na/toluene/105–110°	*p*-$CH_3OC_6H_4C(CH_3)_2COCHOHC(CH_3)_2C_6H_4OCH_3$-*p* (—), [*p*-$CH_3OC_6H_4C(CH_3)_2{-}]_2$ (—), *p*-$CH_3OC_6H_4CH(CH_3)_2$ (—)	51
	1-Adamantyl $CO_2C_2H_5$	Na/xylene/reflux	1-adamantyl COCHOH-1-adamantyl (82)	350

Note: References 184–356 are on pp. 400–403.

[a] The reaction mixture was acidified immediately after the ammonia had been allowed to evaporate.

[b] Identical results were obtained under air or nitrogen. The diketone was not reduced by Na-xylene.

TABLE I. STRAIGHT CHAIN ACYLOINS FROM MONOESTERS OR ACID CHLORIDES (*Continued*)

A. Symmetrical Acyloin Condensations with Monoesters (*Continued*)

	Starting Material	Metal/Solvent/Temp (°C)	Product(s) (% Yield)	Refs.
C_{11}	$CH_3CH{=}CH(CH_2)_7CO_2CH_3$	Na/xylene/reflux	$CH_3CH{=}CH(CH_2)_7COCHOH(CH_2)_7CH{=}CHCH_3$ (48)[c]	309
	$CH_2{=}CH(CH_2)_8CO_2CH_3$	Na/xylene/reflux	$CH_2{=}CH(CH_2)_8COCHOH(CH_2)_8CH{=}CH_2$ (50)[d]	263
	n-$C_{10}H_{21}CO_2CH_3$	Na/xylene/120–125°	n-$C_{10}H_{21}COCHOHC_{10}H_{21}$-$n$ (76)	192
	n-$C_{10}H_{21}CO_2C_2H_5$	Na/xylene/reflux	n-$C_{10}H_{21}COCHOHC_{10}H_{21}$-$n$ (89)	75
C_{12}	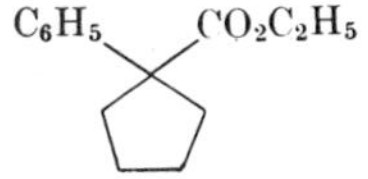 C_6H_5, $CO_2C_2H_5$	Na/toluene/reflux	RCOCHOHR (—), RCOR (—), R—R (13), RH (8), 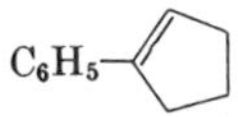 C_6H_5 (trace) (R = 1-phenylcyclopentyl)	51
	1-Adamantyl-$CH_2CO_2C_2H_5$	Na/xylene/100°	1-Adamantyl-$CH_2COCHOHCH_2$-1-adamantyl (86)	258, 200
	n-$C_{11}H_{23}CO_2CH_3$	Na/xylene/110°	n-$C_{11}H_{23}COCHOHC_{11}H_{23}$-$n$ (80–90)	166, 304
		Na/xylene/120–125°	" (89)	192
	n-$C_{11}H_{23}CO_2C_2H_5$	Na/xylene/110–120°	n-$C_{11}H_{23}COCHOHC_{11}H_{23}$-$n$ (>80)	260
C_{13}	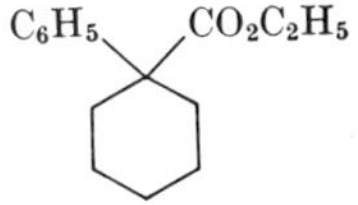 C_6H_5, $CO_2C_2H_5$	Na/toluene/reflux	RCOCHOHR (—), R—R (—), RH 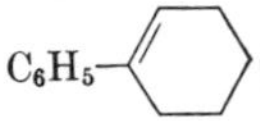 (—), C_6H_5 (Trace) (R = 1-phenylcyclohexyl)	51
	n-$C_{12}H_{25}CO_2CH_3$	Na/xylene/120–125°	n-$C_{12}H_{25}COCHOHC_{12}H_{25}$-$n$ (89)	192
C_{14}	$(C_6H_5)_2CHCO_2C_2H_5$	Na-NH_3	$(C_6H_5)_2CHCOCHOHCH(C_6H_5)_2$ (11), $(C_6H_5)_2CO$ (18), $(C_6H_5)_2CHCH_2OH$ (36), $(C_6H_5)_2CHCO_2H$ (29)	45
	n-$C_{13}H_{27}CO_2CH_3$	Na/xylene/110°	n-$C_{13}H_{27}COCHOHC_{13}H_{27}$-$n$ (80–90)	166
C_{14}	n-$C_{13}H_{27}CO_2CH_3$	Na/xylene/120–125°	" (80)	192
	n-$C_{13}H_{27}CO_2C_2H_5$	Na/xylene/110–120°	" (>80)	260
C_{15}	n-$C_{14}H_{29}CO_2CH_3$	Na/xylene/120–125°	n-$C_{14}H_{29}COCHOHC_{14}H_{29}$-$n$ (65)	192

C_{16}	$(C_6H_5CH_2)_2CHCH_2CO_2C_2H_5$	Na/ether/reflux	$(C_6H_5CH_2)_2CHCOCHOHCH(CH_2C_6H_5)_2$ (83)	190
		K/ether/reflux	" (78)	190
	n-$C_{15}H_{31}CO_2CH_3$	Na/xylene/120–125°	n-$C_{15}H_{31}COCHOHC_{15}H_{31}$-$n$ (76)	192
		Na/xylene/110°	" (80–90)	166
	n-$C_{15}H_{31}CO_2C_2H_5$	Na/xylene/110–120°	" (>80)	260
C_{17}	n-$C_{16}H_{33}CO_2CH_3$	Na/xylene/120–125°	n-$C_{16}H_{33}COCHOHC_{16}H_{33}$-$n$ (80)	192
C_{18}	Methyl linolenate	Na/xylene/115°	$C_2H_5CH{=}CHCH_2CH{=}CHCH_2CH{=}CH(CH_2)_7C{=}O$ | $C_2H_5CH{=}CHCH_2CH{=}CHCH_2CH{=}CH(CH_2)_7CHOH$ (91)	274
	Ethyl oleate	Na/xylene/reflux	$CH_3(CH_2)_7CH{=}CH(CH_2)_7COCHOH(CH_2)_7$-$CH{=}CH(CH_2)_7CH_3$ (~80)	259
	n-$C_{17}H_{35}CO_2CH_3$	Na/xylene/110°	n-$C_{17}H_{35}COCHOHC_{17}H_{35}$-$n$ (80–90)	166
		Na/xylene/120–125°	" (80)	192
	n-$C_{17}H_{35}CO_2C_2H_5$	Na/xylene/110–120°	n-$C_{17}H_{35}COCHOHC_{17}H_{35}$-$n$ (>80)	260
C_{19}	n-$C_{18}H_{37}CO_2CH_3$	Na/xylene/120–125°	n-$C_{18}H_{37}COCHOHC_{18}H_{37}$-$n$ (—)	192
C_{20}	n-$C_{19}H_{39}CO_2CH_3$	Na/xylene/120–125°	n-$C_{19}H_{39}COCHOHC_{19}H_{39}$-$n$ (—)	192
C_{21}	n-$C_{20}H_{41}CO_2CH_3$	Na/xylene/120–125°	n-$C_{20}H_{41}COCHOHC_{20}H_{41}$-$n$ (—)	192
C_{22}	n-$C_8H_{17}CH{=}CH(CH_2)_{11}CO_2C_2H_5$	Na/xylene/reflux	n-$C_8H_{17}CH{=}CH(CH_2)_{11}COCHOH(CH_2)_{11}$—$CH{=}CHC_8H_{17}$-$n$ (80)	259
	n-$C_{21}H_{43}CO_2CH_3$	Na/xylene/120–125°	n-$C_{21}H_{43}COCHOHC_{21}H_{43}$-$n$ (—)	192

B. Symmetrical Condensations of Acid Chlorides

C_4	n-C_3H_7COCl	Na/ether	$[n$-$C_3H_7(n$-$C_3H_7CO_2C{=}]_2$ (—)	3, 4
		Na/neat/reflux	" (—)	1
	i-C_3H_7COCl	Na/ether	$[i$-$C_3H_7(i$-$C_3H_7CO_2)C{=}]_2$ (—)	4
C_5	i-C_4H_9COCl	Na/ether	$[i$-$C_4H_9(i$-$C_4H_9CO_2)C{=}]_2$ (—)	3
		Na/neat	"	2
	$(CH_3)_3CCOCl$	Na/ether/reflux	$(CH_3)_3CCOCOC(CH_3)_3$ (28), $(CH_3)_3CCOCH[O_2CC(CH_3)_3]C(CH_3)_3$ (6)	328

Note: References 184–356 are on pp. 400–403.

[c] The α-diketone was isolated in 1% yield.

[d] The α-diketone was isolated in 2% yield.

TABLE I. STRAIGHT CHAIN ACYLOINS FROM MONOESTERS OR ACID CHLORIDES (*Continued*)

	Starting Material	Metal/Solvent/Temp (°)C	Product(s) (% Yield)	Refs.
	B. Symmetrical Condensations of Acid Chlorides (Continued)			
C_7	C_6H_5COCl	1. Na-Hg/ether 2. KOH, H_2O	$C_6H_5COCHOHC_6H_5$ (—), $C_6H_5CO_2H$ (—)	187
C_{12}	n-$C_{11}H_{23}COCl$	Na/ether/reflux	$[n\text{-}C_{11}H_{23}(n\text{-}C_{11}H_{23}CO_2)C{=}]_2$ (60)	335
C_{14}	n-$C_{13}H_{27}COCl$	Na/ether/reflux	$[n\text{-}C_{13}H_{27}(n\text{-}C_{13}H_{27}CO_2)C{=}]_2$ (64)	335
C_{16}	n-$C_{15}H_{31}COCl$	Na/ether/reflux	$[n\text{-}C_{15}H_{31}(n\text{-}C_{15}H_{31}CO_2)C{=}]_2$ (70)	335
C_{18}	n-$C_{17}H_{35}COCl$	Na/ether/reflux	$[n\text{-}C_{17}H_{35}(n\text{-}C_{17}H_{35}CO_2)C{=}]_2$ (67)	335
	C. Mixed Acyloin Condensations between Esters			
C_2, C_7	$CH_3CO_2C_2H_5$ + $C_6H_5CO_2C_2H_5$	Na/C_6H_6/reflux	$C_6H_5CHOHCOCH_3$ (6)	74
	$(C_2H_5)_2NCOCO_2C_2H_5$ + $C_6H_5CO_2C_2H_5$	Na/C_6H_6/reflux	$(C_2H_5)_2NCOCOCOC_6H_5$ (7)	74
C_3, C_7	$C_2H_5CO_2C_2H_5$ + $C_6H_5CO_2C_2H_5$	Na/ether, C_6H_6, or xylene/reflux	$C_2H_5CHOHCOC_6H_5$ (16–23)	74
C_4, C_7	n-$C_3H_7CO_2C_2H_5$ + $C_6H_5CO_2C_2H_5$	Na/C_6H_6/reflux	n-$C_3H_7CHOHCOC_6H_5$ (18)	74
	$(C_2H_5)_2NCO(CH_2)_2CO_2C_2H_5$ + $C_6H_5CO_2C_2H_5$	Na/C_6H_6/reflux	$(C_2H_5)_2NCO(CH_2)_2COCOC_6H_5$ (1)	74
C_4, C_{11}	n-$C_3H_7CO_2C_2H_5$ + n-$C_{10}H_{21}CO_2C_2H_5$	Na/xylene/reflux	n-$C_3H_7COCHOHC_{10}H_{21}$-n (41)	75
C_5, C_6	$C_2H_5O(CH_2)_4CO_2C_2H_5$ + n-$C_5H_{11}CO_2C_2H_5$	Na/xylene/reflux	$C_2H_5O(CH_2)_4COCHOHC_5H_{11}$-$n$ (—)	212, 280
C_6, C_6	n-$C_5H_{11}CO_2C_2H_5$ + $CH_2{=}CH(CH_2)_3CO_2C_2H_5$	Na/xylene/reflux	n-$C_5H_{11}COCHOH(CH_2)_3CH{=}CH_2$ (—)	212, 280
C_6, C_{11}	$CH_3O(CH_2)_5CO_2C_2H_5$ + $CH_2{=}CH(CH_2)_8CO_2C_2H_5$	Na/xylene/reflux	$CH_3O(CH_2)_5COCHOH(CH_2)_8CH{=}CH_2$ (—)	282
C_7, C_8	n-$C_6H_{13}CO_2C_2H_5$ + $C_2H_5O(CH_2)_7CO_2C_2H_5$	Na/xylene/reflux	n-$C_6H_{13}COCHOH(CH_2)_7OC_2H_5$ (—)	212
C_8, C_9	$C_2H_5O(CH_2)_7CO_2C_2H_5$ + n-$C_8H_{17}CO_2C_2H_5$	Na/xylene/reflux	$C_2H_5O(CH_2)_7COCHOHC_8H_{17}$-$n$ (—)	212

Note: References 184–356 are on pp. 400–403.

TABLE II. Cyclic Acyloins from Diesters

	Starting Material	Metal/Solvent/Temp (°C)	Product(s) (% Yield)	Refs.
A. Carbocyclic Acyloins				
4-Membered Rings				
C_8	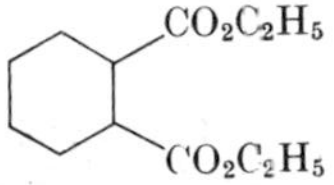 $CO_2C_2H_5$, $CO_2C_2H_5$	Na/xylene/reflux	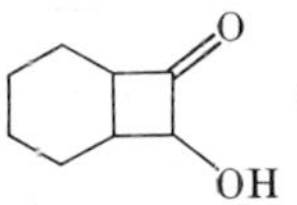 O, OH (12)	81
	$CH_3O_2CC(CH_3)_2C(CH_3)_2CO_2CH_3$	Na-K	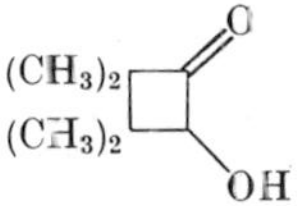 $(CH_3)_2$, $(CH_3)_2$, O, OH (35)	84
C_{10}	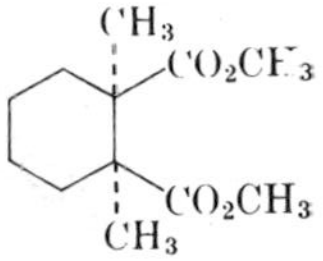 CH_3, CO_2CH_3, CO_2CH_3, CH_3	Na-NH_3/ether/−78°	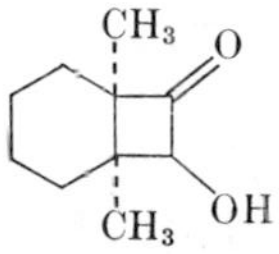 CH_3, O, OH, CH_3 (70)	261
C_{12}	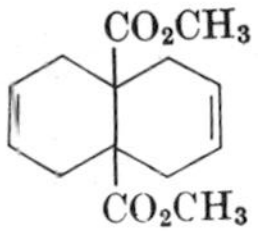 CO_2CH_3, CO_2CH_3	Na-K/C_6H_5/reflux	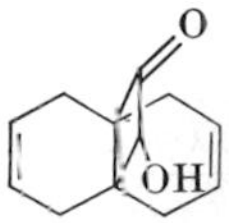 O, OH (39–76)	54
		Na-K/xylene/reflux	" (26–32)	54

Note: References 184–356 are on pp. 400–403.

TABLE II. Cyclic Acyloins from Diesters (*Continued*)

A. Carbocyclic Acyloins (*Continued*)

4-Membered Rings (Continued)

	Starting Material	Metal/Solvent/Temp (°C)	Product(s) (% Yield)	Refs.
		$Na\text{-}NH_3$/ether	" (10), [structure] (36), [structure] (21), [structure] (6).	54
	dl-$C_2H_5O_2CCH[C_4H_9\text{-}t]\text{-}CH[C_4H_9\text{-}t]CO_2C_2H_5$[a]	Na/toluene/reflux	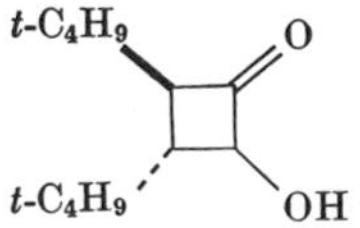 [structure: $t\text{-}C_4H_9$, $t\text{-}C_4H_9$, O, OH]	90
		Na/xylene/reflux	" (30)	90
	5-Membered Rings			
C_5	$CH_3O_2C(CH_2)_3CO_2CH_3$	Na/toluene/reflux	$\overline{(CH_2)_3COCHOH}$ (13–16)	92, 289
		$Na\text{-}NH_3$/ether	" ("Fair")	243
C_7	$CH_3O_2CCH_2C(CH_3)_2CH_2CO_2CH_3$	Na/NH_3/ether	$\overline{CH_2C(CH_3)_2CH_2COCO}$ (52–81)	319, 326, 243
	$C_2H_5O_2CCH_2C(CH_3)_2CH_2CO_2C_2H_5$	$Na\text{-}NH_3$/ether	" (39–52)[b]	318

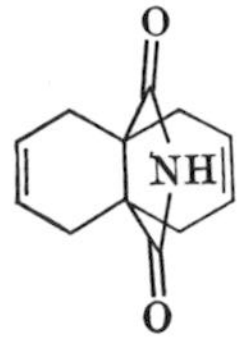

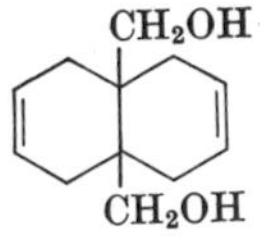

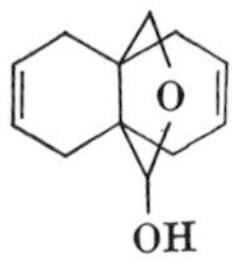

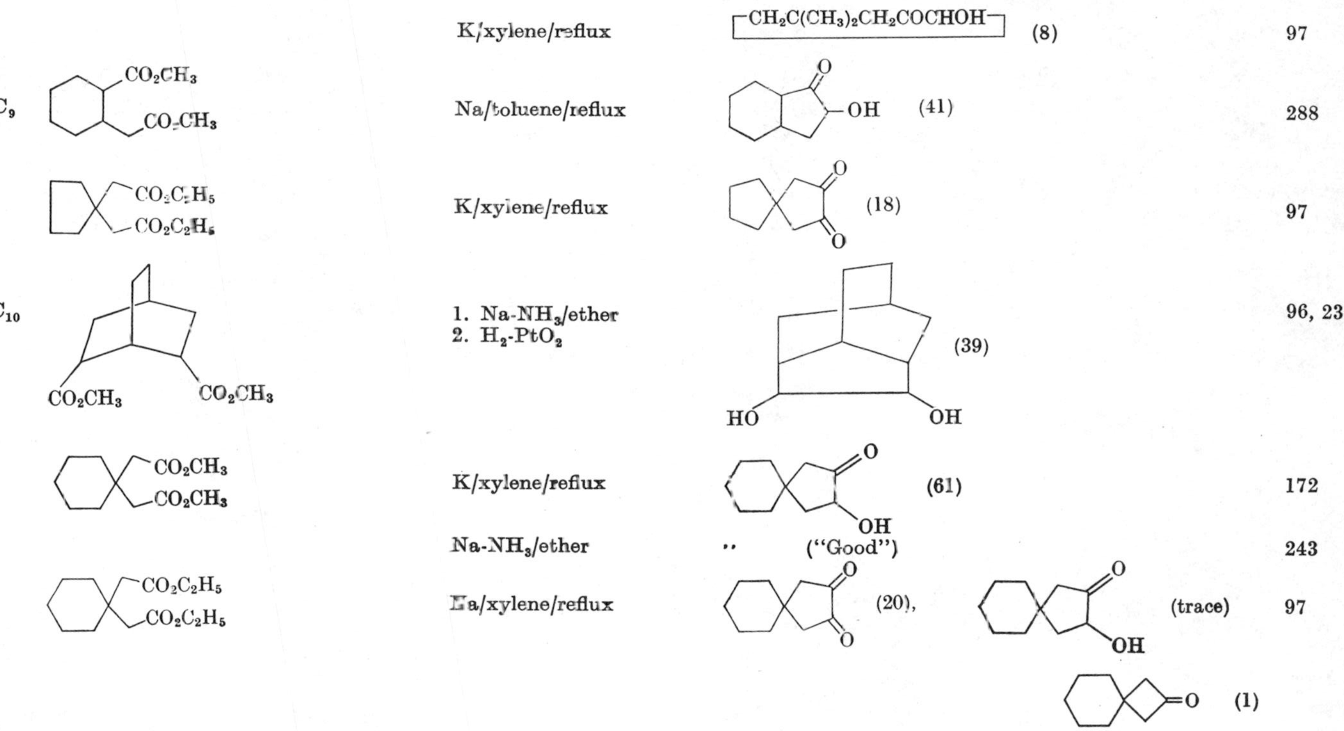

Formula	Reactant	Conditions	Product(s) and Yield(s) (%)	Refs.
		K/xylene/reflux	$\ulcorner CH_2C(CH_3)_2CH_2COCHOH \urcorner$ (8)	97
C_9	[structure; CO_2CH_3, CO_2CH_3]	Na/toluene/reflux	[structure; OH] (41)	288
	[structure; $CO_2C_2H_5$, $CO_2C_2H_5$]	K/xylene/reflux	[structure] (18)	97
C_{10}	[structure; CO_2CH_3, CO_2CH_3]	1. Na-NH_3/ether 2. H_2-PtO_2	[structure; HO, OH] (39)	96, 239
	[structure; CO_2CH_3, CO_2CH_3]	K/xylene/reflux	[structure; OH] (61)	172
		Na-NH_3/ether	" ("Good")	243
	[structure; $CO_2C_2H_5$, $CO_2C_2H_5$]	Na/xylene/reflux	[structure] (20), [structure; OH] (trace), [structure] (1)	97

Note: References 184–356 are on pp. 400–403.

[a] The *meso* form does not react.

[b] The product contained some α-diketone.

TABLE II. CYCLIC ACYLOINS FROM DIESTERS (*Continued*)

A. Carbocyclic Acyloins (Continued)

5-Membered Rings (Continued)

Starting Material	Metal/Solvent/Temp (°C)	Product(s) (% Yield)	Refs.
C_{10} (*Contd.*)			
$CH_2{=}C[(CH_2)_3CO_2R]_2$, R = CH_3, C_2H_5	Na/toluene/-	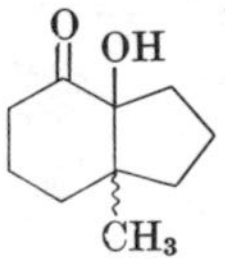 (50)[c,d], 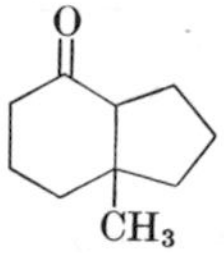 (2), 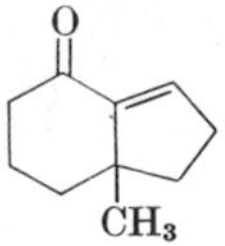 (2)	217, 118
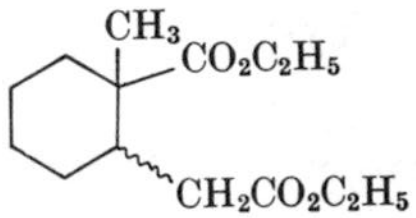	Na-NH_3/ether	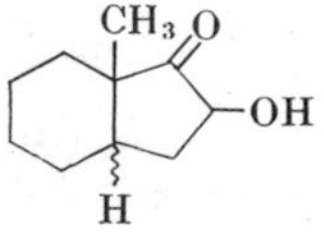 (50), 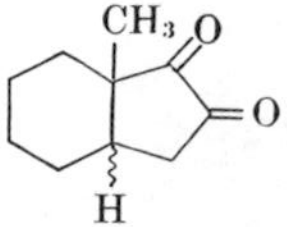 (14)	254
	Na/toluene/reflux	" (30–42), " (12)	254
$CH_3OC{=}[(CH_2)_3CO_2C_2H_5]_2$	Na	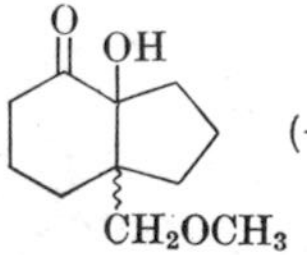 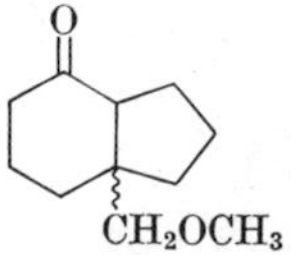 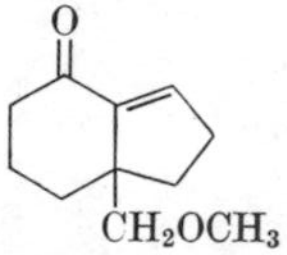 (—), (—), (—)	118
C_{12} 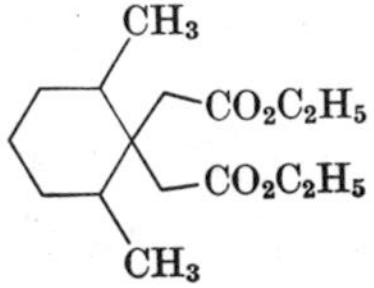	Na-K/xylene/reflux	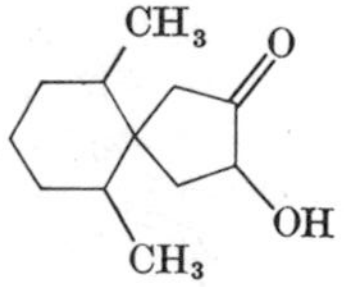 (26), 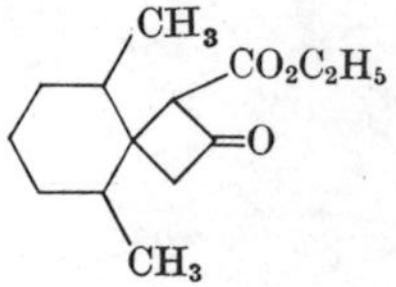 (5), 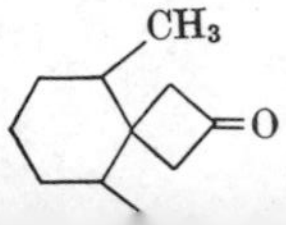 (24)	95

	Reactant	Conditions	Product (Yield, %)	Refs.
C_{13}	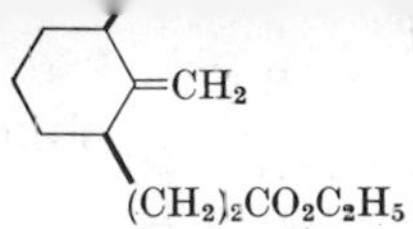 $=CH_2$; $(CH_2)_2CO_2C_2H_5$	Na	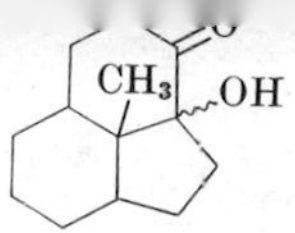 CH_3, OH (30)	118
	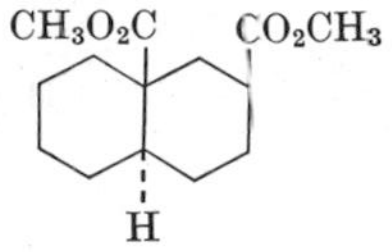 CH_3O_2C, CO_2CH_3, H	Na-NH_3/ether	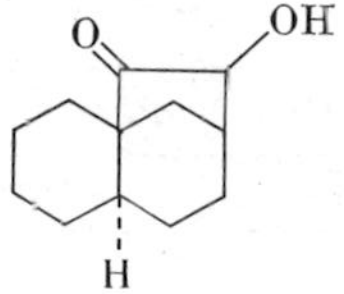 O, OH, H ("Low")	186
C_{18}	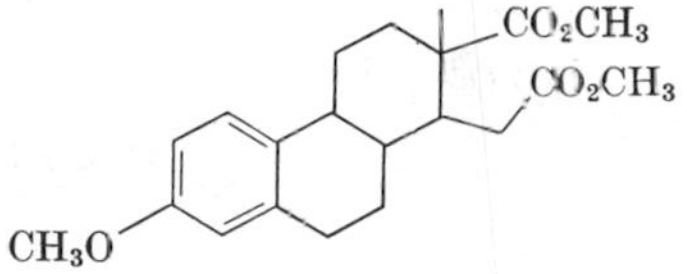 CO_2CH_3, CO_2CH_3, CH_3O	Na-NH_3/ether	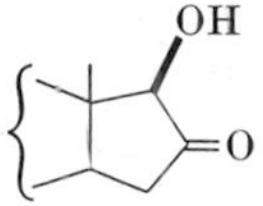 OH, =O (60–96)	169, 181
		,,	,, (91)	249
		1. Na-NH_3/ether 2. $(CH_3CO)_2O$	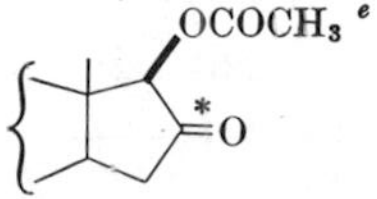 $OCOCH_3$ [e], *, =O (77)	96

Note: References 184–356 are on pp. 400–403.

[c] The original work incorrectly assigned the structure. See ref. 217b.
[d] A nine-membered ring is also formed. See p. 33[illegible].
[e] The primary carboxyl group in the starting material was labeled to provide ^{14}C at position 16.

TABLE II. Cyclic Acyloins from Diesters (*Continued*)

A. Carbocyclic Acyloins (*Continued*)

5-Membered Rings (Continued)

	Starting Material	Metal/Solvent/Temp (°C)	Product(s) (% Yield)	Refs.
C_{19}	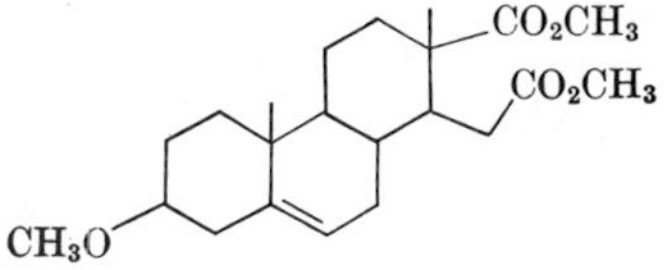 CH_3O, CO_2CH_3, CO_2CH_3	Na-NH_3/ether	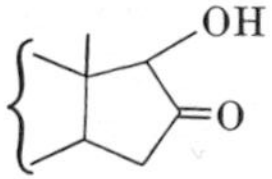 OH, O (–)	244
	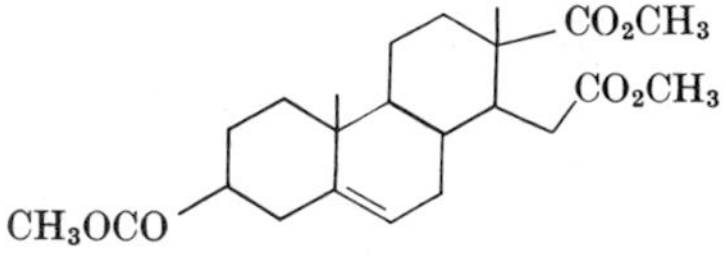 CH_3OCO, CO_2CH_3, CO_2CH_3	Na-NH_3/ether	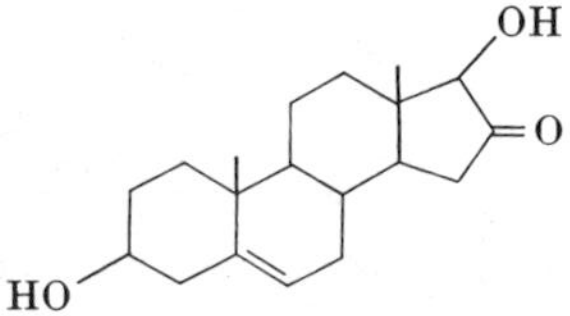 HO, OH, O (–)	244
	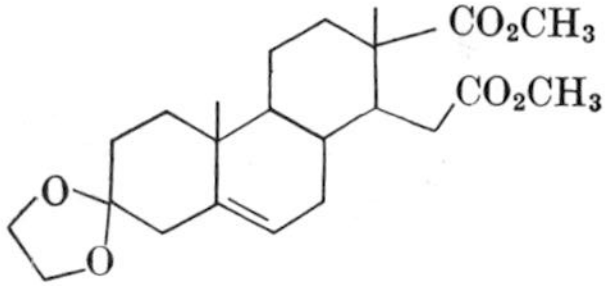 O, O, CO_2CH_3, CO_2CH_3	Na-NH_3/ether	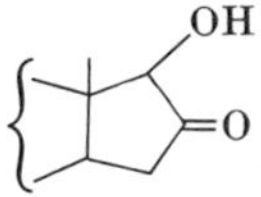 OH, O (–)	244

6-Membered Rings

	Starting Material	Metal/Solvent/Temp (°C)	Product(s) (% Yield)	Refs.
C_6	$CH_3O_2C(CH_2)_4CO_2CH_3$	Na/toluene/reflux	$\ulcorner(CH_2)_4COCHOH\urcorner$ (57)	92
		Na-NH_3/ether	" ("Fair")	243
	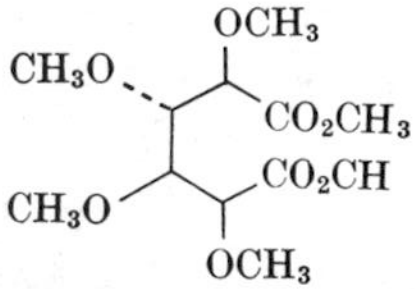 CH_3O, OCH_3, CO_2CH_3, CO_2CH, CH_3O, OCH_3	Na/xylene/reflux	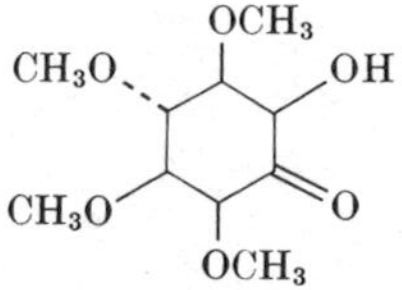 CH_3O, OCH_3, OH, O, CH_3O, OCH_3 (51)	231

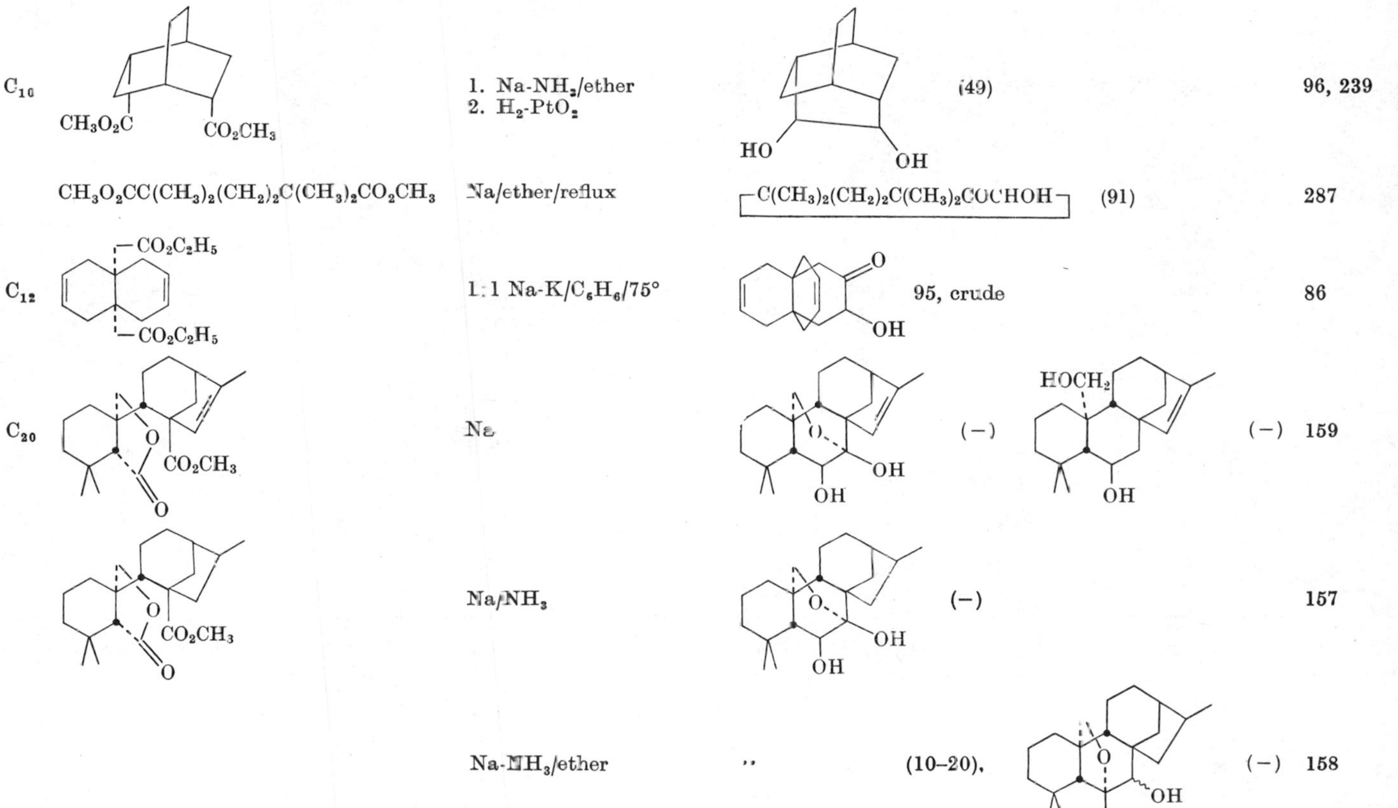

	Reactant	Conditions	Product(s) and Yield(s) (%)	Refs.
C_{10}	CH_3O_2C, CO_2CH_3	1. $Na\text{-}NH_3$/ether 2. $H_2\text{-}PtO_2$	HO, OH (49)	96, 239
	$CH_3O_2CC(CH_3)_2(CH_2)_2C(CH_3)_2CO_2CH_3$	Na/ether/reflux	$C(CH_3)_2(CH_2)_2C(CH_3)_2COCHOH$ (91)	287
C_{12}	$CO_2C_2H_5$, $CO_2C_2H_5$	1:1 $Na\text{-}K/C_6H_6$/75°	O, OH 95, crude	86
C_{20}	O, CO_2CH_3, O	Na	O, OH, OH (−) $HOCH_2$, OH (−)	159
	O, CO_2CH_3, O	Na/NH_3	O, OH, OH (−)	157
		$Na\text{-}NH_3$/ether	" (10–20), O, OH, OH (−)	158

Note: References 184–356 are on pp. 400–403.

TABLE II. Cyclic Acyloins from Diesters (*Continued*)

A. Carbocyclic Acyloins (*Continued*)

6-Membered Rings (Continued)

Starting Material	Metal/Solvent/Temp (°C)	Product(s) (% Yield)	Refs.
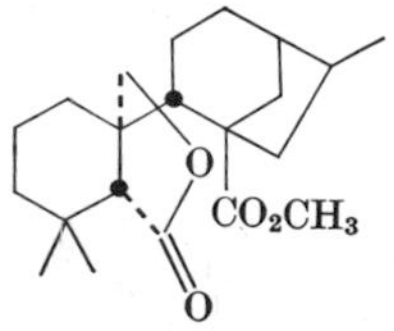 O, CO_2CH_3, O		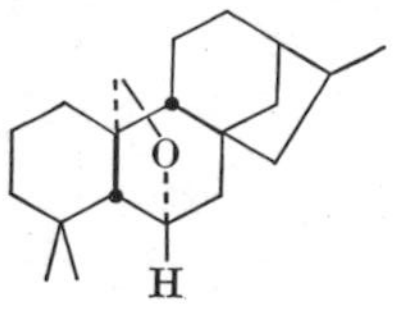 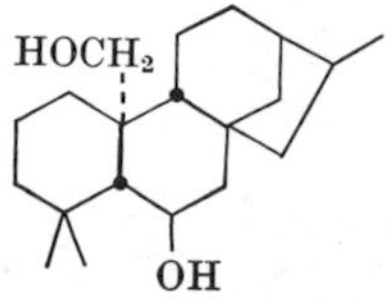 H (−); $HOCH_2$, OH (−)	
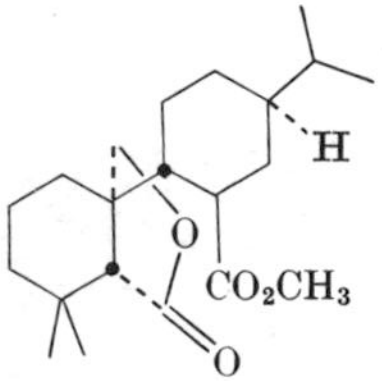 H, O, CO_2CH_3, O	Na-NH_3/ether	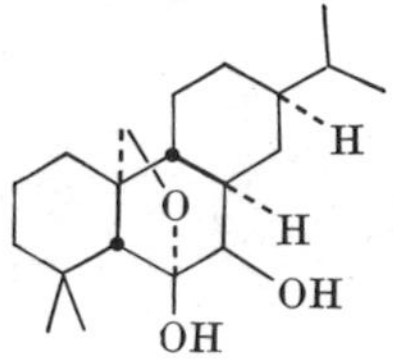 H, H, O, OH, OH (−)	157
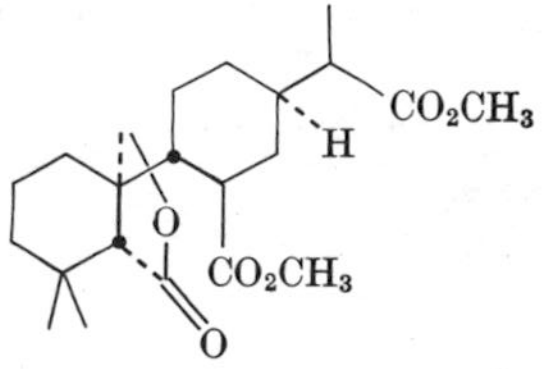 CO_2CH_3, H, O, CO_2CH_3, O	Na-NH_3/ether	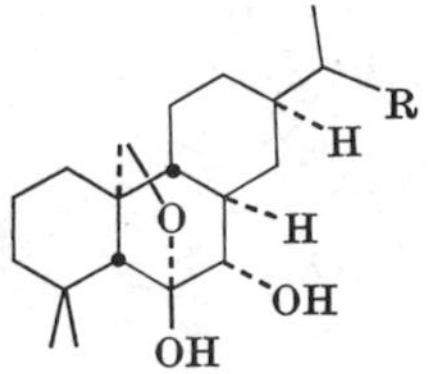 R, H, O, H, OH, OH (−) (R = CO_2CH_3, CH_2OH)	157

	Reactant	Conditions	Product(s) (Yield, %)	Refs.
C_{21}	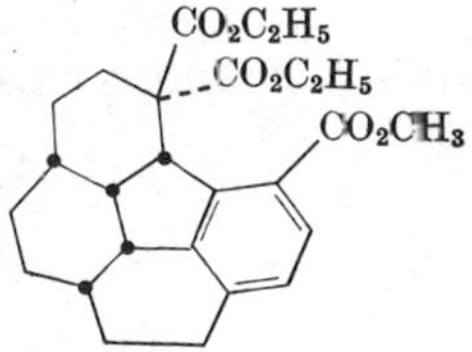 $CO_2C_2H_5$, $CO_2C_2H_5$, CO_2CH_3	$Na\text{-}NH_3$/ether/$-65°$	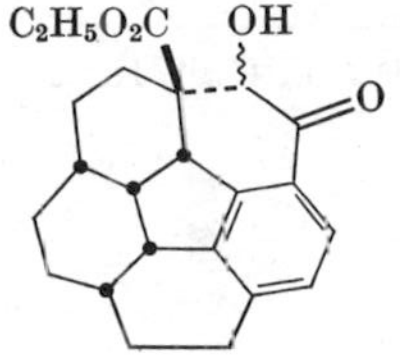 $C_2H_5O_2C$, OH, O (48)[f]	80
C_{24}	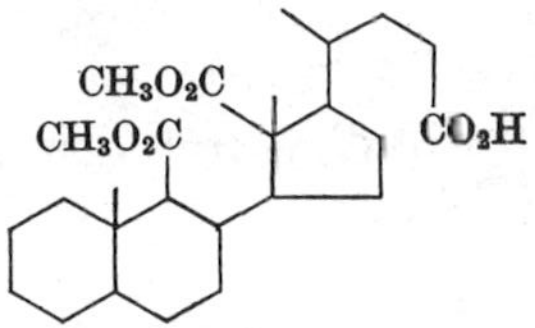 CH_3O_2C, CH_3O_2C, CO_2H	$Na\text{-}NH_3$/ether	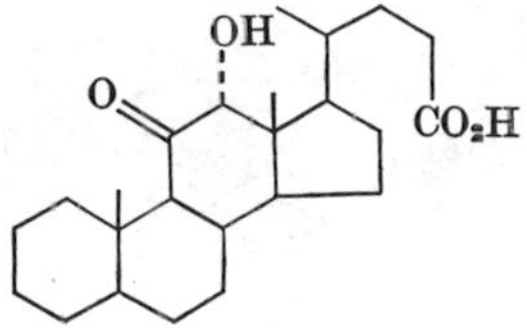 OH, O, CO_2H (80)	169
C_{28}	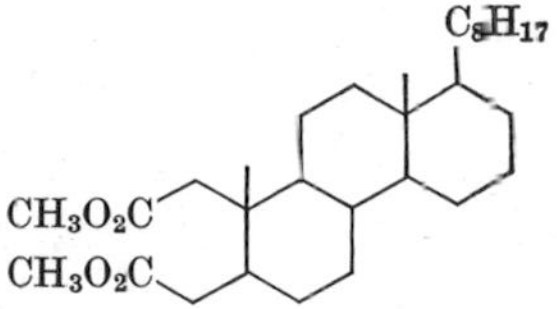 C_8H_{17}, CH_3O_2C, CH_3O_2C	$Na\text{-}NH_3$/ether	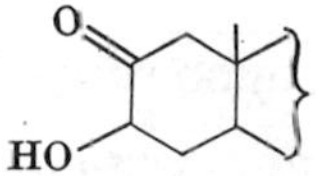 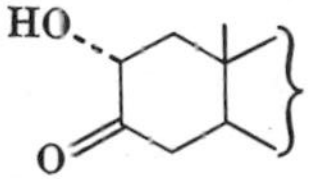 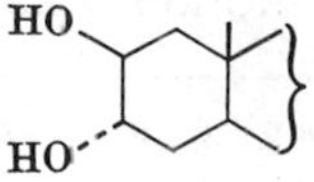 O, HO (82), HO, O (5), HO, HO (1.4)	168

Note: References 184–356 are on pp. 400–403.

[f] The ester group is reduced if excess sodium is used.

TABLE II. Cyclic Acyloins from Diesters (*Continued*)

A. Carbocyclic Acyloins (Continued)

7-Membered Rings

	Starting Material	Metal/Solvent/Temp (°C)	Product(s) (% Yield)	Refs.
C_7	$C_2H_5O_2C(CH_2)_5CO_2C_2H_5$	Na/xylene/reflux	$\ulcorner(CH_2)_5COCHOH\urcorner$ (52)	289
		Na	$\ulcorner(CH_2)_5COCHOH\urcorner$ (—)	102
		Na/xylene/reflux[g]	$\ulcorner(CH_2)_5COCHOH\urcorner$ + $\ulcorner(CH_2)_5CO\dot{C}O\urcorner$ (15) + $\ulcorner(CH_2)_5CO\urcorner$ (13)	98, 216
		Na/xylene/reflux	$\ulcorner(CH_2)_5COCHOH\urcorner$ + $\ulcorner(CH_2)_5COCO\urcorner$ (23)	98
C_8	$CH_2{=}C[(CH_2)_2CO_2C_2H_5]_2$	(—)	C=C1CCC(=O)C(O)CC1 (12) + C=C1CCC(=O)CC1 (27) + C=C1CCC(=O)C(C(=O)OCC)C1 (12)	118
C_{11}	$CH_3O_2CC(CH_3)_2(CH_2)_3C(CH_3)_2CO_2CH_3$	Na/xylene/reflux	$\ulcorner C(CH_3)_2(CH_2)_3C(CH_3)_2COCHOH\urcorner$ (69)	287

	Reactant	Conditions	Product (Yield, %)		Refs.
C_{14}	O O CO_2CH_3 CO_2CH_3	Na/xylene/reflux	O O O OH	(34)	269
C_{16}	CH_3O CH_3O CH_3O O CO_2CH_3 O	Na-NH_3/ether	O OH OH	(9.2)	156
C_{21}	$C_6H_5CH_2$, $CH_2C_6H_5$: $C_2H_5O_2CCH(CH_2)_3CHCO_2C_2H_5$	Na/xylene/reflux	$C_6H_5CH_2$, $CH_2C_6H_5$: $\overline{CH(CH_2)_3CHCOCHOH}$	(87)	223
	8-Membered Rings				
C_8	$CH_3O_2C(CH_2)_6CO_2CH_3$	Na/xylene/reflux	$\overline{(CH_2)_6CO—CHOH}$	(37–56)	100–104
	$C_2H_5O_2C(CH_2)_6CO_2C_2H_5$	Na/toluene-xylene/reflux	"	(35)	101
C_9	$(CH_2)_2CO_2CH_3$ $(CH_2)_2CO_2CH_3$	Na/xylene/reflux	O OH	(75)	105
C_{12}	$CH_3O_2CC(CH_3)_2(CH_2)_4C(CH_3)_2CO_2CH_3$	Na/xylene/105°	$\overline{C(CH_3)_2(CH_2)_4C(CH_3)_2COCHOH}$	(33)	287

Note: References 184–356 are on pp. 400–403.

[g] A commercial 40% sodium dispersion was used.

TABLE II. Cyclic Acyloins from Diesters (*Continued*)

A. Carbocyclic Acyloins (Continued)

9-Membered Rings

	Starting Material	Metal/Solvent/Temp (°C)	Product(s) (% Yield)	Refs.
C_9	$CH_3O_2C(CH_2)_7CO_2CH_3$	Na/xylene/reflux	$\overline{(CH_2)_7COCHOH}$ (9–97)	71, 321, 72, 279, 226, 325, 103, 305, 5
	$C_2H_5O_2C(CH_2)_7CO_2C_2H_5$	Na/xylene/reflux	" (5)	257, 298, 312, 102, 232
	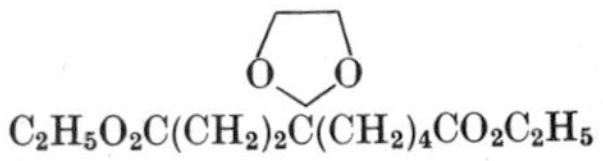 $C_2H_5O_2C(CH_2)_2C(CH_2)_4CO_2C_2H_5$ (C ketal-protected as 1,3-dioxolane)	Na/xylene/reflux	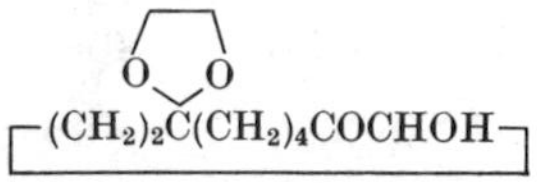 $\overline{(CH_2)_2C(CH_2)_4COCHOH}$ (C as 1,3-dioxolane) ("Low")	232
	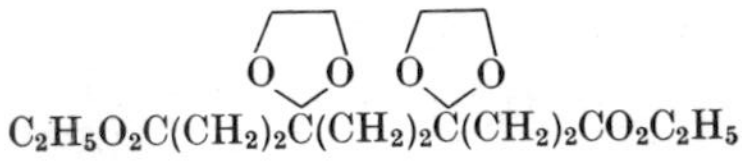 $C_2H_5O_2C(CH_2)_2C(CH_2)_2C(CH_2)_2CO_2C_2H_5$ (both C as 1,3-dioxolanes)	Na/xylene/reflux	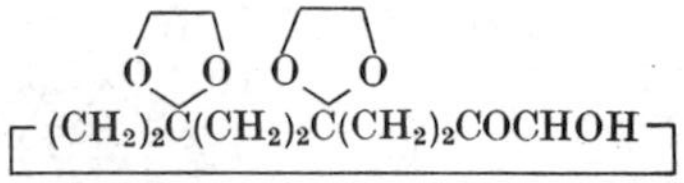 $\overline{(CH_2)_2C(CH_2)_2C(CH_2)_2COCHOH}$ (both C as 1,3-dioxolanes) ("Low")	232
C_{10}	$CH_3O_2CCH_2CH(CH_3)(CH_2)_5CO_2CH_3$	Na/xylene/reflux	$\overline{CH_2CH(CH_3)(CH_2)_5COCHOH}$ (33)	316
	$CH_3O_2C(CH_2)_3CH(CH_3)(CH_2)_3CO_2CH_3$	Na/5:1 xylene-toluene/reflux	$\overline{(CH_2)_3CH(CH_3)(CH_2)_3COCHOH}$ (60)	300
	$CH_2{=}C[(CH_2)_3CO_2C_2H_5]_2$	Na	(10)[h]	118

C_{11}	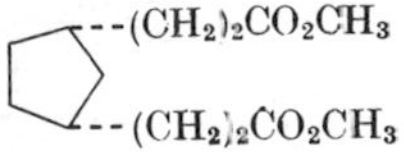 [cyclopentane ring] $-(CH_2)_2CO_2CH_3$, $-(CH_2)_2CO_2CH_3$	Na/dioxane/reflux	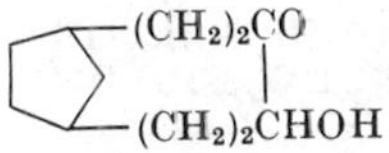 [cyclopentane ring] $-(CH_2)_2CO$ / $-(CH_2)_2CHOH$ (43)	57
	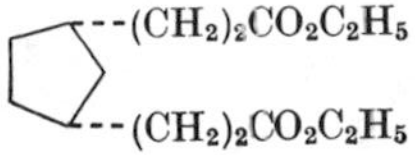 [cyclopentane ring] $-(CH_2)_2CO_2C_2H_5$, $-(CH_2)_2CO_2C_2H_5$	Na/xylene/reflux	" (10)	57
	$C_2H_5O_2C(CH_2)_3C(CH_3)_2(CH_2)_3CO_2C_2H_5$	Na/3:1 xylene-toluene/reflux	$\overline{(CH_2)_3C(CH_3)_2(CH_2)_3COCHOH}$ (66–70)	221
C_{13}	$CH_3O_2CCH_2C(CH_3)_2(CH_2)_3C(CH_3)_2$-$CH_2CO_2CH_3$	Na/toluene/reflux	$\overline{CH_2C(CH_3)_2(CH_2)_3C(CH_3)_2CH_2COCHOH}$ (84)	299
		1. Na 2. $Cu(O_2CCH_3)_2$	$\overline{CH_2C(CH_3)_2(CH_2)_3C(CH_3)_2CH_2COCO}$ (83)	250
	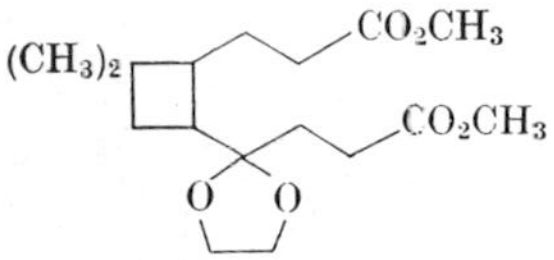 $(CH_3)_2$[cyclobutane] with CO_2CH_3, CO_2CH_3, O O [dioxolane]	—	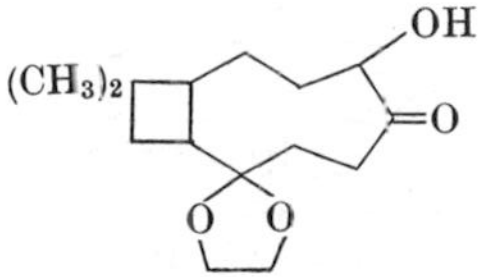 $(CH_3)_2$[cyclobutane-fused ring] OH, =O, O O [dioxolane]	184
C_{21}	$(C_6H_5)_2C[(CH_2)_3CO_2CH_3]_2$	Na/xylene/reflux	$\overline{(CH_2)_3C(C_6H_5)_2(CH_2)_3COCHOH}$ (40)	276
	$(C_6H_5)_2C[(CH_2)_3CO_2C_2H_5]_2$	Na/xylene/reflux	" (35)	296

Note: References 184–356 are on pp. 400–403.

[h] Five-membered ring compounds are also formed. See p. 322.

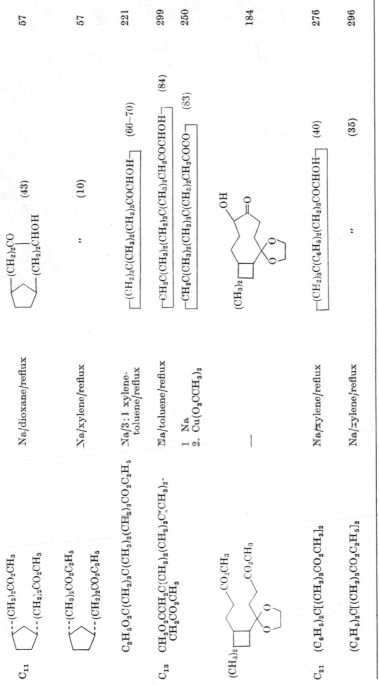

TABLE II. Cyclic Acyloins from Diesters (*Continued*)

A. Carbocyclic Acyloins (*Continued*)

10-Membered Rings

	Starting Material	Metal/Solvent/Temp (°C)	Product(s) (% Yield)	Refs.
C_{10}	*cis*-$CH_3O_2C(CH_2)_3CH{=}CH(CH_2)_3$-$CO_2CH_3$	Na/toluene/reflux	*cis*-$(CH_2)_3CH{=}CH(CH_2)_3COCHOH$ (ring) (78)	35
	trans-$CH_3O_2C(CH_2)_3CH{=}CH(CH_2)_3$-$CO_2CH_3$	Na/xylene/reflux	*trans*-$(CH_2)_3CH{=}CH(CH_2)_3COCHOH$ (ring) (51)	285
	$CH_3O_2C(CH_2)_8CO_2CH_3$	Na/xylene/reflux	$(CH_2)_8COCHOH$ (ring) (24–67)	262, 194, 226, 236, 72, 279, 71, 33, 18, 34, 278, 178, 291, 292, 327, 103, 5
	$C_2H_5O_2C(CH_2)_8CO_2C_2H_5$	Na/xylene/reflux	$(CH_2)_8COCHOH$ (ring) (45–65)	102, 257, 278, 312, 298
		Na/xylene/reflux[g]	" (48)	98
		Na/xylene/reflux[i]	" (<5), $(CH_2)_8COCO$ (ring) (10), $HO_2C(CH_2)_8CO_2H$ (52–62)	236
		Na/xylene/reflux[j]	$(CH_2)_8COCO$ (ring)[k] + $(CH_2)_8CO$ (ring) (14), $HO(CH_2)_{10}OH$ (3), $HO_2C(CH_2)_8CO_2H$ (17), $HO_2C(CH_2)_8CHOH(CH_2)_8CO(CH_2)_8$-$CHOH(CH_2)_8CO_2H$ (49)	58, 59

	Reactant	Conditions	Product (Yield %)		Refs.
	$n\text{-}C_4H_9O_2C(CH_2)_8CO_2C_4H_9\text{-}n$	Na/xylene/reflux	$\ulcorner(CH_2)_8COCHOH\urcorner$ (cyclic)	(60–65)	278
	$CH_3O_2CCH_2CD_2(CH_2)_4CD_2CH_2CO_2CH_3$	Na/xylene/reflux	$\ulcorner CH_2CD_2(CH_2)_4CD_2CH_2COCHOH\urcorner$ (cyclic)	(72)	230
C_{18}	2,2′-Biphenyl with $-(CH_2)_2CO_2CH_3$ on each ring	1. Na/xylene/reflux 2. Bi_2O_3	2,2′-Biphenyl with $-(CH_2)_2CO$ and $-(CH_2)_2CO$ joined (diketone)	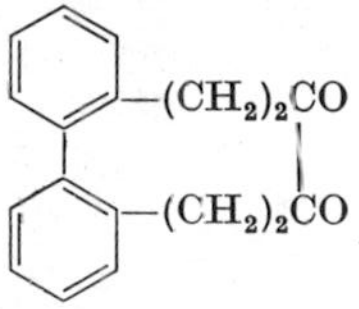 (25)	302
	11-Membered Rings				
C_{11}	$RO_2C(CH_2)_9CO_2R$ ($R = CH_3, C_2H_5, n\text{-}C_4H_9$)	Na/xylene/reflux	$\ulcorner(CH_2)_9COCHOH\urcorner$ (cyclic)	(61–71)	226, 71, 290, 327, 257, 102, 312, 298, 5, 262
C_{15}	$C(CH_3)_2(CH_2)_2CH(CH_3)CO_2CH_3$ | $(CH_2)_3CH(CH_3)CH_2CO_2CH_3$	Na/xylene/reflux	$C(CH_3)_2(CH_2)_2CH(CH_3)CO$ | $(CH_2)_3CH(CH_3)CH_2CHOH$ (cyclic)	(49)	306
	$(CH_2)_2CH(CH_3)CH_2CO_2CH_3$ | $C(CH_3)_2(CH_2)_3CH(CH_3)CO_2CH_3$	Na/xylene/reflux	$(CH_2)_2CH(CH_3)CH_2CHOH$ | $C(CH_3)_2(CH_2)_3CH(CH_3)CO$ (cyclic)	(24)	268
	$C(CH_3)_2(CH_2)_2CH(CH_3)CH_2CO_2CH_3$ | $(CH_2)_3CH(CH_3)CO_2CH_3$	Na/xylene/reflux	$C(CH_3)_2(CH_2)_2CH(CH_3)CH_2CO$ | $(CH_2)_3CH(CH_3)CHOH$ (cyclic)	(37)	306

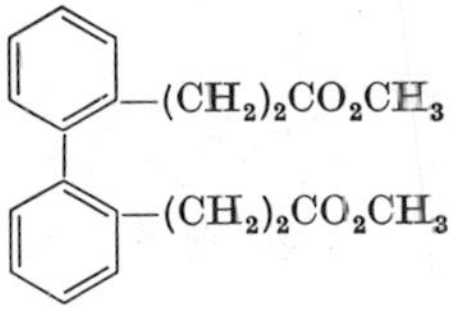

Note: References 184–356 are on pp. 400–403.

[i] The nitrogen atmosphere contained 4% oxygen.
[j] Diester (250 g) dissolved in 350 ml of xylene was refluxed with excess sodium.
[k] Cyclic acyloins of 19 and 20 members are also formed (see p. 340 and 341).

TABLE II. Cyclic Acyloins from Diesters (*Continued*)

A. Carbocyclic Acyloins (Continued)

	Starting Material	Metal/Solvent/Temp (°C)	Product(s) (% Yield)	Refs.
	12-Membered Rings			
C_{12}	$CH_3O_2C(CH_2)_4C{\equiv}C(CH_2)_4CO_2CH_3$	Na/toluene/reflux	$\overline{(CH_2)_4C{\equiv}C(CH_2)_4COCHOH}$ (ring) (73)	35
	$RO_2C(CH_2)_{10}CO_2R$ ($R = CH_3$, C_2H_5)	Na/xylene/reflux	$\overline{(CH_2)_{10}COCHOH}$ (ring) (64–89)	262, 193, 72, 71, 257, 102, 327, 312, 5
	13-Membered Rings			
C_{13}	$RO_2C(CH_2)_{11}CO_2R$ ($R = CH_3$, C_2H_5)	Na/toluene/reflux	$\overline{(CH_2)_{11}COCHOH}$ (ring) (68–82)	262, 257, 102, 327, 312, 5
C_{15}	p-C_6H_4[$(CH_2)_4CO_2CH_3$][$(CH_2)_3CO_2CH_3$]	Na/xylene/reflux	p-C_6H_4[$(CH_2)_4CO$][$(CH_2)_3CHOH$] (CO–CHOH bonded, ring) (35–36)	37, 286, 301
	$CH_3O_2CCH_2$–(cyclohexane-1,4-diyl)–$(CH_2)_6CO_2CH_3$	Na/toluene/reflux	$\overline{CH_2\text{–}(C_6H_{10})\text{–}(CH_2)_6COCHOH}$ (ring) (75)	286

14-Membered Rings

	Substrate	Conditions	Product	Yield (%)	References
C_{14}	$RO_2C(CH_2)_{12}CO_2R$ ($R = CH_3$, C_2H_5)	Na/xylene/reflux	$\ulcorner(CH_2)_{12}COCHOH\urcorner$ (cyclic)	(47–82)	262, 72, 233, 322, 257, 312, 102, 327, 266, 5
C_{16}	p-$C_6H_4[(CH_2)_4CO_2CH_3]_2$	Na/xylene/reflux	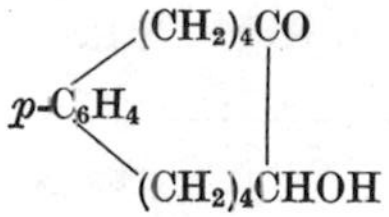 p-C_6H_4 bridged by $(CH_2)_4CO$—$(CH_2)_4CHOH$	(70–76)	308, 270, 286, 225, 283, 293
C_{18}	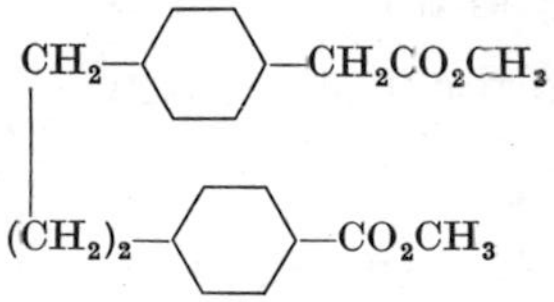 CH_2—C_6H_{10}—$CH_2CO_2CH_3$ / $(CH_2)_2$—C_6H_{10}—CO_2CH_3 (CH₂ and (CH₂)₂ joined)	Na/xylene/reflux	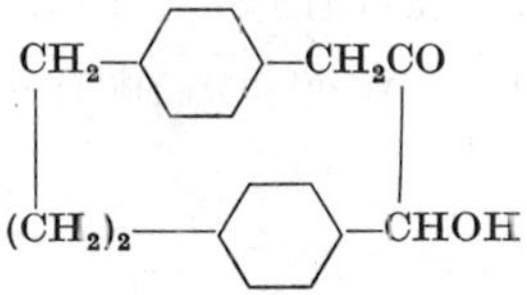 CH_2—C_6H_{10}—CH_2CO / $(CH_2)_2$—C_6H_{10}—$CHOH$ (cyclic)	(~4)	218
	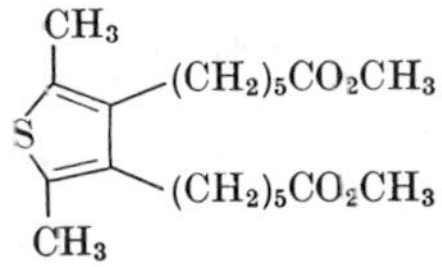 2,5-di-CH_3-thiophene-3,4-di[$(CH_2)_5CO_2CH_3$]	Na-K/xylene-ether/ 60–65°	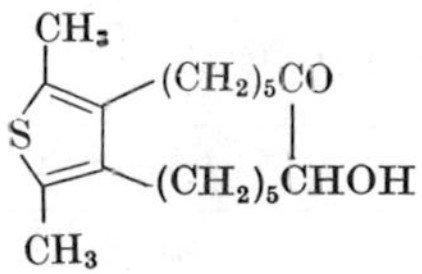 2,5-di-CH_3-thiophene bridged by $(CH_2)_5CO$—$(CH_2)_5CHOH$	(70–72)	117, 171

Note: References 184–356 are on pp. 400–403.

TABLE II. CYCLIC ACYLOINS FROM DIESTERS (*Continued*)

A. Carbocyclic Acyloins (Continued)

14-Membered Rings (Continued)

	Starting Material	Metal/Solvent/Temp (°C)	Product(s) (% Yield)	Refs.
C_{20}	$p\text{-}C_6H_4$ with $(CH_2)_2C(CH_3)_2CH_2CO_2CH_3$ and $(CH_2)_2C(CH_3)_2CH_2CO_2CH_3$	Na/toluene/reflux	$p\text{-}C_6H_4$ with $(CH_2)_2C(CH_3)_2CH_2CO$ and $(CH_2)_2C(CH_3)_2CH_2CHOH$ (ring) [l] (31)	277
	"	Na/xylene/reflux	" (35–38)	208
	$(CH_2)_2CH(CH_3)(CH_2)_3CH(CH_3)CO_2CH_3$ / $CH(C_3H_7\text{-}i)(CH_2)_2CH(CH_3)CH_2CO_2CH_3$	Na/toluene/reflux	$(CH_2)_2CH(CH_3)(CH_2)_3CH(CH_3)CO$ / $CH(C_3H_7\text{-}i)(CH_2)_2CH(CH_3)CH_2CHOH$ (ring) (89)	265, 247

15-Membered Rings

	Starting Material	Metal/Solvent/Temp (°C)	Product(s) (% Yield)	Refs.
C_{15}	$RO_2C(CH_2)_{13}CO_2R$ ($R = CH_3$, C_2H_5)	Na/xylene/reflux	$\overline{(CH_2)_{13}COCHOH}$ (ring) (64–93)	71, 196, 310, 327, 199, 189, 257
	$CH_3O_2C(CH_2)_{13}CO_2CH_3$	Na/xylene-toluene/reflux	$(CH_2)_{13}COCHOH$ (ring) (93)	262
C_{16}	$CH_3O_2CCH(CH_3)(CH_2)_{12}CO_2CH_3$	Na/xylene/reflux	$CH(CH_3)(CH_2)_{12}COCHOH$ (ring) (72)	245, 9
	$C_2H_5O_2CCH(CH_3)(CH_2)_{12}CO_2C_2H_5$	Na/xylene/reflux	" (—)	257
	$C_2H_5O_2CCH_2CH(CH_3)(CH_2)_{11}CO_2C_2H_5$	Na/xylene/reflux	$CH_2CH(CH_3)(CH_2)_{11}COCHOH$ (ring) (74)	272, 9

	ethylene ketal of $CH_2CH(CH_3)(CH_2)_2CO_2C_2H_5$ / $(CH_2)_8CO_2C_2H_5$	Na/xylene/reflux	ethylene ketal of $CH_2CH(CH_3)(CH_2)_2CO$—$(CH_2)_8$—$CHOH$	(63)	251
C_{19}	$(CH_2)_3$ —$CH_2CO_2CH_3$ —$CH_2CO_2CH_3$	Na/xylene/reflux	$(CH_2)_3$ —CH_2CO —CH_2CHOH	(15–26)	209, 320
	16-Membered Rings				
C_{16}	$RO_2C(CH_2)_{14}CO_2R$ ($R = CH_3$, C_2H_5)	Na/xylene/reflux	$(CH_2)_{14}COCHOH$ (cyclic)	(65–95)	262, 257, 71, 195, 72, 327, 102, 5, 324
C_{18}	p-$C_6H_4[(CH_2)_5CO_2CH_3]_2$	Na/xylene/reflux	p-C_6H_4 $(CH_2)_5CO$ $(CH_2)_5CHOH$	(69)	218
C_{20}	CH_2 —$(CH_2)_3CO_2CH_3$ —$(CH_2)_2CO_2CH_3$	Na/xylene/reflux	CH_2 —$(CH_2)_3CO$ —$(CH_2)_2CHOH$	(trace)	284
	$(CH_2)_4$ —$CH_2CO_2CH_3$ —$CH_2CO_2CH_3$	Na/xylene/reflux	$(CH_2)_4$ —CH_2CO —CH_2CHOH	(25)	220

Note: References 184–356 are on pp. 400–403.

[l] A 28-membered diacyloin is also formed (see p. 342).

TABLE II. Cyclic Acyloins from Diesters (*Continued*)

A. Carbocyclic Acyloins (Continued)

17-Membered Rings

	Starting Material	Metal/Solvent/Temp (°C)	Product(s) (% Yield)	Refs.
C_{17}	*cis*-$C_2H_5O_2C(CH_2)_6CH{=}CH(CH_2)_7$-$CO_2C_2H_5$	Na/xylene/reflux	(25) [*cis*-$(CH_2)_6CH{=}CH(CH_2)_7COCHOH$] (ring) (72)	295, 311
	$RO_2C(CH_2)_{15}CO_2R$ ($R = CH_3, C_2H_5$)	Na/xylene/reflux	[$(CH_2)_{15}COCHOH$] (ring) (69–92)	262, 311, 257, 315, 327, 294, 5
	1,3-dioxolane-2,2-bis[$(CH_2)_7CO_2CH_3$]	Na/xylene/reflux	1,3-dioxolane-2,2-diyl ring: $(CH_2)_7CO$ / $(CH_2)_7CHOH$ (77)	240, 7, 9
C_{21}	$CH_2[p\text{-}C_6H_4\text{—}(CH_2)_3CO_2CH_3]_2$	Na/xylene/reflux	CH_2 bridged ring: $p\text{-}C_6H_4\text{—}(CH_2)_3CO$ / $p\text{-}C_6H_4\text{—}(CH_2)_3CHOH$ (17–28)	210, 284
	$(CH_2)_3[p\text{-}C_6H_4\text{—}(CH_2)_2CO_2CH_3]_2$	Na/xylene/reflux	$(CH_2)_3$ bridged ring: $p\text{-}C_6H_4\text{—}(CH_2)_2CO$ / $p\text{-}C_6H_4\text{—}(CH_2)_2CHOH$ (53)	211
	$(CH_2)_5$[cyclohexane-$CH_2CO_2CH_3$]$_2$	Na/xylene/reflux	$(CH_2)_5$ bridged ring: cyclohexane-CH_2CO / cyclohexane-CH_2CHOH (47)	209

18-Membered Rings				
C_{18}	$RO_2C(CH_2)_5C{\equiv}C(CH_2)_9CO_2R$ ($R = CH_3, C_2H_5$)	Na/xylene/reflux	$\ulcorner(CH_2)_5C{\equiv}C(CH_2)_9COCHOH\urcorner$ (70)	303, 314
	$CH_3O_2C(CH_2)_{16}CO_2CH_3$	Na/xylene/reflux	$\ulcorner(CH_2)_{16}COCHOH\urcorner$ (64)	236, 266
		Na/xylene/reflux[i]	$\ulcorner(CH_2)_{16}COCHOH\urcorner$ (32), $HO_2C(CH_2)_{16}CO_2H$ (43)	236
	$C_2H_5O_2C(CH_2)_{16}CO_2C_2H_5$	Na/xylene/reflux	$\ulcorner(CH_2)_{16}COCHOH\urcorner$ (83)	257, 327, 5
C_{22}	$CH_2[p\text{-}C_6H_4(CH_2)_3CO_2CH_3][p\text{-}C_6H_4(CH_2)_4CO_2CH_3]$	Na/xylene/reflux	$CH_2[p\text{-}C_6H_4(CH_2)_3CO][p\text{-}C_6H_4(CH_2)_4CHOH]$ (ring) (46)	284
	$(CH_2)_2[p\text{-}C_6H_4(CH_2)_3CO_2C_2H_5]_2$	Na/xylene/reflux	$(CH_2)_2[p\text{-}C_6H_4(CH_2)_3CO][p\text{-}C_6H_4(CH_2)_3CHOH]$ (ring) (58)	224
	$(CH_2)_4[p\text{-}C_6H_4(CH_2)_2CO_2CH_3]_2$	Na/xylene/reflux	$(CH_2)_4[p\text{-}C_6H_4(CH_2)_2CO][p\text{-}C_6H_4(CH_2)_2CHOH]$ (ring) (37)	210

Note: References 184–356 are on pp. 400–403.

[i] The nitrogen atmosphere contained 4% oxygen.

TABLE II. Cyclic Acyloins from Diesters (*Continued*)

A. Carbocyclic Acyloins (*Continued*)

18-Membered Rings (Continued)

	Starting Material	Metal/Solvent/Temp (°C)	Product(s) (% Yield)	Refs.
C_{22} (*contd.*)	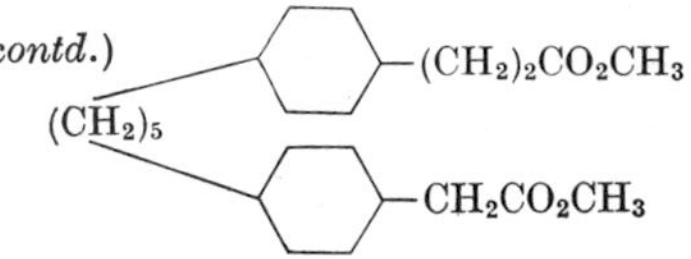 $(CH_2)_5$ bridging two cyclohexane rings bearing $-(CH_2)_2CO_2CH_3$ and $-CH_2CO_2CH_3$	Na/xylene/reflux	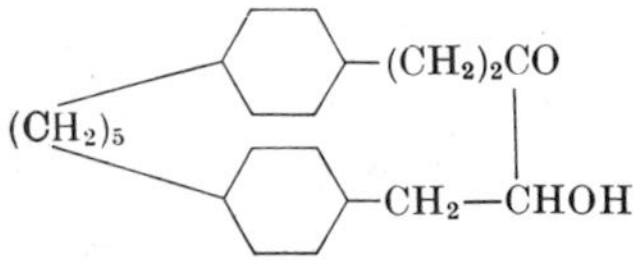 $(CH_2)_5$ bridging two cyclohexane rings joined by $-(CH_2)_2CO$—CH_2—$CHOH$ (62)	209
	$C_2H_5O_2CC(CH_3)_2(CH_2)_{14}C(CH_3)_2$-$CO_2C_2H_5$	Na/xylene/reflux	ring: $C(CH_3)_2(CH_2)_{14}C(CH_3)_2COCHOH$ (–)	287
C_{26}	$C_2H_5O_2C(CH_2)_3CH$—$(CH_2)_8$—$CH(CH_2)_3$-$CO_2C_2H_5$, the two CH also joined by $(CH_2)_8$	Na/xylene/reflux	CH and CH joined by $(CH_2)_3COCHOH(CH_2)_3$, $(CH_2)_8$, $(CH_2)_8$ [m] (–)	107

19-Membered Rings

	Starting Material	Metal/Solvent/Temp (°C)	Product(s) (% Yield)	Refs.
C_{10}	$C_2H_5O_2C(CH_2)_8CO_2C_2H_5$	Na/xylene/reflux[j]	ring: $(CH_2)_8CO(CH_2)_8COCHOH$ [n,o] (15)	58, 59
C_{23}	CH_2 bearing p-C_6H_4—$(CH_2)_4CO_2CH_3$ and p-C_6H_4—$(CH_2)_4CO_2CH_3$	Na/xylene/reflux	CH_2 bearing p-C_6H_4—$(CH_2)_4CO$ and p-C_6H_4—$(CH_2)_4CHOH$, CO–CHOH joined (62)	284
	$(CH_2)_5$ bearing p-C_6H_4—$(CH_2)_2CO_2CH_3$ and p-C_6H_4—$(CH_2)_2CO_2CH_3$	Na/xylene/reflux	$(CH_2)_5$ bearing p-C_6H_4—$(CH_2)_2CO$ and p-C_6H_4—$(CH_2)_2CHOH$, CO–CHOH joined (71)	222

	Reactant	Conditions	Product (Yield, %)		Refs.
	20-Membered Rings				
C_{10}	$C_2H_5O_2C(CH_2)_8CO_2C_2H_5$	Na/xylene/reflux[j]	$\ulcorner(CH_2)_8COCHOH(CH_2)_8COCHOH\urcorner$[p]	(0.5)	58, 59
C_{20}	$CH_3O_2C(CH_2)_{18}CO_2CH_3$	Na/xylene/reflux	$\ulcorner(CH_2)_{18}COCHOH\urcorner$	(96)	72, 5
C_{24}	$p\text{-}C_6H_4\text{—}(CH_2)_2CO_2CH_3$ $(CH_2)_6$ $p\text{-}C_6H_4\text{—}(CH_2)_2CO_2CH_3$	Na/xylene/reflux	$p\text{-}C_6H_4\text{—}(CH_2)_2CO$ $(CH_2)_6$ $p\text{-}C_6H_4\text{—}(CH_2)_2CHOH$	(70)	210
	$p\text{-}C_6H_4\text{—}(CH_2)_5CO_2CH_3$ CH_2 $p\text{-}C_6H_4\text{—}(CH_2)_4CO_2CH_3$	Na/xylene/reflux	$p\text{-}C_6H_4\text{—}(CH_2)_5CO$ CH_2 $p\text{-}C_6H_4\text{—}(CH_2)_4CHOH$	(51)	284
	Rings of 21 or More Members				
C_{25}	$p\text{-}C_6H_4\text{—}(CH_2)_5CO_2CH_3$ CH_2 $p\text{-}C_6H_4\text{—}(CH_2)_5CO_5CH_3$	Na/xylene/reflux	$p\text{-}C_6H_4\text{—}(CH_2)_5CO$ CH_2 $p\text{-}C_6H_4\text{—}(CH_2)_5CHOH$	(80)	284
C_{26}	$p\text{-}C_6H_4\text{—}(CH_2)_5CO_2C_2H_5$ $(CH_2)_2$ $p\text{-}C_6H_4\text{—}(CH_2)_5CO_2C_2H_5$	Na/xylene/reflux	$p\text{-}C_6H_4\text{—}(CH_2)_5CO$ $(CH_2)_2$ $p\text{-}C_6H_4\text{—}(CH_2)_5CHOH$	(51)	224

Note: References 184–356 are on pp. 400–403.

[j] Diester (250 g) dissolved in 350 ml of xylene was refluxed with excess sodium.
[m] The *cis* ester gave *cis* "in, in" isomer. The *trans* ester gave the "in, out" isomer.
[n] A 10-membered α-diketone is also produced (see p. 332).
[o] A 20-membered diacyloin is also produced (See p. 341).
[p] A 10-membered α-diketone and a 19-membered acyloin are also produced (see p. 332 and 340).

TABLE II. Cyclic Acyloins from Diesters (*Continued*)

A. Carbocyclic Acyloins (*Continued*)

Rings of 21 or More Members (Continued)

	Starting Material	Metal/Solvent/Temp (°C)	Product(s) (% Yield)	Refs.
C_{25}	$C_2H_5O_2C(CH_2)_{23}CO_2C_2H_5$	Na/1,6-bis-(dicyclohexylamine)-hexane or 1,10-bis-(dicyclohexylamine)-decane or 1,6-bis-(diisopropylmethoxy)-hexane or 1,1,10,10-tetra-*p*-tolyldecane	cyclic $(CH_2)_{23}COCHOH$ + cyclic $(CH_2)_{23}COCO$ (~50)	345
C_{26}	$C_2H_5O_2C(CH_2)_{24}CO_2C_2H_5$	Na/1,10-bis(dicyclohexylamine)decane or 1,1,10,10-tetra-*p*-tolyldecane	cyclic $(CH_2)_{24}COCHOH$ + cyclic $(CH_2)_{24}COCO$ (~50)	345
C_{20}	p-$C_6H_4[(CH_2)_2C(CH_3)_2CH_2CO_2CH_3]_2$	Na/toluene/reflux	cyclic dimer: p-C_6H_4[$(CH_2)_2C(CH_3)_2CH_2COCHOHCH_2C(CH_3)_2(CH_2)_2$]$_2$$C_6H_4$-$p$ [q] (—)	277
C_{34}	$(CH_2)_2[p$-C_6H_4—$(CH_2)_9CO_2C_2H_5]_2$	Na/xylene/reflux	cyclic $(CH_2)_2$[p-C_6H_4—$(CH_2)_9CO$][p-C_6H_4—$(CH_2)_9CHOH$] (24)	224
C_{36}	$C_2H_5O_2C(CH_2)_{34}CO_2C_2H_5$	Na/1:1 xylene-*cyclo*-$C_{34}H_{65}D_5$/reflux	cyclic $(CH_2)_{32}COCHOH$ (5–20), catenane of cyclo-$C_{34}H_{65}D_5$ and cyclic $(CH_2)_{32}COCHOH$ (small amounts)	109

B. Heterocyclic Acyloins

	Reactant	Conditions	Product (Yield, %)		Refs.
Nitrogen Heterocycles: 6-Membered Rings					
C_9	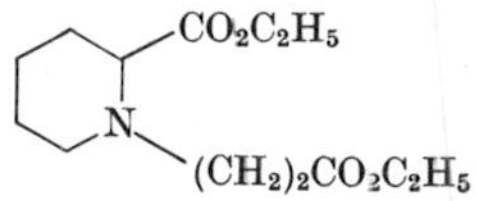 $CO_2C_2H_5$; N; $(CH_2)_2CO_2C_2H_5$	Na/toluene/reflux	O; OH; N	(trace)	110
	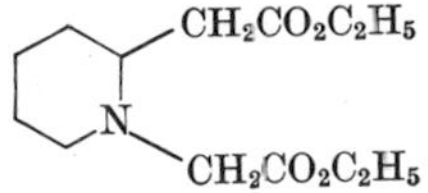 $CH_2CO_2C_2H_5$; N; $CH_2CO_2C_2H_5$	Na/toluene/reflux	O; OH; N	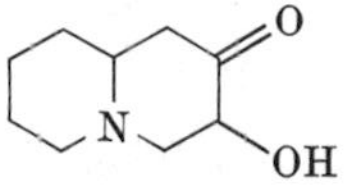 (3)	110
Nitrogen Heterocycles: 7-Membered Rings					
C_{11}	$CH_3N[CH_2C(CH_3)_2CO_2C_2H_5]_2$	Na/toluene/reflux	$C(CH_3)_2CH_2N(CH_3)CH_2C(CH_3)_2COCHOH$ (ring)	(66)	118
C_{12}	$C_6H_5N[(CH_2)_2CO_2C_2H_5]_2$	Na/xylene/reflux	$(CH_2)_2N(C_6H_5)(CH_2)_2COCHOH$ (ring)	(10)	111, 112
C_{14}	$t\text{-}C_4H_9N[CH_2C(CH_3)_2CO_2C_2H_5]_2$	Na/xylene/reflux	$C(CH_3)_2CH_2N(t\text{-}C_4H_9)CH_2C(CH_3)_2COCHOH$ (ring)	(80)	118
	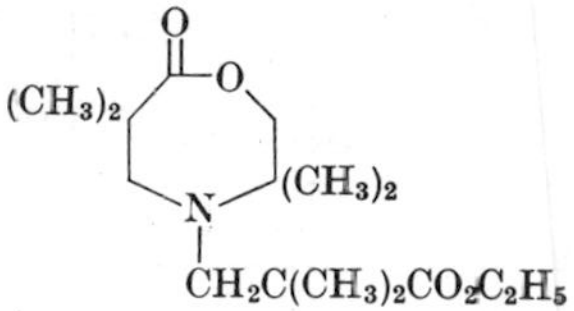 O; O; $(CH_3)_2$; $(CH_3)_2$; N; $CH_2C(CH_3)_2CO_2C_2H_5$	Na/toluene/reflux	OH; O; $(CH_3)_2$; $(CH_3)_2$; N; $C(CH_3)_2CH_2OH$	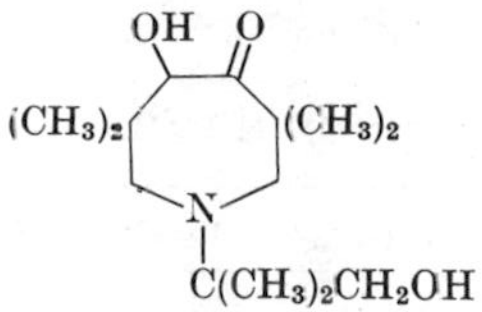 (62)	118

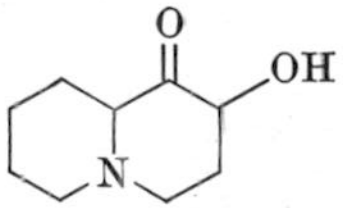

[a] A 14-membered acyloin is also formed (see p. 333).

Note: References 184–356 are on pp. 400–403.

TABLE II. Cyclic Acyloins from Diesters (*Continued*)

B. Heterocyclic Acyloins (*Continued*)

	Starting Material	Metal/Solvent/Temp (°C)	Product(s) (% Yield)	Refs.
	Nitrogen Heterocycles: 7-Membered Rings (Continued)			
C_{15}	$(CH_3)_3CCH_2N[CH_2C(CH_3)_2CO_2C_2H_5]_2$	Na/toluene/reflux	$\overline{C(CH_3)_2CH_2N[CH_2C(CH_3)_3]CH_2C(CH_3)_2COCHOH}$ (66)	118
C_{16}	$C_6H_5N[CH_2C(CH_3)_2CO_2C_2H_5]_2$	Na/toluene/reflux	$\overline{C(CH_3)_2CH_2N(C_6H_5)CH_2C(CH_3)_2COCHOH}$ (83)	118
	C_6H_{11}–$N[CH_2C(CH_3)_2CO_2C_2H_5]_2$ (cyclohexyl)	Na/toluene/reflux	C_6H_{11}–N ring: $CH_2C(CH_3)_2CO$ / $CH_2C(CH_3)_2CHOH$ (84)	118
	Nitrogen Heterocycles: 9-Membered Rings			
C_9	$CH_3N[(CH_2)_3CO_2C_2H_5]_2$	Na/xylene/reflux	$\overline{(CH_2)_3N(CH_3)(CH_2)_3COCHOH}$ (53–75)	111, 112, 14
C_{10}	$C_2H_5N[(CH_2)_3CO_2C_2H_5]_2$	Na/xylene/reflux	$\overline{(CH_2)_3N(C_2H_5)(CH_2)_3COCHOH}$ (60–73)	111, 112
C_{11}	C_3H_5–$N[(CH_2)_3CO_2C_2H_5]_2$ (cyclopropyl)	Na/xylene/reflux	C_3H_5–N ring: $(CH_2)_3CO$ / $(CH_2)_3CHOH$ (45)	113, 14
	i-$C_3H_7N[(CH_2)_3CO_2C_2H_5]_2$	Na/xylene/reflux	$\overline{(CH_2)_3N(i\text{-}C_3H_7)(CH_2)_3COCHOH}$ (60)	116
C_{12}	i-$C_4H_9N[(CH_2)_3CO_2C_2H_5]_2$	Na/xylene/reflux	$\overline{(CH_2)_3N(i\text{-}C_4H_9)(CH_2)_3COCHOH}$ (48)	114
	t-$C_4H_9N[(CH_2)_3CO_2C_2H_5]_2$	Na/xylene/reflux	$\overline{(CH_2)_3N(t\text{-}C_4H_9)(CH_2)_3COCHOH}$ (—)	116

C_{14}	$C_6H_5N[(CH_2)_3CO_2C_2H_5]_2$	Na/xylene/reflux	$\ulcorner(CH_2)_3N(C_6H_5)(CH_2)_3COCHOH\urcorner$ (3)	115
	Nitrogen Heterocycles: 9-Membered Rings (Continued)			
C_{14} (*contd.*)	C_6H_{11}–$N[(CH_2)_3CO_2C_2H_5]_2$ (cyclohexyl)	Na/xylene/reflux	C_6H_{11}–$N\langle(CH_2)_3CO / (CH_2)_3CHOH\rangle$ (ring) (50)	114
C_{15}	p-$CH_3C_6H_4N[(CH_2)_3CO_2C_2H_5]_2$	Na/xylene/reflux	$\ulcorner(CH_2)_3N(C_6H_4CH_3\text{-}p)(CH_2)_3COCHOH\urcorner$ (38)	115
C_{18}	C_4H_8N–C_6H_4–$N[(CH_2)_3CO_2C_2H_5]_2$ (pyrrolidinyl-phenyl)	Na/xylene/reflux	C_4H_8N–C_6H_4–$N\langle(CH_2)_3CO / (CH_2)_3CHOH\rangle$ (ring) (37)	115
	Nitrogen Heterocycles: Rings of 11 or More Members			
C_{11}	$CH_3N[(CH_2)_4CO_2C_2H_5]_2$	Na/xylene/reflux	$\ulcorner(CH_2)_4N(CH_3)(CH_2)_4COCHOH\urcorner$ (47–64)	111, 112, 14
C_{12}	$C_2H_5N[(CH_2)_4CO_2C_2H_5]_2$	Na/xylene/reflux	$\ulcorner(CH_2)_4N(C_2H_5)(CH_2)_4COCHOH\urcorner$ (64–85)	111, 112
C_{13}	$CH_3N[(CH_2)_5CO_2C_2H_5]_2$	Na/xylene/reflux	$\ulcorner(CH_2)_5N(CH_3)(CH_2)_5COCHOH\urcorner$ (86)	112
C_{14}	$C_2H_5N[(CH_2)_5CO_2C_2H_5]_2$	Na/xylene/reflux	$\ulcorner(CH_2)_5N(C_2H_5)(CH_2)_5COCHOH\urcorner$ (77)	112
C_{15}	$CH_3N[(CH_2)_6CO_2C_2H_5]_2$	Na/xylene/reflux	$\ulcorner(CH_2)_6N(CH_3)(CH_2)_6COCHOH\urcorner$ (83)	112

Note: References 184–356 are on pp. 400–403.

TABLE II. CYCLIC ACYLOINS FROM DIESTERS (*Continued*)

B. Heterocyclic Acyloins (*Continued*)

Nitrogen Heterocycles: Rings of 11 or more Members (Continued)

	Starting Material	Metal/Solvent/Temp(°C)	Product(s) (% Yield)		Refs.
C_{16}	$C_2H_5N[(CH_2)_6CO_2C_2H_5]_2$	Na/xylene/reflux	$(CH_2)_6N(C_2H_5)(CH_2)_6COCHOH$ (ring)	(88)	112
C_{17}	$CH_3N[(CH_2)_7CO_2C_2H_5]_2$	Na/xylene/reflux	$(CH_2)_7N(CH_3)(CH_2)_7COCHOH$ (ring)	(84)	112
C_{19}	$CH_3N[(CH_2)_8CO_2C_2H_5]_2$	Na/xylene/reflux	$(CH_2)_8N(CH_3)(CH_2)_8COCHOH$ (ring)	(83)	112
C_{24}	$C_2H_5N[(CH_2)_{10}CO_2C_2H_5]_2$	Na/xylene/reflux	$(CH_2)_{10}N(C_2H_5)(CH_2)_{10}COCHOH$ (ring)	(72)	112

Oxygen and Sulfur Heterocycles: 7-Membered Rings

	Starting Material	Metal/Solvent/Temp(°C)	Product(s) (% Yield)		Refs.
C_{10}	$O[CH_2C(CH_3)_2CO_2C_2H_5]_2$	Na/toluene/reflux	$C(CH_3)_2CH_2OCH_2C(CH_3)_2COCHOH$ (ring)	(25–30)	337, 118
		Na/xylene/reflux	"	(45–50)	238, 118
	$S[CH_2C(CH_3)_2CO_2C_2H_5]_2$	Na/toluene or xylene/ reflux	$C(CH_3)_2CH_2SCH_2C(CH_3)_2COCHOH$ (ring)	(75–80)	227, 118, 297, 338

Oxygen and Sulfur Heterocycles: Rings of 9 or More Members

C_{10}	2,5-[$(CH_2)_2CO_2C_2H_5$]$_2$-tetrahydrofuran	Na/xylene/reflux	tetrahydrofuran-2,5-bridged $(CH_2)_2CO$–$(CH_2)_2CHOH$	(13–52)	213
	$O[(CH_2)_4CO_2CH_3]_2$	Na/xylene/125–130°	$\ulcorner(CH_2)_4O(CH_2)_4COCHOH\urcorner$	(64)	271
C_{12}	$O[(CH_2)_5CO_2CH_3]_2$	Na/xylene/124–130°	$\ulcorner(CH_2)_5O(CH_2)_5COCHOH\urcorner$	(71)	271
	2,5-[$(CH_2)_3CO_2CH_3$]$_2$-thiophene	Na-K/xylene/60–65°	thiophene-2,5-bridged $(CH_2)_3CO$–$(CH_2)_3CHOH$	(41)	171
		Na/xylene-ether/55–60°	"	(29)	171
C_{13}	2-$(CH_2)_3CO_2CH_3$-5-$(CH_2)_4CO_2CH_3$-thiophene	Na-K/xylene/60–65°	thiophene-2,5-bridged $(CH_2)_3CO$–$(CH_2)_4CHOH$	(39)	171
C_{14}	2,5-[$(CH_2)_4CO_2CH_3$]$_2$-thiophene	Na/xylene-ether/55–60°	thiophene-2,5-bridged $(CH_2)_4CO$–$(CH_2)_4CHOH$	(29)	117, 171, 307
		Na-K/xylene-ether/55–60°	"	(41)	117, 171

Note: References 184–356 are on pp. 400–403.

TABLE II. Cyclic Acyloins from Diesters (*Continued*)

B. Heterocyclic Acyloins (Continued)

Oxygen and Sulphur Heterocycles: Rings of 9 or More Members (*Continued*)

	Starting Material	Metal/Solvent/Temp(°C)	Product(s) (% Yield)		Refs.
C_{16}	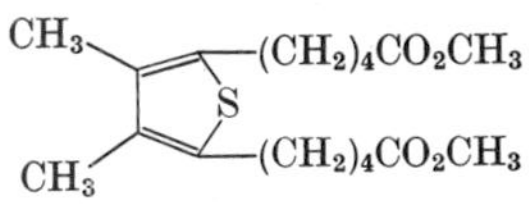 3,4-dimethyl-2,5-bis[$(CH_2)_4CO_2CH_3$]thiophene	Na-K/xylene-ether/55–60°	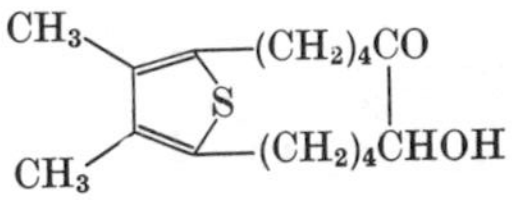 3,4-dimethylthiophene bridged 2,5 by $(CH_2)_4CO$–$(CH_2)_4CHOH$	(42)	117, 171
C_{15}	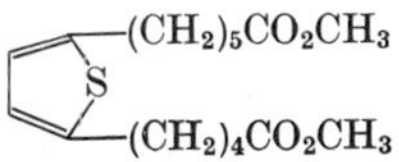 2-[$(CH_2)_5CO_2CH_3$]-5-[$(CH_2)_4CO_2CH_3$]thiophene	Na-K/xylene-ether/55–60°	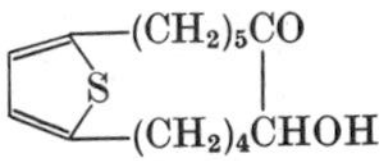 thiophene bridged 2,5 by $(CH_2)_5CO$–$(CH_2)_4CHOH$	(39)	117
C_{22}	$O[(CH_2)_{10}CO_2CH_3]_2$	Na/xylene/120–130°	cyclic $(CH_2)_{10}O(CH_2)_{10}COCHOH$	(56)	271

Silicon and Germanium Heterocycles

	Starting Material	Metal/Solvent/Temp(°C)	Product(s) (% Yield)		Refs.
C_8	$(CH_3)_2Si[(CH_2)_2CO_2CH_3]_2$	Na/toluene/reflux	cyclic $(CH_2)_2Si(CH_3)_2(CH_2)_2COCHOH$	(28)	235
C_9	$(CH_3)_2Si[(CH_2)_3CO_2CH_3][(CH_2)_2CO_2CH_3]$	Na/toluene/reflux	cyclic $(CH_2)_2Si(CH_3)_2(CH_2)_3COCHOH$	(52)	343

C_{10}	$(CH_3)_2Si[(CH_2)_3CO_2CH_3]_2$	Na/toluene/reflux	$\ulcorner(CH_2)_3Si(CH_3)_2(CH_2)_3COCHOH\urcorner$	(45)	343
	$(CH_3)_2Si[(CH_2)_3CO_2C_2H_5]_2$	"	"	(47)	235
C_{11}	$(CH_3)_2Si<(CH_2)_3CO_2CH_3, (CH_2)_4CO_2CH_3$	Na/toluene/reflux	$\ulcorner(CH_2)_3Si(CH_3)_2(CH_2)_4COCHOH\urcorner$	(49)	343
C_{12}	$(CH_3)_2Si[(CH_2)_4CO_2CH_3]_2$	Na/toluene/reflux	$\ulcorner(CH_2)_4Si(CH_3)_2(CH_2)_4COCHOH\urcorner$	(38)	343
			"	(65)	119
C_{14}	$(C_2H_5)_2Ge[(CH_2)_4CO_2C_2H_5]_2$	Na/xylene/reflux	$\ulcorner(CH_2)_4Ge(C_2H_5)_2(CH_2)_4COCHOH\urcorner$	(60)	119

C. Ferrocene Acyloins

C_{14}	$CH_3O_2CCH_2(C_5H_4\text{-}\pi)Fe(C_5H_4\text{-}\pi)CH_2\text{-}CO_2CH_3$	Na/xylene/reflux	$\ulcorner CH_2(C_5H_4\text{-}\pi)Fe(C_5H_4\text{-}\pi)CH_2COCHOH\urcorner$	(20)	121
C_{16}	$CH_3O_2C(CH_2)_2(C_5H_4\text{-}\pi)Fe(C_5H_4\text{-}\pi)\text{-}(CH_2)_2CO_2CH_3$	Na/xylene/reflux	$\ulcorner(CH_2)_2(C_5H_4\text{-}\pi)Fe(C_5H_4\text{-}\pi)(CH_2)_2COCHOH\urcorner$	(50)	121
C_{18}	$CH_3O_2C(CH_2)_3(C_5H_4\text{-}\pi)Fe(C_5H_4\text{-}\pi)\text{-}(CH_2)_3CO_2CH_3$	Na/xylene/reflux	$\ulcorner(CH_2)_3(C_5H_4\text{-}\pi)Fe(C_5H_4\text{-}\pi)(CH_2)_3COCHOH\urcorner$	(58)	121
C_{19}	$CH_3O_2C(CH_2)_3(C_5H_4\text{-}\pi)Fe(C_5H_4\text{-}\pi)\text{-}(CH_2)_4CO_2CH_3$	Na/xylene/reflux	$\ulcorner(CH_2)_3(C_5H_4\text{-}\pi)Fe(C_5H_4\text{-}\pi)(CH_2)_4COCHOH\urcorner$	(55)	121
C_{20}	$CH_3O_2C(CH_2)_4(C_5H_4\text{-}\pi)Fe(C_5H_4\text{-}\pi)\text{-}(CH_2)_4CO_2CH_3$	Na/xylene/reflux	$\ulcorner(CH_2)_4(C_5H_4\text{-}\pi)Fe(C_5H_4\text{-}\pi)(CH_2)_4COCHOH\urcorner$	(75)	121

Note: References 184–356 are on pp. 400–403.

TABLE III. STRAIGHT-CHAIN ACYLOIN CONDENSATIONS CARRIED OUT WITH TRIMETHYLCHLOROSILANE (TMCS)*

	Starting Material	Solvent[a]	Product(s) (% Yield)	Refs.
			A. From Esters	
C_2	$CH_3CO_2CH_3$	Toluene	$(CH_3)_3SiOC(CH_3){=}C(CH_3)OSi(CH_3)_3$ (—)	61
	$CH_3CO_2C_2H_5$	Ether	" (65–76)	16, 160, 203, 76
	$CH_3CO_2C_4H_9$-n	Ether	" (73)	16, 160, 203
	$CH_3CO_2Si(CH_3)_3$	Ether	" (73–76)	203
	$ClCH_2CO_2CH_3$	Ether	$(CH_3)_3SiCH_2C[OSi(CH_3)_3]{=}C[OSi(CH_3)_3]CH_2Si(CH_3)_3$ (21)	16
	$(CH_3)_3SiCH_2CO_2C_2H_5$	Ether, toluene, or xylene	$(CH_3)_3SiOC[CH_2Si(CH_3)_3]{=}C[CH_2Si(CH_3)_3]OSi(CH_3)_3$ (90)	16
C_3	$C_2H_5CO_2CH_3$	Ether	$(CH_3)_3SiOC(C_2H_5){=}C(C_2H_5)OSi(CH_3)_3$ (69)	160
	$C_2H_5CO_2C_2H_5$	Ether	" (63–91)	16, 160, 203, 76
	$C_2H_5CO_2C_4H_9$-n	Ether	" (56)	16, 160
	$C_2H_5O(CH_2)_2CO_2C_2H_5$	Ether, toluene, or xylene	$(CH_3)_3SiOC[(CH_2)_2OC_2H_5]{=}C[(CH_2)_2OC_2H_5]OSi(CH_3)_3$ (47)	16, 85
	$C_2H_5S(CH_2)_2CO_2C_2H_5$	Ether, toluene, or xylene	$(CH_3)_3SiOC[(CH_2)_2SC_2H_5]{=}C[(CH_2)_2SC_2H_5]OSi(CH_3)_3$ (69)	16, 85
	$(CH_3)_3Si(CH_2)_2CO_2CH_3$	Ether	$(CH_3)_3SiOC[(CH_2)_2Si(CH_3)_3]{=}C[(CH_2)_2Si(CH_3)_3]OSi(CH_3)_3$ (65)	16, 160
	$(C_2H_5)_2N(CH_2)_2CO_2C_2H_5$	Ether, toluene, or xylene	$(CH_3)_3SiOC[(CH_2)_2N(C_2H_5)_2]{=}C[(CH_2)_2N(C_2H_5)_2]OSi(CH_3)_3$ (62)	16, 85
C_4	n-$C_3H_7CO_2C_2H_5$	Ether	$(CH_3)_3SiOC(C_3H_7$-$n){=}C[C_3H_7$-$n)OSi(CH_3)_3$ (92)	16, 160, 203
		—	"	346
		Ether, $ClSi(C_6H_5)_3$	$(C_6H_5)_3SiOC(C_3H_7$-$n){=}C(C_3H_7$-$n)OSi(C_6H_5)_3$ (61)	76
	i-$C_3H_7CO_2C_2H_5$	Ether	$(CH_3)_3SiOC(C_3H_7$-$i){=}C(C_3H_7$-$i)OSi(CH_3)_3$ (81)	16, 160, 203
C_5	n-$C_4H_9CO_2C_2H_5$	Ether	$(CH_3)_3SiOC(C_4H_9$-$n){=}C(C_4H_9$-$n)OSi(CH_3)_3$ (64–77)	16, 160, 203, 76
	$(CH_3)_2CHCH_2CO_2C_2H_5$	Ether	$(CH_3)_3SiOC[CH_2CH(CH_3)_2]{=}C[CH_2CH(CH_3)_2]OSi(CH_3)_3$ (73)	16, 160, 203
	$(CH_3)_3CCO_2C_2H_5$	Ether	$(CH_3)_3SiOC[C(CH_3)_3]{=}C[C(CH_3)_3]OSi(CH_3)_3$ (42–68)	31, 16
	$CH_3C(OCH_2CH_2O)(CH_2)_2CO_2CH_3$	—	$CH_3C(OCH_2CH_2O)(CH_2)_2C[OSi(CH_3)_3]{=}C[OSi(CH_3)_3](CH_2)_2C(OCH_2CH_2O)CH_3$ (65)	347
C_6	$(CH_3)_2CH(CH_2)_2CO_2C_2H_5$	Ether	$(CH_3)_3SiOC[(CH_2)_2CH(CH_3)_2]{=}C[(CH_2)_2CH(CH_3)_2]OSi(CH_3)_3$ (72)	16, 160, 203
C_7	$C_6H_5CO_2C_2H_5$	Ether, toluene, or xylene	$(CH_3)_3SiOC(C_6H_5){=}C(C_6H_5)OSi(CH_3)_3$ (39)	16
	$C_6H_5CO_2Si(CH_3)_3$	Ether, toluene or xylene	" (41)	16

	$p\text{-}CH_3C_6H_4CO_2C_2H_5$	Ether, toluene or xylene	$(CH_3)_3SiOC[C_6H_4CH_3\text{-}p]{=}C(C_6H_4CH_3\text{-}p)OSi(CH_3)_3$ (41)	16
	$p\text{-}CH_3C_4H_4CO_2Si(CH_3)_3$	Ether, toluene or xylene	" (86)	16
	$C_6H_5CH_2CO_2C_2H_5$	Ether	$(CH_3)_3SiOC(CH_2C_6H_5){=}C(CH_2C_6H_5)OSi(CH_3)_3$ (I, 40), $C_6H_5CH{=}C(OC_2H_5)OSi(CH_3)_3$ (II, 11), $C_6H_5CH_2Si(CH_3)_3$ (III, 26)	16, 203
		Toluene[b]	I (48), II (0), III (16)	16, 203
		Toluene[c]	I (37), II (0), III (0)	16, 203
		C_6H_5	I (30), II (0), III (34)	16, 203
	$C_6H_5CH_2CO_2Si(CH_3)_3$	Ether	$(CH_3)_3SiOC[(CH_2C_6H_5]{=}C(CH_2C_6H_5)OSi(CH_3)_3$ (I, 21), $C_6H_5CH{=}C[OSi(CH_3)_3]_2$ (II, 15), $C_6H_5CH_2Si(CH_3)_3$ (III, 17), $(C_6H_5CH_2)_2CO$ (IV, 15)	16
		THF[b]	I (3), II (5), III (70), IV (10)	16
		THF	I (0), II (0), III (64), IV (0)	16
		Toluene[b]	I (47), II (2), III (9), IV (4)	16
		Toluene	I (15), II (5), III (21), IV (0)	16
	$n\text{-}C_7H_{15}CO_2C_2H_5$	Ether	$(CH_3)_3SiOC(C_7H_{15}\text{-}n){=}C(C_7H_{15}\text{-}n)OSi(CH_3)_3$ (80)	16, 160, 203
C_{11}	1-Adamantyl-$CO_2C_2H_5$	Ether	(1-adamantyl)$C(OSi(CH_3)_3){=}C(OSi(CH_3)_3)$(adamantyl-1) [1-adamantyl / OSi(CH₃)₃ above; (CH₃)₃SiO / adamantyl-1 below] (55)	31

B. From Acid Chlorides, Anhydrides and Lactones

C_2	CH_3COCl	Ether	$(CH_3)_3SiOC(CH_3){=}C(CH_3)OSi(CH_3)_3$ (11)	76
	$(CH_3CO)_2O$	Xylene	$(CH_3)_3SiOC(CH_3){=}C(CH_3)OSi(CH_3)_3$ (20)	77
C_3	C_2H_5COCl	Ether	$(CH_3)_3SiOC(C_2H_5){=}C(C_2H_5)OSi(CH_3)_3$ (15)	76
C_4	$n\text{-}C_3H_7COCl$	Ether	$(CH_3)_3SiOC(C_3H_7\text{-}n){=}C(C_3H_7\text{-}n)OSi(CH_3)_3$ (16)	76
	γ-Butyrolactone	Xylene	$(CH_3)_3SiO(CH_2)_3C(OSi(CH_3)_3){=}C(OSi(CH_3)_3)(CH_2)_3OSi(CH_3)_3$ (35)	77
C_5	$n\text{-}C_4H_9COCl$	Ether	$(CH_3)_3SiOC(C_4H_9\text{-}n){=}C(C_4H_9\text{-}n)OSi(CH_3)_3$ (17)	76

* *Caution*: See warning on pp. 305–306.
Note: References 184–356 are on pp. 400–403.

[a] Condensations were brought about by sodium and TMCS and were carried out at reflux unless otherwise noted.
[b] The reaction temperature was 36°.
[c] The reaction temperature was 50°.

TABLE IV. CYCLIC ACYLOIN CONDENSATIONS CARRIED OUT WITH TRIMETHYLCHLOROSILANE (TMCS)*

	Starting Material	Metal/Solvent/Temp (°C)	Product(s) (% Yield)	Refs.
		A. Carbocycles from Diesters		
		3-Membered Rings		
C_5	$(CH_3)_2C(CO_2CH_3)$	1. $Na\text{-}NH_3$/ether/−34° 2. TMCS	*cis*- (structure: cyclopropane with CH_3, CH_3, H, H, $OSi(CH_3)_3$, $OSi(CH_3)_3$) (I, 25), $(CH_3)_2C{=}C(OCH_3)[OSi(CH_3)_3]$ (II, 6),[a] $(CH_3)_2C[CH_2OSi(CH_3)_3]CONHSi(CH_3)_3$ (III, 25), $(CH_3)_2C[CH_2OSi(CH_3)_3]_2$ (IV, 3), $(CH_3)_2CHCONHSi(CH_3)_3$ (V, 25)	56
		1. $Na\text{-}NH_3$/ether/−78° 2. TMCS	I (25), II (57), IV (3)	56
		1. $K\text{-}NH_3$/ether/−78° 2. TMCS	I (25), II (38), III (7), IV (5)	56
		1. $Na\text{-}NH_3/CH_3OH$, ether/−34° 2. TMCS	I (22), IV (3), V (10), $(CH_3)_2CHCH_2OSi(CH_3)_3$ (55)	56
		4-Membered Rings		
C_4	$C_2H_5O_2C(CH_2)_2CO_2C_2H_5$	Na/toluene TMCS/reflux	(structure: cyclobutene with two $OSi(CH_3)_3$ groups) (88)	77
		Na-K/ether, TMCS/reflux	" (80–93)	86
		Na/toluene, TMCS/reflux	" (75–85)	83, 86, 87, 183
		1. $Na\text{-}NH_3$/ether 2. TMCS	" + $(CH_3)_3SiO(CH_2)_4OSi(CH_3)_3$ (1:4 mixture), succinimide (30)	77

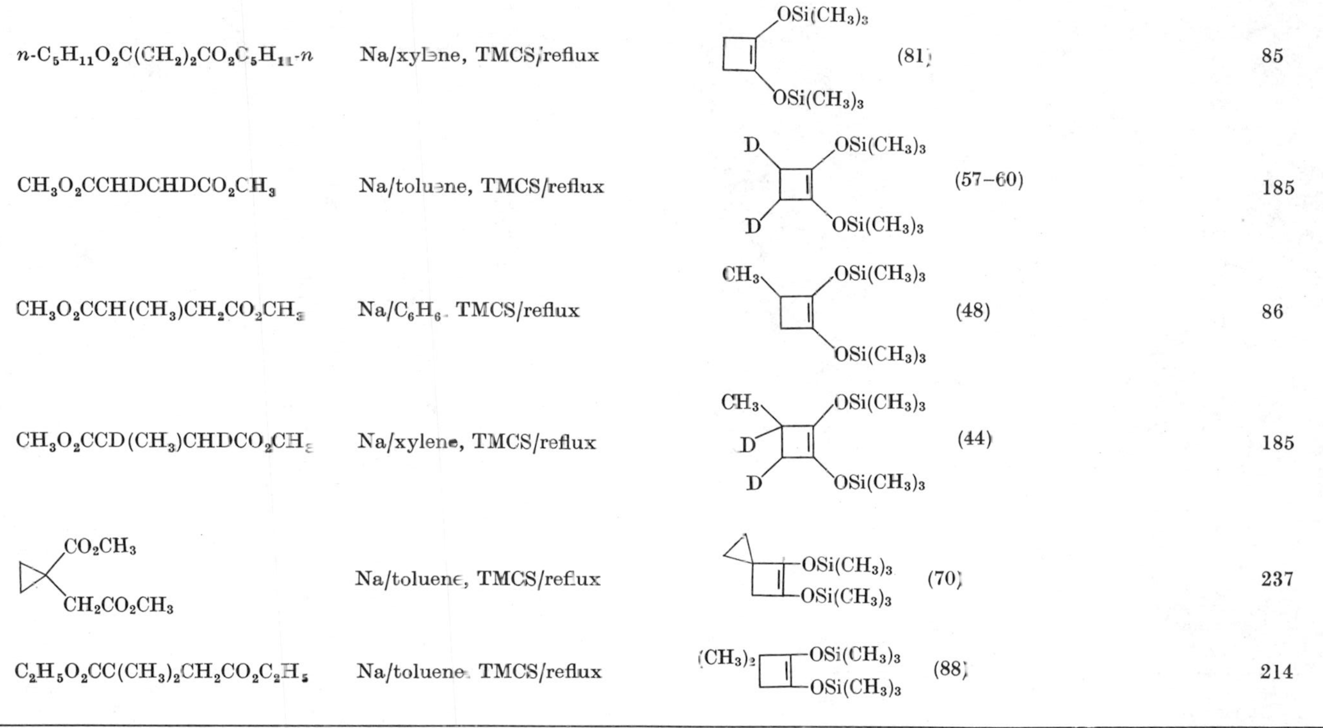

	Reactant	Conditions	Product(s) and Yield(s) (%)	Refs.
	$n\text{-}C_5H_{11}O_2C(CH_2)_2CO_2C_5H_{11}\text{-}n$	Na/xylene, TMCS/reflux	(81)	85
	$CH_3O_2CCHDCHDCO_2CH_3$	Na/toluene, TMCS/reflux	(57–60)	185
C_5	$CH_3O_2CCH(CH_3)CH_2CO_2CH_3$	Na/C_6H_6, TMCS/reflux	(48)	86
	$CH_3O_2CCD(CH_3)CHDCO_2CH_3$	Na/xylene, TMCS/reflux	(44)	185
C_6	(cyclopropane-1-CO_2CH_3-1-$CH_2CO_2CH_3$)	Na/toluene, TMCS/reflux	(70)	237
	$C_2H_5O_2CC(CH_3)_2CH_2CO_2C_2H_5$	Na/toluene, TMCS/reflux	(88)	214

* *Caution:* See warning on pp. 305–306.
Note: References 184–356 are on pp. 400–403.

[a] This is the exclusive product when the reduction is carried out in hydrocarbon solvents. See Table VIA.

TABLE IV. CYCLIC ACYLOIN CONDENSATIONS CARRIED OUT WITH TRIMETHYLCHLOROSILANE (TMCS)* (*Continued*)

A. Carbocycles from Diesters (Continued)

4-Membered Rings (Continued)

	Starting Material	Metal/Solvent/Temp (°C)	Product(s) (% Yield)	Refs.
C_7	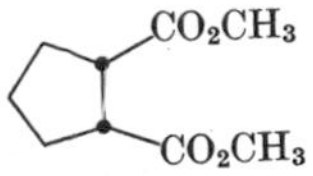 CO_2CH_3, CO_2CH_3	Na/toluene, TMCS/reflux	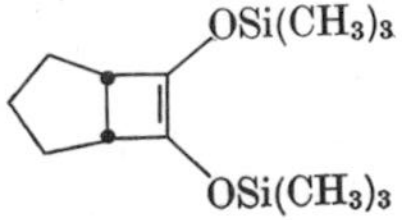 $OSi(CH_3)_3$, $OSi(CH_3)_3$ (77)	86
C_8	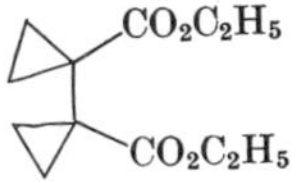 $CO_2C_2H_5$, $CO_2C_2H_5$	Na/toluene, TMCS/reflux	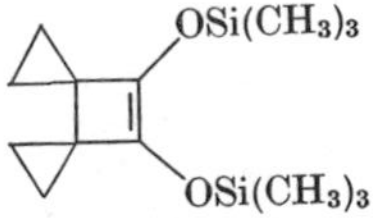 $OSi(CH_3)_3$, $OSi(CH_3)_3$ (42)	89
	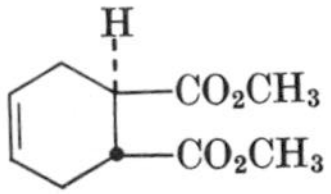 H, CO_2CH_3, CO_2CH_3	Na-K (excess)/ether, TMCS/−5 to 0°	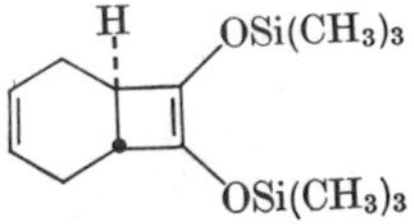 H, $OSi(CH_3)_3$, $OSi(CH_3)_3$ (65)	86
	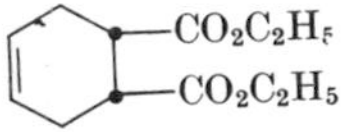 $CO_2C_2H_5$, $CO_2C_2H_5$	Na/toluene, TMCS/reflux	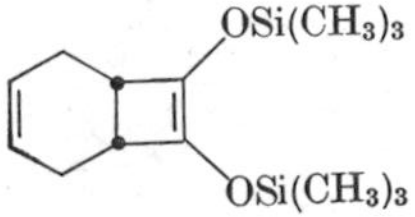 $OSi(CH_3)_3$, $OSi(CH_3)_3$ (80–86)	83, 267
	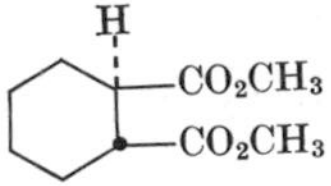 H, CO_2CH_3, CO_2CH_3	Na-K (excess)/ether, TMCS/ −-5 to 0°	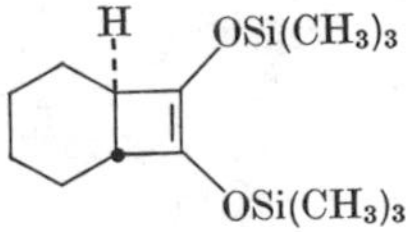 H, $OSi(CH_3)_3$, $OSi(CH_3)_3$ (57–65)	86

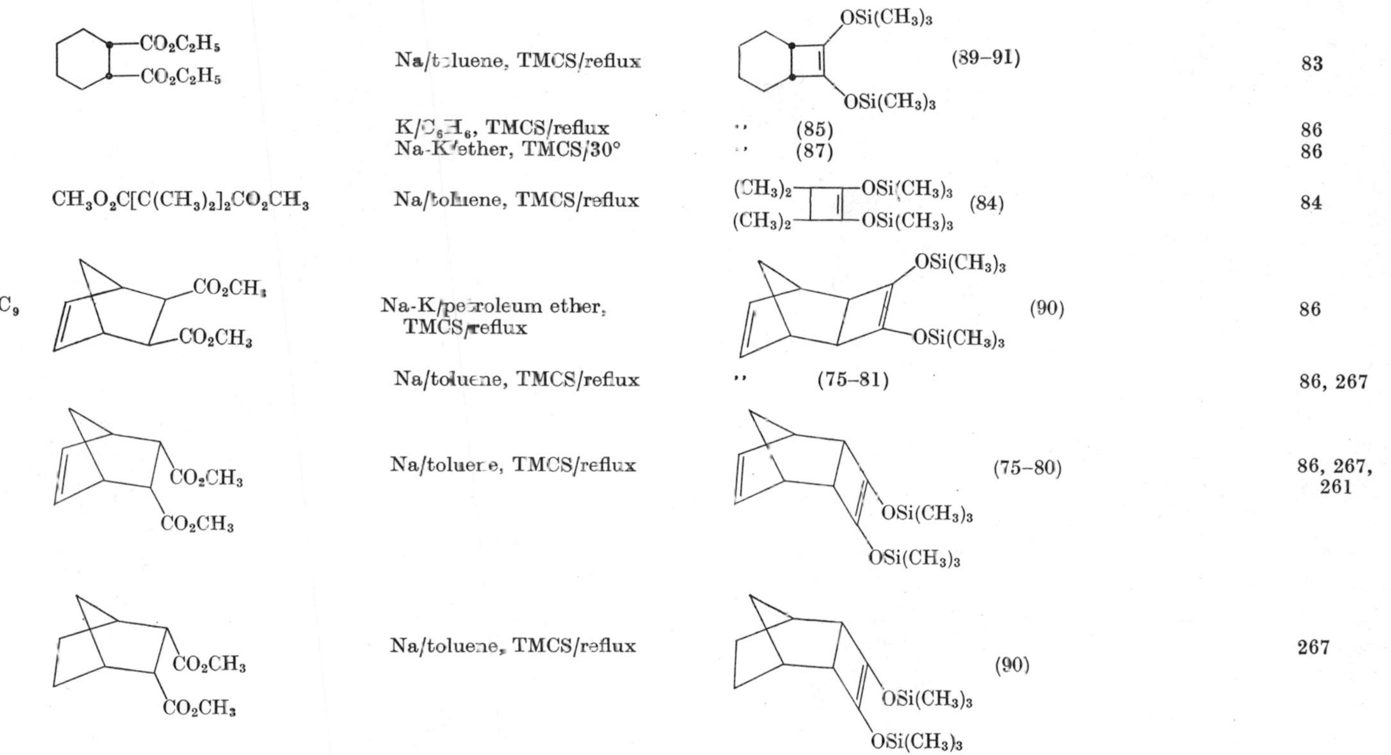

	Reactant	Conditions	Product (Yield, %)	Refs.
	1,2-$(CO_2C_2H_5)_2$-cyclohexane	Na/toluene, TMCS/reflux	$OSi(CH_3)_3$, $OSi(CH_3)_3$ (89–91)	83
		K/C_6H_6, TMCS/reflux	" (85)	86
		Na-K/ether, TMCS/30°	" (87)	86
	$CH_3O_2C[C(CH_3)_2]_2CO_2CH_3$	Na/toluene, TMCS/reflux	$(CH_3)_2$, $(CH_3)_2$, $OSi(CH_3)_3$, $OSi(CH_3)_3$ (84)	84
C_9	CO_2CH_3, CO_2CH_3	Na-K/petroleum ether, TMCS/reflux	$OSi(CH_3)_3$, $OSi(CH_3)_3$ (90)	86
		Na/toluene, TMCS/reflux	" (75–81)	86, 267
	CO_2CH_3, CO_2CH_3	Na/toluene, TMCS/reflux	$OSi(CH_3)_3$, $OSi(CH_3)_3$ (75–80)	86, 267, 261
	CO_2CH_3, CO_2CH_3	Na/toluene, TMCS/reflux	$OSi(CH_3)_3$, $OSi(CH_3)_3$ (90)	267

* *Caution:* See warning on pp 305–306.

Note: References 184–356 are on pp. 400–403.

TABLE IV. Cyclic Acyloin Condensations Carried out with Trimethylchlorosilane (TMCS)* (*Continued*)

Starting Material	Metal/Solvent/Temp (°C)	Product(s) (% Yield)	Refs.
A. Carbocycles from Diesters (Continued)			
4-Membered Rings (Continued)			
C_9 (*contd.*) 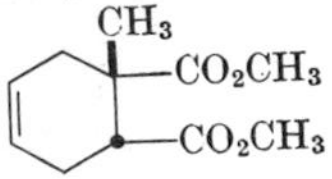 CH_3, CO_2CH_3, CO_2CH_3	Na/toluene, TMCS/reflux	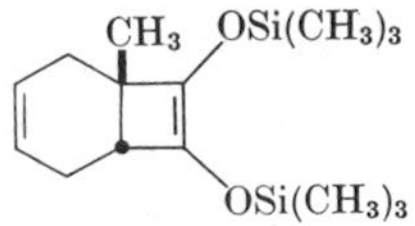 CH_3, $OSi(CH_3)_3$, $OSi(CH_3)_3$ (84–94)	83, 86
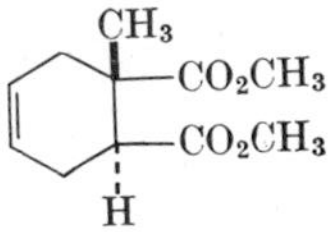 CH_3, CO_2CH_3, CO_2CH_3, H	Na-K (excess)/ether, TMCS/−5 to 0°	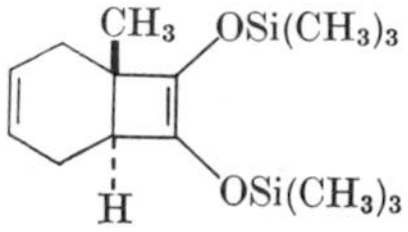 CH_3, $OSi(CH_3)_3$, $OSi(CH_3)_3$, H (83)	86
	Na/toluene, TMCS/reflux	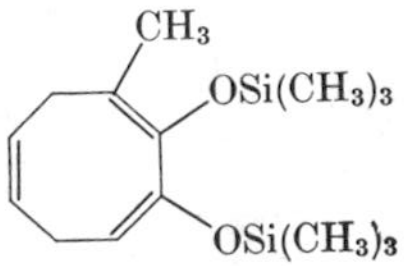 CH_3, $OSi(CH_3)_3$, $OSi(CH_3)_3$ (86–92)	83
C_{10} 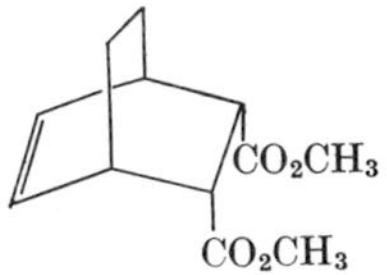 CO_2CH_3, CO_2CH_3	Na/toluene, TMCS/reflux	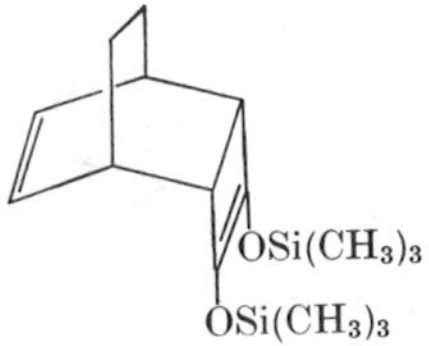 $OSi(CH_3)_3$, $OSi(CH_3)_3$ (83)	267

Reactant	Conditions	Product(s) and Yield(s) (%)	Refs.
$CH_3O_2CCH(C_6H_5)CH_2CO_2CH_3$	Na-K (excess)/ether-THF, TMCS/−3°	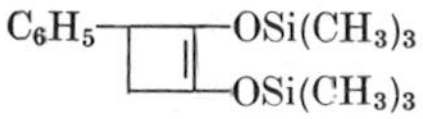 C_6H_5-cyclobutene-1,2-bis[$OSi(CH_3)_3$] (60)	86
	Na-K (excess)/C_6H_6, TMCS/25°	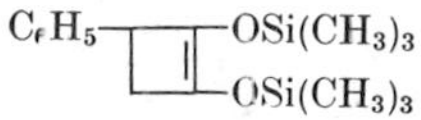 C_6H_5-cyclobutene-1,2-bis[$OSi(CH_3)_3$] (—)	86
	Na/toluene, TMCS/reflux	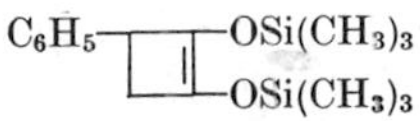 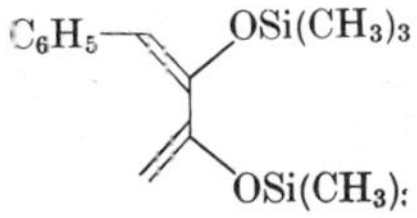 C_6H_5-cyclobutene-1,2-bis[$OSi(CH_3)_3$] + $C_6H_5CH{=}C[OSi(CH_3)_3]C[OSi(CH_3)_3]{=}CH_2$ (Total: 62)	86
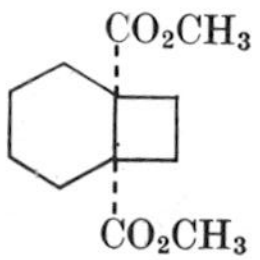 bicyclo[4.2.0]octane-1,6-bis(CO_2CH_3)	Na/toluene, TMCS/25° or reflux	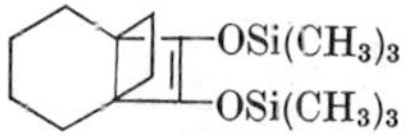 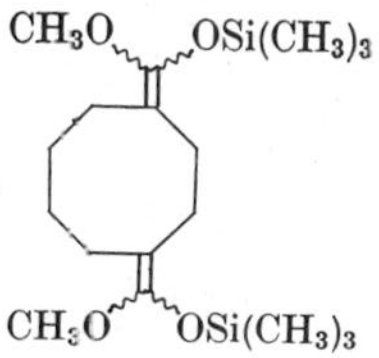 bridged bicyclic cyclobutene-bis[$OSi(CH_3)_3$] (24–40), cyclooctane-1,5-diylidene-bis[$C(OCH_3)OSi(CH_3)_3$] (40–58)	69
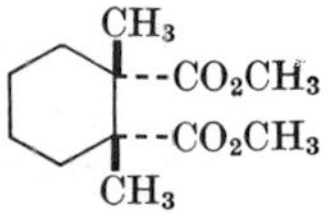 1,2-di-CH_3-cyclohexane-1,2-bis(CO_2CH_3)	Na/toluene, TMCS/reflux	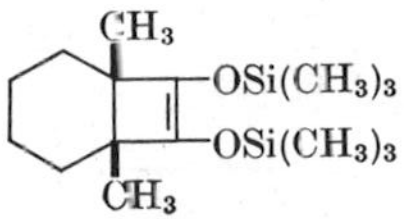 1,6-di-CH_3-bicyclo[4.2.0]oct-7-ene-7,8-bis[$OSi(CH_3)_3$] (88)	261

* *Caution:* See warning on pp. 305–306.

Note: References 184–356 are on pp. 400–403.

TABLE IV. Cyclic Acyloin Condensations Carried out with Trimethylchlorosilane (TMCS)* (*Continued*)

Starting Material	Metal/Solvent/Temp (°C)	Products (% Yield)	Refs.
A. Carbocycles from Diesters (Continued)			
4-Membered Rings (Continued)			
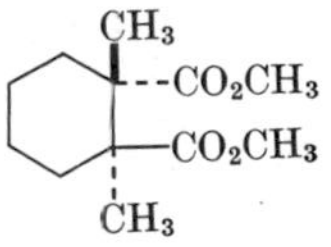 CH_3, CO_2CH_3, CO_2CH_3, CH_3	Na/toluene, TMCS/reflux	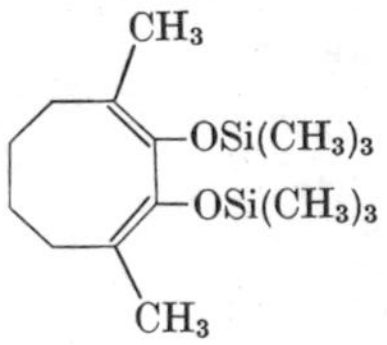 CH_3, $OSi(CH_3)_3$, $OSi(CH_3)_3$, CH_3 (88)	70, 261
	Na-K/C_6H_6, TMCS/25°	" (57)	261
C_{11} 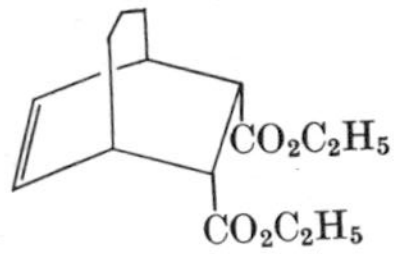 $CO_2C_2H_5$, $CO_2C_2H_5$	Na/toluene, TMCS/reflux	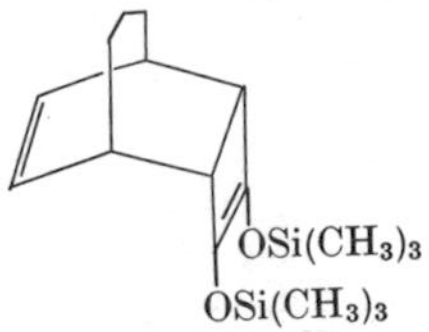 $OSi(CH_3)_3$, $OSi(CH_3)_3$ (81)	267
C_{12} 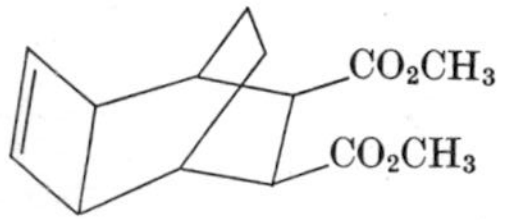 CO_2CH_3, CO_2CH_3	Na/xylene, TMCS/40–80	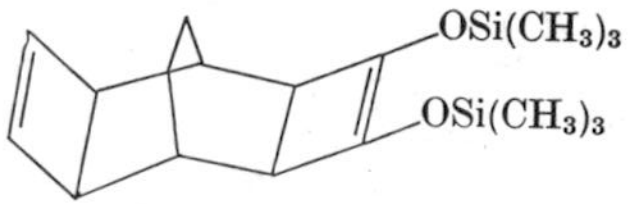 $OSi(CH_3)_3$, $OSi(CH_3)_3$ (—)	344
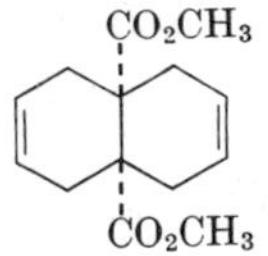 CO_2CH_3, CO_2CH_3	Na/toluene, TMCS/reflux	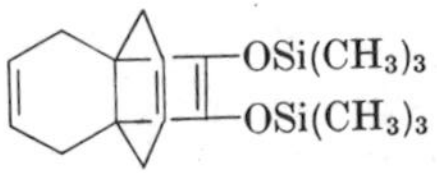 $OSi(CH_3)_3$, $OSi(CH_3)_3$ (88–91)	83, 29

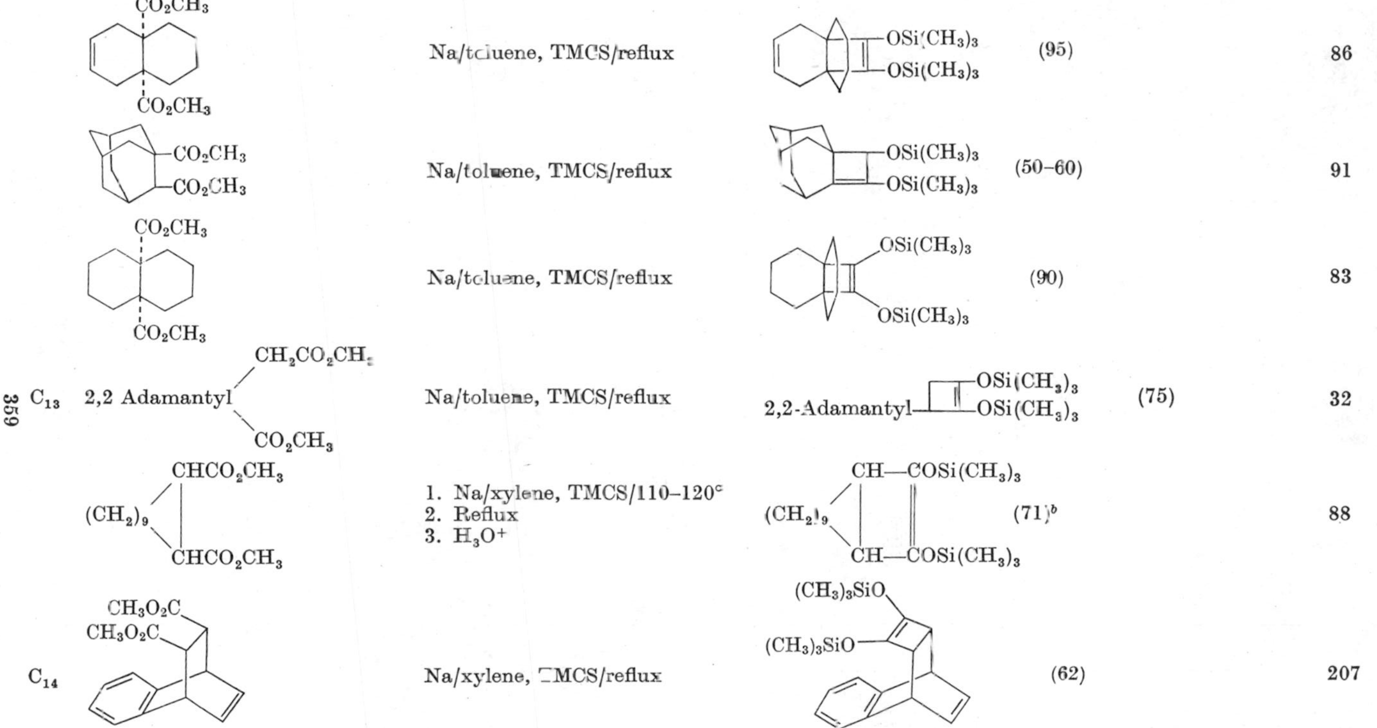

	Reactant	Conditions	Product (Yield, %)		Refs.
	CO_2CH_3 / CO_2CH_3	Na/toluene, TMCS/reflux	$OSi(CH_3)_3$ / $OSi(CH_3)_3$	(95)	86
	CO_2CH_3 / CO_2CH_3	Na/toluene, TMCS/reflux	$OSi(CH_3)_3$ / $OSi(CH_3)_3$	(50–60)	91
	CO_2CH_3 / CO_2CH_3	Na/toluene, TMCS/reflux	$OSi(CH_3)_3$ / $OSi(CH_3)_3$	(90)	83
C_{13}	2,2 Adamantyl $CH_2CO_2CH_3$ / CO_2CH_3	Na/toluene, TMCS/reflux	2,2-Adamantyl $OSi(CH_3)_3$ / $OSi(CH_3)_3$	(75)	32
	$(CH_2)_9$ $CHCO_2CH_3$ / $CHCO_2CH_3$	1. Na/xylene, TMCS/110–120° 2. Reflux 3. H_3O^+	$(CH_2)_9$ $CH—COSi(CH_3)_3$ / $CH—COSi(CH_3)_3$	(71)[b]	88
C_{14}	CH_3O_2C / CH_3O_2C	Na/xylene, TMCS/reflux	$(CH_3)_3SiO$ / $(CH_3)_3SiO$	(62)	207

* *Caution:* See warning on pp. 305–306.
Note: References 184–356 are on pp. 400–403.
[b] The yield is based on the diketone obtained after thermal ring opening to the bis-(trimethylsilyloxy)cycloalkadiene and hydrolysis.

TABLE IV. Cyclic Acyloin Condensations Carried out with Trimethylchlorosilane (TMCS)* (*Continued*)

A. Carbocycles from Diesters (*Continued*)

4-Membered Rings (Continued)

	Starting Material	Metal/Solvent/Temp (°C)	Products (% Yield)		Refs.
	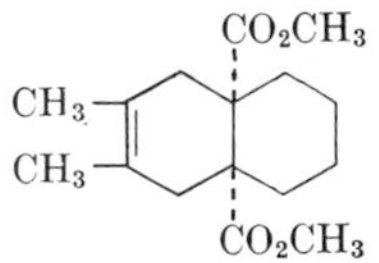 CH_3, CH_3 (bicyclic diester with two CO_2CH_3 groups)	Na/toluene, TMCS/reflux	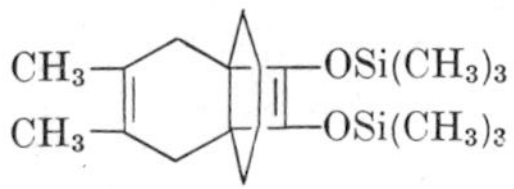 CH_3, CH_3 (tricyclic product with $-OSi(CH_3)_3$, $-OSi(CH_3)_3$)	(80)	86
	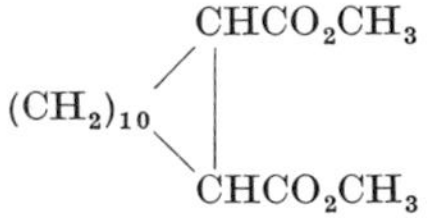 $(CH_2)_{10}$ (ring with $CHCO_2CH_3$, $CHCO_2CH_3$)	1. Na/xylene, TMCS/110–120° 2. Reflux 3. H_3O^+	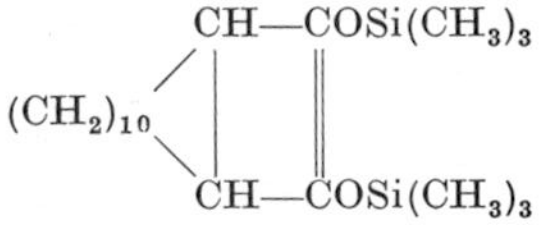 $(CH_2)_{10}$ (ring with $CH—COSi(CH_3)_3$, $CH—COSi(CH_3)_3$)	$(73)^b$	88
C_{15}	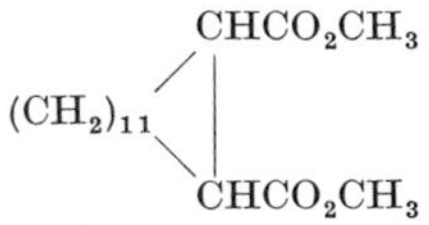 $(CH_2)_{11}$ (ring with $CHCO_2CH_3$, $CHCO_2CH_3$)	1. Na/xylene, TMCS/110–120° 2. Reflux 3. H_3O^+	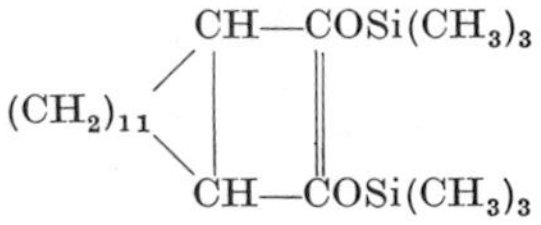 $(CH_2)_{11}$ (ring with $CH—COSi(CH_3)_3$, $CH—COSi(CH_3)_3$)	$(74)^b$	88
C_{18}	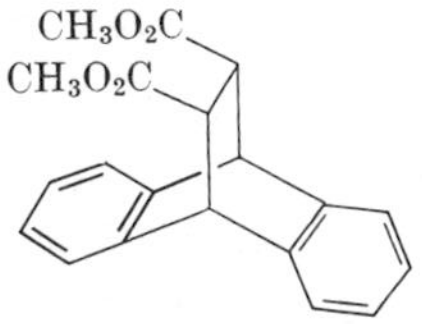 CH_3O_2C, CH_3O_2C (dibenzobicyclic diester)	Na/xylene, TMCS/reflux	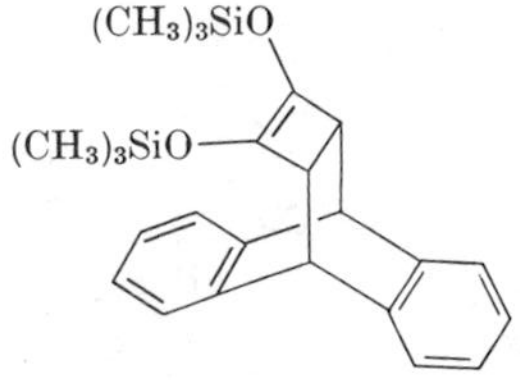 $(CH_3)_3SiO$, $(CH_3)_3SiO$ (cyclobutene-fused dibenzobicyclic product)	(26)	85, 207

		5-Membered Rings			
C_5	$C_2H_5O_2C(CH_2)_3CO_2C_2H_5$	Na/toluene, TMCS/reflux	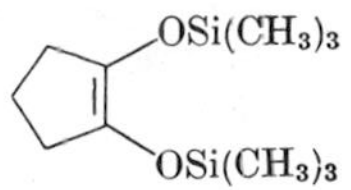 1,2-bis($OSi(CH_3)_3$)cyclopentene	(91–93)	93, 203
		Na/ether TMCS/reflux	"	(—)	61
C_7	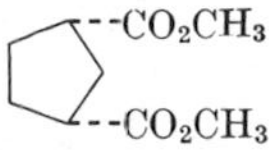 cyclopentane-1,2-di-CO_2CH_3	Na/toluene, TMCS/reflux	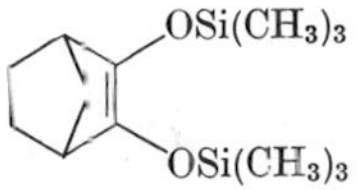 bicyclic ene-bis($OSi(CH_3)_3$)	(89)	261
		1. Na-NH_3/ether 2. TMCS	"	(74)	77
C_8	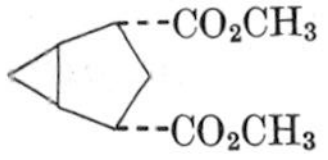 cyclopropane-fused cyclopentane-di-CO_2CH_3	Na/toluene, TMCS/reflux	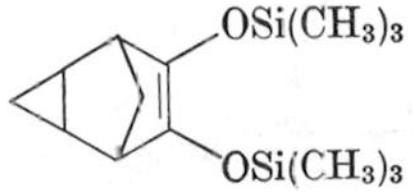 tricyclic ene-bis($OSi(CH_3)_3$)	(84)	261
	 cyclobutane-$(CH_2CO_2C_2H_5)_2$	Na/toluene, TMCS/reflux	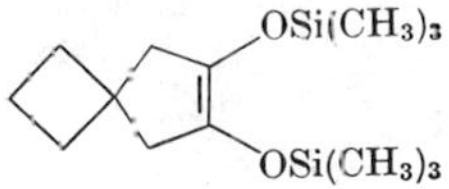 spiro[3.4]octene-bis($OSi(CH_3)_3$)	(84)	334
C_9	 cyclopentane-$(CH_2CO_2CH_3)_2$	Na/toluene, TMCS/reflux	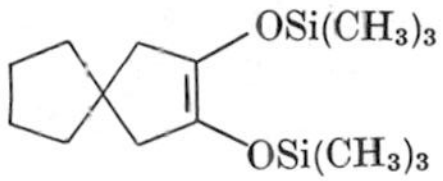 spiro[4.4]nonene-bis($OSi(CH_3)_3$)	(87)	48

* *Caution:* See warning on pp. 305–306.

Note: References 184–356 are on pp. 400–403.

TABLE IV. CYCLIC ACYLOIN CONDENSATIONS CARRIED OUT WITH TRIMETHYLCHLOROSILANE (TMCS)* (*Continued*)

Starting Material	Metal/Solvent/Temp (°C)	Product(s) (% Yield)	Refs.
A. Carbocycles from Diesters (Continued)			
5-Membered Rings (Continued)			
Cyclopentylidene $(CH_2CO_2C_2H_5)_2$	Na/ether, TMCS/reflux	Spiro[4.4]non-2-ene-2,3-bis[$OSi(CH_3)_3$] (59)	30
$C(CH_2CO_2CH_3)_4$	1. $Na\text{-}NH_3$/ether 2. TMCS	Spiro[4.4]nona-2,7-diene with four $OSi(CH_3)_3$ groups [two $(CH_3)_3SiO$, two $OSi(CH_3)_3$] (88)	153, 77
	Na/xylene, TMCS/reflux	" (Low)	153
C_{10} Cyclopentane with CO_2CH_3, $(CH_3)_2$, CO_2CH_3, CH_3	$Na/C_6H_5C_2H_5$, TMCS/reflux	Bicyclic ene with CH_3, CH_3, CH_3, and two $OSi(CH_3)_3$ (67)	77
$CH_2{=}C[(CH_2)_3CO_2R]_2$, $R = CH_3, C_2H_5$	Na/(—), TMCS/-	Bicyclic (hydrindene) ene with two $OSi(CH_3)_3$ and CH_3 (—)	217a

6-Membered Rings

	Reactant	Conditions	Product (Yield %)	Refs.
C_6	$CH_3O_2C(CH_2)_4CO_2CH_3$	Na/toluene, TMCS/reflux	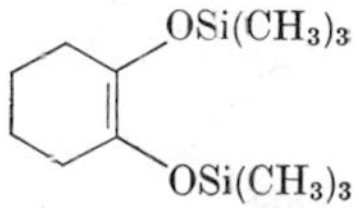 $OSi(CH_3)_3$, $OSi(CH_3)_3$ (85–99)	86, 203
	$C_2H_5O_2C(CH_2)_4CO_2C_2H_5$	Na/toluene, TMCS/reflux	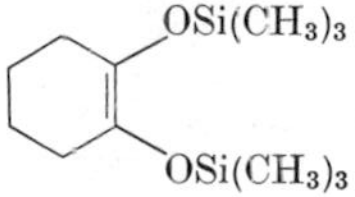 $OSi(CH_3)_3$, $OSi(CH_3)_3$ (55–89)	93
	$CH_3O_2C[CHOSi(CH_3)_3]_4CO_2CH_3$	Na/toluene, TMCS/reflux	o-$C_6H_4[OSi(CH_3)_3]_2$ (18)	234
C_8	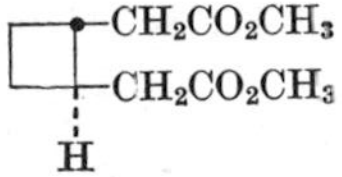 $CH_2CO_2CH_3$, $CH_2CO_2CH_3$, H	Na/C_6H_6, TMCS/reflux	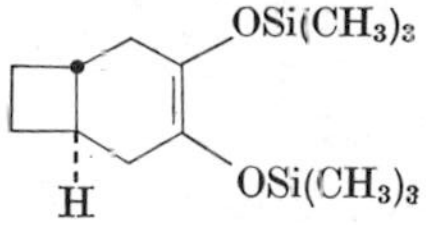 $OSi(CH_3)_3$, $OSi(CH_3)_3$, H (22)	241
C_{14}	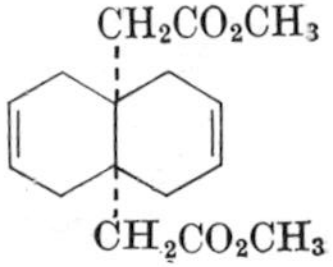 $CH_2CO_2CH_3$, $CH_2CO_2CH_3$	Na-K/ether/TMCS/reflux	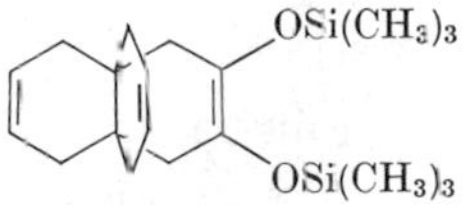 $OSi(CH_3)_3$, $OSi(CH_3)_3$ (82)	99
		Na-K (excess)/toluene, TMCS/reflux	" (89–92)	86
	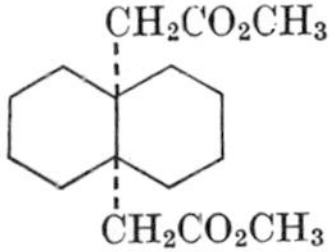 $CH_2CO_2CH_3$, $CH_2CO_2CH_3$	Na-K (excess)/ether, TMCS/reflux	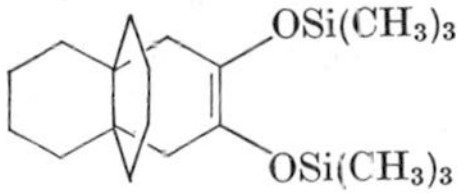 $OSi(CH_3)_3$, $OSi(CH_3)_3$ (88–90)	99, 86

* *Caution*: See warning on pp. 305–306.
Note: References 184–356 are on pp. 400–403.

TABLE IV. Cyclic Acyloin Condensations Carried out with Trimethylchlorosilane (TMCS)* (*Continued*)

A. Carbocycles from Diesters (Continued)

7-Membered Rings

	Starting Material	Metal/Solvent/Temp (°C)	Product(s) (% Yield)	Refs.
C_7	$C_2H_5O_2C(CH_2)_5CO_2C_2H_5$	Na/toluene, TMCS/reflux or K/C_6H_6, TMCS/reflux	$(CH_2)_5$ ring, $COSi(CH_3)_3$=$COSi(CH_3)_3$ (75–81)	16, 99
	1,3-dioxolane-2,2-$[(CH_2)_2CO_2C_2H_5]_2$	Na/toluene, TMCS/reflux	1,3-dioxolane spiro: $(CH_2)_2COSi(CH_3)_3$=$(CH_2)_2COSi(CH_3)_3$ (85)	99
C_8	$CH_2{=}C[(CH_2)_2CO_2C_2H_5]_2$	Na/(—), TMCS/	CH_2= cycloheptene with two $OSi(CH_3)_3$ (—)	118
C_{10}	$RO_2C(CH_2)_4CH(i\text{-}C_3H_7)CO_2R^c$	Na/(—), TMCS	i-C_3H_7 cycloheptene with two $OSi(CH_3)_3$ (—)	336
	$RO_2C(CH_2)_3CH(C_3H_7\text{-}i)CH_2$-$CO_2R^c$	Na/(—), TMCS	i-C_3H_7 cycloheptene with two $OSi(CH_3)_3$ (—)	336

	$RO_2C(CH_2)_2CH(C_3H_7\text{-}i)(CH_2)_2CO_2R$[c]	Na/(—), TMCS	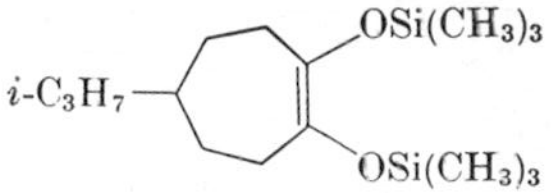 $i\text{-}C_3H_7$ cycloheptene, $OSi(CH_3)_3$, $OSi(CH_3)_3$	(–)	336
C_{11}	$C_2H_5O_2CC(CH_3)_2(CH_2)_3C(CH_3)_2CO_2C_2H_5$	Na/toluene, TMCS/reflux	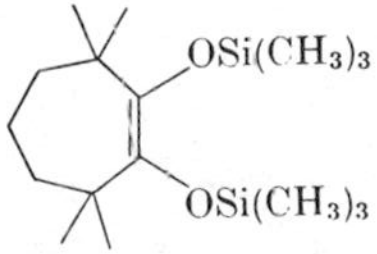 $OSi(CH_3)_3$, $OSi(CH_3)_3$	(~50)	31
		Rings of 8 or More Members			
C_8	$CH_3O_2C(CH_2)_6CO_2CH_3$	Na/toluene, TMCS/reflux	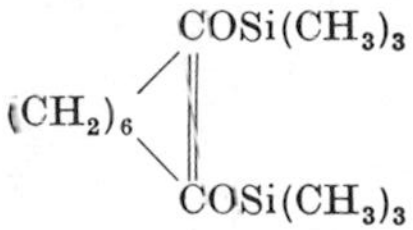 $(CH_2)_6$, $COSi(CH_3)_3$, $COSi(CH_3)_3$	(85)[d]	86
		Na/methylcyclohexane, TMCS/reflux	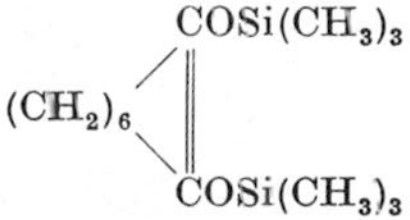 $(CH_2)_6$, $COSi(CH_3)_3$, $COSi(CH_3)_3$	(72)[e]	86
C_{12}	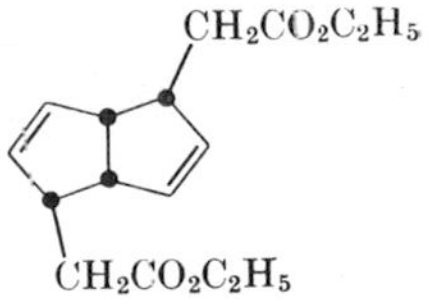 $CH_2CO_2C_2H_5$, $CH_2CO_2C_2H_5$	Na/toluene/TMCS/60°	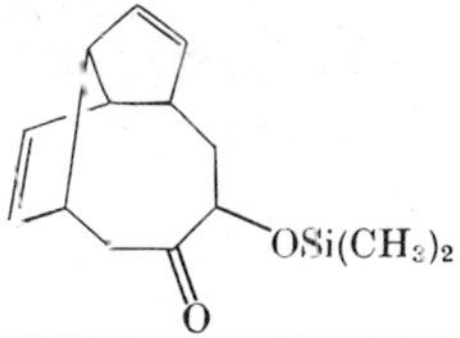 $OSi(CH_3)_2$, O	(5)	106

* *Caution:* See warning on pp. 305–306.
Note: References 184–356 are on pp. 400–403.

[c] The R group of the ester was not specified.
[d] High-dilution cycle was used; 0.2 mol of ester was added over 68 hours.
[e] High-dilution cycle was used; 0.3 mol of ester was added over 90 hours.

TABLE IV. CYCLIC ACYLOIN CONDENSATIONS CARRIED OUT WITH TRIMETHYLCHLOROSILANE (TMCS)* (*Continued*)

A. Carbocycles from Diesters (Continued)

Rings of 8 or More Members (Continued)

	Starting Material	Metal/Solvent/Temp (°C)	Product(s) (% Yield)	Refs.
	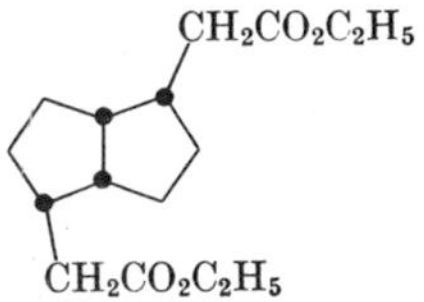 [bicyclic diester: $CH_2CO_2C_2H_5$, $CH_2CO_2C_2H_5$]	Na/toluene, TMCS/60°	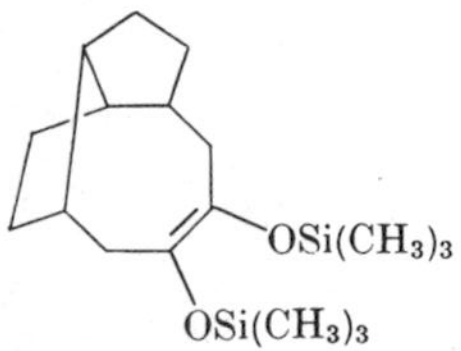 [tricyclic bis-$OSi(CH_3)_3$ enediol ether] (–)	106
C_9	$CH_3O_2C(CH_2)_7CO_2CH_3$	Na/toluene, TMCS/reflux[f]	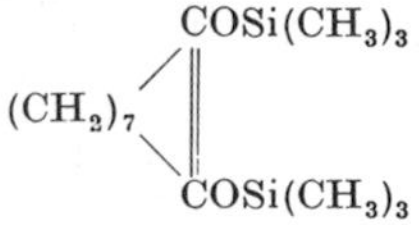 $(CH_2)_7$ ring with $C(OSi(CH_3)_3)=C(OSi(CH_3)_3)$ (68)	86
	$C_2H_5O_2C(CH_2)_7CO_2C_2H_5$	Na/toluene, TMCS/reflux	" (22)[g]	16
C_{10}	$CH_3O_2C(CH_2)_8CO_2CH_3$	Na/toluene, TMCS/reflux[h]	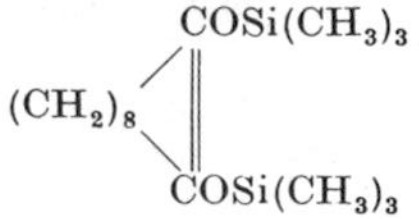 $(CH_2)_8$ ring with $C(OSi(CH_3)_3)=C(OSi(CH_3)_3)$ (69)	86, 203
		Na/toluene, TMCS/reflux[i]	" (58)	86

	$C_2H_5O_2C(CH_2)_8CO_2C_2H_5$	Na/xylene, TMCS/reflux	$(CH_2)_8$ ring with $C(OSi(CH_3)_3)=C(OSi(CH_3)_3)$	(53)[j]	16, 93
		Na/toluene, TMCS/reflux	"	(22)[j]	16, 93
C_{11}	$CH_3O_2C(CH_2)_9CO_2CH_3$	Na/methylcyclohexane, TMCS/reflux[k]	$(CH_2)_9$ ring with $C(OSi(CH_3)_3)=C(OSi(CH_3)_3)$	(48)	86
C_{12}	$CH_3O_2C(CH_2)_{10}CO_2CH_3$	Na/toluene, TMCS/reflux[l]	$(CH_2)_{10}$ ring with $C(OSi(CH_3)_3)=C(OSi(CH_3)_3)$	(68)	86
C_{13}	$CH_3O_2C(CH_2)_{11}CO_2CH_3$	Na/toluene, TMCS/reflux[m]	$(CH_2)_{11}$ ring with $C(OSi(CH_3)_3)=C(OSi(CH_3)_3)$	(84)	86
C_{14}	$CH_3O_2C(CH_2)_{12}CO_2CH_3$	Na/methylcyclohexane, TMCS/reflux[n]	$(CH_2)_{12}$ ring with $C(OSi(CH_3)_3)=C(OSi(CH_3)_3)$	(67)	86

* *Caution*: See warning on pp. 305–306.

Note: References 184–356 are on pp. 400–403.

[f] High-dilution cycle was used: 0.2 mol of ester was added over 55 hours.
[g] An 18-membered ring dimer is also formed. See below (p. 368).
[h] High-dilution cycle was used; 0.2 mol of ester was added over 90 hours.
[i] High-dilution cycle was used; 0.3 mol of ester was added over 90 hours.
[j] A 20-membered ring dimer is also formed. See below (p. 368).
[k] High-dilution cycle was used; 0.05 mol of ester was added over 36 hours.
[l] High-dilution cycle was used; 0.2 mol of ester was added over 63 hours.
[m] High-dilution cycle was used; 0.2 mol of ester was added over 45 hours.
[n] High-dilution cycle was used; 0.075 mol of ester was added over 54 hours.

TABLE IV. CYCLIC ACYLOIN CONDENSATIONS CARRIED OUT WITH TRIMETHYLCHLOROSILANE (TMCS)* (*Continued*)

	Starting Material	Metal/Solvent/Temp (°C)	Product(s) (% Yield)		Refs.
		A. Carbocycles from Diesters (Continued) Rings of 8 or More Members (*Continued*)			
C_9	$C_2H_5O_2C(CH_2)_7CO_2C_2H_5$	Na/xylene, TMCS/reflux	$(CH_3)_3SiOC(CH_2)_7COSi(CH_3)_3$ $(CH_3)_3SiOC(CH_2)_7COSi(CH_3)_3$	(62)[o]	16, 93
C_{10}	$C_2H_5O_2C(CH_2)_8CO_2C_2H_5$	Na/xylene, TMCS/reflux	$(CH_3)_3SiOC(CH_2)_8COSi(CH_3)_3$ $(CH_3)_3SiOC(CH_2)_8COSi(CH_3)_3$	(20)[p]	93, 203
		Na/toluene, TMCS/reflux	"	(22–73)[p]	16, 93
		B. Cyclic Bis(trimethylsilyloxy)alkanes from Anhydrides			
C_4	Succinic anhydride	Na/xylene, TMCS/reflux	$OSi(CH_3)_3$ $OSi(CH_3)_3$	(30)	77
		C. Nitrogen Heterocycles			
C_7	$CH_3N[(CH_2)_2CO_2CH_3]_2$	Na/ether, toluene, or xylene, TMCS/reflux	CH_3N — $OSi(CH_3)_3$ $OSi(CH_3)_3$	(67)	85
C_{12}	$C_6H_5N[(CH_2)_2CO_2C_2H_5]_2$	Na/ether, toluene, or xylene, TMCS/reflux	C_6H_5N — $OSi(CH_3)_3$ $OSi(CH_3)_3$	(74)	85
		D. Oxygen and Sulfur Heterocycles			
C_6	$S[(CH_2)_2CO_2CH_3]_2$	Na/ether, toluene, or xylene, TMCS/reflux	S — $OSi(CH_3)_3$ $OSi(CH_3)_3$	(26)	85

	Reactant	Conditions	Product(s) and Yield(s) (%)	Refs.
C_{10}	$O[CH_2C(CH_3)_2CO_2C_2H_5]_2$	1. Na/toluene, TMCS/reflux 2. H_3O^+	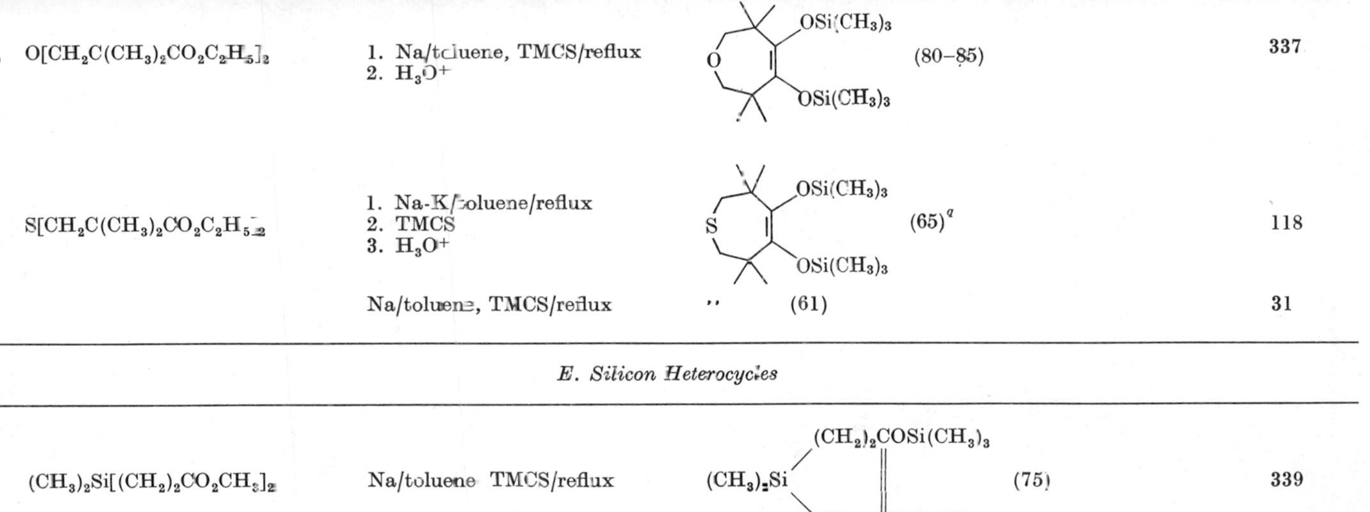 (80–85)	337
	$S[CH_2C(CH_3)_2CO_2C_2H_5]_2$	1. Na-K/toluene/reflux 2. TMCS 3. H_3O^+	(65)[q]	118
		Na/toluene, TMCS/reflux	" (61)	31
	E. Silicon Heterocycles			
C_8	$(CH_3)_2Si[(CH_2)_2CO_2CH_3]_2$	Na/toluene TMCS/reflux	(75)	339
C_9	$CH_3O_2C(CH_2)_3Si(CH_3)_2(CH_2)_2$-$CO_2CH_3$	Na/toluene, TMCS/reflux	(60)	339
C_{18}	$(C_6H_5)_2Si[(CH_2)_2CO_2CH_3]_2$	Na/toluene, TMCS/reflux	(72)	120

Caution: see warning on pp. 305–306.
Note: References 184–356 are on pp. 400–403.

[o] A 9-membered ring is also formed. See p. 366.
[p] A 10-membered ring is also formed. See p. 367.
[q] The acyloin was isolated after hydrolysis of the crude product.

TABLE V. Condensations between Esters and Ketones

A. Alkali Metal Reductions

	Starting Materials	Metal/Solvent/Temp (°C)	Product(s) (% Yield)	Refs.
C_2, C_3	$CH_3CO_2C_2H_5 + (CH_3)_2CO$	Na/ether, CH_3CO_2H/reflux	$(CH_3)_2C—C(CH_3)C(CH_3)_2$ (—), $(CH_3)_2C(OH)COCH_3$ (—) HO OH OH (I)	134
		Na/ether/reflux	$(CH_3)_2C(OH)COCH_3$ (Low)	215
		Na/C_6H_6, CH_3CO_2H/reflux	(I) (—)	135
C_2, C_4	$CH_3CO_2C_2H_5 + C_2H_5COCH_3$	Na/C_6H_6, CH_3CO_2H/reflux	$C_2H_5C(CH_3)C(CH_3)C(CH_3)(C_2H_5$ (—), OH OH OH $C_2H_5(CH_3)C(OH)COCH_3$ (—)	134
C_3, C_3	$C_2H_5CO_2C_2H_5 + (CH_3)_2CO$	Na/ether/reflux	$C_2H_5CHOHCOC_2H_5$ (Low)	215
		Na/C_6H_6, CH_3CO_2H/reflux	" (—), $C_2H_5COC(CH_3)_2OH$ (—), $(CH_3)_2C(OH)C(C_2H_5)(OH)C(CH_3)_2OH$ (—)	135
C_4, C_3	$n\text{-}C_3H_7CO_2C_2H_5 + (CH_3)_2CO$	Na/C_6H_6, CH_3CO_2H/reflux	$n\text{-}C_3H_7COC(CH_3)_2OH$ (—), $n\text{-}C_3H_7COCHOHC_3H_7\text{-}n$ (—), $(CH_3)_2C(OH)C(C_3H_7\text{-}n)(OH)C(CH_3)_2OH$ (—)	134, 135
	$i\text{-}C_3H_7CO_2C_2H_5 + (CH_3)_2CO$	Na/C_6H_6, CH_3CO_2H/reflux	$i\text{-}C_3H_7COCHOHC_3H_7\text{-}i$ (—), $(CH_3)_2C(OH)C(C_3H_7\text{-}i)(OH)C(CH_3)_2OH$ (—)	134, 135
C_4, H_7	$n\text{-}C_3H_7CO_2C_2H_5 +$ $C_6H_5COCH_3$	Na/toluene/reflux	$n\text{-}C_3H_7COC(CH_3)(C_6H_5)OH$ (—)	197
C_7, C_3	$C_6H_5CO_2C_2H_5 + (CH_3)_2CO$	Na/C_6H_6, CH_3CO_2H/reflux	$(CH_3)_2C(OH)C(C_6H_5)(OH)C(CH_3)_2OH$ (—)	134
C_7, C_4	$C_6H_5CO_2C_2H_5 + C_2H_5COCH_3$	Na/C_6H_6, CH_3CO_2H/reflux	$C_2H_5C(CH_3)C(C_2H_5)C(CH_3)C_2H_5$ (—) OH OH OH	134
C_8	O $CO_2C_2H_5$	Na-$C_{10}H_8$/THF	HO O (2)	130

	Reactant	Conditions	Product(s) and Yield(s) (%)	Refs.
C_9	CO_2CH_3	Na-NH_3/ether	HO O (I, 5.5 parts), O (II, 1.5 parts).	130
			O (III, trace), O (IV, 1 part)	
			O (V, 1 part),	
		Na-NH_3/ether (high dilution)	I (1 part), III (4.3 parts), OH (1.1 part)	130
		Na-$C_{10}H_8$/THF	I (10.9 parts), III (2.4 parts)	130
		1. Li-NH_3/ether 2. CrO_3-H_2SO_4/$(CH_3)_2CO$	I (8), IV (22), V (60),	256
			O O (8), O O O (2)	
C_{10}	O CO_2CH_3	Na-NH_3/ether	O (45)	129
		Na-$C_{10}H_8$/THF	HO O (20)	129

Note: References 184–356 are on pp. 400–403.

TABLE V. CONDENSATIONS BETWEEN ESTERS AND KETONES (*Continued*)

Starting Materials	Metal/Solvent/Temp (°C)	Product(s) (% Yield)	Refs.
A. Alkali Metal Reductions (Continued)			
C_{10} (*contd.*) CO_2CH_3	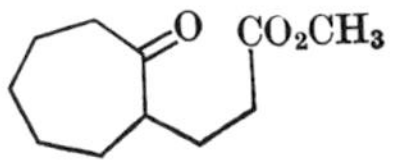 Na-NH_3/ether	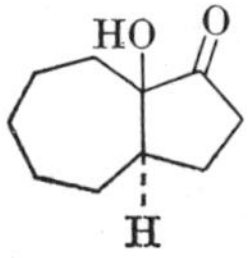 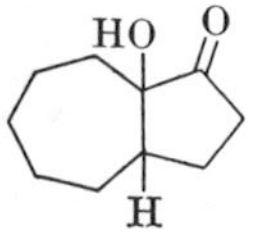 HO O + HO O H H (5.5 parts), O (trace), O (1.6 parts) OH (1 part), O (1.6 parts)	130

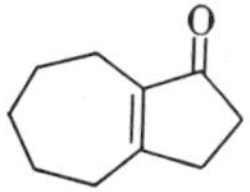

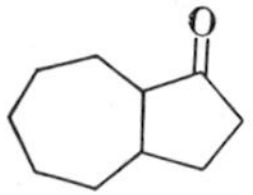

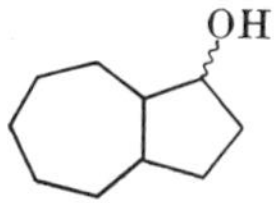

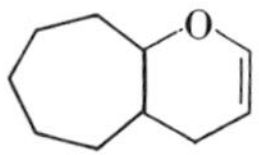

	Reactant	Conditions	Product(s) and Yield(s) (%)	Refs.
C_{10} (*contd.*)	$CO_2C_2H_5$, O	Na-$C_{10}H_8$/THF	HO, O (11)	130
C_{12}	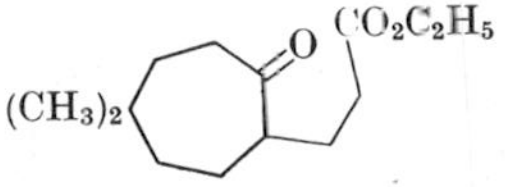 $(CH_3)_2$, O, $CO_2C_2H_5$	Na-$C_{10}H_8$/THF	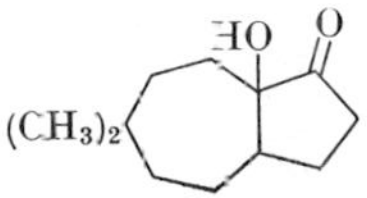 $(CH_3)_2$, HO, O (27)	130
C_{19}	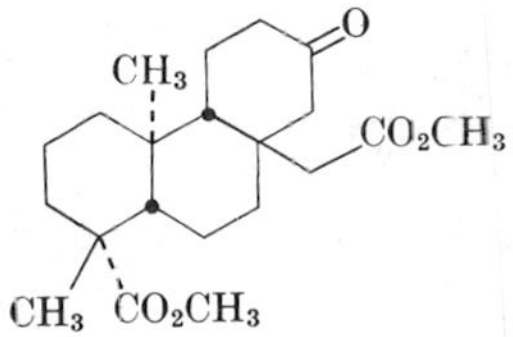 CH_3, O, CO_2CH_3, CH_3 CO_2CH_3	Na-NH_3/THF	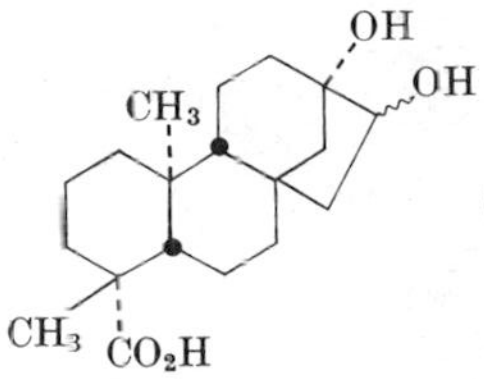 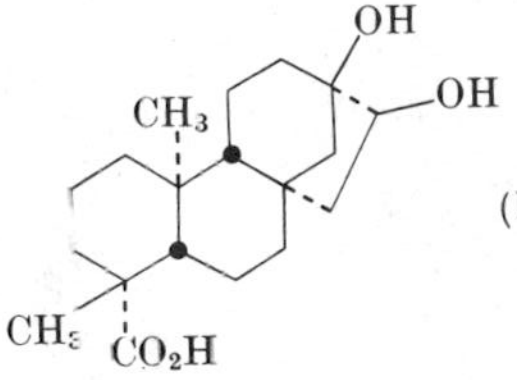 CH_3, OH, OH, CH_3 CO_2H (30), CH_3, OH, OH, CH_3 CO_2H (10)	133
	B. Electrochemical Reductions			
C_2, C_7	$(CH_3CO)_2O + C_6H_5CHO$	e^- (Hg)/CH_3CN/25°	$C_6H_5CH(OCOCH_3)CH(OCOCH_3)C_6H_5$ (—) (1:1 *meso* + *dl*)	132
C_2, C_8	$(CH_3CO)_2O + C_6H_5COCH_3$	e^- (Hg)/CH_3CN/25°	$C_6H_5C(OCOCH_3)(CH_3)COCH_3$ (15)	132
C_2, C_{13}	$(CH_3CO)_2O + (C_6H_5)_2CO$	e^- (Hg)/CH_3CN/25°	$(C_6H_5)_2C(OCOCH_3)COCH_3$ (50)	132

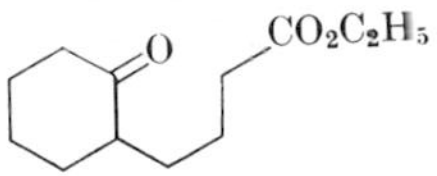

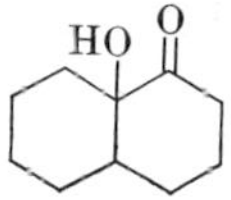

Note: References 180–356 are on pp. 400–403.

TABLE VI. Esters which Fail To Undergo the Acyloin Condensation

	Starting Material	Metal/Solvent/Temp (°C)	Product(s) (% Yield)	Refs.
	A. Malonic Esters which Fragment by Dealkoxycarbonylation			
C_5	$(CH_3)_2C(CO_2CH_3)_2$	Na/xylene, TMCS/reflux	$(CH_3)_2C{=}C(OCH_3)OSi(CH_3)_3$ (87)	78, 56
		Na-K/ether, TMCS	" (60)	56
	$(CH_3)_2C(CO_2C_2H_5)_2$	Na-K/ether, TMCS/reflux	$(CH_3)_2C{=}C(OC_2H_5)OSi(CH_3)_3$ (86)	86
		Na/petroleum ether, TMCS/reflux	" (85)	86
		Na/xylene/reflux	$(CH_3)_2CHCO_2C_2H_5$ (36)	79
		K/xylene/reflux	" (30)	79
C_6	$\overline{(CH_2)_3—C}(CO_2C_2H_5)_2$ (ring)	Na-K/petroleum ether, TMCS/reflux	$\overline{(CH_2)_3—C}{=}C(OC_2H_5)OSi(CH_3)_3$ (ring) (80)	86
		Na/(—), TMCS	" (60)	78
C_7	$(C_2H_5)_2C(CO_2C_2H_5)_2$	Na/(—), TMCS	$(C_2H_5)_2C{=}C(OC_2H_5)OSi(CH_3)_3$ (88)	78
		K/xylene/reflux	$(C_2H_5)_2CHCO_2C_2H_5$ (46)	79
C_9	$(n\text{-}C_3H_7)_2C(CO_2C_2H_5)_2$	Na/xylene/reflux	$(n\text{-}C_3H_7)_2CHCO_2C_2H_5$ (61)	79
		K/xylene/reflux	" (37)	79
C_{11}	$C_6H_5(C_2H_5)C(CO_2C_2H_5)_2$	Na/(—), TMCS	$C_6H_5(C_2H_5)C{=}C(OC_2H_5)OSi(CH_3)_3$ (90)	78
C_{15}	$(C_6H_5)_2C(CO_2CH_3)_2$	Na/(—), TMCS	$(C_6H_5)_2C{=}C(OCH_3)OSi(CH_3)_3$ (85)	78
C_{17}	$(C_6H_5CH_2)_2C(CO_2C_2H_5)_2$	Na/xylene/reflux	$(C_6H_5CH_2)_2CHCO_2C_2H_5$ (38)	79
		K/xylene/reflux	" (32)	79
	B. 1,2-Diesters which Fragment by Breaking the 1,2-Carbon-Carbon Bond			
C_5	[cyclopropane structure] CO_2CH_3, CO_2CH_3	Na/toluene, TMCS	$CH_2(CH{=}C(OCH_3)[OSi(CH_3)_3])_2$ (47)	70, 261
		Na/C_6H_6/TMCS/reflux	" (39)	341
		$Na\text{-}NH_3/-78°$	$CH_3O_2C(CH_2)_3CO_2CH_3$ (22)	70, 261
	[cyclopropane structure] CO_2CH_3, CO_2CH_3	$Na\text{-}NH_3/-78°$	$CH_3O_2C(CH_2)_3CO_2CH_3$ (25)	70, 261

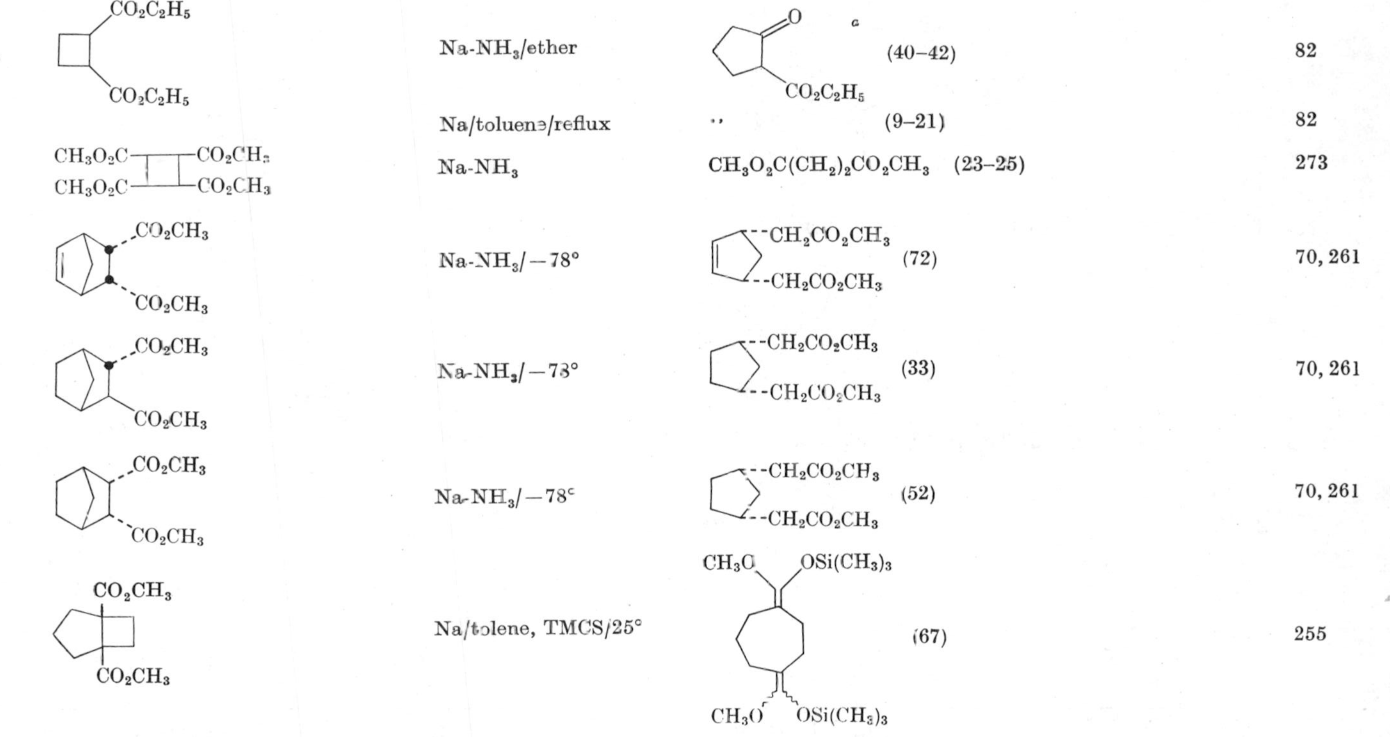

Carbon	Reactant	Conditions	Product(s) (% yield)	Refs.
C_6		Na-NH_3/ether	[a] (40–42)	82
		Na/toluene/reflux	" (9–21)	82
C_8		Na-NH_3	$CH_3O_2C(CH_2)_2CO_2CH_3$ (23–25)	273
C_9		Na-NH_3/−78°	(72)	70, 261
		Na-NH_3/−78°	(33)	70, 261
		Na-NH_3/−78°	(52)	70, 261
		Na/tolene, TMCS/25°	(67)	255

Note: References 184–356 are on pp. 400–403.

[a] Ring opening was followed by a Dieckmann cyclization.

TABLE VI. ESTERS WHICH FAIL TO UNDERGO THE ACYLOIN CONDENSATION (*Continued*)

B. 1,2-Diesters which Fragment by Breaking the 1,2-Carbon-Carbon Bond (Continued)

	Starting Material	Metal/Solvent/Temp (°C)	Product(s) (% Yield)	Refs.
C_{10}	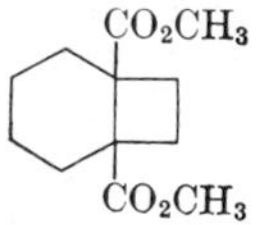 CO_2CH_3, CO_2CH_3	Na-K/toluene, TMCS/25°	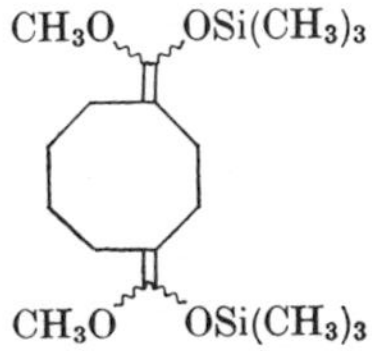 CH_3O, $OSi(CH_3)_3$, CH_3O, $OSi(CH_3)_3$ (88)	86
	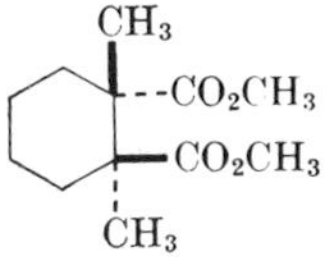 CH_3, CO_2CH_3, CO_2CH_3, CH_3	Na-NH$_3$/−78°	$CH_3O_2CCH(CH_3)(CH_2)_4CH(CH_3)CO_2CH_3$ (54)	70, 261
	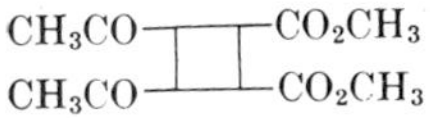 CH_3CO, CO_2CH_3, CH_3CO, CO_2CH_3	Na-NH$_3$	$CH_3O_2C(CH_2)_2CO_2CH_3$ (4), $CH_3CO(CH_2)_2COCH_3$ (24), $CH_3CO(CH_2)_2CO_2CH_3$ (12)	273
C_{11}	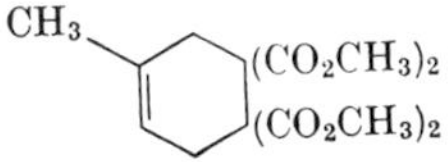 CH_3, $(CO_2CH_3)_2$, $(CO_2CH_3)_2$	Na-NH$_3$/THF/ −70 to −30°	*cis*-$(CH_3O_2C)_2CHCH_2C(CH_3){=}CHCH_2CH(CO_2CH_3)_2$ (87)	122
C_{14}	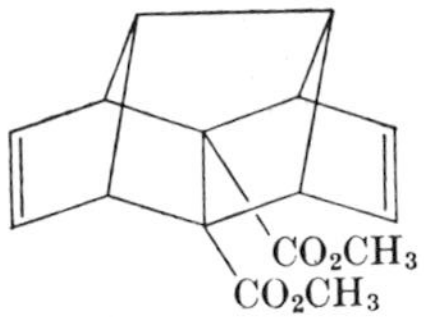 CO_2CH_3, CO_2CH_3	Na/toluene, TMCS/-	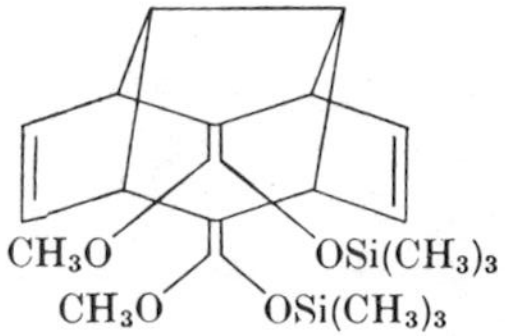 CH_3O, $OSi(CH_3)_3$, CH_3O, $OSi(CH_3)_3$ (—)	342

C. α,β-Unsaturated Esters which Couple Tail-to-Tail and then Undergo Dieckmann Cyclization

C_9	$C_6H_5CH{=}CHCO_2R$ ($R = CH_3$, C_2H_5, $CH_2C_6H_5$)	Na/xylene/reflux	C_6H_5, C_6H_5, $=O$	(—)	125, 126
	$C_6H_5CH{=}CHCO_2R$[b]	Na/ether/reflux	C_6H_5, C_6H_5, $=O$, CO_2R	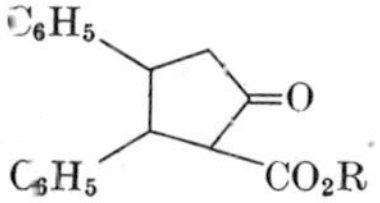 (10–41)	123, 124
	o-$ClC_6H_4CH{=}CHCO_2C_2H_5$	Na/ether/reflux	$CO_2C_2H_5$, o-ClC_6H_4, $=O$, o-ClC_6H_4	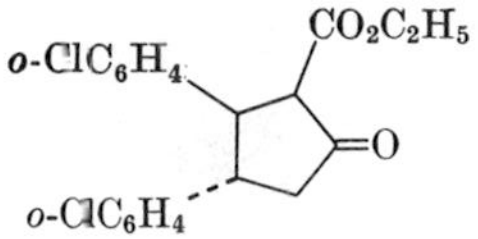 (9)	123
	p-$ClC_6H_4CH{=}CHCO_2C_2H_5$	Na/ether/reflux	$CO_2C_2H_5$, p-ClC_6H_4, $=O$, p-ClC_6H_4	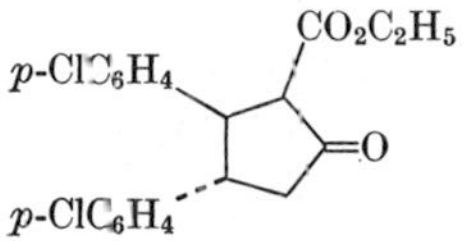 (9)	123
C_{10}	p-$CH_3C_6H_4CH{=}CHCO_2C_2H_5$	Na/ether/reflux	$CO_2C_2H_5$, p-$CH_3C_6H_4$, $=O$, p-$CH_3C_3H_4$	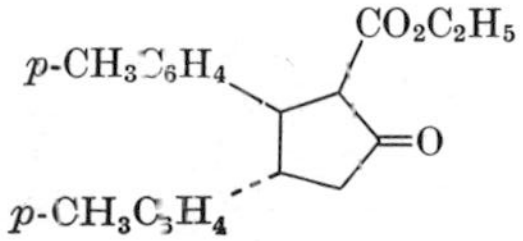 (25)	123

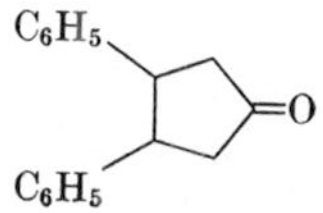

Note: References 184–356 are on pp. 400–403.

[b] R = C_2H_5, n-C_3H_7, i-C_3H_7, n-C_4H_9, i-C_4H_9, sec-C_4H_9, t-C_4H_9, n-C_5H_{11}, $CH_2CH(CH_3)C_2H_5$, cyclohexyl, $CH_2C_6H_5$, $(CH_2)_2C_6H_5$.

TABLE VI. ESTERS WHICH FAIL TO UNDERGO THE ACYLOIN CONDENSATION (*Continued*)

	Starting Material	Metal/Solvent/Temp (C°)	Products (% Yield)	Refs.
	D. Esters which Give Dieckmann Condensation Products under Acyloin Conditions			
C_7	$C_2H_5O_2C(CH_2)_5CO_2C_2H_5$	Na/xylene/reflux[c]	$(CH_2)_5$—C=O (13), $(CH_2)_5$COCHOH[d] (15)	98, 216
	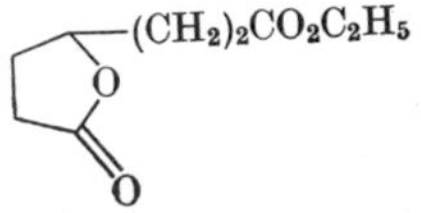 $(CH_2)_2CO_2C_2H_5$ (lactone)	Na/xylene/reflux	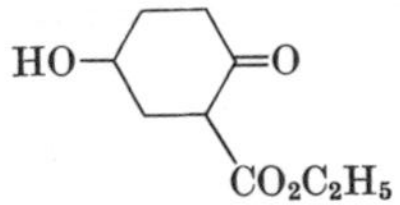 HO—, =O, $CO_2C_2H_5$ (7)	98, 216
	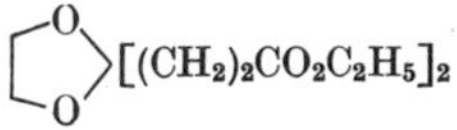 $[(CH_2)_2CO_2C_2H_5]_2$ (dioxolane)	Na/xylene/reflux	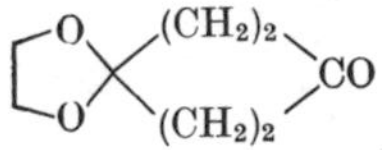 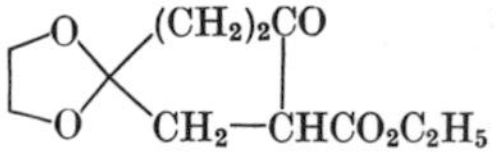 $(CH_2)_2$, CO, $(CH_2)_2$ (12), $(CH_2)_2CO$, CH_2—$CHCO_2C_2H_5$ (15)	98, 216
C_8	$CH_3OCH(CH_2)_2CO_2CH_3$ $CH_3OCH(CH_2)_2CO_2CH_3$	Na/xylene/reflux	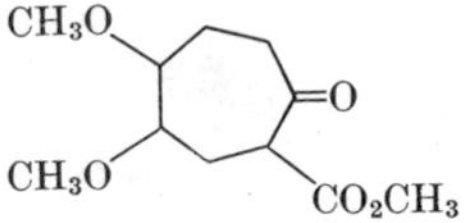 CH_3O, CH_3O, =O, CO_2CH_3 (31)	281
C_9	$CO(CH_2CH_2CH_2CO_2(CH_5)_2$	Na/toluene-naphthalene/ reflux	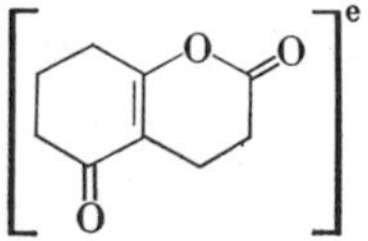 [ ][e] (70)	118
C_9	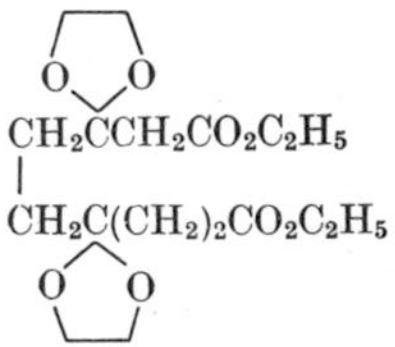 $CH_2CCH_2CO_2C_2H_5$ $CH_2C(CH_2)_2CO_2C_2H_5$	Na-NH_3/ether	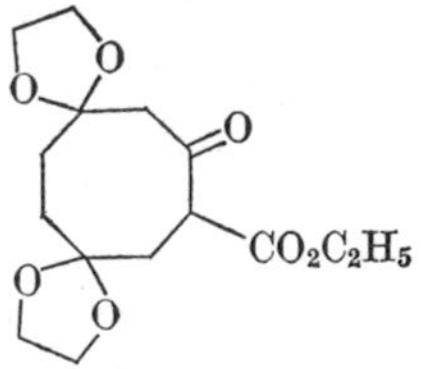 =O, $CO_2C_2H_5$ (—)	232

C_{10}	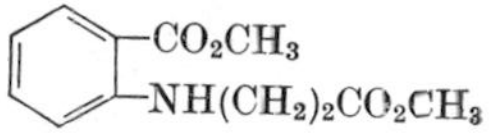 o-$C_6H_4(CO_2CH_3)NH(CH_2)_2CO_2CH_3$	Na/xylene/reflux	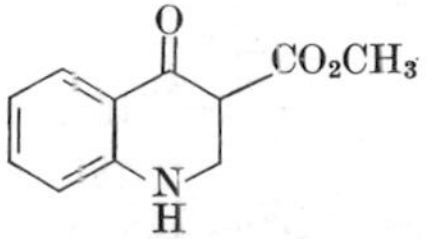 3-carbomethoxy-4-oxo-1,2,3,4-tetrahydroquinoline (–)	252
C_{14}	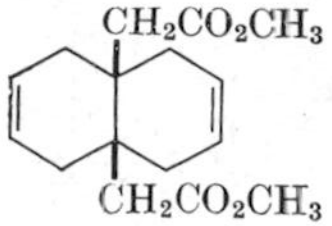 $CH_2CO_2CH_3$, $CH_2CO_2CH_3$ (angular hexahydronaphthalene diester)	Na/dioxane/reflux	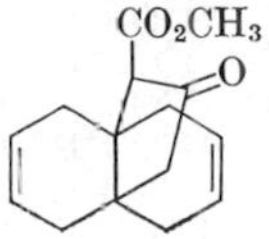 CO_2CH_3, O (bridged keto ester) (86)	242
C_{28}	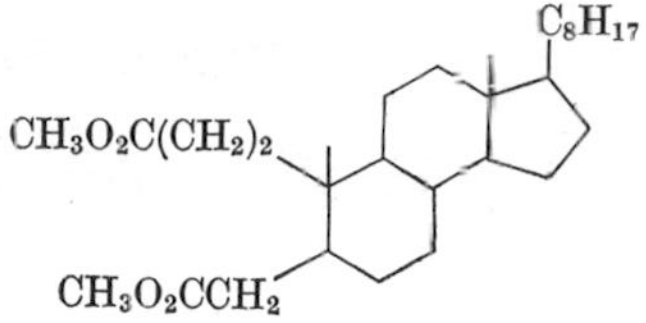 $CH_3O_2C(CH_2)_2$, $CH_3O_2CCH_2$, C_8H_{17} (steroid diester)	Na/xylene/reflux	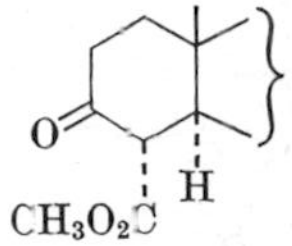 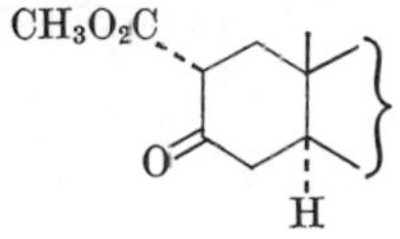 CH_3O_2C, O, H (20), CH_3O_2C, O, H (52)	323

E. Esters which are Reduced but do not Give Acyloins

C_2	CH_3CO_2R ($R = n$-C_3H_7, n-C_4H_9)	Na-NH_3/ether/$-70°$	CH_3CHO, C_2H_5OH, CH_3CONH_2 (—)	55
	$C_2H_5O_2CCO_2C_2H_5$	Na-K (excess)/ether, TMCS/reflux	$(C_2H_5O)[(CH_3)_3SiO]C{=}C[OSi(CH_3)_3](OC_2H_5)$ (65)	86, 78
		K/C_6H_6, TMCS/25°	" (Poor)[f]	86
	i-$C_3H_7O_2CCO_2C_3H_7$-i	Na-K (excess)/ether, TMCS/reflux	$(i$-$C_3H_7O)[(CH_3)_3SiO]C{=}C[OSi(CH_3)_3](OC_3H_7$-$i)$ (66)	78

Note: References 184–356 are on pp. 400–403.

[c] A commercial 40% sodium dispersion was used.

[d] The product was contaminated with 1,2-cyclohexanedione.

[e] This is not strictly a Dieckmann reaction but is an internal Claisen condensation. The product actually isolated was not the lactone but the free acid obtained on hydrolytic work-up of the reaction mixture. No reaction occurs in the absence of naphthalene, Table VIF.

[f] Benzene is reduced to give 1,4-bis-(trimethylsilyl)-2,5-cyclohexadiene.

TABLE VI. Esters which Fail To Undergo the Acyloin Condensation (*Continued*)

E. Esters which are Reduced but do not Give Acyloins (*Continued*)

	Starting Material	Metal/Solven/Temp (°C)	Product(s) (% Yield)	Refs.
C_3	$Cl(CH_2)_2CO_2C_2H_5$	Na/ether, TMCS/reflux	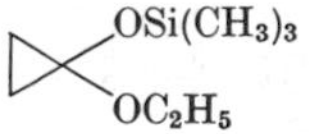 1-$OSi(CH_3)_3$-1-OC_2H_5-cyclopropane (78)	16
C_4	$CH_3CHBrCO_2CH_3$	Na/ether, TMCS/reflux	$CH_3CH{=}C(OCH_3)OSi(CH_3)_3$ (75)	16
	i-$C_3H_7CO_2C_2H_5$	Na-NH_3	i-$C_3H_7CH_2OH$ (21), i-$C_3H_7CONH_2$ (41), i-$C_3H_7CO_2H$ (4)	45
C_5	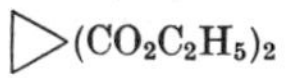 cyclopropane-$(CO_2C_2H_5)_2$	1. Na-NH_3/ether 2. TMCS	$C_2H_5CH(CO_2C_2H_5)_2$ (53)	77
	$CH_3O_2C(CH_2)_2CO_2CH_3$	Na/conc. soln. in toluene/reflux	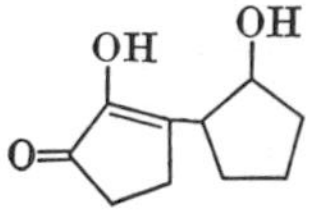 2-hydroxy-3-(2-hydroxycyclopentyl)cyclopent-2-en-1-one (O=, OH, OH) (42)	127
C_6	$C_6H_5CO_2CH_3$	Na-NH_3/C_6H_6-ether/$-70°$	C_6H_5CHO, $C_6H_5CH_2OH$, $C_6H_5CONH_2$, (—) $C_6H_5CONH{-})_2$ (—)	55
	$C_6H_5CO_2CH_3$	Na/toluene/reflux[g]	$C_6H_5CO_2H$ (78), CH_4 (—)	73
	$C_6H_5CO_2C_2H_5$	Na/toluene/reflux[g]	$C_6H_5CO_2H$ (—), C_2H_6 (—)	73
	$C_6H_5CO_2CH_2CH{=}CH_2$	Na/toluene/reflux[g]	$C_6H_5CO_2H$ (—), $CH_2{=}CHCH_2{-})_2$ (34)	73
	$C_6H_5CO_2CH_2C_6H_5$	Na/toluene/reflux[g]	$C_6H_5CO_2H$ (—), $C_6H_5CH_2{-})_2$ (38)	73
	$C_6H_5CO_2CH(CH_3)C_6H_5$	Na/toluene/reflux[g]	$C_6H_5CO_2H$ (—), *meso*-$C_6H_5CH(CH_3){-})_2$ (30)	73
	$C_6H_5CO_2CH_2CH{=}CHC_6H_5$	Na/toluene/reflux[g]	$C_6H_5CO_2H$ (—), $C_6H_5CH{=}CHCH_2{-})_2$ (39)	73
	$C_6H_5CO_2CH(C_6H_5)_2$	Na/toluene/reflux[g]	$C_6H_5CO_2H$ (—), $(C_6H_5)_2CH{-})_2$ (48)	73
C_{10}	2,4,6-$(CH_3)_3C_6H_2CO_2CH_3$	Li-NH_3/THF/$-34°$	2,4,6-$(CH_3)_3C_6H_2CH_2OH$ (49)	53
C_{11}	 $RO_2C(CH_2)_2$-cyclopentane-$(CH_2)_2CO_2R$ ($R = CH_3$, C_2H_5)	Na/dioxane/reflux	 $HO(CH_2)_3$-cyclopentane-$(CH_2)_3OH$ (–)	57

	Substrate	Conditions	Product(s) and Yield(s) (%)	Refs.
C_{13}	[structure: $CO_2C_2H_5$, H]	$Li\text{-}NH_3/THF/-34°$	[structure: R, H] $R = CO_2H$ (40); $R = CH_2OH$ (30)	123
C_{17}	[structure: OH, CO_2CH_3]	$Li\text{-}NH_3/THF/-34°$	[structure: R] $R = CO_2H$ (77) $R = CH_2OH$ (23)	53
C_{20}	[structure: CO_2CH_3]	$Li\text{-}NH_3/THF/-34°$	[structure: R] $R = CO_2H$ (3) $R = CH_2OH$ (62)	53
C_{28}	[structure: HO, CO_2CH_3]	$Li\text{-}NH_3/THF/-34°$	[structure: R] $R = CO_2H$ (68) $R = CH_2OH$ (28)	53

Note: References 180–356 are on pp. 400–403.

[f] Benzene is reduced to give 1,4-bis-(trimethylsilyl)-2,5-cyclohexadiene.

[g] The sodium was not highly dispersed.

TABLE VI. Esters which Fail To Undergo the Acyloin Condensation (*Continued*)

F. Esters which Are not Reduced under the Given Conditions

	Starting Material	Metal/Solvent/Temp (°C)	Product(s) (% Yield)	Refs.
C_2	$CH_3O_2CCO_2CH_3$	Na-K (excess)/ether, TMCS/reflux		86
	$(CH_3)_2NCOCON(CH_3)_2$	Na-K (excess)/ether, TMCS/reflux		86
C_3	$C_2H_5CON(C_2H_5)_2$	Na/C_6H_6/reflux		74
C_4	$C_2H_5O_2C(CH_2)_2CO_2C_2H_5$	Li/ether, TMCS/reflux		86
C_6	*trans*-$(CH_2)_2CH(CO_2CH_3)CH(CO_2CH_3)$ (ring)	Na-K (excess)/ether, TMCS/reflux		86
	$CH_3O_2CCH_2CH(CO_2CH_3)CH_2CO_2CH_3$	Na or K/toluene or C_6H_6, TMCS/reflux	(Not reduced in the presence of esters which are reduced!)	176
C_7	C_6H_{11}—$CON(C_2H_5)_2$ (cyclohexyl)	Na/C_6H_6/reflux		74
	trans-$CH_2CH(CH_2CO_2CH_3)CH(CH_2CO_2CH_3)$ (ring)		Na or Na-K/C_6H_6 toluene, TMCS/25° or reflux	341
	trans-$(CH_2)_3CH(CO_2CH_3)CH(CO_2CH_3)$ (ring)	Na-K (excess)/C_6H_6, TMCS/reflux		86
	n-$C_4H_9CH(CO_2C_2H_5)_2$	Na-K/ether or petroleum ether, TMCS/reflux	n-$C_4H_9C(CO_2C_2H_5)$=$C(OC_2H_5)[OSi(CH_3)_3]$	86
C_8	*cis*-$(CH_2)_4CH(CO_2CH_3)CH(CO_2CH_3)$ (ring)	Na(Hg)/toluene, TMCS/25°		86
	trans-$CH_3O_2C(CH_2)_2CH$=$CH(CH_2)_2CO_2CH_3$			341
C_9	p-$O_2NC_6H_4CH$=$CHCO_2C_2H_5$	Na/C_6H_6, toluene, or xylene/reflux		124
	$CO(CH_2CH_2CH_2CO_2C_2H_5)_2$	Na/toluene/reflux[h]		118
C_{10}	$CH_3O_2C(CH_2)_3C{\equiv}C(CH_2)_3CO_2CH_3$	Na/toluene/reflux		35
	bicyclo[4.2.0]oct-7-ene-1,6-di(CO_2CH_3) (structure)	Na/petroleum ether, TMCS/reflux		86

	Reactant	Conditions	Refs.
	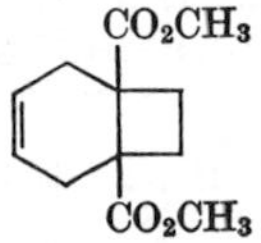 (bicyclic diester: CO_2CH_3, CO_2CH_3)	Na/(—), TMCS	275
C_{12}	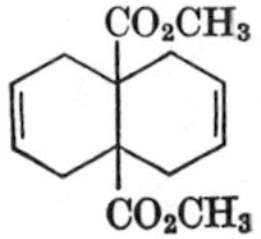 (bicyclic diester: CO_2CH_3, CO_2CH_3)	Na/toluene/reflux	54
	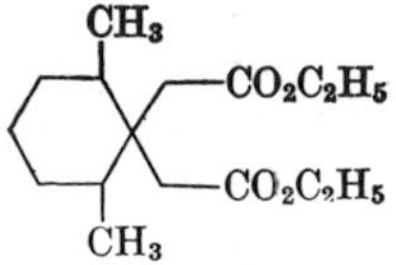 (dimethylcyclohexane: CH_3, $CO_2C_2H_5$, $CO_2C_2H_5$, CH_3)	Na/xylene/reflux	95
C_{15}	$CH_3O_2C(CH_2)_m$-p-$C_6H_4(CH_2)_nCO_2CH_3$ ($m = 1, n = 6, m = 2, n = 5$)	Na/xylene/reflux	286
	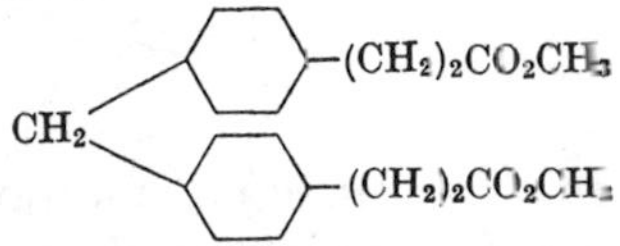 CH_2 (bis-cyclohexyl: $(CH_2)_2CO_2CH_3$, $(CH_2)_2CO_2CH_3$)	Na/xylene/reflux	284
C_{16}	CH_2 (p-$C_6H_4(CH_2)_2CO_2CH_3$, p-$C_6H_4(CH_2)_3CO_2CH_3$)	Na/xylene/reflux	284

Note: References 184–356 are on pp. 400–403.

[h] In the presence of naphthalene an internal Claisen condensation occurs, Table VID.

TABLE VI. Esters which Fail To Undergo the Acyloin Condensation (*Continued*)

G. Esters which Give Either Polymers or Uncharacterized Mixtures

	Starting Material	Metal/Solvent/Temp (°C)	Product(s) (% Yield)	Refs.
C_4	$CH_3CH{=}CHCO_2C_2H_5$	Na/xylene/reflux		125
	$CH_3O_2CCH_2OCH_2CO_2CH_3$	Na/(—), TMCS		86
C_8	$p\text{-}NH_2C_6H_4CH_2CO_2C_2H_5$	Na		191
	trans-$(CH_2)_4CH(CO_2CH_3)CH(CO_2CH_3)$	Na-K/C_6H_6 or ether, TMCS/25–30°		86
C_{11}	$CH_3O_2C(CH_2)_3C{\equiv}C(CH_2)_4CO_2CH_3$	Na/xylene/reflux		285
C_{12}	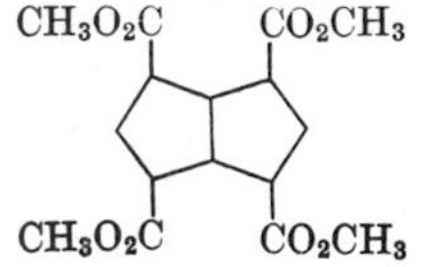 CH_3O_2C, CO_2CH_3, CH_3O_2C, CO_2CH_3	Na-NH_3/ether		313
C_{13-16}, C_{18}, C_{19}, C_{21}	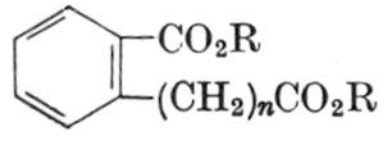 CO_2R, $(CH_2)_nCO_2R$ (n = 5–8, 10, 11, 13; R = CH_3, C_2H_5)	Na-NH_3/ether or Na/xylene/reflux		206
C_{16}	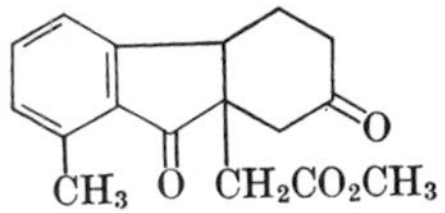 O, CH_3, O, $CH_2CO_2CH_3$	Na/(—)		241
C_{17}, C_{20-} C_{22}	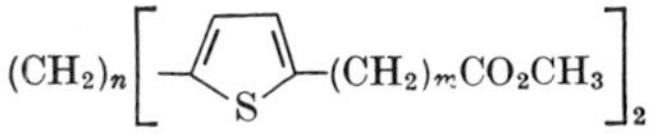 $(CH_2)_n[\text{—}\text{(S)}\text{—}(CH_2)_mCO_2CH_3]_2$ (m = 3, n = 1; m = 5, n = 1; m = 2, n = 6; m = 3, n = 6)	Na/xylene/reflux		202
C_{20-22}	$(CH_2)_m[N(R)\text{-}p\text{-}C_6H_4(CH_2)_nCO_2CH_3]_2$ (R = H, C_6H_5CO, $C_6H_5CH_2$; m = 4, n = 1, 2; m = 3, n = 2)	Na		219
C_{21}	$ROCH[p\text{-}C_6H_4(CH_2)_3CO_2CH_3]_2$ (R = H, tetrahydropyranyl)	Na/xylene/reflux		317

Note: References 184–356 are on pp. 400–403.

TABLE VII. SEMIDIONES

	Starting Material	Method[a]	Semidiones	Refs.
	Cyclobutane Semidiones			
C_4	$CH_3O_2C(CH_2)_2CO_2CH_3$	A	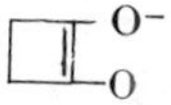 O^-, O	136
C_7	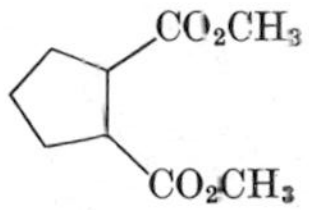 CO_2CH_3, CO_2CH_3	B	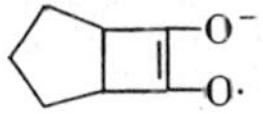 O^-, $O\cdot$	41
C_8	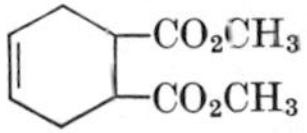 CO_2CH_3, CO_2CH_3	C	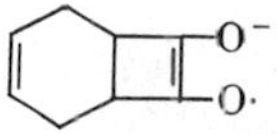 O^-, $O\cdot$	41, 340
	CO_2CH_3, CO_2CH_3	C	O^-, $O\cdot$	41, 340
	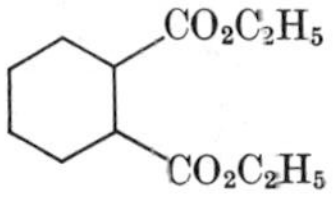 $CO_2C_2H_5$, $CO_2C_2H_5$	A	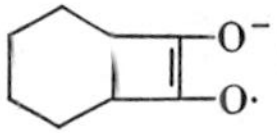 O^-, $O\cdot$	136
C_9	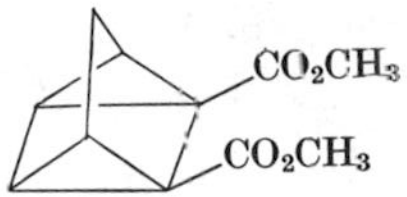 CO_2CH_3, CO_2CH_3	B	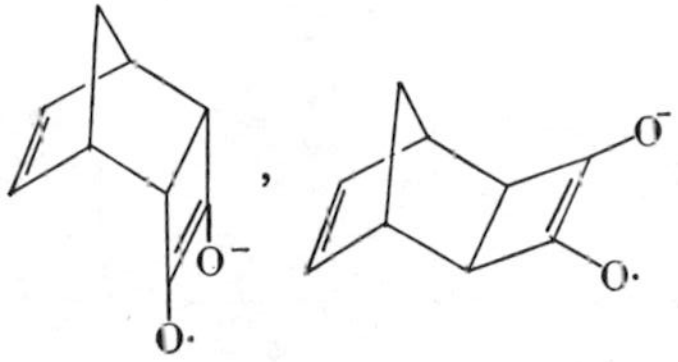 O^-, $O\cdot$, O^-, $O\cdot$	41

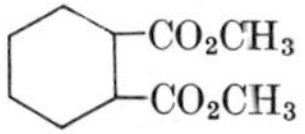

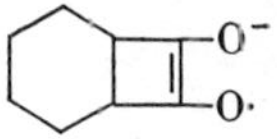

Note: References 184–356 are on pp. 400–403.

[a] Method A = Na-K/glyme. Method B = 1. Na-K/glyme or ether, TMCS; and 2. t-C_4H_9OK/DMSO. Method C = 1. Na-K/glyme and 2. t-C_4H_9OK/DMSO.

TABLE VII. SEMIDIONES (*Continued*)

Cyclobutane Semidiones

	Starting Material	Method[a]	Semidiones	Refs.
C_9 (*contd.*)	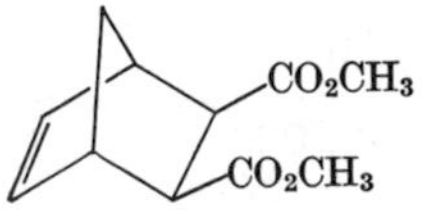 CO_2CH_3, CO_2CH_3	A	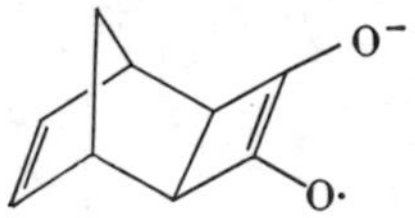 O^-, O·	136
		C	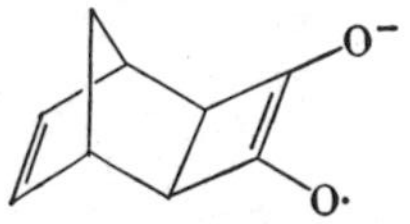 O^-, O·	41
	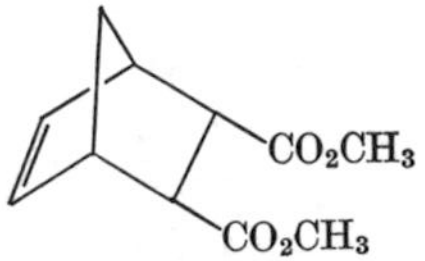 CO_2CH_3, CO_2CH_3	A or C	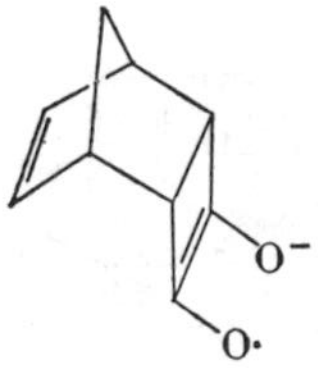 O^-, O·	136, 41
	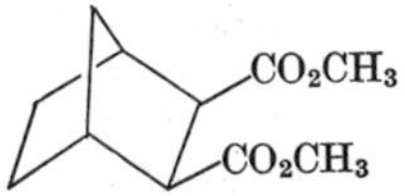 CO_2CH_3, CO_2CH_3	A or C	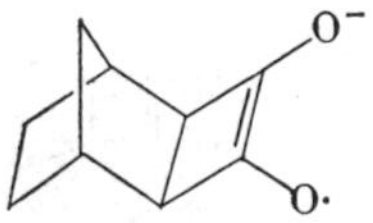 O^-, O·	136, 41
	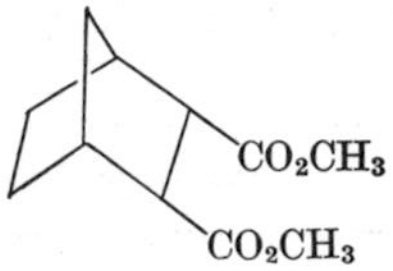 CO_2CH_3, CO_2CH_3	A or C	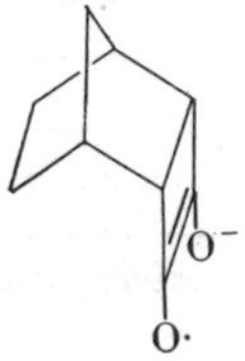 O^-, O·	136, 41

		B		41, 340
		B		41, 340
C_{10}		C	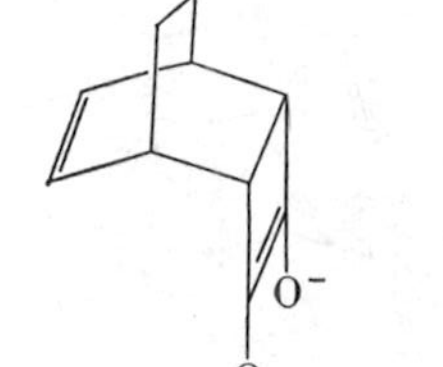	41
		C	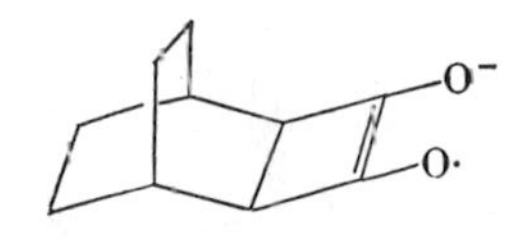	41
C_{11}		B		41

Note: References 184–356 are on pp. 400–403.

[a] Method A = Na-K/glyme. Method B = 1. Na-K/glyme or ether, TMCS; and 2. $t\text{-}C_4H_9OK$/DMSO. Method C = 1. Na-K/glyme and 2. $t\text{-}C_4H_9OK$/DMSO.

TABLE VII. SEMIDIONES (*Continued*)

	Starting Material	Method[a]	Semidiones	Refs.
	Cyclobutane Semidiones			
C_{11} (*contd.*)	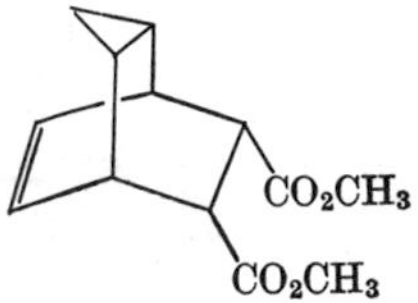 CO_2CH_3, CO_2CH_3	B	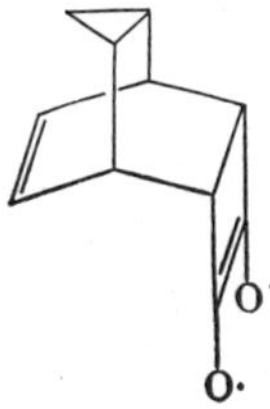 O^-, O·	41
	CO_2CH_3, CO_2CH_3	C	O^-, O·	41
	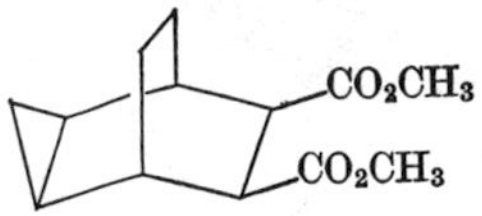 CO_2CH_3, CO_2CH_3	B	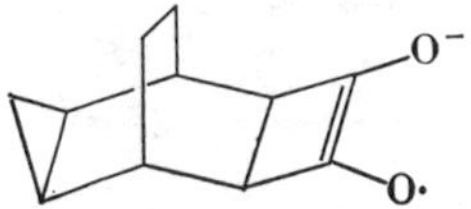 O^-, O·	41
C_{12}	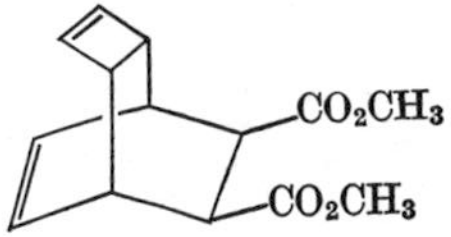 CO_2CH_3, CO_2CH_3	C	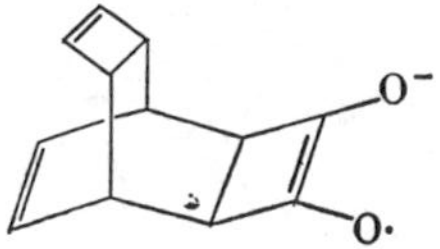 O^-, O·	41
	CO_2CH_3, CO_2CH_3	B	O^-, O·	41

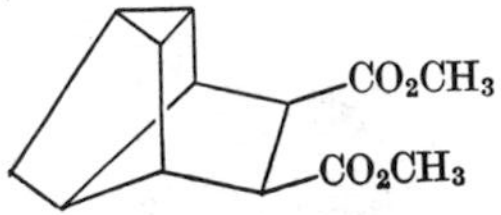

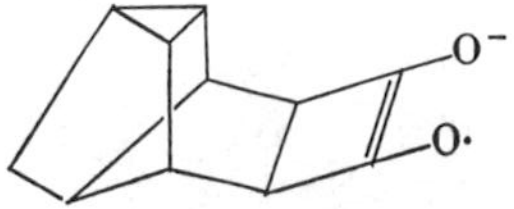

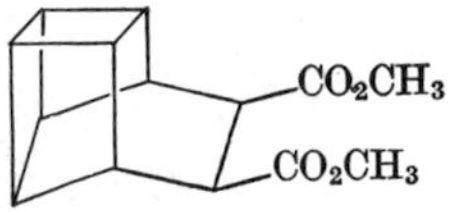

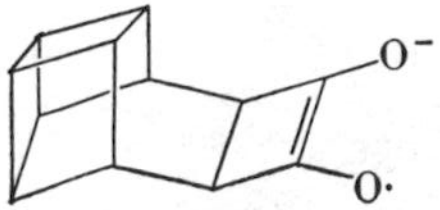

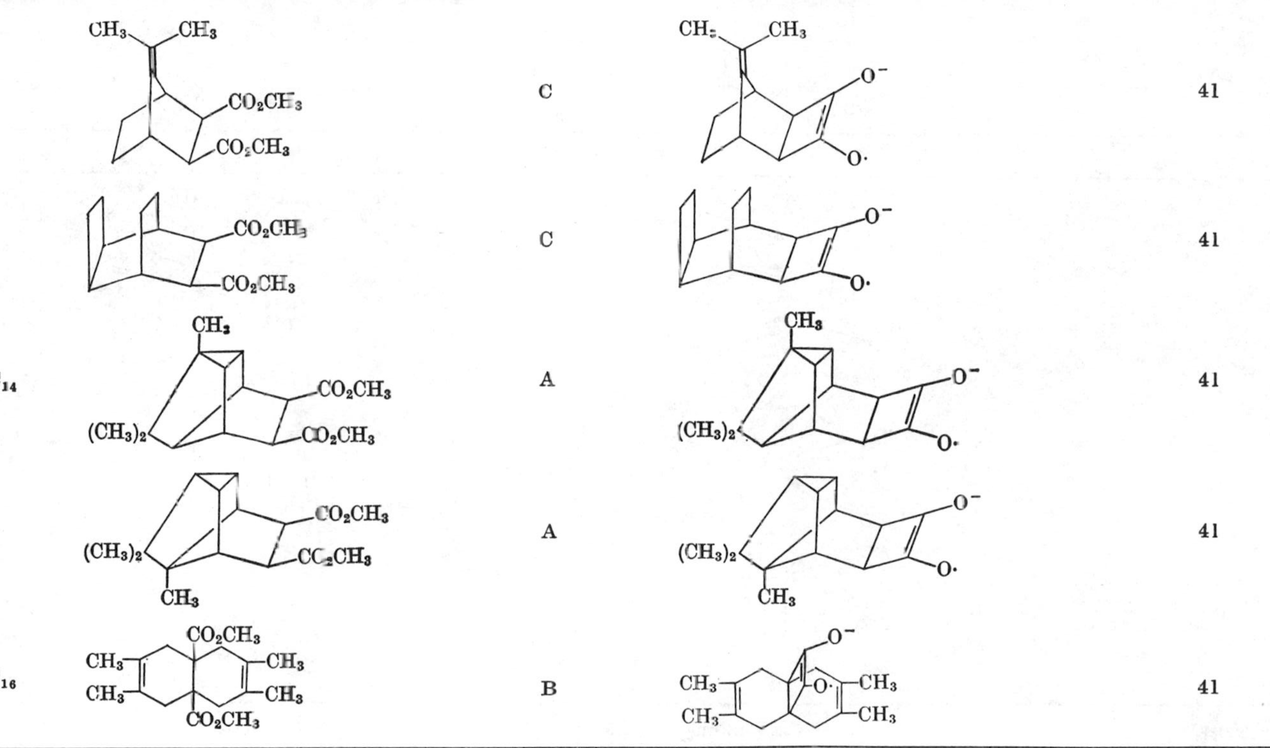

		Method[a]		Refs.
		C		41
		C		41
C_{14}		A		41
		A		41
C_{16}		B		41

Note: References 184–356 are on pp. 400–403.

[a] Method A = Na-K/glyme. Method B = 1. Na-K/glyme or ether, TMCS; and 2. t-C_4H_9OK/DMSO. Method C = 1. Na-K/glyme and 2. t-C_4H_9OK/DMSO.

TABLE VII. SEMIDIONES (*Continued*)

	Starting Material	Method[a]	Semidiones	Refs.
Cyclobutane Semidiones (Continued)				
C_{18}	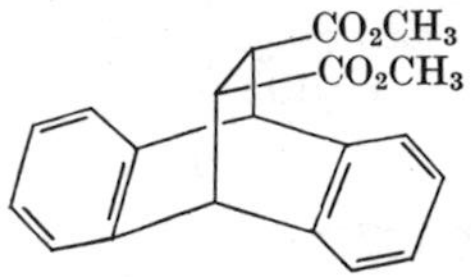 CO_2CH_3, CO_2CH_3	—	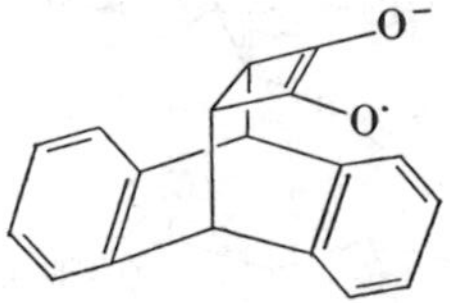 O^-, $O^\cdot$	41
C_{19}	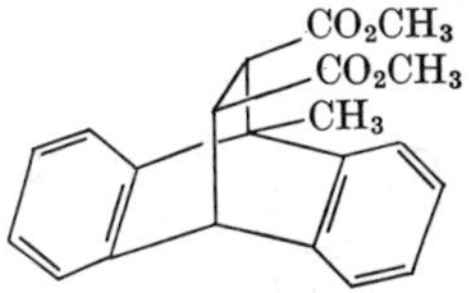 CO_2CH_3, CO_2CH_3, CH_3	—	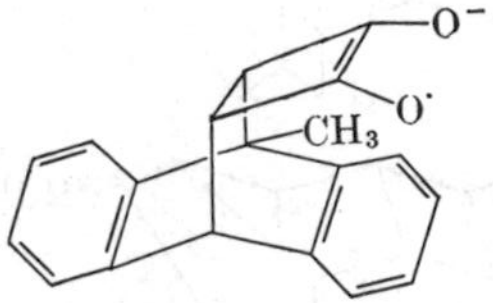 O^-, $O^\cdot$, CH_3	41
C_{20}	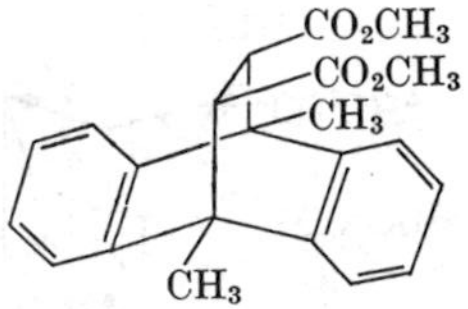 CO_2CH_3, CO_2CH_3, CH_3, CH_3	—	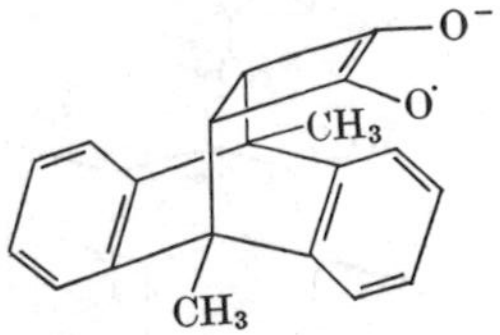 O^-, $O^\cdot$, CH_3, CH_3	41
Cyclopentane Semidiones				
C_5	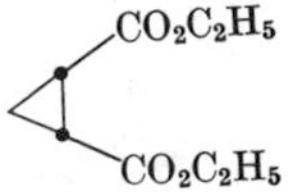 $CO_2C_2H_5$, $CO_2C_2H_5$	B	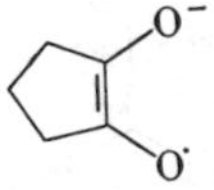 O^- [b], $O^\cdot$	137

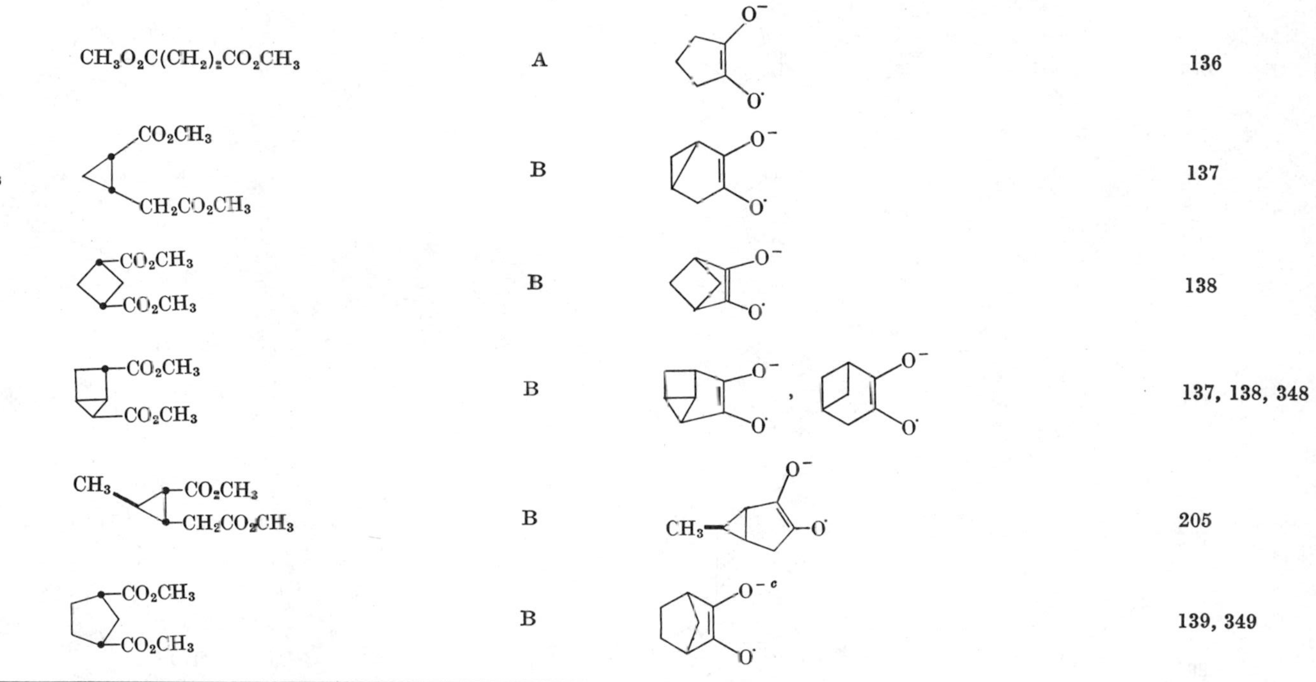

	Reactant	Method[a]	Product	Refs.
	$CH_3O_2C(CH_2)_{[illegible]}CO_2CH_3$	A	O^-, $O^\cdot$	136
C_6	CO_2CH_3, $CH_2CO_2CH_3$	B	O^-, $O^\cdot$	137
	CO_2CH_3, CO_2CH_3	B	O^-, $O^\cdot$	138
C_7	CO_2CH_3, CO_2CH_3	B	O^-, $O^\cdot$, O^-, $O^\cdot$	137, 138, 348
	CH_3, CO_2CH_3, $CH_2CO_2CH_3$	B	CH_3, O^-, $O^\cdot$	205
	CO_2CH_3, CO_2CH_3	B	O^{-} [c], $O^\cdot$	139, 349

Note: References 184–356 are on pp. 400–403.

[a] Method A = Na-K/glyme. Method B = 1. Na-K/glyme or ether, TMCS; and 2. t-C_4H_9OK/DMSO. Method C = 1. Na-K/glyme and 2. t-C_4H_9OK/DMSO.

[b] It has been pointed out that this product is probably the result of fragmentation followed by cyclization.[341]

[c] A series of deuterated analogs of this compound were prepared.

TABLE VII. SEMIDIONES (*Continued*)

Cyclopentane Semidiones (Continued)

	Starting Material	Method[a]	Semidiones	Refs.
C_8	CO_2CH_3, CO_2CH_3	B	O^-, $O^\cdot$	138
	CO_2CH_3, CO_2CH_3	B	O^-, $O^\cdot$	204
	CO_2CH_3, CO_2CH_3	B	O^-, $O^\cdot$	204
	CO_2CH_3, CH_3, CO_2CH_3	B	CH_3, O^-, $O^\cdot$	205
	CO_2CH_3, CH_3, CO_2CH_3	B	CH_3, O^- [d], $O^\cdot$	349

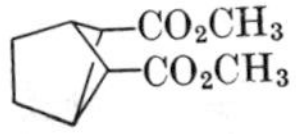

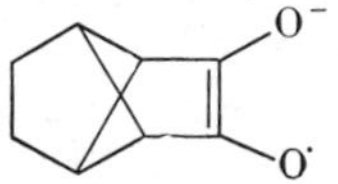

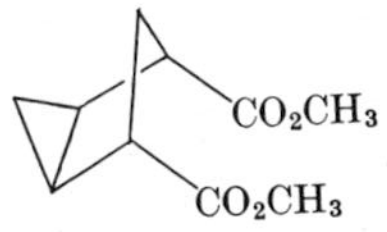

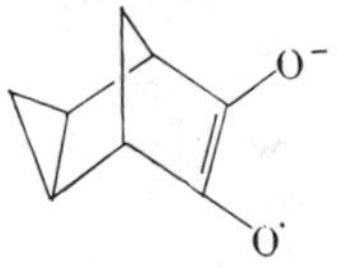

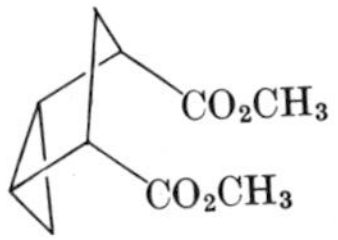

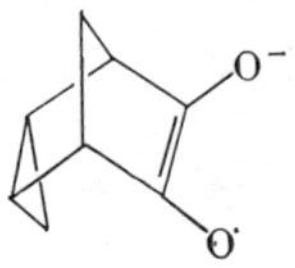

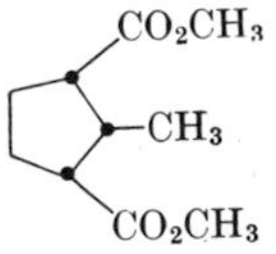

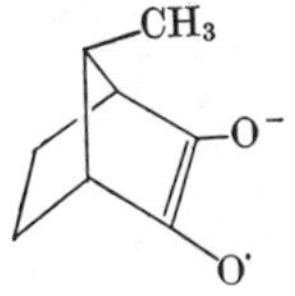

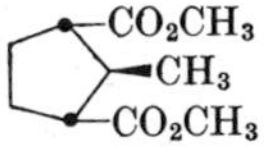

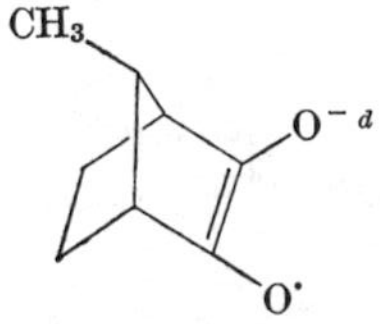

	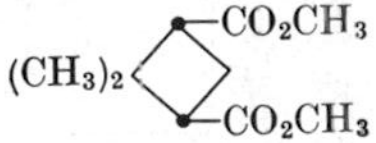	(—)	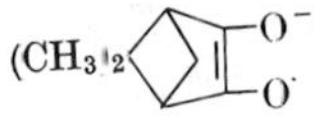	138
	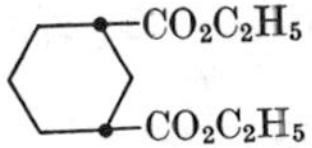	A	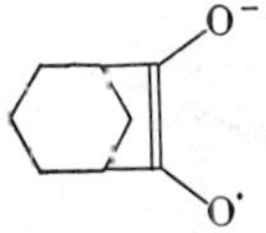	136
C_9	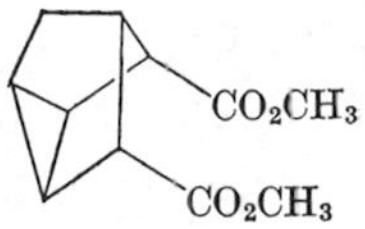	B	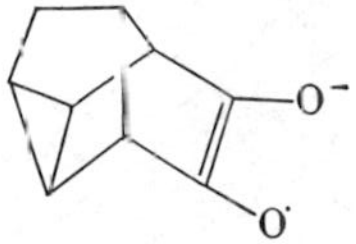	204
	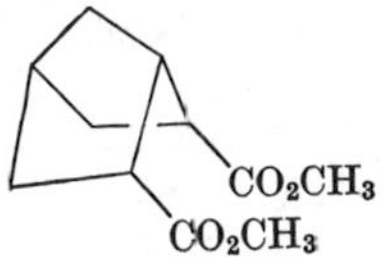	B	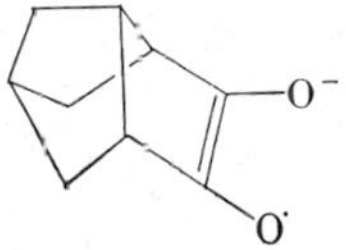	204
	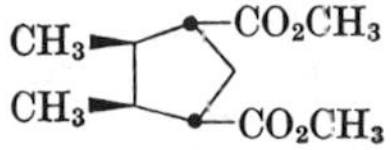	B	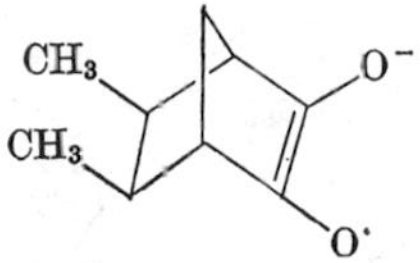	349
	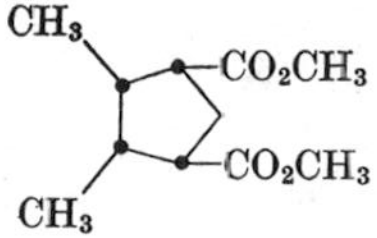	B	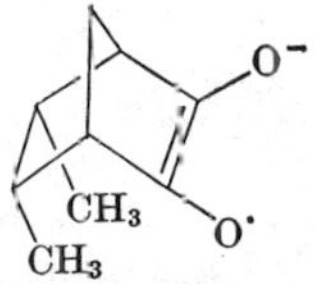	349

Note: References 184–356 are on pp. 400–403.

[a] Method A = Na-K/glyme. Method B = 1. Na-K/glyme or ether, TMCS; and 2. t-C_4H_9OK/DMSO. Method C = 1. Na-K/glyme and 2. t-C_4H_9OK/DMSO.

[d] A trideutero-methyl analog was also prepared.

TABLE VII. SEMIDIONES (*Continued*)

Cyclopentane Semidiones (Continued)

	Starting Material	Method[a]	Semidiones	Refs.
C_{10}	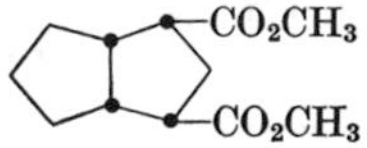 CO_2CH_3, CO_2CH_3	B	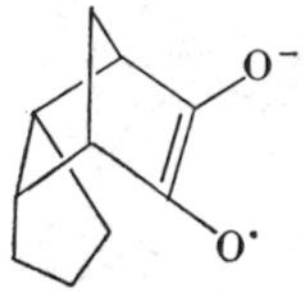 O^-, $O^{\cdot}$	349
	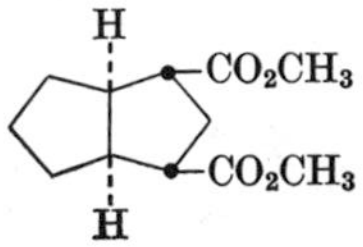 H, H, CO_2CH_3, CO_2CH_3	B	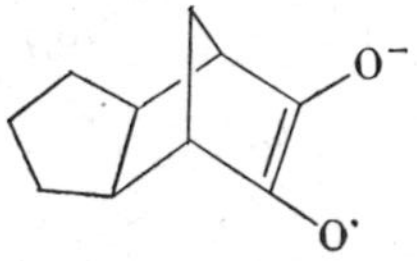 O^-, $O^{\cdot}$	349
C_{11}	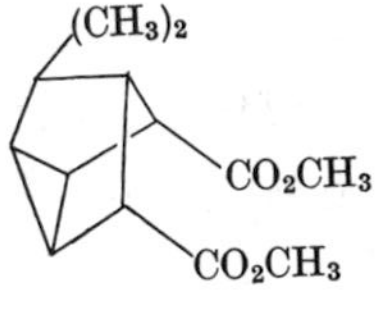 $(CH_3)_2$, CO_2CH_3, CO_2CH_3	B	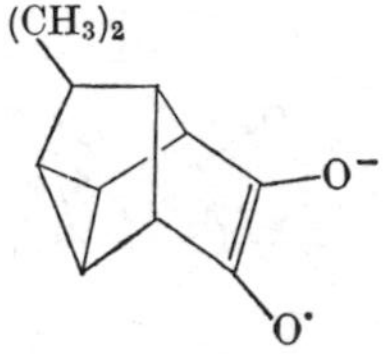 $(CH_3)_2$, O^-, $O^{\cdot}$	204
C_{12}	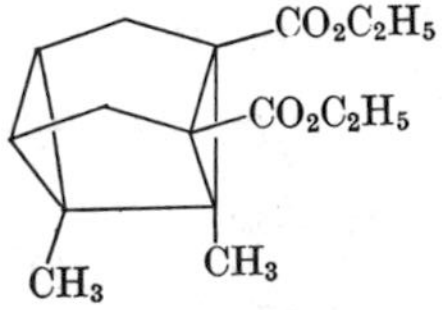 $CO_2C_2H_5$, $CO_2C_2H_5$, CH_3, CH_3	B	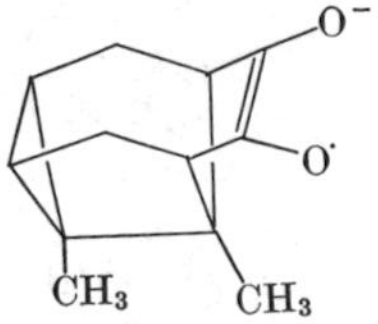 O^-, $O^{\cdot}$, CH_3, CH_3	137

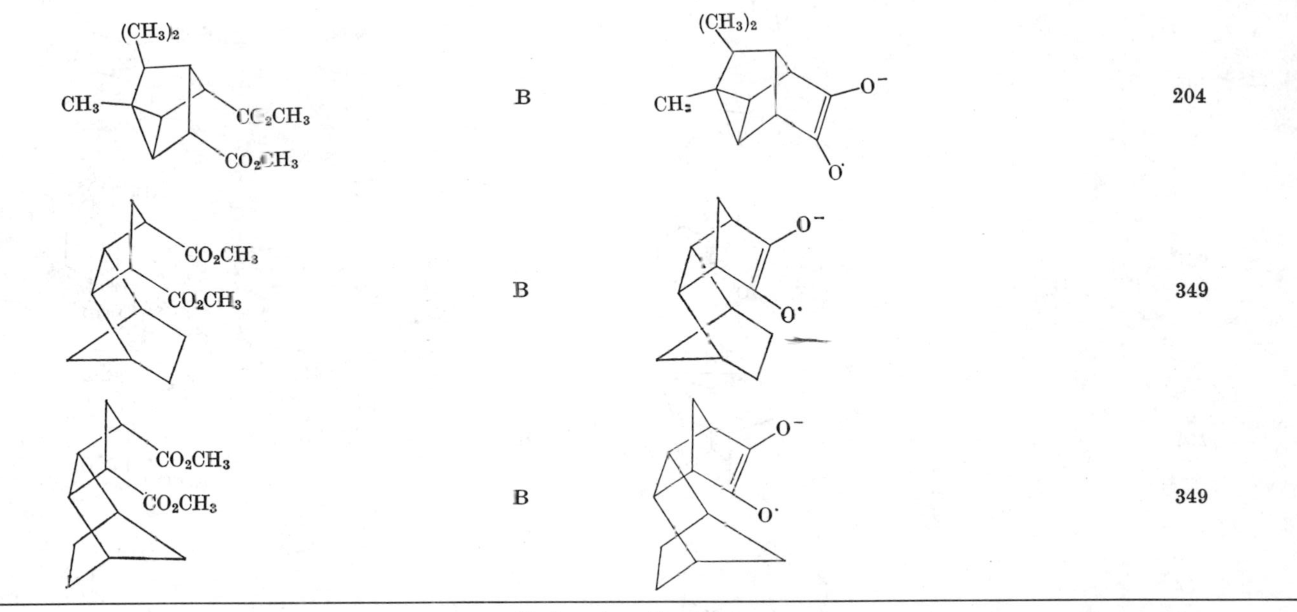

Carbons	Starting material	Method[a]	Semidione	Refs.
	$(CH_3)_2$, CH_3, CO_2CH_3, CO_2CH_3	B	$(CH_3)_2$, CH_3, O^-, $O^{\cdot}$	204
	CO_2CH_3, CO_2CH_3	B	O^-, $O^{\cdot}$	349
	CO_2CH_3, CO_2CH_3	B	O^-, $O^{\cdot}$	349
	Cyclopentane Semidiones			
C_6	$CH_3O_2C(CH_2)_4CO_2CH_3$	A	O^-, $O^{\cdot}$	136

Note: References 184–356 are on pp. 400–403.

[a] Method A = Na-K/glyme. Method B = 1. Na-K/glyme or ether, TMCS; and 2. t-C_4H_9OK/DMSO. Method C = 1. Na-K/glyme and 2. t-C_4H_9OK/DMSO.

TABLE VII. SEMIDIONES (*Continued*)

Cyclohexane Semidiones (*Continued*)

	Starting Material	Method[a]	Semidiones	Refs.
C_7	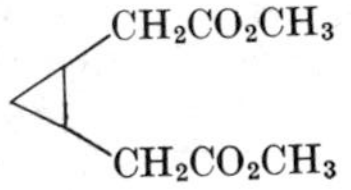 $CH_2CO_2CH_3$ / $CH_2CO_2CH_3$	B	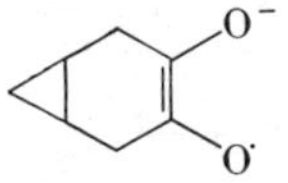 O^- / $O^\cdot$	204
C_8	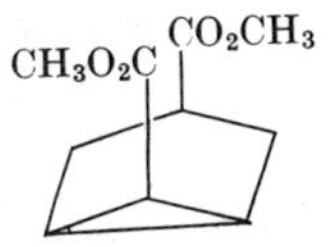 CH_3O_2C / CO_2CH_3	B	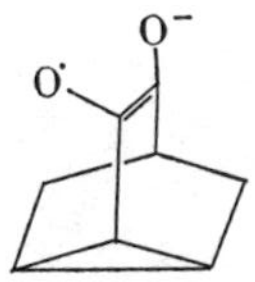 $O^\cdot$ / O^-	137
	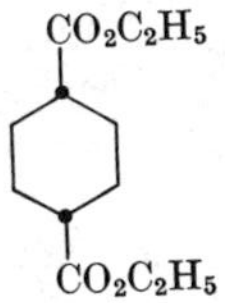 $CO_2C_2H_5$ / $CO_2C_2H_5$	A	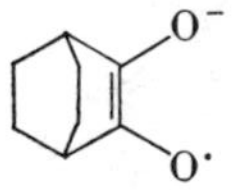 O^- / $O^\cdot$	136
C_9	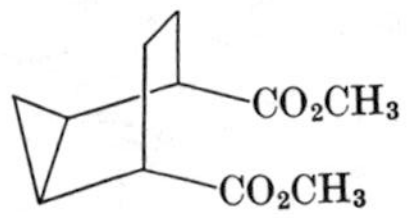 CO_2CH_3 / CO_2CH_3	B	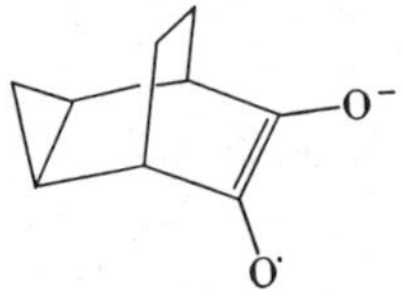 O^- / $O^\cdot$	204

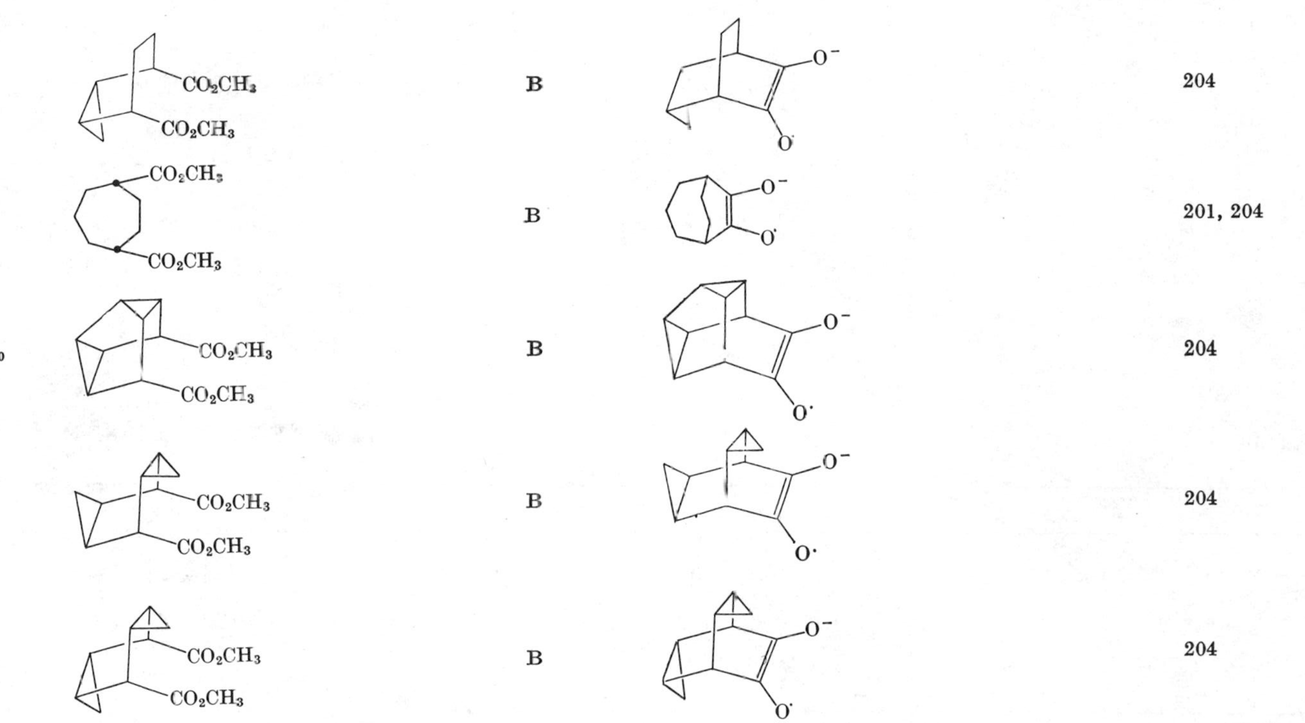

	Method	References
	B	204
	B	201, 204
C_{10}	B	204
	B	204
	B	204

Note: References 184–356 are on pp. 400–403.

[a] Method A = Na-K/glyme. Method B = 1. Na-K/glyme or ether, TMCS; and 2. t-C_4H_9OK/DMSO. Method C = 1. Na-K/glyme and 2. t-C_4H_9OK/DMSO.

TABLE VII. SEMIDIONES (*Continued*)

	Starting Material	Method[a]	Semidiones	Refs.
	Cyclohexane Semidiones (Continued)			
C_{11}	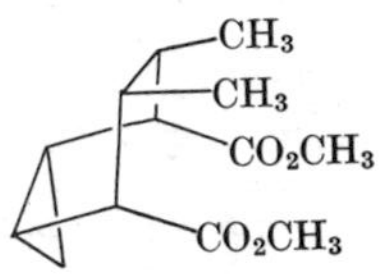	B	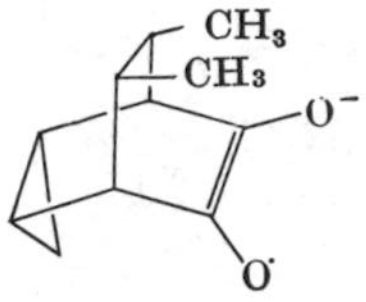	204
	Cycloheptane Semidiones			
C_7	$CH_3O_2C(CH_3)_5CO_5CH_3$	A	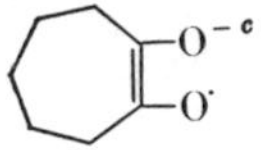 [c]	136, 204
C_8	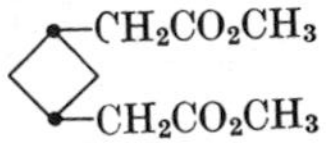	B	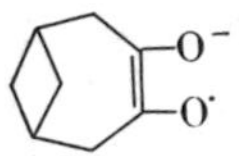	138
	$CH_3CH[(CH_2)_2CO_2CH_3]_2$	B	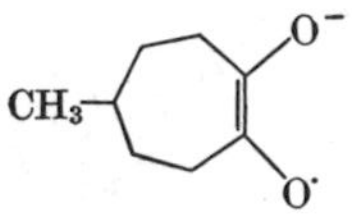	204
	$CH_3O_2C(CH_2)_3CH(CH_3)CH_2CO_2CH_3$	B	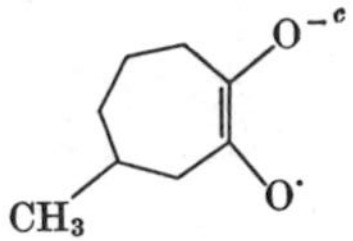 [c]	204

C_9	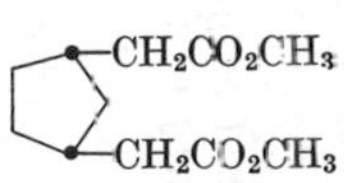 $CH_2CO_2CH_3$, $CH_2CO_2CH_3$	B	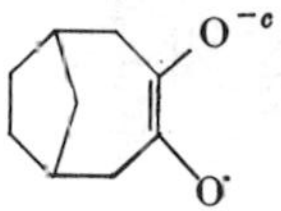 O^{-c}, O^-	204
	$CH_2[CH(CH_3)CH_2CO_2CH_3]_2$	B	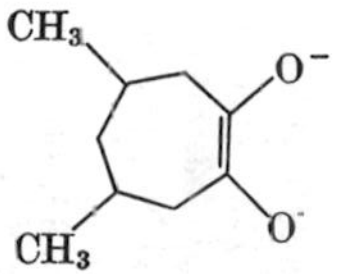 CH_3, CH_3, O^-, O^-	204
C_{10}	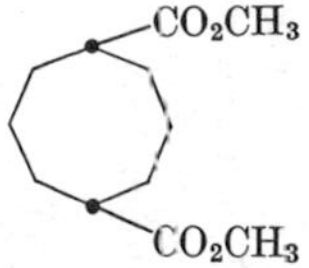 CO_2CH_3, CO_2CH_3	B	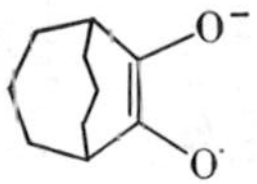 O^-, O^-	201, 204
C_{11}	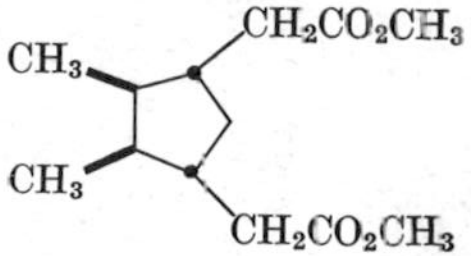 CH_3, CH_3, $CH_2CO_2CH_3$, $CH_2CO_2CH_3$	B	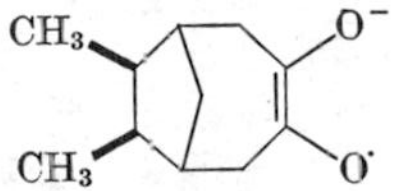 CH_3, CH_3, O^-, O^-	204

Note: References 184–356 are on pp. 400–403.

[a] Method A = Na-K/glyme. Method B = 1. Na-K/glyme or ether, TMCS; and 2. t-C_4H_9OK/DMSO. Method C = 1. Na-K/glyme and 2. t-C_4H_9OK/DMSO.

[c] A series of deuterated analogs of this compound were also prepared.

REFERENCES TO TABLES

[184] A personal communication from Prof. E. J. Corey as footnote 31 in ref. 15.

[185] R. E. K. Winter and M. L. Honig, *J. Amer. Chem. Soc.*, **93,** 4616 (1971).

[186] R. B. Turner, R. E. Lee, Jr., and E. G. Hildenbrand, *J. Org. Chem.*, **26,** 4800 (1961).

[187] H. Klinger and O. Sandke, *Ber.*, **24,** 1264 (1891).

[188] B. B. Corson, W. L. Benson, and T. T. Goodwin, *J. Amer. Chem. Soc.*, **52,** 3988 (1930).

[189] B. B. Ghatge, U. G. Nayak, K. K. Chakravarti, and S. C. Bhattacharyya, *Chem. Ind.* (*London*), **1960,** 1334.

[190] H. Scheibler and F. Emden, *Ann.*, **434,** 265 (1923).

[191] H. R. Usala, Ph.D. Dissertation, Purdue University, 1959 [*Diss. Abstr.*, **20,** 2570 (1960)].

[192] A. Khalique, *Sci. Res.* (Dacca, Pakistan), **4,** 129 (1967) [*C.A.*, **68,** 39009p (1968)].

[193] W. Küng and V. Prelog, *Croat. Chem. Acta*, **29,** 357 (1957) [*C.A.*, **53,** 16022e (1959)].

[194] F. M. Panizo, *An. Real Soc. Espan. Fis. Quim., Ser. B*, **46,** 727 (1950) [*C.A.*, **48,** 12,014d (1954)].

[195] P. Teisseire and B. Corbier, *Rech.* (*Paris*), **No. 11,** 27 (1961) [*C.A.*, **58,** 4417e (1963)].

[196] B. B. Ghatge, U. G. Nayak, K. K. Chakravarti, and S. C. Bhattacharyya, Brit. Pat. 857,163 (1960) [*C.A.*, **55,** 14315e (1961)].

[197] J. Wiemann and L. Mamlok, Fr. Pat. 1,099,096 (1965) [*C.A.*, **52,** 20062a (1958)].

[198] M. Hudlický, *Chem. Listy*, **45,** 506 (1951) [*C.A.*, **46,** 7530i (1952)].

[199] H. Yonetani and M. Kubo, *Koryo*, **48,** 22 (1958) [*C.A.*, **53,** 21717c (1959)].

[200] G. I. Danilenko, *Vestn. Kiev. Politekh. Inst., Ser. Khim.-Tekhnol.*, **1966,** 18 [*C.A.*, **68,** 86879g (1968)].

[201] G. A. Russell and R. G. Keske, *J. Amer. Chem. Soc.*, **92,** 4460 (1970).

[202] R. D. Schuetz and R. A. Baldwin, *J. Org. Chem.*, **27,** 2841 (1962).

[203] K. Rühlmann, *Int. Symp. Organosilicon Chem., Sci. Commun., Prague*, **1965,** 5 [*C.A.*, **65,** 7208c (1966)].

[204] R. G. Keske, Ph.D. Dissertation, Iowa State University, 1970 [*Diss. Abstr.*, **31B,** 5263 (1971)].

[205] G. A. Russell and R. G. Keske, *J. Amer. Chem. Soc.*, **92,** 4458 (1970).

[206] I. Ugi, R. Huisgen, and D. Pawalleck, *Ann.*, **641,** 63 (1961).

[207] I. Murata, Y. Sugihara, and N. Ueda, *Tetrahedron Lett.*, **1973,** 1183.

[208] A. T. Blomquist and B. H. Smith, *J. Org. Chem.*, **32,** 1684 (1967).

[209] N. L. Allinger and D. J. Cram, *J. Amer. Chem. Soc.*, **76,** 2362 (1954).

[210] J. Abell and D. J. Cram, *J. Amer. Chem. Soc.*, **76,** 4406 (1954).

[211] D. J. Cram and H. Steinberg, *J. Amer. Chem. Soc.*, **73,** 5691 (1951).

[212] P. Baudart, *C. R. Acad. Sci.*, **220,** 404 (1945).

[213] G. M. LeClerq, Ph.D. Dissertation, University of Washington, 1956 [*Diss. Abstr.*, **17,** 747 (1957)].

[214] M. Stiles and K. Sundaresan, personal communication.

[215] J. Wiemann, *C. R. Acad. Sci.*, **220,** 606 (1945).

[216] G. R. Haynes, Ph.D. Dissertation, University of Texas, 1957 [*Diss. Abstr.*, **17,** 2814 (1957)].

[217a] P. Y. Johnson and M. A. Priest, *J. Amer. Chem. Soc.*, **96,** 5618 (1974).

[217b] J. Panzer, Ph.D. Dissertation, Cornell University, 1956 [*Diss. Abstr.*, **17,** 39 (1957)].

[218] D. J. Cram, N. L. Allinger, and H. Steinberg, *J. Amer. Chem. Soc.*, **76,** 6132 (1954).

[219] R. C. Fuson and R. Jaunin, *J. Amer. Chem. Soc.*, **76,** 1171 (1954).

[220] D. J. Cram and N. L. Allinger, *J. Amer. Chem. Soc.*, **76,** 726 (1954).

[221] A. T. Blomquist, E. S. Wheeler, and Y. Chu, *J. Amer. Chem. Soc.*, **77,** 6307 (1955).

[222] H. Steinberg and D. J. Cram, *J. Amer. Chem. Soc.*, **74,** 5388 (1952).

[223] N. J. Leonard and G. C. Robinson, *J. Amer. Chem. Soc.*, **75,** 2143 (1953).

[224] R. C. Fuson and G. P. Speranza, *J. Amer. Chem. Soc.*, **74,** 1621 (1952).

[225] D. J. Cram and H. U. Daeniker, *J. Amer. Chem. Soc.*, **76,** 2743 (1954).

[226] H. C. Brown and M. Borkowski, *J. Amer. Chem. Soc.*, **74,** 1894 (1952).

[227] A. Wahl, *Bull. Soc. Chim. Fr.*, **3** [4], 946 (1908).

[228] A. Wohl and B. Mylo, *Ber.*, **45**, 322 (1912).
[229] H. Wynberg and Ae. de Groot, *Chem. Commun.*, **1965**, 171.
[230] G. Binsch and J. D. Roberts, *J. Amer. Chem. Soc.*, **87**, 5157 (1965).
[231] V. Prey and F. Stadler, *Ann.*, **660**, 155 (1962).
[232] C. Kashima, K. Kuo, Y. Omote, and N. Sugiyama, *Bull. Chem. Soc. Jap.*, **38**, 255 (1965).
[233] A. J. Hubert and J. Dale, *J. Chem. Soc.*, **1963**, 4091.
[234] D. Detert and B. Lindbergh, *Acta Chem. Scand.*, **23**, 690 (1969).
[235] R. A. Benkeser and R. F. Cunico, *J. Org. Chem.*, **32**, 395 (1967).
[236] M. Stoll and J. Hulstkamp, *Helv. Chim. Acta*, **30**, 1815 (1947).
[237] W. Hartmann, L. Schrader, and D. Wendisch, *Chem. Ber.*, **106**, 1076 (1973).
[238] A. Krebs and G. Burgdörfer, *Tetrahedron Lett.*, **1973**, 2063.
[239] M. Tichý and J. Sicher, *Coll. Czech. Chem. Commun.*, **37**, 3106 (1972).
[240] M. Stoll, J. Hulstkamp, and A. Rouvé, *Helv. Chim. Acta*, **31**, 543 (1948).
[241] H. O. House and R. Darms, *J. Org. Chem.*, **30**, 2528 (1965).
[242] R. L. Dunn, Ph.D. Dissertation, University of Florida, 1967 [*Diss. Abstr.*, **29B**, 935 (1968)].
[243] G. Foster, Ph.D. Dissertation, University of Pennsylvania, 1956 [*Diss. Abstr.*, **16**, 1582 (1956)].
[244] W. J. Adams, D. K. Patel, V. Petrow, and I. A. Stuart-Webb, *J. Chem. Soc.*, **1956**, 297.
[245] M. Stoll, *Helv. Chim. Acta*, **31**, 1082 (1948).
[246] L. A. Higley, *Amer. Chem. J.*, **37**, 293 (1907).
[247] I. D. Entwistle and R. A. W. Johnstone, *Chem. Commun.*, **1966**, 136.
[248] J. Wegmann and H. Dann, *Helv. Chim. Acta*, **29**, 101 (1946).
[249] J. C. Sheehan, W. F. Erman, and P. A. Cruickshank, *J. Amer. Chem. Soc.*, **79**, 147 (1957).
[250] A. T. Blomquist and R. D. Miller, *J. Amer. Chem. Soc.*, **90**, 3233 (1968).
[251] M. S. R. Nair, H. H. Mathur, and S. C. Bhattacharyya, *J. Chem. Soc.*, **1964**, 4154.
[252] G. R. Proctor and R. H. Thompson, *J. Chem. Soc.*, **1957**, 2312.
[253] J. M. Snell and S. M. McElvain, *J. Amer. Chem. Soc.*, **53**, 750 (1931).
[254] J. C. Sheehan and R. C. Coderre, *J. Amer. Chem. Soc.*, **75**, 3907 (1953).
[255] D. C. Owsley and J. J. Bloomfield, unpublished results.
[256] R. G. Carlson and R. G. Blecke, *J. Org. Chem.*, **32**, 3538 (1967).
[257] M. Stoll and A. Rouvé, *Helv. Chim. Acta*, **30**, 1822 (1947).
[258] F. N. Stepanov and G. I. Danilenko, *J. Org. Chem. USSR*, **3**, 510 (1967).
[259] F. Bouquet and C. Paquot, *Bull. Soc. Chim. Fr.*, **1949**, 110.
[260] F. Bouquet and C. Paquot, *Bull. Soc. Chim. Fr.*, **1948**, 1165.
[261] P. G. Gassman and A. Creary, personal communication.
[262] J. Coops, H. Van Kamp, W. A. Lambregts, B. J. Visser, and H. Dekker, *Rec. Trav. Chim. Pays-Bas*, **79**, 1226 (1960).
[263] L. Ruzicka, Pl. A. Plattner, and W. Widmer, *Helv. Chim. Acta*, **25**, 604 (1942).
[264] V. Prelog, H. J. Urech, A. A. Bothner-By, and J. Würsch, *Helv. Chim. Acta*, **38**, 1095 (1955).
[265] I. D. Entwistle and R. A. W. Johnstone, *J. Chem. Soc.*, *C*, **1968**, 1818.
[266] R. B. Ingraham, D. M. MacDonald, and K. Wiesner, *Can. J. Res.*, **B28**, 453 (1950).
[267] R. D. Miller, personal communication.
[268] R. W. Fawcett and J. O. Harris, *J. Chem. Soc.*, **1954**, 2669.
[269] T. N. Wheeler, personal communication.
[270] K. Wiesner, D. M. MacDonald, R. B. Ingraham, and R. B. Kelly, *Can. J. Res.*, **B28**, 561 (1950).
[271] V. Prelog, M. Fausy El-Neweihy and O. Häfliger, *Helv. Chim. Acta*, **33**, 1937 (1950).
[272] M. Stoll and A. Commarmont, *Helv. Chim. Acta*, **31**, 1435 (1948).
[273] G. W. Griffin and R. B. Hager, *Rev. Chim., Acad. Rep. Pop. Roum.*, **7**, 901 (1962).
[274] E. J. Gauglitz, Jr., and D. C. Malins, *J. Amer. Oil Chem. Soc.*, **37**, 425 (1960).
[275] D. Ginsburg, *Accounts Chem. Res.*, **7**, 249 (1972).
[276] A. T. Blomquist and B. F. Hallam, *J. Amer. Chem. Soc.*, **81**, 676 (1959).
[277] A. T. Blomquist and F. Jaffe, *J. Amer. Chem. Soc.*, **80**, 3405 (1958).

[278] E. A. Braude and B. F. Gofton, *J. Chem. Soc.*, **1957**, 4720.
[279] A. C. Cope, D. C. McLean, and N. A. Nelson, *J. Amer. Chem. Soc.*, **77**, 1628 (1955).
[280] P. Baudart, *Bull. Soc. Chim. Fr.*, [5], **13**, 87 (1946).
[281] A. C. Cope and A. S. Mehta, *J. Amer. Chem. Soc.*, **86**, 1268 (1964).
[282] P. Baudart, *C. R. Acad. Sci.*, **221**, 205 (1945).
[283] D. J. Cram and M. Cordon, *J. Amer. Chem. Soc.*, **77**, 4090 (1955).
[284] D. J. Cram and M. F. Antar, *J. Amer. Chem. Soc.*, **80**, 3103 (1958).
[285] D. J. Cram and L. K. Gaston, *J. Amer. Chem. Soc.*, **82**, 6386 (1960).
[286] D. J. Cram and M. Goldstein, *J. Amer. Chem. Soc.*, **85**, 1063 (1963).
[287] N. J. Leonard and P. M. Mader, *J. Amer. Chem. Soc.*, **72**, 5388 (1950).
[288] J. C. Sheehan and R. C. O'Neill, *J. Amer. Chem. Soc.*, **72**, 4614 (1950).
[289] J. D. Knight and D. J. Cram, *J. Amer. Chem. Soc.*, **73**, 4136 (1951).
[290] K. Ziegler, H. Sauer, L. Bruns, H. Froitzheim-Kühlhorn, and J. Schneider, *Ann.*, **589**, 122 (1954).
[291] N. L. Allinger, *J. Amer. Chem. Soc.*, **79**, 3443 (1957).
[292] A. T. Blomquist and A. Goldstein, *J. Amer. Chem. Soc.*, **77**, 998 (1955).
[293] A. T. Blomquist, R. E. Stahl, Y. C. Meinwald, and B. H. Smith, *J. Org. Chem.*, **26**, 1687 (1961).
[294] S. C. Bhattacharyya, K. K. Chakravarti, and U. G. Nayak, *Chem. Ind.* (*London*), **1960**, 588.
[295] S. C. Bhattacharyya and H. H. Mathur, *Chem. Ind.* (*London*), **1960**, 1087.
[296] A. T. Blomquist and C. J. Buck, *J. Amer. Chem. Soc.*, **81**, 672 (1959).
[297] Ae. de Groot and H. Wynberg, *J. Org. Chem.*, **31**, 3954 (1966).
[298] A. T. Blomquist and F. W. Schlaefer, *J. Amer. Chem. Soc.*, **83**, 4547 (1961).
[299] A. T. Blomquist and G. A. Miller, *J. Amer. Chem. Soc.*, **83**, 243 (1961).
[300] A. T. Bloomquist and Y. C. Meinwald, *J. Amer. Chem. Soc.*, **80**, 630 (1958).
[301] N. L. Allinger, L. A. Freiberg, R. B. Hermann, and M. A. Miller, *J. Amer. Chem. Soc.*, **85**, 1171 (1963).
[302] K. Mislow, S. Hyden, and H. Schaefer, *Tetrahedron Lett.*, **1961**, 410.
[303] I. Marsźak, J.-P. Guermont, and R. Epsztein, *Bull. Soc. Chim. Fr.*, **1960**, 1807.
[304] V. L. Hansley, *Ind. Eng. Chem.*, **43**, 1759 (1951).
[305] S. I. Khromov, E. S. Bálenkova, O. E. Lishenok, and B. A. Kazanskii, *Proc. Acad. Sci., USSR, Chem. Sect.*, **135**, 1337 (1960).
[306] F. Šorm, M. Streibel, V. Jarolím, L. Novotný, L. Dolejš, and V. Herout, *Coll. Czech. Chem. Commun.*, **19**, 570 (1954).
[307] Ya. L. Gol'dfarb, S. Z. Taits, and L. I. Belen'kii, *Bull. Acad. Sci. USSR*, **1957**, 1287.
[308] R. Kelly, D. M. MacDonald, and K. Wiesner, *Nature*, **166**, 225 (1950).
[309] L. Ruzicka, Pl. A. Plattner, and W. Widmer, *Helv. Chim. Acta*, **25**, 1086 (1942).
[310] H. H. Mathur and S. C. Bhattacharyya, *J. Chem. Soc.*, **1963**, 3505.
[311] H. H. Mathur and S. C. Bhattacharyya, *J. Chem. Soc.*, **1963**, 114.
[312] N. J. Leonard and F. H. Owens, *J. Amer. Chem. Soc.*, **80**, 6039 (1958).
[313] E. R. Hanna, K. T. Finley, W. H. Saunders, Jr., and V. Boekelheide, *J. Amer. Chem. Soc.*, **82**, 6342 (1960).
[314] R. Epsztein and I. Marsźak, *C. R. Acad. Sci.*, **243**, 283 (1956).
[315] V. V. Dhekne, B. B. Ghatge, U. G. Nayak, K. K. Chakravarti, and S. C. Bhattacharyya, *J. Chem. Soc.*, **1962**, 2348.
[316] C. Djerassi and G. W. Krakower, *J. Amer. Chem. Soc.*, **81**, 237 (1959).
[317] D. J. Cram and L. A. Singer, *J. Amer. Chem. Soc.*, **85**, 1084 (1963).
[318] J. Meinwald and P. C. Lee, *J. Amer. Chem. Soc.*, **82**, 699 (1960).
[319] R. S. Rouse and W. E. Tyler, III, *J. Org. Chem.*, **26**, 3525 (1961).
[320] D. J. Cram, W. J. Wechter, and R. W. Kierstead, *J. Amer. Chem. Soc.*, **80**, 3126 (1958).
[321] V. Prelog, H. H. Kagi, and E. H. White, *Helv. Chim. Acta*, **45**, 1658 (1962).
[322] V. Prelog and S. Polyák, *Helv. Chim. Acta*, **40**, 816 (1957).
[323] N. A. Nelson and R. N. Schut, *J. Amer. Chem. Soc.*, **80**, 6630 (1958).
[324] H. Bredereck and G. Theilig, *Chem. Ber.*, **86**, 88 (1953).
[325] A. T. Blomquist, L. H. Liu, and J. C. Bohrer, *J. Amer. Chem. Soc.*, **74**, 3643 (1952).

[326] H. Kwart and J. A. Ford, Jr., *J. Org. Chem.*, **24**, 2060 (1959).
[327] M. Stoll, U.S. Pat. 2,529,825 (1950) [*C.A.*, **45**, 2976h (1951)].
[328] V. I. Egorova, *J. Russ. Phys.-Chem. Soc.*, **60**, 1199 (1928) [*C.A.*, **23**, 2935 (1929)].
[329] L. Bouveault and R. Locquin, *C. R. Acad. Sci.*, **140**, 1593 (1905).
[330] L. Bouveault and R. Locquin, *Bull. Soc. Chim. Fr.*, [3], **35**, 641 (1906).
[331] L. Bouveault and R. Locquin, *Bull. Soc. Chim. Fr.*, [3], **35**, 637 (1906).
[332] L. Bouveault and R. Locquin, *Bull. Soc. Chim. Fr.*, [3], **35**, 633 (1906).
[333] L. Bouveault and R. Locquin, *Bull. Soc. Chim. Fr.*, [3], **35**, 629 (1906).
[334] R. D. Miller, M. Schneider, and D. L. Dolce, *J. Amer. Chem. Soc.*, **95**, 8468 (1973).
[335] A. W. Ralston and W. M. Selby, *J. Amer. Chem. Soc.*, **61**, 1019 (1939).
[336] H. D. Durst and L. Liebeskind, Abstr. 163rd National ACS Meeting, Organic Division, April 1972.
[337] P. Y. Johnson, J. Zitsman, and C. E. Hatch, *J. Org. Chem.*, **38**, 4087 (1973).
[338] J. Kooi, H. Wynberg, and R. M. Kellogg, *Tetrahedron*, **29**, 2135 (1973).
[339] W. P. Weber, R. A. Felix, A. K. Willard, and H. G. Boettger, *J. Org. Chem.*, **36**, 4060 (1971).
[340] G. A. Russell and P. R. Whittle, *J. Amer. Chem. Soc.*, **91**, 2813 (1969).
[341] C. U. L. Delbaere and G. H. Whitman, *J. Chem. Soc. Perkin I*, **1974**, 879.
[342] D. McNeil, B. R. Vogt, J. J. Sudol, S. Theodopulus, and E. Hedaya, *J. Amer. Chem. Soc.*, **96**, 4673 (1974).
[343] J. V. Swisher and H. H. Chen, *J. Organometal. Chem.*, **69**, 93 (1974).
[344] M. Avram, I. E. Dinelescu, F. Chiraleu and C. D. Nenitezscu, *Rev. Roum. Chim.*, **18**, 863 (1973). [*C.A.*, **79**, 65885t (1973)].
[345] G. Porzi, C. Concilio and A. Bongini, *Gazz. Chem. Ital.*, **103**, 393 (1973).
[346] T. Wakamatsu, K. Akasaka and Y. Ban, *Tetrahedron Lett.*, **1974**, 3879.
[347] T. Wakamatsu, K. Akasaka and Y. Ban, *Tetrahedron Lett.*, **1974**, 3883.
[348] G. A. Russell, P. R. Whittlo, C. S. C. Chung, Y. Kosugi, K. Schmitt and E. Goettert, *J. Amer. Chem. Soc.*, **96**, 7053 (1974).
[349] G. A. Russell, G. W. Holland, K.-W. Chang, R. G. Keske, J. Mattox, C. S. C. Chung, K. Stanley, K. Schmitt, R. Blankespoor and Y. Kosugi, *J. Amer. Chem. Soc.*, **96**, 7237 (1974).
[350] H. Stotter and E. Rauscher, *Chem. Ber.*, **93**, 1161 (1960).
[351] S. Danilow and E. Venus-Danilowa, *Ber.*, **62**, 2563 (1929).
[352] J. H. Stocker, *J. Org. Chem.*, **29**, 3593 (1964).
[353] H. Shubert and H. Ladish, *J. Prakt.* [4], **18**, 203 (1962).
[354] A. O. Hellwig and A. O. Shubert, *Zeitschrift für Chemie*, **4**, 227 (1964).
[355] H. H. Inhoffen, K. Radscheit, U. Stache and V. Koppe, *Ann.*, **684**, 24 (1965).
[356] F. E. Deatherage and H. S. Olcott, *J. Amer. Chem. Soc.*, **61**, 630 (1939).

CHAPTER 3

ALKENES FROM TOSYLHYDRAZONES

Robert H. Shapiro

University of Colorado, Boulder, Colorado

CONTENTS

INTRODUCTION

In 1952 Bamford and Stevens[1] observed that tosylhydrazones* of aliphatic ketones yielded alkenes when treated with the sodium salt of ethylene glycol in boiling ethylene glycol (EG).† In addition to the alkene, molecular nitrogen and *p*-toluenesulfinate anion were produced.

$$R_1C(=NNHTs)CHR_2R_3 \xrightarrow[\text{Heat}]{\text{Na/EG}} R_1CH=CR_2R_3 + N_2 + Ts^-$$

$$R_1, R_2, R_3 = \text{alkyl or hydrogen; Ts} = p\text{-}CH_3C_6H_4SO_2)$$

Tosylhydrazones of unbranched ketones (*e.g.*, phenylacetone) and unstrained cyclic ketones (*e.g.*, cyclohexanone) gave the corresponding alkenes (propenylbenzene and cyclohexene, respectively) in good to excellent yield. In contrast, tosylhydrazones of branched ketones (*e.g.*, pinacolone) and strained cyclic ketones (*e.g.*, camphor) decomposed under the reaction conditions to yield rearranged alkenes (2,3-dimethyl-2-butene and camphene, respectively).

$$(CH_3)_3CC(=NNHTs)CH_3 \xrightarrow{\text{Na/EG}} (CH_3)_2C=C(CH_3)_2$$

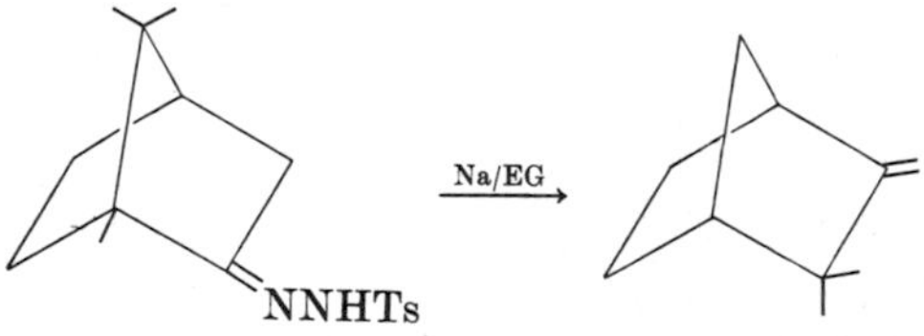

NNHTs $\xrightarrow{\text{Na/EG}}$

This chapter reviews such syntheses of alkenes from tosylhydrazones and the formation of cyclopropanes and acetylenes from tosylhydrazones, because the several reactions involved are so closely interrelated. The conversions of tosylhydrazones to alkenes, carbene plus double-bond cycloaddition products, and alkanes are treated in the section on Related

* Tosylhydrazone is the accepted contracted name for the *p*-toluenesulfonylhydrazone of an aldehyde or ketone.

† The following abbreviations are used throughout the chapter: ethylene glycol, EG; diethylene glycol, DEG; diethylene glycol dimethyl ether, diglyme; diethylene glycol diethyl ether, DEC; N-methylpyrrolidone, NMP; tetrahydrofuran, THF; unspecified heating or reflux, heat.

[1] W. R. Bamford and T. S. Stevens, *J. Chem. Soc.*, **1952**, 4735.

and Anomalous Reactions. Some duplication of material previously reviewed is unavoidable.[2,3] No attempt is made here to cover all aspects of carbene generation from tosylhydrazones, since this subject has received excellent treatment recently.[3] Other methods for generating the reaction intermediates, except where they seem relevant to considerations of reaction mechanism, are neglected.

The four major reaction types reviewed are, in order of presentation: (1) base-induced decomposition of tosylhydrazones in protic solvents; (2) base-induced decomposition of tosylhydrazones in aprotic solvents; (3) reaction of tosylhydrazones with alkyllithium reagents; and (4) fragmentation reactions of α,β-epoxytosylhydrazones and analogous systems. The first two types of reactions are simply called the protic and aprotic Bamford-Stevens reaction, respectively, and are discussed together with the third type, since all are base induced.

The Bamford-Stevens Reaction

The protic reaction is defined as the reaction of a tosylhydrazone with a strong base in a protic solvent, *i.e.*, the original Bamford-Stevens conditions. The solvent most commonly employed is ethylene glycol (EG), but higher-boiling glycols such as diethylene glycol (DEG) have also been used. The base is prepared by dissolving metallic sodium in the solvent before the tosylhydrazone is added, or commercial sodium methoxide may be added to the tosylhydrazone dissolved or suspended in the solvent.

The "aprotic" reaction employs a solvent of little or no proton-donating ability.* Diglyme (diethylene glycol dimethyl ether) is typical, but other high-boiling ethers, such as diethyl carbitol (diethylene glycol diethyl ether, DEC), hydrocarbons (*e.g.*, decalin) and N,N-disubstituted amides (*e.g.*, N-methylpyrrolidone, NMP) have been used with success. Even acetamide is a reliable "aprotic" solvent. In reactions employing acetamide as the solvent it is convenient to prepare the base *in situ* by the dissolution of metallic sodium, as is done with ethylene glycol. The most common base employed in the aprotic reaction is sodium methoxide, but other alkoxides as well as sodium and lithium hydride have been used occasionally. Lithium aluminium hydride[4] and sodium amide[5] have been

* In one modification of the reaction the dry sodium or lithium salt of the tosylhydrazone is pyrolyzed. This technique is especially useful for the preparation of volatile products where solvent separation is inconvenient. This modification gives results which are nearly identical to those obtained using an aprotic solvent (see Table II).

[2] (a) E. Chinoporos, *Chem. Rev.*, **63**, 235 (1963); (b) L. Caglioti, *Ric. Sci., Riv.* (Rome), **34,** 41 (1964) [*C.A.*, **61,** 1781d (1964)].

[3] (a) W. Kirmse, *Carbene Chemistry*, 2nd ed., Academic Press, New York, 1971; (b) M. Jones, Jr., and R. A. Moss, Eds., *Carbenes*, John Wiley and Sons, New York, 1973, Ch. 1.

[4] L. Caglioti and M. Magi, *Tetrahedron Lett.*, **1962,** 1261.

[5] W. Kirmse, B.-G. von Bülow, and H. Schepp, *Ann.*, **691,** 41 (1966).

tested and found to be less satisfactory; the former may cause reductive elimination, and both may lead to products expected from reactions employing alkyllithium reagents.

Protic and "aprotic" solvents frequently give rise to totally different reaction products. For example, under protic conditions, pinacolone tosylhydrazone gives only the rearranged alkene tetramethylethylene.[1] Under aprotic conditions it decomposes to an unrearranged alkene and a cyclopropane.[6] On the other hand, cyclohexanone tosylhydrazone yields only cyclohexene under either protic[1] or aprotic[7] conditions.

$$(CH_3)_3CC(=NNHTs)CH_3 \xrightarrow{\text{NaOMe, DEC}} (CH_3)_3CCH{=}CH_2 + \text{(1,1,2-trimethylcyclopropane)}$$

It is not always obvious by inspection of its structure whether a solvent will be "aprotic" or "protic" in a Bamford-Stevens reaction. Acetamide acts as an aprotic solvent in the Bamford-Stevens reaction,[7] as does triethylcarbinol.[8] One study showed that ethanol may be a poor proton donor in this reaction![9]

Solvent effects on the course of the reaction led early workers to the conclusion that under protic conditions a diazonium ion and/or a carbonium ion is an intermediate, but when no important proton source is present the intermediate is a carbene.[6,7] These intermediates would lead to the observed products. Later studies, however, revealed that some products rationalized as stemming from carbene intermediates were probably derived from cationic intermediates; these details are discussed in the Mechanism section.

The proton-donating ability of the solvent, or lack of it, is not the only important criterion in determining the reaction pathway and thus the products. Even under aprotic conditions, products derived from cationic intermediates can predominate when less than one equivalent of base is used. Since the hydrogen on nitrogen is fairly acidic (the pK_a of acetone mesylhydrazone is about 8.5),[7] the tosylhydrazone is almost entirely in the form of its salt when one or more equivalents of base are present. With a deficiency of base, the free tosylhydrazone can act as a proton donor leading to cationic products.[10]

[6] L. Friedman and H. Shechter, *J. Amer. Chem. Soc.*, **81**, 5512 (1959).

[7] J. W. Powell and M. C. Whiting, *Tetrahedron*, **7**, 305 (1959).

[8] J. H. Bayless, L. Friedman, F. B. Cook, and H. Shechter, *J. Amer. Chem. Soc.*, **90**, 531 (1968).

[9] P. Clarke, M. C. Whiting, G. Papenmeier, and W. Reusch, *J. Org. Chem.*, **27**, 3356 (1962).

[10] J. A. Smith, H. Shechter, J. Bayless, and L. Friedman, *J. Amer. Chem. Soc.*, **87**, 659 (1965).

A third effect, a minor one, is related to the nature of the base. Aluminum- and boron-containing bases, such as aluminum isopropoxide and sodium borohydride, and to a lesser extent lithium-containing bases. may lead to cationic products even when used in large excess in aprotic media.[11] Apparently the metal ion can complex with the diazo intermediate to form a cationoid species which can lead to molecular rearrangements like a carbonium ion species.

The Reaction Of Tosylhydrazones with Alkyllithium Reagents

A different course is followed when a tosylhydrazone bearing an α-hydrogen atom is allowed to react with an alkyllithium reagent or, at least in a few cases, with lithium aluminum hydride or sodium amide. In such reactions an unrearranged, less substituted alkene is almost always the exclusive product.[12] To illustrate the utility of this reaction, and to contrast it with the protic and aprotic Bamford-Stevens reactions, the behavior of camphor tosylhydrazone (**1**) toward the three sets of reaction conditions is shown in Eq. 1.* Clearly the formation of 2-bornene

1 NNHTs

Na/EG[1,7] Heat → 2

NaOMe/Diglyme[6] → 3

MeLi/Et_2O[12] Room temp → 4

(Eq. 1)

(**4**) is not consistent with a cationic or carbenic intermediate. Another example contrasting with the aprotic Bamford-Stevens reaction is shown

* Camphor tosylhydrazone and lithium aluminum hydride yield a mixture of camphene (**2**) and tricyclene (**3**); the ratio depends on the amount of base.[11]

[11] (a) R. H. Shapiro, J. H. Duncan, and J. C. Clopton, *J. Amer. Chem. Soc.*, **89**, 1442 (1967); (b) R. H. Shapiro, J. H. Duncan, and J. C. Clopton, *ibid.*, **89**, 471 (1967).

[12] R. H. Shapiro and M. J. Heath, *J. Amer. Chem. Soc.*, **89**, 5734 (1967).

NNHTs

NaOMe/NMP[13], Heat → (63) + (27) + other products (10)

MeLi/Et_2O[12], Room temp → (<2) (>98) (−)

(Eq. 2)

in Eq. 2. 2-Methylcyclohexanone tosylhydrazone yields the more substituted alkene, 1-methylcyclohexene, as the major product upon treatment with sodium methoxide in N-methylpyrrolidone;[13] the less substituted alkene, 3-methylcyclohexene, is dominant with methyllithium in ether. The first observation of different products from a tosylhydrazone resulted from the treatment of the tosylhydrazones of two 17-ketosteroids with lithium aluminum hydride to give unrearranged Δ^{16}-steroid alkenes in about 70% yield.[4] Later it was shown that a carbene intermediate leads to a cyclopropane (a 17, 18-cyclosteroid);[14] it had been previously demonstrated that a cationic intermediate leads to rearranged products (a mixture of 17-methyl-18-norsteroid alkenes).[15] With the aid of deuterium-labeled materials,[16] a reasonable anionic mechanism could be proposed to explain these earlier observations.[4] Actually, alkene formation from tosylhydrazones promoted by lithium aluminum hydride is anomalous; this reagent usually induces reductive elimination leading to alkanes (p. 430).

Additionally, it was discovered that sodium amide or sodium hydride can convert tosylhydrazones containing α-hydrogen atoms to unrearranged alkenes.[5] Sodium hydride, however, appears to be somewhat unreliable, since it sometimes leads to carbenic products.[5,11] With acetylcyclopropane tosylhydrazone, one can see the contrast between the aprotic Bamford-Stevens reaction and sodium amide reaction.[5] The details of the formation of the cyclobutene by a carbenic process are discussed in the Mechanism section.

Lithium aluminum hydride, sodium hydride, and sodium amide possess the ability to direct tosylhydrazones to unrearranged alkenes, but alkyllithium reagents have replaced them.[12,17] The alkyllithiums are commercially available, easy to handle, and induce elimination at low temperatures (−78 to +25°). Two or more equivalents of alkyllithium must be

[13] J. W. Wilt and W. J. Wagner, *J. Org. Chem.*, **29**, 2788 (1964).

[14] L. Caglioti, P. Grasselli, and A. Selva, *Gazz. Chim. Ital.*, **94**, 537 (1964).

[15] W. F. Johns, *J. Org. Chem.*, **26**, 4583 (1961).

[16] M. Fischer, Z. Pelah, D. H. Williams, and C. Djerassi, *Chem. Ber.*, **98**, 3236 (1965).

[17] G. Kaufman, F. Cook, H. Shechter, J. Bayless, and L. Friedman, *J. Amer. Chem. Soc.*, **89**, 5736 (1967).

=NNHTs (cyclopropyl)

NaOMe / Diglyme → (1) + (92) + other products (7)

$NaNH_2$ / Decalin → (99.7) (0.3) (−)

used, since the first equivalent is consumed by the acidic hydrogen on nitrogen in the tosylhydrazone. Herz has shown that, with a large excess of alkyllithium, nucleophilic substitution competes very effectively with elimination.[18] Nucleophilic substitution had been previously observed as a competitive process,[19] and as the exclusive reaction of tosylhydrazones lacking α-hydrogen atoms.[12]

MECHANISM

The Bamford-Stevens Reaction

The mechanism of the Bamford-Stevens reaction is believed to involve initial proton abstraction by base and subsequent rate-determining thermal elimination of *p*-toluenesulfinyl anion to give a diazo compound.[6,7] The net result of these two steps is α elimination of *p*-toluenesulfinic acid. The fact that diazo compounds can be isolated under some circumstances supports this mechanism.[20–22]

$$R_1R_2C{=}NNHTs \xrightarrow[-H^+]{\text{Base}} R_1R_2C{=}N\bar{N}Ts \xrightarrow[Ts]{} R_1R_2C{-}N_2$$

The fate of diazo compound in the presence of excess base is largely dependent on the nature of the solvent. If no proton source is available, or if available proton sources react too slowly with the diazo intermediate, molecular nitrogen will be eliminated and a carbene generated. If, however, proton donation is faster than nitrogen elimination, a diazonium ion is formed. The diazonium ion can lose nitrogen, giving a carbonium ion, which expels a proton with or without previous rearrangement. Protonation of the carbene is not thought to be an important route.[23]

[18] (a) J. E. Herz and C. V. Ortiz, *J. Chem. Soc., C,* **1971,** 2294. (b) J. E. Herz and E. Gonzalez, *Chem. Commun.*, **1969,** 1395.

[19] J. Meinwald and F. Uno, *J. Amer. Chem. Soc.*, **90,** 800 (1968).

[20] D. G. Farnum, *J. Org. Chem.*, **28,** 870 (1963).

[21] G. L. Closs and R. A. Moss, *J. Amer. Chem. Soc.*, **86,** 4042 (1964).

[22] G. M. Kaufman, J. A. Smith, G. G. Vander Stouw, and H. Shechter, *J. Amer. Chem. Soc.*, **87,** 935 (1965).

[23] (a) A. Nickon and N. H. Werstiuk, *J. Amer. Chem. Soc.*, **49,** 7081 (1972); (b) *ibid.*, **88,** 4543 (1966).

$$\begin{array}{ccccc} R_2C{=}N_2 & \xrightarrow{-N_2} & R_2C: & \longrightarrow & \text{carbenic products} \\ \downarrow{\scriptstyle SH} & & \not\downarrow{\scriptstyle H^+} & & \\ R_2CHN_2^+ & \xrightarrow{-N_2} & R_2CH^+ & \xrightarrow{-H^+} & \text{cationic products} \end{array}$$

Many of the important mechanistic studies of the Bamford-Stevens reaction have been conducted on the tosylhydrazones of camphor (**1**) and cyclopropanecarboxaldehyde (**5**). Both compounds give different major products under protic and aprotic reaction conditions. Camphor tosylhydrazone yields camphene (**2**) under protic conditions and tricyclene (**3**) under aprotic conditions (Eq. 1, p. 409).* Cyclopropanecarboxaldehyde tosylhydrazone (**5**) behaves anomalously in the Bamford-Stevens reaction, giving bicyclo[1.1.0]butane[10,24] under protic conditions and cyclobutene under aprotic conditions.[25]

▷—CH=NNHTs (**5**) —Na/EG→ bicyclo[1.1.0]butane (76%)

▷—CH=NNHTs (**5**) —NaOMe/DEC→ cyclobutene (60%)

The formation of camphene (**2**) from camphor tosylhydrazone (**1**) appears to be a straightforward example of a bornyl cation undergoing rearrangement with subsequent loss of a proton.[26] It is clear that the proton source effecting the conversion of diazo compound **6** to a diazonium ion is the solvent. With a fortyfold excess of deuterium oxide in the reaction mixture, the camphene produced is an 80% d_1 and 20% d_0 mixture.[11] This corresponds to a deuterium isotope effect (k_H/k_D) of about 8, since the original proton from the tosylhydrazone was still in the reaction mixture.

The formation of tricyclene (**3**) appears to result from intramolecular insertion of the carbene. Although this mechanism may be the dominant route, it is not the exclusive pathway to tricyclene. When a deficiency of base is employed in a fortyfold excess of deuterium oxide, the tricyclene formed is 64% d_1 and 36% d_0. The possibility that exchange of the α-hydrogen atom was occurring to this extent was precluded by the fact

* Reproduction of the original conditions of Bamford and Stevens[1] showed that the camphene contained 20% of tricyclene.[7]

[24] H. M. Frey and I. D. R. Stevens, *Proc. Chem. Soc.*, **1964,** 144.

[25] L. Friedman and H. Shechter, *J. Amer. Chem. Soc.*, **82,** 1002 (1960).

[26] P. Beltramé, C. A. Bunton, A. Dunlop, and D. Whittaker, *J. Chem. Soc.*, **1964,** 658; and references contained therein.

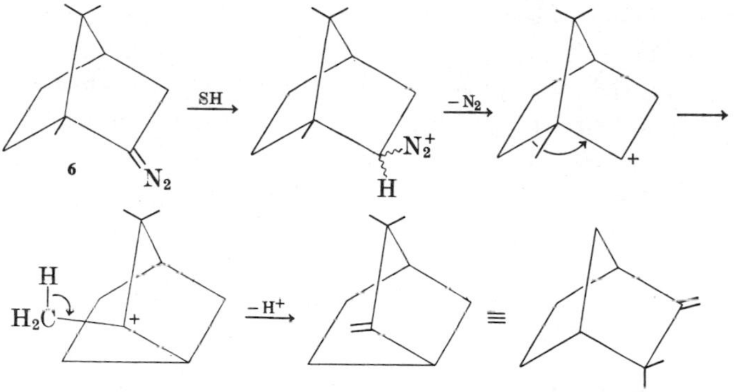

6 N_2 SH N_2^+ H $-N_2$ + H H_2C + $-H^+$ ≡

that with excess base the isotopic composition of tricyclene was found to be 6% d_1 and 94% d_0. Since no external proton is incorporated in a simple

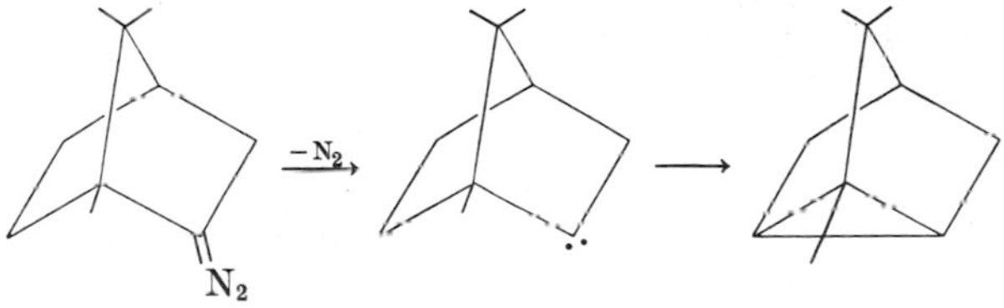

$-N_2$ N_2

intramolecular insertion, it appears that tricyclene can also be formed by way of a poorly solvated cationic intermediate. This information together with that obtained by using various amounts of base[10] led to the mechanism shown in the accompanying scheme.[11] If an equilibrium between diazo compound and diazonium ion competes effectively with nitrogen expulsion from either, then a large excess of base (B^-) will favor the diazo species and carbenic products; low concentrations of base will favor the diazonium ion and the resulting cationic products.

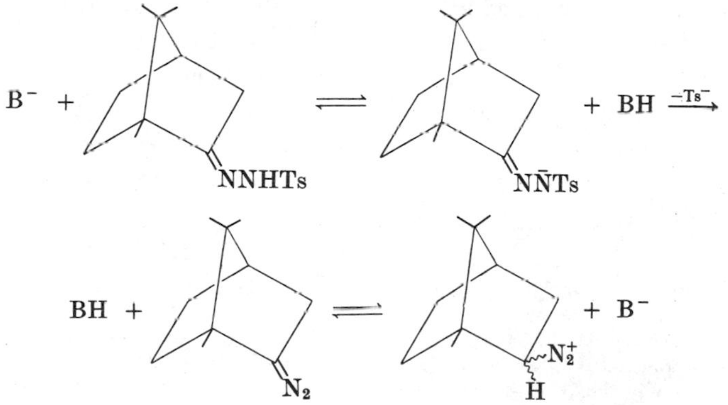

B^- + ⇌ + BH $\xrightarrow{-Ts^-}$

NNHTs $N\bar{N}Ts$

BH + ⇌ + B^-

N_2 N_2^+ H

Similar conclusions were reached in a related study with norbornanone tosylhydrazone.[23] This tosylhydrazone decomposes almost exclusively under both protic[27] and aprotic[28,29] conditions to nortricyclene. In a protic solvent such as ethylene glycol a cationic path may compete with a carbenic path to products, but the key intermediate is not identical with norbornyl cation generated through solvolysis reactions.[23] If the normal

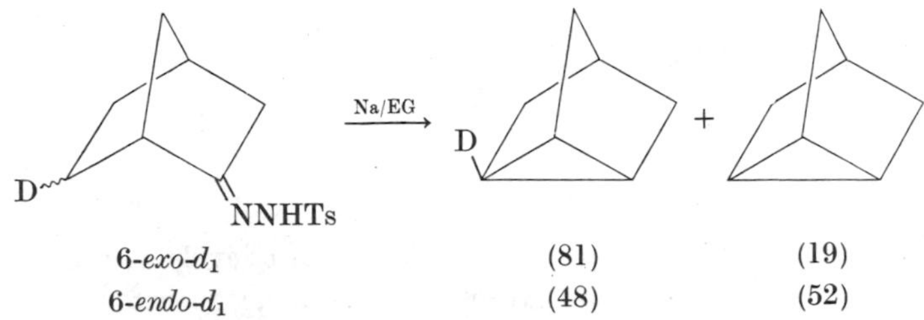

norbornyl cation were the only intermediate, then the 6-*exo* and 6-*endo* protons would become equivalent and, starting from either substrate, would be lost to the same extent. Although several mechanisms may be operating in these conversions, an important one is thought to be the collapse of a diazonium ion directly to product, but carbene and carbonium ion mechanisms are thought to compete.

It seems clear, however, that diazonium ion collapse does not always represent a major route to product. For example, studies have revealed that the products obtained in a protic Bamford-Stevens reaction are consistent with an E_1 mechanism.[30,31] In the first study, only the Δ^7-alkene (Saytzeff product) was obtained from a 7-ketosteroid tosylhydrazone;[30] in the second, there was seen a high *cis/trans* ratio in the 2-butenes obtained from butanone tosylhydrazone.[31] These observations are more consistent with carbonium ion intermediates than with diazonium ions undergoing E_2 eliminations.

[27] A. Nickon, J. L. Lambert, S. J., and J. E. Oliver, *J. Amer. Chem. Soc.*, **88**, 2787 (1966).
[28] L. Friedman and H. Shechter, *J. Amer. Chem. Soc.*, **83**, 3159 (1961)
[29] P. K. Freeman, D. E. George, and V. N. M. Rao, *J. Org. Chem.*, **29**, 1682 (1964).
[30] E. J. Corey and R. A. Sneen, *J. Amer. Chem. Soc.*, **78**, 6269 (1956).
[31] C. H. DePuy and D. H. Froemsdorf, *J. Amer. Chem. Soc.*, **82**, 634 (1960).

C_8H_{17} HO NNHTs $\xrightarrow{\text{Na/EG}}$ C_8H_{17} HO

$$C_2H_5C(=NNHTs)CH_3 \xrightarrow{\text{Na/EG}} CH_3CH=CHCH_3 + C_2H_5CH=CH_2$$
(*cis/trans*, 1.00 : 1.24)

Perhaps the most conclusive experiment involved the formation of *cis*-cyclodecene from ^{14}C-labeled cyclodecanone tosylhydrazone in a protic Bamford-Stevens reaction.[32a] No appreciable hydride shifts occurred during the course of the reaction, in contrast to previous reports of a 17% label scrambling in the deamination of cyclodecylamine[32b] and more than 30% label scrambling in the solvolysis of cyclodecyl tosylate.[32c,d] If it can be assumed that amine deaminations proceed by more than one mechanism, *i.e.*, concerted and stepwise collapse of intermediate diazonium ions,[33] and that tosylate solvolyses give solvated carbonium ions, then it may be deduced that cyclodecanone tosylhydrazone decomposes by concerted collapse of a diazonium ion. It must not, however, be assumed that all protic Bamford-Stevens reactions follow this route exclusively, as the structure of the diazonium ion must be taken into consideration.[33]

The decomposition of cyclopropanecarboxaldehyde tosylhydrazone (**5**), formulated on p. 412, has also been studied in great detail. The protic Bamford-Stevens reaction appears to be a good method of preparing bicyclobutane, and the aprotic reaction has been used to prepare numerous substituted cyclobutenes (see Table II, p. 446). The mechanism of the formation of bicyclobutane from **5** was primarily investigated through deuterium labeling techniques.[34,35] The original aldehyde hydrogen becomes the *exo* hydrogen in the bicyclobutane, and the proton donated by the solvent becomes the *endo* hydrogen. Some hydrogen-deuterium exchange of the original aldehyde hydrogen was noted; most likely this

[32] (a) V. Prelog and S. Smolinski, *Helv. Chim. Acta*, **49,** 2275 (1966); (b) V. Prelog, H. J. Urech, A. A. Bothner-By, and J. Würsch, *ibid.*, **38,** 1095 (1955); (c) H. J. Urech and V. Prelog, *ibid.*, **40,** 477 (1957); (d) V. Prelog, W. Küng, and T. Tomljenovic, *ibid.*, **45,** 1352 (1962).

[33] L. Friedman, *Carbonium Ions*, Vol. II., G. A. Olah and P. R. von Schleyer, Eds., Wiley-Interscience, New York, 1970, pp. 655–713.

[34] (a) K. B. Wiberg and J. M. Lavanish, *J. Amer. Chem. Soc.*, **88,** 5272 (1966); (b) *ibid.*, **88,** 365 (1966).

[35] (a) F. Cook, H. Shechter, J. Bayless, L. Friedman, R. L. Foltz, and R. Randall, *J. Amer. Chem. Soc.*, **88,** 3870 (1966); (b) J. Bayless, L. Friedman, J. A. Smith, F. B. Cook, and H. Shechter, *ibid.*, **87,** 661 (1965).

exchange occurred on the intermediate diazo compound. Again it is probable that the intermediate diazonium ion collapses directly to product,[34a,35] or that it loses nitrogen to give a bicyclobutonium ion.[33] A

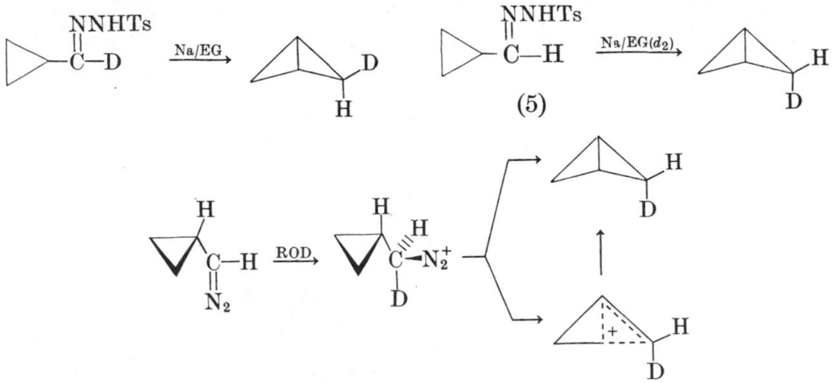

(5)

normal cyclopropylcarbinyl cation yields a mixture consisting mainly of unrearranged and ring-expanded products.[33,36]

The aprotic reactions of substituted cyclopropanecarboxaldehyde tosylhydrazones have become important for the preparation of cyclobutene derivatives.[37–41]

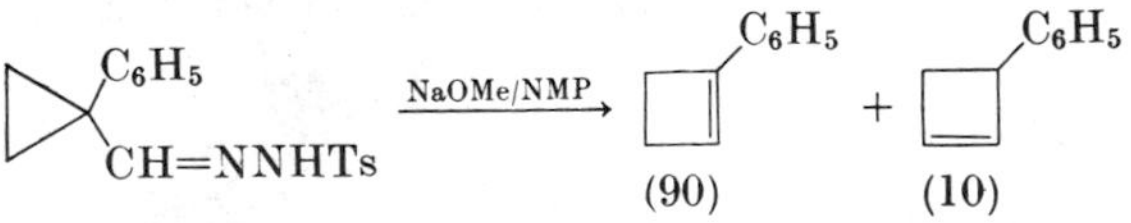

(90) (10) (Ref. 37)

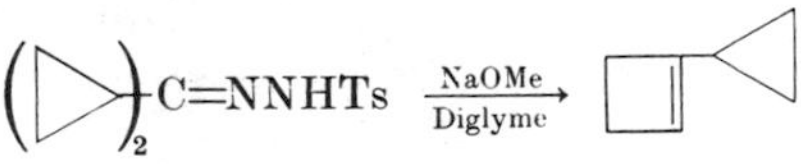

(Ref. 38)

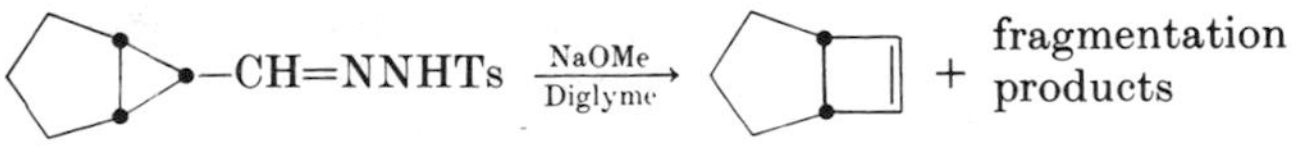

fragmentation products (Ref. 39)

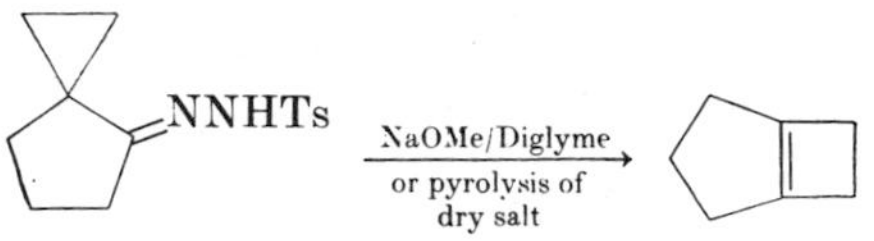

(Refs. 40, 41)

[36] J. D. Roberts and R. H. Mazur, *J. Amer. Chem. Soc.*, **73,** 2509 (1951).

[37] J. W. Wilt, J. M. Kosturik, and R. C. Orlowski, *J. Org. Chem.*, **30,** 1052 (1965).

[38] H. M. Ensslin and M. Hanack, *Angew. Chem., Int. Ed. Engl.*, **6,** 702 (1967).

[39] W. Kirmse and K.-H. Pook, *Chem. Ber.*, **98,** 4022 (1965).

[40] W. Kirmse and K.-H. Pook, *Angew. Chem., Int. Ed. Engl.*, **5,** 594 (1966).

[41] (a) K. B. Wiberg, G. J. Burgmaier, and P. Warner, *J. Amer. Chem. Soc.*, **93,** 246 (1971); (b) K. B. Wiberg, J. E. Hiatt, and G. J. Burgmaier, *Tetrahedron Lett.*, **1968,** 5855.

The formation of the ring-expanded product is consistent with alkyl migration to the divalent center of a carbene intermediate.[25] Such a migration is also important in the cyclobutylcarbinyl system[37,42] but diminishes with increasing ring size.[37,43]

$$\text{(cyclobutyl)}-CH{=}N_2 \xrightarrow{-N_2} \text{(cyclobutyl)}-\ddot{C}H \longrightarrow \text{(cyclopentene)}$$

Tosylhydrazones have also been used in other systems to study relative migratory aptitudes in carbene rearrangements.[44,45] Coupling the results from these studies with those obtained with carbenes generated from other precursors[46] has led to the conclusion that hydrogen migrates faster than phenyl, which migrates faster than methyl. Obviously, cyclopropylcarbene is an exception. In the substituted phenyl series the following order of migration aptitudes is obeyed: $o\text{-}CH_3O > p\text{-}CH_3O > p\text{-}CH_3 > p\text{-}Cl > H > m\text{-}Cl > p\text{-}O_2N$.[44,45] Alkylthio and arylthio groups have also been shown to be excellent migrating groups,[47,48] but oxygen and nitrogen groups are not.[47]

The Reaction of Tosylhydrazones with Alkyllithium Reagents

The first observations on the reaction of tosylhydrazones with alkyllithium reagents were that an unrearranged alkene is produced (Eq. 1, p. 409)[12,17] and the less substituted alkene is formed almost exclusively.[12] The first experiment to shed any light on the mechanism was one demonstrating that an α-hydrogen atom was eliminated in the reaction,[12] thus precluding a hydride shift mechanism and an intermediate carbene. Similar results had been previously reported with sodium hydride[5] and lithium aluminum hydride[16] in special cases.

(bicyclic tosylhydrazone, D, D, NNHTs) $\xrightarrow[\text{Room temp}]{\text{MeLi/Et}_2\text{O}}$ (bicyclic alkene, D)

The anionic mechanism for the lithium aluminum hydride reaction (*cf.* p. 418) is consistent with the elimination of an α-hydrogen atom and

[42] D. H. Paskovich and P. W. N. Kwok, *Tetrahedron Lett.*, **1967**, 2227.
[43] J. W. Wilt, J. F. Zawadzki, and D. G. Schultenover, *J. Org. Chem.*, **31**, 876 (1966).
[44] P. B. Sargeant and H. Shechter, *Tetrahedron Lett.*, **1964**, 3957.
[45] J. A. Landgrebe and A. G. Kirk, *J. Org. Chem.*, **32**, 3499 (1967).
[46] H. Phillip and J. Keating, *Tetrahedron Lett.*, **1961**, 523.
[47] J. H. Robson and H. Shechter, *J. Amer. Chem. Soc.*, **89**, 7112 (1967).
[48] I. Ojima and K. Kondo, *Bull. Chem. Soc. Jap.*, **46**, 2571 (1973).

with the preferential elimination of the most acidic α-hydrogen atom to give the less substituted alkene.[16] This mechanism would predict deuterium incorporation in the alkene if the reaction mixture were quenched with deuterium oxide. The tosylhydrazone of a 17-ketosteroid does, in fact, yield 81% d_1 alkene when allowed to react with lithium aluminum hydride and subsequently quenched with deuterium oxide.[16] Deuterium incorporation in the methyllithium reaction, however, is not so reliable.* Treatment of the reaction mixture from camphor tosylhydrazone (**1**) with deuterium oxide led to only about 10% deuterium incorporation; 60% incorporation was realized in the case of a 17-ketosteroid.[12] Apparently the ether solvent can act as a proton donor to varying extents which are dependent on the system. Evidence for solvent participation is supported by the fact that ethanol is found in some reaction mixtures.

The anionic mechanism referred to above[5,16] included, as its initial step, a concerted 1,4 elimination of *p*-toluenesulfinic acid from the tosylhydrazone anion, yielding a vinyl diimide anion. Alternatively, one can view this elimination as a two-step process in which the α-proton is abstracted to give a dianion intermediate which undergoes subsequent expulsion of *p*-toluenesulfinate anion.[17]

$$\mathrm{>C(-C(H)<)=N-\bar{N}-Ts\ Li^+} \xrightarrow{-H^+} \mathrm{>C(-\bar{C}<)=N-\bar{N}-Ts\ Li^+} \xrightarrow{-Ts^-} \mathrm{>C(=C<)-N=\bar{N}\ Li^+} \xrightarrow{-N_2}$$

$$\mathrm{>\bar{C}=C<\ Li^+} \xrightarrow[\text{Workup}]{H^+} \mathrm{>C(H)=C<}$$

Unpublished experiments by the author gave strong support for the two-step elimination of *p*-toluenesulfinic acid, at least in cases where the α-proton is primary or secondary.† For example, pinacolone tosylhydrazone (primary α-proton) and cyclohexanone tosylhydrazone (secondary α-proton) can be methylated (or deuterated) at the α position by treatment with 2 equivalents of methyllithium in tetrahydrofuran followed by treatment with 1 equivalent of methyl iodide (or deuterium oxide). The optimum temperature for α methylation (or deuteration) for primary

* Note added in proof: The vinyl anion intermediate has been trapped using deuterium oxide in tetramethylenediamine (J. E. Stemke and F. T. Bond, *Tetrahedron Lett.*, *1975*, 1815). The author has also found this same result with other reagents.

† Note added in proof: Some details of dianion and vinyl trapping experiments have now been published (R. H. Shapiro, M. F. Lipton, K. J. Kolonko, R. L. Buswell, and L. A. Capuano, *Tetrahedron Lett.*, *1975*, 1811).

systems is $-78°$ and for secondary systems is 0°. At these temperatures the decomposition to alkene is slow, and yields of 80–90% have been realized. Dialkylation is readily achieved by adding 2 equivalents of methyl iodide at the optimum temperature and slowly heating the reaction to 55°. In addition, monoalkylation on nitrogen occurs with 1 equivalent of methyllithium followed by the addition of 1 (or more) equivalents of methyl iodide at the optimum temperature and then warming to 55°. To date no such successes have been observed with tertiary α-protons. For example, isobutyrophenone tosylhydrazone has not been successfully alkylated or deuterated on the α-carbon atom when subjected to these conditions; unchanged starting material is isolated nearly quantitatively.* Although the experiments have obvious synthetic implications, for now we can only say that they support the dianionic mechanism.

$$(CH_3)_3CC(=NNHTs)CH_3 \xrightarrow[\text{2. MeI, } -78° \quad \text{3. } H_2O]{\text{1. 2MeLi/THF, } -78°} (CH_3)_3CC(=NNHTs)C_2H_5$$

$$\text{Cyclohexanone tosylhydrazone (=NNHTs)} \xrightarrow{\text{Same sequence at 0°}} \text{2-Methylcyclohexanone tosylhydrazone (=NNHTs)}$$

$$C_6H_5C(=NNHTs)CH(CH_3)_2 \xrightarrow{\text{Same sequence at 0°}} \text{No reaction}$$

After formation of the vinyl diimide anion, elimination of molecular nitrogen would give a vinyl anion which may abstract a proton from solvent or external sources in the workup. When open-chain tosylhydrazones were treated with methyllithium, however, abnormally high *cis/trans* ratios were observed, as shown in the accompanying equations.[49]

$$C_6H_5C(=NNHTs)C_2H_5 \xrightarrow{\text{MeLi/Et}_2\text{O}} C_6H_5CH=CHCH_3 \quad cis/trans\ 76:24$$

$$C_6H_5C(=NNHTs)C_2H_5 \xrightarrow{\text{NaOMe/Diglyme}} C_6H_5CH=CHCH_3 \quad cis/trans\ 20:80$$

$$C_6H_5CH(OH)C_2H_5 \xrightarrow[\text{AcOH}]{H_2SO_4} C_6H_5CH=CHCH_3 \quad cis/trans\ 5:95$$

If the *cis*- and *trans*-alkenes were generated by way of vinyl anions, would the high *cis/trans* ratio be expected? Such anions have been reported to equilibrate.[50] Perhaps, at least in some cases, the vinyl diimide

* The reaction of alkyllithium reagents with tosylhydrazones containing only tertiary α-protons has not been the subject of any detailed study. See discussion in the section on Scope and Limitations of the reaction.

[49] R. H. Shapiro, *Tetrahedron Lett.*, **1968**, 345.

[50] D. Y. Curtin and J. W. Crump, *J. Amer. Chem. Soc.*, **80,** 1922 (1958); and references contained therein.

anion abstracts a proton from solvent to give a vinyl diimide which collapses to product. This type of mechanism is not without precedent; for example, the collapse of a vinyl diimide to alkene is believed to occur in the Kishner reductive elimination,[51,52] and alkyl diimides are known to give alkenes with retention of configuration.[53] In any event, it seems reasonable that at least two mechanisms are operative: one in which a vinyl anion is a discrete intermediate, and a second in which a vinyl diimide is produced. The amount of each route depends on structure and reaction conditions.

$$>C(=C<)-N=\bar{N} \xrightarrow[(-C_2H_4,\ -EtO^-)]{Et_2O} >C(=C<)-N=NH \xrightarrow{-N_2} >C(H)=C<$$

SCOPE AND LIMITATIONS

The Protic Reaction

Tosylhydrazones are easily prepared from aldehydes and ketones and are highly crystalline, conveniently handled, and indefinitely stable at ordinary laboratory temperatures.

Both the protic Bamford-Stevens reaction and ketone reduction followed by dehydration of the corresponding alcohol involve two steps from the ketone. Although the yield of alkene from tosylhydrazone is frequently poor,[31] in some cases it is still the more convenient and hence preferred route.

$$R_1C(=O)CHR_2R_3 \xrightarrow{[H]} R_1CHOHCHR_2R_3 \xrightarrow[-H_2O]{H^+} R_1CH{=}CR_2R_3$$

$$R_1C(=O)CHR_2R_3 \xrightarrow{TsNHNH_2} R_1C(=NNHTs)CHR_2R_3 \xrightarrow[EG]{Na} R_1CH{=}CR_2R_3$$

The relationship between the two routes is similar to that between the Clemmensen and Wolff-Kishner reductions: one is carried out in acidic, the other in basic media, and the stability of other functional groups in the molecule must be considered when selecting one or the other procedure. The intermediates in the protic Bamford-Stevens reaction (diazonium ions and/or poorly solvated carbonium ions) can lead to product ratios

[51] P. S. Wharton, S. Dunny, and L. S. Krebs, *J. Org. Chem.*, **29**, 958 (1964).

[52] N. J. Leonard and S. Gelfand, *J. Amer. Chem. Soc.*, **77**, 3272 (1955).

[53] D. J. Cram and J. S. Bradshaw, *J. Amer. Chem. Soc.*, **85**, 1108 (1963).

different from those produced in alcohol dehydrations (well-solvated cations). A side reaction competing occasionally with alkene formation gives azines. Except for the convenience it offers in some cases, the protic Bamford-Stevens reaction has no clear advantage over other methods of alkene synthesis. Examples of this reaction appear in Table I (p. 435).

The Aprotic Reaction

The aprotic Bamford-Stevens reaction, on the other hand, generates carbene intermediates most conveniently. When the carbene intermediates undergo 1,2-hydride migrations instead of intramolecular insertions, the aprotic Bamford-Stevens reaction gives alkenes in useful yields. For open-chain tosylhydrazones with both α- and β-hydrogen atoms, a competition between hydride migration and insertion occurs.[6] The rate of alkene formation by the former route is usually faster, as the two accompanying examples illustrate.[54]

$$s\text{-}C_4H_9CH{=}NNHTs \xrightarrow[\text{Diglyme}]{\text{NaOMe}} \underset{(63.5)}{C_2H_5C({=}CH_2)CH_3} + \underset{(21.5)}{\overset{C_2H_5}{\triangle}} + \underset{(cis\text{ and }trans,\ 15.0)}{\overset{CH_3}{\triangle}\text{—}CH_3}$$

$$(C_2H_5)_2CHCH{=}NNHTs \xrightarrow[\text{Diglyme}]{\text{NaOMe}} \underset{(73.5)}{(C_2H_5)_2C{=}CH_2} + \underset{(cis\text{ and }trans,\ 26.5)}{\overset{C_2H_5}{\triangle}\text{—}CH_3}$$

Tosylhydrazones of cyclopentanones and cyclohexanones seem to undergo hydrogen migration exclusively upon reaction with base in aprotic media.[13,24,55–63] With larger rings intramolecular insertion again competes with hydrogen migration.[28,64–69] The formation of bicyclic compounds is

[54] W. Kirmse and G. Wächtershäuser, *Tetrahedron*, **22**, 63 (1966).

[55] J. W. Powell and M. C. Whiting, *Tetrahedron*, **12**, 168 (1961).

[56] C. Swithenbank and M. C. Whiting, *J. Chem. Soc.*, **1963**, 4573.

[57] R. W. Alder and M. C. Whiting, *J. Chem. Soc.*, **1963**, 4595.

[58] A. P. Krapcho and R. Donn, *J. Org. Chem.*, **30**, 641 (1965).

[59] N. C. G. Campbell, J. R. P. Clark, R. R. Hill, P. Oberhänsli, J. H. Parish, R. M. Southam, and M. C. Whiting, *J. Chem. Soc., B*, **1968**, 349.

[60] J. M. Coxon, M. P. Hartshorn, D. N. Kirk, and M. A. Wilson, *Tetrahedron*, **25**, 3107 (1969).

[61] R. M. Coates and E. F. Bertram, *J. Org. Chem.*, **36**, 3722 (1971).

[62] S. Oida and E. Ohki, *Chem. Pharm. Bull.* (Tokyo), **17**, 1405 (1969).

[63] W. Kirmse and L. Ruetz, *Ann.*, **726**, 36 (1969).

[64] A. C. Cope, M. Brown, and G. L. Woo, *J. Amer. Chem. Soc.*, **87**, 3107 (1965).

[65] A. C. Cope and S. S. Hecht, *J. Amer. Chem. Soc.*, **89**, 6920 (1967).

[66] W. Kirmse and G. Münscher, *Ann.*, **726**, 42 (1969).

[67] M. R. Vegar and R. J. Wells, *Tetrahedron Lett.*, **1969**, 2565.

[68] A. P. Krapcho and J. Diamanti, *Chem. Ind.* (London), **1965**, 847.

[69] J. Casanova and B. Waegell, *Bull. Soc. Chim. Fr.*, **1971**, 1289.

dominant with nine- and ten-membered rings[28,64,65] and almost disappears with twelve-membered rings.[68,69] Six-membered rings which exist in fixed boat conformations, such as those found in the norbornyl system (*e.g.*, Eq. 1, p. 409) decompose almost exclusively by intramolecular insertion.[6,7,11,12,17–19,70–75] It should also be noted that hydrogen migration in a carbene intermediate may be somewhat indiscriminate. For example, migrations of secondary and tertiary hydrogen atoms compete in the decomposition of 2-methylcyclohexanone tosylhydrazone as illustrated in Eq. 2 (p. 410).[13] Open-chain tosylhydrazones can also show the same type of nonselectivity;[6] even with methyl isopropyl ketone tosylhydrazone,[37] 5% of the terminal alkene, 3-methyl-1-butene, is produced.[76]

$$i\text{-}C_3H_7C(=NNHTs)CH_3 \longrightarrow \underset{(90)}{(CH_3)_2C{=}CHCH_3} + \underset{(5)}{(CH_3)_2CHCH{=}CH_2} + \underset{(5)}{\text{1,2-dimethylcyclopropane}}$$

In spite of these limitations, the aprotic Bamford-Stevens reaction has found numerous applications. For example, the 12-ketosteroid tosylhydrazone formulated below gives exclusively unrearranged olefin when heated with sodium methoxide in diglyme,[60] but suffers extensive rearrangement under protic conditions.[77,78]

TsHNN R′ RO —NaOMe / Diglyme→ R′ RO

In both the protic and aprotic reactions base-labile functional groups in the substrate will be altered. Esters, for example, do not survive the reaction conditions, except when lithium hydride in toluene is used to decompose the tosylhydrazone.[14] (This modification of the aprotic

[70] A. Nickon, H. Kwansik, T. Swartz, R. O. Williams, and J. B. DiGiorgio, *J. Amer. Chem. Soc.*, **87**, 1613 (1965).

[71] S. Julia and C. Gueremy, *Bull. Soc. Chim. Fr.*, **1965**, 3002.

[72] H. Krieger, S.-E. Masar, and H. Ruotsalainen, *Suom. Kemistilehti*, **39**, 237 (1966) [*C.A.*, **66**, 65170n (1967)].

[73] T. Sasaki, S. Eguchi, and T. Kiryama, *J. Amer. Chem. Soc.*, **91**, 212 (1969).

[74] C. A. Grob and J. Hostynek, *Helv. Chim. Acta*, **46**, 1676 (1963).

[75] J. A. Berson and R. G. Bergman, *J. Amer. Chem. Soc.*, **89**, 2569 (1967).

[76] A. M. Mansoor and I. D. R. Stevens, *Tetrahedron Lett.*, **1966**, 1733.

[77] (a) R. Hirschmann, C. S. Snoddy, Jr., C. F. Hiskey, and N. L. Wendler, *J. Amer. Chem. Soc.*, **76**, 4013 (1954); (b) C. F. Hiskey, R. Hirschmann, and N. L. Wendler, *ibid.*, **75**, 5135 (1953).

[78] J. Elks, G. H. Phillips, D. A. H. Taylor, and L. J. Wyman, *J. Chem. Soc.*, **1954**, 1739.

Bamford-Stevens reaction has not received much attention, but it appears to have excellent potential in alkene syntheses). Table II lists the reported examples of the aprotic Bamford-Stevens reaction (p. 446).

The Alkyllithium Reaction

The reaction of tosylhydrazones with alkyllithium reagents is an extremely useful synthesis of difficultly obtainable, unrearranged, and less substituted alkenes. The yields are good to excellent. Although almost all reported examples involve tosylhydrazones of cyclic ketones, unpublished experiments from the author's laboratory show that open-chain tosylhydrazones behave similarly. For example, treatment of the tosylhydrazone of phenylacetone with methyllithium in ether at room temperature yields allylbenzene as the only alkene. To date, however, no detailed study of the reactions of unsymmetrical, open-chain tosylhydrazones, which bear the same number of α-hydrogen atoms (*e.g.*, 3-hexanone tosylhydrazone) with alkyllithium reagents has been reported. The phenylacetone tosylhydrazone decomposition indicates that steric factors may have more important influences on product ratios than electronic factors, since benzylic hydrogens should be more acidic than methyl hydrogens. Additional study on open-chain systems is clearly warranted.

$$C_6H_5CH_2C(=NNHTs)CH_3 \xrightarrow{MeLi/Et_2O} C_6H_5CH_2CH{=}CH_2$$

As mentioned in the Mechanism section, the scope of the reaction between alkyllithium reagents and tosylhydrazones containing only tertiary α protons has not been tested.* The author has recently examined the behavior of isobutyrophenone (and to a lesser extent diisopropyl ketone) tosylhydrazone in the presence of methyllithium (unpublished work). Although the optimum reaction conditions for production of isobutenylbenzene have not yet been determined, it is clear that the abstraction of the α proton is very slow. In fact, elimination is so slow that nucleophilic substitution competes extremely effectively with it, even when only 2 equivalents of methyllithium are employed. The accompanying equation illustrates the fate of isobutyrophenone tosylhydrazone in the presence of methyllithium; the label was introduced in order to ascertain whether the α-hydrogen atom could act as a proton source (it does not).

$$C_6H_5C(=NNHTs)CD(CH_3)_2 \xrightarrow[Et_2O]{MeLi} \underset{(25\%)}{C_6H_5CH{=}C(CH_3)_2} + \underset{(25\%)}{C_6H_5CH(CH_3)CD(CH_3)_2}$$

The ability of the alkyllithium reaction to produce difficultly obtainable alkenes is illustrated through additional examples below. Bis- and even

* Two such cases have been studied. The unrearranged product was obtained in unreported[79] or poor[190] yield.

[79] Y. Chretien-Bessiere and J.-P. Bras., *C.R. Acad. Sci., Ser. C*, **268**, 2221 (1969).

tris-tosylhydrazones yield unrearranged products, as do cyclopropylcarbinyl systems. Neopentyl systems, such as camphor tosylhydrazone (**1**) also yield unrearranged alkenes (Eq. 1, p. 409), and α,β-unsaturated tosylhydrazones that contain an α-hydrogen atom on the saturated side of the carbon-nitrogen double bond lead cleanly to conjugated dienes.[80]

NNHTs $\xrightarrow{\text{MeLi}}$ (Ref. 19)

NNHTs $\xrightarrow{\text{MeLi}}$ (Ref. 79)

NNHTs, C_6H_5, C_6H_5 $\xrightarrow{n\text{-BuLi}}$ C_6H_5, C_6H_5 (Ref. 81)

NNHTs $\xrightarrow{\text{MeLi}}$ (Ref. 82)

NNHTs, TsHNN $\xrightarrow{n\text{-BuLi}}$ (Ref. 83)

TsHNN, TsHNN, NNHTs $\xrightarrow{\text{MeLi}}$ (Ref. 84)

[80] W. G. Dauben, M. E. Lorber, N. D. Vietmeyer, R. H. Shapiro, J. H. Duncan, and K. Tomer, *J. Amer. Chem. Soc.*, **90,** 4762 (1968).

[81] J. S. Swenton, J. A. Hyatt, T. J. Walker, and A. L. Crumrine, *J. Amer. Chem. Soc.*, **93,** 4808 (1971).

[82] B. D. Cuddy, D. Grant, and M. A. McKervey, *J. Chem. Soc., C*, **1971,** 3173.

[83] T. Tsuji, S. Nishida, and H. Tsubomura, *J. Chem. Soc., Chem. Commun.*, **1972,** 284.

[84] J. Font, F. Lopez, and F. Serratosa, *Tetrahedron Lett.*, **1972,** 2589.

Most examples of this reaction deal with tosylhydrazones containing no other functional group, but a few show that remote cyclic ethers and amines survive the reaction conditions, as expected. Certain cyclic ethers undergo fragmentation (see next section). As mentioned earlier, substitution occasionally competes with elimination.[12,18,19,79] For example, when 8 equivalents of alkyllithium, such as *t*-butyllithium, are used on cholestanone tosylhydrazone, substitution rather than elimination results.[18]

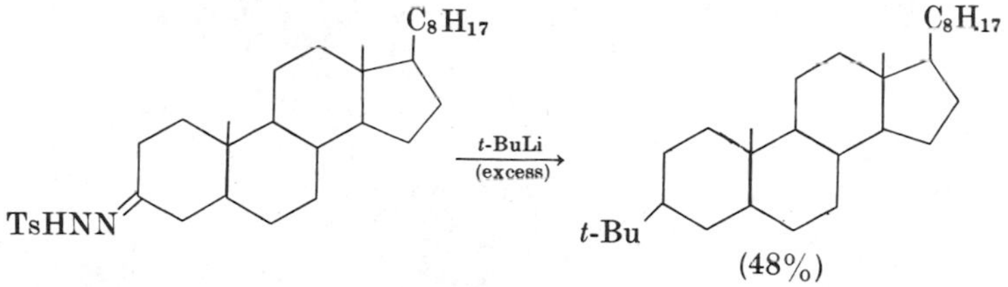

C_8H_{17} C_8H_{17}

TsHNN $\xrightarrow[\text{(excess)}]{t\text{-BuLi}}$ *t*-Bu (48%)

FRAGMENTATION REACTIONS OF TOSYLHYDRAZONES

In 1967 two groups of investigators discovered that tosylhydrazones of many α,β-epoxy ketones are quite unstable and spontaneously decompose at room temperature to give acetylenic ketones.[85–88] This reaction may be formally viewed as a fragmentation of the Grob type.[89] For example, isophorone oxide (**7**) reacts with tosylhydrazine at 0° and, when the reaction mixture is allowed to warm to room temperature, nitrogen is evolved and an acetylenic ketone is produced in 74% yield.[88]

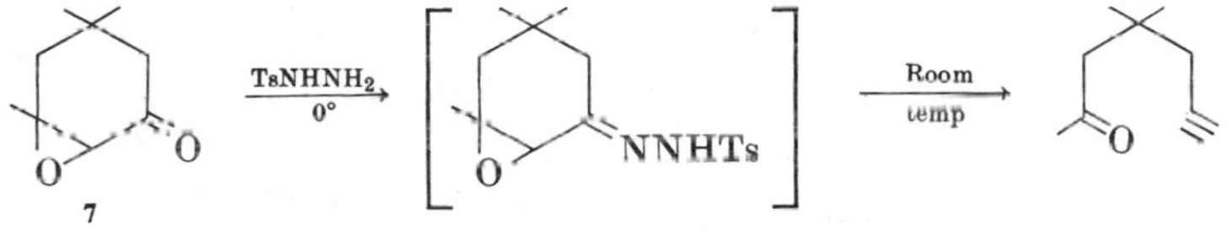

O O **7** $\xrightarrow[0°]{TsNHNH_2}$ [O NNHTs] $\xrightarrow[\text{temp}]{\text{Room}}$ O

A formal mechanism to rationalize the fragmentation has been postulated.[85–88]

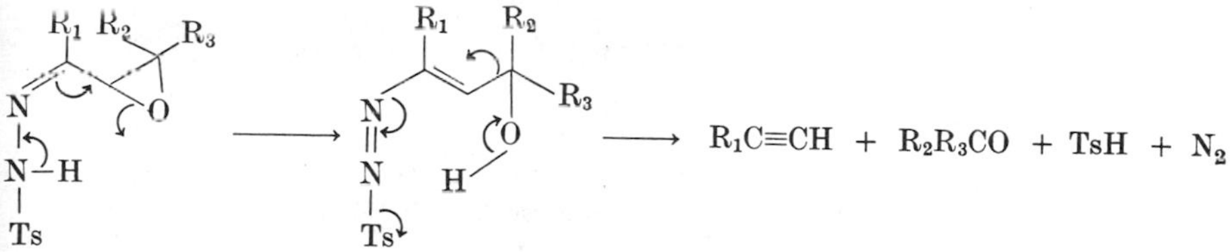

R_1 R_2 R_3 N O N–H Ts ⟶ R_1 R_2 R_3 N ‖ N O H Ts ⟶ $R_1C{\equiv}CH + R_2R_3CO + TsH + N_2$

[85] (a) A. Eschenmoser, D. Felix, and G. Ohloff, *Helv. Chim. Acta*, **50**, 708 (1967); (b) D. Felix, J. Schreiber, G. Ohloff, and A. Eschenmoser, *ibid.*, **54**, 2896 (1971).

[86] J. Schreiber, D. Felix, A. Eschenmoser, M. Winter, F. Gautschi, K. H. Schulte-Elte, E. Sundt, G. Ohloff, J. Kalvoda, H. Kaufmann, P. Wieland, and G. Anner, *Helv. Chim. Acta*, **50**, 2101 (1967).

[87] M. Tanabe, D. F. Crowe, R. L. Dehn, and G. Detre, *Tetrahedron Lett.*, **1967**, 3739.

[88] M. Tanabe, D. F. Crowe, and R. L. Dehn, *Tetrahedron Lett.*, **1967**, 3943.

[89] C. A. Grob and P. W. Schiess, *Angew. Chem., Int. Ed. Engl.*, **6**, 1 (1967).

If R_1 and R_2 in the epoxytosylhydrazone are connected to form a ring, an acetylenic ketone is the product. If R_1 and R_2 are unconnected, two products are formed.[86]

$$(CH_2)_{11}(CH_2)_2\text{-epoxide-}C_6H_5 \xrightarrow{TsNHNH_2} (CH_2)_{11}(CH_2C{\equiv}CCH_2) + C_6H_5CHO$$

The sodium salts of the stable tosylhydrazones of furfural and related systems undergo fragmentation on pyrolysis.[90] Fragmentation of saturated analogs of furfural tosylhydrazone can be induced with butyllithium, the products being mainly allenes.[91]

$$\text{2-furyl-}CH{=}N\bar{N}Ts\ Na^+ \xrightarrow{250^\circ\ (0.5\ mm)} HC{\equiv}CCH{=}CHCHO$$

(66%) (*cis*/*trans*, 81:19)

$$\text{2-methyltetrahydrofuran-2-yl-}C(CH_3){=}NNHTs \xrightarrow{n\text{-BuLi}} HO(CH_2)_3C(CH_3){=}C{=}CH_2$$

(48%)

Molecules having such leaving groups as hydroxy,[92] acetoxy,[92] benzoyloxy,[92], mesyloxy,[93] and fluorine[94] adjacent to the imino carbon of the tosylhydrazone function will also undergo fragmentation in the presence of base to give acetylenes. Two possible mechanisms for decomposition are shown at the top of page 427.

Although fragmentation reactions of tosylhydrazones have not been fully studied, the process shows great promise in the synthesis of acetylenic compounds. A few failures, however, have been reported.[85,86]

RELATED AND ANOMALOUS REACTIONS

The formation of allenes and alkynes by reactions not covered in the preceding section, and of alkanes is the main subject of this section. In addition, the formation of cyclopropenes and other products from intramolecular cycloaddition reactions is discussed.

[90] R. V. Hoffman and H. Shechter, *J. Amer. Chem. Soc.*, **93**, 5940 (1971).

[91] A. M. Foster and W. C. Agosta, *J. Org. Chem.*, **37**, 61 (1972).

[92] T. Iwadare, I. Adachi, M. Hayashi, A. Matsunaga, and T. Kitai, *Tetrahedron Lett.*, **1969**, 4447.

[93] P. Wieland, *Helv. Chim. Acta*, **53**, 171 (1970).

[94] S. J. Cristol and J. K. Harrington, *J. Org. Chem.*, **28**, 1413 (1963).

R$_2$ R$_1$ C H C −HX N N−Ts −TsH −N$_2$ R$_2$ R$_1$ C H C X N H N Ts R$_1$C=CR$_2$ −H$^+$, N$_2$ R$_2$ R$_1$ C H C X −TsH +N N −X$^-$ R$_1$ R$_2$ C=C H +N N

Allenes and Alkynes

The first report of alkyne formation concerned the aprotic decomposition of cyclopropanecarboxaldehyde tosylhydrazone where acetylene was obtained in 10–13% yield.[25] A few years later it was noted that an attempt to prepare quadricyclene from nortricyclenone tosylhydrazone gave only an alkyne and an allene.[94]

NNHTs NaOMe Diglyme, heat C≡CH +

Ambient-temperature photolysis[95] of the sodium salt of the same tosylhydrazone led not to fragmentation, however, but to sulfone formation by recombination of the expelled p-toluenesulfinate with the intermediate carbene.[96] A few other examples of strained cyclopropylcarbinyl

[95] W. G. Dauben and F. G. Willey, *J. Amer. Chem. Soc.*, **84**, 1497 (1962).

[96] D. M. Lemal and A. J. Fry, *J. Org. Chem.*, **29**, 1673 (1964).

tosylhydrazones undergoing similar fragmentations have been reported. Three are shown in the accompanying equations.

Na^+ $N\bar{N}Ts$ → Heat → (64) + (36) (Ref. 97)

NNHTs → NaOMe / Diglyme, heat → (13) + (12) + other products (Ref. 98)

=NNHTs → NaOMe / Tetraglyme, heat → —C≡CH (Ref. 99)

Tetramethylallene[100] and a cumulene[101,102] were prepared by the pyrolyses shown in the accompanying equations.

Na^+ $N\bar{N}Ts$ S O_2 → 110–116° → >=C=<

Na^+ $N\bar{N}Ts$ $N\bar{N}Ts$ Na^+ → Heat[101] or heat Cu[102] → >=C=C=<

A promising method for producing cycloalkynes has been discovered.[103]

[97] J. W. Wheeler, R. H. Chung, V. N. Vaishnav, and C. C. Shroff, *J. Org. Chem.*, **34,** 545 (1969).

[98] P. K. Freeman and D. G. Kuper, *J. Org. Chem.*, **30,** 1047 (1965).

[99] R. G. Bergman and V. J. Rajadhyaksha, *J. Amer. Chem. Soc.*, **92,** 2163 (1970).

[100] R. Kalish and W. G. Pirkle, *J. Amer. Chem. Soc.*, **89,** 2781 (1967).

[101] F. T. Bond and D. E. Bradway, *J. Amer. Chem. Soc.*, **87,** 4977 (1965).

[102] G. Maier, *Tetrahedron Lett.*, **1965,** 3603.

[103] H. Meier and I. Menzer, *Synthesis*, **1971,** 215.

In several examples, yields of 36–55% were obtained by photolyzing *gem*-bistosylhydrazones.

$(CH_2)_n$ C=NNHTs / C=NNHTs $\xrightarrow[h\nu]{NaOH/H_2O—MeOH}$ $(CH_2)_n$ C≡C

(n = 6, 10, 13)

Cyclopropenes

It has been shown that some tosylhydrazones of open-chain α,β-unsaturated carbonyl compounds can undergo base-induced decomposition to give cyclopropenes by way of a vinylcarbene intermediate that undergoes intramolecular cycloaddition.[104] Similar findings have been reported by others.[105–107]

R_1, R_3, R_2, =NNHTs, R_4 $\xrightarrow[Diglyme]{NaOMe}$ R_1, R_3, R_2, :, R_4 ⟶ R_1, R_2, R_4, R_3

Longer-range intramolecular cycloaddition of carbenes generated from unsaturated tosylhydrazones has been attempted in a few cases. The natural product thujopsene (**8**) was formed in 51% yield by such a reaction, along with 10% of a cyclopropene.[108] A similar example is known,[109] but an attempt to make bicyclo[2.1.0]pentane by this technique gave only products from 1,2- and 1,3-intramolecular insertion.[110]

CH=NNHTs $\xrightarrow{NaH}$ **8** +

Li$^+$ CH=N$\bar{N}$Ts $\xrightarrow{250°}$ (92%) + (4%)

[104] G. L. Closs, L. E. Closs, and W. A. Böll, *J. Amer. Chem. Soc.*, **85,** 3796 (1963).
[105] H.-H. Stechl, *Chem. Ber.*, **97,** 2681 (1964).
[106] H. Dürr, *Chem. Ber.*, **103,** 369 (1970).
[107] H. Dürr, *Angew. Chem., Int. Ed. Engl.*, **6,** 1084 (1967).
[108] G. Büchi and J. D. White, *J. Amer. Chem. Soc.*, **86,** 2884 (1964).
[109] W. Kirmse and D. Grassmann, *Chem. Ber.*, **99,** 1746 (1966).
[110] D. H. White, P. B. Condit, and R. G. Bergman, *J. Amer. Chem. Soc.*, **94,** 1348 (1972).

Alkanes

In 1963 Caglioti and Magi reported that tosylhydrazones of some aromatic aldehydes and steroidal ketones react with lithium aluminum hydride to give alkanes.[111] 17-Ketosteroid tosylhydrazones gave alkenes rather than reductive elimination products.[4,111,112] Other systems which do not give reductive elimination include camphor tosylhydrazone (**1**), formulated on p. 409, which gives tricyclene (**3**) by way of a carbene,[11] hydroxycaryolanone tosylhydrazone (**9**),[113] and a few steroidal examples.[16]

$$R_1R_2C{=}NNHTs \xrightarrow[2.\ H_2O]{1.\ LiAlH_4} R_1R_2CH_2 + N_2 + TsH$$

HO— ... =NNHTs (**9**) $\xrightarrow[2.\ H_2O]{1.\ LiAlH_4}$ HO— ...

9

An investigation of the Caglioti reaction led to the conclusion that the hydride attacks the original carbonyl carbon of the tosylhydrazone salt.[16] The anionic intermediate remains in the reaction mixture until a proton source is added. Thus the reaction may be a useful alternative to the Wolff-Kishner procedure and an excellent method for introducing deuterium, sometimes stereoselectively, into organic compounds.[16,114–121] By use of lithium aluminum deuteride and/or deuterium oxide, one or two deuterium atoms can be incorporated.

$$H^- \longrightarrow (R_1)(R_2)C{=}N{-}\bar{N}{-}Ts \longrightarrow H{-}C(R_1)(R_2){-}N{=}\bar{N} \xrightarrow{H_2O}$$

$$R_1R_2CH{-}N{=}NH \xrightarrow{-N_2} R_1CH_2R_2$$

[111] L. Caglioti and M. Magi, *Tetrahedron*, **19**, 1127 (1963).

[112] (a) L. Caglioti, *Tetrahedron*, **22**, 487 (1966); (b) L. Caglioti and P. Graselli, *Chem. Ind.* (London), **1964**, 153; *Chim. Ind.* (Milan); **46**, 799, 1492 (1964); **47**, 62 (1965) [*C.A.*, **61**, 8365b (1964); **62**, 7669h, 10388g (1965)].

[113] F. Y. Edamura and A. Nickon, *J. Org. Chem.*, **35**, 1509 (1970).

[114] C. Djerassi and L. Tokes, *J. Amer. Chem. Soc.*, **88**, 536 (1966).

[115] L. Tokes, G. Jones, and C. Djerassi, *J. Amer. Chem. Soc.*, **90**, 5465 (1968).

[116] L. Tokes and C. Djerassi, *J. Amer. Chem. Soc.*, **91**, 5017 (1969).

[117] R. T. Gray and C. Djerassi, *Org. Mass Spectrom.*, **3**, 245 (1970).

[118] R. T. LaLonde, J. Ding, and M. A. Tobias, *J. Amer. Chem. Soc.*, **89**, 6651 (1967).

[119] I. Elphimoff-Felkin and M. Verrier, *Tetrahedron Lett.*, **1968**, 1515.

[120] A. Buchs, *Helv. Chim. Acta*, **51**, 688 (1968).

[121] C. Asselineau, H. Montrozier, and J.-C. Promé, *Bull. Soc. Chim. Fr.*, **1969**, 1911.

Recently, the Caglioti reaction has been modified by using sodium cyanoborohydride to obtain excellent yields of reduced products; many of the compounds tried contained reducible functional groups that would not have survived treatment with lithium aluminum hydride.[122]

$$CH_3CO(CH_2)_3CO_2C_8H_{17}\text{-}n \xrightarrow[\text{DMF/Sulfolane}]{\text{TsNHNH}_2} \xrightarrow[\text{TsOH}]{\text{NaBH}_3\text{CN}} CH_3(CH_2)_4CO_2C_8H_{17}\text{-}n$$

Since the tosylhydrazone reduction is carried out on the acid side, many base-sensitive groups (esters, amides, cyano, nitro, and halo) have been shown to be stable to the reaction conditions. Moreover, under these acidic reaction conditions it is the protonated tosylhydrazone that is reduced with sodium cyanoborohydride, and thus the mechanism differs considerably from that of the Caglioti reaction. The proposed mechanism, shown below, resembles that of the Wolff-Kishner reduction after the elimination of *p*-toluenesulfinic acid.

$$R_1R_2C{=}NNHTs \xrightarrow{H^+} R_1R_2C{=}\overset{+}{N}HNHTs \xrightarrow{\text{NaBH}_3\text{CN}} R_1R_2CHNHNHTs \xrightarrow{\text{Base}}$$
$$R_1R_2CHN{=}NH \xrightarrow{-N_2} R_1R_2CH_2$$

EXPERIMENTAL CONDITIONS AND PROCEDURES

The protic Bamford-Stevens reaction is ordinarily conducted by dissolving metallic sodium in ethylene glycol, adding the tosylhydrazone, and heating the mixture under reflux for 30 minutes to 2 hours. Commercially available sodium methoxide may be used in place of sodium. In both protic and aprotic reactions, distilled solvents often increase yields. Several liters of diglyme at a time may be dried over lithium aluminum hydride, distilled from sodium, and stored under nitrogen in sealed 100-ml brown bottles. Traces of mositure do not affect the protic reaction but may produce lower yields in the aprotic reactions.

There have been reports (see the tables) of decompositions of sulfonylhydrazones other than tosylhydrazones, such as mesylhydrazones, benzenesulfonylhydrazones, and 2,4-dichlorobenzenesulfonylhydrazones. These derivatives offer no advantage over tosylhydrazones; under some reaction conditions the mesylhydrazones may suffer metalation α to sulfur and lead to additional products. Ethylene glycol (protic) and diglyme (aprotic) are the preferred solvents. Some of the solvents recommended, such as acetamide, are less convenient and lack compensating advantages.

[122] (a) R. O. Hutchins, C. A. Milewski, and B. E. Maryanoff, *J. Amer. Chem. Soc.*, **95**, 3662 (1973); (b) *ibid.*, **93**, 1973 (1971).

The literature indicates that the choice of methyllithium or butyllithium in the alkyllithium reaction seems to be a matter of preference or availability; both have been used with success. Preliminary experiments with phenyllithium showed that it was inferior. Obviously, the reaction with alkyllithium reagents requires dry equipment, reagents, and solvents.

The progress of the alkyllithium reactions, which are commonly run at temperatures below the boiling points of the solvents, can occasionally be monitored when the solvent acts as a proton donor. In these cases the reaction mixtures develop a very bright orange to red color which disappears when the reaction is complete. On the other hand, the evolution of nitrogen (after the initial evolution of methane when methyllithium is used) is difficult to follow, since it appears to depend on a proton source which is at least patially the reaction solvent.

Tosylhydrazine. Tosylhydrazine is conveniently prepared by the reaction of tosyl chloride with hydrazine hydrate[123] or is purchased from commercial suppliers.

Tosylhydrazones (General Procedure). Ketones and aldehydes are readily converted to tosylhydrazones with tosylhydrazine in acidic, and occasionally neutral, media. Solvents commonly employed are acidified ethanol[1] and acetic acid.[20] The following general procedure is typical.[20]

The carbonyl compound (1 equiv), dissolved in the minimal volume of hot acetic acid, was mixed with a solution of 1 equiv of tosylhydrazine in hot acetic acid. (The latter solution is prepared by dissolving 4 parts of tosylhydrazine in 5 parts of hot acetic acid.) The reaction mixture was heated to boiling or until crystallization began and then cooled to 5°. The crystalline product was filtered, washed with cold acetic acid, cold aqueous acetic acid, water, and finally air-dried in a warm place. Yields of 60–96% were reported. The Bamford-Stevens procedure with acidic ethanol is quite similar, except that the reaction mixture is usually heated under reflux for ½ to 1 hour. The preparation of tosylhydrazones under neutral conditions, a procedure useful for acid-sensitive carbonyl compounds, is illustrated in the next paragraph.

Cyclopropanecarboxaldehyde Tosylhydrazone.[34a] Cyclopropanecarboxaldehyde (5.1 g, 73 mmol) was added to a solution of 13.5 g (73 mmol) of tosylhydrazine in 25 ml of 60% aqueous methanol at 60°. The solution was placed in a refrigerator. After the reaction mixture had stood overnight, the solid which had separated was collected and washed with 60% methanol, giving 15.8 g (91%) of the tosylhydrazone, mp 95–97°. Recrystallization from methanol raised the melting point to 96.5–98°.

[123] L. Friedman, R. L. Little, and W. R. Reichle, *Org. Syntheses*, **40**, 93 (1960).

Cyclohexene (The Protic Bamford-Stevens Reaction.)[1] The following procedure can be used to generate many volatile alkenes. In an apparatus equipped for distillation, 5 g (18.6 mmol) of cyclohexanone tosylhydrazone was added to 50 ml of 1.3 *N* sodium β-hydroxyethoxide in ethylene glycol (prepared by adding metallic sodium to the solvent). The mixture was heated to 190–200° and the cyclohexene (1.5 g, 18.3 mmol, 98%) was isolated by distillation.

For the production of nonvolatile alkenes the reaction mixture was heated for 10 minutes to 1 hour at 160–190°, cooled to room temperature, and diluted with water. The alkene was isolated by extraction with ether or chloroform and purified by recrystallization or chromatography.

Alkenes (The Aprotic Bamford-Stevens Reaction). The same procedure as above is used except that commercial sodium methoxide is the base and diglyme is the solvent. Since diglyme is miscible with water, the workup is the same except that several water washings are necessary to remove all traces of diglyme through extraction.

***exo, exo*-Tricyclo[6.2.1.0^{2,7}]undec-3-ene**[57] **(Decomposition of a Mesylhydrazone in Sodium and Acetamide).** Sodium (620 mg) was dissolved in molten acetamide (80°, reduced pressure) and the solution was cooled. *exo, exo*-Tricyclo[6.2.1.0^{2,7}]undec-3-one mesylhydrazone (2 g) was added and the mixture was heated to 156° in a bromobenzene-vapor bath and maintained there until nitrogen evolution virtually ceased. After cooling to 65°, water (100 ml) was added and the solution was extracted with pentane (6 × 15 ml); the organic phase was washed with water (2 × 10 ml) and the solvent was evaporated. Distillation yielded the alkene (248 mg, 28%), bp 76–77° (8 mm).

2-Bornene (The Alkyllithium Reaction).[124] In a dry 1-l, three-necked flask protected from moisture were placed camphor tosylhydrazone (32 g, 0.1 mol) and 400 ml of dry ether. The flask was immersed in a water bath (20–25°) and the contents magnetically stirred. To the stirred solution was added 150 ml (0.24 mol) of 1.6 *N* methyllithium in ether during 1 hour. Stirring was continued for 8 or 9 hours, during which time the reaction mixture developed a deep red-orange color. A small amount of water was added to destroy the excess methyllithium and then an additional 200 ml was added. The layers were separated and the organic layer was washed with water and dried. The solvent was removed by distillation and the product purified by distillation. The yield of 2-bornene was 8.5–8.8 g (63–65%), mp 110–111°.

[124] R. H. Shapiro and J. H. Duncan, *Org. Syntheses*, **51**, 66 (1971).

Cyclopentadec-5-yn-1-one (Fragmentation Reaction).[85b] A solution of 708 mg (3.0 mmol) of 1,12-epoxybicyclo[10.3.0]pentadecan-3-one in 7 ml of 1:1 methylene chloride: acetic acid standing at −24° was mixed with a precooled (−24°) solution of tosylhydrazine (614 mg, 3.3 mmol) in 7 ml of the same solvent. After the resulting solution had remained at −24° for 36 hours, it was allowed to warm to 0°, held there for 2 hours, allowed to come to room temperature, and held there for 4 hours. The reaction mixture was poured onto ice water and extracted with ether. The ether extract was washed with aqueous carbonate and ice water and dried over anhydrous sodium sulfate. The solvent was removed with a rotary evaporator and the product was purified by distillation [bp 95–100° (0.03 mm)]. The yield of cyclopentadec-5-yn-1-one (mp 38–40°) was 555 mg (84%).

TABULAR SURVEY

The four tables that follow contain examples of the reactions covered in the chapter. The literature survey covers articles appearing up to April 1973, but several later references that appeared in leading journals are also included. Table I and Table II list examples of the protic and aprotic Bamford-Stevens reaction, respectively, and are arranged according to increasing carbon number. The reactions of tosylhydrazones with alkyllithium reagents (Table III) are sorted roughly according to compound type; similar compounds are grouped so that the general trends of the reaction can be seen at a glance. Table IV, the fragmentation reactions, is arranged according to increasing carbon number.

Acknowledgements

The author wishes to thank Nancee Bershoff, University of Colorado, who helped with the literature search, and Professor Anders Kjaer of the Technical University of Denmark, who provided time, space, and encouragement.

TABLE I. The Protic Bamford-Stevens Reaction

Carbon Atoms in Starting Ketone	Reactant	Conditions	Product(s)—Distribution ( ) or Yield(s) (%)	Refs.
		A. Monotosylhydrazones		
3	$(CH_3)_2C{=}NNHTs$	Na/ED, heat	$CH_3CH{=}CH_2$	1
4	$C_2H_5C({=}NNHTs)CH_3$	Na/EG, 170°	$CH_2{=}CHC_2H_5$ (28) + $CH_3CH{=}CHCH_3$ (*cis*, 37; *trans*, 30) + n-C_4H_{10} (5) (30%)	31
	i-$C_3H_7CH{=}NNHTs$	NaOMe/DEG, heat	$(CH_3)_2C{=}CH_2$ (65) + $CH_2{=}CHC_2H_5$ (10) + $CH_3CH{=}CHCH_3$ (*cis*, 4; *trans*, 8)  + [methylcyclopropane structure] (12)	6
		NaOMe/various protic solvents, heat	Same products in varying ratio depending on solvent	8
	[cyclopropyl structure]–CH=NNHTs	NaOMe/EG, heat	[cyclobutene structure] (16) + [bicyclobutane structure] (79) + $CH_2{=}CH{-}CH{=}CH_2$ (4)	10
	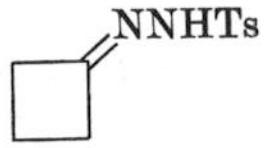 [cyclobutylidene structure]=NNHTs	Na/EG, heat	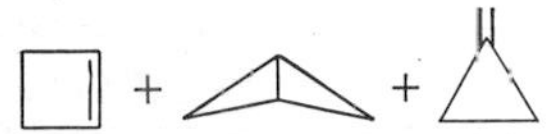 [cyclobutene structure] (12%) + [bicyclobutane structure] (60%) + [methylenecyclopropane structure] (6%)	35b

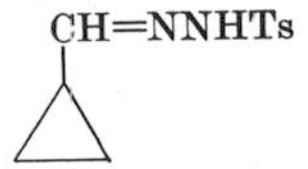

Note: References 125–210 are on pp. 506–507.

TABLE I. THE PROTIC BAMFORD-STEVENS REACTION (*Continued*)

Carbon Atoms in Starting Ketone	Reactant	Conditions	Product(s)—Distribution ( ) or Yield(s) (%)	Refs.
	A. Monotosylhydrazones			
4 (*contd.*)	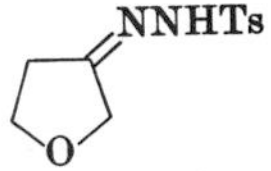 NNHTs, O	Na/EG, heat	  O (75%) + O (5%)	137
	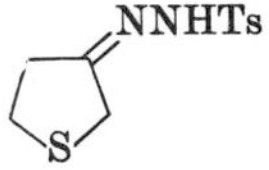 NNHTs, S	Na/EG, heat	  S (43%) + S (23%)	137
5	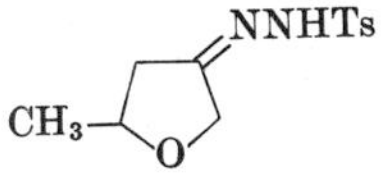 CH_3, NNHTs, O	Na/EG, heat	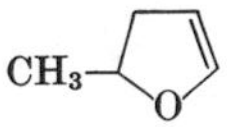 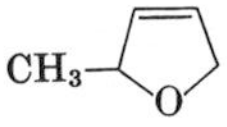 CH_3, O (70%) + CH_3, O (3%)	137
	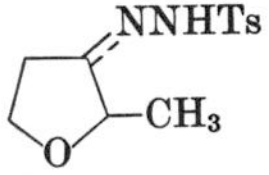 NNHTs, CH_3, O	Na/EG, heat	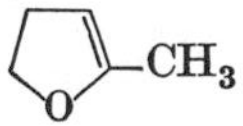 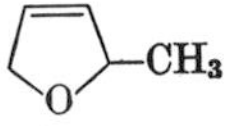 O, CH_3 (52%) + O, CH_3 (6%)	137
	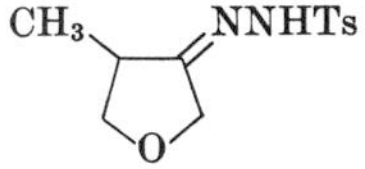 CH_3, NNHTs, O	Na/EG, heat	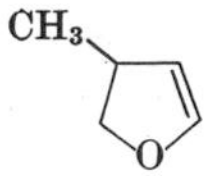 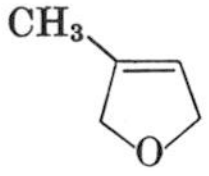 CH_3, O (49%) + CH_3, O (24%)	137
	$i\text{-}C_3H_7C(=NNHTs)CH_3$	Na/EG, 170°	$(CH_3)_2C{=}CHCH_3$ I (56) + $C_2H_5C(=CH_2)CH_3$ II (12) + $i\text{-}C_3H_7CH{=}CH_2$ (24) + $i\text{-}C_5H_{12}$ (8) (35%)	31

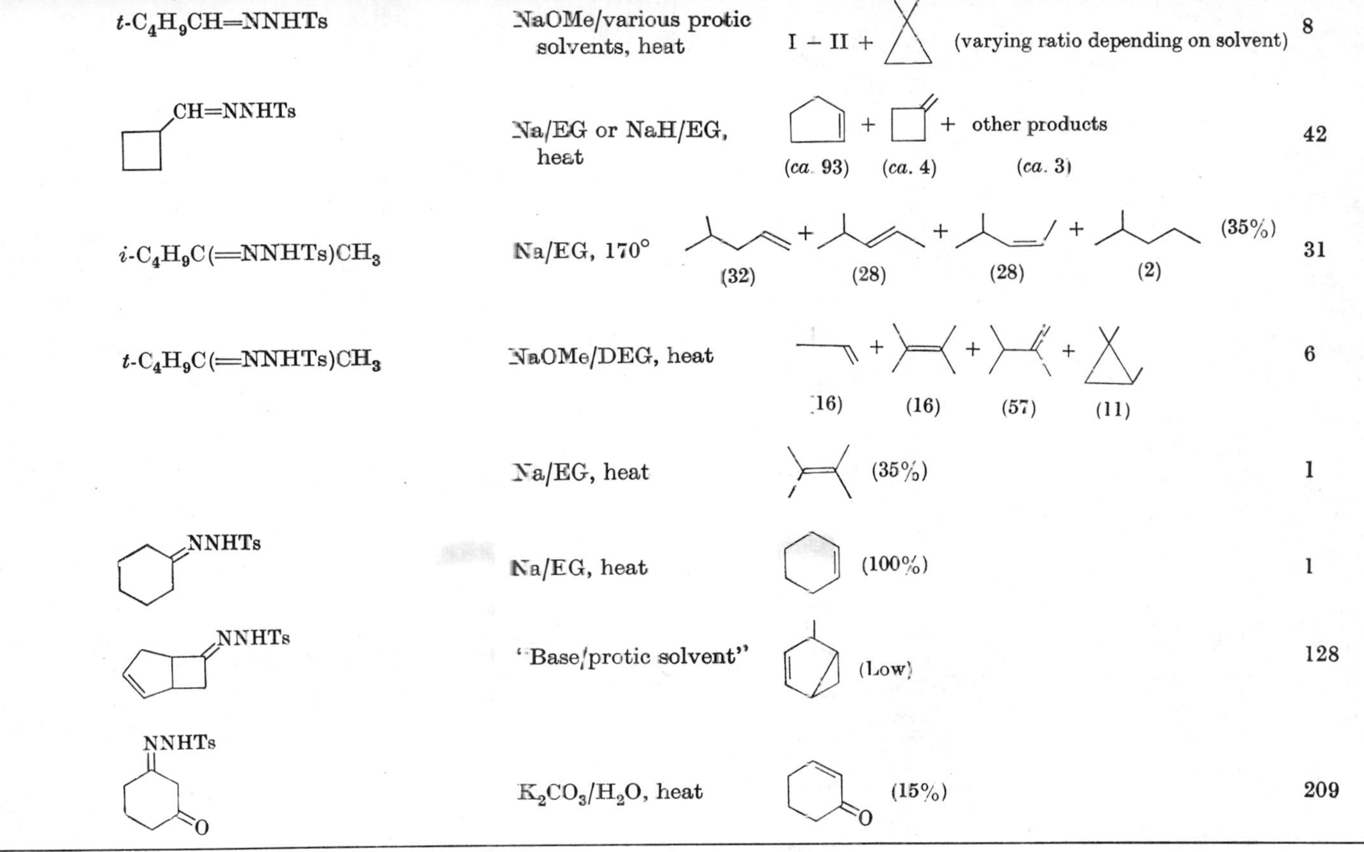

	t-$C_4H_9CH{=}NNHTs$	NaOMe/various protic solvents, heat	I − II + (varying ratio depending on solvent)	8
	CH=NNHTs	Na/EG or NaH/EG, heat	+ + other products (ca. 93) (ca. 4) (ca. 3)	42
6	i-$C_4H_9C({=}NNHTs)CH_3$	Na/EG, 170°	+ + + (35%) (32) (28) (28) (2)	31
	t-$C_4H_9C({=}NNHTs)CH_3$	NaOMe/DEG, heat	+ + + (16) (16) (57) (11)	6
		Na/EG, heat	(35%)	1
	NNHTs	Na/EG, heat	(100%)	1
	NNHTs	"Base/protic solvent"	(Low)	128
	NNHTs	K_2CO_3/H_2O, heat	(15%)	209

Note: References 125–210 are on pp. 506–507.

TABLE I. THE PROTIC BAMFORD-STEVENS REACTION (*Continued*)

Carbon Atoms in Starting Ketone	Reactant	Conditions	Product(s)—Distribution ( ) or Yield(s) (%)	Refs.
		A. Monotosylhydrazones		
7	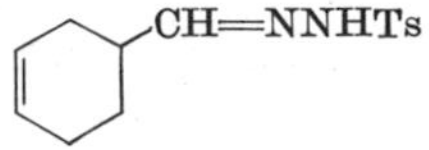 CH=NNHTs	Na/EG, heat	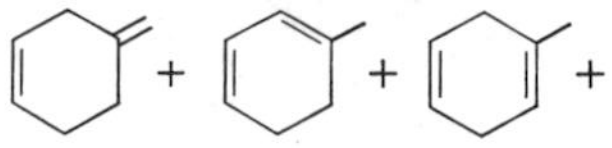 (1) + (1) + (1) + other products (30%)	129, 130
	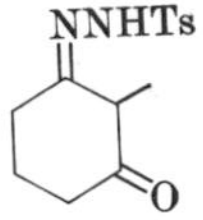 NNHTs, O	K_2CO_3/H_2O, heat	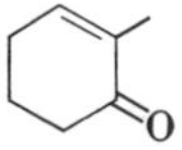 O (41%)	209
	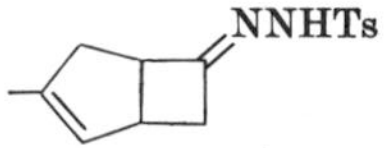 NNHTs	"Base/protic solvent"	 (Low)	128
8	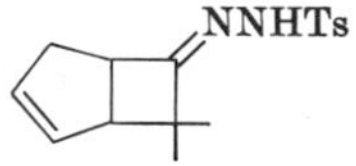 NNHTs	"Base/protic solvent"	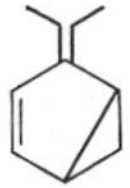 (Low)	128
	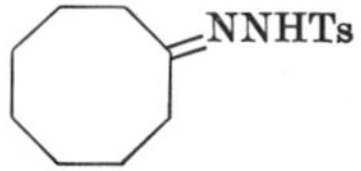 NNHTs	Na/EG, heat	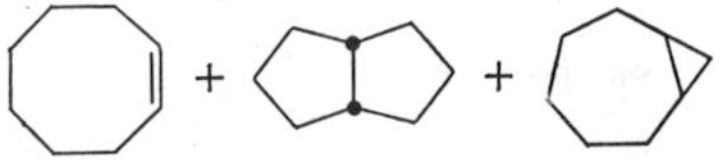 (83.3) + (6.4) + (10.4) (40%)	64

No. of C Atoms	Reactant	Conditions	Product(s) and Yield(s) (%)	Refs.
	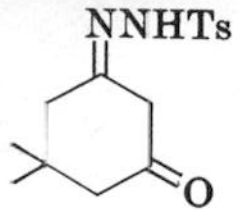 NNHTs, O	K_2CO_3/H_2O, heat	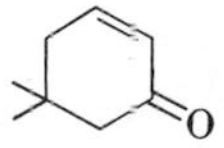 O (53%)	209
	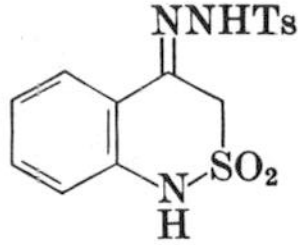 NNHTs, N, H, SO_2	NaOMe/EtOH—H_2O,	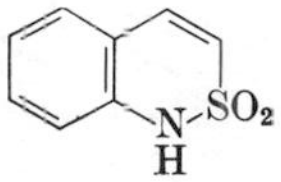 N, H, SO_2 (67%)	139
	$C_6H_5CH_2CH{=}NNHTs$	Na/EG, heat	$C_6H_5CH{=}CH_2$ (20%)	143
9	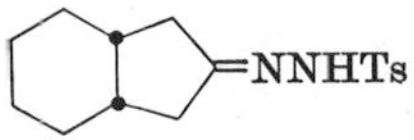 =NNHTs	Na/EG, heat	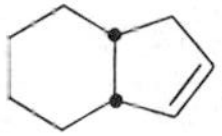 (—)	142
	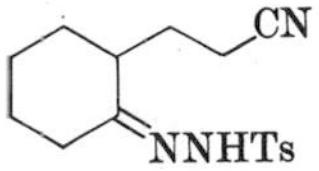 CN, NNHTs	Na/ethoxyethanol, heat	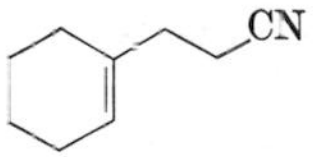 CN (60%)	144
	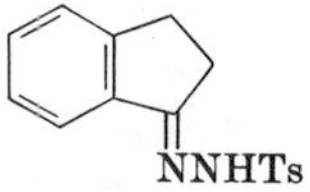 NNHTs	Na/EG, heat	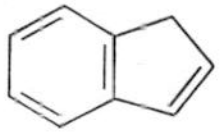 (92%)	1
	$C_6H_5CH_2C(=NNHTs)CH_3$	Na/EG, heat	$C_6H_5CH{=}CHCH_3$ (80%)	1
	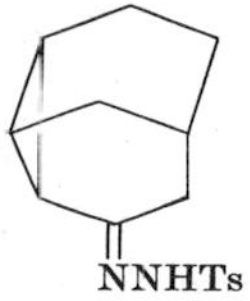 NNHTs	Na/EG, heat	 (85%)	126

Note: References 125–210 are on pp. 506–507.

TABLE I. THE PROTIC BAMFORD-STEVENS REACTION (*Continued*)

Carbon Atoms in Starting Ketone	Reactant	Conditions	Product(s)—Distribution ( ) or Yield(s) (%)	Refs.
			A. Monotosylhydrazones	
9 (*contd.*)	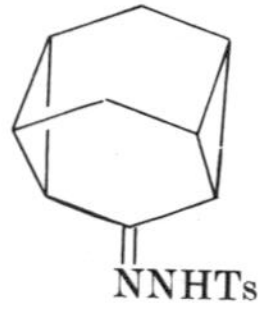 NNHTs	Na/EG, heat	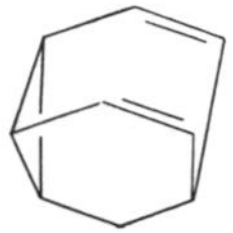 (56%)	126
	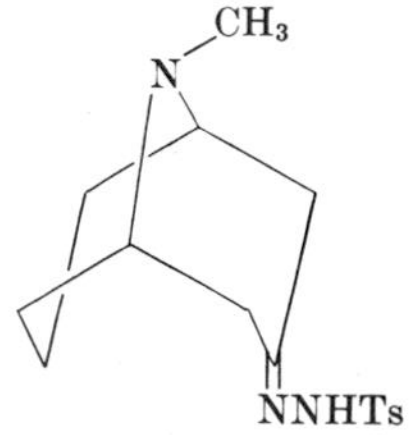 CH_3 N NNHTs	Na/EG, heat	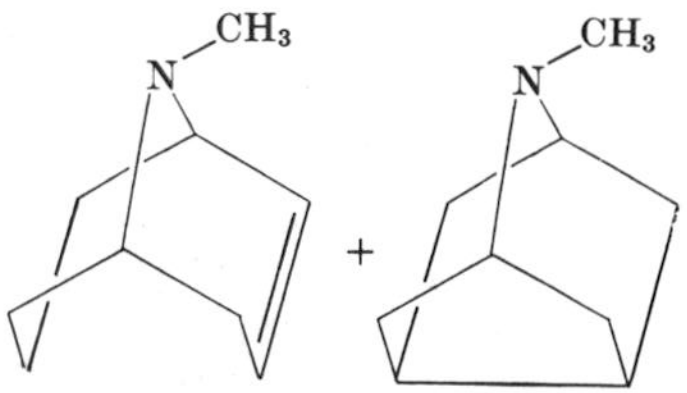 CH_3 N (76.5) + CH_3 N (23.5) (46%)	73
	$C_6H_5C(=NNHTs)CH_2OCH_3$	Na/EG, heat	$C_6H_5CH=CHOCH_3$ (54%)	143
10	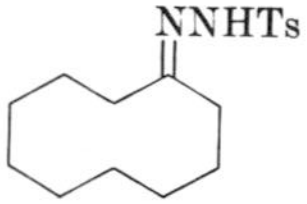 NNHTs	Na/EG, heat	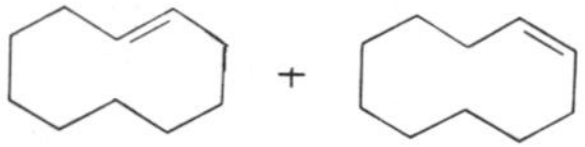 (30.6 *trans*) + (69.4 *cis*) (36%)	64
	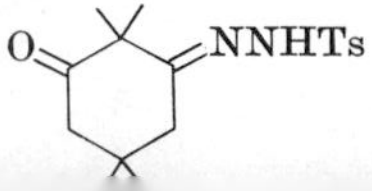 O NNHTs	Na/EG, heat	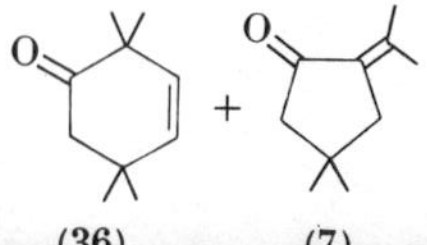 O (36) + O (7) (75%)	136

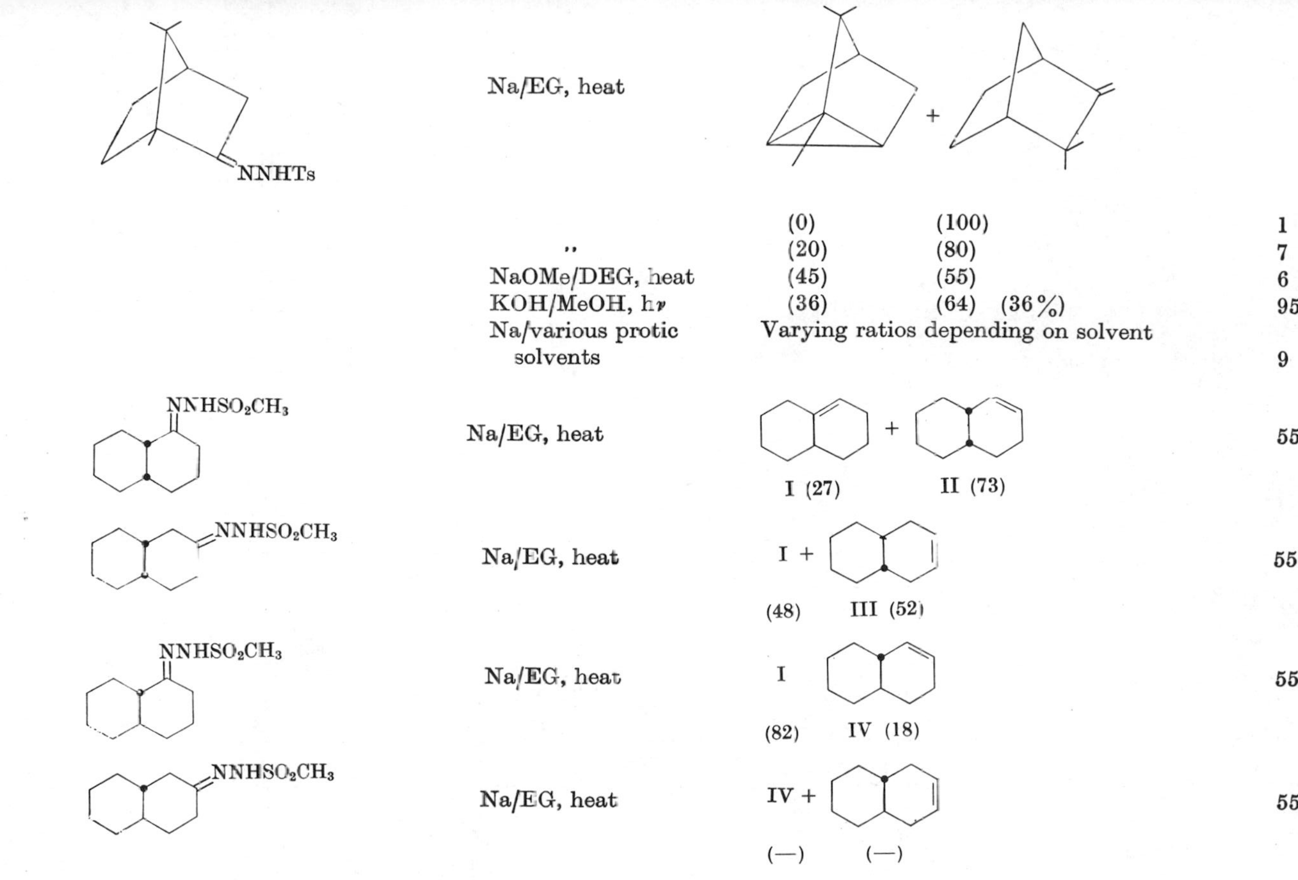

Reactant	Conditions	Product(s) and Yield(s) (%)	Refs.
NNHTs	Na/EG, heat	+	
		(0) (100)	1
	"	(20) (80)	7
	NaOMe/DEG, heat	(45) (55)	6
	KOH/MeOH, $h\nu$	(36) (64) (36%)	95
	Na/various protic solvents	Varying ratios depending on solvent	9
$NNHSO_2CH_3$	Na/EG, heat	+ I (27) II (73)	55
$NNHSO_2CH_3$	Na/EG, heat	I + (48) III (52)	55
$NNHSO_2CH_3$	Na/EG, heat	I (82) IV (18)	55
$NNHSO_2CH_3$	Na/EG, heat	IV + (—) (—)	55

Note: References 125–210 are on pp. 505–507.

TABLE I. THE PROTIC BAMFORD-STEVENS REACTION (*Continued*)

Carbon Atoms in Starting Ketone	Reactant	Conditions	Product(s)—Distribution ( ) or Yield(s) (%)	Refs.
		A. Monotosylhydrazones		
10 (*contd.*)	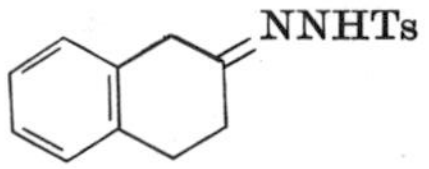 NNHTs	Na/EG, heat	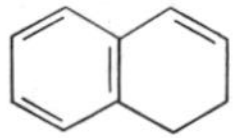 (60%)	143
	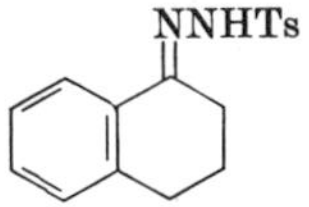 NNHTs	Na/EG, heat	Product is azine (75%)	143
	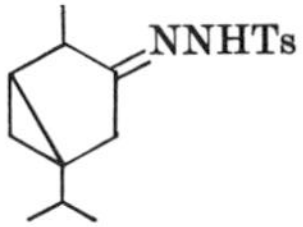 NNHTs	Na/EG, heat	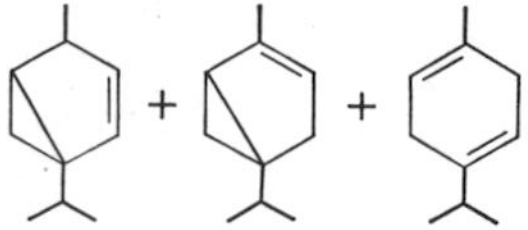 + + (42%) (16%) (13%)	127
11	$C_6H_5CH_2C(CH_3)_2CH{=}NNHTs$	Na/EG, heat	$C_6H_5CH_2C(CH_3)_2{=}CHCH_3$ (46) + $C_6H_5CH_2CH{=}C(CH_3)_2$ (33) 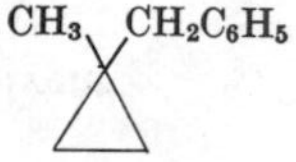 + CH_3 $CH_2C_6H_5$ (20)	138
	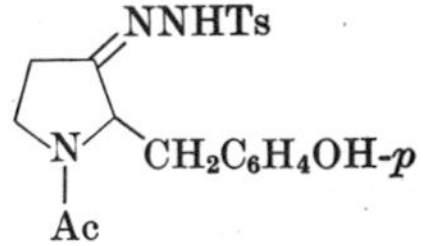 NNHTs, N, Ac, $CH_2C_6H_4OH$-*p*	KOH/DEG, 135–140	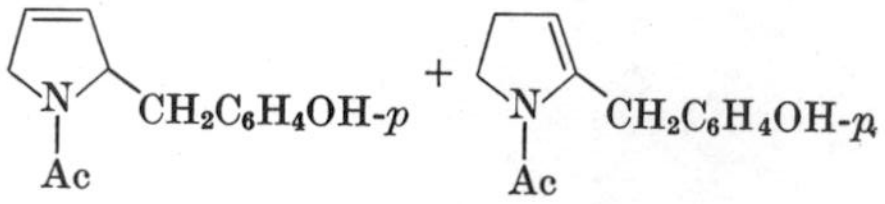 N, Ac, $CH_2C_6H_4OH$-*p* (55%) + N, Ac, $CH_2C_6H_4OH$-*p* (30%)	62
12	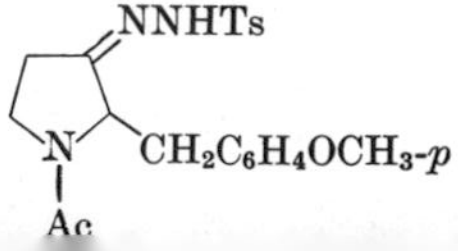 NNHTs, N, Ac, $CH_2C_6H_4OCH_3$-*p*	KOH/DEG, 135–140°	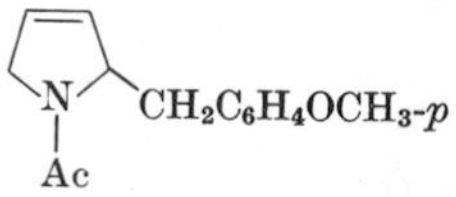 N, Ac, $CH_2C_6H_4OCH_3$-*p*	62

			(cis/trans, 54%/9%)	69
14	$C_6H_5C(=NNHTs)CHOHC_6H_5$	Na/EG, heat	$C_6H_5COCH_2C_6H_5$ (65%)	1
	$C_6H_5C(=NNHTs)CH_2OC_6H_5$	Na/EG, heat	$C_6H_5CH{=}CHOC_6H_5$ (40%)	143
15	$(C_6H_5CH_2)_2C{=}NNHTs$	Na/EG, heat	$C_6H_5CH_2CH{=}CHC_6H_5$ (95%)	1
	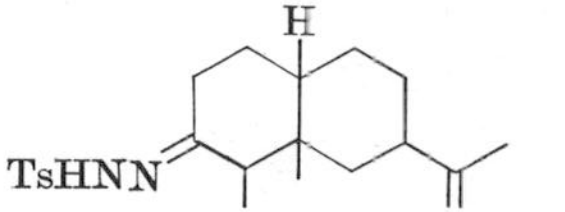	Na/EG, heat	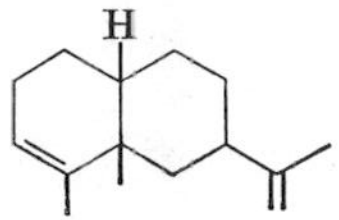 (95%)	140
	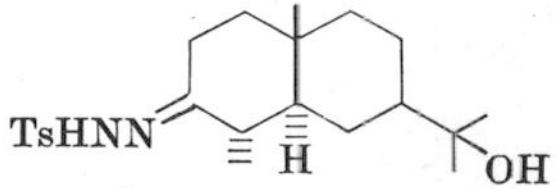	Na/EG, heat	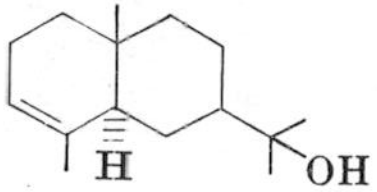	141
19	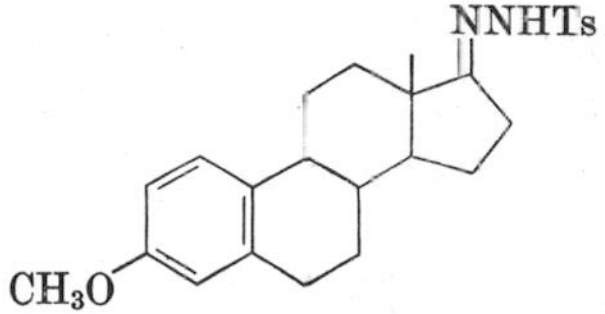	Na/EG, heat	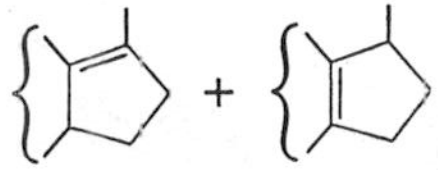 (15%) + (15%)	15
	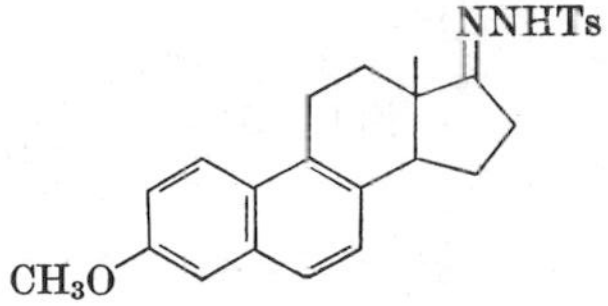	Na/EG, heat	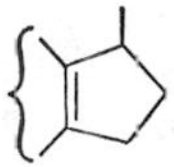 (89%)	134
27	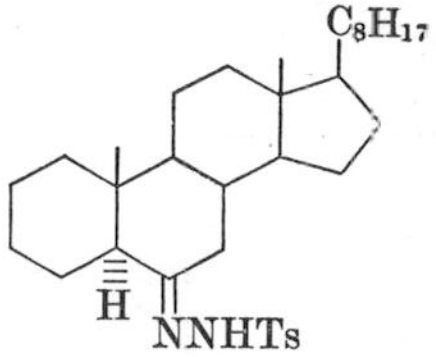	Na/EG, heat	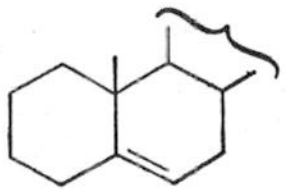	125, 132

Note: References 125–210 are on pp. 506–507.

TABLE I. THE PROTIC BAMFORD-STEVENS REACTION (*Continued*)

Carbon Atoms in Starting Ketone	Reactant	Conditions	Product(s)—Distribution ( ) or Yield(s) (%)	Refs.
		A. Monotosylhydrazones		
27 (*contd.*)	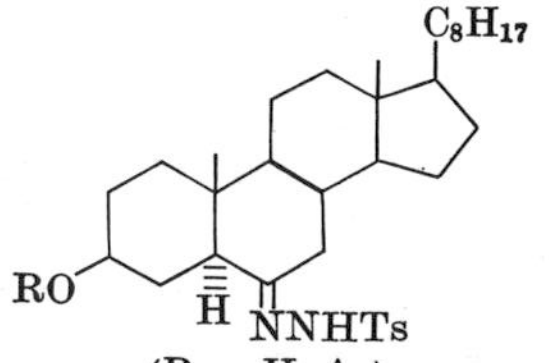 C_8H_{17}; RO; H; NNHTs (R = H, Ac)	Na/EG, heat	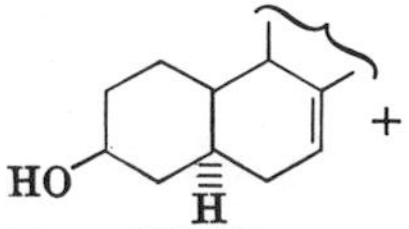 HO; H; + 3α-alcohol (58%) (13%)	133
			(43%) (0%)	30
29	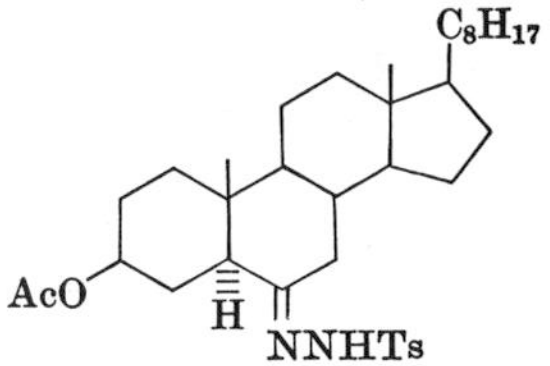 C_8H_{17}; AcO; H; NNHTs	1. Na/EG, heat; 2. Ac_2O	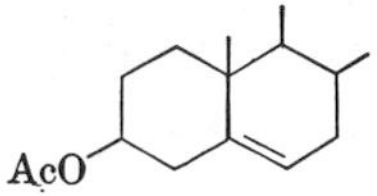 AcO (28–47%)	125, 132
	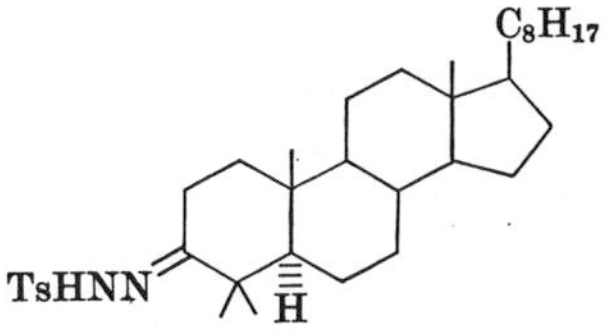 C_8H_{17}; TsHNN; H	Na/EG, heat	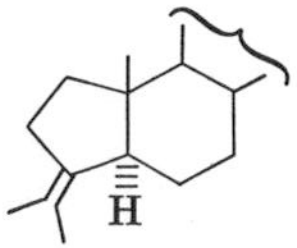 H ("Quant.")	131
	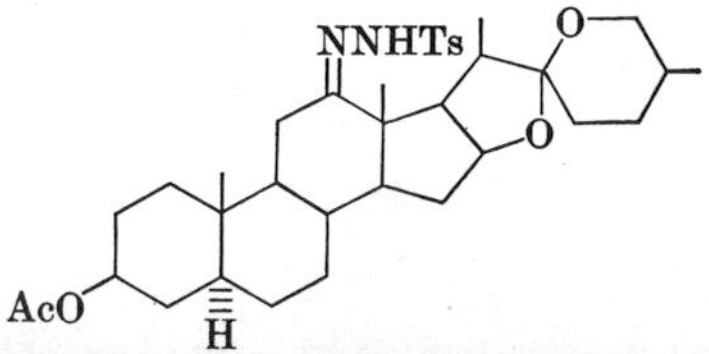 NNHTs; O; O; AcO; H	Various bases and protic solvents, heat	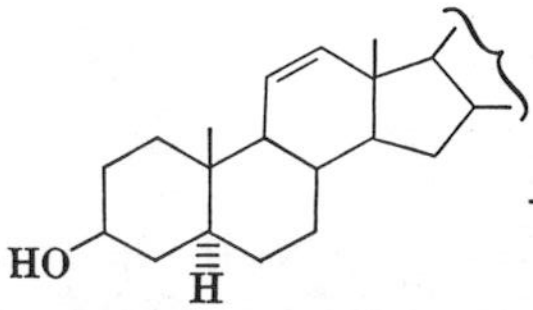 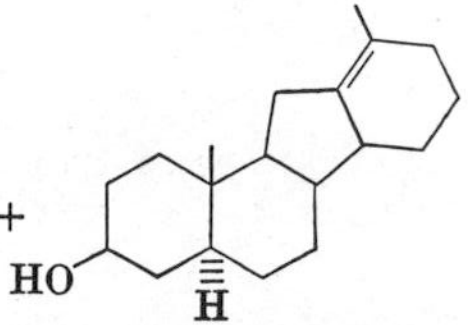 HO; H + HO; H	60, 77, 78

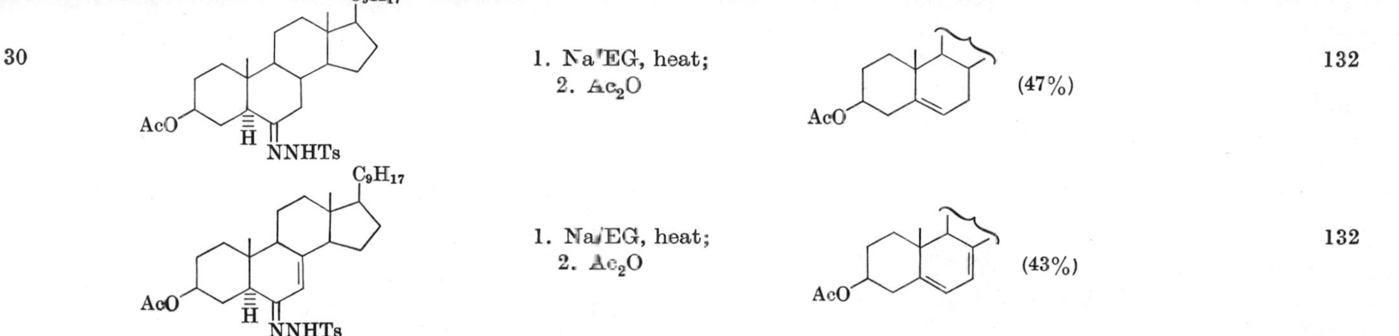

No. of C Atoms	Reactant	Conditions	Product(s) and Yield(s) (%)	Refs.
30		1. Na/EG, heat; 2. Ac_2O	(47%)	132
		1. Na/EG, heat; 2. Ac_2O	(43%)	132
		B. Bistosylhydrazones		
6		Na/EG, heat	(40%)	143
9		Na/EG, 120–150° (Cu)	(6.5%) + (2.7%) + (0.5%)	145
14	$C_6H_5C(=NNHTs)C(=NNHTs)C_6H_5$	Na/EG, heat	$C_6H_5C{\equiv}CC_6H_5$ (73%)	1[a]
16		Na/EG, heat	$C_6H_5CH{=}CHCH{=}CHC_6H_5$ (27%)	143

Note: References 125–210 are on pp. 506–507.

[a] Several other 1,2-bistosyldrazones are reported to give 1,2,3-triazole derivatives.

TABLE II. THE APROTIC BAMFORD-STEVENS REACTION

Carbon Atoms in Starting Ketone	Reactant	Conditions	Product(s)—Distribution ( ) or Yield(s) (%)	Refs.
		A. Monotosylhydrazones		
3	$C_2H_5CH{=}NNHTs$	NaOMe/DEC, heat	$CH_3CH{=}CH_2$ (90)  + [cyclopropane] (10)	6
4	*n*-$C_3H_7CH{=}NNHTs$	NaOMe/DEC, heat	$C_2H_5CH{=}CH_2$ (92) + $CH_3CH{=}CHCH_3$ (*cis*, 1,2; *trans*, 2.3)  + [methylcyclopropane] (4.6)	6
	i-$C_3H_7CH{=}NNHTs$	NaOMe/DEC, heat	$(CH_3)_2C{=}CH_2$ (62)  + [methylcyclopropane] (38)	6
	$C_2H_5C({=}NNHTs)CH_3$	NaOMe/DEC, heat	$CH_3CH{=}CHCH_3$ (*cis*, 28; *trans*, 67) + $C_2H_5CH{=}CH_2$ (5)  + [methylcyclopropane] (0.5)	6
	$CH_3CH{=}CHCH{=}NNHTs$ (*trans*)	NaOMe/diglyme, 160°	 [structure] (3%)	104
	 [cyclopropyl]$CH{=}NNHTs$	NaOMe (1.1 equiv)/ DEC, 180°	[cyclobutene] (83)  + [bicyclobutane] (3) + C_2H_2 + C_2H_4 (10)	10

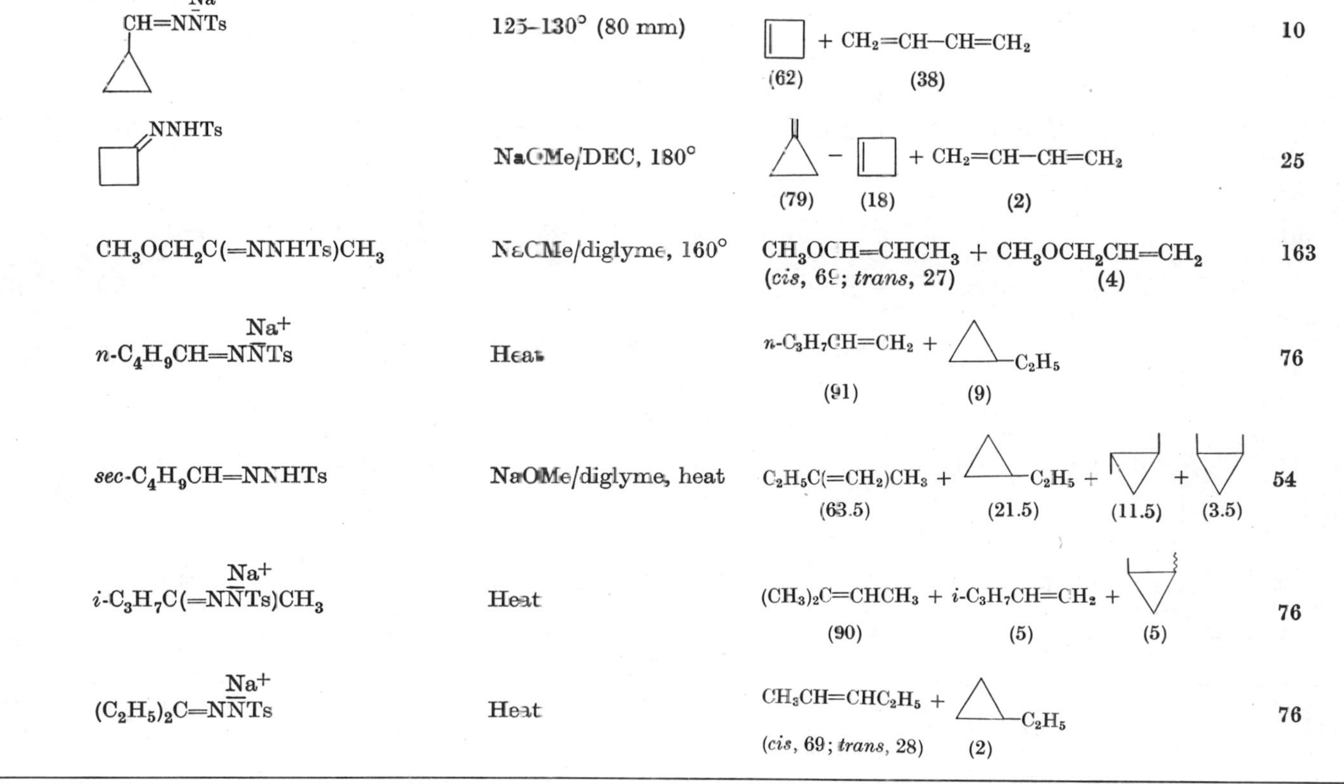

	Reactant	Conditions	Product(s) and Yield(s) (%)	Refs.
	Cyclopropyl-$CH{=}N\bar{N}Ts$ Na^+	125–130° (80 mm)	cyclobutene (62) + $CH_2{=}CH{-}CH{=}CH_2$ (38)	10
	Cyclobutanone tosylhydrazone (=NNHTs)	NaOMe/DEC, 180°	methylenecyclopropane (79) – cyclobutene (18) + $CH_2{=}CH{-}CH{=}CH_2$ (2)	25
	$CH_3OCH_2C({=}NNHTs)CH_3$	NaOMe/diglyme, 160°	$CH_3OCH{=}CHCH_3$ (*cis*, 69; *trans*, 27) + $CH_3OCH_2CH{=}CH_2$ (4)	163
5	$n\text{-}C_4H_9CH{=}N\bar{N}Ts$ Na^+	Heat	$n\text{-}C_3H_7CH{=}CH_2$ (91) + ethylcyclopropane (9)	76
	$sec\text{-}C_4H_9CH{=}NNHTs$	NaOMe/diglyme, heat	$C_2H_5C({=}CH_2)CH_3$ (63.5) + ethylcyclopropane (21.5) + 1,2-dimethylcyclopropane (11.5) + 1,2-dimethylcyclopropane (3.5)	54
	$i\text{-}C_3H_7C({=}N\bar{N}Ts)CH_3$ Na^+	Heat	$(CH_3)_2C{=}CHCH_3$ (90) + $i\text{-}C_3H_7CH{=}CH_2$ (5) + 1,2-dimethylcyclopropane (5)	76
	$(C_2H_5)_2C{=}N\bar{N}Ts$ Na^+	Heat	$CH_3CH{=}CHC_2H_5$ (*cis*, 69; *trans*, 28) + ethylcyclopropane (2)	76

Note: References 125–210 are on pp. 506–507.

TABLE II. THE APROTIC BAMFORD-STEVENS REACTION (*Continued*)

Carbon Atoms in Starting Ketone	Reactant	Conditions	Product(s)—Distribution ( ) or Yield(s) (%)	Refs.
		A. Monotosylhydrazones (Continued)		
(5 *contd.*)	$t\text{-}C_4H_9CH{=}NNHTs$	NaOMe/DEC, heat	$(CH_3)_2C{=}CHCH_3$ (7) + $C_2H_5C({=}CH_2)CH_3$ (1) +  (92)	6
	$CH_2{=}CHCH_2CH_2CH{=}N\bar{N}Ts$ Na^+	250°	$CH_2{=}CHCH_2CH{=}CH_2$ (92.2) 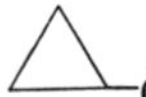 + $CH{=}CH_2$ (3.8)	110
	$(CH_3)_2C{=}CHCH{=}NNHTs$	NaOMe/diglyme, 160°	 (50%)	104
	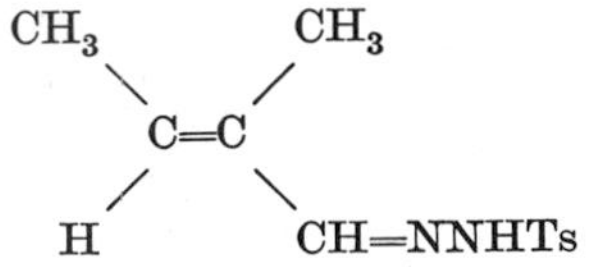 CH_3, CH_3 C=C H, $CH{=}NNHTs$	NaOMe/diglyme, 160°	 (4%)	104
	$CH_3OCH_2C({=}NNHTs)C_2H_5$	NaOMe/diglyme, heat	$CH_3OCH{=}CHC_2H_5$ (*cis*, 61; *trans*, 23) + $CH_3OCH_2CH{=}CHCH_3$ (*cis*, 5; *trans*, 11)	163
	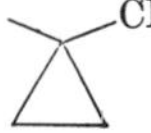 $CH{=}NNHTs$	NaOMe/diglyme, 160–180°	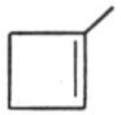 (>90%)	147

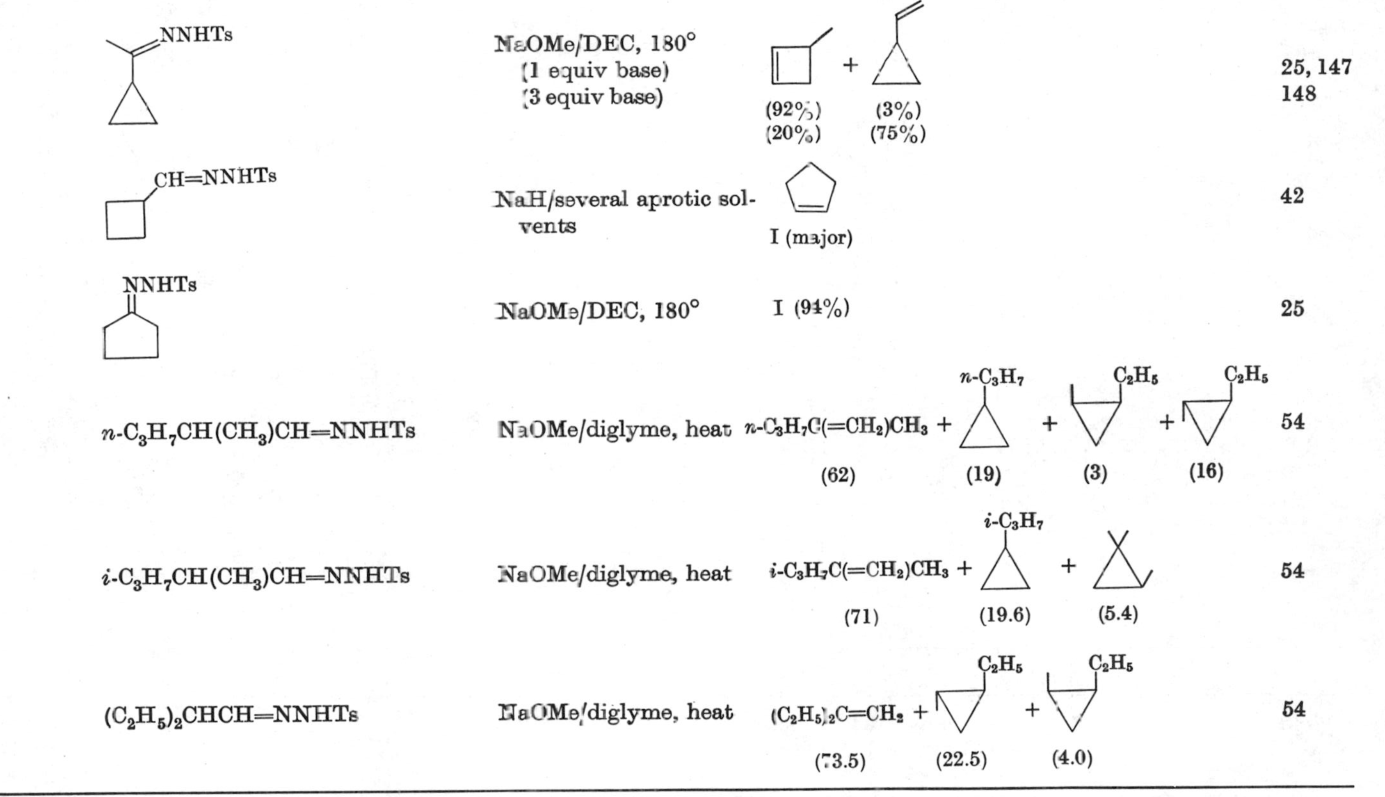

	(1-cyclopropylethylidene) NNHTs	NaOMe/DEC, 180° (1 equiv base) (3 equiv base)	[methylcyclobutene] (92%) (20%) + [vinylcyclopropane] (3%) (75%)	25, 147 148
	(cyclobutyl) CH=NNHTs	NaH/several aprotic solvents	[cyclopentene] I (major)	42
	(cyclopentylidene) NNHTs	NaOMe/DEC, 180°	I (94%)	25
6	n-$C_3H_7CH(CH_3)CH{=}NNHTs$	NaOMe/diglyme, heat	n-$C_3H_7C({=}CH_2)CH_3$ (62) + n-C_3H_7-cyclopropane (19) + C_2H_5, methyl-cyclopropane (3) + C_2H_5, methyl-cyclopropane (16)	54
	i-$C_3H_7CH(CH_3)CH{=}NNHTs$	NaOMe/diglyme, heat	i-$C_3H_7C({=}CH_2)CH_3$ (71) + i-C_3H_7-cyclopropane (19.6) + trimethylcyclopropane (5.4)	54
	$(C_2H_5)_2CHCH{=}NNHTs$	NaOMe/diglyme, heat	$(C_2H_5)_2C{=}CH_2$ (73.5) + C_2H_5, methyl-cyclopropane (22.5) + C_2H_5, methyl-cyclopropane (4.0)	54

Note: References 125–210 are on pp. 506–507.

TABLE II. The Aprotic Bamford-Stevens Reaction (*Continued*)

Carbon Atoms in Starting Ketone	Reactant	Conditions	Product(s)—Distribution ( ) or Yield(s) (%)	Refs.
		A. Monotosylhydrazones (Continued)		
6 (*contd.*)	$(CH_3)_2C{=}C(CH_3)CH{=}NNHTs$	NaOMe/diglyme, 160°	 I (72%)	104
	$(CH_3)_2C{=}CHCH({=}NNHTs)CH_3$	NaOMe/diglyme, 160°	I (39%)	104
		NaOMe/diglyme, $h\nu$	I + $(CH_3)_2C{=}CHCH{=}CH_2$ (36–41%) (29–33%)	106, 107
	$CH_3CH{=}C(CH_3)C({=}NNHTs)CH_3$	NaOMe/diglyme, 160°	 (1.5%)	104
	$t\text{-}C_4H_9C({=}NNHTs)CH_3$	NaOMe/DEC, heat	$t\text{-}C_4H_9CH{=}CH_2$ 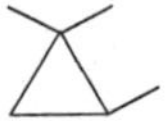 + (52) (47)	6
	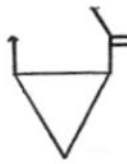 =NNHTs	NaOMe/diglyme, 135–140°	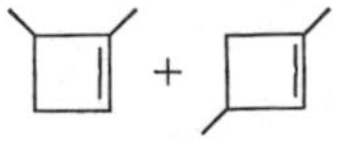 + I (4.4) II (95.6)	150
	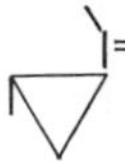 =NNHTs	NaOMe/diglyme, 135–140°	I (29.7) II (72.1)	150

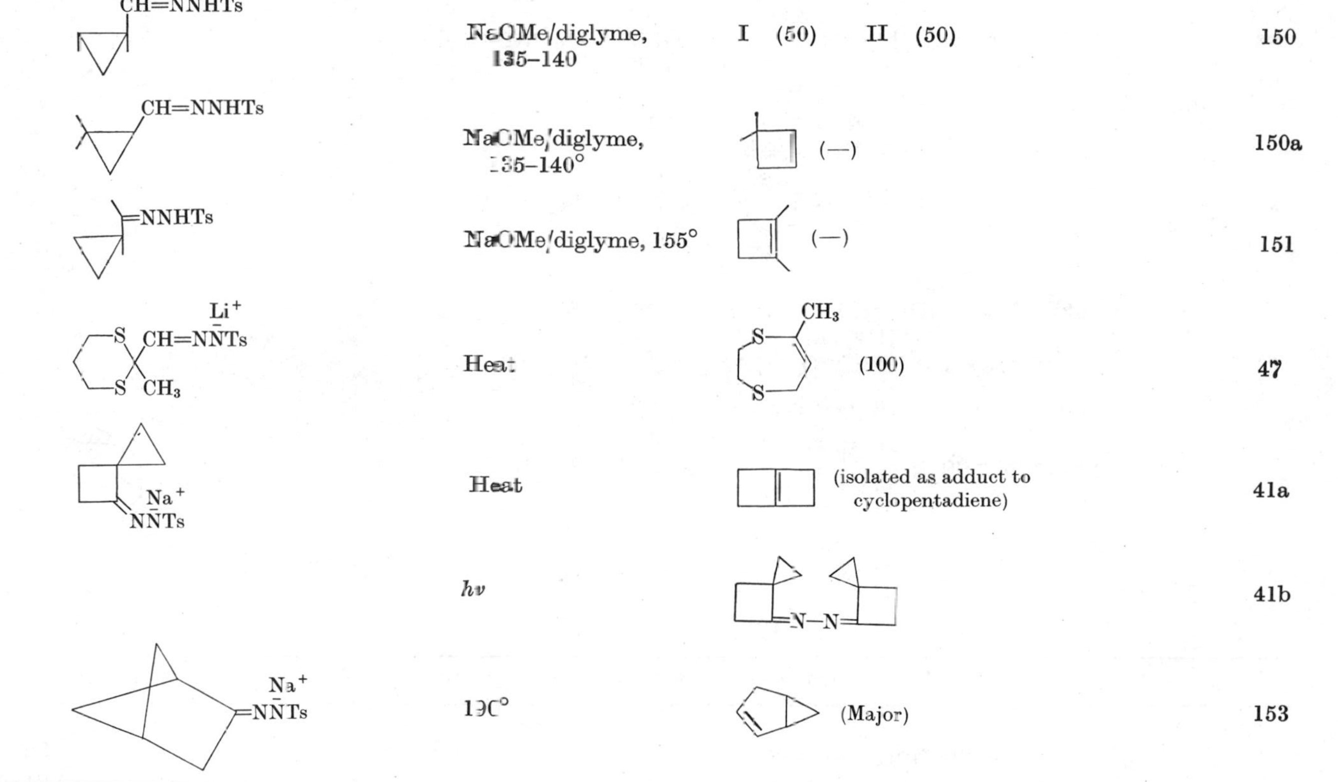

Reactant	Conditions	Product(s) and Yield(s) (%)	Refs.
CH=NNHTs	NaOMe/diglyme, 135–140	I (50) II (50)	150
CH=NNHTs	NaOMe/diglyme, 135–140°	(—)	150a
=NNHTs	NaOMe/diglyme, 155°	(—)	151
Li^+ CH=N$\bar{N}$Ts, S, S, CH_3	Heat	CH_3, S, S (100)	47
Na^+ N$\bar{N}$Ts	Heat	(isolated as adduct to cyclopentadiene)	41a
	$h\nu$	=N–N=	41b
Na^+ =N$\bar{N}$Ts	130°	(Major)	153

Note: References 125–210 are on pp. 506–507.

TABLE II. THE APROTIC BAMFORD-STEVENS REACTION (*Continued*)

Carbon Atoms in Starting Ketone	Reactant	Conditions	Product(s)—Distribution ( ) or Yield(s) (%)	Refs.
		A. Monotosylhydrazones (*Continued*)		
6 (*contd.*)	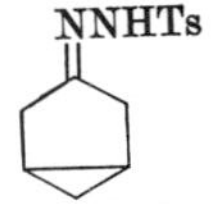 NNHTs	NaOMe/diglyme, 160°	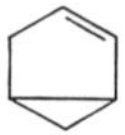 (91%)	98
	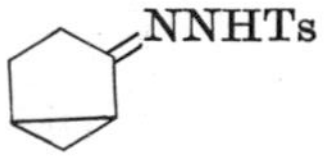 NNHTs	NaOMe/diglyme, 160°	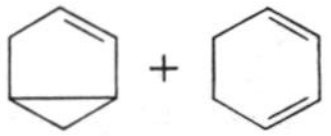 (19) + (14) + $CH\equiv CCH_2CH_2CH=CH_2$ (13) + $CH_2=C=CHCH_2CH=CH_2$ (12) + $CH_2=CHCH=CHCH=CH_2$ (42) (49%)	98
	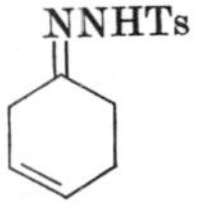 NNHTs	NaOMe/diglyme, 160°	 (65) + (17.5) + (17.5) (18%)	98
	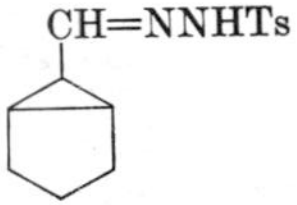 CH=NNHTs	NaOMe/diglyme, 160°	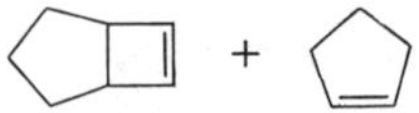 (71) + (29) (52%)	98

7	$(CH_3)_2C{=}C(CH_3)C({=}NNHTs)CH_3$	NaOMe/diglyme, 160°	 I (22%)	105
		NaOMe/diglyme, $h\nu$	I (25%)	106, 107
	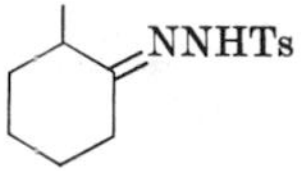 =NNHTs	NaOMe/NMP, 180°	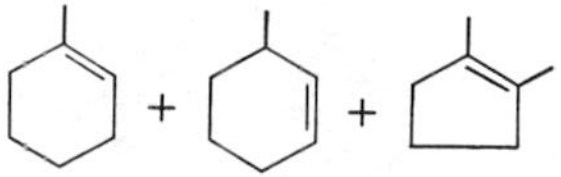 + + (54–64%) (63) (27) (4)	13
	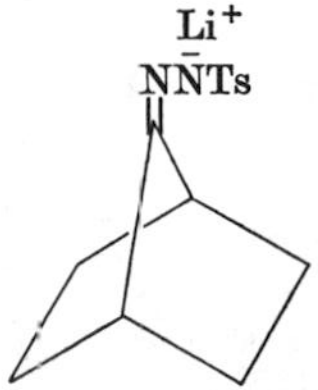 Li^+ $N\bar{N}Ts$	180° (20 mm)	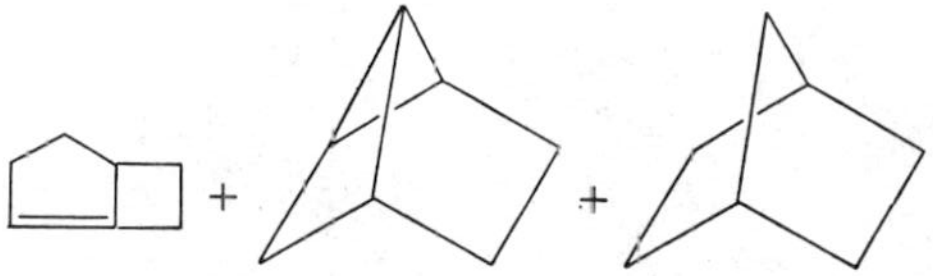 + + (80%) (6.2) (1.0) (1.2)	178
	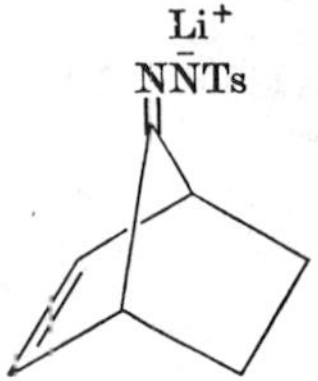 Li^+ $N\bar{N}Ts$	190° (0.05 mm)	 + five minor components (67%)	166
	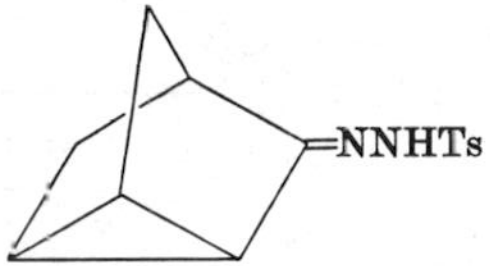 =NNHTs	NaOMe/diglyme, 160°	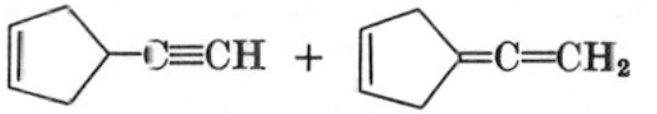 —C≡CH + =C=CH_2 (19%) (69) (29)	94

Note: References 125–210 are on pp. 506–507.

TABLE II. THE APROTIC BAMFORD-STEVENS REACTION (*Continued*)

Carbon Atoms in Starting Ketone	Reactant	Conditions	Product(s)—Distribution ( ) or Yield(s) (%)	Refs.
		A. Monotosylhydrazones (*Continued*)		
7 (*contd.*)	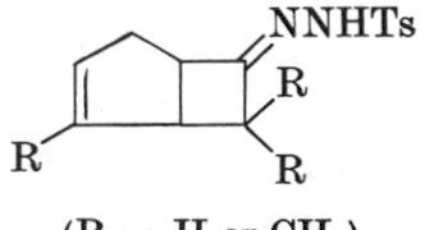 NNHTs, R, R, R (R = H or CH_3)	Base, aprotic conditions	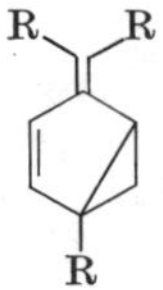 R, R, R (Low yield)	128
	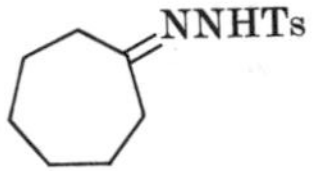 NNHTs	NaOMe/DEC, heat	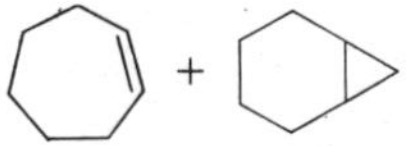 (82) + (18)	28
	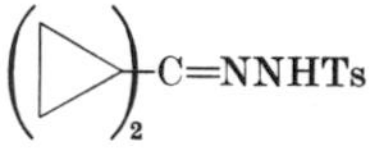 ( )$_2$C=NNHTs	NaOMe/diglyme, 85–100	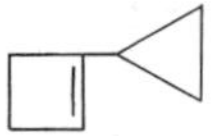 (41%)	38, 150
	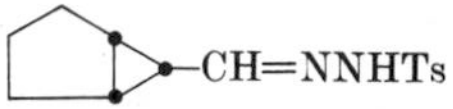 CH=NNHTs	NaOMe/diglyme, 160–170°	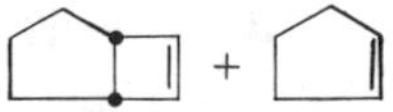 I (76) + II (24) (30%)	39
	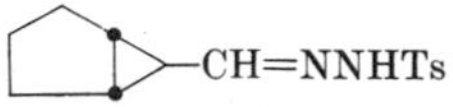 CH=NNHTs	NaOMe/diglyme, 160–170°	I (70) II (30) (60%)	39
	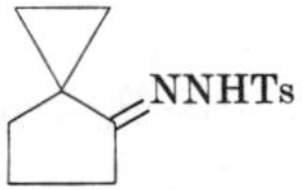 NNHTs	NaOMe/diglyme, heat	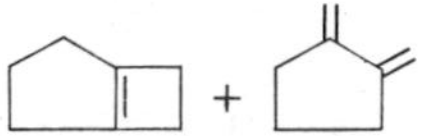 (50) + (50) (65%)	40, 41

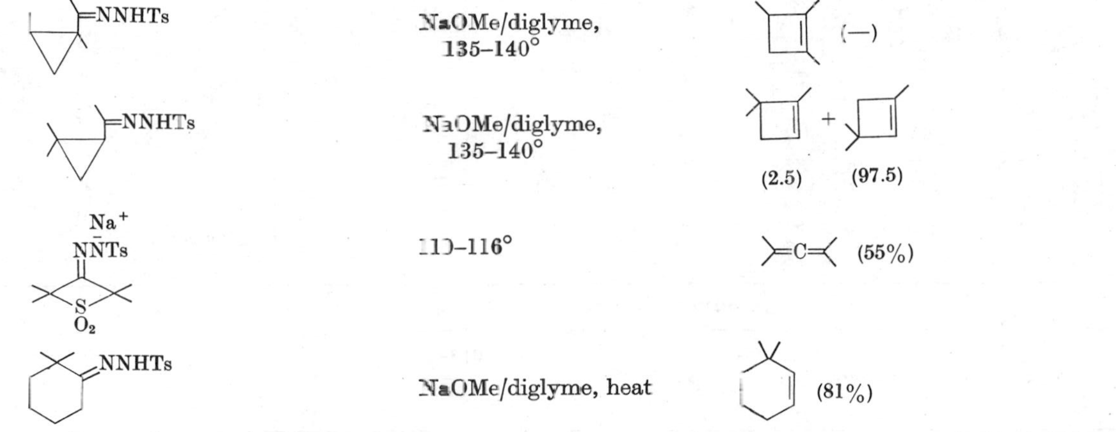

	Reactant	Conditions	Products (Yield, %)	Refs.
	CH=NNHTs	NaOMe/diglyme, heat	(65) + (24) + (6) + (5) (60–80%)	129
	$CH_2{=}CHCH_2C(CH_3)_2C{=}NNHTs$	NaOMe/diglyme, 170° With added Cu	$CH_2{=}CHCH{=}C(CH_3)C_2H_5$ (87) (80) + CH=CH_2 (13) (20) + $CH_2CH{=}CH_2$ (<5%) (60–70%)	109
	C_2H_5 CH=NNHTs	NaOMe/diglyme, 135–140°	C_2H_5 (—)	150a
	=NNHTs	NaOMe/diglyme, 135–140°	(—)	150a
	=NNHTs	NaOMe/diglyme, 135–140°	(2.5) + (97.5)	150
	Na^+ $\bar{N}NTs$ S O_2	110–116°	C (55%)	100
8	NNHTs	NaOMe/diglyme, heat	(81%)	58

Note: References 125–210 are on pp. 506–507.

TABLE II. The Aprotic Bamford-Stevens Reaction (*Continued*)

Carbon Atoms in Starting Ketone	Reactant	Conditions	Product(s)—Distribution ( ) or Yield(s) (%)	Refs.
		A. Monotosylhydrazones (Continued)		
8 (*contd.*)	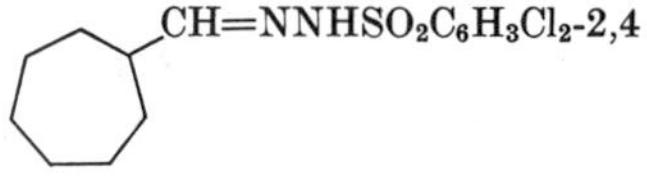 CH=$NNHSO_2C_6H_3Cl_2$-2,4	NaOMe/diglyme, heat	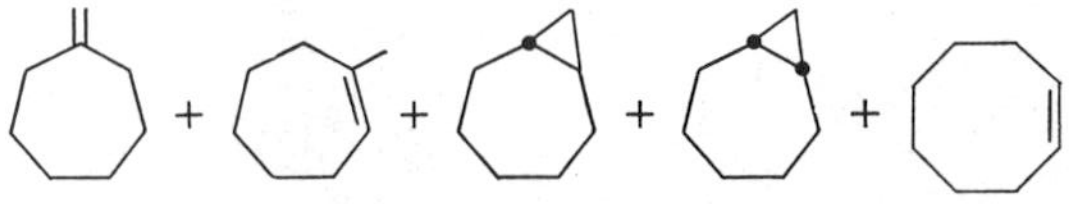 (48) + (2) + (12) + (37) + (1)	160
	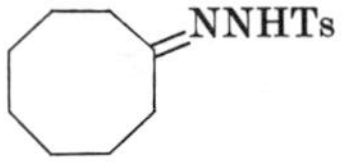 NNHTs	NaOMe/DEC, heat NaOMe/diglyme, heat	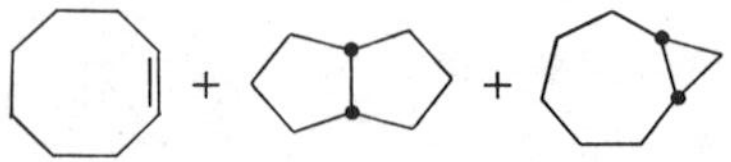 (45) + (46) + (9) (43.5) + (50) + (6.5) (52.5%)	28 64
	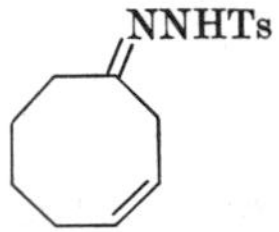 NNHTs	NaOMe/diglyme, heat	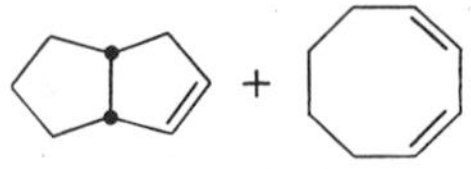 I (60) + II (40)	66
	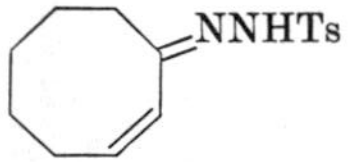 NNHTs	NaOMe/diglyme, 160°	I (51–55) + II (35–45) + minor products (7–12)	66

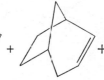

Reactant	Conditions	Products (Yield, %)	Refs.
NNHTs (cyclooctenone tosylhydrazone)	NaOMe/diglyme, 160°	+ + II (20) (35) (36)	66
NNHTs (bicyclooctanone tosylhydrazone)	NaOMe/diglyme, heat	I + + (97) (3) (<1)	63
CH=NNHTs (norbornyl)	NaOMe/tetraglyme, 130–140°	+ + (19.5) (9.5)  +  (60) (11) (65%)	146; also 176

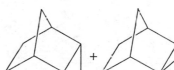

Note: References 125–210 are on pp. 506–507.

TABLE II. THE APROTIC BAMFORD-STEVENS REACTION (*Continued*)

Carbon Atoms in Starting Ketone	Reactant	Conditions	Product(s)—Distribution ( ) or Yield(s) (%)	Refs.
		A. Monotosylhydrazones (Continued)		
8 (*contd.*)	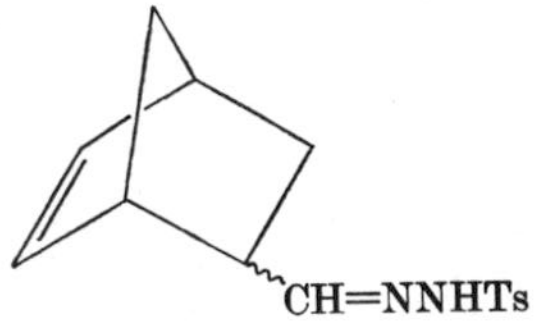 CH=NNHTs	NaOMe/tetraglyme, 135–140°	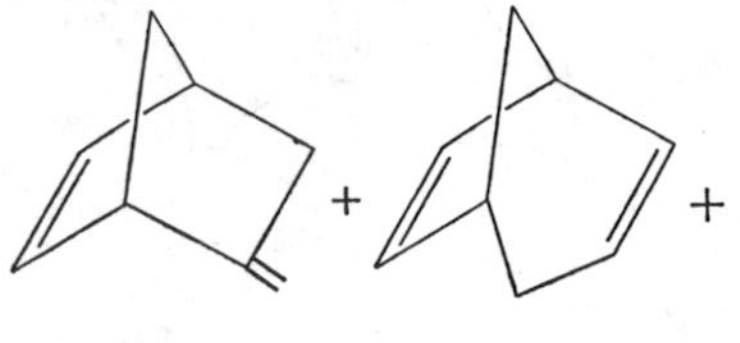 (40) + (7) 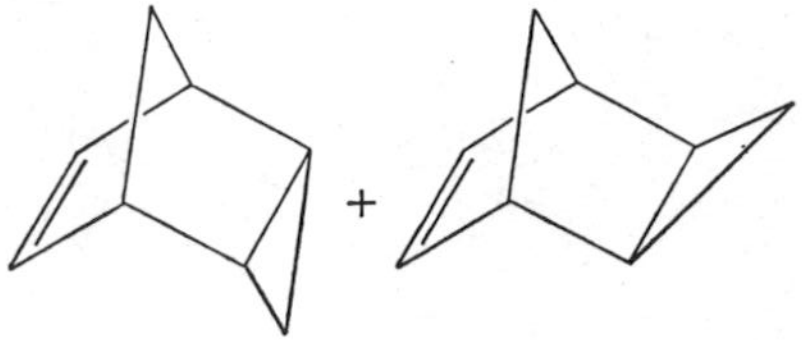 + (42) + (11)	146; see also 176
	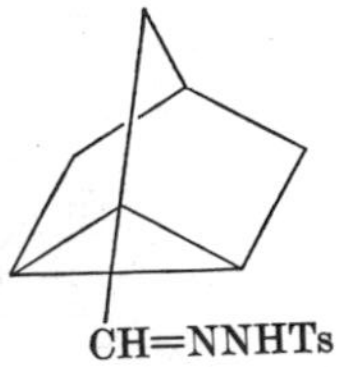 CH=NNHTs	1. NaOMe/NMP, heat; 2. H_2O	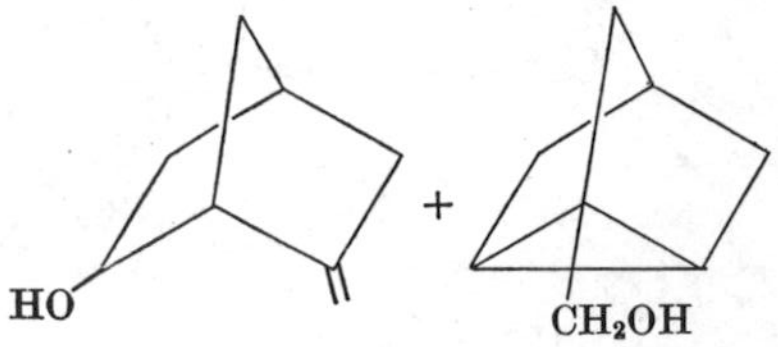 HO (28%) + CH_2OH (14%)	154

Reactant	Conditions	Product(s)	Refs.
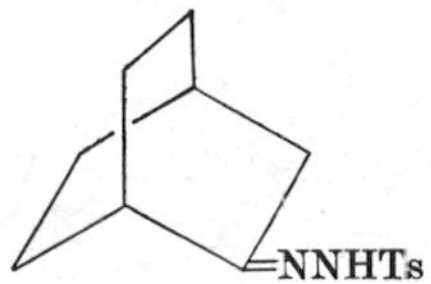 =NNHTs	Na/CH_3CONH_2, heat	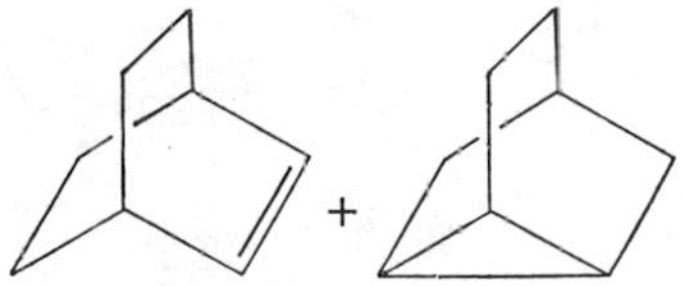 + (30) (70)	74
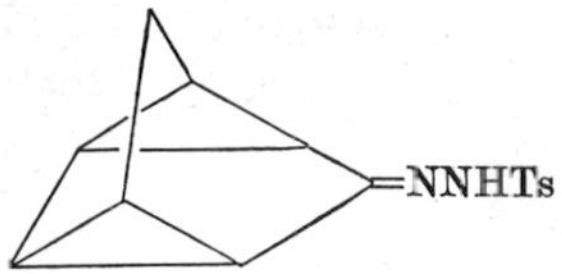 =NNHTs	NaOMe/tetraglyme, heat	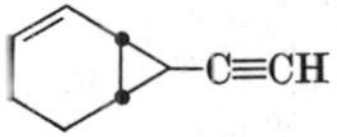 $-C\equiv CH$ (38%)	99
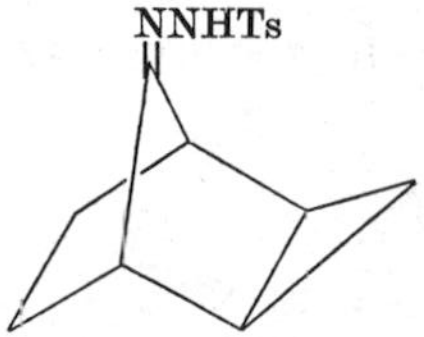 NNHTs	NaOMe/decalin, 160°	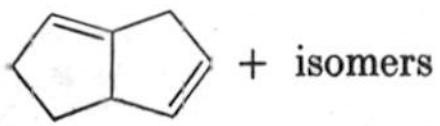 + isomers (13%) (7%)	177
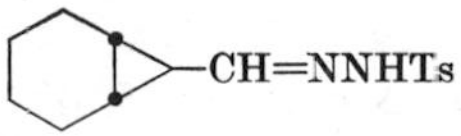 $-CH=NNHTs$	NaOMe/diglyme, 160–170°	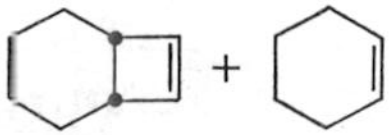 + (60%) (67) (33)	39
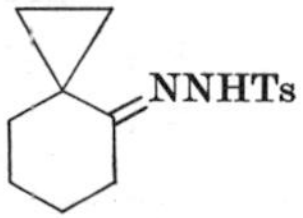 NNHTs	NaOMe/diglyme, 160–170°	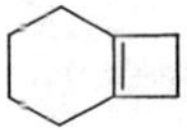 (65%)	40

Note: References 125–210 are on pp. 506–507.

TABLE II. THE APROTIC BAMFORD-STEVENS REACTION (*Continued*)

Carbon Atoms in Starting Ketone	Reactant	Conditions	Product(s)—Distribution ( ) or Yield(s) (%)	Refs.
		A. Monotosylhydrazones (Continued)		
8 (*contd.*)	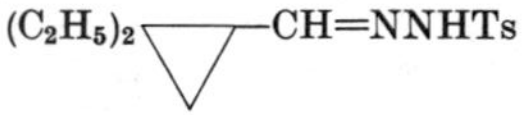 $(C_2H_5)_2$-cyclopropyl-CH=NNHTs	NaOMe/diglyme, 135–140°	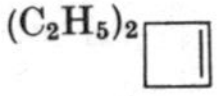 $(C_2H_5)_2$-cyclobutene (—)	150a
	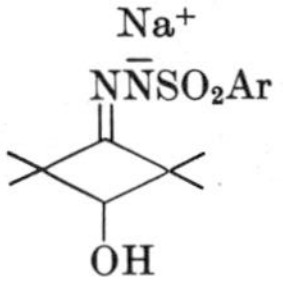 Na⁺ salt: $N\bar{N}SO_2Ar$, OH (Ar = C_6H_5 or p-$CH_3C_6H_4$)	Heat	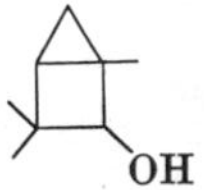 OH (76%)	171, 172[a]
		$h\nu$, CH_3OH	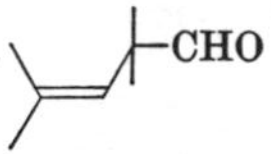 CHO (44%)	168
	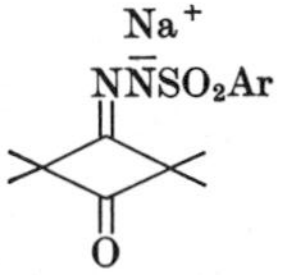 Na⁺ salt: $N\bar{N}SO_2Ar$, O (Ar = C_6H_5 or p-$CH_3C_6H_4$)	Heat, CH_3OH	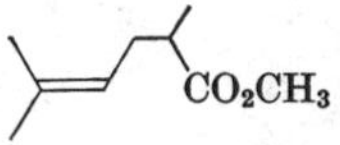 CO_2CH_3 (—)	171
		$h\nu$, CH_3OH	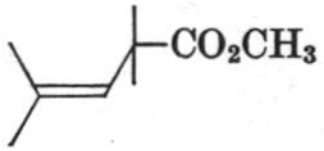 CO_2CH_3	168

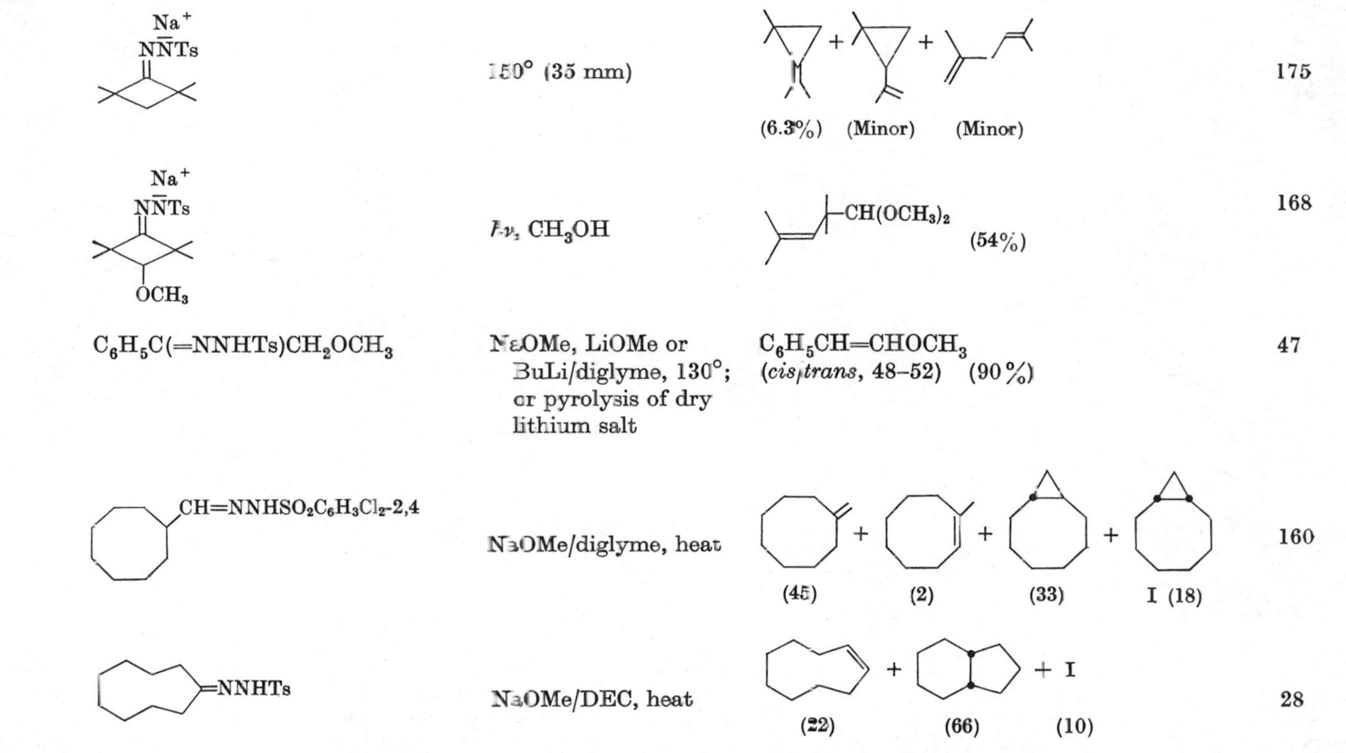

	Reactant	Conditions	Product(s) and Yield(s) (%)	Refs.
	Na$^+$ $\bar{N}$NTs	150° (35 mm)	(6.3%) + (Minor) + (Minor)	175
9	Na$^+$ $\bar{N}$NTs, OCH_3	$h\nu$, CH_3OH	$-CH(OCH_3)_2$ (54%)	168
	$C_6H_5C(=NNHTs)CH_2OCH_3$	NaOMe, LiOMe or BuLi/diglyme, 130°; or pyrolysis of dry lithium salt	$C_6H_5CH{=}CHOCH_3$ (*cis*/*trans*, 48–52) (90%)	47
	$CH{=}NNHSO_2C_6H_3Cl_2$-2,4	NaOMe/diglyme, heat	(45) + (2) + (33) + I (18)	160
	=NNHTs	NaOMe/DEC, heat	(22) + (66) + I (10)	28

[a] See Refs. 173 and 174 for reaction of the corresponding propionates.

Note: References 125–210 are on pp. 506–507.

TABLE II. THE APROTIC BAMFORD-STEVENS REACTION (*Continued*)

Carbon Atoms in Starting Ketone	Reactant	Conditions	Product(s)—Distribution ( ) or Yield (%)	Refs.
		A. Monotosylhydrazones (Continued)		
9 (*contd.*)	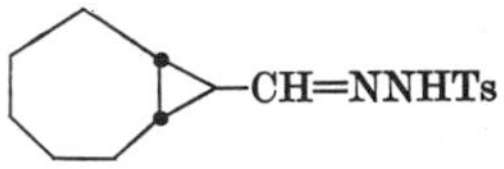 CH=NNHTs	NaOMe/diglyme, 160–170°	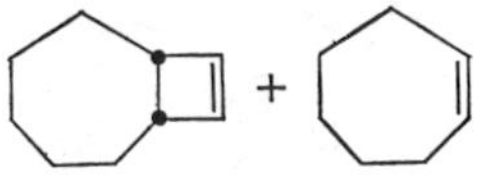 (72) + (28) (33%)	39
	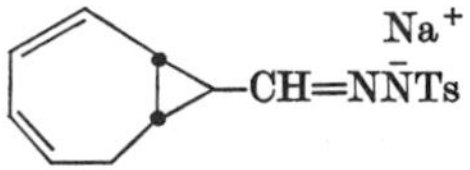 CH=N$\bar{N}$Ts Na$^+$	Heat	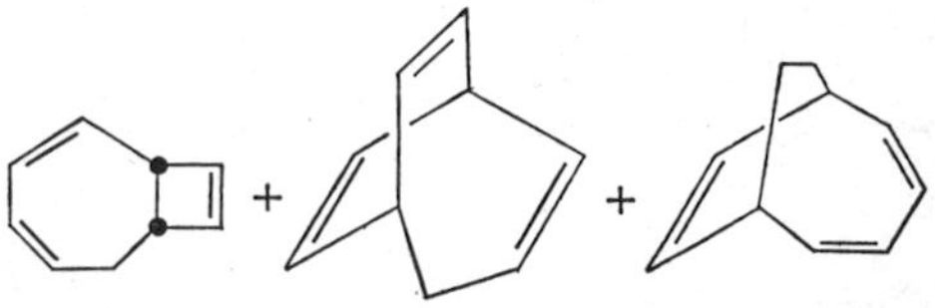 (27) + (12) + (37)	152
	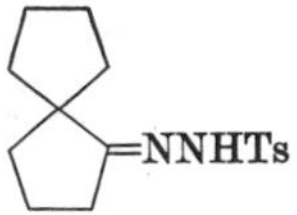 =NNHTs	NaOMe/diglyme, heat	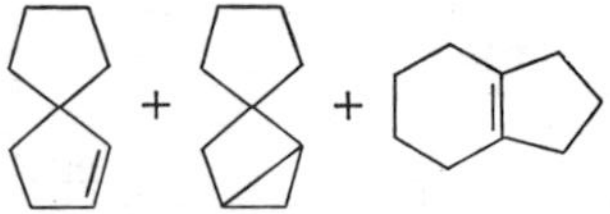 (53) + (32) + (12) + unknown (3)	58
	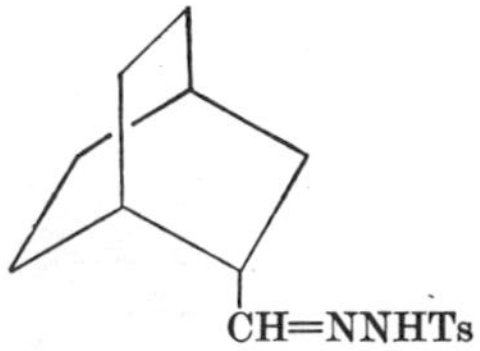 CH=NNHTs	NaOMe/tetraglyme, 130–140°	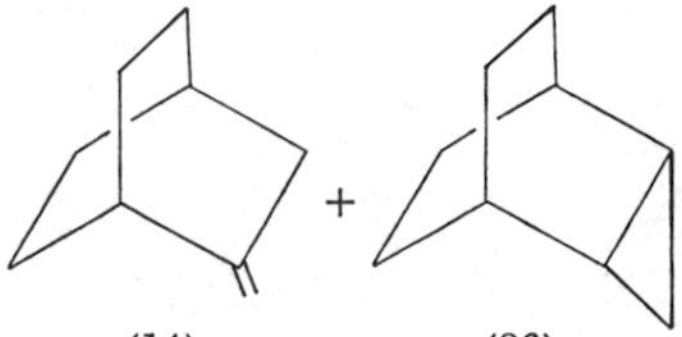 (14) + (86) (70%)	146

Reactant	Conditions	Product(s) and Yield(s) (%)	Refs.
CH=NNHTs	NaOMe/tetraglyme, 130–140°	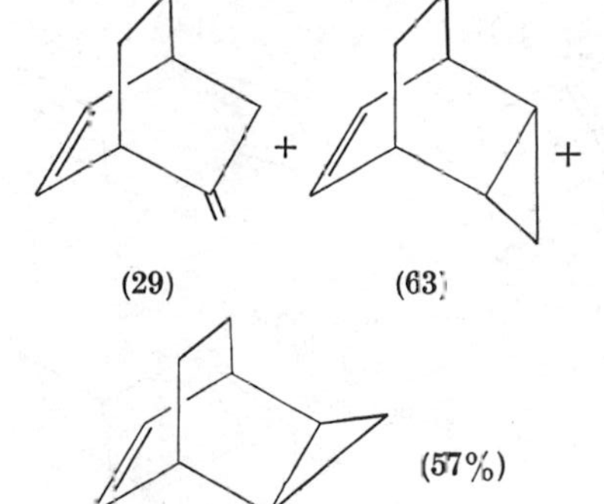 (29) + (63) + (8) (57%)	146
Na⁺ $N\bar{N}Ts$	Heat	$C{\equiv}CH$ (64) + $CH{=}C{=}CH_2$ (36)	97
Na⁺ $N\bar{N}Ts$	Heat	(35) + (65) (90%)	67

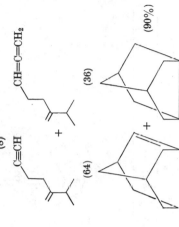

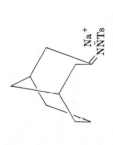

Note: References 125–210 are on pp. 506–507.

TABLE II. The Aprotic Bamford-Stevens Reaction (*Continued*)

Carbon Atoms in Starting Ketone	Reactant	Conditions	Product(s)—Distribution ( ) or Yield (%)	Refs.
		A. Monotosylhydrazones (*Continued*)		
9 (*contd.*)	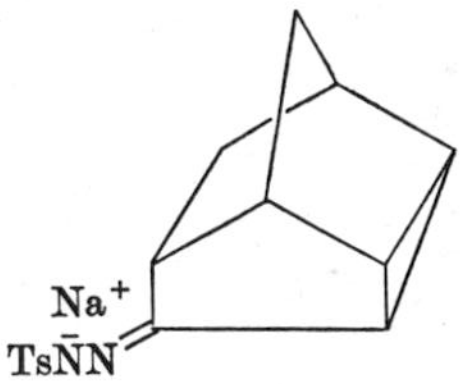 Na^+ $Ts\bar{N}N$=	180–190°, reduced pressure	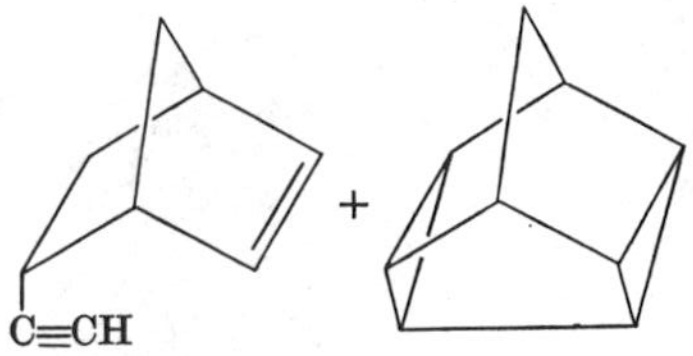 C≡CH (29) + (63)	156
	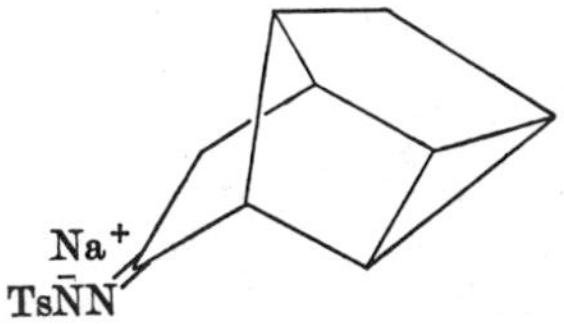 Na^+ $Ts\bar{N}N$=	180–190°, reduced pressure	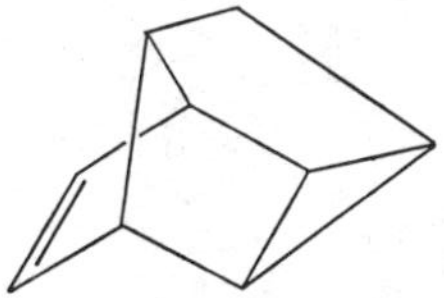 (56%)	156
	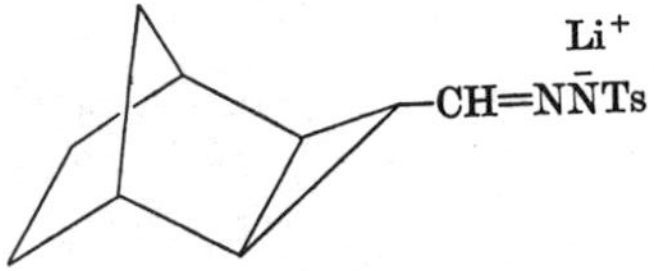 —CH=N$\bar{N}$Ts Li^+	95–110° (0.25 mm)	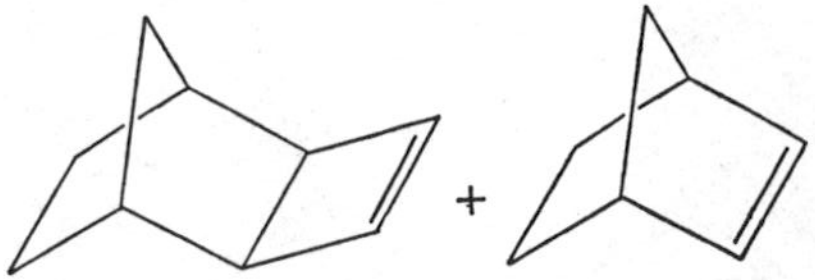 (3.4) + (1)	170

No. of C Atoms	Reactant	Conditions	Product(s) and Yield(s) (%)	Refs.
10	Li^+ $\bar{N}NTs$	95–110° (0.25 mm)	(87) + (13)	170
	NNHTs	Na/CH_3CONH_2, heat	I (80) + (20) (97%)	127
		Pyrolysis of sodium salt	I + (70%) (50) (50)	97
	Na^+ $\bar{N}NTs$	Heat	+ —C≡CC_3H_7-*i* (31) (69)	97
	$C_6H_5C(=NNHTs)CH_2\overset{+}{N}H(CH_3)_2Cl^-$	NaOMe, LiOMe or *n*-BuLi/diglyme, 130° or pyrolysis of dry sodium salt	$C_6H_5CH{=}CHN(CH_3)_2$ (—) (*cis* and *trans*)	47

Note: References 125–210 are on pp. 506–507.

TABLE II. The Aprotic Bamford-Stevens Reaction (*Continued*)

Carbon Atoms in Starting Ketone	Reaction	Conditions	Product(s)—Distribution ( ) or Yield(s) (%)	Refs.
		A. Monotosylhydrazones (*Continued*)		
10 (*contd.*)	$C_6H_5C(=NNHTs)CH_2C_2H_5$	NaOMe, LiOMe or *n*-BuLi/diglyme, 130°; or pyrolysis of dry sodium salt	$C_6H_5CH{=}CHSC_2H_5$ (9–15) + $C_6H_5C(=CH_2)SC_2H_5$ (91–85)	47
	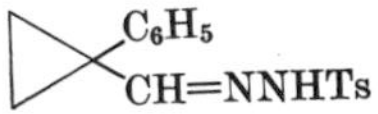 1-C_6H_5-1-(CH=NNHTs)cyclopropane	NaOMe/NMP, 180° 5 min	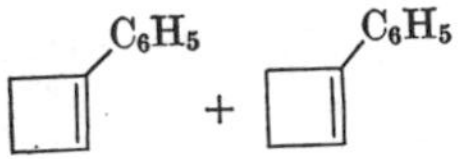 1-phenylcyclobutene (71%) + 3-phenylcyclobutene (8%)	37
	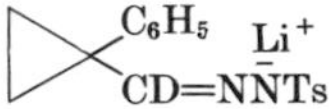 1-C_6H_5-1-(CD=N$\bar{N}$Ts Li^+)cyclopropane	125–135° (80 mm)	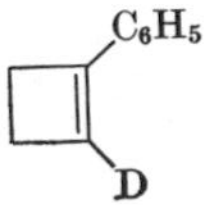 1-phenyl-2-deuteriocyclobutene (78%)	10
	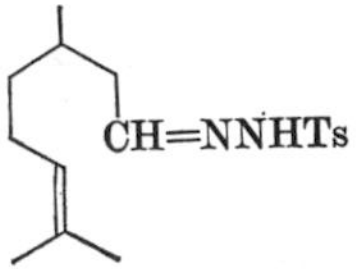 citronellal tosylhydrazone (CH=NNHTs)	NaOMe/diglyme, 165–170° NaOMe/diglyme, 100–110° NaOMe/diglyme, Cu, 165–170° NaOMe/diglyme, Cu, 100–110°	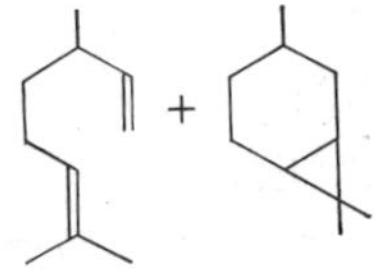 diene + bicyclic (carane) product (96) (4) (15%) (89) (11) (17%) (84) (16) (17%) (77) (23) (19%)	109

	Reactant	Conditions	Product(s) and Yield(s) (%)	Refs.
10	Na^+ $N\bar{N}Ts$	Heat	$C{\equiv}CH$ (76) + $C{=}CH_2$ (2) +	97
			(12) + (10) (50%)	
	Na^+ $Ts\bar{N}N$	Heat	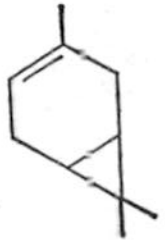 (—)	97
	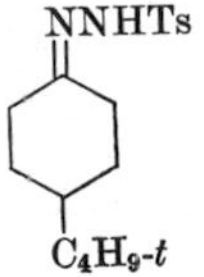 NNHTs, C_4H_9-*t*	Na/CH_3CONH_2, 160°	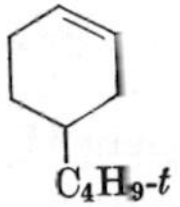 C_4H_9-*t* (61%)	59

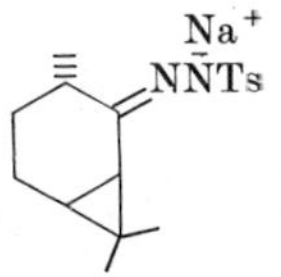

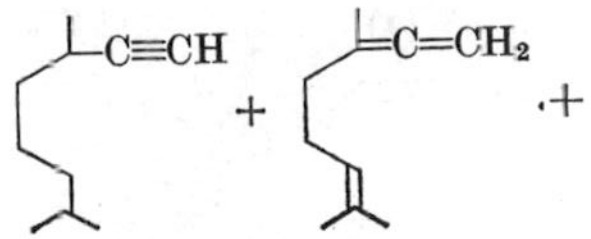

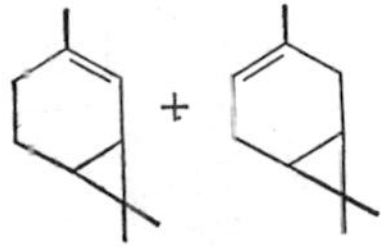

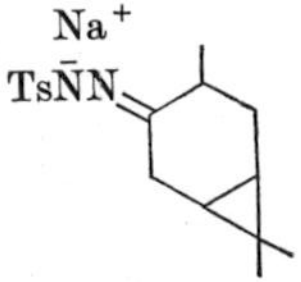

Note: References 125–210 are on pp. 506–507.

TABLE II. THE APROTIC BAMFORD-STEVENS REACTION (*Continued*)

Carbon Atoms in Starting Ketone	Reaction	Conditions	Product(s)—Distribution ( ) or Yields() (%)	Refs.
		A. Monotosylhydrazones (Continued)		
10 (*contd.*)	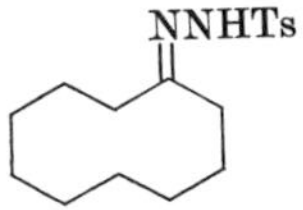 NNHTs	NaOMe/DEC, heat NaOMe/diglyme, 165–170°	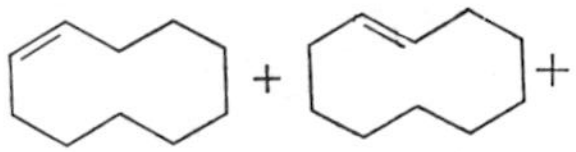 + + (14) (6) (15.4) (3.5) 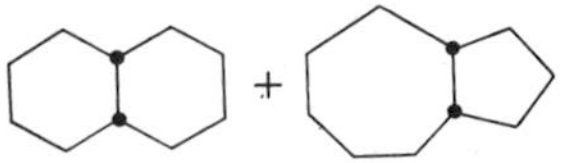 + (18) (62) (18) (63) (57%)	28 64
	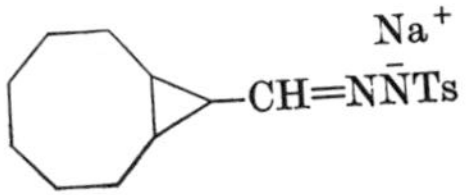 Na^+ —CH=N$\bar{N}$Ts	Heat	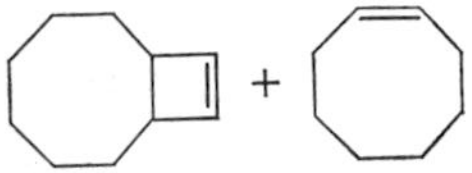 + (4.5) (1)	152
	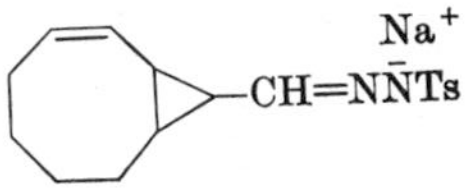 Na^+ —CH=N$\bar{N}$Ts	Heat	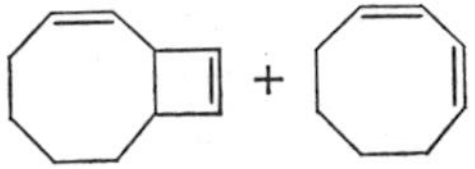 + (3.7) (1)	152

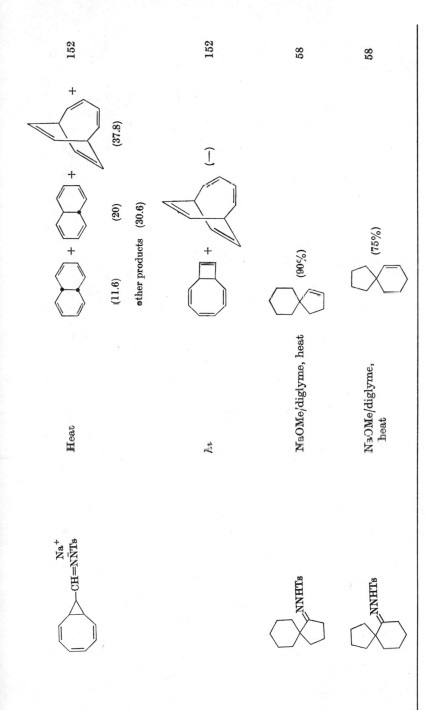

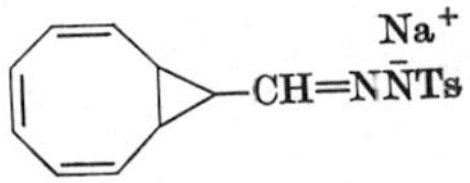	Heat	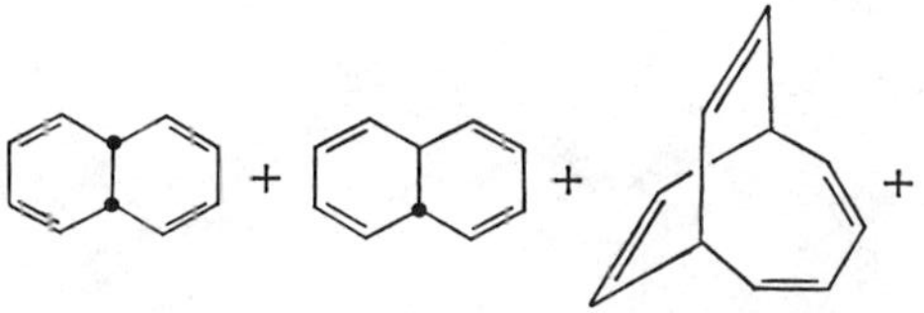 (11.6) (20) (37.8) other products (30.6)	152
	$h\nu$	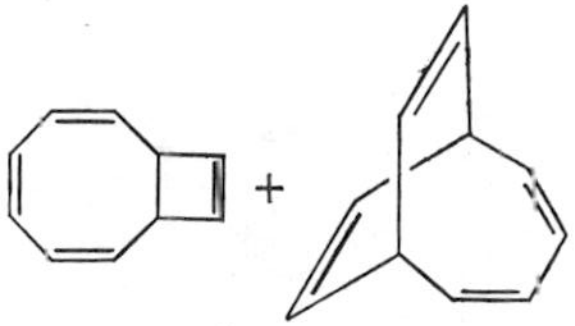 (—)	152
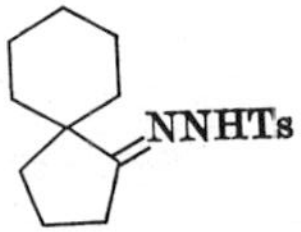	NaOMe/diglyme, heat	 (90%)	58
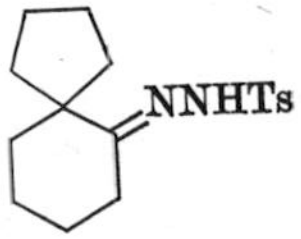	NaOMe/diglyme, heat	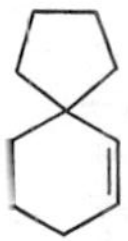 (75%)	58

Note: References 125–210 are on pp. 506–507.

TABLE II. THE APROTIC BAMFORD-STEVENS REACTION (*Continued*)

Carbon Atoms in Starting Ketone	Reaction	Conditions	Product(s) —Distribution ( ) or Yield(s) (%)	Refs.
		A. Monotosylhydrazones (*Continued*)		
10 (*contd.*)	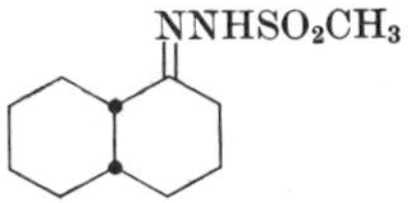 $NNHSO_2CH_3$	Na/CH_3CONH_2, 156°	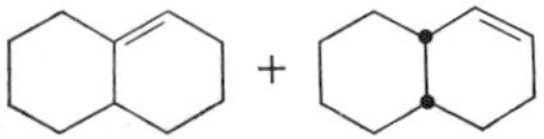 I (25) + II (75)	55
	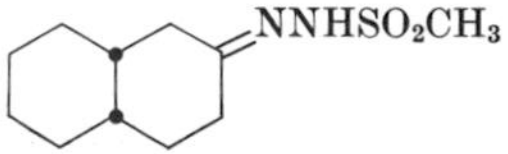 $NNHSO_2CH_3$	Na/CH_3CONH_2, 156°	II 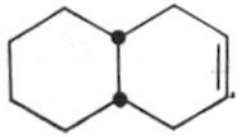 (51) + III (49)	55
	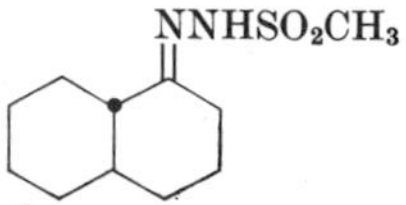 $NNHSO_2CH_3$	Na/CH_3CONH_2, 156°	I 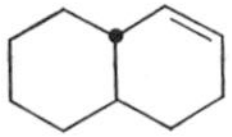 (74) + IV (26)	55
	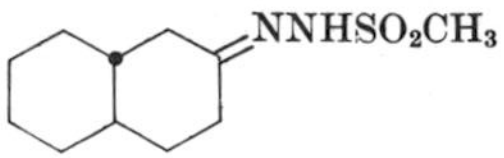 $NNHSO_2CH_3$	Na/CH_3CONH_2, 156°	IV  (35) + (65)	55

	Reactant	Conditions	Product(s) and Yield(s) (%)	Refs.
	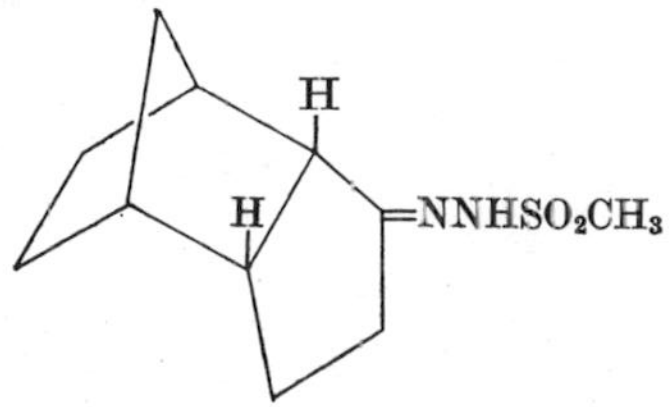 =$NNHSO_2CH_3$ (H, H)	Na/CH_3CONH_2, 156°	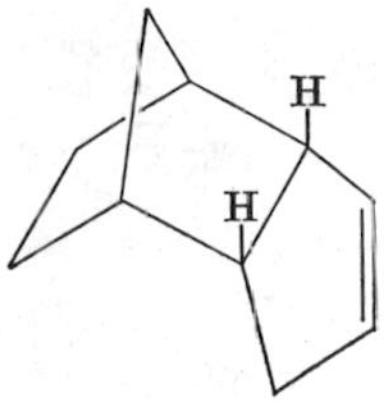 (H, H) (49%)	57
11	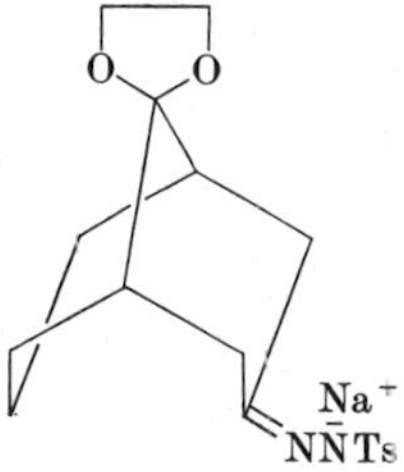 $\overset{-}{N}Ts$ Na^+ (ketal, O, O)	Heat	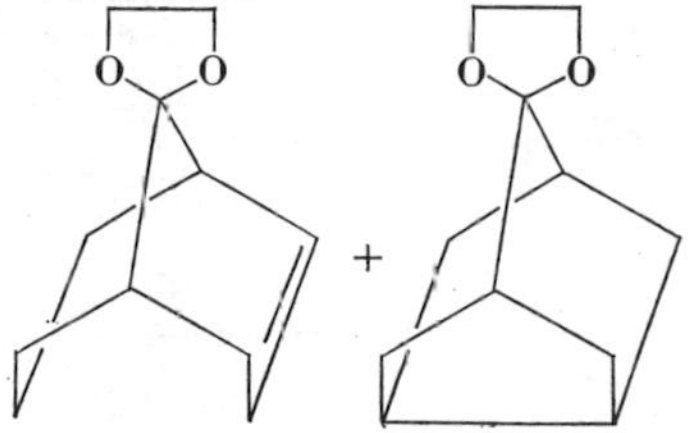 (O, O) + (O, O) (40) (60) (35%)	67
	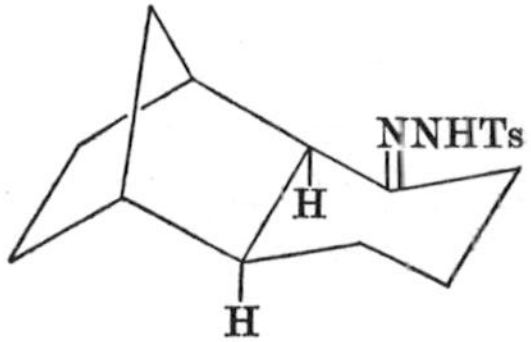 NNHTs (H, H)	Na/CH_3CONH_2, 156°	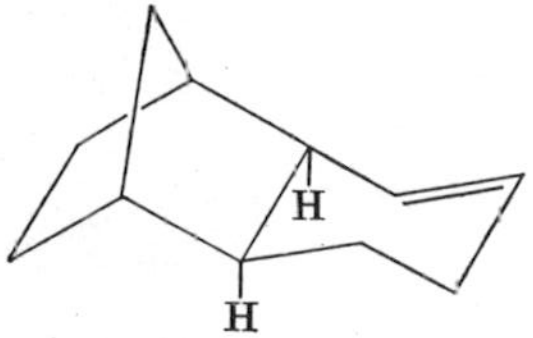 (H, H) (—)	56
	 NNHTs	170–180° (0.1 mm)	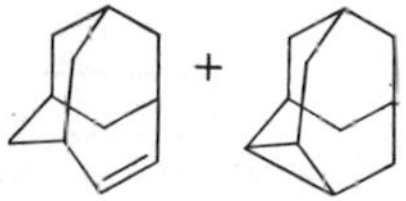 + (55) (45) (98%)	167

Note: References 125–210 are on pp. 506–507.

TABLE II. The Aprotic Bamford-Stevens Reaction (*Continued*)

Carbon Atoms in Starting Ketone	Reactant	Conditions	Product(s)—Distribution ( ) or Yield(s) (%)	Refs.
		A. Monotosylhydrazones (*Continued*)		
11 (*contd.*)	Br; Li^+ $\overset{-}{N}NTs$	$h\nu$	Br (6.5–8.5%)	155
	C_6H_5; CH=NNHTs	NaOMe/NMP, 180°	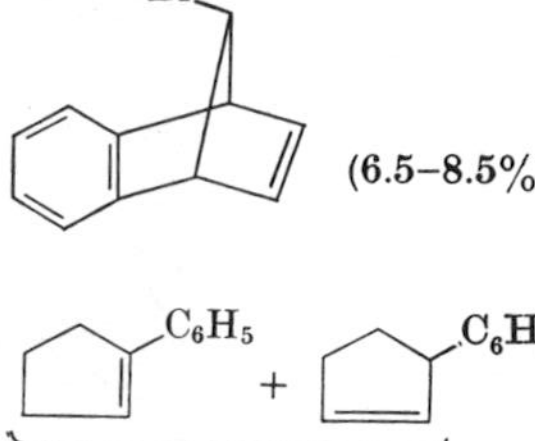 (69%) + $CH_2C_6H_5$ + CHC_6H_5 (6%)	37
	NNHTs	NaOMe/diglyme, heat	(82%)	58

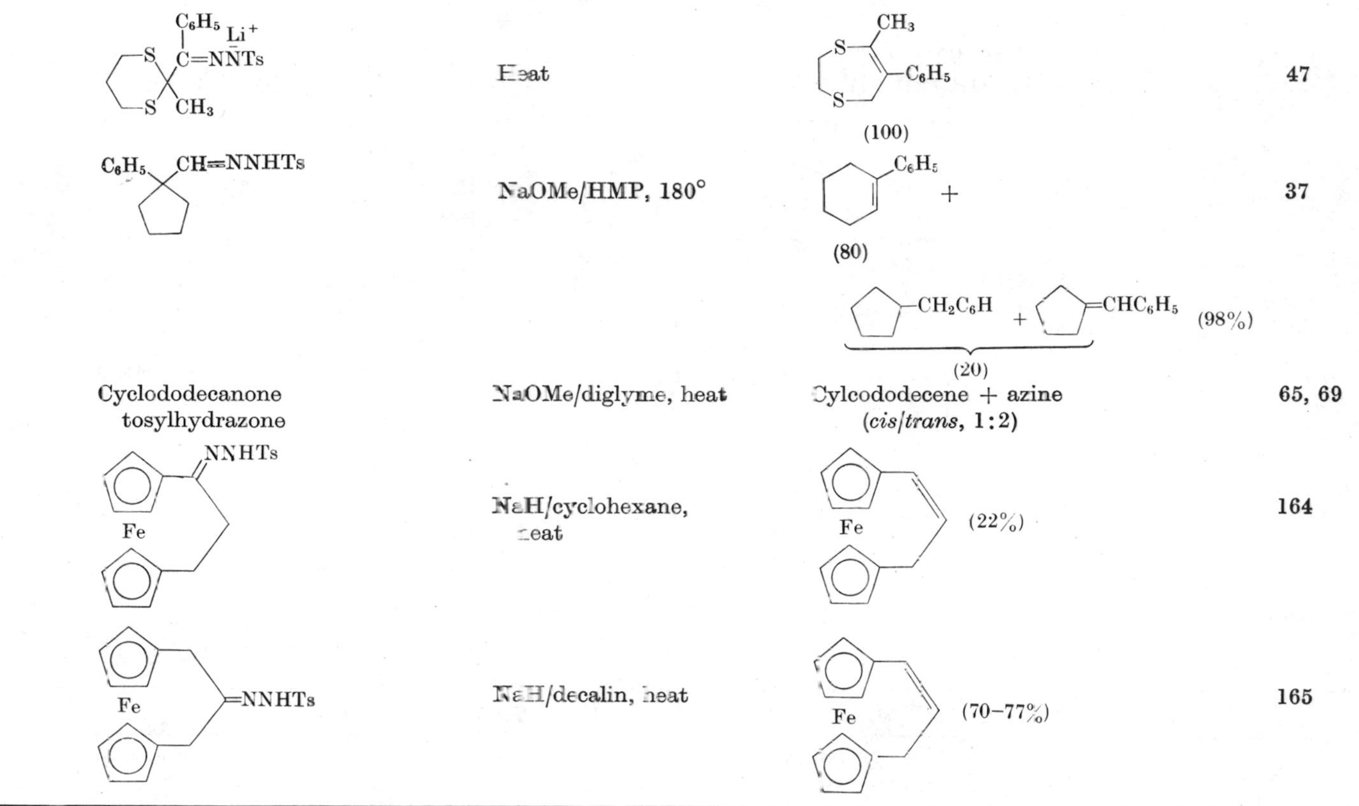

12	[structure: 2-methyl-1,3-dithiane bearing C_6H_5–C=N$\bar{N}$Ts Li^+]	Heat	[structure: dihydrodithiepin with CH_3, C_6H_5] (100)	47
	[structure: 1-phenylcyclopentyl CH=NNHTs]	NaOMe/HMP, 180°	[structure: 1-phenylcyclohexene] (80) + [structure: cyclopentyl–CH_2C_6H] + [structure: cyclopentylidene=CHC_6H_5] (20) (98%)	37
	Cyclododecanone tosylhydrazone	NaOMe/diglyme, heat	Cylcododecene + azine (*cis*/*trans*, 1:2)	65, 69
13	[structure: bridged ferrocene ketone NNHTs]	NaH/cyclohexane, heat	[structure: bridged ferrocene alkene] (22%)	164
	[structure: bridged ferrocene =NNHTs]	NaH/decalin, heat	[structure: bridged ferrocene alkene] (70–77%)	165

Note: References 125–210 are on pp. 506–507.

TABLE II. The Aprotic Bamford-Stevens Reaction (*Continued*)

Carbon Atoms in Starting Ketone	Reactant	Conditions	Product(s)—Distribution ( ) or Yield(s) (%)	Refs.
		A. Monotosylhydrazones (Continued)		
13 (*contd.*)	1-C_6H_5-1-(CH=NNHTs)cyclohexane	NaOMe/NMP, 180°	1-($CH_2C_6H_5$)cyclohexene (31) + (CHC_6H_5=)cyclohexane (10) + 1-C_6H_5-cycloheptene (59) (95%)	37
14	1-C_6H_5-1-(CH=NNHTs)cyclooctane	NaOMe/NMP, 180°	1-($CH_2C_6H_5$)cyclooctene (30) + (CHC_6H_5=)cyclooctane (13) + 1-C_6H_5-cyclononene (57) (93%)	43
	$C_6H_5C(=NNHTs)CH_2OC_6H_5$	NaOMe, LiOMe, or BuLi/diglyme, 130° or pyrolysis of dry sodium salt	$C_6H_5CH=CHOC_6H_5$ (90%) (*cis* and *trans*)	47

No. of C Atoms	Reactant	Conditions	Product(s) and Yield(s) (%)	Refs.
	$C_6H_5C(=NNHTs)CH_2SC_6H_5$	As above	$C_6H_5CH{=}CHSC_6H_5$ (0–8) + $C_6H_5C(=CH_2)SC_6H_5$ (100–92)	47
	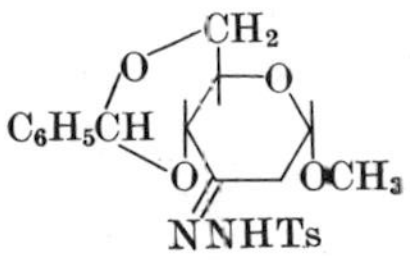 [structure: C_6H_5CH acetal, NNHTs, OCH_3]	NaOMe/NMP, 180° 7 min	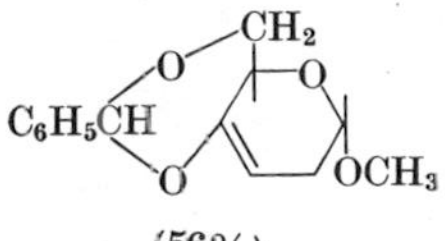 [structure: C_6H_5CH acetal, OCH_3] (56%)	135
	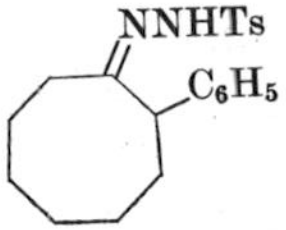 [structure: NNHTs, C_6H_5 cyclooctanone tosylhydrazone]	NaOMe/diglyme, 165–170°	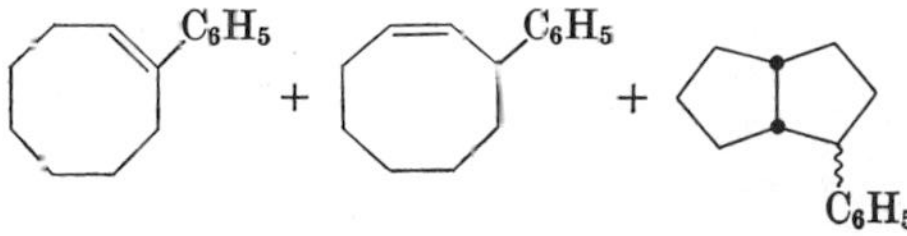 [structure: C_6H_5] (18) + [structure: C_6H_5] (42) + [structure: C_6H_5] (19)	65
	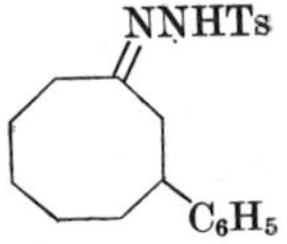 [structure: NNHTs, C_6H_5 cyclooctanone tosylhydrazone]	NaOMe/diglyme, 165–170°	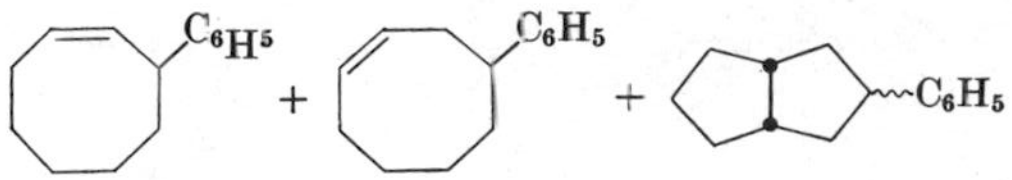 [structure: C_6H_5] (≤0) + [structure: C_6H_5] (38) + [structure: C_6H_5] (17)	65
15	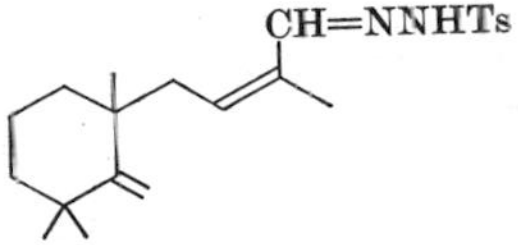 [structure: CH=NNHTs]	NaH/i-C_8H_{18}, 5% monoglyme, $h\nu$	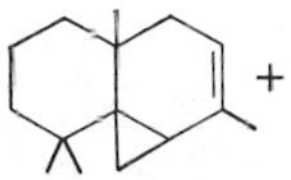 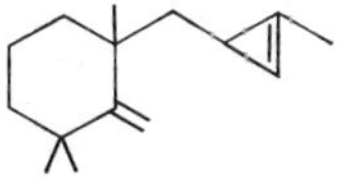 [structure] (51%) + [structure] (10%)	108

Note: References 125–210 are on pp. 506–507.

TABLE II. THE APROTIC BAMFORD-STEVENS REACTION (*Continued*)

Carbon Atoms in Starting Ketone	Reactant	Conditions	Product(s)—Distribution ( ) or Yield(s) (%)	Refs.
		A. Monotosylhydrazones (Continued)		
15 (*contd.*)	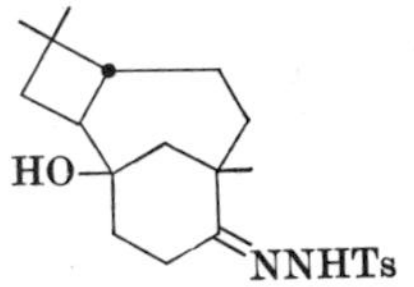	NaOMe/DEC, heat $LiAlH_4$/THF, heat	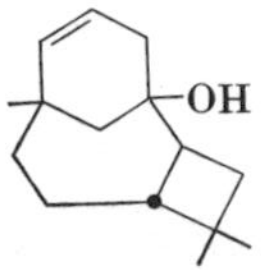 (50%) (69%)	113
	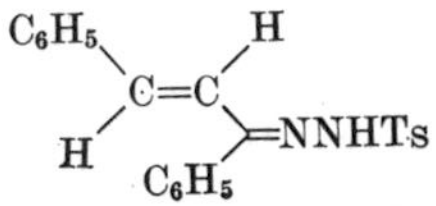	NaOMe/diglyme, *hν*	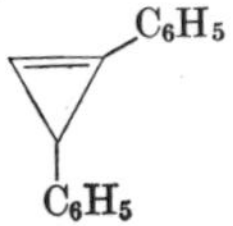 (19%)	106, 107
	Cyclopentadecanone tosylhydrazone	NaOMe/diglyme, heat	*trans*-Cyclopentadecene (94%)	68
16	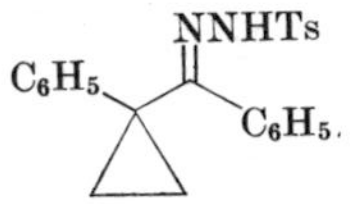	NaOMe/NMP, 120°	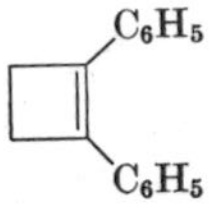 (76%)	149
17	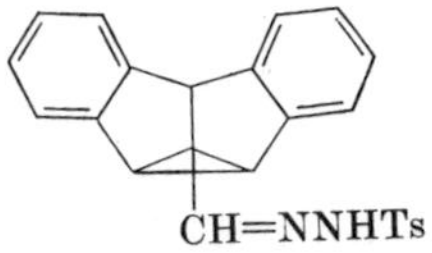	Diverse conditions	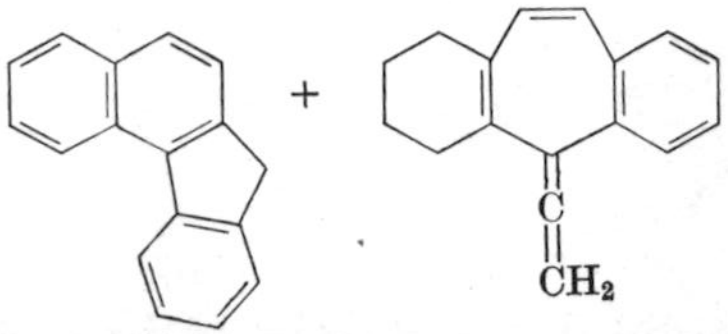	158

No. of C Atoms	Reactant	Conditions	Product(s) and Yield(s) (%)	Refs.
	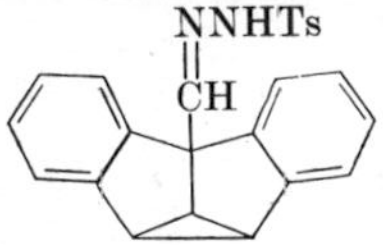 NNHTs=CH, Na$^+$	Diverse conditions	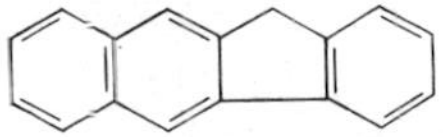	159
18	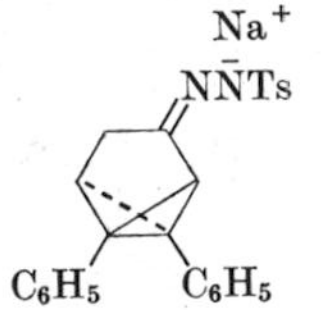 $N\bar{N}Ts$, C_6H_5, C_6H_5	Heat	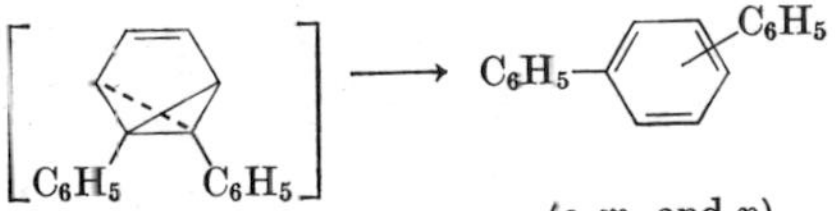 [C_6H_5 C_6H_5] → C_6H_5–C_6H_4–C_6H_5 (o, m, and p)	157
	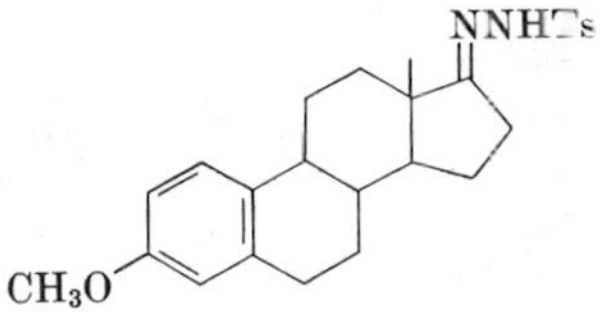 NNHTs, CH_3O	$LiAlH_4$/THF, heat	 (60–70%)	4
	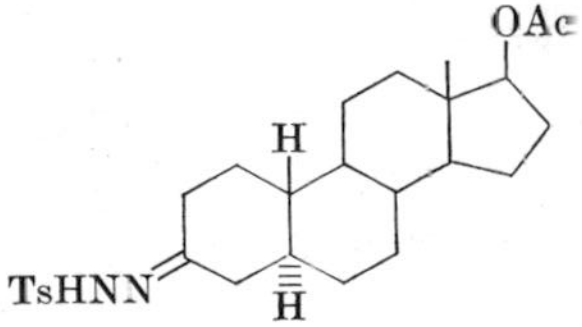 OAc, H, H, TsHNN	$LiH/C_6H_5CH_3$, heat	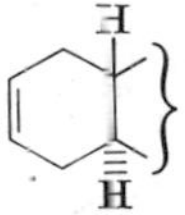 H, H (95%)	14
	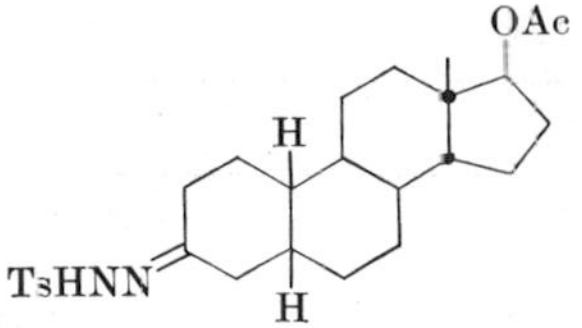 OAc, H, H, TsHNN	$LiH/C_6H_5CH_3$, heat	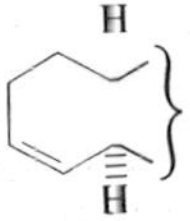 H, H (70%)	14

Note: References 125–210 are on pp. 506–507.

TABLE II. THE APROTIC BAMFORD-STEVENS REACTION (*Continued*)

Carbon Atoms in Starting Ketone	Reactant	Conditions	Product(s)—Distribution ( ) or Yield(s) (%)	Refs.
		A. Monotosylhydrazones (Continued)		
19	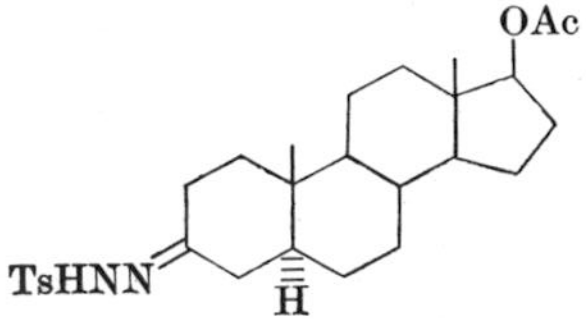 TsHNN, OAc, H	LiH/$C_6H_5CH_3$, heat	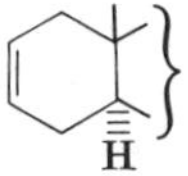 H (95%)	14
	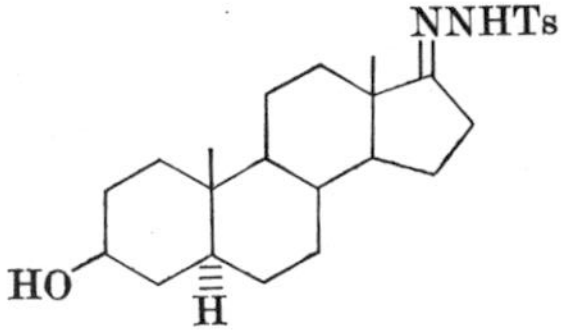 NNHTs, HO, H	$LiAlH_4$/THF, heat	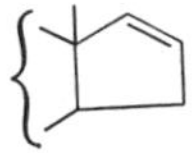 (70%)	4
	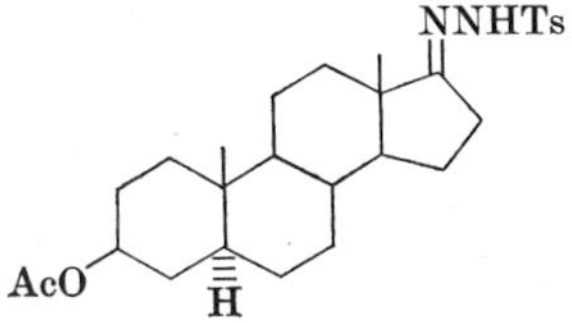 NNHTs, AcO, H	LiH/$C_6H_5CH_3$	 (82%)	14
20	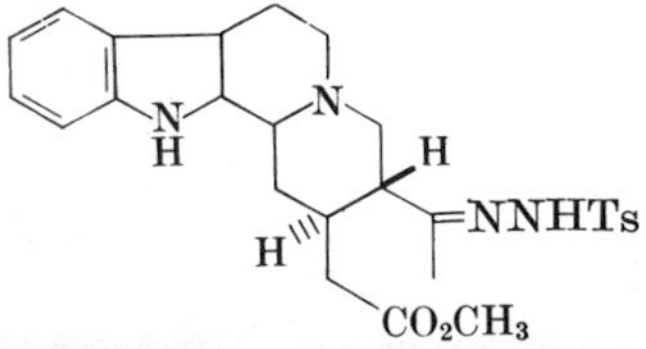 N, H, N, H, =NNHTs, H, CO_2CH_3	NaOMe/diglyme, heat	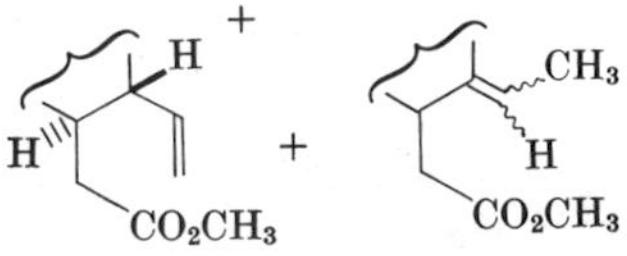 H, H, CO_2CH_3 (5.6%) + CH_3, H, CO_2CH_3 (14.3%) +	162

21	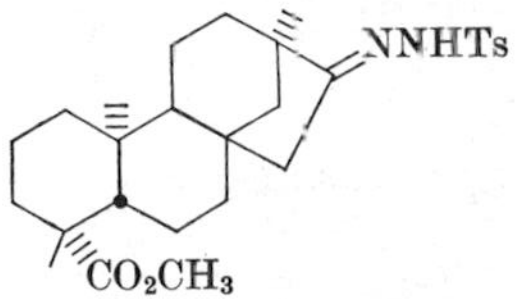 =NNHTs; CO_2CH_3	NaOMe/diglyme, heat	 (Main product)	61
	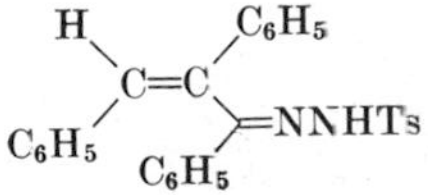 H, C_6H_5, C=C, =NNHTs, C_6H_5, C_6H_5	NaOMe/diglyme, $h\nu$	1,2,3/Triphenylcyclopropene (21%)	106, 107
	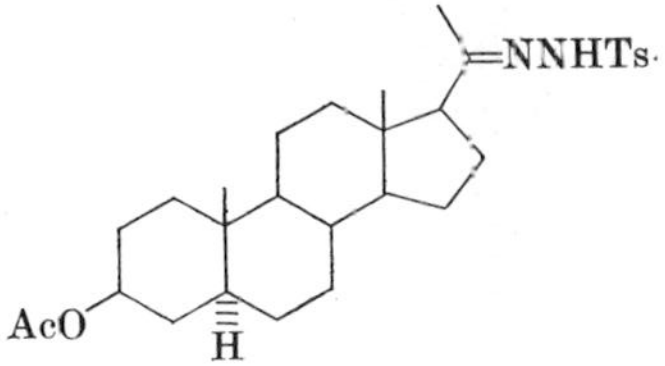 =NNHTs; AcO; H	LiH/$C_6H_5CH_3$, heat	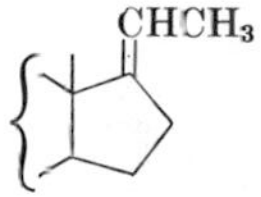 $CHCH_3$ (75%)	14
24	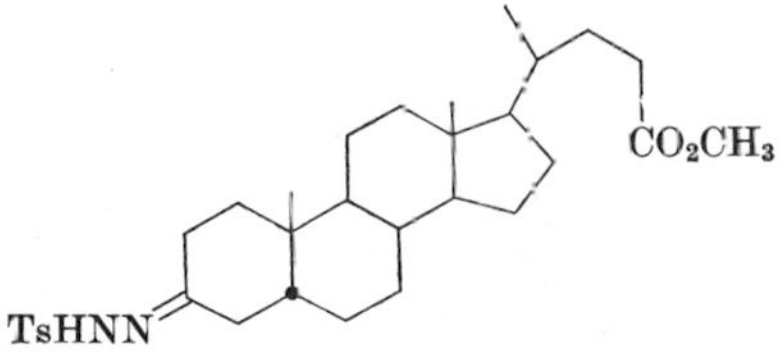 CO_2CH_3; TsHNN=	LiH/$C_6H_5CH_3$, heat	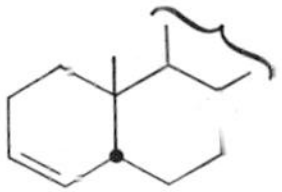 (70%)	14
27	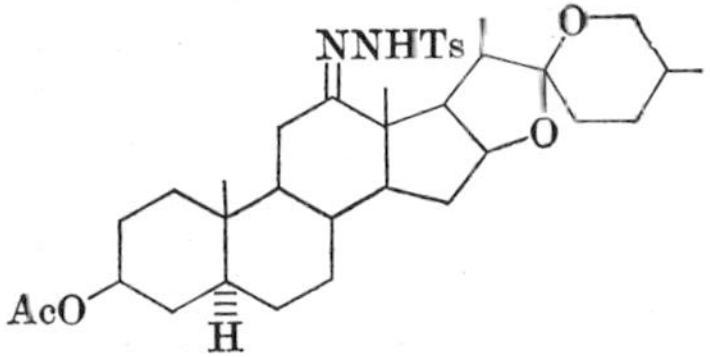 NNHTs; O; O; AcO; H	NaOMe/diglyme, 155°	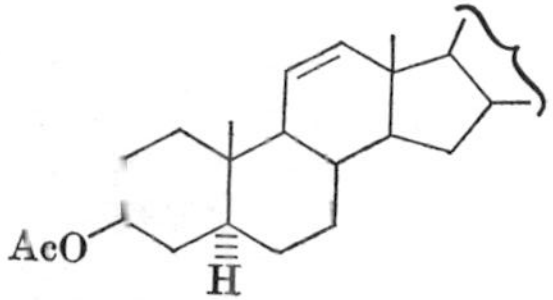 AcO; H (60% after acetylation)	60

Note: References 125–210 are on pp. 506–507.

TABLE II. THE APROTIC BAMFORD-STEVENS REACTION (*Continued*)

Carbon Atoms in Starting Ketone	Reactant	Conditions	Product(s)—Distribution ( ) or Yield(s((%)	Refs.
		A. Monotosylhydrazones (*Continued*)		
27 (*contd.*)	TsHNN= ; C_8H_{17}	LiH/diglyme, 160°	(66) (34) (20%)	
		$LiNH_2$/diglyme, 160°	(65) (35) (17%)	
		NaOMe/diglyme, 160°	(63) (37) (21%)	161
		KOMe/diglyme, 160°	(44) (56) (15%)	
		CaH_2/diglyme, 160°	(33) (67) (22.5%)	
		KOBu-*t*/diglyme, 160°	(20) (80) (33%)	
	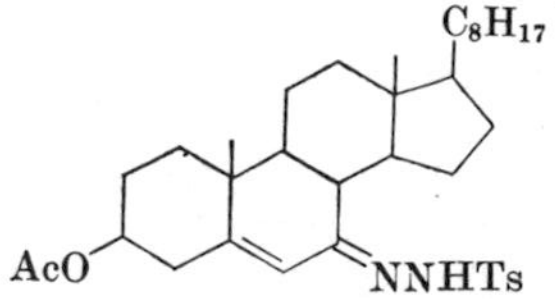 AcO ; =NNHTs ; C_8H_{17}	LiH/$C_6H_5CH_3$, heat 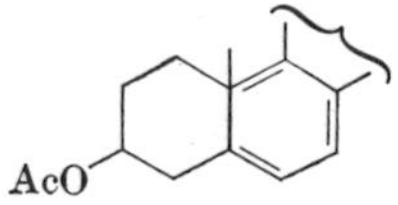	AcO (86%)	186
	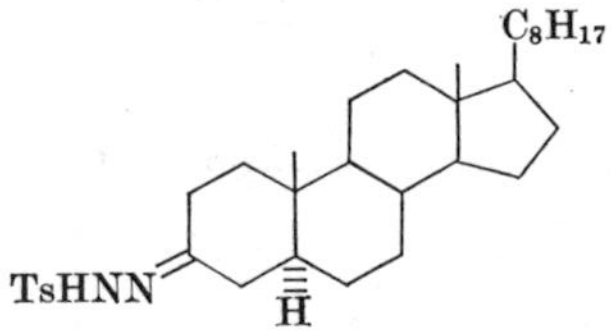 TsHNN= ; H ; C_8H_{17}	LiH/$C_6H_5CH_3$, heat 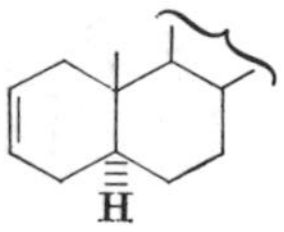	H (92%)	14

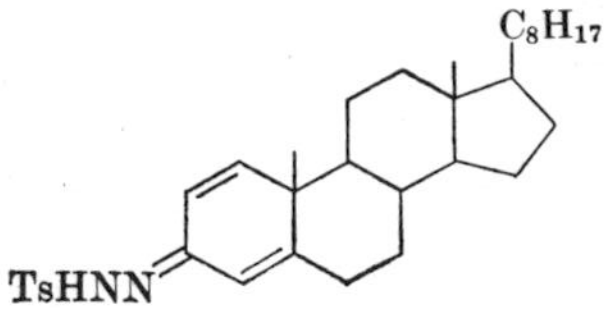

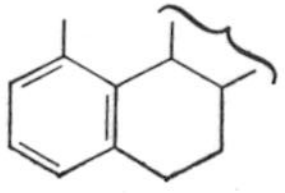

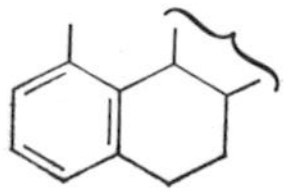

No. of C atoms	Tosylhydrazone	Conditions	Product(s) and Yield(s) (%)	Refs.
29	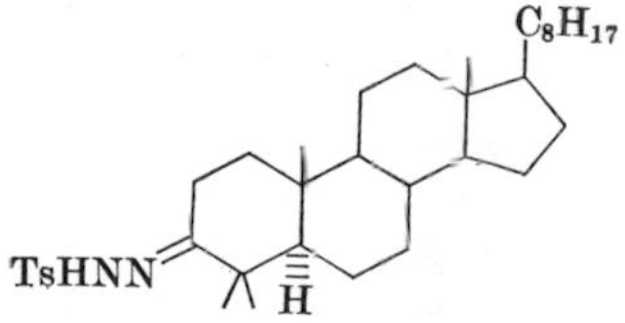 TsHNN= (steroid, C_8H_{17} side chain) [structure]	$NaH/C_6H_5CH_3$, heat	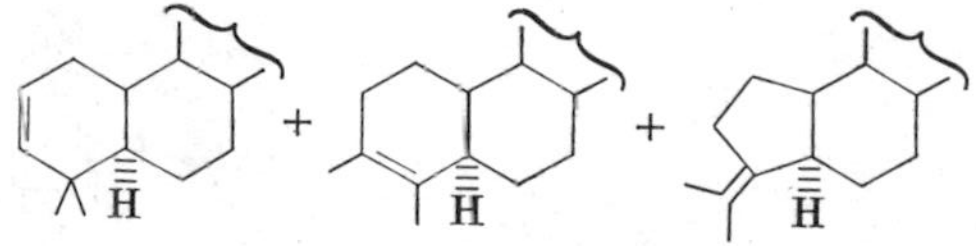 [structure] (100) + [structure] (0) + [structure] (0)	181
		$NaOC_5H_{11}$-t/C_6H_6, heat	(91) (—) (—)	
		$LiH/C_6H_5CH_3$, heat	(27) (6) (16)	
		$BuLi/C_6H_5CH_3$, heat	(50) (—) (—)	
		NaH/DMSO, heat	(36) (—) (—)	

B. α-Oxophosphonic Acid Tosylhydrazones—Aprotic decomposition of Diazo compound

No. of C atoms	Tosylhydrazone	Conditions	Product(s) and Yield(s) (%)	Refs.
5	$CH_3OCH_2C(=NNHTs)P(O)(OCH_3)_2$	1. Na_2CO_3/H_2O; 2. Cu/C_6H_6	*trans*-$CH_3OCH{=}CHP(O)(OCH_3)_2$ (87%)	182
6	i-$C_3H_7C(=NNHTs)P(O)(OCH_3)_2$	1. Na_2CO_3/H_2O; 2. Cu/C_6H_6	$(CH_3)_2{=}CHP(O)(OCH_3)_2$ (87%)	182
6	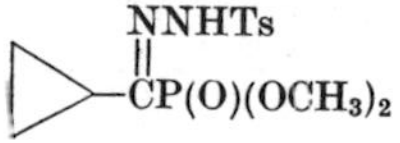 cyclopropyl-$C(=NNHTs)P(O)(OCH_3)_2$	1. Na_2CO_3/H_2O; 2. Cu/C_6H_6	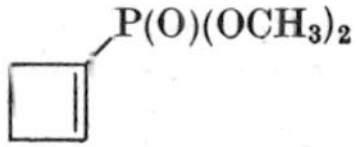 cyclobutenyl-$P(O)(OCH_3)_2$ (71%)	182
7	t-$C_4H_9C(=NNHTs)P(O)(OCH_3)_2$	1. Na_2CO_3/H_2O; 2. Cu/C_6H_6	$(CH_3)_2C{=}C(CH_3)P(O)(OCH_3)_2$ (81%) 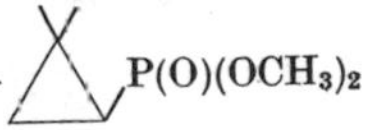 + 2,2-dimethylcyclopropyl-$P(O)(OCH_3)_2$ (9%)	182

Note: References 125–210 are on pp. 506–507.

TABLE II. THE APROTIC BAMFORD-STEVENS REACTION (*Continued*)

Carbon Atoms in Starting Ketone	Reactant	Conditions	Product(s)—Distribution ( ) or Yield(s) (%)	Refs.
		C. Bistosylhydrazones		
7 (*contd.*)	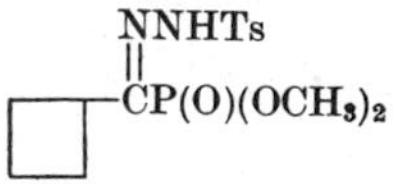 NNHTs ‖ $CP(O)(OCH_3)_2$	1. Na_2CO_3/H_2O; 2. Cu/C_6H_6	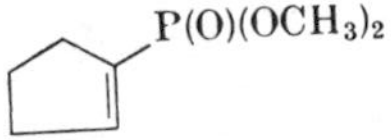 $P(O)(OCH_3)_2$ (96%)	182
4	$[TsNHN{=}CHCH{=}]_2$	NaOMe/diglyme, heat	$CH_2{=}C{=}C{=}CH_2$ (>3%)	179
	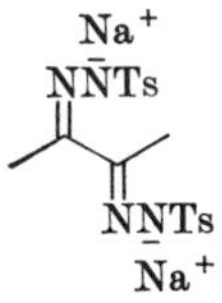 Na^+ $N\bar{N}Ts$ $N\bar{N}Ts$ Na^+	$h\nu$, MeOH	$CH_3C{\equiv}CCH_3$ (—)	169
5	$(CH_3)_2C(CH{=}NNHTs)_2$	NaOMe/diglyme, 160–180°	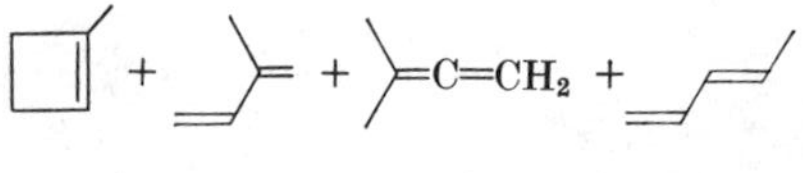 + + $C{=}CH_2$ + (—)	147
		NaOMe/diglyme, heat	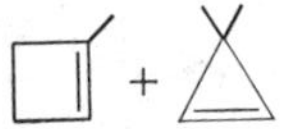 + (5) (6)	179

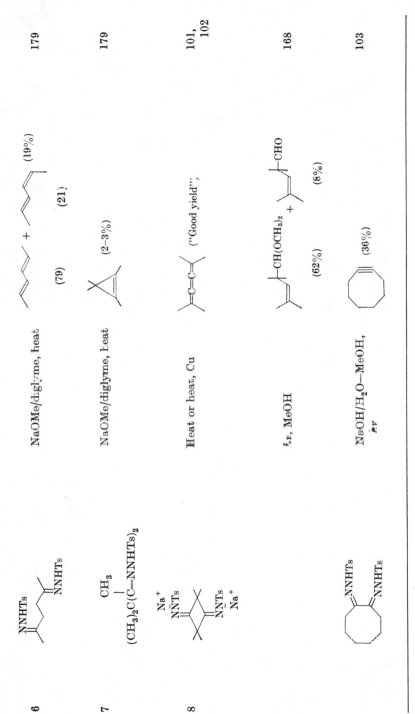

6	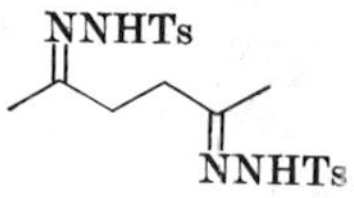	NaOMe/diglyme, heat	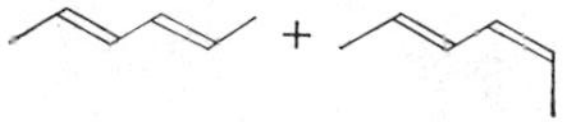 (19%) (79) (21)	179
7	$(CH_3)_2C(\overset{\overset{CH_3}{\vert}}{C}{=}NNHTs)_2$	NaOMe/diglyme, heat	 (2–3%)	179
8	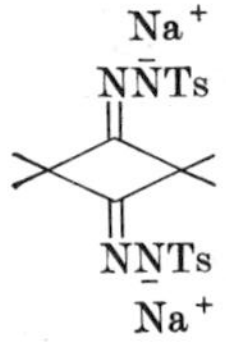	Heat or heat, Cu	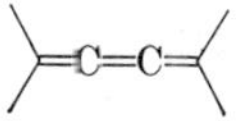 ("Good yield")	101, 102
		hν, MeOH	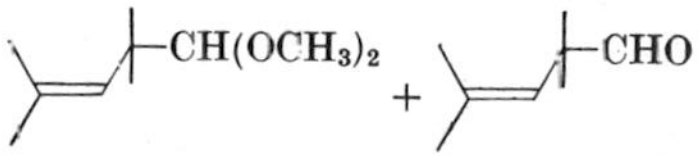 (62%) (8%)	168
	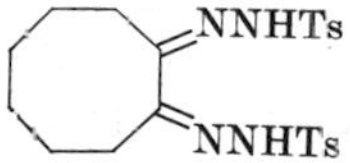	NaOH/H_2O—MeOH, hν	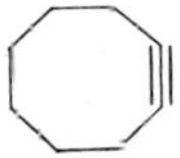 (36%)	103

Note: References 125–210 are on pp. 506–507.

TABLE II. THE APROTIC BAMFORD-STEVENS REACTION (*Continued*)

Carbon Atoms in Starting Ketone	Reactant	Conditions	Product(s)—Distribution ( ) or Yield(s) (%)	Refs.
		C. Bistosylhydrazones (Continued)		
10	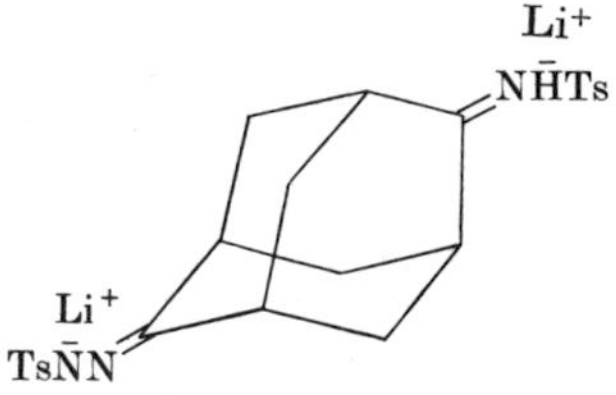 Li$^+$ $N\bar{H}Ts$ … $Ts\bar{N}N$ Li$^+$	155°	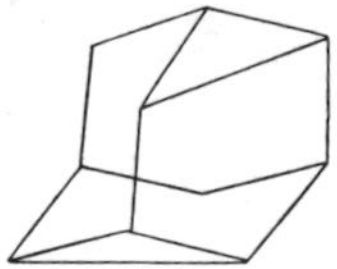 (12%) 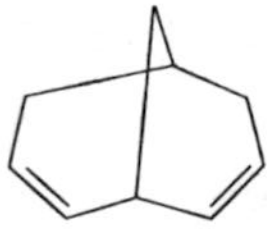 + (6%) "$C_{10}H_{12}$ alkene" (3%)	180
12	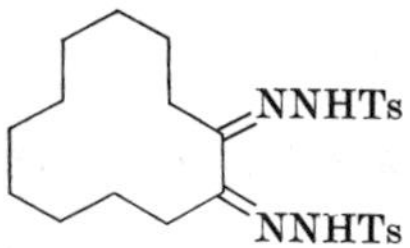 NNHTs NNHTs	NaOH/H_2O—MeOH, *hν*	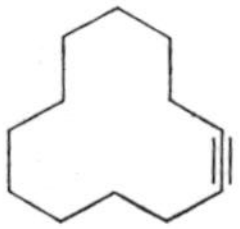 (53%)	103
14	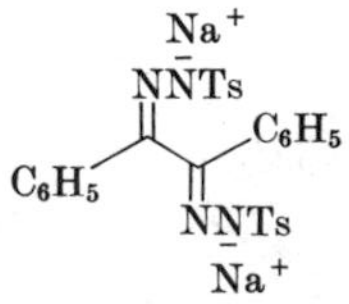 C_6H_5, $N\bar{N}Ts$ Na$^+$, C_6H_5, $N\bar{N}Ts$ Na$^+$	*hν*, MeOH	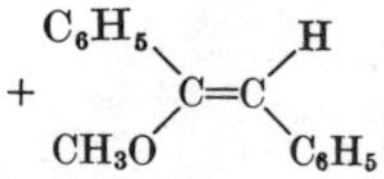 $C_6H_5C{\equiv}CC_6H_5$ (20%) 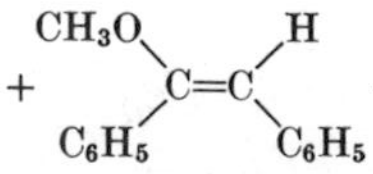 + $CH_3O(C_6H_5)C{=}C(H)C_6H_5$ (28%) + $C_6H_5(CH_3O)C{=}C(H)C_6H_5$ (45%)	169

14	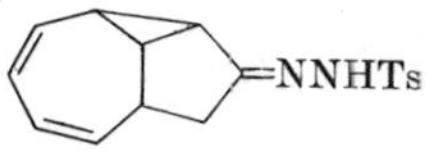 =NNHTs	n-BuLi	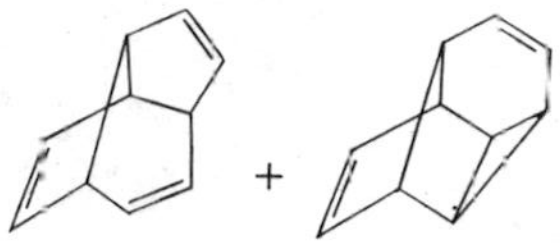 I + II	
		2 equiv	(15%) (1–2%)	210
		6 equiv	(0°) (20%)	
	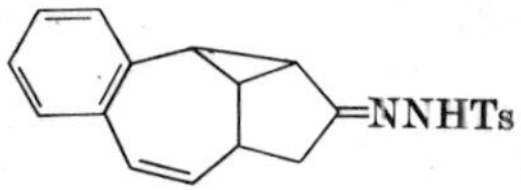 =NNHTs	n-BuLi (excess)	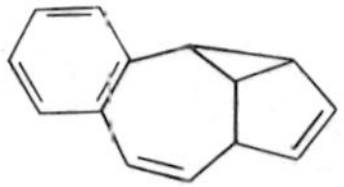 (60%)	210
	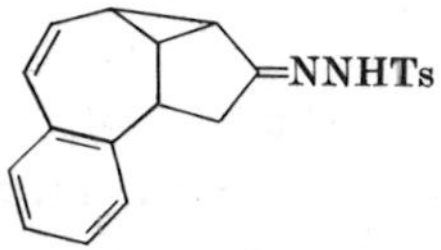 =NNHTs	n-BuLi (excess)	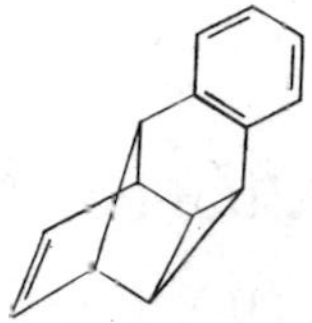 (35%)	210
15	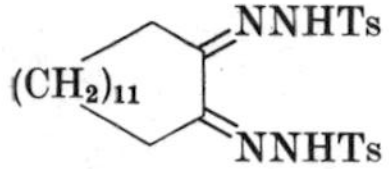 $(CH_2)_{11}$ NNHTs NNHTs	NaOH/H_2O–MeOH, [illegible]	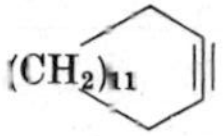 $(CH_2)_{11}$ (55%)	103

Note: References 125–210 are on pp. 506–507.

TABLE III. THE ALKYLLITHIUM REACTION

Reactant	Conditions	Product(s)—Distribution ( ) or Yield(s) (%)	Refs.
n-$C_3H_7C(=NNHSO_2C_6H_5)CH_3$	n-BuLi/decalin, 70–110°	n-Pentane (100)	17
i-$C_3H_7C(=NNHSO_2C_6H_5)CH_3$	n-BuLi/decalin, 70–110°	i-$C_3H_7CH{=}CH_2$ (100)	17
t-$C_4H_9C(=NNHTs)CH_3$	n-BuLi/decalin, 70–110°	t-$C_4H_9CH{=}CH_2$ (100)	17
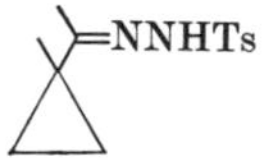 =NNHTs	n-BuLi/Et_2O—C_6H_{14}	 (—)	91
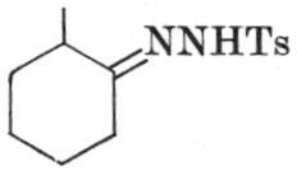 NNHTs	MeLi/Et_2O	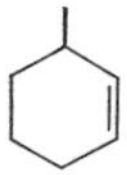 + (98) 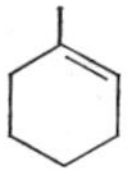 (2)	12
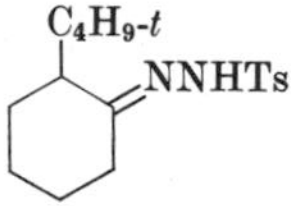 C_4H_9-t NNHTs	n-BuLi/Et_2O	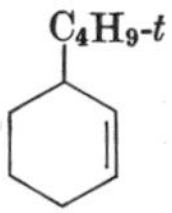 C_4H_9-t + C_4H_9-t (99%) 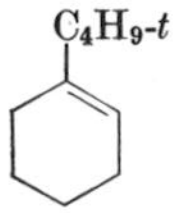 (1%)	192
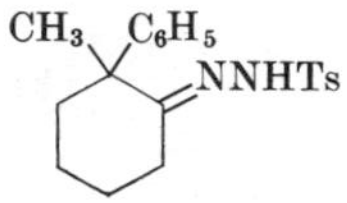 CH_3 C_6H_5 NNHTs	n-BuLi/Et_2O	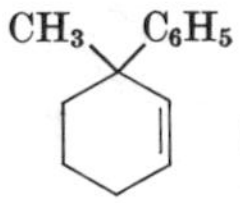 CH_3 C_6H_5 (45%)	202

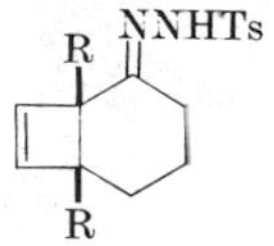 NNHTs, R, R (R = H or D)	MeLi/Et_2O	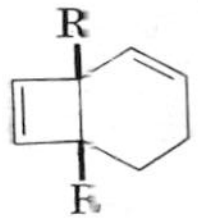 R, R (32.5%)	184
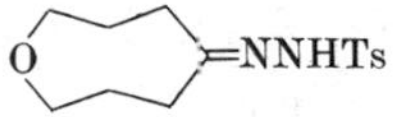 O, =NNHTs	n-BuLi/Et_2O—C_6H_{12}, 0°	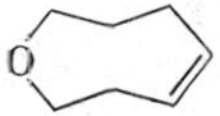 O (83%)	191
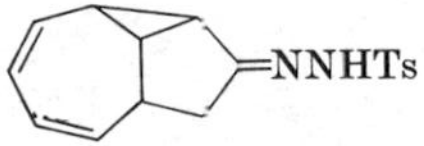 =NNHTs	n-BuLi	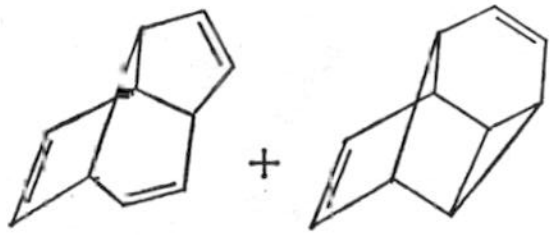 I + II	210
	2 equiv	(15%) (1–2%)	
	6 equiv	(0°) (20%)	
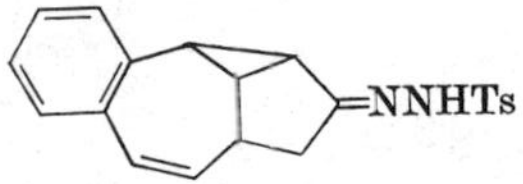 =NNHTs	n-BuLi (excess)	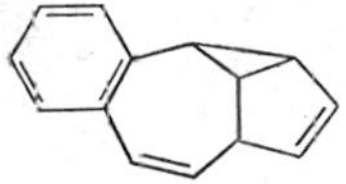 (60%)	210
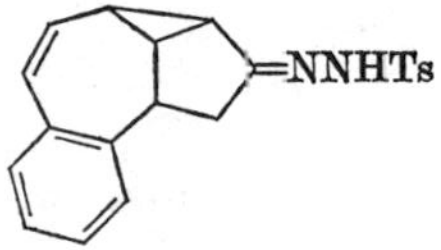 =NNHTs	n-BuLi (excess)	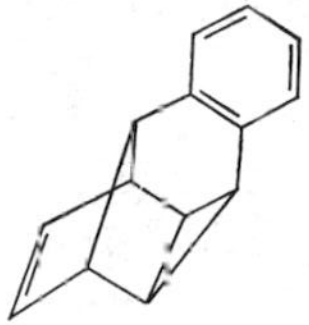 (35%)	210

Note: References 125–210 are on pp. 506–507.

TABLE III. THE ALKYLLITHIUM REACTION (*Continued*)

Reactant	Conditions	Product(s)—Distribution ( ) or Yields (%)	Refs.
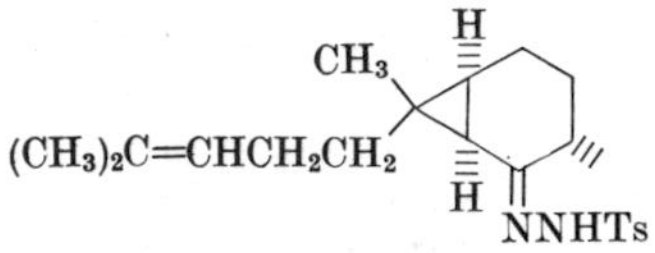 $(CH_3)_2C{=}CHCH_2CH_2$, CH_3, H, H, NNHTs	n-BuLi/Et_2O—C_6H_6, 0–5°	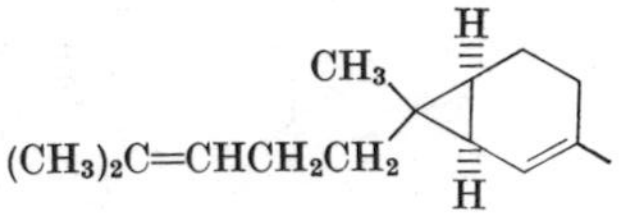 CH_3, $(CH_3)_2C{=}CHCH_2CH_2$, H, H	190
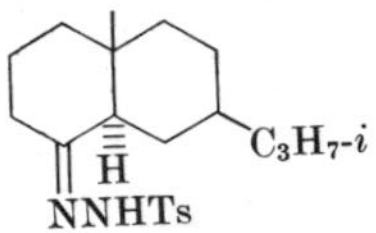 H, C_3H_7-i, NNHTs	n-BuLi/n-C_6H_{14}, 10–15°	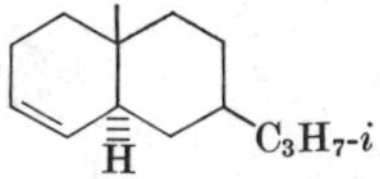 H, C_3H_7-i (97.5%)	188
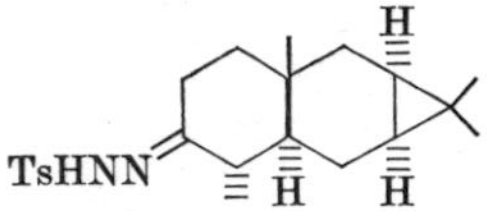 TsHNN, H, H, H	n-BuLi/Et_2O—n-C_6H_{14}, 10–15°	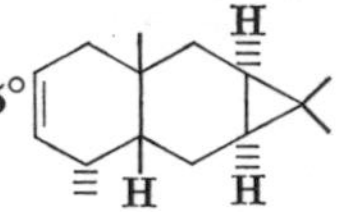 H, H, H (76%)	189
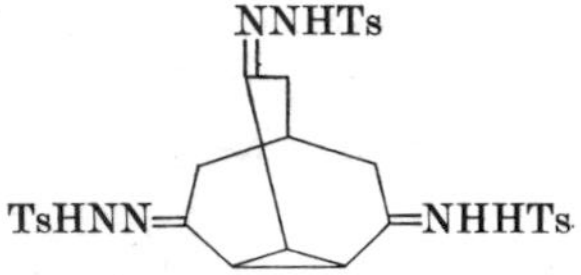 NNHTs, TsHNN=, =NHHTs	MeLi/Et_2O	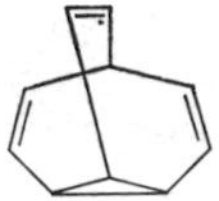 (20 ± 5%)	84
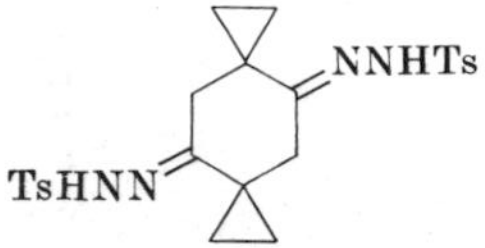 NNHTs, TsHNN	MeLi	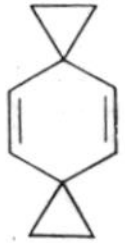 (64%)	83

Reactant	Conditions	Product(s) (Yield)	Refs.
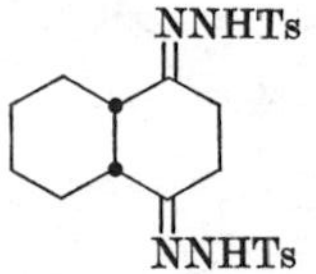 NNHTs / NNHTs	MeLi/Et_2O—C_6H_6	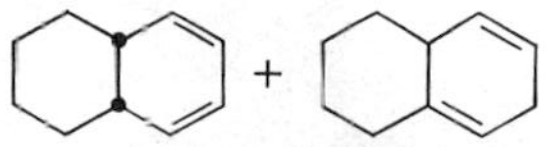 (30%) + (5%)	199
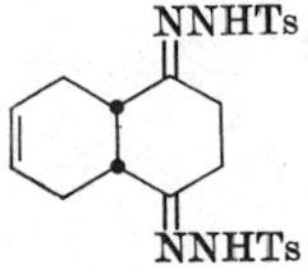 NNHTs / NNHTs	MeLi/Et_2O—C_6H_6	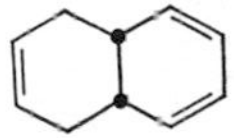 (30%)	199
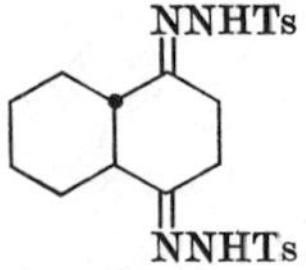 NNHTs / NNHTs	MeLi/Et_2O—C_6H_6	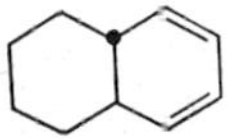 (50%)	199
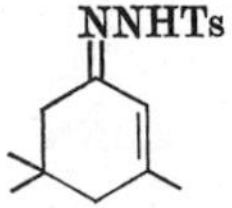 NNHTs	MeLi/Et_2O—C_6H_6	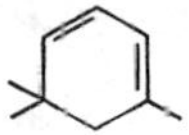 (Sole product)	80
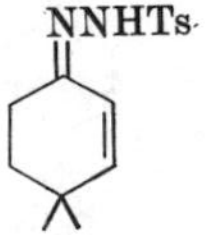 NNHTs	MeLi/Et_2O—C_6H_6	 (85%)	80
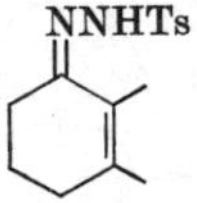 NNHTs	MeLi/Et_2O—C_6H_6	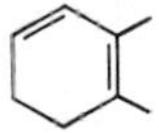 (87%)	80

Note: References 125–210 are on pp. 506–507.

TABLE III. THE ALKYLLITHIUM REACTION (*Continued*)

Reactant	Conditions	Product(s)—Distribution ( ) or Yield(s) (%)	Refs.
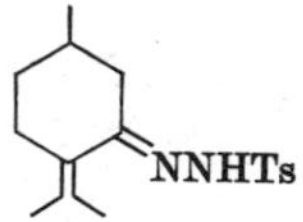 NNHTs	MeLi/Et_2O—C_6H_6	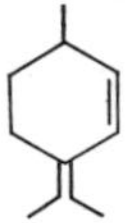 (Sole product)	80
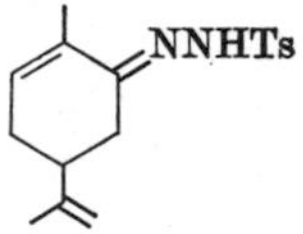 NNHTs	MeLi/Et_2O—C_6H_6	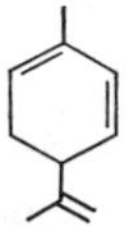 (80%)	80
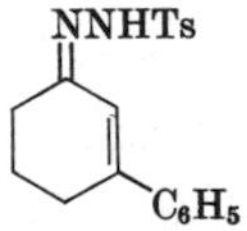 NNHTs, C_6H_5	*n*-BuLi/Et_2O—C_6H_6	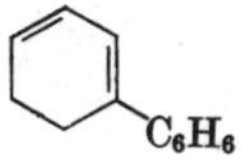 C_6H_6 (40%)	80
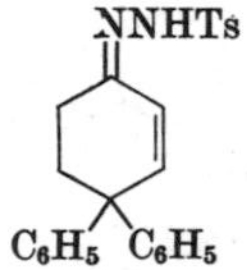 NNHTs, C_6H_5 C_6H_5	*n*-BuLi/Et_2O—C_6H_6	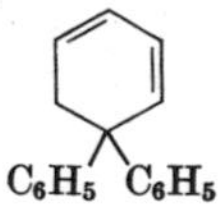 C_6H_5 C_6H_5 (55%)	80, 81
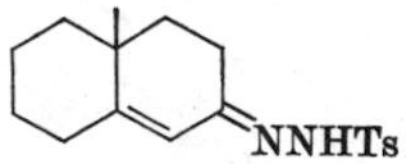 NNHTs	MeLi/Et_2O—C_6H_6	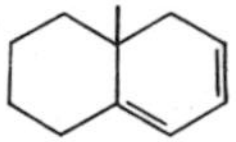 (85%)	80
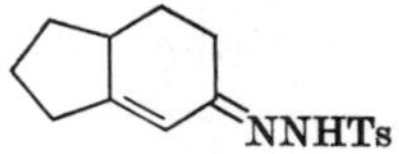 NNHTs	MeLi/Et_2O—C_6H_6	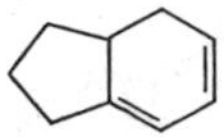 (60%)	80
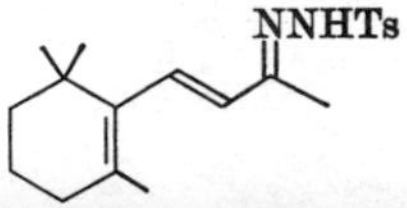 NNHTs	MeLi/Et_2O—C_6H_6	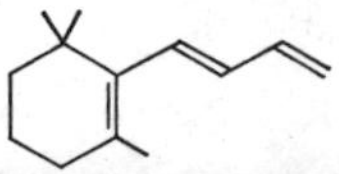 (75–80%)	80

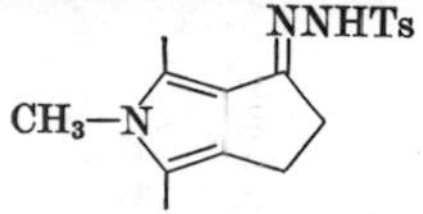 CH_3–N, NNHTs	n-BuLi	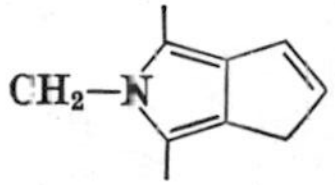 CH_2–N (60%)	204
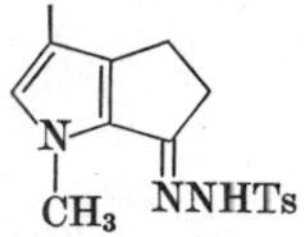 N, CH_3, NNHTs	n-BuLi	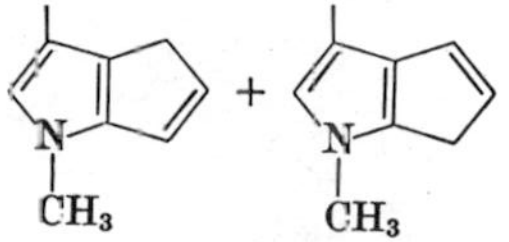 N, CH_3 + N, CH_3 (87%)	205
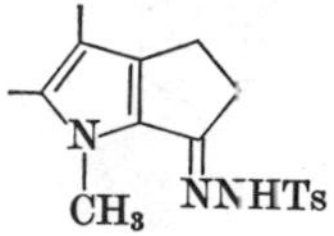 N, CH_3, NNHTs	n-BuLi	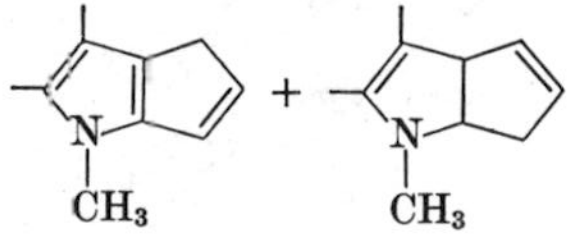 N, CH_3 + N, CH_3 (53%)	206
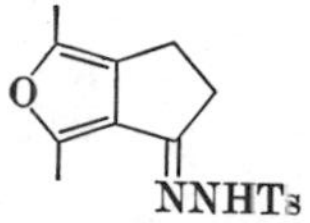 O, NNHTs	$MeLi/Et_2O$, 5°	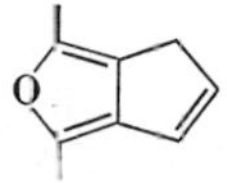 O (61–75%)	187
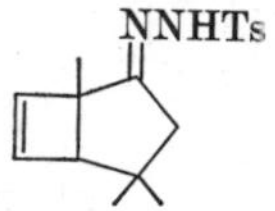 NNHTs	$MeLi/Et_2O$	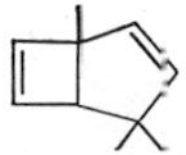 (—)	185
$C_6H_5C(=NNHTs)CH_2C_6H_5$	$MeLi/Et_2O$	$C_6H_5CH=CHC_6H_5$ (*cis*/*trans*, 14:86)	49
$C_6H_5C(=NNHTs)C_2H_5$	$MeLi/Et_2O$	$C_6H_5CH=CHCH_3$ (*cis*/*trans*, 76:24)	49
$C_6H_5C(=NNHTs)CH_2CH_2C_6H_5$	$MeLi/Et_2O$	$C_6H_5CH=CHCH_2C_6H_5$ I (*cis*/*trans*, 48:52)	49
$(C_6H_5CH_2)_2C=NNHTs$	$MeLi/Et_2O$	I (*cis*/*trans*, 86:14)	49
$C_6H_5C(=NNHTs)CH_2SC_2H_5$	n-BuLi/diglyme, heat	$C_6H_5CH=CHSC_2H_5$ (—)	47

Note: References 125–210 are on pp. 506–507.

TABLE III. THE ALKYLLITHIUM REACTION (*Continued*)

Reactant	Conditions	Product(s)—Distribution ( ) or Yield(s) (%)	Refs.
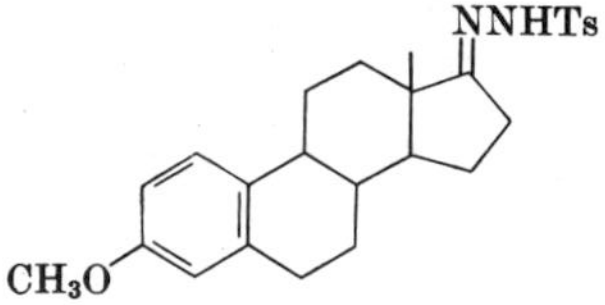 NNHTs, CH_3O	MeLi/Et_2O	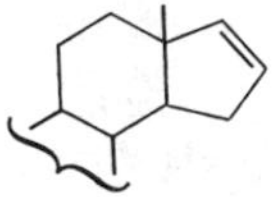 (86%)	200, 201
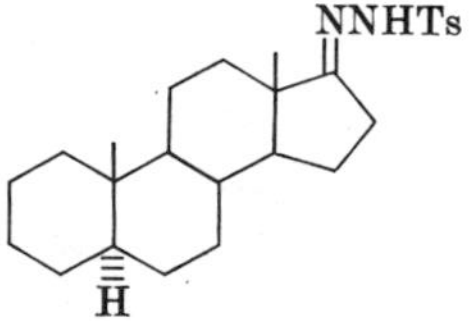 NNHTs, H	MeLi/Et_2O	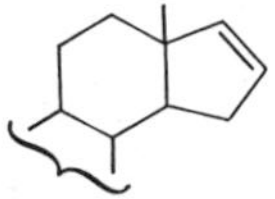 (Sole product)	12
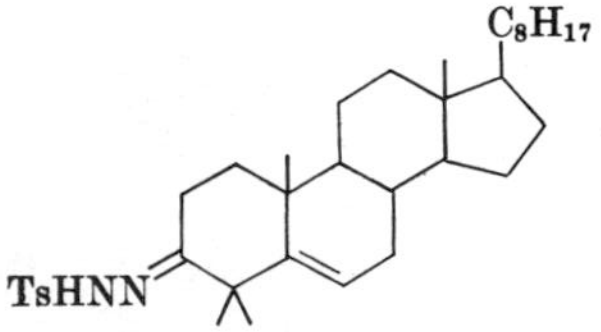 C_8H_{17}, TsHNN	MeLi/Et_2O	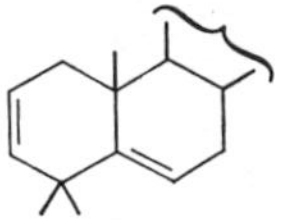 (Sole product)	12
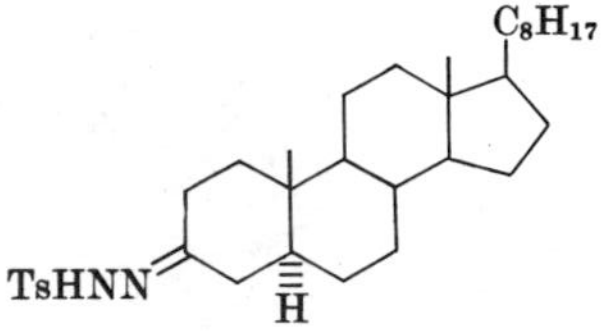 C_8H_{17}, TsHNN, H	MeLi or n-BuLi (<3 equiv)	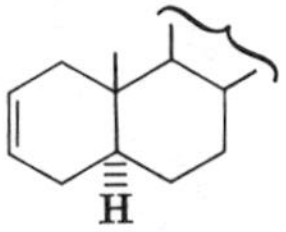 H (78%)	203

Substrate	Conditions	Product(s) and Yield(s) (%)	Refs.
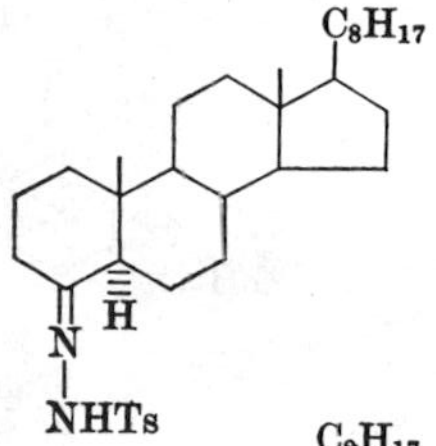 C_8H_{17}; H; N–NHTs	MeLi or n-BuLi (<3 equiv)	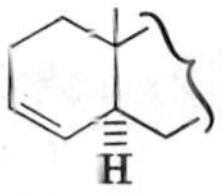 H (73%)	203
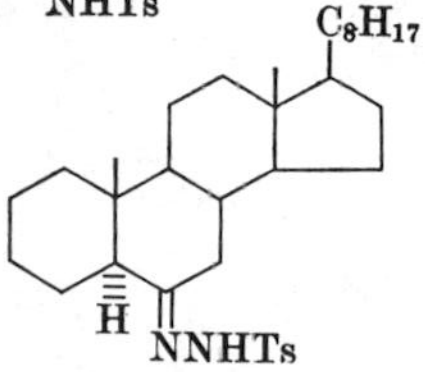 C_8H_{17}; H; NNHTs	MeLi or n-BuLi (<3 equiv)	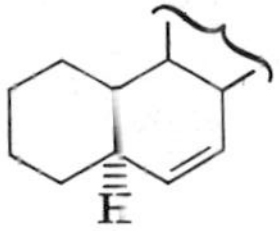 H (70%)	203
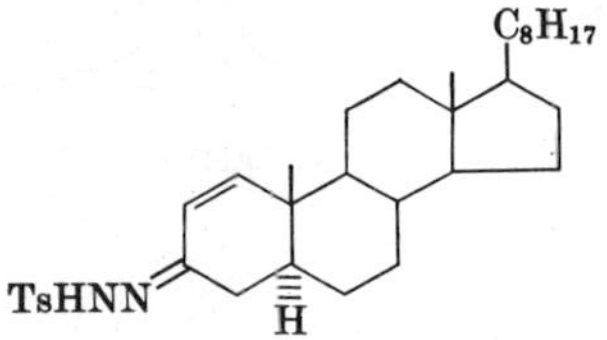 C_8H_{17}; TsHNN; H	MeLi or n-BuLi (<3 equiv)	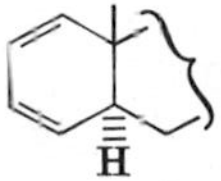 H (60%)	203
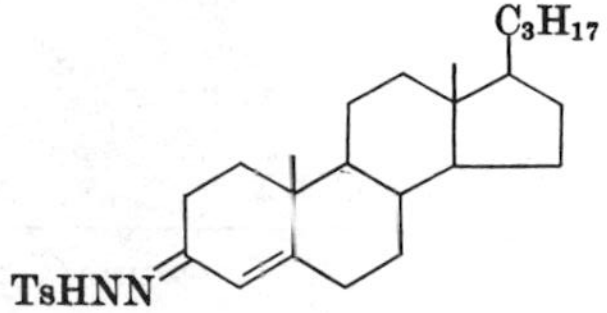 C_8H_{17}; TsHNN	$MeLi/Et_2O$	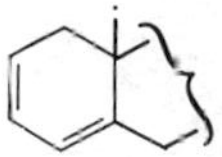 (Sole product)	80

Note: References 125–210 are on pp. 506–507.

TABLE III. The Alkyllithium Reaction (*Continued*)

Reactant	Conditions	Product(s)—Distribution ( ) or Yield(s) (%)	Refs.
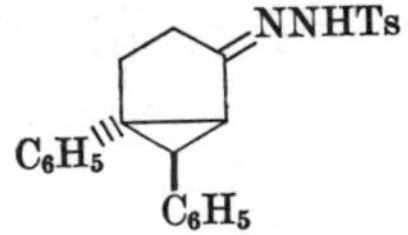 NNHTs; C_6H_5; C_6H_5	n-BuLi/C_6H_6-hexane	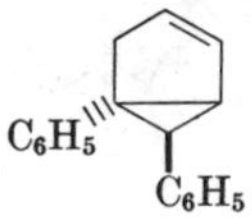 C_6H_5; C_6H_5 (53%)	81
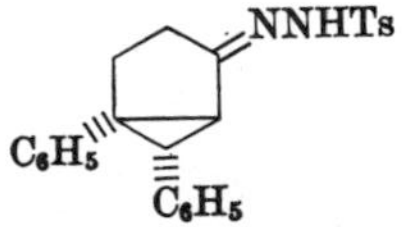 NNHTs; C_6H_5; C_6H_5	n-BuLi/C_6H_6—n-C_6H_{14}	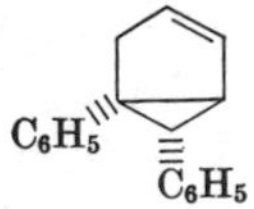 C_6H_5; C_6H_5 (43%)	81
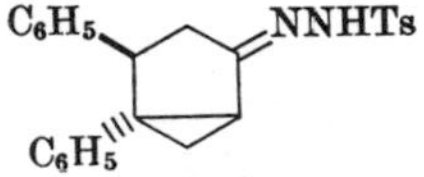 C_6H_5; NNHTs; C_6H_5	n-BuLi/C_6H_6—n-C_6H_{14}	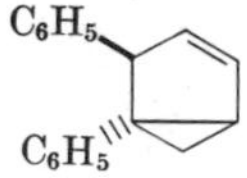 C_6H_5; C_6H_5 (43%)	81
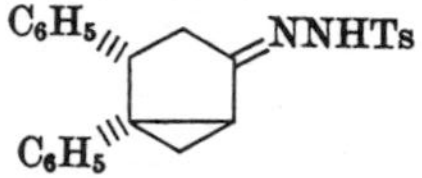 C_6H_5; NNHTs; C_6H_5	n-BuLi/C_6H_6—n-C_6H_{14}	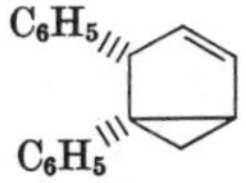 C_6H_5; C_6H_5 (35%)	81
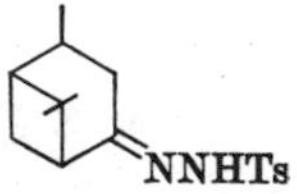 NNHTs	MeLi	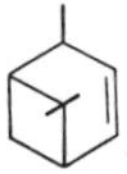 (—)	79
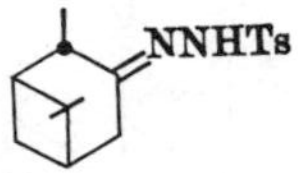 NNHTs	MeLi	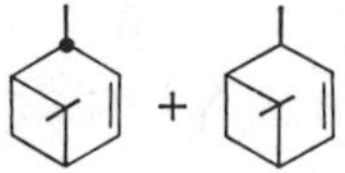 + (—)	79
 NNHTs	MeLi	 (—)	79

Reactant	Conditions	Product(s) and Yield(s) (%)	Refs.
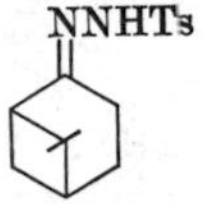 NNHTs	MeLi	 (—)	79
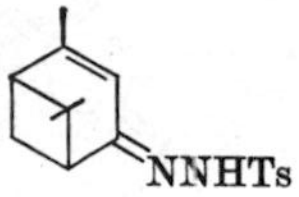 NNHTs	MeLi	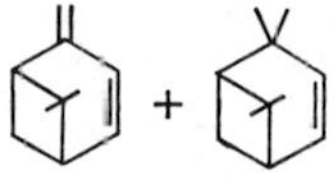 + (1) (1)	79
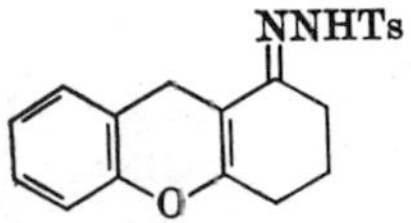 NNHTs O	MeLi	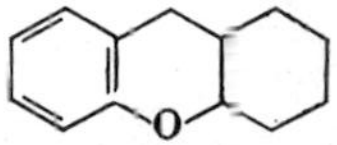 O (—)	207
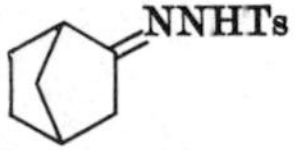 NNHTs	MeLi/Et_2O, 20°	 (—)	195
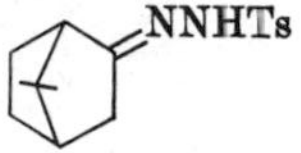 NNHTs	MeLi/Et_2O, 20°	 (—)	195
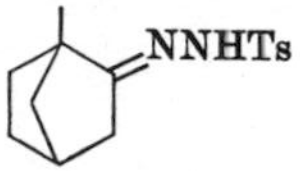 NNHTs	MeLi/Et_2O, 20°	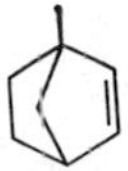 (—)	197
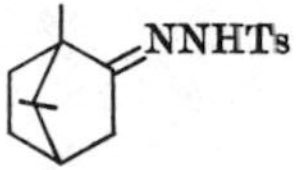 NNHTs	MeLi/Et_2O, 20°	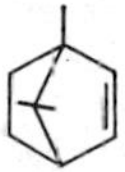 (65%)	12, 124, 196, 197
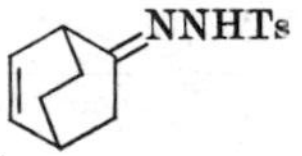 NNHTs	MeLi/ Et_2O, 20°	 (—)	183

Note: References 125–210 are on pp. 506–507.

TABLE III. THE ALKYLLITHIUM REACTION (*Continued*)

Reactant	Conditions	Product(s)—Distribution ( ) or Yield(s) (%)	Refs.
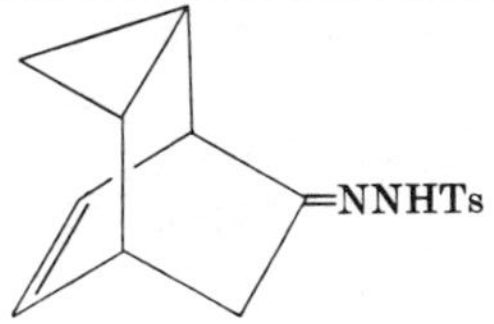 =NNHTs	$MeLi/Et_2O$, 20°	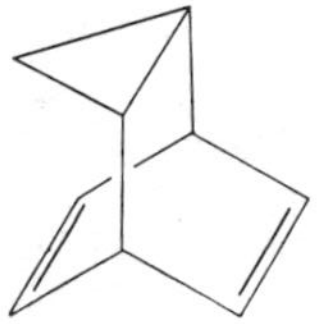 (77%)	208
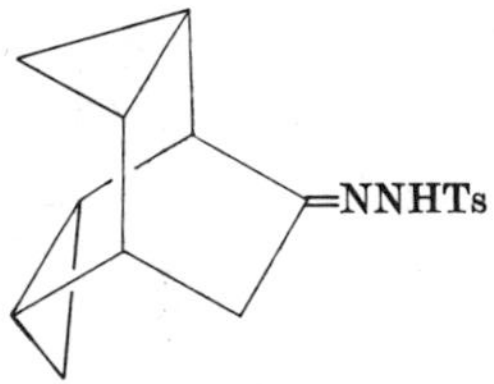 =NNHTs	$MeLi/Et_2O$, 20°	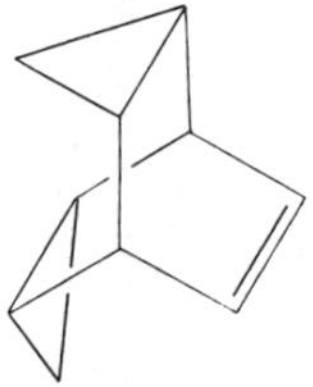 (72%)	208
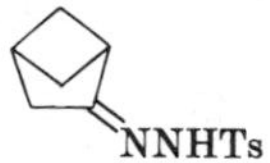 NNHTs	$MeLi/Et_2O$	 (25%)	19
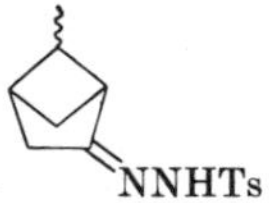 NNHTs	$MeLi/Et_2O$	 (—)	198
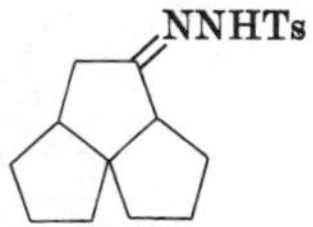 NNHTs	$MeLi/Et_2O$	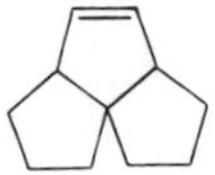 (41%)	194
 NNHTs	$MeLi/Et_2O$	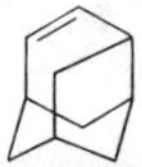 (72%)	82

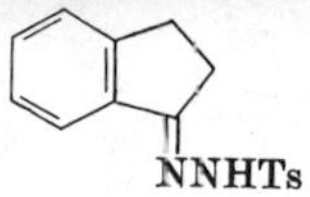 NNHTs	MeLi/Et_2O	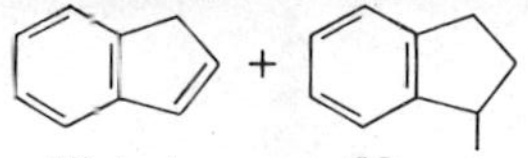 + (Major) (Minor)	12
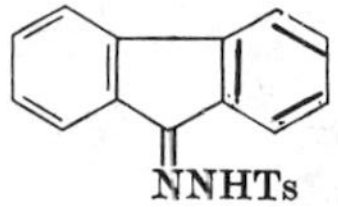 NNHTs	MeLi/Et_2O	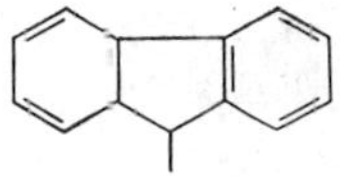 (100)	12
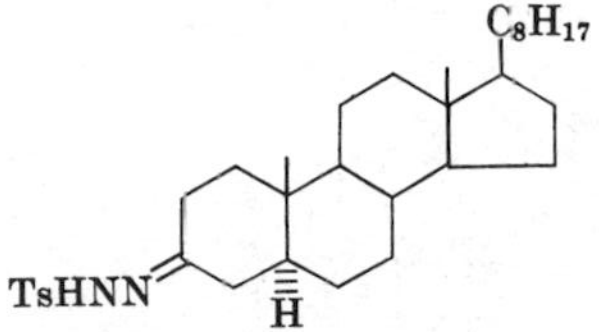 C_8H_{17} TsHNN H	RLi (8 equiv)	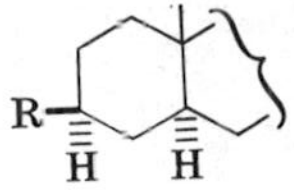 R H H (R = n-C_4H_9, 55%; R = sec-C_4H_9, 50%; R = t-C_4H_9, 48%; R = i-C_3H_7, 30%.)	18
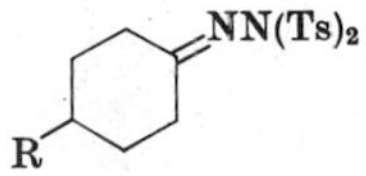 $NN(Ts)_2$ R	t-BuLi/THF, −78°	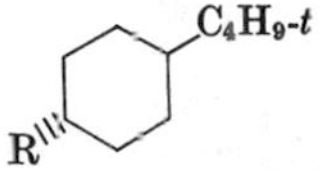 C_4H_9-t R (R = H, 35%; R = t-C_7H_9, 56%)	193
	MeLi/THF, 0°	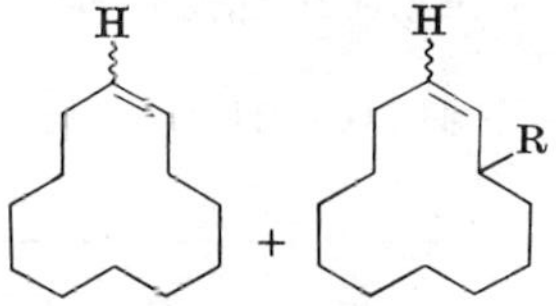 H H R + (cis + trans, 34%)	193
Cyclododecanone tosylhydrazone	t-BuLi/THF, −78°	(R = CH_3, cis + trans 30%) (cis/trans = 41:59, 50%) (R = t-C_7H_9, trans 29%)	193
$(C_6H_5CH_2)_2C{=}NNHTs$	MeLi/THF, 0°	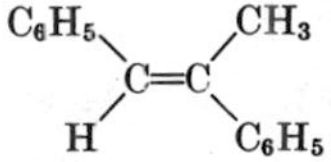 C_6H_5 CH_3 C=C H C_6H_5 (79%)	193

Note: References 125–210 are on pp. 506–507.

TABLE IV. FRAGMENTATION REACTIONS OF TOSYLHYDRAZONES

Carbon Atoms in Starting Ketone	Reactant	Conditions	Product(s)—Distribution ( ) or Yield(s) (%)	Refs.
		A. Tosylhydrazones		
5	2-Furyl-$CH{=}N\bar{N}Ts$ Na^+	250° (0.5 mm)	$HC{\equiv}CCH{=}CHCHO$ (66%; *cis/trans*, 81:19)	90
6	5-CH_3-2-furyl-$CH{=}N\bar{N}Ts$ Na^+	250° (0.5 mm)	$HC{\equiv}CCH{=}CHCOCH_3$ (43%; *cis/trans*, 97:13)	90
	2-Furyl-$C(CH_3){=}N\bar{N}Ts$ Na^+	250° (0.5 mm)	$CH_3C{\equiv}CCH{=}CHCHO$ (36%; *cis/trans*, 73:27)	90
7	2-Furyl-$C(C_2H_5){=}N\bar{N}Ts$ Na^+	250° (0.5 mm)	$C_2H_5C{\equiv}CCH{=}CHCHO$ (47%; *cis/trans*, 68:32)	90
	2-CH_3-tetrahydrofuran-2-yl-$C(=NNHTs)CH_3$	n-BuLi/Et_2O—n-C_6H_{14} 25°	$HO(CH_2)_3C(CH_3){=}C{=}CH_2$ (48%) + $HO(CH_2)_3CH(CH_3)C{\equiv}CH$ (—)	91

No.	Reactant	Conditions	Product(s) and Yield(s) (%)	Refs.
8	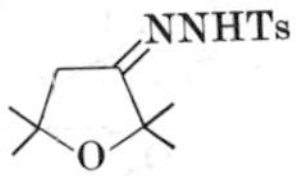 2,2,5,5-Tetramethyltetrahydrofuran-3-one tosylhydrazone (=NNHTs)	n-BuLi/Et_2O—n-C_6H_{14}, 25°	$(CH_3)_2C{=}C{=}CHC(CH_3)_2OH$ (58%) 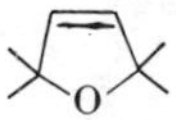 + 2,2,5,5-tetramethyl-2,5-dihydrofuran (3%)	91
9	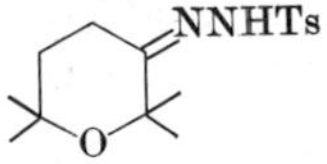 2,2,6,6-Tetramethyltetrahydropyran-3-one tosylhydrazone (=NNHTs)	n-BuLi/Et_2O—n-C_6H_{14}, 25°	$(CH_3)_2C{=}C{=}CHCH_2C(CH_3)_2OH$ (49%) 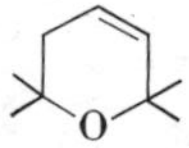 + 2,2,6,6-tetramethyl-3,6-dihydro-2H-pyran (—)	91
11	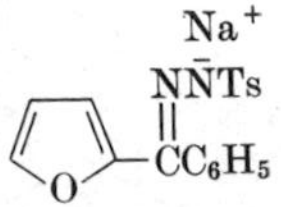 2-Furyl-$C(=N\bar{N}Ts\ Na^+)C_6H_5$	250° (0.5 mm)	$C_6H_5C{\equiv}CCH{=}CHCHO$ (43%; *cis/trans*, 53:47)	90
14	$C_6H_5C(=NNHTs)CHOHC_6H_5$	Na/EG, heat	$C_6H_5C{\equiv}CC_6H_5$ (I) + $C_6H_5CH_2COC_6H_5$ (II)	92
		NaOMe/diglyme, heat	I (13%) II (72%)	
16	$C_6H_5C(=NNHTs)CH(OHC)C_6H_5$	Na/EG, heat	I (53%) II (43%)	92
		NaOMe/diglyme, heat	I (95%) II (3%)	
21	$C_6H_5C(=NNHTs)CH(OBz)C_6H_5$	Na/EG, heat	I (42%) II (42%)	92
		NaOMe/diglyme, heat	I (98%) II (0%)	
	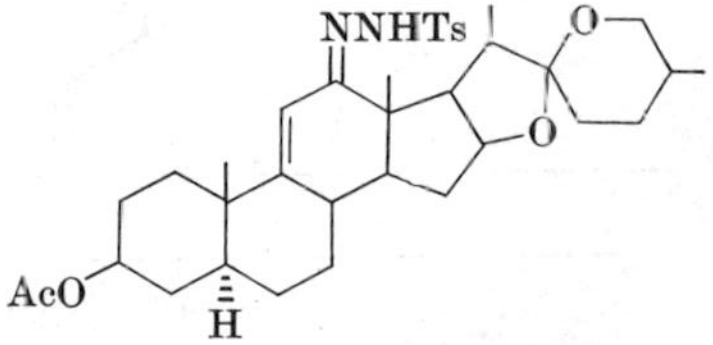 Steroidal spiroketal 12-tosylhydrazone (AcO, =NNHTs, H)	m-$ClC_6H_4CO_2H$	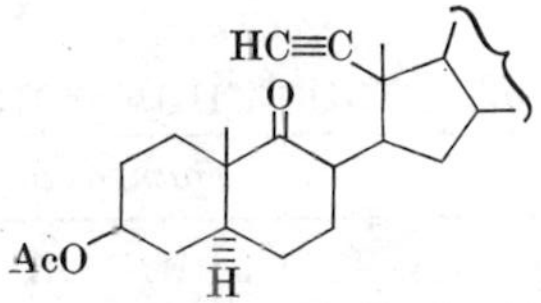 Steroidal ethynyl ketone (AcO, O, HC≡C, H) (5%)	87

Note: References 125–210 are on pp. 506–507.

TABLE IV. FRAGMENTATION REACTIONS OF TOSYLHYDRAZONES (*Continued*)

Carbon Atoms in Starting Ketone	Reactant	Conditions	Product(s)—Distribution ( ) or Yield(s) (%)	Refs.
B. Epoxyketones and Related Compounds				
7	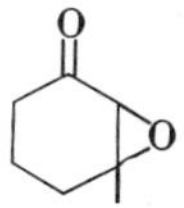	TsNNNH$_2$/ CH_2Cl_2—AcOH, 0°, 14 hr; 25°, 3 hr	$HC{\equiv}C(CH_2)_3COCH_3$ (63%)	86
	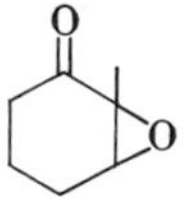	TsNHNH$_2$/ CH_2Cl_2—AcOH, 0°, 1.5 hr	$CH_3C{\equiv}C(CH_2)_3CHO$ (38%)	86
9	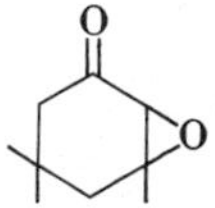	TsNHNH$_2$/ CH_2Cl_2—AcOH, 0°, 25°, 12 hr	$HC{\equiv}CCH_2C(CH_3)_2CH_2COCH_3$ (74%)	86, 88
10	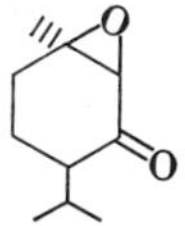	TsNHNH$_2$/ CH_2Cl_2—AcOH, 2°, 14 hr; 25°, 1 hr, TsOH	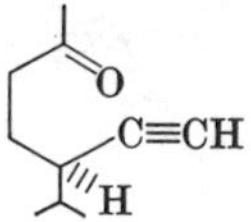 O, C≡CH, H (57%)	86
	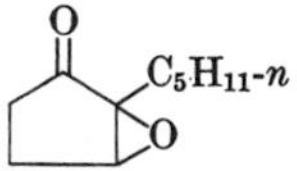 C_5H_{11}-*n*	TsNHNH$_2$/diglyme-AcOH, 0°, 15 min	*n*-$C_5H_{11}C{\equiv}C(CH_2)_2CHO$ (27%)	86
	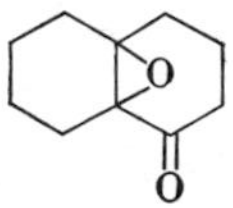	TsNHNH$_2$/AcOH, 25°, 45 min	 I (81%)	86, 88

No.	Reactant	Conditions	Product(s) and Yield(s) (%)	Refs.
		$TsNHNH_2$/MeOH-AcOH-H_3PO_4, 25°, 30 min	I (86%)	86
11	C_5H_{11}-*n*	$TsNHNH_2$/CH_2Cl_2—AcOH, 0–20°, 4 hr	*n*-$C_5H_{11}C{\equiv}C(CH_2)_2COCH_3$ (55%)	86
		$TsNHNH_2$/CH_2Cl_2-AcOH, 0°, 45 min; 25°, 1 hr	$(CH_2)_2C{\equiv}CH$ (67%)	86, 88
13		$TsNHNH_2$/CH_2Cl_2-AcOH, −35, 50 hr; −10 to +25°, 3 hr	(70%)	86
	$COCH_3$	$TsNHNH_2$/CH_2Cl_2-AcOH, −18°, 5 days; 25°, 2 hr; 35°, 4 hr	OH O (30%) + OH $C{\equiv}CCH_3$ (10%)	
15	$(CH_2)_9$	$TsNHNH_2$/AcOH-H_3PO_4, 25°, 36 hr	$(CH_2)_9$ $C{\equiv}C$ (63%)	85, 86

Note: References 125–210 are on pp. 506–507.

TABLE IV. Fragmentation Reactions of Tosylhydrazones (*Continued*)

Carbon Atoms in Starting Ketone	Reactant	Conditions	Product(s)—Distribution ( ) or Yield(s) (%)	Refs.
		B. Epoxyketones and Related Compounds (*Continued*)		
15 (*contd.*)	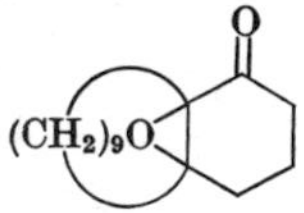 $(CH_2)_9$ O, O	$TsNHNH_2$/MeOH, 25°, 3 days	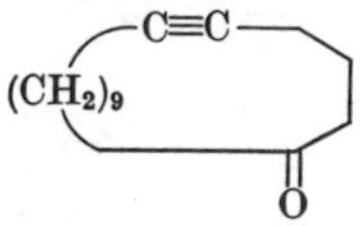 $(CH_2)_9$, C≡C, O (58%)	85, 86
16	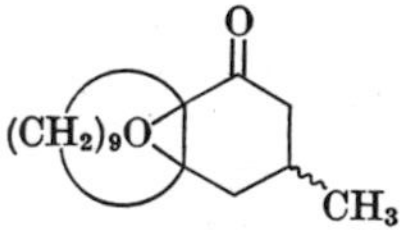 $(CH_2)_9$ O, O, CH_3	$TsNHNH_2$/AcOH-H_3PO_4, 25°, 24 hr	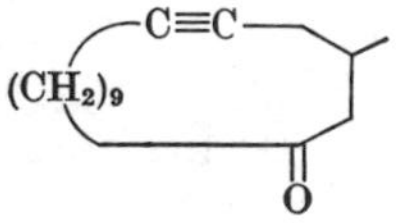 $(CH_2)_9$, C≡C, O (76%)	85, 86
18	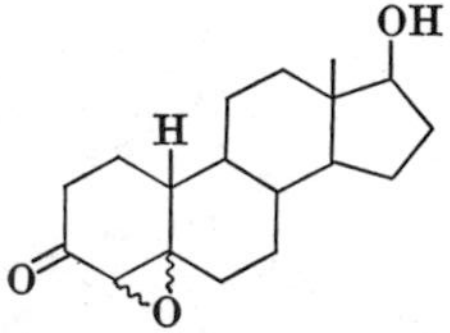 OH, H, O, O	$TsNHNH_2$/EtOH, 50°	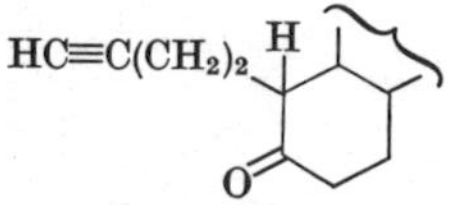 HC≡C$(CH_2)_2$, H, O (65%)	87
19	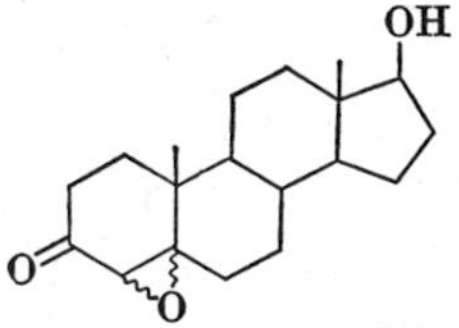 OH, O, O	$TsNHNH_2$/EtOH, 50°	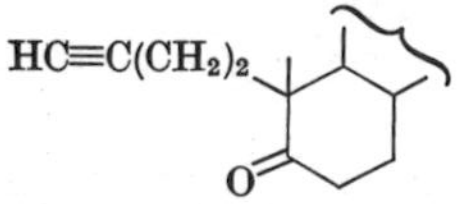 HC≡C$(CH_2)_2$, O (85%)	87
20	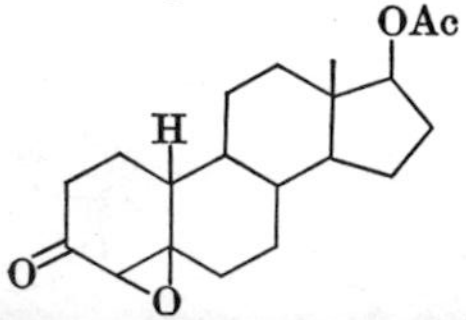 OAc, H, O, O	$TsNHNH_2$, EtOH 25°, 16 hr	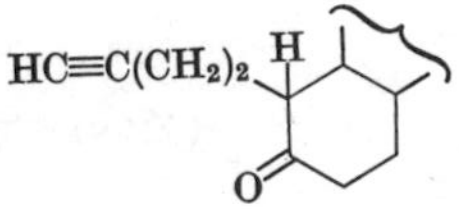 HC≡C$(CH_2)_2$, H, O (72%)	86

	Reactant	Conditions	Product(s) and Yield(s) (%)	Refs.
21	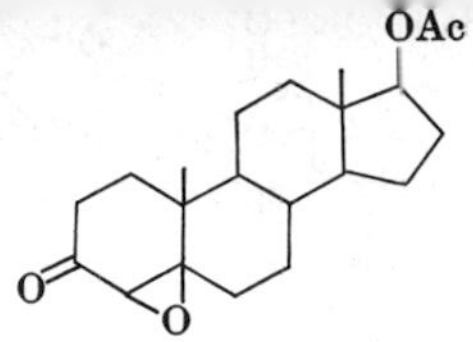 OAc; O; O	$TsNHNH_2/CH_2Cl_2$-AcOH, $-18°$, 20 hr; 25°, 1 hr	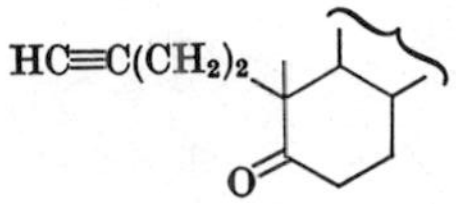 $HC{\equiv}C(CH_2)_2$; O (83%)	86
21	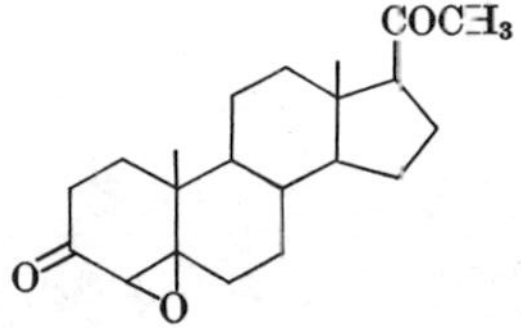 $COCH_3$; O; O	$TsNHNH_2$/EtOH, 25°, 2 hr	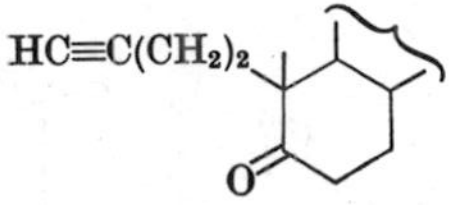 $HC{\equiv}C(CH_2)_2$; O (70%)	86
	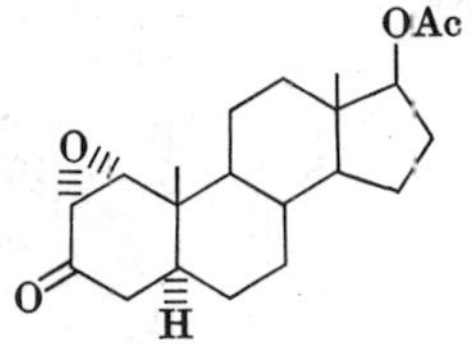 OAc; O; O; H	$TsNHNH_2/CH_2Cl_2$-AcOH, $-18°$, 15 hr; 25°, 2 hr	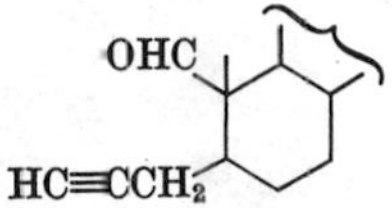 OHC; $HC{\equiv}CCH_2$ (79%)	86
	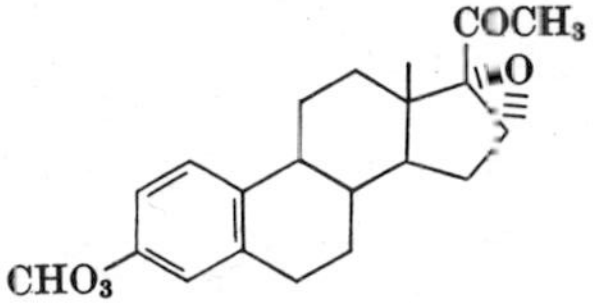 $COCH_3$; O; CH_3O	$TsNHNH_2/CH_2Cl_2$-AcOH, $-20°$, 15 hr; 25°, 5 hr	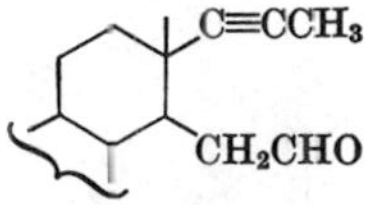 $C{\equiv}CCH_3$; CH_2CHO (30% as dimethyl acetal)	86
	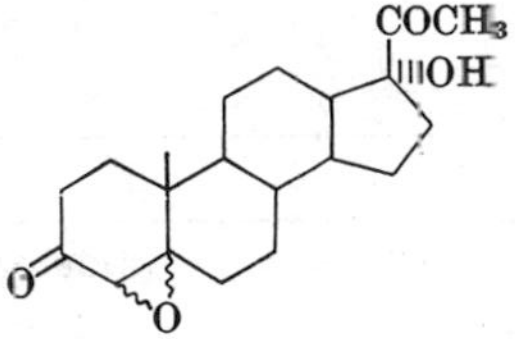 $COCH_3$; OH; O; O	$TsNHNH_2$/EtOH, 50°	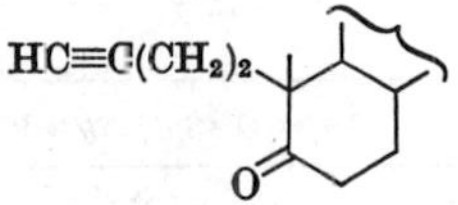 $HC{\equiv}C(CH_2)_2$; O (76%)	87

Note: References 125–210 are on pp. 506–507.

TABLE IV. FRAGMENTATION REACTIONS OF TOSYLHYDRAZONES (*Continued*)

Carbon Atoms in Starting Ketone	Reactant	Conditions	Product(s)—Distribution ( ) or Yield(s) (%)	Refs.
		B. Epoxyketones and Related Compounds (Continued)		
22	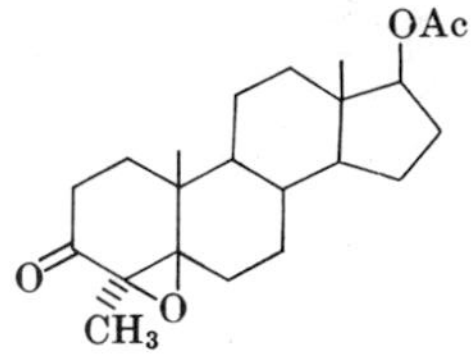 OAc; O; CH_3; O	$TsNHNH_2/CH_2Cl_2$-AcOH, −18°, 20 hr; 25°, 1 hr	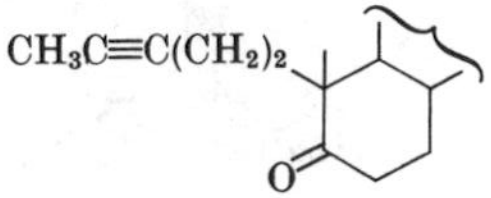 $CH_3C{\equiv}C(CH_2)_2$; O (84%)	86
	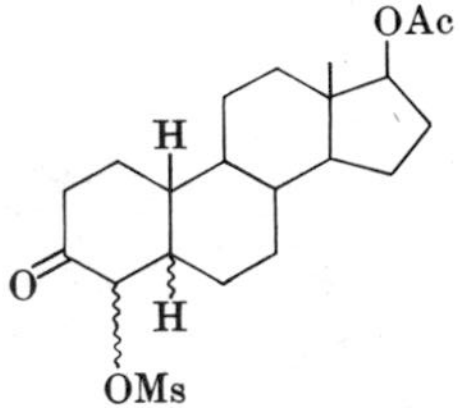 OAc; H; O; H; OMs	$TsNHNH_2/CH_2Cl_2$-AcOH, KOAc, 25°, 18 hr	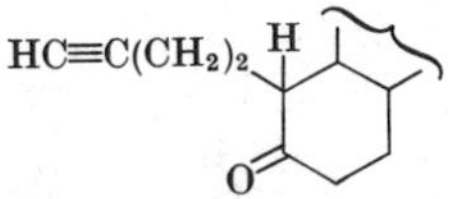 $HC{\equiv}C(CH_2)_2$; H; O (44%)	86
	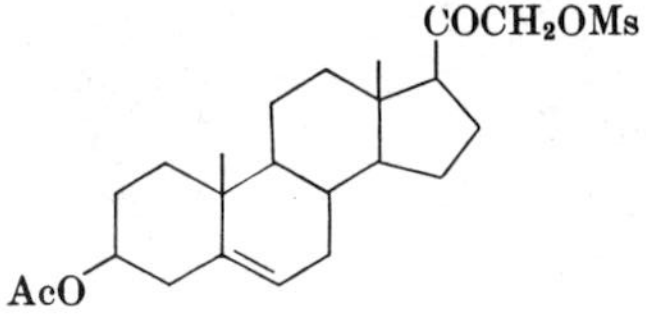 $COCH_2OMs$; AcO	$TsNHNH_2/CH_2Cl_2$-AcOH, KOAc	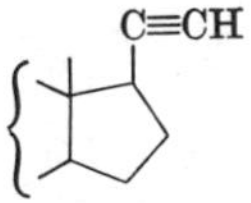 C≡CH (50%)	93
	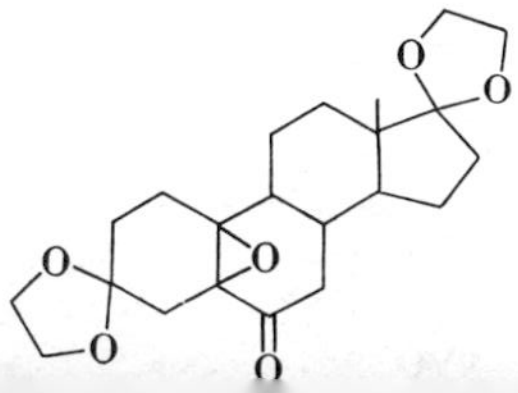 O; O; O; O; O; O	$TsNHNH_2$/EtOH, 25°, 40 hr	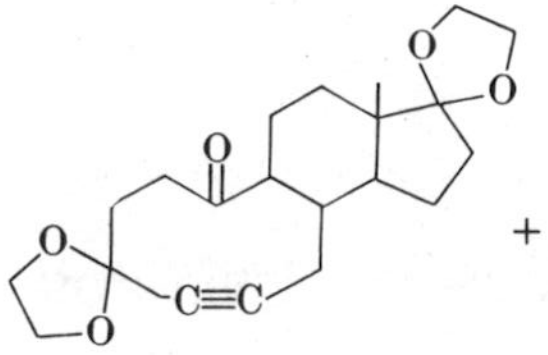 O; O; O; O; O; C≡C (40%) + 17-ketone (40%)	86

No.	Reactant	Conditions	Product(s) and Yield(s) (%)	Refs.
22	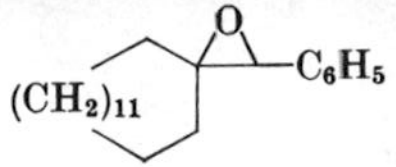 $(CH_2)_{11}$ (C$_6$H$_5$ epoxide)	$TsNHNH_2/CH_2Cl_2$-AcOH, 25°, 24 hr; 40°, 6 hr	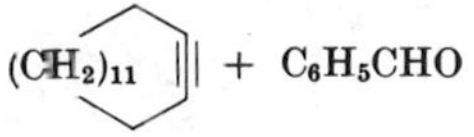 $(CH_2)_{11}$ (cycloalkyne) + C_6H_5CHO (52%)	86
23	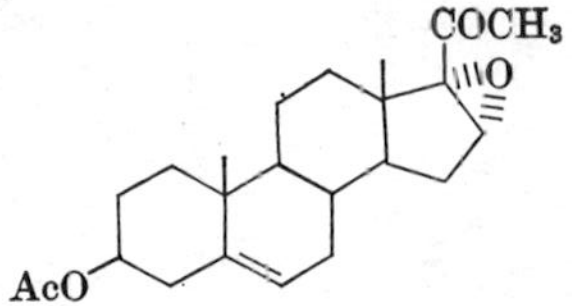 AcO, $COCH_3$ (steroid epoxide)	$TsNHNH_2/CH_2Cl_2$-AcOH, −18°, 18 hr; 25°, 1 hr	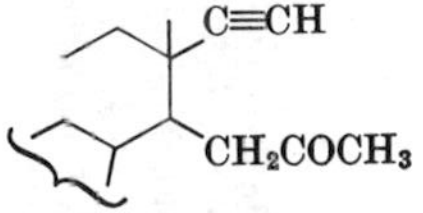 C≡CH, CH_2COCH_3 (63%)	86
24	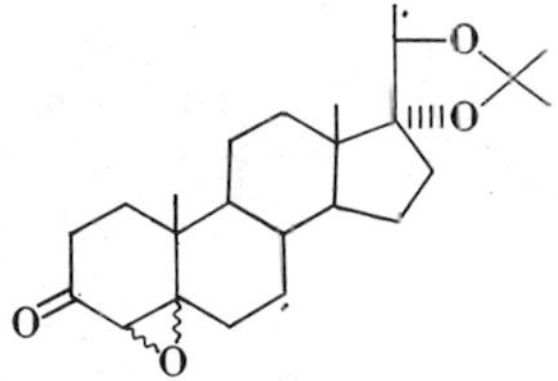 (steroid epoxy ketone)	$TsNHNH_2$/EtOH, 50°	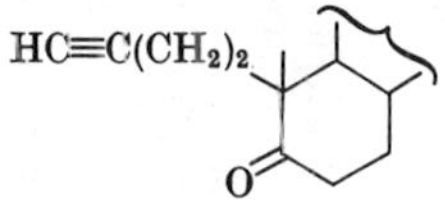 $HC{\equiv}C(CH_2)_2$ (70%)	87
25	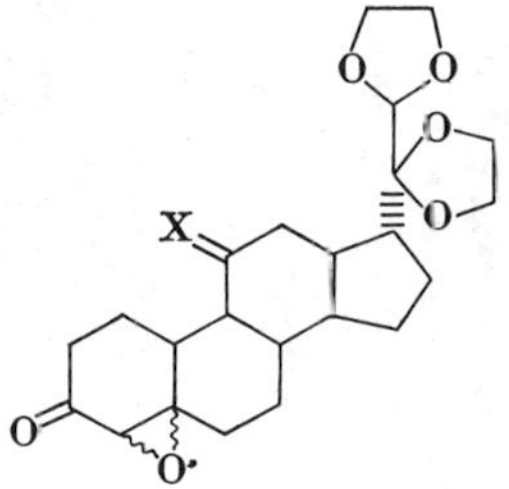 (steroid epoxy ketone) (X = H_2, O)	$TsNHNH_2$/EtOH, 50°	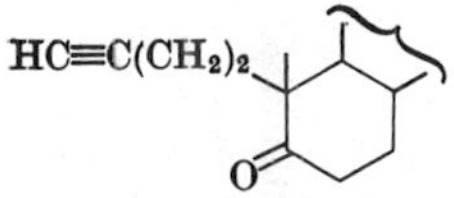 $HC{\equiv}C(CH_2)_2$ (X = H_2, 66%, X = 0, 57%)	87

Note: References 125–210 are on pp. 506–507.

REFERENCES TO TABLES

[125] M. S. Ahmad and R. P. Sharma, *Aust. J. Chem.*, **21,** 527 (1968).

[126] U. Biethan, H. Klusacek, and H. Musso, *Angew Chem., Int. Ed.*, **6,** 176 (1967).

[127] J. E. Baldwin and H. C. Krauss, Jr., *J. Org. Chem.*, **35,** 2426 (1970).

[128] M. Rey, U. A. Huber, and A. S. Dreiding, *Tetrahedron Lett.*, **1968,** 3583.

[129] M. Rey, R. Begrich, W. Kirmse, and A. S. Dreiding, *Helv. Chim. Acta*, **51,** 1001 (1968).

[130] H. Babad, W. Flemon, and J. B. Wood, III, *J. Org. Chem.*, **32,** 2871 (1967).

[131] G. Bancroft, Y. M. Y. Haddad, and G. H. R. Summers, *J. Chem. Soc.*, **1961,** 3295.

[132] D. H. R. Barton and C. H. Robinson, *J. Chem. Soc.*, **1954,** 3045.

[133] D. E. Evans and G. H. R. Summers, *J. Chem. Soc.*, **1956,** 4821.

[134] J. T. Edward, N. E. Lawson, and D. L'Anglais, *Can. J. Chem.*, **50,** 766 (1972).

[135] R. J. Ferrier, *J. Chem. Soc.*, **1964,** 5443.

[136] Y. Gaoni and E. Wenkert, *J. Org. Chem.*, **31,** 3809 (1966).

[137] M. A. Gianturco, P. Friedel, and V. Flanagan, *Tetrahedron Lett.*, **1965,** 1847.

[138] W. Kirmse, H. J. Schladetsch, and H.-W. Bücking, *Chem. Ber.*, **99,** 2579 (1966).

[139] B. Loev, M. F. Kormendy, and K. M. Snader, *J. Org. Chem.*, **31,** 3532 (1966).

[140] (a) E. Piers and R. J. Keziere, *Can. J. Chem.*, **47,** 137 (1969); (b) *Tetrahedron Lett.*, **1968,** 583.

[141] (a) D. C. Humber, A. R. Pinder, and R. A. Williams, *J. Org. Chem.*, **32,** 2335 (1967); (b) D. C. Humber and A. R. Pinder, *Tetrahedron Lett.*, **1966,** 4985.

[142] K. M. Shumate and G. J. Fonken, *J. Amer. Chem. Soc.*, **88,** 1073 (1966).

[143] R. K. Bartlett and T. S. Stevens, *J. Chem. Soc., C*, **1967,** 1964.

[144] R. R. Wittekind, C. Weissman, S. Farber, and R. I. Meltzer, *J. Heterocycl. Chem.*, **4,** 143 (1967).

[145] H. Musso and U. Biethan, *Chem. Ber.*, **100,** 119 (1967).

[146] K. Geibel, *Chem. Ber.*, **103,** 1637 (1970).

[147] W. Kirmse and H.-W. Bücking, *Ann.*, **711,** 31 (1968).

[148] M. Jones, Jr., and M. B. Sohn, unpublished work quoted in Ref. 5, p. 34.

[149] M. A. Battiste and M. E. Burns, *Tetrahedron Lett.*, **1966,** 523.

[150] (a) C. L. Bird, H. M. Frey, and I. D. R. Stevens, *Chem. Commun.*, **1967,** 707; (b) I. D. R. Stevens, H. M. Frey, and C. L. Bird, *Angew. Chem., Int. Ed. Engl.*, **7,** 646 (1968).

[151] M. Julia, Y. Noël, and R. Guégan, *Bull. Soc. Chim. Fr.*, **1968,** 3742.

[152] (a) M. Jones, Jr., S. D. Reich, and L. T. Scott, *J. Amer. Chem. Soc.*, **92,** 3118 (1970); (b) M. Jones, Jr., and L. T. Scott, *ibid*, **89,** 150 (1967).

[153] F. T. Bond and L. Scerbo, *Tetrahedron Lett.*, **1968,** 2789.

[154] J. W. Wilt, C. A. Schneider, H. F. Dabek, Jr., J. F. Kraemer, and W. J. Wagner, *J. Org. Chem.*, **31,** 1543 (1966).

[155] J. W. Wilt and P. J. Chenier, *J. Org. Chem.*, **35,** 1562 (1970).

[156] P. K. Freeman and D. M. Balls, *J. Org. Chem.*, **32,** 2354 (1967).

[157] S. Masamune, K. Fukumoto, Y. Yasunari, and D. Darwish, *Tetrahedron Lett.*, **1966,** 193.

[158] L. A. Paquette and G. V. Mechan, *J. Amer. Chem. Soc.*, **93,** 3039 (1970).

[159] L. A. Paquette and G. H. Birnberg, *J. Amer. Chem. Soc.*, **94,** 164 (1972).

[160] W. Kirmse and C. Hase, *Angew. Chem., Int. Ed. Engl.*, **7,** 891 (1968).

[161] H. Dannerberg and H. J. Bross, *Tetrahedron*, **21,** 1611 (1965).

[162] E. E. van Tamelen and I. G. Wright, *J. Amer. Chem. Soc.*, **91,** 7349 (1969).

[163] W. Kirmse and M. Buschhoff, *Chem. Ber.*, **100,** 1491 (1967).

[164] M. N. Applebaum, R. W. Fish, and M. Rosenblum, *J. Org. Chem.*, **29,** 2452 (1964).

[165] A. Sonoda and I. Moritani, *J. Organometal. Chem.*, **26,** 133 (1971).

[166] R. A. Moss, U.-H. Dolling, and J. R. Whittle, *Tetrahedron Lett.*, **1971,** 931.

[167] Z. Majerski, S. H. Liggero, and P. von Schleyer, *Chem. Commun.*, **1970,** 949.

[168] P. K. Freeman and R. C. Johnson, *J. Org. Chem.*, **34,** 1751 (1969).

[169] P. K. Freeman and R. C. Johnson, *J. Org. Chem.*, **34,** 1746 (1969).

[170] R. R. Sauers, S. B. Schlosberg, and P. E. Pfeffer, *J. Org. Chem.*, **33**, 2175 (1968).

[171] A. H. Rees and M. C. Whiting, *J. Org. Chem.*, **35**, 4167 (1970).

[172] J. R. Chapman, *Tetrahedron Lett.*, **1966**, 113.

[173] B. Singh, *J. Org. Chem.*, **31**, 181 (1966).

[174] G. Maier and M. Strasser, *Tetrahedron Lett.*, **1966**, 6453.

[175] J. Meinwald, J. W. Wheeler, A. A. Nimetz, and J. S. Liu, *J. Org. Chem.*, **30**, 1038 (1965).

[176] P. K. Freeman and K. B. Desai, *J. Org. Chem.*, **36**, 1554 (1971).

[177] P. K. Freeman, R. S. Raghavan, and D. G. Kuper, *J. Amer. Chem. Soc.*, **93**, 5288 (1971).

[178] R. A. Moss and J. R. Whittle, *Chem. Commun.*, **1969**, 341.

[179] K. Geibel and H. Mäder, *Chem. Ber.*, **103**, 1645 (1970).

[180] (a) H. W. Geluk and T. J. deBoer, *Tetrahedron*, **28**, 3351 (1972); (b) *Chem. Commun.*, **1972**, 3.

[181] J.-F. Biellman and J.-P. Péte, *Bull. Soc. Chim. Fr.*, **1967**, 675.

[182] R. S. Marmor and D. Seyferth, *J. Org. Chem.*, **36**, 128 (1971).

[183] E. L. Allred and A. L. Johnson, *J. Amer. Chem. Soc.*, **93**, 1300 (1971).

[184] (a) J. E. Baldwin and M. S. Kaplan, *J. Amer. Chem. Soc.*, **93**, 3969 (1971); (b) *Chem. Commun.*, **1970**, 1560.

[185] J. E. Baldwin and M. S. Kaplan, *J. Amer. Chem. Soc.*, **94**, 668 (1972).

[186] L. Caglioti, P. Grasselli, and G. Maina, *Chim. Ind.* (Milan), **45**, 559 (1963) [*C.A.*, **60**, 10749g (1963)].

[187] T. S. Cantrell and B. L. Harrison, *Tetrahedron Lett.*, **1969**, 1299.

[188] F. Fringuelli and A. Taticci, *J. Chem. Soc., C*, **1971**, 1809.

[189] F. Fringuelli and A. Taticci, *J. Chem. Soc., C*, **1971**, 756.

[190] (a) K. Mori and M. Matsui, *Tetrahedron*, **26**, 2801 (1970); (b) *Tetrahedron Lett.*, **1969**, 2729.

[191] L. A. Paquette and M. K. Scott, *J. Amer. Chem. Soc.*, **94**, 6751 (1972).

[192] J.-C. Richer and C. Freppel, *Can. J. Chem.*, **46**, 3709 (1968).

[193] J. F. W. Keana, D. P. Dolata, and J. Ollerenshaw, *J. Org. Chem.*, **38**, 3815 (1973).

[194] R. L. Cargill and A. M. Foster, *J. Org. Chem.*, **35**, 1971 (1970).

[195] K.-T. Liu and R. H. Shapiro, *J. Chin. Chem. Soc.* (*Tapei*), **16**, 30 (1969) [*C.A.*, **71**, 90548c (1969)].

[196] M. R. Willcott, III, and C. J. Boriack, *J. Amer. Chem. Soc.*, **93**, 2354 (1971).

[197] J. B. Grutzner, M. Jautelat, J. B. Dence, R. A. Smith, and J. D. Roberts, *J. Amer. Chem. Soc.*, **92**, 7107 (1970).

[198] W. R. Roth and A. Friedrich, *Tetrahedron Lett.*, **1969**, 2607.

[199] B. M. Jacobson, *J. Amer. Chem. Soc.*, **95**, 2597 (1973).

[200] B. Schönecker, K. Ponsold, and P. Neuland, *Z. Chem.*, **10**, 221 (1970).

[201] I. R. Trehan, S. C. Narang, V. K. Sekhri, and K. C. Gupta, *Indian J. Chem.*, **9**, 287 (1971).

[202] A. G. Brook, H. W. Kucera, and D. M. MacRae, *Can. J. Chem.*, **48**, 819 (1970).

[203] J. E. Herz, E. Gonzales, and B. Mandel, *Aust. J. Chem.* **23**, 857 (1970).

[204] H. Volz and B. Messner, *Tetrahedron Lett.*, **1969**, 4111 (1969).

[205] H. Volz, U. Zirngibl, and B. Messner, *Tetrahedron Lett.*, **1970**, 3593.

[206] H. Volz and R. Draese, *Tetrahedron Lett.*, **1970**, 4917.

[207] D. J. Bichan and P. Yates, *J. Amer. Chem. Soc.*, **94**, 4773 (1972).

[208] A. de Meijere, *Tetrahedron Lett.*, **1973**, 3483.

[209] G. A. Hiegel and P. Burk, *J. Org. Chem.*, **38**, 3637 (1973).

[210] E. Vedejs and R. P. Steiner, *J. Chem. Soc., Chem. Commun.*, **1973**, 599.

AUTHOR INDEX, VOLUMES 1–23

CHAPTER AND TOPIC INDEX, VOLUMES 1–23

Many chapters contain brief discussions of reactions and comparisons of alternative synthetic methods which are related to the reaction that is the subject of the chapter. These related reactions and alternative methods are not usually listed in this index. In this index the volume number is in BOLDFACE, the chapter number in ordinary type.

SUBJECT INDEX, VOLUME 23

Since the table of contents provides a quite complete index, only those items not readily found from the contents pages are listed here. Numbers in BOLDFACE type refer to experimental procedures.